China's Economic Gene Mutations

Zhaoguang Hu • Jian Zhang • Ning Zhang

China's Economic Gene Mutations

By Electricity Economics and Multi-agent

Springer

Zhaoguang Hu
State Grid Energy Research Institute
Beijing, China

Jian Zhang
Beijing Jiaotong University
Beijing, China

Ning Zhang
Beijing Jiaotong University
Beijing, China

ISBN 978-3-662-47297-2 ISBN 978-3-662-47298-9 (eBook)
DOI 10.1007/978-3-662-47298-9

Library of Congress Control Number: 2015946242

Springer Heidelberg New York Dordrecht London

Printed on acid-free paper

Springer-Verlag GmbH Berlin Heidelberg is part of Springer Science+Business Media (www.springer.com)

Preface

A gene has four characteristics as replication, mutation, evolution, and uniqueness. The human body is very complex, and a gene can be used to diagnose health problems. The economic system is very complex too; is there something like a gene in the economy? It has been found that production function with electricity in electricity economics has the four characteristics of a gene.[1]

Electricity plays a crucial role in the economic activities because it is widely used for commercial purposes, in agriculture, industry, and our daily life. It is clear that the more output of production, the more electricity must be used in the production. Thus, there is a positive correlation between electricity use and production output. On the other hand, electricity data can serve as an accurate measure for economic productions in nearly all kinds of economic activities since there are many meters measuring the electricity from the power generation side to demand side. The electric power data has the features of real time, reliability, accuracy, comprehensiveness, and the synchronism with economic activity. And also, it is very easy to get these data. A mechanism of collecting power (power generation, power consumption) data has been established in China, which ensures that the electric power data can be obtained accurately and reliably every month.

Production function with electricity, putting electricity as a representative production factor, has been studied in electricity economics.[2] Millions of production functions with electricity from a firm's level, to sectoral level, to industrial level, and to national economy level can form a gene mapping of the economy. Economic gene mutations on the four levels can depict the big changes of the marginal representative factor productivity in the economy, and it can be used to diagnose the economic problems.

[1] Zheng Hu and Zhaoguang Hu (2013), "Production function with electricity consumption and its applications" [J], *Energy Economics* (39, 171–192)

[2] Zhaoguang Hu and Zheng Hu (2013), *Electricity Economics: Production Functions with Electricity* [M], Springer

In this book, all these concepts are introduced in detail, and they are applied to review China's economy. We hope that readers can have a deep view and diagnose the economy for other countries from the perspective of gene mutations.

In order to simulate the operation of China's economy, authors proposed the Agents Response Equilibrium (ARE) model after studying the strengths and weaknesses of Computable General Equilibrium (CGE) model and Agent-Based Computational Economics (ACE) model. The ARE model is developed on the basis of the multi-agent technique. Authors build the intelligent agent model with 42 sectors in the input-output table of China. As agents, they have their goal, knowledge base, learning ability, the capability to communicate with others, and decision making on their business. Besides, the overall economic activities of the agents, interaction mechanism, and markets are designed according to the economy in China. Then, it can be used to simulate the operation of the national economy and conduct certain policy experiments. Based on the input-output table and some assumptions of fiscal and monetary policies, the economic activities can be simulated and result in new input-output tables and prices of the products of the 42 sectors year by year. Thus, it is easy to study the impacts of the policies, just like playing games on a computer. China's economy has been simulated from 2008 to 2014 to verify the accuracy and reliability of the ARE model, while the simulation from 2015 to 2025 serves as the economic development with the results of input-output tables. It is also the foundation to extend the research on economic gene mutations in China.

This book can be divided into three parts. The first part describes the concepts and applications of electricity economics and analyzes the gene mutations and gene mapping in China's economy. It contains three chapters. In Chap. 1, a typical theory of electricity economics is introduced, and economic development in China is analyzed from the perspective of electricity economics. In Chap. 2, electricity and energy supply in China is reviewed, and the concept of mutation is expressed through case studies in the USA, Japan, and China. Chapter 3 presents the concept of gene mapping and studies the situation in China until 2025 in both business as usual scenario and policy scenario. The second part introduces our ARE model which acts as the economic simulation tool for policy study. It consists of four chapters. In Chap. 4, economic modeling techniques that are widely used and discussed nowadays are reviewed. In Chap. 5, the basic content of intelligent engineering is introduced. In Chap. 6, the foundations, framework, features, and the overall design of the ARE model are introduced. In Chap. 7, each kind of agent in the ARE model is described one by one, and some details about our computer program are shown. The third part acts as the verification of the ARE model and presents some simulation results such as the annual input-output tables. Three chapters are included in this part. In Chap. 8, the simulation results of China's economy during the period of 2008–2010 are analyzed, while Chap. 9 focuses on the simulation from 2011 to 2014. In Chap. 10, all the annual input-output tables of China in 2011–2025 obtained by our simulation are listed, since they are very useful data foundation to conduct various researches but mostly not available by published materials.

The authors are the delegates of our team that has been committed in the area of economic modeling by intelligent agents since 2003. As the team leader, Prof. Zhaoguang Hu studies and directs the process of research in detail. Typically, principal ideas in this book are mostly proposed by him. Dr. Jiahai Yuan is the first member of the team to study the intelligent agent theory. He learned a lot about the multi-agent system and introduced this theory to our team. He began to try to apply this technique to the modeling and simulation of the economic system, especially the electricity market. Dr. Minjie Xu summarized the methodology of intelligent engineering and developed the generalized model in the intelligent space. He utilized the agent-based model to simulate residential electric power consumption and the impacts of macroeconomic policies on electricity consumption. The electricity price, income level, and government expenditure can be simulated to observe the influences. Dr. Meng Li carried out the analysis and empirical study of electric power supply and demand systematically. Through the discovery of the quantitative relationship between electrical load rate and coal consumption, he developed the simulation and analysis model to research the influences of relevant policies. Dr. Xiandong Tan studied the input-output table updating approach using electricity data. He also participated in the design of response rules of sector agents, government agent, and resident agent and the software system of multi-agent model. Besides, he executed the simulation of the demand side management compensation mechanism in China. Dr. Xiaoyou Jiao explored the theory extension from system engineering to intelligent engineering. A modified intelligent inference model to deal with the analysis of complicated issues was proposed by him. He executed the intelligent simulation of the influences of macroeconomic policy on electricity consumption. Dr. Jianwei Tian creatively integrated intelligent engineering theory and multi-agent modeling technology. His applications focused on the simulation of smart grid operation. He put forward an agent-based hierarchical hybrid coordination model and algorithm to simulate the grid-consumer interoperability among related stakeholders under smart grid framework. Dr. Xiao Xiao started to build a multi-agent system model that was a complex giant system which contains control agents, coordinate agents, and functional agents, in order to reflect the production, exchange, distribution, consumption, etc. in the economic system. By means of the multi-agent model, he attempted to simulate the asymmetric effects of the monetary policies and underlined the significant intermediary role of depository financial institutions. He developed the computer program of the ARE model in JAVA on the platform of Swarm. Mr. Wei Duan updated China's 2010 input-output table with the intelligent agent technique. Then, he reviewed the fiscal policies, such as carbon tax and other sorts of taxes, and their impacts on the economy, and the method to simulate the fiscal policy. In the computer program, diverse entities in the practical economy were designed as intelligent agents who could negotiate with each other and change their behaviors dynamically according to the environment circumstances. Mr. Mingtao Yao made some modifications on the ARE model. He devoted himself to build the module of the labor market in the model and code-relevant program, in order to embody the effects of labor issues on the national economy. Besides, he executed some experiments of fiscal policies and monetary policies to test the consequences of various policy packages by virtue of the model. Mr. Jian Zhang improved the ARE model and related computer program, especially in the field of international commodity market to make it more reasonable and realistic and the module of inventory in all the production sector agents to adjust their behavior process. He perfected the approach to calculate the energy consumption and electricity consumption in the model. All the simulations from 2008 to 2025 displayed in this book were conducted by him. Mr. Ning Zhang made some corrections on the computer program to consummate the mechanism and interfaces of the fiscal policies and monetary policies and revised the design of the computer program window and some ideas on improving the program. As the last team member, he was committed to collect and analyze the policy experiment results and summarize all the previous work in the team. And also Dr. Jing Wang, Dr. Wei Ding, Dr. Junjie Kang, Dr. Jinghong Zhou, Ms. Yan Xu, Ms. Chuning Na, and Dr. Yanan Zheng have contributed a lot on the theory and the model when they were doctoral students and postdoctoral from 2003 to 2014.

We are very grateful for the help from a lot of researchers and scholars in a series of discussions on electricity economics and the ARE model. And also Dr. Zhaoguang Hu's presentations on the above issues at Tsinghua University, Massachusetts Institute of Technology, Lawrence Berkeley National Laboratory, Argonne National Laboratory, International Energy Agency, World Bank, and other institutions have been discussed in detail. They have provided us with many beneficial suggestions for our research. We appreciate their time and support a lot. They are listed as follows:

Prof. Huijiong Wang (Development Research Center, the State Council of China), Prof. Xiliang Zhang (Tsinghua University, China), Prof. Yi Wang, Prof. Junyong Wu, Prof. Jinghan He, Prof. Yuhui Zhou (Beijing Jiaotong University, China), Prof. Dongxiao Niu, Prof. Zhongfu Tan, Prof. Huiru Zhao (North China Electric Power University, China), Dr. Dong Wang (Rio Tinto, China), Dr. Bo Shen (Lawrence Berkeley National Laboratory, USA), Dr. Valerie J. Karplus, Mr. Loren C. Cox, Dr. John E. Parsons (Massachusetts Institute of Technology, USA), Dr. Jianhui Wang (Argonne National Laboratory, USA), Dr. Ming Yang (GEF, World Bank), Max Dupuy (Regulatory Assistance Project, USA), Dr. Fuqiang Yang and Ms. Finamore Barbara (Natural Resources Defense Council, USA), Mr. Kevin JianjunTu (International Energy Agency), Prof. Hameed Nezhad (Metropolitan State University, USA), Mr. Zheng Hu (University of Delaware, USA), Mr. Froylan E. Sifuentes (University of California at Berkeley, USA), Mr. Baoguo Shan, and Mr. Xinyang Han (State Grid Energy Research Institute, China).

We would like to express our sincere appreciation to all of them.

Since the knowledge scope of our team is limited, we know that our research cannot be perfect. Accordingly, there may be some mistakes in this book, although we have tried our best to avoid them. It is our will that readers can gain something after reading the book. More importantly, we look forward to the discussion and even criticisms from the readers, because we believe that truth develops in criticism, while fallacy grows in praise.

Beijing, China
28 February 2015

Prof. Dr. Zhaoguang Hu
Mr. Jian Zhang
Mr. Ning Zhang

Contents

About the Authors

Dr. Zhaoguang Hu is the vice president of consultant and chief energy specialist and the head of the Intelligent Laboratory for Economy-Energy-Electricity-Environment at the State Grid Energy Research Institute, State Grid Corporation of China. He is also editor in chief of *Electric Power* (Chinese journal) and chairman of the Electricity Economics Committee of the Chinese Society for Electric Engineering and serves as a professor at Beijing Jiaotong University and North China Electric University. He has studied electricity economics by using electricity consumption data. It has been used to review and study China's economy monthly by production functions with electricity. He has extended power planning model to integrated resource strategic planning (IRSP) to promote energy efficiency and renewable generation on both power supply and demand sides. It has been used as a policy study tool on low carbon electricity in China. He has studied intelligent engineering, as a methodology for solving complex problems, by using artificial intelligence, fuzzy systems, and other advanced technologies. It has been used in economic modeling. Many of his studies and policy suggestions have been accepted or referenced by the Chinese government. He has received a Special Contribution Award from the Chinese government for greatly contributing to China's science development and a lot of awards on science and technology from the National Energy Administration, Chinese Society for Electrical Engineering, State Grid Corporation of China, and other institutions. He has published more than 140 papers in journals and 10 books. He received his bachelor of science degree at Huadong Petroleum Institute in 1982 and master's and Ph.D. at Electric Power Research Institute of China in 1984 and 1989, respectively.

Jian Zhang Ph.D. Candidate at Beijing Jiaotong University

Ning Zhang Ph.D. Candidate at Beijing Jiaotong University

Part I
Economic Gene Mapping in China by Electricity Economics

Chapter 1
China's Electricity Economy

Mutation is a word to show a big change in gene of a living body, and it will result in some diseases usually. It looks like an accident because it is very difficult to know when it will happen. However, it is inevitability of living things in the historical developing process. For a long period, living environments have many big changes; how do they meet the challenge of the changes? Mutation is the only way for living things to do so. Therefore, a series of mutations will make gene changes to guaranty living things to be alive in the changing environments. This is the evolution of the living things generation by generation.

What is mutation of an economy? Is there any gene in the economy? Production functions with electricity have the characteristics of gene, and it can reflect the mutation of the economy by electricity economics [1]. People are used to reviewing national economy by some indicators such as Gross Domestic Product (GDP), value added, new jobs, etc. However, the accuracy of the indicators is still worried by some experts and scholars, since they are collected by the statistical method. Is it possible to calculate them by accurate data such as electricity consumption? Yes, it is one study area of electricity economics. This book typically tries to provide a new perspective of study on China's economy by electricity economics. As one of the necessary inputs, electricity has been commonly used in almost all types of productions, and *electricity use is a representative of inputs of production factors; it can be called as representative production factor. Electricity data can also serve as an accurate measure for economic productions in nearly all kinds of economic activities*. On the other hand, with the smart meters, it is convenient to access real-time electricity data even on per 15 min bases for every electricity user. It is a useful data source for reviewing economic activities that require accurate and reliable electricity data from firm to sector, to industry, and to national economy.

By means of electricity economics, economic development in China will be reviewed in this chapter by electricity data. It will provide some new scenarios on economic activities for some sectors, primary industry, secondary industry, tertiary industry, and national economy. For the convenience of the readers, the basic principle of electricity economics will be introduced first in this chapter.

Z. Hu et al., *China's Economic Gene Mutations*,
DOI 10.1007/978-3-662-47298-9_1

1.1 Electricity Economics

One function of electricity economics is to review and study economic issues by electricity consumption data, thus, it is a cross area of electric engineering and economics.

1.1.1 The Characteristics of Electricity Data

In our daily life, when we turn the switch on, the lamp will light. The power units generate electricity, and then, the electricity can be transmitted to the users through the power grid. Power system consists of power generation, transmission, distribution, and utilization. There are a few different technologies on power generation, such as thermal power generation, hydropower generation, nuclear power generation, wind power generation, and solar power generation. Thermal power plant uses coal, oil, gas, and biomass as the fuel to generate electricity. Its equipment mainly includes boilers, turbines, generators, and the auxiliary production system. Hydropower plant is driven by water. Its equipment mainly includes diversion pipes, turbine generators, and the auxiliary production system. Nuclear power plant consists of nuclear reactors, turbines, generators, and the auxiliary production system. Wind farm mainly includes blades, fans, and the auxiliary production system to transform wind power to electric power. In order to deliver the electric energy to users with long distance, large scale, and low loss, it is necessary to raise the voltage level by the transformer on the generation side and then transport the electricity to the electric load center via high-voltage power transmission lines. On electric power demand side, the high voltage must be step down to the 110/220/380 voltage by the transformers. Then, the power is supplied to the manufacturing firms, shops, schools, and thousands of households through the low-voltage distribution system [2]. Power consumers use a variety of electrical equipment to convert electrical energy into power, heat, light, and other forms of energy, serving for the economic production and our life.

Electricity data is very accurate. In terms of current technology, electricity cannot be stored in large scale. The production, transmission, distribution, and consumption of electricity must be completed in an instant. There are many meters measuring the electricity from power generation side to demand side. The electric power data measured by the meters has the features of real time, reliability, accuracy, comprehensiveness, and the synchronism of economic operation. It is very easy to get these data. A mechanism of collecting power (power generation, power consumption) data has been established in China, which ensures that the electric power data can be obtained accurately and reliably. Monthly electricity data is published in the middle of the next month, while the GDP data is published once every quarter. Smart Grid technology can make the data collecting system more complete, and we can even get these data weekly, daily, or even hourly.

Electricity has been widely used in almost all economic activities such as firms' production, shops' sale service, hotels' service, and farmers' watering except transport vehicles and crafts using gasoline. However, with the technology development of electric vehicles, the electricity will be further used in transportation sector. It is a necessary element in the production, and it can be regarded as one of production factors. Since it is the reading from electric meter to be the base of electricity bill, it must be very accurate in quantity. On the other hand, it is clear that the more electricity is used, the more amount of product will be produced. Therefore, the data of electricity consumption can be used to measure the output of the production at different production levels, i.e., from firms to sectors, from sectors to industries, and from industries to the national economy.

Usually, the productions in agriculture, forestry, animal husbandry, and fishery are classified as primary industry. The productions in mining, manufacturing, and construction are called as secondary industry. The other economic activities such as commercial business are called as tertiary industry. Primary industry, secondary industry, and tertiary industry can be regarded as three industries of the economy. Thus, an industry is consisted of many sectors, and a sector contains many firms and companies. The electricity used in primary industry is expressed by E_1, the electricity consumption in secondary industry is expressed by E_2, and the electricity use in tertiary industry is denoted by E_3. The electricity consumption for the three industries is called as whole industrial electricity consumption E ($E = E_1 + E_2 + E_3$). The total electricity consumption is denoted by TE, in which the electricity consumption of household, RE, is considered ($TE = E + RE$).

Table 1.1 shows the details of electricity data in Nov. 2014 in China. The total electricity consumption (TE) of China was 463.22 billion kWh (kilowatt-hours) and grew by 3.28 % year on year (YoY), while the electricity consumption by the three industries (E) is 411.93 billion kWh, accounting for 88.92 % of the total electricity consumption, and grew by 3.05 % YoY. Besides, the annual growth rate of the electricity consumed by primary industry (E_1), secondary industry (E_2), and tertiary industry (E_3) is −6.47 %, 2.71 %, and 6.96 %, respectively. The residential electricity consumption RE is 51.29 billion kWh, accounting for 11.08 % of the total electricity consumption, and increases 5.2 % year on year. For the first 11 months in 2014, total electricity use TE was 5011.61 billion kWh, which grew by 3.74 % YoY, and industrial electricity use E was 4373.97 TWh, grew by 4 %. Other more than 100 detailed sectorial electricity uses and growth rates in November 2014 are not shown in Table 1.1 since it will take a big space.

The shares of primary industry, secondary industry, and tertiary industry in the total electricity consumption were 1.55 %, 76.22 %, and 11.15 %, respectively, in Nov. 2014. It can be easily seen from the data above that the electricity consumption by secondary industry has an absolute advantage in quantity in China and the electricity consumption by tertiary industry has to be further enhanced. Because electricity consumption data is published every month with 2 weeks delay usually in China, it is a very important data source to review sectorial economy and study the trends of the sectors by government, policy officers, business people, stock market analyzers, and other persons who are interested in it.

Table 1.1 Electricity consumption in China in November 2014

	In November		From January to November	
	Electricity (TWh)	Growth (%)	Electricity (TWh)	Growth (%)
Total electricity use	**463.22373**	**3.28**	**5011.60740**	**3.74**
Industrial electricity use	411.93425	3.05	4373.97465	4.00
Primary industry	7.17052	−6.47	92.46115	0.42
Secondary industry	353.11885	2.71	3674.37532	3.75
Tertiary industry	51.64488	6.96	607.13818	6.08
Household electricity use	51.28948	5.20	637.63275	2.01
Town	28.81732	5.72	361.18426	1.64
Farmer	22.47216	4.55	276.44849	2.50

Note: 1 TWh = 1 billion kWh

Table 1.2 Electricity consumption in the USA in 2014 (million kWh)

Period	Residential	Commercial	Industrial	Transportation	All sectors
January	146,435	114,230	77,616	724	339,006
February	130,478	104,662	73,135	723	308,997
March	114,158	106,873	78,081	645	299,756
April	92,188	102,403	77,638	634	272,863
May	95,507	109,713	82,174	655	288,049
June	117,630	118,776	82,282	615	319,302
July	136,239	126,080	84,179	653	347,151
August	135,247	126,527	85,597	642	348,014
Sept	120,118	120,693	81,717	628	323,157
October	97,570	113,553	81,299	630	293,052

Data Source: U.S. Energy Information Administration, November 2012 Monthly Energy Review, Table 7.6. Electricity End Use

Monthly electricity consumption data of the USA is published with about 2-month delay; it consists of residential, commercial, industrial, and transportation electricity consumption as listed in Table 1.2. Although it is in no more details of sectorial electricity use as published in China and takes around 2 months on data publication, it is still very useful to review and study employments and the US economy monthly.

Smart meters can firstly record and show different sorts of information such as the type of electricity consumption, the amount of electricity consumption, and the load curve of users' various electrical equipment. And then, the meters would store and transmit them to the electric power company. Therefore, some people regard electricity meter as the barometer of economic operation. The electricity meter can act as the oscilloscope of the economic activity of the electricity customer. It accurately records the tracks of almost all the production of the firms, sectors, and industries and the national economy. The electricity consumption data is not only the unique and important basic data of electricity economics but also the important basis of economic review of government offices and business people. It has certain reference significance of grasping the economic (especially the real economy) trends and studying impacts of macroeconomic policy [3].

All in all, the electricity has the following basic features that make the electricity data accurate and meaningful.

Electricity cannot be stored in large scale. It is very difficult to store electric energy massively with current technology. Therefore, the electricity generation is equal to the electricity consumption plus line loss at each moment.

The power supply and demand have to be balanced in the power system at any second. Since the electric power cannot be stored in large scale, it is required that the power generation and consumption must be completed at every instant. If the power generation is less or greater than the power consumption, the power system will be unstable.

Electricity is closely interlinked to all level of economic activities from firms to sectors and to the national economy. Electricity is widely used in agriculture, industry, commercial, and our daily life, because various types of equipment and machines cannot work without electricity. Therefore, the electricity data would be the accurate measure of the economic activities and residents' lives.

Electricity use is a representative of inputs of production factors. It is clear that the more inputs of production, the more electricity must be used in the production. Thus, there is a positive correlation between electricity use and production inputs. Electricity data can serve as an accurate measure for economic productions in nearly all kinds of economic activities.

1.1.2 Production Functions with Electricity

Production function depicts the relationship between the output and all kinds of inputs in the production. If a, b, c, d, e, etc. represent the inputs and y represents the output of the production, the production function can be expressed as

$$y = f(a, b, c, d, e, \ldots) \tag{1.1}$$

Production function indicates the maximum amount of product that any particular collection of inputs is capable of producing [4]. Usually, the economists consider two inputs, labor and capital, as the main inputs in the production function. The famous Cobb-Douglas production function is

$$y = AL^{\alpha}K^{\beta} \tag{1.2}$$

Here, A, α, β are constant. Formula (1.2) reflects the relationship between the labor L, the capital K, and the output y if the technical level A remains unchanged. A is also named as total factor of productivity. α and β are the output elasticity of labor and capital, respectively.

Electricity is also a typical input in the production. With a certain level of production technology, the electricity consumption in a firm's production corresponds to the inputs of all raw materials, labors, investments, and other factors required for production. It is clear that the more inputs are put into the production, the more electricity will be used to produce the product. If e denotes electricity consumption, then there is a map between e and other inputs as follows:

$$e = g(a, b, c, d, \ldots) \quad (1.3)$$

This is the relationship between the production inputs and the electricity consumption. Since the more electricity is used, the more amount of product will be produced. Therefore, *the formula* (1.3) *is a positive correlation between the electricity consumption and the inputs of production factors*. Moreover, electricity consumption is positively correlated with the output of production. In this case, if we treat *electricity consumption as representative production factor (RPF), i.e., the indication of all other inputs of production factors*, the relationship between electricity consumption, representative production factor, and the overall output can be named as production function with electricity. The output y will be the function of electricity consumption. The production function with electricity is shown as follows:

$$y = F(e) \quad (1.4)$$

This is a one-variable function of the relationship between e and y. Production output y can be the amount of product, sales revenue, profits, value added, etc. Apart from a certain firm, the production function with electricity can also be used for various sectors, industries, and the national economy. Therefore, it is both simple and practical to adopt this method for finding the relationship between electricity consumption and the value added. At the sectoral level, if output V refers to the value added of the sector and Es depicts the electricity consumption of the sector, then Es-V function, the production function with electricity of the sector, can be written as follows [5]:

$$V = fv(Es) \quad (1.5)$$

For the three industries, i.e., primary industry, secondary industry, and tertiary industry, the industrial production function with electricity, $Ei-Vi$ $(i = 1,2,3)$ function, can be written as follows [5]:

$$Vi = fv(Ei) \quad (i = 1, 2, 3) \quad (1.6)$$

At the national level, if the whole industrial electricity consumption E indicates the inputs and GDP indicates the output of the national economy, then the production function with electricity, the E-GDP function, is

$$GDP = F(E) \quad (1.7)$$

As a case study, the *E-GDP* function for Japan is shown in the following equation:

$$GDP = 689.4E + 21546 \tag{1.8}$$

The GDP is calculated in constant local currency Yen in the year of 2000. The *E-GDP* function is a linear function with the correlation coefficient being 0.9935. The slope of formula (1.8) means the increase of national GDP resulted by a one-unit increase of the whole industrial electricity consumption, which is called as marginal GDP of electricity (*MGDP*) [5]. As mentioned above, *E* is the representative production factor. Thus, *MGDP* can also be called as marginal representative factor productivity (MRFP). MRFP shows the increase of a unit of RPF can result in the increase of output of the production. MRFP depends on technology innovation and improvement, energy efficiency, electrification, and also the economic mix of the national economy. In the formula (1.8), MRFP is 689.4 Yen/kWh in Japan. It shows that the increase of a unit of RPF, one kWh industrial electricity consumption, can produce 689.4 Yen of GDP in Japan. The positive intercept indicates that the average GDP of the electricity is declining.

What about the production functions with electricity of other countries in the world? In order to do so, it is necessary to get the data of the whole industrial electricity consumption and GDP for each country. The GDP in constant currency 2005 USD during 1980–2009 can be found on the website of the World Bank [6]; the total electricity consumption *TE* during 1980–2009 can be accessed from the US Energy Information Administration website [7]. Since it is very difficult to get the data of the whole industrial electricity consumption *E* for each country in the world, total electricity consumption *TE* is used to replace *E* in our study. Then, the production function with electricity is the *TE-GDP* function for some countries and the world as shown in Table 1.3. All the *TE-GDP* functions are in the form of linear function. We can see that the largest slope of all the *TE-GDP* functions is 9.623 for the UK. It means the marginal GDP of electricity, marginal representative factor productivity, is 9.623$/kWh. The main reason is that the share of tertiary industry in the GDP is relatively higher in the UK. It is clear that the output of electricity in tertiary industry is greater than the output of electricity in secondary industry. The low share of secondary industry will result in more imports of secondary industrial goods in the UK. The minimum slope of the *TE-GDP* functions occurs is China. It shows the MRFP is only 1.06$/kWh since the share of secondary industry is higher than tertiary. It means China exports more secondary products and imports more tertiary products (service). On the other hand, the positive intercept of the *TE-GDP* function means the average GDP of electricity consumption is declining with the increase of electricity use. And the negative intercept means the average GDP of electricity is rising with the increase of electricity use. The countries with average GDP of electricity rising include the USA, Canada, Germany, the UK, and India in Table 1.3. It shows that the high share of tertiary industry will result in the large slope of the *E-GDP* function for the national economy. It means the fast growth of the tertiary industry will increase the MRFP of the economy. Usually, the

Table 1.3 The *E-GDP* functions for some countries (2005USD)

Country			Slope			Intercept
World	GDP	=	2.724	TE	+	2092.7
The USA	GDP	=	3.947	TE	–	2699.6
Canada	GDP	=	2.706	TE	–	362.81
Brazil	GDP	=	1.734	TE	+	245.32
39 European countries	GDP	=	3.906	TE	+	2185.6
France	GDP	=	4.249	TE	+	220.62
Germany	GDP	=	7.139	TE	–	1006.1
UK	GDP	=	9.623	TE	–	1123.7
Japan	GDP	=	4.314	TE	+	320.56
Australia	GDP	=	3.336	TE	+	20.302
Taiwan region	GDP	=	1.736	TE	+	30.069
South Korea	GDP	=	2.03	TE	+	154.62
China Mainland	GDP	=	1.06	TE	+	2.3846
India	GDP	=	1.664	TE	–	4.3041

Source: Zhaoguang Hu and Zheng Hu [5, p 276–321]

share of the tertiary industry in the GDP depends on the economic stage. The tertiary share tends to be lower in the industrialization stage.

Studying the economy in this simple way allows us to uncover its intrinsic characters. Through the *E-GDP* function, we can excavate and understand these characteristics. We can also analyze their replication and mutation, diagnose economic operations, discover problems, and provide timely warnings beforehand. We can also conduct in-depth research and suit the remedy to each specific case. In this way, we will be able to realize healthy economic operation.

1.1.3 Stages of Economic Development in Terms of Electricity Consumption

There are four features for the economic development in a country/region as (1) GDP keeps constantly growing generally; (2) technological innovation and progress play an important role in the GDP growth; (3) the structure of the three industries is optimized gradually; and (4) the market mechanism is in perfecting. The economist Hollis B. Chenery has studied the economic development stages that contain primary product stage, industrialization stage, and developed economy stage in terms of per capita income as shown in Table 1.4. However, the monetary value is changing with time since money has inflation. And money exchange rate for foreign currency will be different with purchasing power parity (PPP) and market exchange rate.

There are no such problems if we use electricity data. As mentioned in Sect. 1.1.1, electricity data is accurate and available for almost all the economic

Table 1.4 The division standard of economic development stages (Chenery's model)

Development stage		Per capita income (in 1982 US dollars)
Primary products stage		260–364
		364–728
Industrialization stage	Early	728–1456
	Middle	1456–2912
	Late	2912–5460
Developed economy stage		5460–8736
		8736–13104

Data Source: Hollis B Chenery [11] and Shi Donghui [12]

Table 1.5 Electricity consumptions per capita in the economic development stages (unit: kWh/person)

Development stage		Per capita electricity consumption	Per capita residential electricity consumption
Stage of primary commodity		<300	<20
Industrialization stage	Early	300–1000	20–80
	Middle	1000–2400	80–240
	Late	2400–4500	240–810
Stage of industrialization completion		4500–5000	810–900
Developed economic stage	Early	5000–6000	900–1500
	Middle	6000–8000	1500–2400
	Late	>8000	>2400

Source: Zhaoguang Hu and Zheng Hu [5, p 232]

activities. The more electricity consumption per capita is, the more income per capita will be. How to use electricity consumption per capita to replace the income per capita in Table 1.4? It means the economic stage can be measured by per capita electricity consumption. Since the household electricity consumption can reflect the income level, the per capita residential electricity consumption is also the index in measuring economic stages. Maybe the per capita electricity consumption is higher because of the high share of heavy industry in the national economy. The per capita residential electricity consumption will be useful to know the per capita income of the country. As shown in Table 1.5, if per capita of electricity consumption is less than 300 kWh and household electricity consumption is less than 20 kWh/person annually, the economy is in the stage of primary commodity. There are three periods in the industrialization stage: early, middle, and late. The per capita electricity is around 300–4500 kWh and the per capita residential electricity is 20–810 kWh in the three periods. The finish of the industrialization stage in a country is marked by the electricity consumption in 4500–5000 kWh/p and residential electricity consumption in 810–900 kWh/p. In the developed economic stage, the two indexes can also indicate the three periods of the stage.

1.1.4 The Characteristics of Gene in an Economy

As we all know, every person has its own special gene. There are four characteristics of gene, i.e., replication, mutation, evolution, and uniqueness. What about the economic system? Is there anything like gene in an economy? We find that there are the four characteristics of gene from the production functions with electricity, such as *E-GDP* function of the economy [5]. The *E-GDP* function is a linear function as listed in Table 1.3. The slope of it is the marginal GDP of electricity consumption, which means the increase of one-unit electricity consumption will result in the increase of GDP; it is also marginal representative factor productivity of the economy. Thus, it is the replication with the constant GDP increase resulting from the addition of one unit of electricity consumption in a period. It shows that the economic growth is in the replication by the slope of the function, i.e., it is in the replication by the MRFP of the economy.

What is an economic mutation? A mutation of the economy means that there will be a big change of the marginal GDP of electricity. The intercept of the linear function can reflect the change. If the intercept is positive, the average GDP of electricity will experience a decline process when electricity consumption is increasing, and it is bigger than marginal GDP of electricity. On the other hand, if the intercept of the function is negative, the average GDP of electricity will increase when the electricity consumption is growing, and it is less than the marginal GDP of electricity. Therefore, the change of the intercept from positive to negative and from negative to positive will result in the big change of the marginal GDP of electricity, the change of average GDP trend, and also the change of relationship between average GDP of electricity and marginal GDP of electricity. Therefore, the intercept of the *E-GDP* function will determine the mutation of the economy.

With the technology innovation and technical improvement, including energy efficiency and electrification growth by using electricity to replace other primary energy, the slope of the *E-GDP* function will be changed gradually. In order to meet these changes, technology progress and innovations are required. If the impacts of them were bigger and bigger gradually, the MRFP would change. It shows the changes of interrelationships among firms, sectors, and industries of the economy to meet the outside changes of the economy. This is the evolution process in a long time. Each economy has its own *E-GDP* function. It is the uniqueness of the economy.

To make it easier to understand the concept of mutation, let us go through an ideal example. Figure 1.1 illustrates a typical case of mutation in a supposed country. It is clear that mutation occurs three times in this case. As a consequence, the *E-GDP* function is divided into four parts as the four lines in Fig. 1.1. It is the gene line of the economy.

The equation of *E-GDP* function in each part can be calculated. Formula (1.9) shows the function of the first part, a period. Evidently, the intercept is positive. Formula (1.10) lists the function of the second period. The intercept is changed from positive to negative, which means the mutation happens. The intercept of the

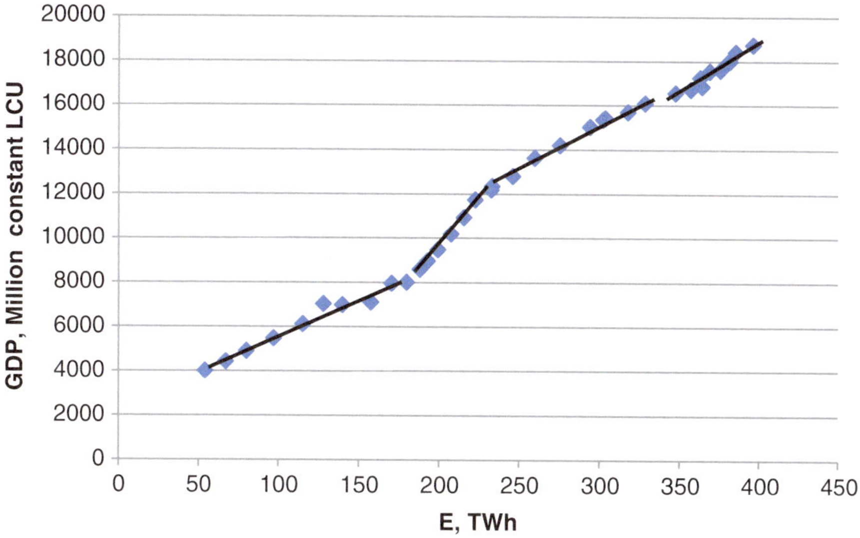

Fig. 1.1 A typical example of mutations in a supposed nation (Data source: World Bank Data [9]. http://data.worldbank.org/, and statistical material of China Electricity Council over years)

function of the third period is positive, while the intercept in the fourth period is negative, as shown in Formula (1.11) and Formula (1.12).

$$\mathrm{GDP} = 32.54E + 2327 \tag{1.9}$$

$$\mathrm{GDP} = 90.07E - 8415 \tag{1.10}$$

$$\mathrm{GDP} = 41.26E + 2749 \tag{1.11}$$

$$\mathrm{GDP} = 49.74E - 937.2 \tag{1.12}$$

Because of the mutations, the relationships between electricity consumption and GDP are different within different period, which are reflected by the *E-GDP* function. More specifically, the slope and intercept of the function represent the intrinsic characteristics of the national economy during a certain stage. If we use a straight line segment to depict the *E-GDP* function of each period in the figure, we can have a line like as shown in Fig. 1.1.

The reasons for mutation are diverse. Particularly, they have something to do with the technology level, economic structure, infrequent accident, electrification, and so on. As for the example displayed in Fig. 1.1, the first mutation and the third mutation are probably caused by the following issues: Firstly, the production technology is improved, which leads to the promotion of energy efficiency, and, secondly, the economic structure is changed, rendering less proportion for energy-intensive sectors in the national economy. In any case, less electricity is needed to gain the same GDP. Oppositely, the second mutation may be triggered by the relatively rapid development of energy-intensive sectors in the country, the speedy improvement of electrification level resulting from certain technologies and policies, and some other uncommon accidents such as the supply shortage or price increase of other kinds of energy sources, which makes more electricity be utilized. What is particularly worth mentioning is that we should distinguish between necessity and contingency. The necessity actually refers to the mutation with the slope and intercept of the *E-GDP* function being changed. For example, true mutation occurs from the tenth point from the left in Fig. 1.1, which implies that the country has entered into another period with a brand new *E-GDP* function. On the contrary, the contingency means the irregular fluctuation of one or two certain years around the *E-GDP* function for the period. For instance, the sixth point and eighth point in Fig. 1.1 stand for the contingency. The point of a single year may be higher or lower than the expected level determined by the *E-GDP* function of the period. However, it does not violate the regular pattern from the perspective of the entire stage.

Therefore, if we find that the point of 1 year departs from the *E-GDP* line, it is necessary to observe the situations in the next few years. If the subsequent points return to the original trend, it belongs to the scope of contingency. Oppositely, it is the consequence of necessity given a new trend is established by the points corresponding to the following years. This is just like the diagnosis of disease to some extent. When we find some abnormal signs of our bodies, we need to keep observing to make sure whether it is a true change of our condition or just an occasional fluctuation.

Let us have a brief view on an actual example of mutation. Figure 1.2 displays the economic mutation line of Canada during the period of 1961–2010. It is very evident that the period can be divided into two parts. The relationships between GDP and electricity consumption are quite different in the two stages.

We can build *E-GDP* function for each part. The function for the period of 1961–1992 is shown in Formula (1.13), and the function for the period of 1993–2010 is shown in Formula (1.14). It shows a mutation happened because the intercept of the first stage is positive, while that of the second stage is negative. The importance of the mutation is that it reflects the big increment of the slope of the *E-GDP* function. According to the theory mentioned previously in this section, a mutation happens, rendering the national economy with a different intrinsic property.

$$\mathrm{GDP} = 1.474E + 114.0 \tag{1.13}$$

$$\mathrm{GDP} = 4.758E - 1434 \tag{1.14}$$

It can be concluded that the economic mutation has two sides: one is the opportunity of the economy in health development resulting from technology innovation and improvement like formula (1.14) with negative intercept, and another side is the challenge of the sick economy like formula (1.13) with positive intercept. Perhaps, it can be used to act as the diagnostic tool of the economy to see if it is healthy or sick. Every country has its own gene line of the national economic mutation.

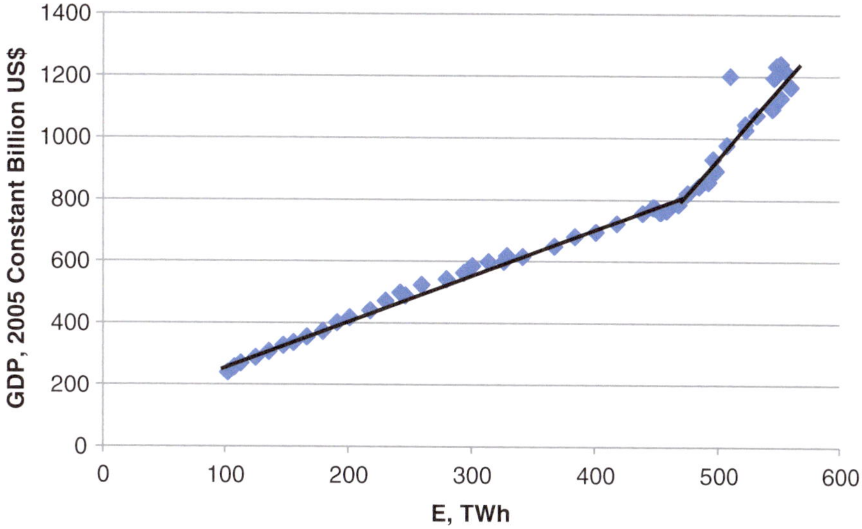

Fig. 1.2 Economic mutation line of Canada in the period 1961–2010. (Data source: World Bank Data (2014). http://data.worldbank.org/, and statistical material of China Electricity Council over years [9])

In summary, the economy is a very complicated system. Economists research and describe it from many aspects to depict the whole picture of it. The electricity economics discussed in this section is to observe and study the economic system from the perspective of electricity consumption, which is believed to be accurate and timely. Although it has different theoretical foundations from traditional economics, it can be regarded as a new approach to enrich our researches. In this book, some economic issues will be expressed in terms of the electricity consumption according to the principle of the electricity economics.

1.2 Review of the Economic Development in China by Electricity Economics

The People's Republic of China, located in Asia, is one of the biggest countries in the world with 1.36 billion people and a land area of around 9.6 million square kilometers. China has the largest population in the world. Besides, China is No. 1 in terms of electricity consumption and is No. 2 according to GDP, following the USA. China's economic growth makes great contributions to the global economy. The economic activities of China such as the import and export in the international market will affect economic growth and employment of some countries, and also international market prices of some goods. China's economic growth will contribute greatly in the global economy. Therefore, a lot of people in the world pay much attention to China's economic activities.

1.2.1 Economic Activities of Three Industries

China has undergone dramatic policy reforms on transforming from planned economy towards the market economy since 1978, which brings about rapid economic growth and electricity consumption growth. Chinese mode has achieved an extraordinary success and attracted wide attention globally [8].

In the three industries in China, the secondary industry takes the first place according to the amount of electricity consumption. The number was 218.2 billion kWh in 1978. In the early industrialization stage, it was 526.12 billion kWh in 1991 and 1064.64 billion kWh in 2001. In the middle industrialization stage, it was 2147.42 billion kWh in 2006. In the late industrialization stage, it was 3145 billion kWh in 2010 and reached 4065 billion kWh in 2014 (Fig. 1.3). The average annual growth rate was 8.5 % in 36 years, while it was 10.6 % during 1999–2014.

As for the tertiary industry, electricity consumption was 12.29 billion kWh in 1978; it was 100.87 billion kWh in 1996, 210.88 billion kWh in 2003, 318.46 billion kWh in 2007, 447.9 billion kWh in 2010, and 666 billion kWh in 2014. The average growth rate was 11.89 % during 1978–2014 annually. It was 11.36 % in the period of 1999–2014. The electricity consumption growth of tertiary industry is the highest in the three industries.

The electricity consumption of the primary industry is quite lower compared with secondary and tertiary industries. It was 10.917 billion kWh in 1978, 101.5 billion kWh in 2011, and 99.4 billion kWh in 2014. The average annual growth rate was 6.45 % in 36 years. The electricity consumptions of primary industry, secondary industry, and tertiary industry are shown in Fig. 1.3.

From the electricity consumptions of the three industries, we can get the industrial production outputs such as value added and new job positions by electricity economics. Since the share of value added of primary industry in GDP is very lower, we only introduce production functions with electricity of the secondary and tertiary industries in this chapter.

The production function with electricity for secondary industry E_2–V_2 in China is shown as follows [5]:

$$V_2(E) = 6.09E_2 - 21.81 \tag{1.15}$$

It is clear that the marginal value added of electricity for secondary industry MV_2 is 6.09CNY/kWh. The V_2 was 1307.156 billion CNY (Chinese Yuan) in constant 2010 CNY in 1978, 11389.58 billion CNY in 2005, 21470.67 billion CNY, and 24736.3 billion CNY in 2014 (Fig. 1.4). The average annual growth rate was 8.59 % in 36 years. However, the yearly growth rate was quite different as shown in Fig. 1.4. The highest growth rate was 16.72 % in 2004, and the lowest one was

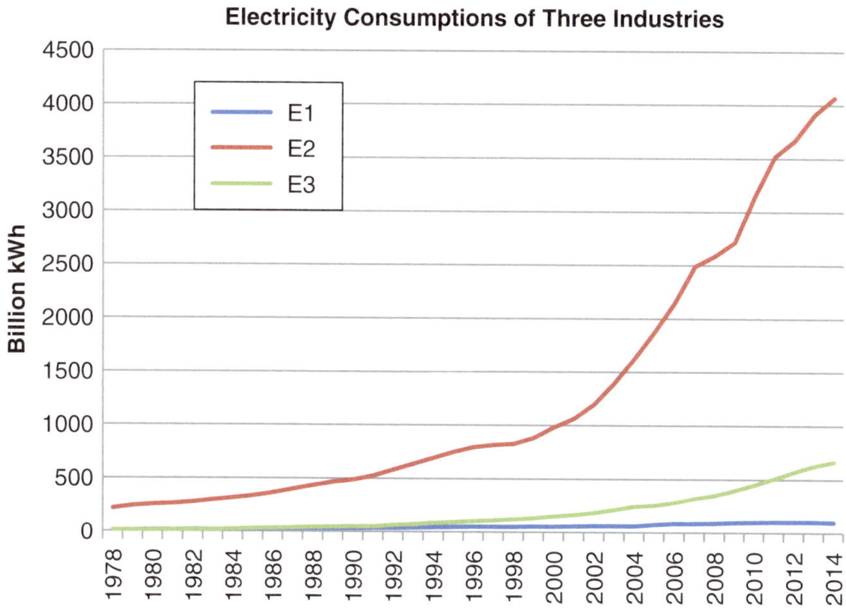

Fig. 1.3 Electricity consumptions of the three industries in China (Data source: statistical material of China Electricity Council over years)

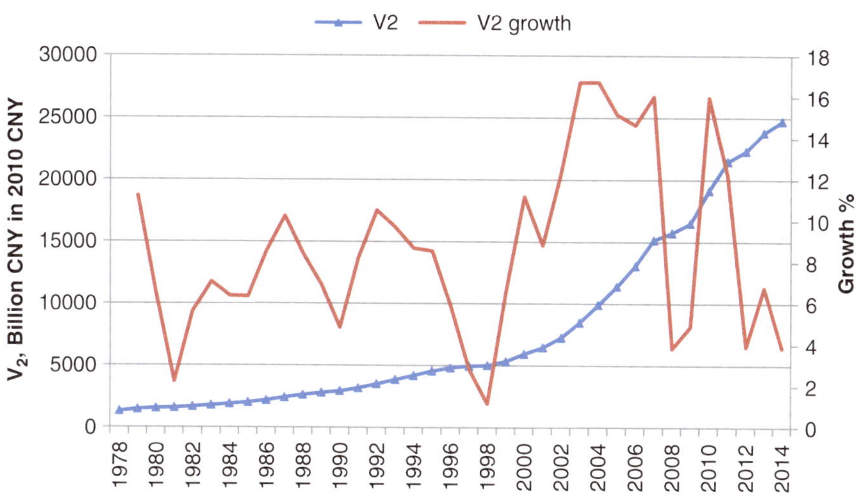

Fig. 1.4 Value added and the growth of secondary industry in China (Data source: National Bureau of Statistics of China, 2014 China Statistical Yearbook [10])

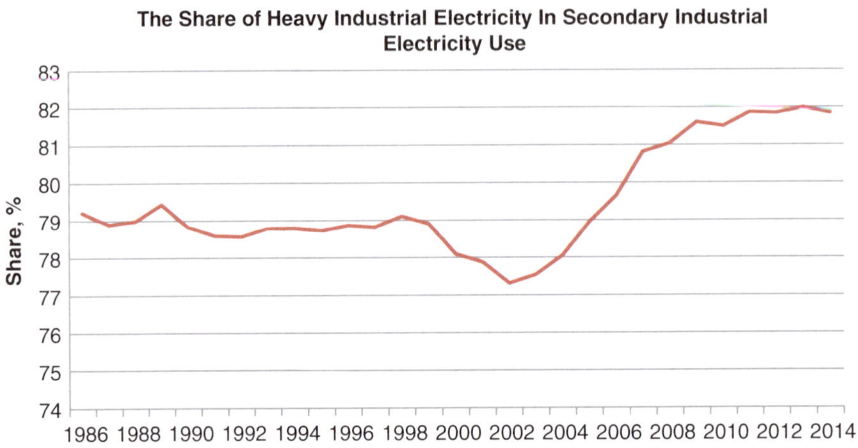

Fig. 1.5 Share of heavy industrial electricity consumption in secondary industrial electricity consumption (Data source: statistical material of China Electricity Council over years)

1.16 % in 1998, the year of Asian crisis. It was 2.22 % in 1981. During the period of global financial crisis, it was 3.84 % in 2008 and 4.88 % in 2009. In 2010 and 2011, the growth rate was 15.97 % and 12.22 %, respectively. The high growth after 2009 was driven by great investment of Chinese government with the aim of restoring the economic growth. However, it decreased to 3.92 % in 2012, which verifies that the effect of the investment is very limited in the economic recovery in China.

The share of heavy industrial electricity consumption in the secondary electricity consumption decreased from 79.2 % in 1986 to 77.3 % in 2002. However, it increased to 81.85 % in 2014 due to the huge investment to the secondary industry as shown in Fig. 1.5.

There is a strong positive relationship between investment and heavy industrial electricity consumption as shown in Fig. 1.6. The slope of the linear function is 1.7558, which means the increase of one USD (in constant 2000$) in investment will result in the increase of 1.7558 kWh of heavy industrial electricity consumption in China, i.e., the marginal heavy industrial electricity consumption of investment (*MHIEI*) is 1.7558 kWh/$. And the intercept of the function is positive, which means average heavy industrial electricity consumption of investment (*AHIEI*) is larger than the marginal heavy industrial electricity consumption of investment. However, *AHIEI* will be lower and lower with the growing gross fixed capital formation. It was 3.2393 kWh/$ in 1986, declined to 2.1229 kWh/$ in 2000, and then decreased to 1.84 kWh/$ in 2011 as shown in Fig. 1.7. Elasticity, the growth rate of investment over the growth rate of heavy industrial electricity consumption, has the same trend as *AHIEI*, from 1.8449 to 1.048 during the period of 1986–2011 as shown in Fig. 1.7.

On the other hand, let us see the growth rates of the investment and electricity consumption of heavy industry. Chinese government intended to promote the economic growth by virtue of investment. As shown in Fig. 1.8, the gross fixed

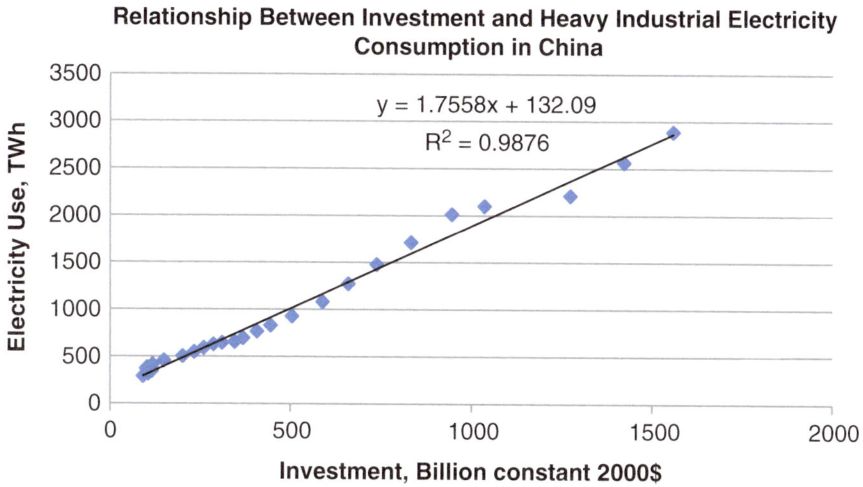

Fig. 1.6 Relationship between investment and heavy industrial electricity consumption in China (Data source: World Bank Data [9]. http://data.worldbank.org/, and statistical material of China Electricity Council over years)

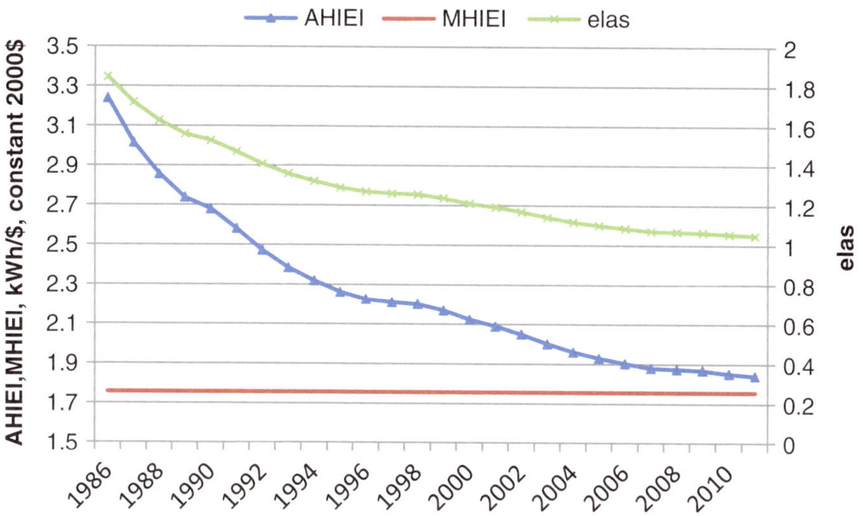

Fig. 1.7 Average heavy industrial electricity consumption of the investment, marginal heavy industrial electricity consumption of investment, and the elasticity in China (Data source: World Bank Data [9]. http://data.worldbank.org/, and statistical material of China Electricity Council over years)

capital formation growth rate in 1991–1993 was 15.1 %, 26.12 %, and 34.61 %, respectively, which reveals the investment was huge. Accordingly, heavy industrial electricity consumption level in 1991–1993 was relatively high. When the Asian crisis appeared, the growth rate of heavy industrial electricity consumption was

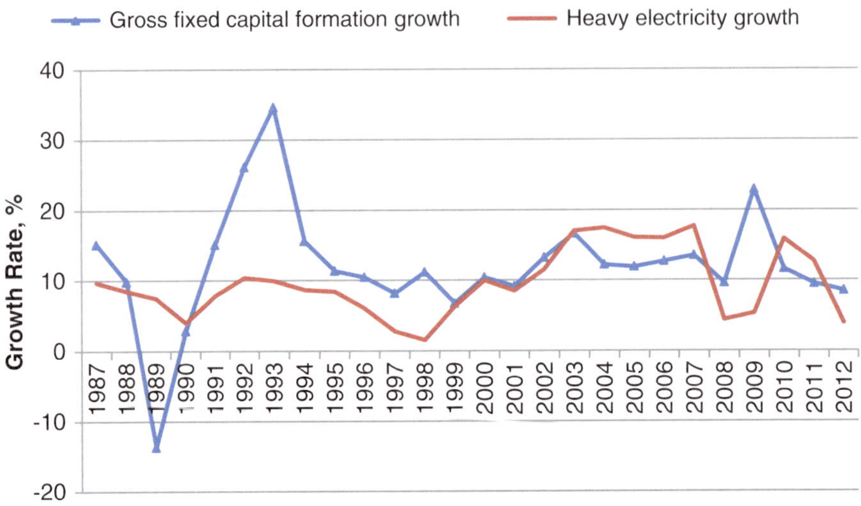

Fig. 1.8 Gross fixed capital formation growth rate and heavy industrial electricity consumption growth rate (Data source: World Bank Data [9]. http://data.worldbank.org/, and statistical material of China Electricity Council over years)

only 1.53 % in 1998, the lowest growth rate in the decades. Meanwhile, the growth rate of gross fixed capital formation was very low, even 8.16 % in 1997. During the period of 2002–2007, the growth rate of heavy industrial electricity consumption was in two digits, due to the high gross fixed capita growth rate, i.e., 13.21 % in 2002, 16.66 % in 2003, 12.19 % in 2004, 11.9 % in 2005, 12.7 % in 2006, and 13.5 in 2007 [12]. The growth rate of heavy industrial electricity consumption was higher than the growth rate of gross fixed capital formation during 2003–2007. The main reason was that China's export grew fast after joining WTO in 2002. China was called as world factory because of the considerably low labor cost and energy cost in the period. Growth rate of heavy industrial electricity consumption was also high in 2010. It reached 15.88 % as a result of the huge investment growth rate of 22.9 % to push economic growth decided by Chinese government in 2009. Therefore, we can conclude that gross fixed capital formation is the main reason to drive heavy industrial electricity consumption, and export and other factors will affect the heavy industrial electricity consumption weakly.

The products of light industry can be consumed directly or exported to other countries. Therefore, the final consumption expenditure and export of goods will have positive relationship with light industrial electricity consumption. It is a linear function between them as illustrated in Fig. 1.9. The slope of the function is 0.1803 and the intercept is 31.402. The slope is marginal light industrial electricity use of the final consumption and export (*MLIEFCE*), which shows an increase of 1$ (constant 2000$) in consumption, and export will result in an increase of 0.1803 kWh in light industrial electricity use in China. The positive intercept means the average light industrial electricity consumption of the final consumption

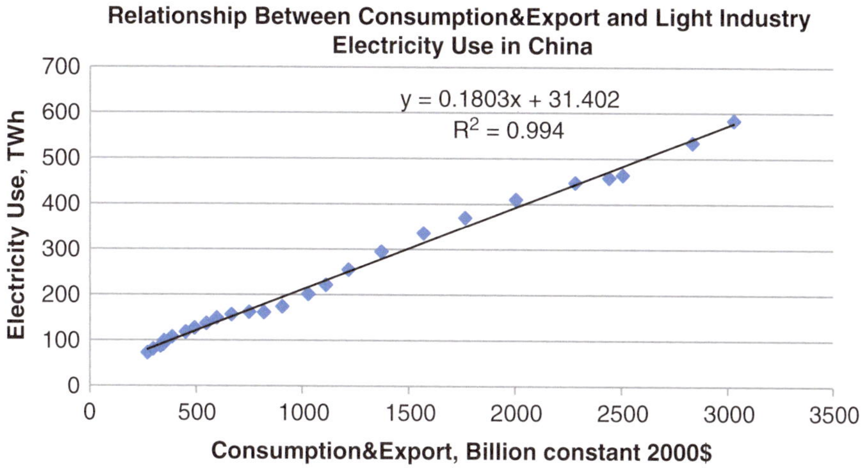

Fig. 1.9 Relationship between consumption and export and light industry electricity consumption in China (Data source: World Bank Data (2014). http://data.worldbank.org/, and statistical material of China Electricity Council over years)

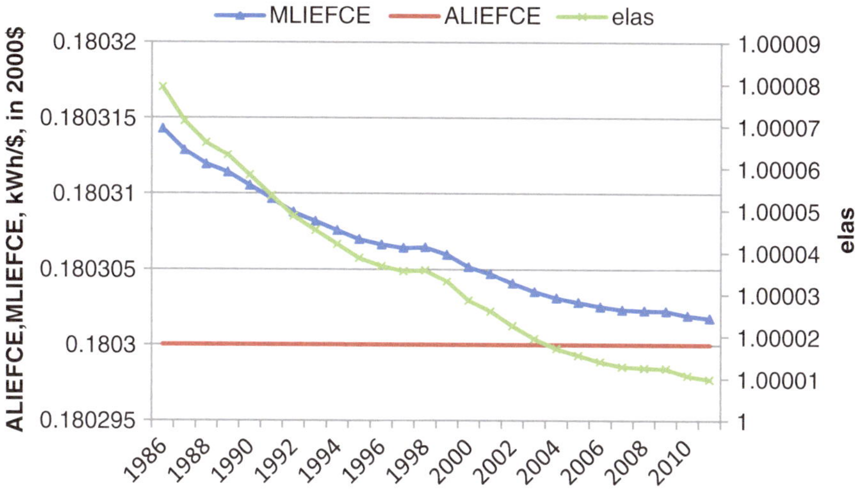

Fig. 1.10 Average light industrial electricity consumption of final consumption and export, marginal light industrial electricity consumption of final consumption and export, and elasticity in China (Data source: World Bank Data (2014). http://data.worldbank.org/, and statistical material of China Electricity Council over years)

and export (*ALIEFCE*) will fall with the growing final consumption expenditure and export. Figure 1.10 shows that *ALIEFCE* was 0.1803128 kWh/$ in 1986 and 0.1803017 kWh/$ in 2011, and the elasticity decreased in the same way from 1.00008 to 1.00001.

Based on the classification of industrial sectors in China, there are 20 sectors that belong to manufacturing as follows:

1. Food, beverages, and tobacco sector (light)
2. Textile sector (light)
3. Garments, leather, and down products sector (light)
4. Wood processing and furniture manufacturing
5. Paper and paper products sector (light)
6. Printing and media reproduction (light)
7. Sports goods manufacturing (light)
8. Petroleum processing, coking, and nuclear fuel processing sector
9. Chemical materials and chemical products manufacturing
Include:
Chlor-alkali
Calcium carbide
Yellow phosphorus
Fertilizer manufacturing
10. Pharmaceutical manufacturing (light)
11. Chemical fiber manufacturing (light)
12. Rubber and plastic products sector
13. Nonmetallic mineral products sector
Include:
Cement manufacturing
14. Ferrous metals smelting and rolling processing sector
Include:
Ferroalloy smelting
15. Nonferrous metals smelting and rolling processing sector
Include:
Aluminum smelting
16. Fabricated metal products
17. General and special equipment manufacturing
18. Transportation, electrical, electronic equipment manufacturing
Include:
Transportation equipment manufacturing
19. Artwork and other manufacturing (light)
20. Exhaust gas resources and materials recycling and processing

Electricity consumption of manufacturing was 2851.86 TWh in 2013. It accounts for around 70 % of the secondary electricity consumption in China, and thus, it can reflect the production growth of the secondary industry. It accounts for 88.9 % of the heavy industrial electricity consumption, and it has the features of heavy industry such as it has strong positive relationship with investment. As shown in Fig. 1.11, it grew fast in the past few decades, especially during the period of 2000–2007 and 2010–2011. The low growth rates indicate some economic recessions. It was 4.4 % in 1990 because of the political storm that happened in 1989 in

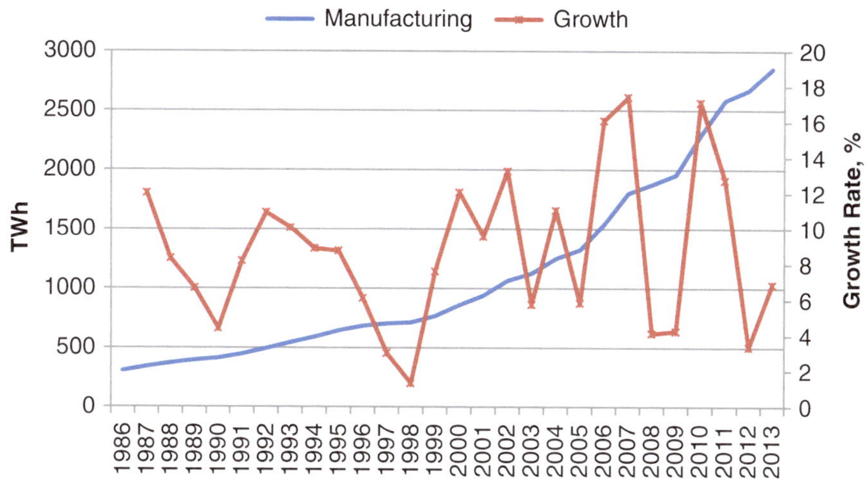

Fig. 1.11 Electricity consumption and growth rate of manufacturing in China (Data source: statistical material of China Electricity Council over years)

China. It was 1.31 % in 1998 due to the Asian crisis. It was 4.12 % and 4.25 % in 2008 and 2009, respectively, since global crisis happened. It was 3.35 % in 2012 because the effect of the huge investment in 2009 is very weak in stimulating and restoring the economy.

Electricity consumption in construction sector also has the feature of positive relationship with investment. The more investment, the more building, highway, railway, airport, and other construction projects will be constructed. Then, electricity consumption of construction sector will grow fast. Figure 1.12 shows that investment growth is highly correlated with electricity consumption growth of construction in China.

Since the share of heavy industrial electricity consumption is quite high in E_2, it is also meaningful to analyze the relationship between the investment and E_2. The growth rates of investment and E_2 are depicted in Fig. 1.13. We can see that the investment played an important role in the E_2 growth from 1979 to 2013. The high investment would result in the high E_2 growth and V_2 growth. Similarly, low investment growth would result in low E_2 and V_2 growth.

We can also observe the employment by electricity consumption in the secondary industry. It has been studied that the labor positions are related with electricity consumption in the secondary industry in China as follows [5]:

$$L_2 = 0.0307E_2 + 122.2 \tag{1.16}$$

The interception of formula (1.16) is 0.0307, which means the increase of one billion kWh in electricity consumption of secondary industry will lead to the increase of 30,700 in labor positions in the secondary industry. It is also called as the marginal electricity employee for secondary industry in China.

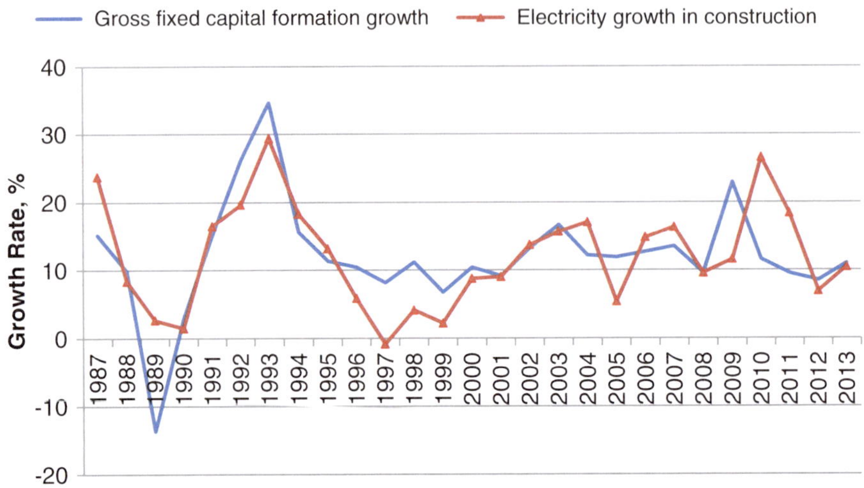

Fig. 1.12 Investment growth and electricity consumption growth of construction in China (Data source: World Bank Data (2014). http://data.worldbank.org/, and statistical material of China Electricity Council over years)

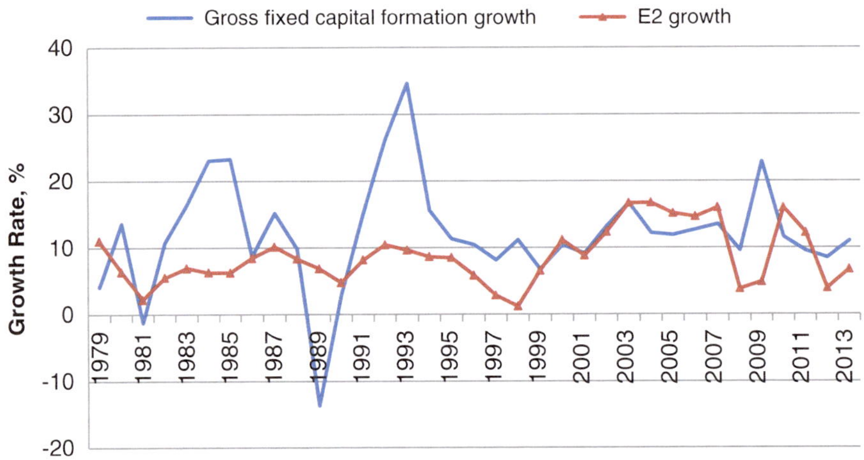

Fig. 1.13 Growths of investment and secondary industrial electricity use (Data source: World Bank Data (2014). http://data.worldbank.org/, and statistical material of China Electricity Council over years)

What about the tertiary industry? Its production function with electricity in China can be depicted as E_3–V_3 [5]:

$$V_3(E) = 39.48E_3 - 186.1 \tag{1.17}$$

Evidently, the marginal value added of electricity consumption of tertiary industry MV_3 is 39.48 CNY/kWh, which is much higher than that of the secondary industry.

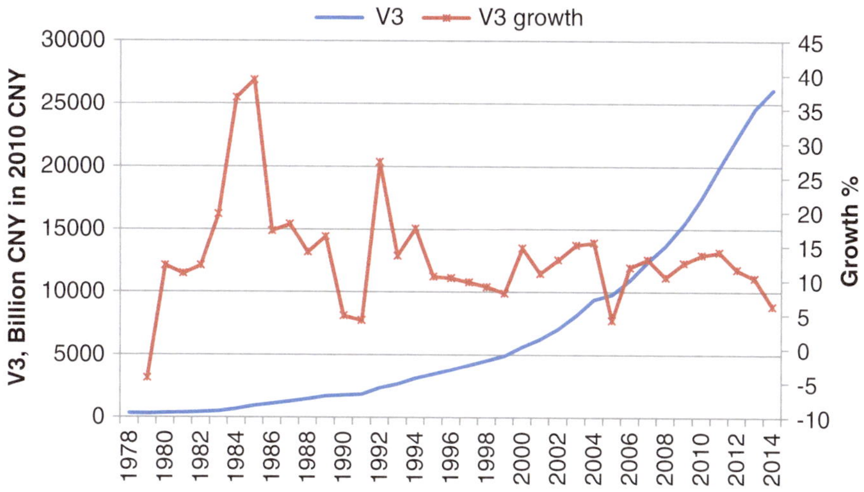

Fig. 1.14 Value added and the growth of tertiary industry in China (Data source: National Bureau of Statistics of China, 2014 China Statistical Yearbook [10])

The value added of tertiary industry was 298.92 billion CNY in 1978 and 26,107.5 billion CNY in 2014 as shown in Fig. 1.14. The average annual growth rate was 13.48 % from 1978 to 2014. The highest growth rate was 39.29 % in 1985 and the second highest was 27.28 % in 1992. However, the growth rate still reached 9.07 and 10.45 % in the Asian crisis in 1998 and global crisis in 2008, respectively. It means the financial crisis would affect the value added of tertiary industry weakly in the middle stage of industrialization for China.

Usually, the electricity consumption in tertiary industry is driven by the final consumption. The growth curves of the final consumption and tertiary industrial electricity consumption have the same trends as illustrated in Fig. 1.15. It reveals that if the growth rate of the final consumption rises, the growth rate of E_3 rises too, and vice versa. Thus, the electricity consumption of tertiary industry can be used as a meter to measure the final consumption of the economy.

The growth of electricity consumption in shopping center and hotel and restaurant is sensitive to economic growth. The electricity growth rate in commercial, accommodation, and catering services was more than 20 % during 1992–1994, around 12 % during the period of Asian crisis, and 8.82 % in global crisis in 2008 as shown in Fig. 1.16. However, the growth rate is declining with the increasing electricity consumption in the sector. The amount of electricity consumption in commercial, accommodation, and catering services reached 187.69 TWh in 2013 from 7.7 TWh in 1990 as shown in Fig. 1.16, increasing 24.38 times in 23 years. It reflects the business of commercial, accommodation, and catering services has grown fast in China, and there is still a big growth space in the future since the share of electricity consumption in the service sectors is still relatively low compared with some other countries.

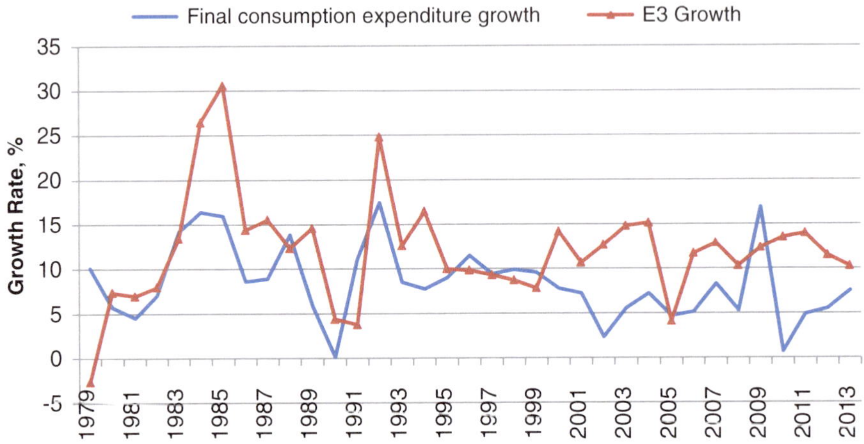

Fig. 1.15 Growths of final consumption and tertiary electricity use in China (Data source: statistical material of China Electricity Council over years)

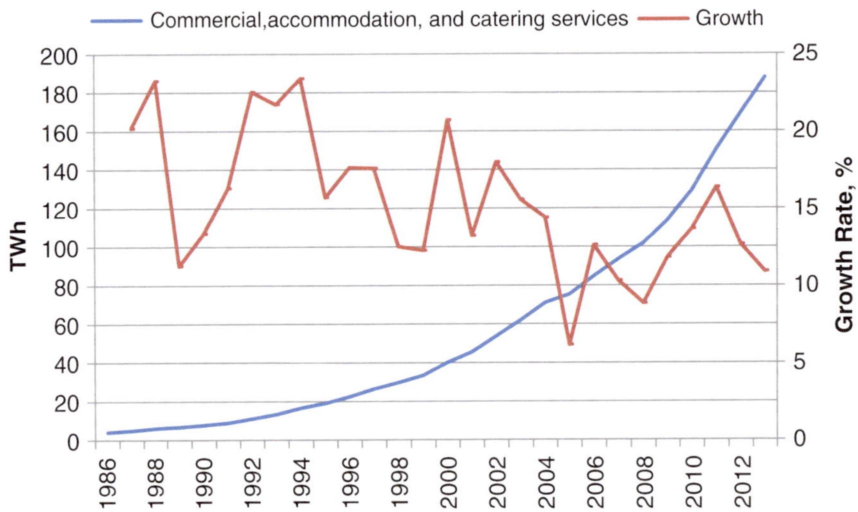

Fig. 1.16 Electricity use and growth rate for shop and hotel and restaurant in China (Data source: statistical material of China Electricity Council over years)

In tertiary industry, transportation, warehousing, and postal services used to consume more electricity than commercial, accommodation, and catering services before the middle of the 1990s. However, the electricity consumption of transportation, warehousing, and postal services was 18.23 TWh, while the electricity consumption of commercial, accommodation, and catering services was 19.06 TWh in 1995, and the two indicators were 91.42 TWh and 187.69 TWh, respectively, in 2013 as shown in Fig. 1.17. In terms of the average annual growth

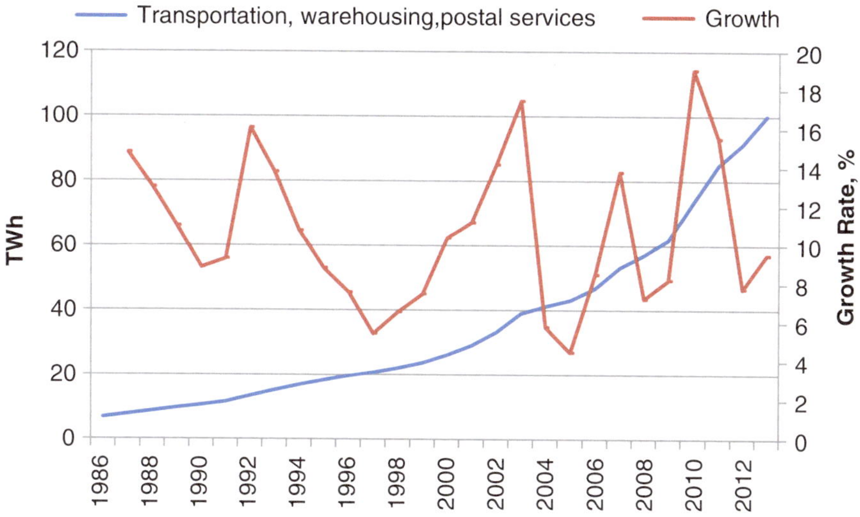

Fig. 1.17 Electricity use and growth rate for transportation, warehousing, and postal services in China (Data source: statistical material of China Electricity Council over years)

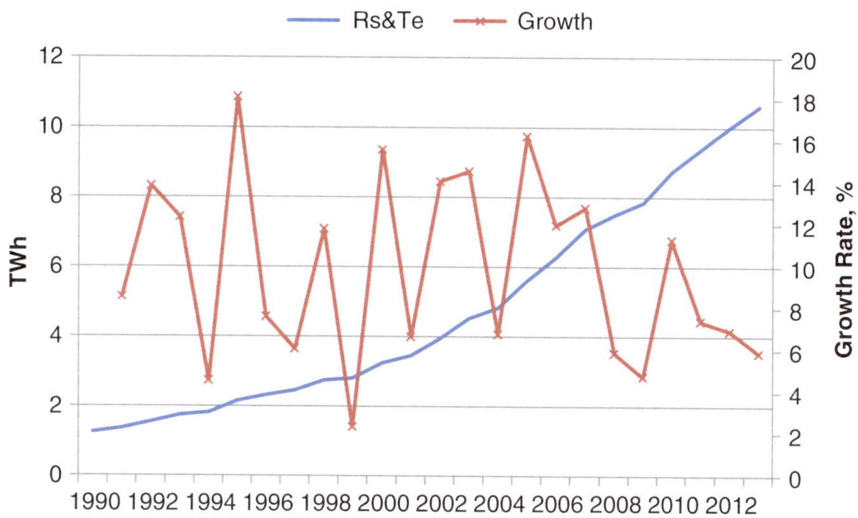

Fig. 1.18 Electricity use and growth rate for research and technical service in China (Data source: statistical material of China Electricity Council over years)

rate of electricity consumption, it was 10.54 % as for transportation, warehousing, and postal services and 15.19 % as for commercial, accommodation, and catering services during 1986–2013, which reflects the growth of the final consumption in China's economy.

Electricity consumption of research and technical service (Rs&Te) grew from 1.25 TWh in 1990 to 10.61 TWh in 2013 as exhibited in Fig. 1.18. The average

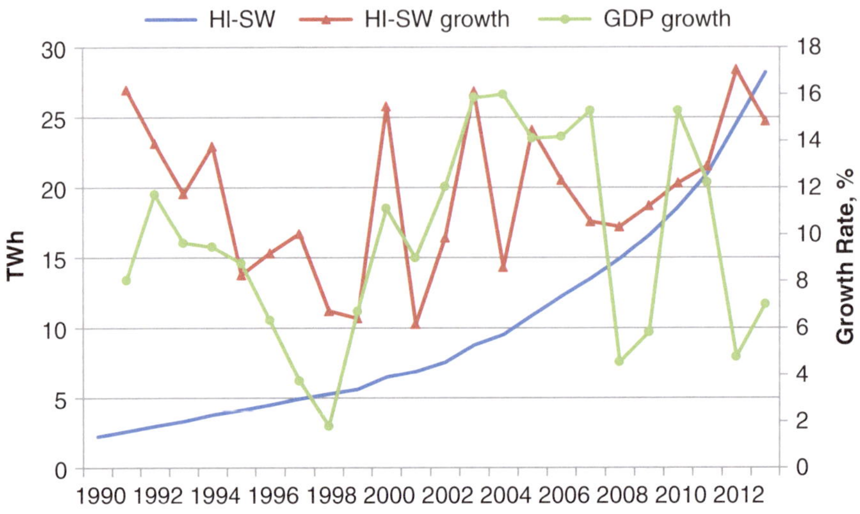

Fig. 1.19 Electricity use and growth rate for health and social welfare service and GDP growth rate (Data source: statistical material of China Electricity Council over years)

annual growth rate was 9.74 % in 23 years. However, it was 14.9 % as for commercial, accommodation, and catering services in the same period. It reveals that the research and technical improvement needs to be further stimulated now. Otherwise, it is very difficult for the sustainable development of the economy because it will have great impacts on the progress of both secondary industry and tertiary industry. This is also a caution for China's economy.

Electricity consumption of health and social welfare service (HI-SW) can reflect the business progress as shown in Fig. 1.19. It was 2.26 TWh in 1990 and reached 24.57 TWh in 2012, with the average annual growth rate equaling 11.46 % in 22 years. Since the electricity consumption in health service such as hospital occupy a large share in the health and social welfare service, the high electricity growth rate means hospital electricity consumption grows fast, i.e., more and more people go to see doctor in the period. Evidently, the level of people's health is decreasing. The growth rate 11.46 % is higher than that of the research and technical service, 9.92 %, which gives us another caution that the growth is unreasonable for the economy. On the other hand, the growth rate of the whole industrial electricity consumption was only 9.7 % and GDP was 9.75 % from 1990 to 2012, while it was more than 11.46 % for hospital electricity growth. Figure 1.19 also reveals that there has been a larger difference between economic growth rate and HI-SW electricity growth rate since 2008. We can get the trend that the HI-SW electricity growth rate will be larger than GDP growth rate. There is a question: Why do we have to develop our economy? For higher income? For better health?

The tertiary industry can provide more job positions for each unit of electricity consumption. The relationship between the electricity consumption and employment is shown as follows [5]:

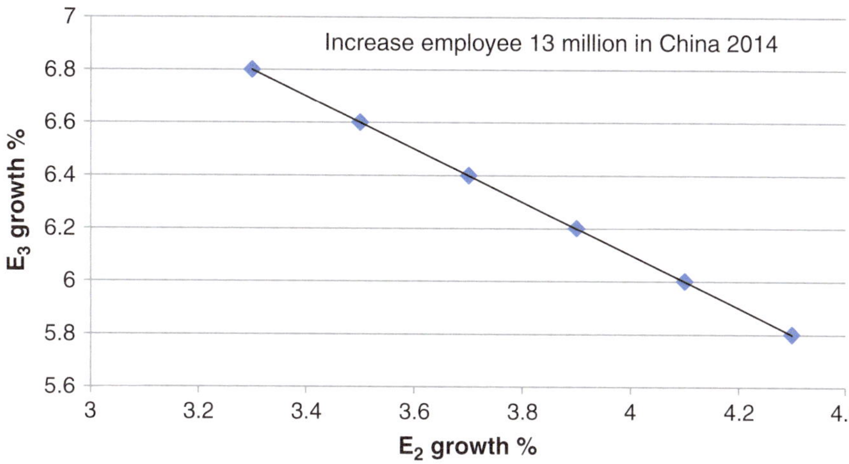

Fig. 1.20 Alternative of E_2 and E_3 growths for 13 million employees in 2014

$$L_3 = 0.2008E_3 + 176.96 \tag{1.18}$$

It is clear that the marginal electricity employee for tertiary industry is 0.2008, which is 6.54 times greater than that of the secondary industry 0.0307. Therefore, the final consumption would provide more labor positions. The increase of new labor is always the key issue of the government. There are 7.27 million undergraduate and graduate students looking for jobs in 2014 in China. The government worries about the employment for the students since the lower electricity growth rate in the first half of 2014. It has been calculated that 13 million job positions can be provided with few alternatives of the electricity growth in secondary and tertiary industries. It is shown in Fig. 1.20 that it will create positions for 13 million employees if E_2 increases by 3.3 % and E_3 increases by 6.8 %, if E_2 increases by 3.5 % and E_3 increases by 6.6 %, if E_2 increases by 3.7 % and E_3 increases by 6.4 %, or if E_2 increases by 3.9 % and E_3 increases by 6.2 %. Based on the published data, the growth rate of E_2 and E_3 was 3.7 % and 6.4 % (YoY) in 2014; the total new jobs were 13.05 million in which secondary industry provided 4.58 million and tertiary industry 8.47 million. New jobs accumulated each month in 2014 are shown in Fig. 1.21; we can see that tertiary industry has created great job positions in China.

1.2.2 National Economic Development

As we know that the total electricity consumption *TE* consists of the whole industrial electricity consumption *E* and residential electricity consumption *RE*. The residential electricity consumption has nothing to do with economic activities, and it would not produce value added calculated in GDP. However, it can show the

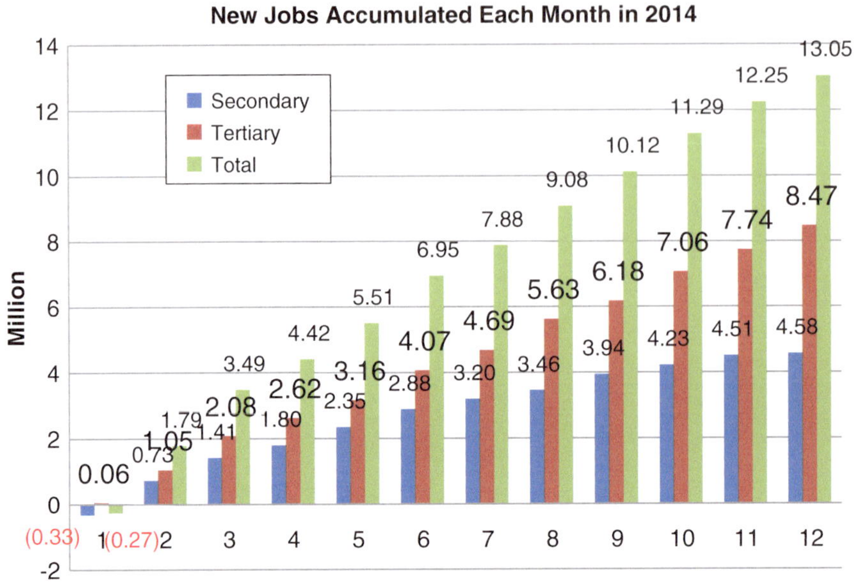

Fig. 1.21 New job positions accumulated each month in China in 2014

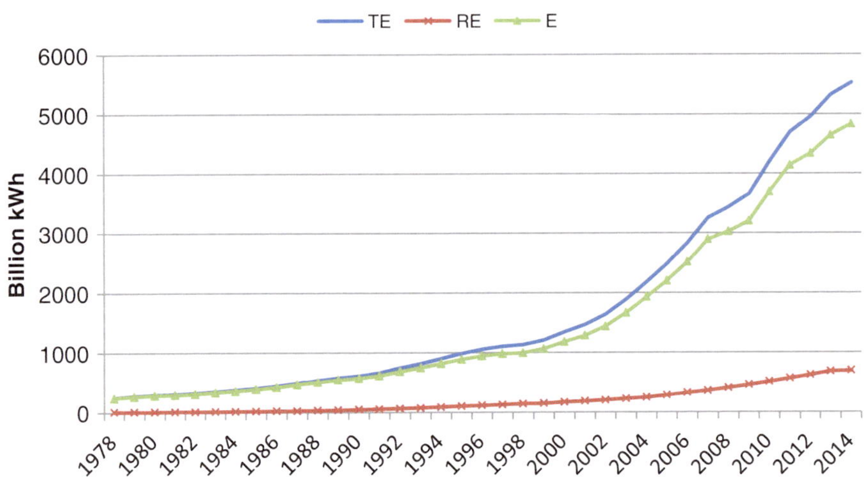

Fig. 1.22 Total residential electricity consumption and the whole industrial electricity consumption in China (Data source: statistical material of China Electricity Council over years)

people's income. China's economic development in the period of 1978–2014 can be revealed from the significant increase in the electricity consumption as shown in Fig. 1.22. The total electricity consumption in 1978 was 249.8 billion kWh, and it was 5523.3 billion kWh in 2014. The average growth rate of the total electricity consumption was 9.04 % in 36 years annually. The whole industrial electricity consumption increased from 241.4 billion kWh in 1978 to 4830.5 billion kWh in 2014, with the average growth rate equaling 8.74 % annually. As for household electricity consumption, it was 8.41 billion kWh in 1978 and 692.8 billion kWh in 2014 with average growth rate of 13.12 % annually.

TE was higher than 1000 billion kWh in 1996, when *TE*, *RE*, and *E* were 1057 billion kWh, 113.3 billion kWh, and 943.72 billion kWh, respectively. The average annual growth rate of *TE*, *RE*, and *E* was 8.34 %, 15.55 %, and 7.87 %, respectively, in 18 years. The growth rate of residential electricity consumption was much higher than the growth rate of the whole industrial electricity consumption. It means that the growth rate of residential income was higher than the growth rate of industrial production. This is the meaningful feature in the economic growth in China.

TE got higher than 2000 billion kWh in 2004, where *TE*, *RE*, and *E* were 2181.5 billion kWh, 245.3 billion kWh, and 1930.53 billion kWh, respectively. The average annual growth rate of *TE*, *RE*, and *E* was 9.48 %, 10.14 %, and 9.36 %, respectively, in the period of 1996–2004. It is also in line with the feature that residential growth rate was higher than industrial growth rate.

TE increased to 3000 billion kWh in 2007, where *TE*, *RE*, and *E* were 3256.53 billion kWh, 360.83 billion kWh, and 2895.7 billion kWh, respectively. The average annual growth rate of *TE*, *RE*, and *E* was 14.29 %, 13.73 %, and 14.47 %, respectively, in the period of 2004–2007.

As the largest electricity consumer in the world in 2010, China's total electricity consumption exceeded 4000 billion kWh. *TE*, *RE*, and *E* were 4199.9 billion kWh, 509.4 billion kWh, and 3690.5 billion kWh, respectively. The average annual growth rate of *TE*, *RE*, and *E* was 8.85 %, 12.18 %, and 8.42 %, respectively, in 2007–2010.

TE got higher than 5500 billion kWh in 2014, when *TE*, *RE*, and *E* were 5523.3 billion kWh, 692.8 billion kWh, and 4830.4 billion kWh, respectively. The average annual growth rate of *TE*, *RE*, and *E* was 7.09 %, 7.99 %, and 6.96 %, respectively, in 2010–2014.

It is clear that the more these three industries consume electricity, the more value added or GDP they will produce. What is the relationship between the industrial electricity consumption and *GDP* in China? China's *E-GDP* function is shown as follows [17]:

$$GDP(E) = 10.989E - 74.162 \tag{1.19}$$

It can be calculated from formula (1.17) that the GDP was 2578.65 billion CNY in 1978 as shown in Fig. 1.23 and it was 53008.20 billion CNY (about 8414 billion USD, exchange rate as 1USD = 6.3CNY in 2010, as shown in Fig. 1.24) in 2014. The annual growth rate of GDP was 8.76 %, and the annual growth rate of the whole industrial electricity consumption *E* was 8.68 % in 36 years.

The growth rate of the industrial electricity consumption *E* is shown in Fig. 1.25. The elasticity of electricity in China was 0.99 in the period of 1978–2014.

The slope of formula (1.17) is 10.989 CNY/kWh in China which means that the increase of one kWh in the whole industrial electricity consumption in China would

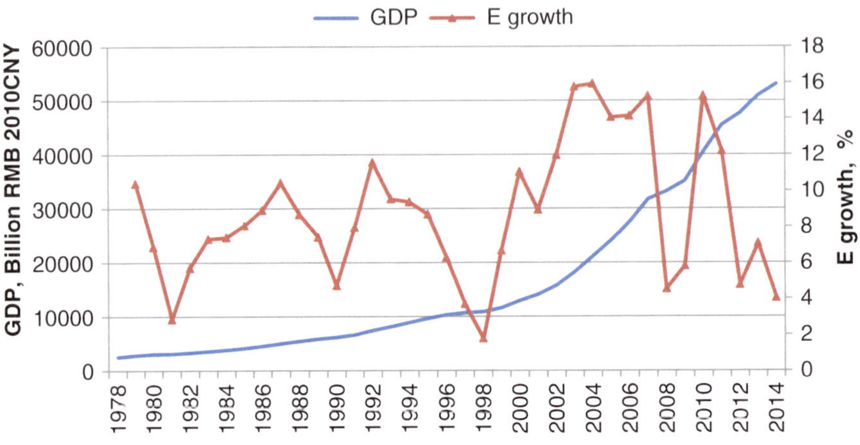

Fig. 1.23 China's GDP and *E* growth rate (Data source: National Bureau of Statistics of China, 2014 China Statistical Yearbook [10])

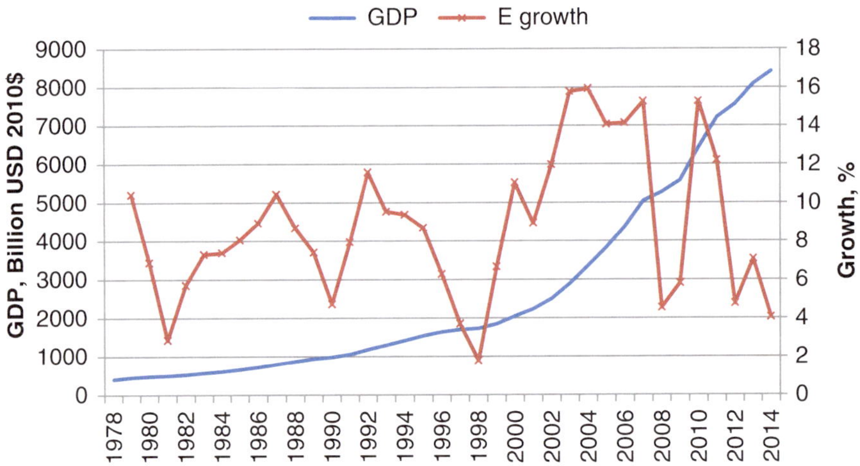

Fig. 1.24 China's GDP in USD and *E* growth rate (Data source: National Bureau of Statistics of China, 2014 China Statistical Yearbook [10])

produce 10.989 CNY, about 1.74USD (1USD = 6.3CNY in 2010), in GDP. Since the marginal value added of electricity is quite different for the three industries and GDP as shown in formulas (1.15), (1.16), and (1.17), their growth rates are quite different too. Based on formulas (1.15), (1.16), and (1.17), Fig. 1.26 depicts the growth of GDP, V_1, V_2, and V_3 with index in the period of 1978–2014. The GDP, V_1, V_2, and V_3 in 1978 are assumed to be 100. The highest growth was the value added of tertiary industry that is 8733.89 in 2014; the next one was GDP that is 2055.66 in 2014; then it comes that V_2 is 1892.38 in 2014; and the lowest one was primary industry that is 218.77. It shows that the V_3 has increased by 87.34 times, GDP has

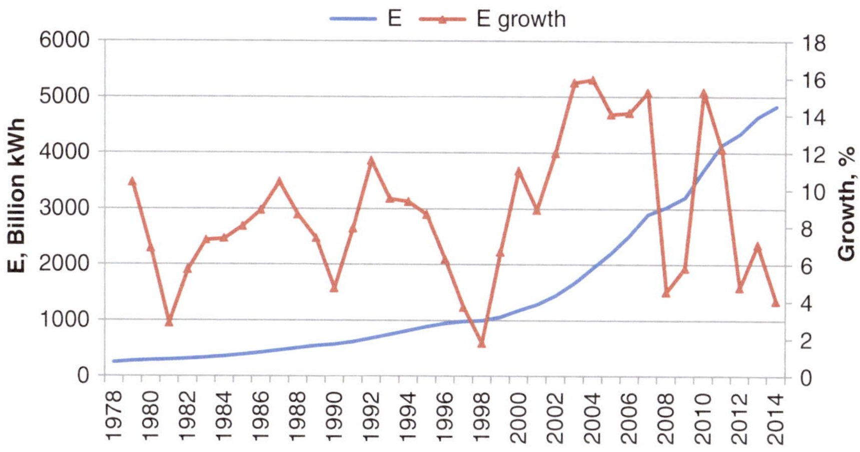

Fig. 1.25 Whole industrial electricity consumption and its growth rate in China 1978–2014 (Data source: statistical material of China Electricity Council over years)

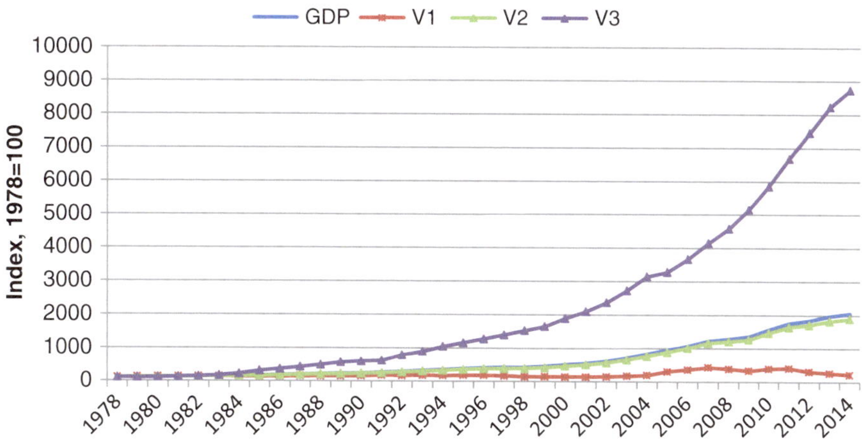

Fig. 1.26 Index of GDP and three industries in China 1978–2014 (Data source: National Bureau of Statistics of China, 2014 China Statistical Yearbook [10])

increased by 20.56 times, V_2 has increased by 18.92 times, and V_1 has increased by 2.19 times in 36 years.

As shown in Fig. 1.27, the share of primary industry was decreasing sharply, the share of secondary industry was declining slowly, while the share of tertiary industry was increasing greatly during 1978–2014 in China. However, the share of primary increased by over 10 % in 2005–2008. The main reason is that the products of primary industry are exported to other countries due to the competitive prices in the global market. With the gradual growth of tertiary industry, its share reached 26.37 % in 1988, which means it was higher than that of primary industry,

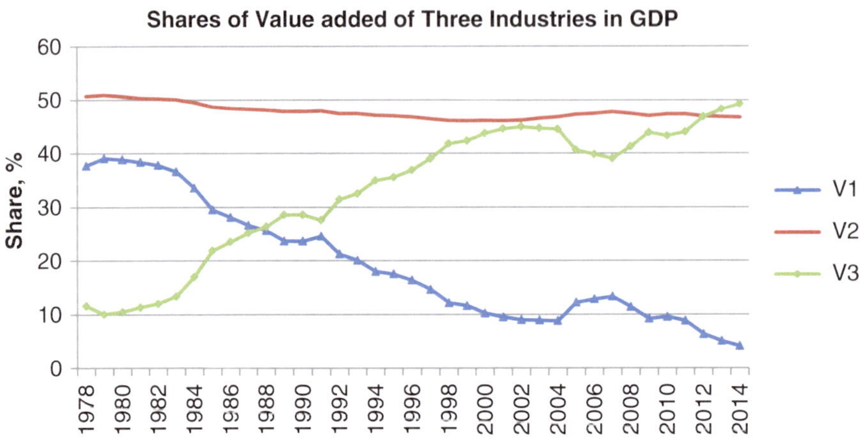

Fig. 1.27 Shares of V_1, V_2, and V_3 in GDP in China (Data source: National Bureau of Statistics of China, 2014 China Statistical Yearbook [10])

25.57 %. It reached 49.25 % in 2014, which means it exceeds the share of secondary industry, 46.66 %.

On the other hand, what is mixture of GDP by expenditure approach in China? As shown in Fig. 1.28, the final consumption occupied quite great share before 2000, and it was lower and lower, while gross capital formation was higher and higher in the first decade in the twenty-first century.

1.2.3 *Stages of Economic Development in China*

The population of China was 962.59 million in 1978 and 1.3678 billion in 2014 [7]. The GDP per capita was 2678.9 CNY in 1978, and it was 38753.55 CNY in 2014. The GDP per capita in constant 2010 USD grew by 6151.35$/p in 2014 from 425.2$/p in 1978 as expressed in Fig. 1.29. The average annual growth rate was 7.7 % in 36 years.

Total electricity consumption per capita and residential electricity consumption per capita are shown in Fig. 1.30. From Table 1.5 in Sect. 1.1.3, we can see the economic stages in China as follows:

Primary Commodity Stage (−1982)

In this stage, electricity consumption per capita is less than 300 kWh and household electricity consumption per capita is less than 20 kWh. *TE*/*p* was 293 kWh and *RE*/*p* was 12.66 kWh in 1982. Thus, China's economy was in the primary commodity stage.

Early Period of Industrialization (1983–2000)

In this period, electricity consumption per capita is in the range of 300–1000 kWh and residential electricity consumption per capita is in the range of

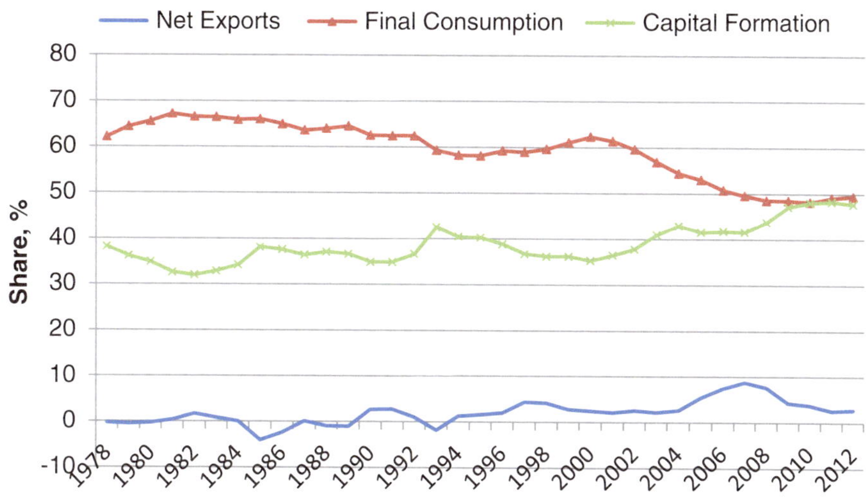

Fig. 1.28 Mixture of GDP by expenditure approach in China (Data source: National Bureau of Statistics of China, 2014 China Statistical Yearbook [10])

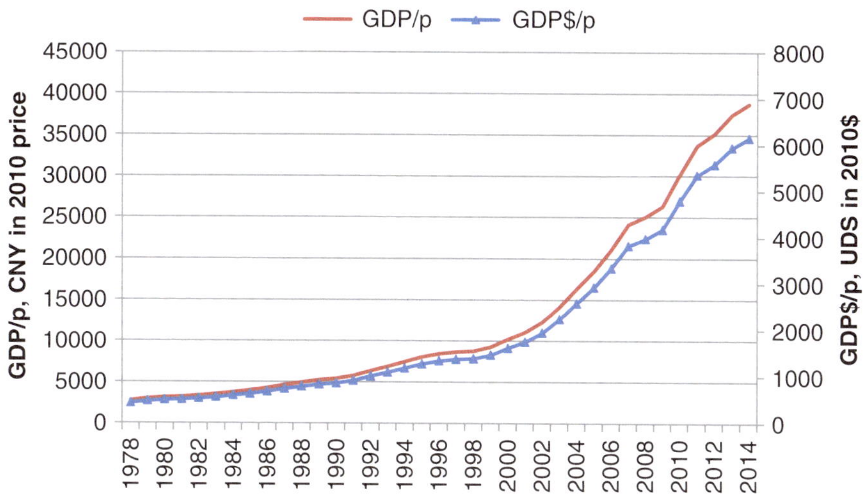

Fig. 1.29 GDP per capita in CNY and USD in China (Data source: National Bureau of Statistics of China, 2014 China Statistical Yearbook [10])

20–80 kWh. *TE/p* was 308.9 kWh and *RE/p* was 14 kWh in 1983. Although *RE/p* was only 14 kWh, it would be over 20 kWh. In this period, there were lots of household in the area of secondary industry, and the electricity consumption was put to the whole industrial electricity use. In 2000, *TE/p* was 980.67 kWh and *RE/p* was 131.92 kWh. It belongs to the early period of industrialization.

Middle Period of Industrialization (2001–2007)

In this period, electricity consumption per capita is in the 1000–2400 kWh range, and residential electricity consumption per capita is in the 80–240 kWh range. *TE/p*

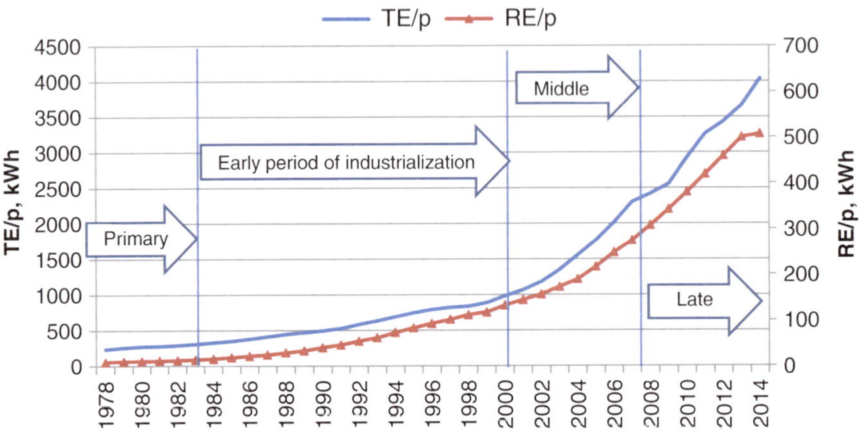

Fig. 1.30 *TE/p* and *RE/p* in China (Data source: statistical material of China Electricity Council over years)

was 1063.56 kWh and *RE/p* was 143.78 kWh in 2001. It belongs to the middle period of industrialization.

Late Period of Industrialization (2008–)

In this period, electricity consumption per capita is in the 2400–4500 kWh range, and residential electricity consumption per capita is in the range of 240–810 kWh. *TE/p* was 2415.74 kWh and RE/p was 307.08 kWh in 2008. It reveals that China's economy was entering to the middle period of industrialization in 2008.

The shares of E_1, E_2, and E_3 in the whole industrial electricity consumption in China are shown in Fig. 1.31. In the primary commodity stage, the share of E_2 in E was around 90 %, while E_1 and E_2 were only around 5 %. The main reason is the history of planed economy in China since 1949. With the economic model of former Soviet Union, heavy industry was developed quickly to promote military production. Then, steel and iron were produced as the action of great growth, "*DA LIAN GONG TIE*" and "*DA YUE JIN*."

In the early period of industrialization, 1983–2000, China was transferring the planned economy to market economy. The microeconomic activities were very active. People had more way to do business and got more incomes. The demand of products for heavy industry was lower. Therefore, the share of E_2 decreased to around 83 %. On the other hand, the light industry grew fast to meet the demand of human life. Some appliances such as TV, washing machine, refrigerator, and air conditioner were in shortage in the market. Thus, the share of E_3 increased fast to 12.5 %.

In the middle period of industrialization during 2001–2007, the rich people were interested in buying big house or apartment. Thus, the great growth of building construction led to high demand on steel and iron, cement, chemical products, and others, which rendered the share of E_2 to increase. On the other hand, the products

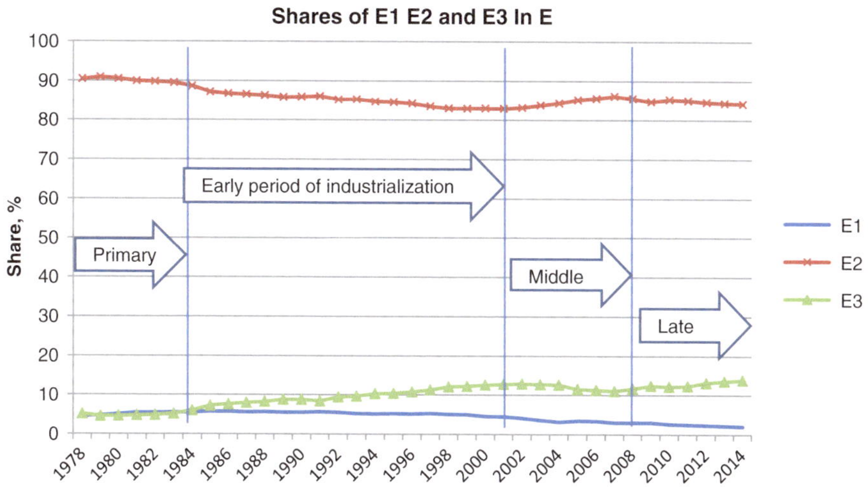

Fig. 1.31 Shares of E_1, E_2, and E_3 in the whole industrial electricity consumption in China (Data source: statistical material of China Electricity Council over years)

exported increased so fast that China was named as the World Factory. The export also promoted the increase of E_2. While the share of E_3 was increasing in 2001–2004, it was then decreasing during the period 2005–2007.

In the late period of industrialization, the export and secondary industrial production have declined in global financial crisis since 2008. And the share of E_2 has decreased too. Based on the experience in Japan, the share of E_2 decreased fast and E_3 increased quickly in the late period of industrialization in 1969–1979, and it was true in the stage of industrialization completion in 1980–1986 as shown in Fig. 1.32.

The shares of the four parts E_1, E_2, E_3, and *RE* in the total electricity consumption *TE* during 1978–2014 are depicted in Fig. 1.33. The share of primary industry in the total electricity consumption decreased from 4.37 to 1.8 %. The share of secondary industry declined from 87.35 % in 1978 to 72.5 % in 2001 and then up and down to 73.6 % in 2014. However, it always occupied the largest proportion. The share of tertiary industry increased from 4.92 % in 1978 to 4.33 % in 1979 and then rose to 12.06 % in 2014, maintaining a gradual rising. The share of residential electricity consumption was increasing from 3.36 to 12.54 % with a growth rate higher than the tertiary industry during 1978–2014. Since 1991, the share of residential electricity consumption has been more than the tertiary industry. The share changes of E_1, E_2, and E_3 can reflect production share changes of the primary industry, secondary industry, and tertiary industry. And the increasing share of *ER* reveals the rising income of the Chinese people. The per capita electricity consumption increased to 4038.1 kWh in 2014 from 259.52 kWh in 1978, with an average annual growth of 8.18 %.

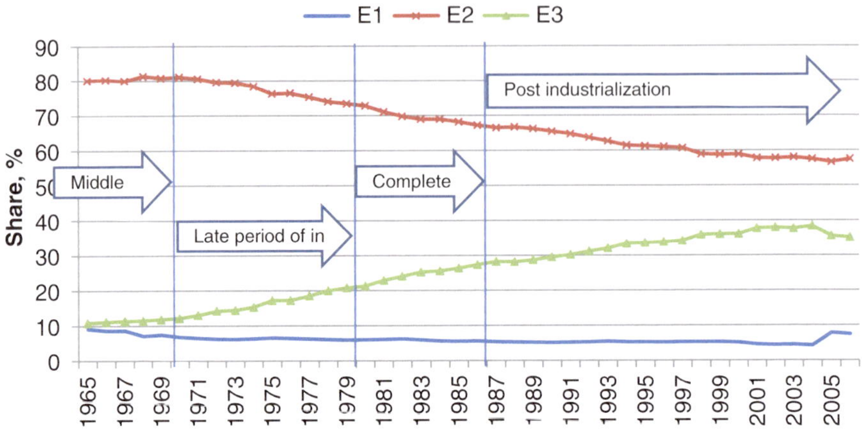

Fig. 1.32 Shares of E_1, E_2, and E_3 in the whole industrial electricity consumption in Japan (Data source: The Institute of Energy, Economics, Japan: Handbook of Energy & Economic Statistics in Japan, 2014)

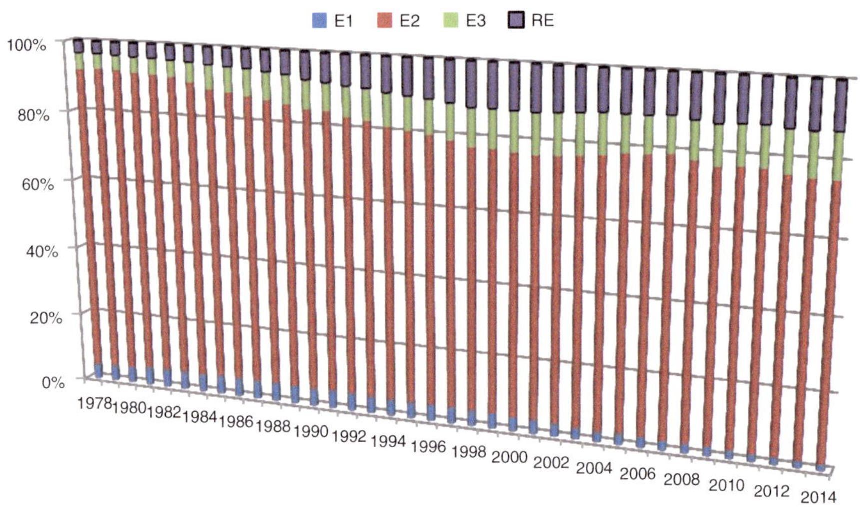

Fig. 1.33 Mixture of the total electricity consumption in China (Data source: statistical material of China Electricity Council over years)

1.3 Review of China's Economy in Global Crisis

Economic growth has been affected greatly by the global financial crisis since 2008, which is reflected by electricity consumptions. With the challenge, Chinese government decided to put huge investment to stimulate the economy in April 2009. The secondary industrial electricity use E_2, tertiary industrial electricity use E_3, and, in detail, some sectoral electricity consumptions will be analyzed to review the economy. Based on that, the impacts of the huge investment on new jobs in 2009–2011 have been studied in this section.

1.3.1 Electricity Growths of Secondary and Tertiary Industries

In order to analyze the trend of growth change, Fig. 1.34 shows the monthly growth of E, E_2, and E_3 from Jan–Feb of 2007 to December of 2008. Electricity consumptions are quite different in January and February due to more than 7 days of public holiday for Chinese New Year. If the Chinese New Year is in January, the electricity consumption in this month will be very low because the production would be influenced around 15–30 days; and sometimes the Chinese New Year would be in February; then the electricity consumption will be lower in February. Thus, the growth (YoY) will be not comparable in January and February, respectively. Therefore, electricity consumptions in the first 2 months of a year are put together as one, so that there is 11 growth data each year. As exhibited in Fig. 1.34, the monthly growths of E, E_2, and E_3 were declining in 2007, though the yearly growth rate of E was 15.24 %. Although the growth rate of E_2 was over 10 % in the first 5 months in 2008, some firms producing export products found difficulty in getting new contracts from abroad. The declining growth rate of E_2 went on to 4.4 % in September of 2008 and decreased sharply to be negative in the last quarter and to be −18.59 % in November. The growth rate of E is similar to E_2 due to the high share of secondary electricity consumption. Growth rate of E_3 is waved around

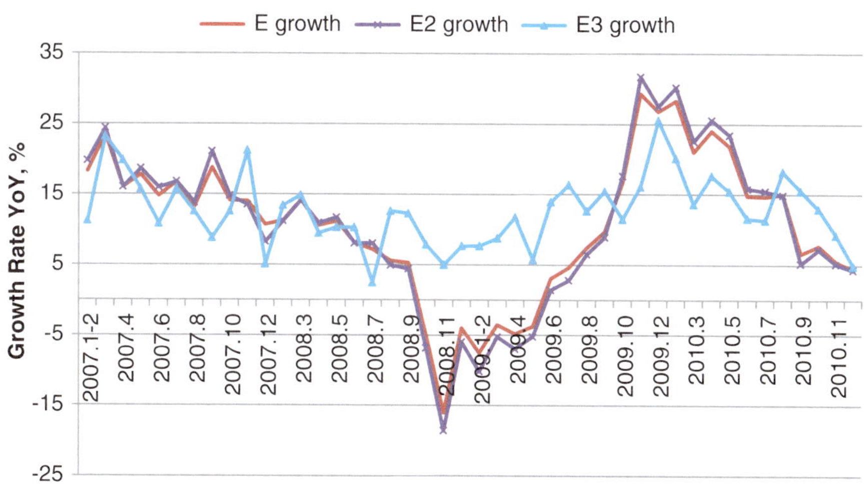

Fig. 1.34 Monthly growth rates of E, E_2, and E_3 in 2007–2010 in China (Data source: statistical material of China Electricity Council over years)

the growth rate of E with the declining trend. However, there was no negative growth for E_3.

In order to stimulate economy, the Chinese government quickly decided to put around four trillion CNY as investment to construct highway, railway, airport, and other constructions in early 2009. A lot of projects had been proved by the central government. Furthermore, local governments had also taken some measures to stimulate the local economy by investment in 2009. The total investment has been more than four trillion CNY since 2009. The projects on construction have stimulated the demands on iron and steel, cement, aluminum, and machine manufactures. Therefore, after 8 months of negative growth, there was a positive E_2 growth rate of 1.4 % in June 2009, and then it reached 31.69 % in November of the same year as shown in Fig. 1.33. Within 12 months, E_2 growth rate changes from −18.59 to 31.69 %, which verifies the effect of stimulating economy by investment. However, it takes another 12 months to decline to 5.13 % which shows the effect of the investment is gradually weak. The growth amplitude of vibration has recorded the swing of the economy in 3 years of global crisis in China.

As for tertiary industry, the growth rate of E_3 was relatively stable compared with that of E_2, and it kept rising and arrived at 25.53 % in December 2009 and then declined to 4.88 % in December 2010. It reveals that the effect of the investment is weak for tertiary industry in China.

1.3.2 Electricity Growths for Some Sectors

Now, let us pay attention to manufacturing sectors. The electricity consumption growth rates of textile sector, garments sector, and furniture sector are shown in Fig. 1.35. The products of those sectors used to contribute greatly to export. It seems that the textile sector changes beforehand compared with other sectors. Its electricity consumption grew by −0.52 % in July 2008. It is a signal of global economic recession. On contrast, the electricity consumption growth rates of the other two sectors turned negative in the last quarter of 2008. The growth rate of textile sector became positive in August 2009, 2 months later than the other two sectors. After a short swing in the first quarter of 2010, the growth rates of the three sectors arrived at 5–9 % by the end of 2010. The process of changes in electricity consumption growth rate reflects the influences of the crisis. Figures 1.35, 1.36, 1.37, 1.38, 1.39, 1.40, and 1.41 illustrate the monthly production of the manufacturing sectors by the electricity consumption growth rate. It is clear that their growth rates share the same trend.

The ferrous metal smelting and rolling processing (iron and steel) sector is very sensitive to investment. Figure 1.40 depicts the results of investment clearly for the three sectors. Electricity consumption growth rate of Iron&Steel changed from −24.55 % in November 2008 to 31.69 % in November 2009 and, then, decreased to −2 % in December 2010, which is the record of the results of investment during the global crisis.

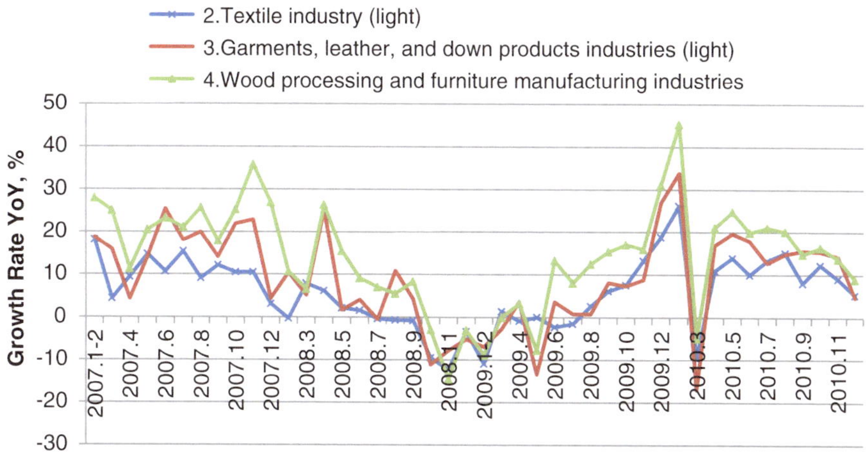

Fig. 1.35 Monthly electricity growth rates of textile, garments, and furniture in 2007–2010 in China (Data source: statistical material of China Electricity Council over years)

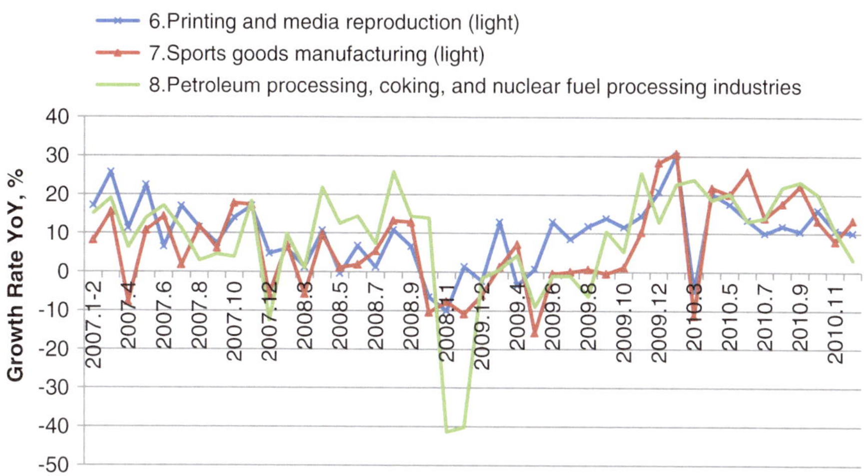

Fig. 1.36 Monthly electricity growth rates of printing, sports products, and petroleum products in 2007–2010 in China (Data source: statistical material of China Electricity Council over years)

As for tertiary industry, the influence of the recession in 2007–2008 was not as evident as the secondary industry in China. Electricity consumption growth rates of computer service, shopping center, and hotel and restaurant are illustrated in Fig. 1.42. There were some negative growth rates for hotel and restaurant service in 2007 and 2008. The business of those sectors is driven by domestic final consumption in China.

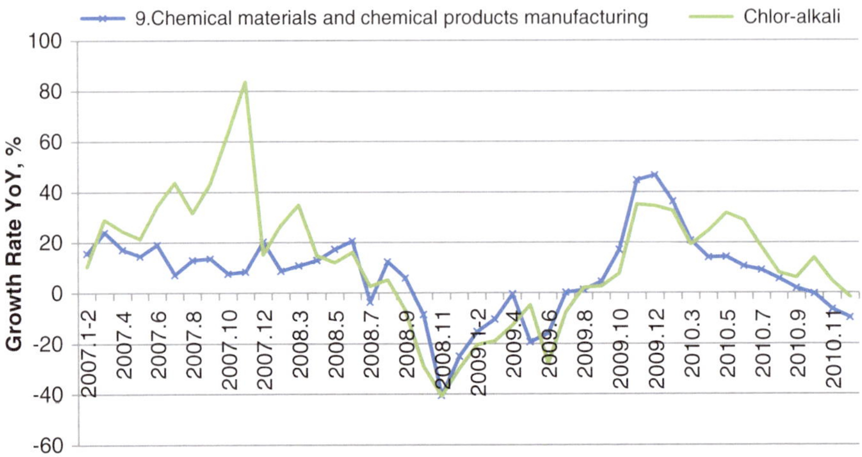

Fig. 1.37 Monthly electricity growth rates of chemical products in 2007–2010 in China (Data source: statistical material of China Electricity Council over years)

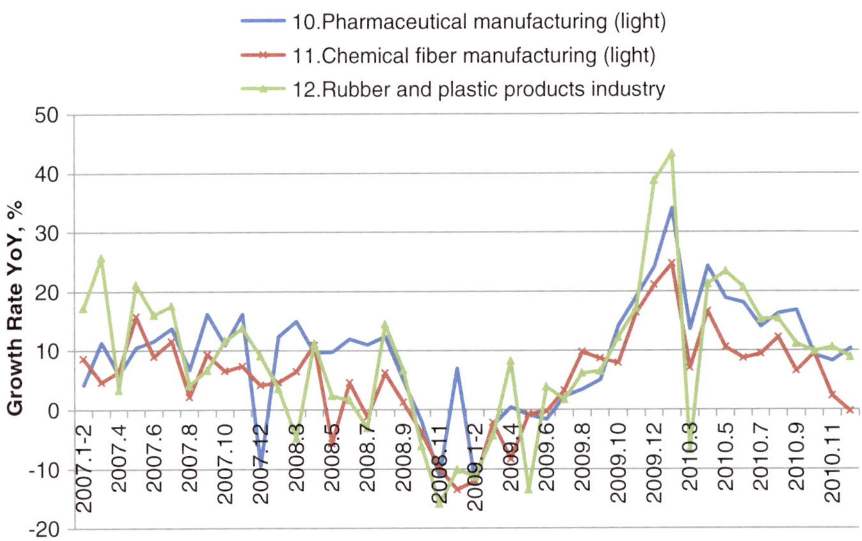

Fig. 1.38 Monthly electricity growth rates of pharmaceutical, chemical, and rubber products in 2007–2010 in China (Data source: statistical material of China Electricity Council over years)

1.3.3 Electricity Growth Trends

From the above discussion, we can find that E_2 growth can reflect the economic operation effected by both the global crisis and Chinese government actions to recover the economy by huge investment. The trend of the E_2 growth can also reveal the economic trend during 2007–2013 (Fig. 1.43). The trend curve shows

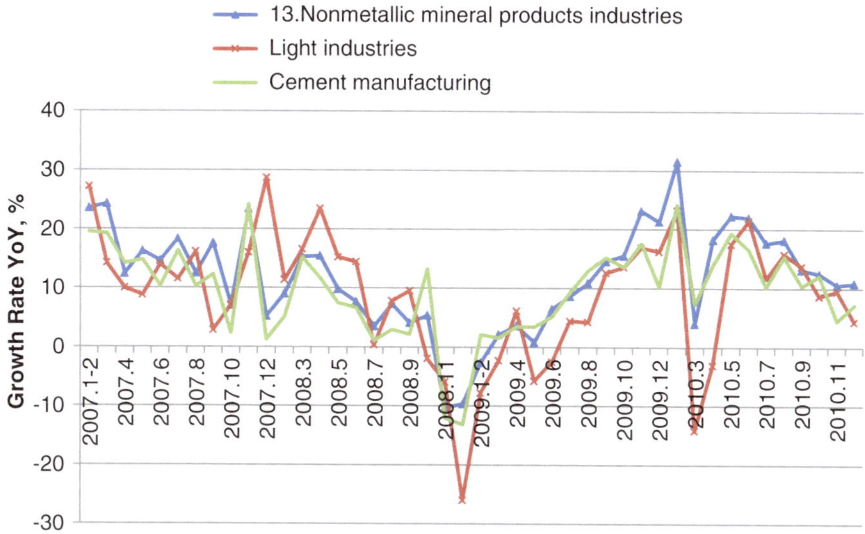

Fig. 1.39 Monthly electricity growth rates of nonmetallic (including light and cement products) in 2007–2010 in China (Data source: statistical material of China Electricity Council over years)

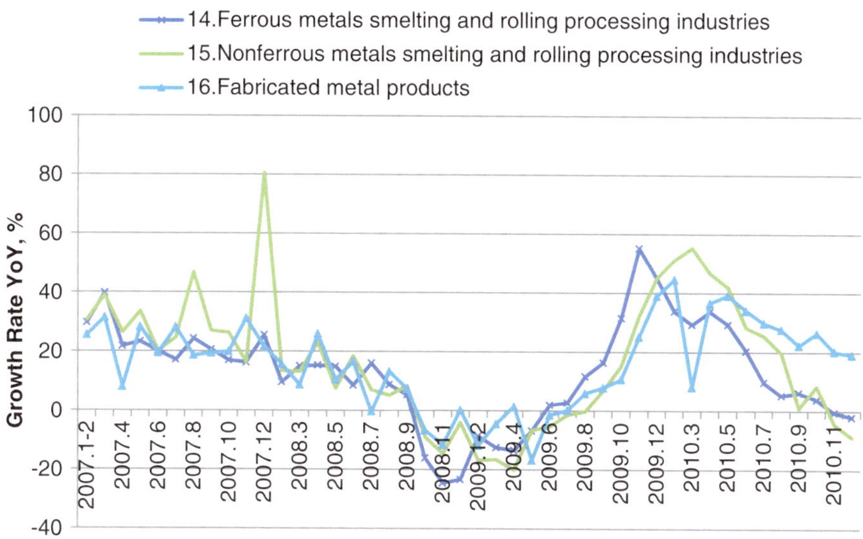

Fig. 1.40 Monthly electricity growth rates of ferrous, nonferrous, and fabricated products in 2007–2010 in China (Data source: statistical material of China Electricity Council over years)

that the growth valleys were in October 2008 and June 2012, and the peak of the growth was in June 2007, September 2010, and August 2013. The cycle is around 1.5–2 years. This is the result of the combined effects of both global crisis and government investment on China's economy.

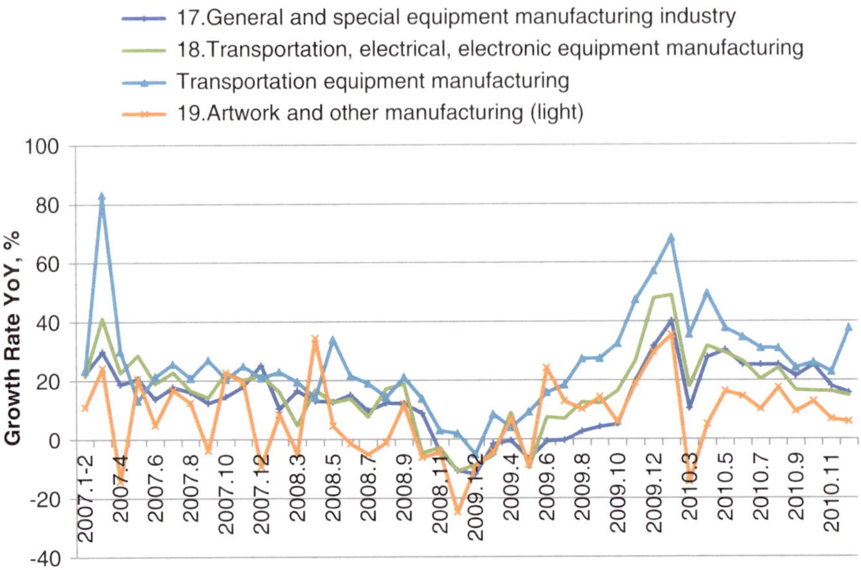

Fig. 1.41 Monthly electricity growth rates of equipment, transportation (including equipment), and artwork products in 2007–2010 in China (Data source: statistical material of China Electricity Council over years)

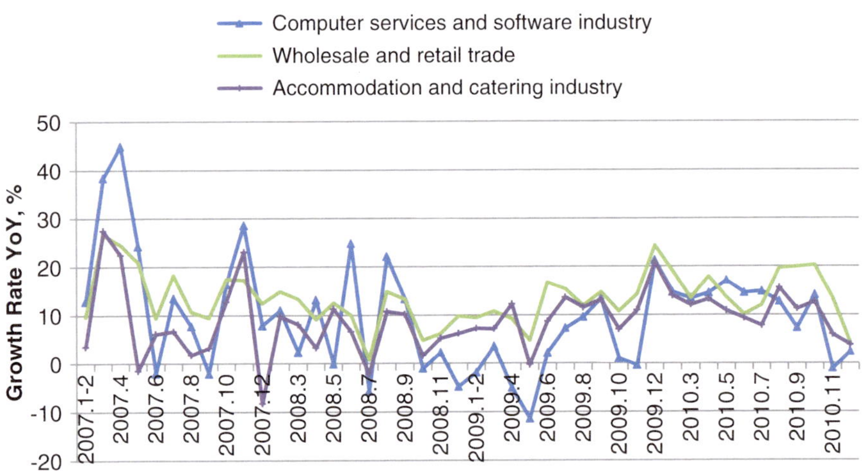

Fig. 1.42 Monthly electricity growth rates for computer, sales, and hotel and restaurant in 2007–2010 in China (Data source: statistical material of China Electricity Council over years)

As discussed above in this section, the investment has less effect on the electricity consumption of tertiary industry. Thus, E_3 growth can reflect the trend of China's economy without the huge investment during the global crisis. As shown in Fig. 1.44, the valleys of the growth were in July 2008 and September 2012, and the

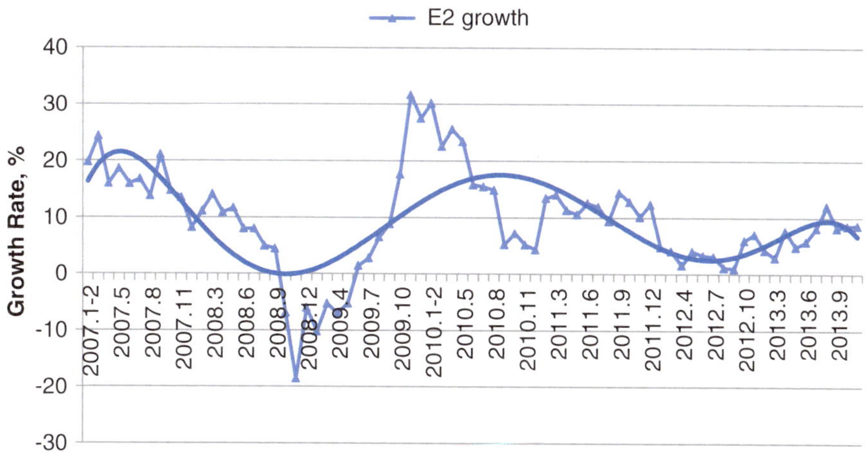

Fig. 1.43 E_2 growth rate and its trend from 2007 to 2013 in China

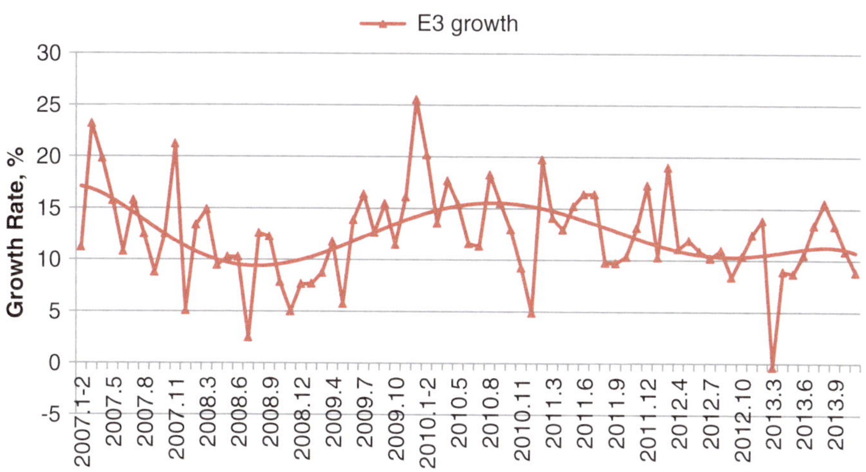

Fig. 1.44 E_3 growth rate and its trend from 2007 to 2013 in China

peaks of the growth were in August 2010 and September 2013. The cycle is around 3–4 years.

There are some who criticize the huge investment to recover China's economy. What the economy would be if the investment did not exist in China? Since the trend of E_2 growth could be the result of both global crisis and investment decided by the Chinese government, and the trend of E_3 growth could be affected by the investment weakly, then the difference of the two trends would be the effect of the investment in 2007–2013. It is evident from Fig. 1.45 that there will be no big growth amplitude of vibration by the swing of the economy in the global crisis in China. The effect of the investment last from June 2009 to May 2012, only about 3 years.

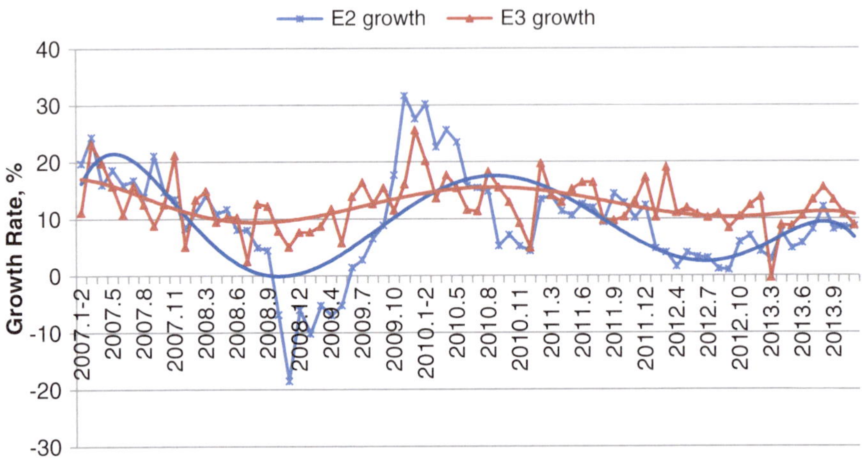

Fig. 1.45 Growth rates and the trends of E_2 and E_3 from 2007 to 2013 in China

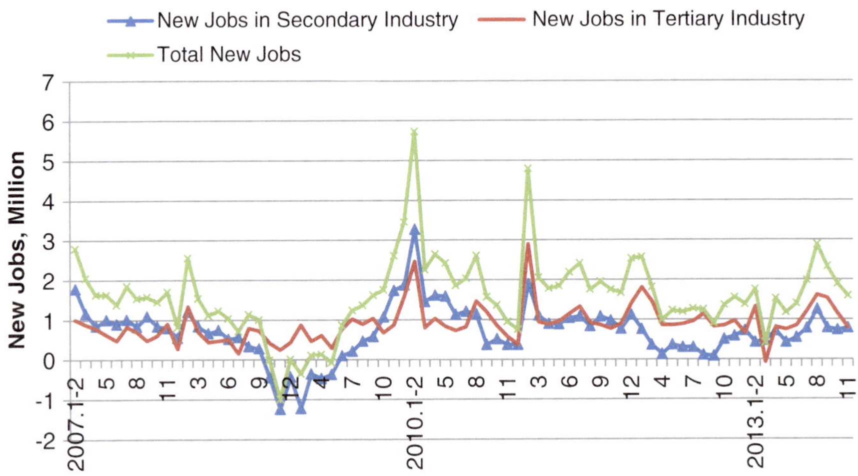

Fig. 1.46 Monthly total new employment in secondary industry and tertiary industry in 2007–2013 in China

1.3.4 Employment Analysis

On the other hand, what about employment influenced by the investment? Figure 1.46 depicts the monthly real employment in 2007–2013. For the first 2 months of 2007, total new jobs were 2.79 million in which 1.785 million from the secondary industry and one million from the tertiary industry, and the total new jobs were 18.42 million in 2007. It was 9.3 and 18.43 million, respectively, in 2008 and 2009. We can find that the effect of investment on new employment lasts about 2 years. Suppose E_2 growth was in the same trend of E_3 growth which can be regarded as the

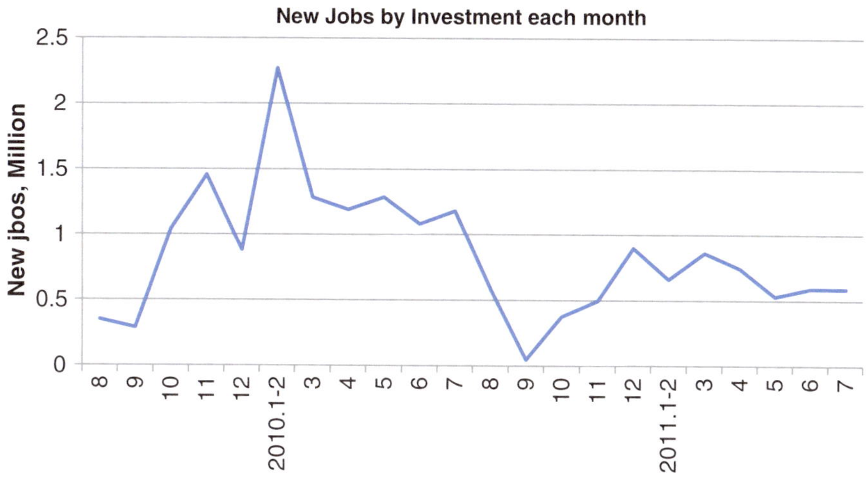

Fig. 1.47 New jobs in secondary industry resulting from investment from August 2009 to July 2011

case without investment from August 2009 to July 2011; then there were a total of 18.67 million new jobs created by the investment in the secondary industry. The monthly new jobs provided by the investment are shown in Fig. 1.47.

References

1. Hu Zheng, Zhaoguang Hu (2013) Production function with electricity consumption and its applications [J]. Energy Econ 39:171–192
2. Encyclopedia of China electric power (2001) China Electric Power Press (2nd edn), , p 114
3. Hu Zhaoguang, Tan Xiandong, Xu Zhaoyuan (2011) 2050 Chinese economic development and demand for electric power- based on the electric power supply and demand research laboratory (ILE4) simulation [M]. China Electric Power Press, Beijing
4. Baumol WJ, Blinder AS (2002) Economics principles and policy (translated version) [M]. Peking University Press, Beijing, p 151
5. Zhaoguang Hu, Zheng Hu (2013) Electricity economics: production functions with electricity [M]. Springer, Heidelberg, pp 70, 111, 140, 116, 125–144, 431
6. Data Source: World Bank World Development Indicators, International Financial Statistics of the IMF, Global Insight, and Oxford Economic Forecasting, as well as estimated and projected values developed by the Economic Research Service all converted to a 2005 base year
7. Independent Statistics & Analysis US Energy Information Administration, http://tonto.eia.doe.gov/cfapps/ipdbproject/IEDIndex3.cfm
8. Huijiong Wang (2003) Integrated study of China's development and reform [M]. Foreign Languages Press, Beijing, p 7
9. World Bank Data (2014) http://data.worldbank.org/
10. National Bureau of Statistics of China (2014) China statistical yearbook [M], p 4
11. Chenery HB, Comparative study of industry and economic growth, p 71
12. Shi Donghui (1999) Introduction of the industrialization of later-rising countries [M]. Shanghai University Finance and Economics Press, Shanghai

Chapter 2
Challenge–Opportunity and Mutations in China's Economy

WEI JI(危机) in Chinese means challenge and opportunity are the two sides of a coin, and it also shows a challenge would become an opportunity by smart people. Although the economic growth is higher in the last decade, the marginal representative factor productivity is declining. It is a challenge in economic development in China. Is there any opportunity to increase productivity in China? On the other hand, with high energy demand, the supplies of electric power, coal, oil, and gas have been reviewed in this chapter. The great amount of energy consumption, especially the high share of coal use in the total energy consumption and chemical production, has resulted in a big amount of emissions and wasted water pollutions which have seriously damaged people's health and living conditions. How do we meet the challenges? Should the energy demand keep fast growth to meet the economic growth in the future? Is it determined by economic gene? Are there any relations with economic mutation? These issues have been discussed in this chapter. Some challenges/opportunities have been studied on the mutation of the national economy by comparing the mutations of the national economy for the USA, Japan, and China. Our study findings show that the high energy consumption appears with a positive mutation and low energy demand coexists with a negative mutation; positive mutation and negative mutation will happen in turn alternatively in an economy. For China's economy, it was in the high energy consumption period that showed a positive mutation which happened in 2000. And the next economic mutation will be a negative one which means that the economy will enter into a lower energy demand period. If the mutation were powerful enough to increase marginal representative factor productivity greatly and take a longer period, it would be an opportunity for China's economy to grow healthily.

Z. Hu et al., *China's Economic Gene Mutations*,
DOI 10.1007/978-3-662-47298-9_2

2.1 Review of Electricity and Energy Supply in China

As secondary energy, electricity is produced by primary energy, such as coal, oil, gas, nuclear, hydropower, wind, solar, biomass, and others. Electricity plays an important role in the economic production and human life because it is convenient to be used in the economic activities and household, and it will take more and more share of the primary energy. Electrification is defined as the ratio between electricity consumption and the total end energy use excluding electricity. The world's electrification was 19.8 %, and those of the USA were 21.7 %; Japan, 25.7 %; and China, 23.1 % in 2011 [1]. "*Electricity is a driving force in the changing economic landscape. A 'great electrification' is taking place as growth in emerging economies and changing technologies puts air conditioners, computers, and much more at the disposal of billions*" [2]. It is true, and the growth of electricity demand was, is, and will be higher than other kinds of energy in China.

2.1.1 *Electric Power Supply*

Power sector is one of the high capital intensity sectors in the secondary industry. Power generation, transmission, and distribution need a great amount of capital for construction. As fundamental construction, the investment in power sector can also result in national economic growth. Total generation capacity in China was only 57.12 GW with serious power shortage in 1978 (Fig. 2.1), since the investment in the power sector was limited from the government budget, while other investment and foreign investment were not allowed in the planned economy in that period in China. With the economic reform, Chinese government had realized the power

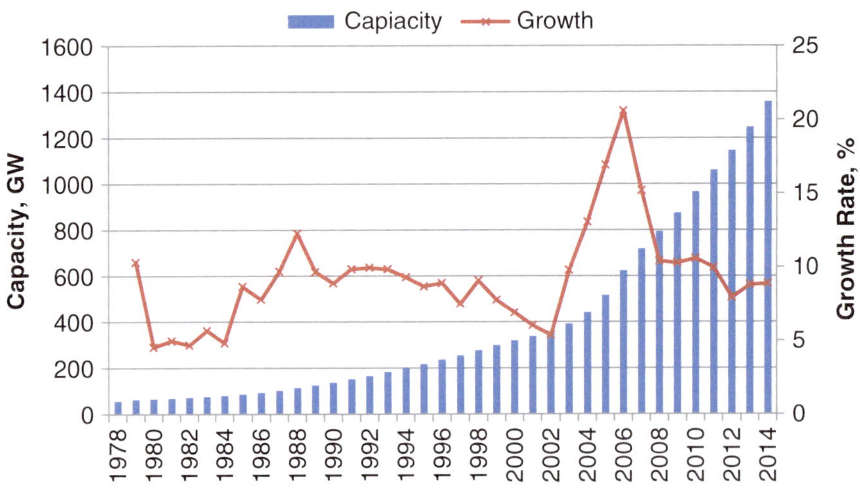

Fig. 2.1 Total power generation capacity and growth in China (Data source: statistical material of China Electricity Council over years)

shortage would limit the economic growth in the early 1980s. Thus, foreign investment in power sector had been allowed in 1984. Therefore, the generation capacity had a high growth from 8.6 % in 1985 to 12.25 % in 1988, and the total generation capacity reached 115.5 GW by the end of 1988.

In the Asian economic crisis during 1998–2000, the electricity demand was lower and there were power surplus. On the other hand, the power deregulation had been studied, and the investors found some risks in power generation. Thus, the growth of power investment was lower. However, after joining WTO since 2002, the electricity demand growth was fast and there were power shortages in China in 2003–2006. The power generation had been separated from grid in the power restructuring in 2003. The state-owned power generation companies have paid more attention on investment to meet the great electricity demand since 2003. In this case, the generation capacity has increased greatly. The new added generation capacity was 106.5 GW in 2006, and it was 110.56 GW in 2014. And then, the total power generation capacity reached 1357.95 GW, which means China became the biggest electric power country in the world in 2014. It has been shown in Fig. 2.1 that the highest growth was 12.25 % in 1998 and average annual growth was 7.93 % during 1978–2002 and the highest growth was 20.59 % in 2005 and average annual growth was 11.97 % during 2003–2014.

The coal-fired power generator has contributed a great share in China since 1978, and it will be the same case in the next 10 years because of the great availability of coal resource and lower price in China. The generation capacity of the coal-fired power plant was 39.84 GW in 1978 and reached 825.24 GW in 2014. The gas power was 55.67 GW, and coal fired and gas together as thermal power was 915.69 GW in 2014 (see Fig. 2.2). The share of coal-fired power capacity in the total generation capacity was 60.77 % in 2014, the biggest one in the world. In order to increase energy efficiency, more than 75 GW of coal-fired power plants with the unit less than 100 MW had been closed during 2006–2009. Therefore, the coal-fired

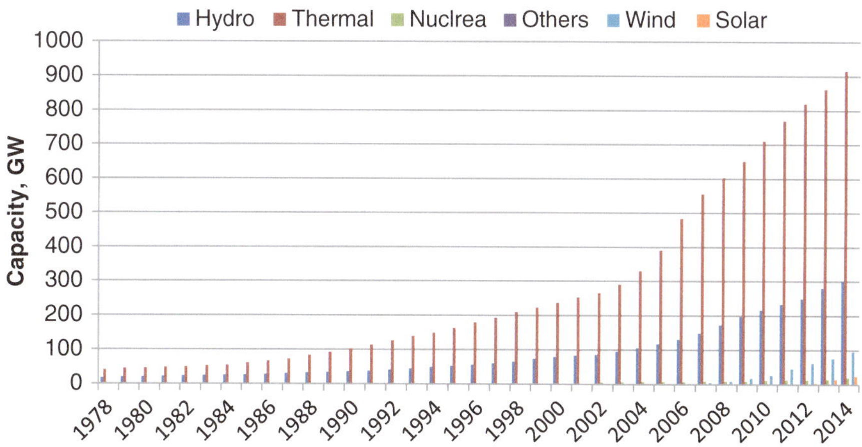

Fig. 2.2 Power generation capacity with different technologies in China (Data source: statistical material of China Electricity Council over years)

power units that are higher than 200 MW except cogeneration units play a crucial role to produce heat and electricity. The generation units that are less than 200 MW will be closed in the next few years in China.

Hydropower takes the second share in the total power capacity. It was 17.28 GW in 1978, 105.24 in 2004, 216.05 in 2010, and 301.83 GW in 2014 (Fig. 2.2). The growth rate is 8.27 % for 36 years annually. The share of hydropower in the total generation capacity was 22.22 % in 2014.

China's nuclear power station started to generate electricity in 2000, and the capacity was only in 2.1 GW by the end of 2000. The capacity was 19.88 GW in 2014. The share of nuclear power in the total generation capacity was only 1.46 % in 2014, while it was 21 % in Japan and 10 % in the USA [3]. We can see that there is enormous space to develop nuclear power in the future in China.

As the green energy, wind power is promoted with series of incentive policies and laws in China. Wind power has grown fast since 2005. The capacity was 1.06 GW in 2005. As the biggest wind power country in the world, China's wind power capacity was 95.81 GW in 2014. The new added capacity was 20.25 GW, 15.2 GW, 14.06 GW, and 20.33GW in 2011, 2012, 2013, and 2014, respectively.

Solar power generation is also promoted in China. The capacity of solar power was only 0.025 GW in 2009, 0.27 GW in 2010, 2.12 GW in 2011, 3.41 GW in 2012, and 14.79 GW in 2013. However, it experienced great growth by 64 % and reached 24.28 GW in 2014. With series of incentive policies, Chinese government plays an important role in promoting solar energy.

The others generation capacity illustrated in Fig. 2.2 consists of biomass and other power sources with very little share in the total power generation capacity.

People pay more and more attentions to electric power generation. As the biggest electric power country in the world, China's total electricity generation was 5545.88 TWh in 2014 (Fig. 2.3). However, it was only 256.55 TWh in 1978,

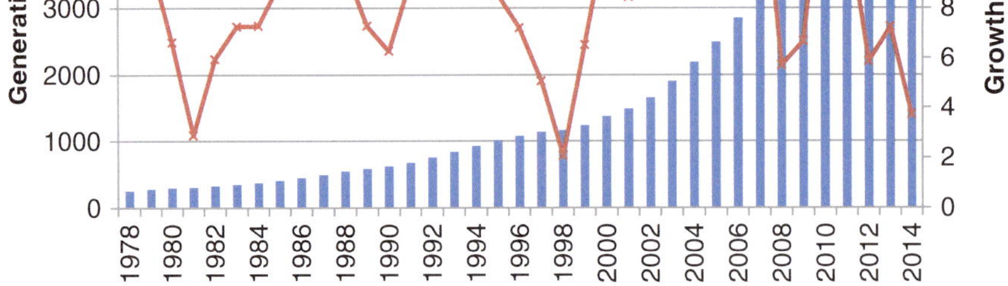

Fig. 2.3 Total power generation and growth in China (Data source: statistical material of China Electricity Council over years)

545.06 TWh in 1988, 1006.95 TWh in 1995, 2194.35 TWh in 2004, 3264.4 TWh in 2007, and 4227.8 TWh in 2010. As for the growth of power generation, the lowest one was 2.07 % in 1998 due to the Asian crisis which damaged China's economy seriously. The second lowest was 2.88 % in 1981 because of the transferring from the planned economy to market economy. The third one was 3.71 % in 2014 due to the decline of the impacts of investment around four trillion CNY, about 634.92 billion USD, in fundamental construction such as highway, high speed electric railway, airport, etc., during the global financial crisis in 2009. The great investment had stimulated the growths of 14.85 % and 11.41 % in electricity generation in 2010 and 2011, respectively. However, it got to its lowest in 2014. We can see in Fig. 2.3 that the annual growth of power generation was 7.76 % during 1978–1999 and 11.14 % in the 2000–2014 period. It reflects that the global economy has provided the great opportunities to China's economic growth. After joining WTO, China's comparative factors on lower labor cost, high working efficiency (Chinese people working hard), energy price, etc., have promoted the export. However, the labor cost has increased due to the tremendously rising house price since 2008.

Coal-fired power generation takes great share in the total power generation and more than 90 % in thermal power generation in China due to the lower coal price. As shown in Fig. 2.4, thermal power generation reached 83.34 % in 2007 and 83.3 % in 2006. Usually, the power shortage will make the coal-fired power generation generate more electricity to meet the demand. Moreover, the dry weather with few rains will also cause the same result—the hydropower station cannot generate more power without enough water. Anyway, the share of coal-fired power generation was around 75.43–70.46 % during 1978–2014. It is the highest level in the world.

The share of hydropower generation in the total power generation takes around 14.5–25 % in China. The hydropower generation is limited by water. If the weather

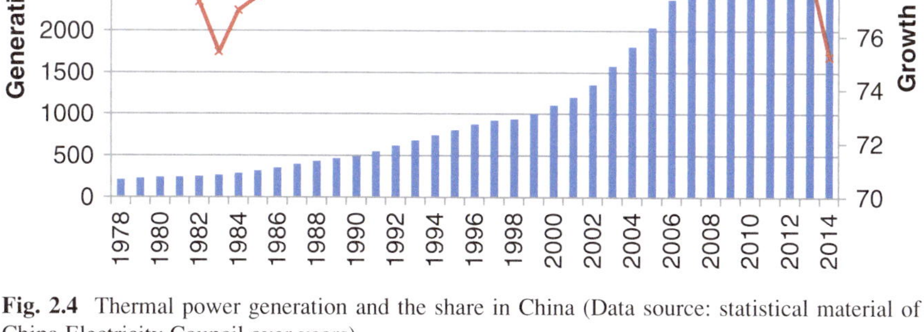

Fig. 2.4 Thermal power generation and the share in China (Data source: statistical material of China Electricity Council over years)

can provide more rains, the share will be higher. As depicted in Fig. 2.5, the highest share was 24.57 % in 1983, and the lowest one was 14.18 % in 2011. The hydropower generation was 1006.14 TWh with the share of 19.23 % in 2014.

China's nuclear power generation takes very little share around 2 % as shown in Fig. 2.6, while it was 23 % in Japan and 19 % in the USA [3]. It is clear that the share will grow fast in the next few decades due to the pressure of pollutions. The nuclear power generation was 126.21 TWh with a share of 2.28 % in total generation in 2013. The average annual growth was 15.53 % in the 2000–2014 period.

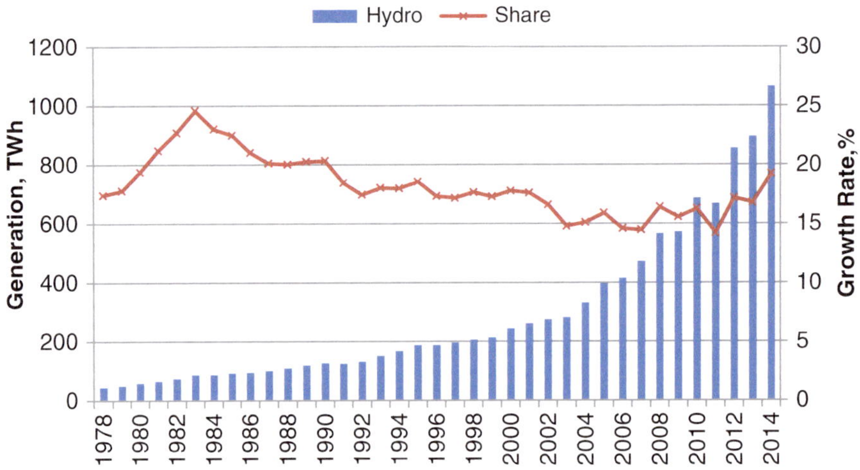

Fig. 2.5 Hydropower generation and the share in China (Data source: statistical material of China Electricity Council over years)

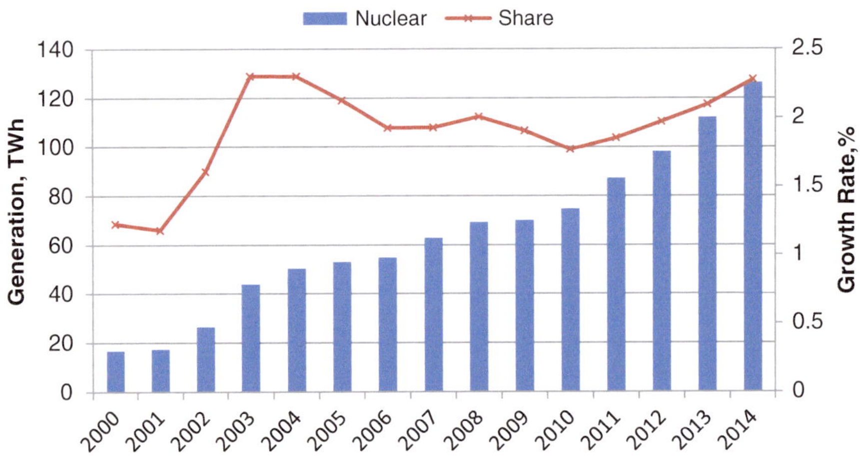

Fig. 2.6 Nuclear power generation and the share in China (Data source: statistical material of China Electricity Council over years)

China has started to develop wind power since 2005 and solar power since 2011. However, the growth of wind power generation was fast from 1.64 TWh in 2005 to 179.47 TWh in 2014. Solar power generation was 0.6 TWh in 2011, 3.6 TWh in 2012, 8.67 TWh in 2013, and 24.28 TWh in 2014 as shown in Fig. 2.7. Apart from wind and solar generation, their share in the total power generation is displayed in Fig. 2.7; it was 3.23 % in 2014. There will be substantial growth in the next decades with the promotion of incentive policies and laws in China.

In summary, nonfossil energy generation and its share in the total power generation have been shown in Fig. 2.8. It was 1371.92 TWh with the share of 24.74 % in total power generation in 2014, as the highest share since 1978 in China.

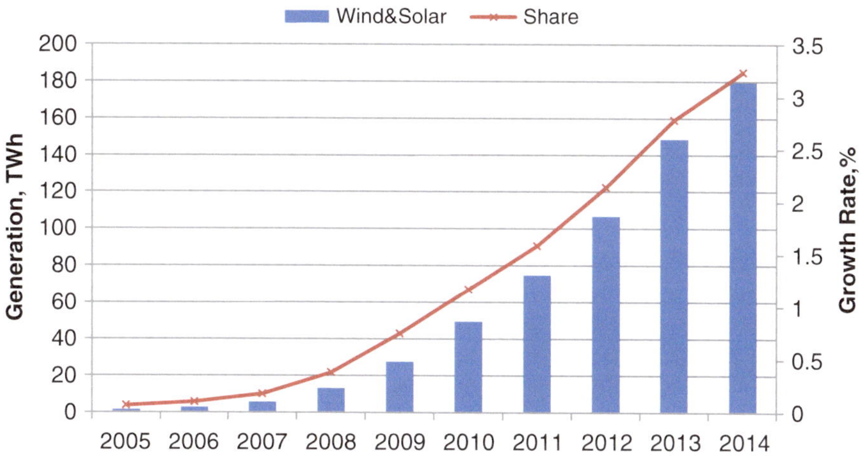

Fig. 2.7 Wind and solar power generation and the share in China (Data source: statistical material of China Electricity Council over years)

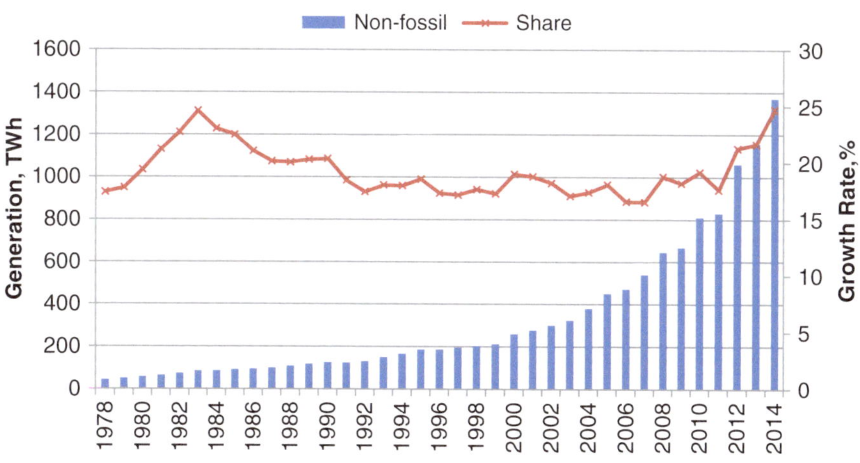

Fig. 2.8 Nonfossil energy generation and the share in total generation China (Data source: statistical material of China Electricity Council over years)

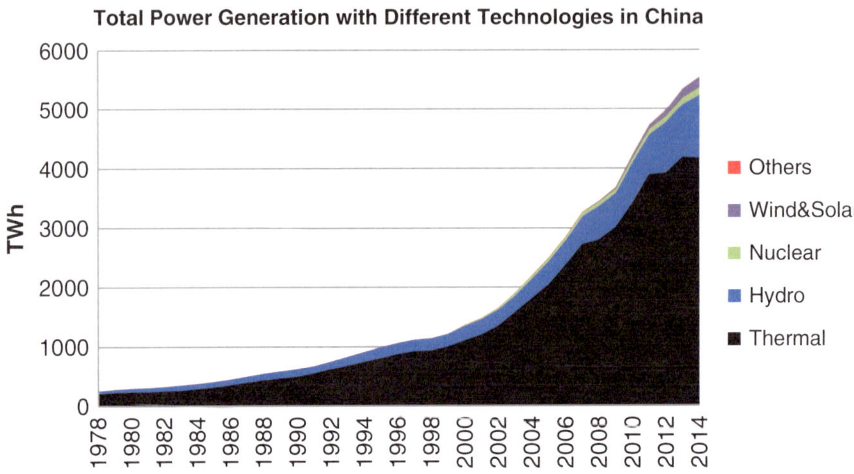

Fig. 2.9 Total power generation with different technologies in China (Data source: statistical material of China Electricity Council over years)

As exhibited in Fig. 2.9, the power generations with different technologies are quite different. The thermal power generation, mainly coal-fired power generation, has contributed a great share, followed by hydropower generation. However, the coal-fired power generation will be retailed due to pollution, and hydropower is also limited by the availability of resources. The nuclear, wind, and solar power will grow fast in the next few decades. Therefore, the nonfossil energy generation will have a big space in the future in China.

2.1.2 *Coal Production and Consumption*

There is rich coal resource in China. During 1995–2012, the investments in coal mining and processing for state-owned firms are shown in Fig. 2.10. It was 28 billion CNY in 1995 and declined to 20 billion in 2000 because of the Asian crisis, it reached 62 billion in 2005 since the coal demand grow fast, and it was 178 billion in 2012. The coal production can be reflected by electricity consumption in coal production sector (Fig. 2.11). It was 30.92 TWh in 1990 and 91.21 TWh in 2012; the average annual growth was 5.04 % for 22 years. The lowest growth was −4.41 % during the Asian crisis in 1998, and the highest growth was 14.58 % in 2005.

The share of coal in the total energy production is around 69.4–77.8 % as displayed in Fig. 2.12. Coal production was 441.3 Mtce (million tons of coal equivalent) in 1978, 771.71 Mtce in 1990, 988.6 Mtce in 2000, and 2583.6 Mtce in 2012. However, comparing with Figs. 2.11 and 2.12, the average coal production by using one GWh electricity was quite different from 23.5 tce/GWh in 1992 to

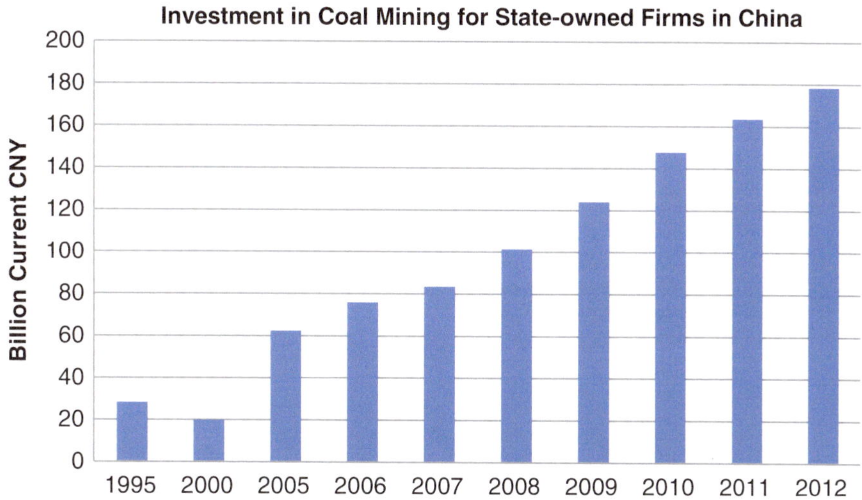

Fig. 2.10 Investment in coal mining and processing for state-owned firms (Data source: Department of Industry and Transport Statistics & National Energy Administration of China, China Energy Statistical Year Book 2013, China Statistics Press, Beijing, China

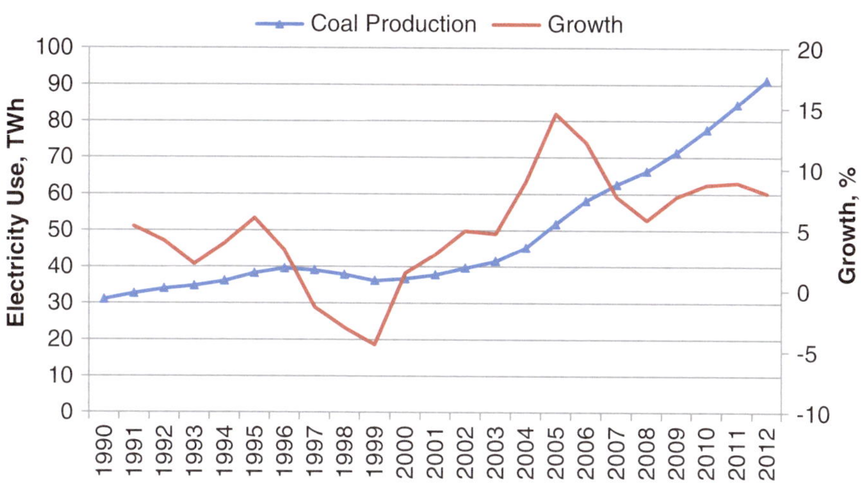

Fig. 2.11 Electricity use and the growth in coal production in China (Data source: China Statistics Press, Beijing, China)

33.57 tce/GWh in 2004 (Fig. 2.13). It would increase with technology improvement. Maybe there are some errors in the coal production published in statistic way, since the data of electricity use in the coal production sector is quite accurate. Based on the electricity economics, the revised coal production in the 1990–2012 period is

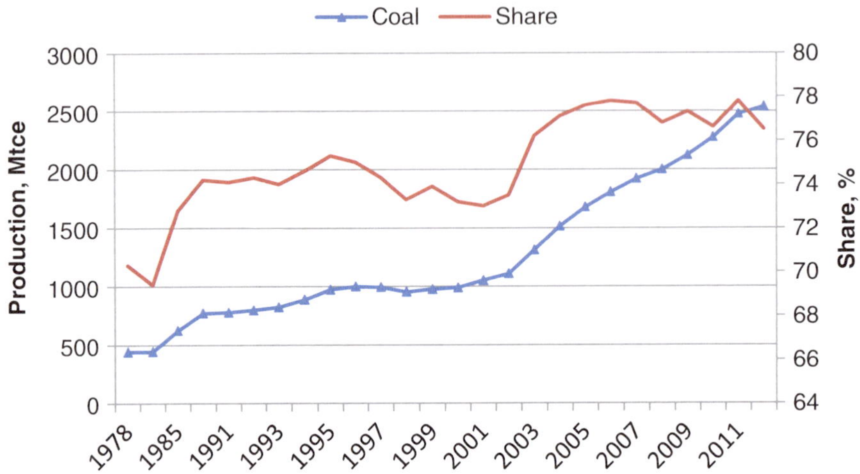

Fig. 2.12 Coal production and the share in total energy production in China (Data source: Department of Industry and Transport Statistics & National Energy Administration of China, China Energy Statistical Year Book 2013)

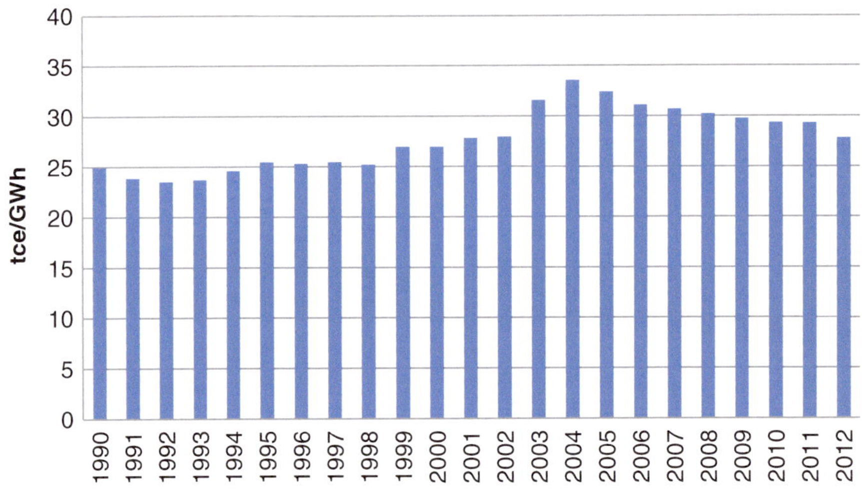

Fig. 2.13 Coal productions by using one GWh electricity in coal sector in China (Data source: Department of Industry and Transport Statistics & National Energy Administration of China, China Energy Statistical Year Book 2013, China Statistics Press, Beijing, China)

shown in Fig. 2.14. We can see that the revised coal production would be 2740.81 Mtce in 2012, which is 202.17 Mtce more than the published one.

Coal consumption is illustrated in Fig. 2.15. The share of coal consumption in the total energy consumption is around 66.6–76.2 %. The coal consumption was 404 Mtce with the share of 70.7 % in 1978, 435.19 Mtce with the share of 72.2 % in

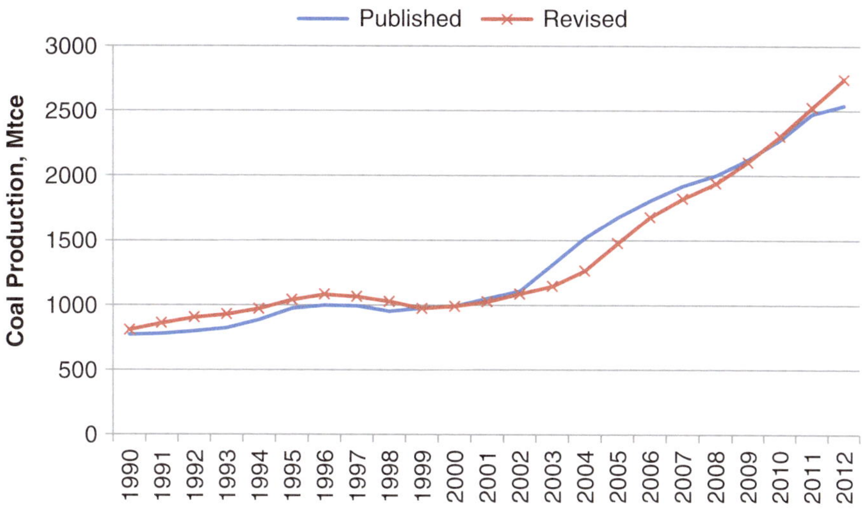

Fig. 2.14 Revised and published coal production in China

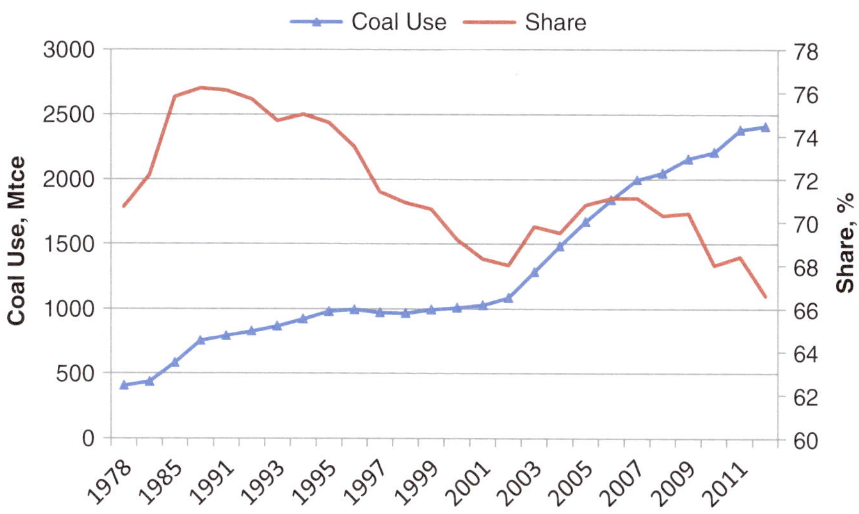

Fig. 2.15 Coal consumption and the share in total energy consumption in China (Data source: Department of Industry and Transport Statistics & National Energy Administration of China, China Energy Statistical Year Book 2013, China Statistics Press, Beijing, China)

1980, 581.25 Mtce with the share of 75.8 % in 1985, 752.12 Mtce with the share of 76.2 % in 1990, 978.57 Mtce with the share of 75 % in 1995, 1027.27 Mtce with the share of 68.3 % in 2001, and 2409.16 Mtce with the share of 66.6 % in 2012.

However, since the coal production has been revised, the coal consumption ought to be revised too. Based on the coal balance, the revised coal consumption during 1990–2012 is shown in Fig. 2.16. It is revised as 2611.31 Mtce in 2012.

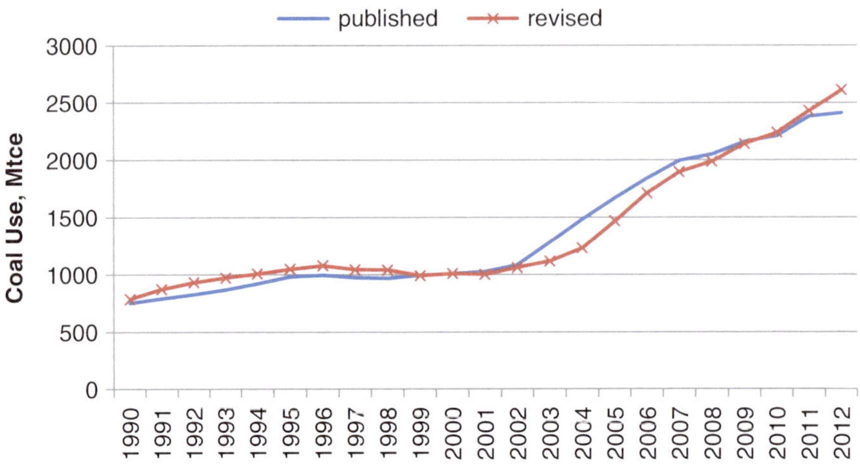

Fig. 2.16 Revised and published coal use in China

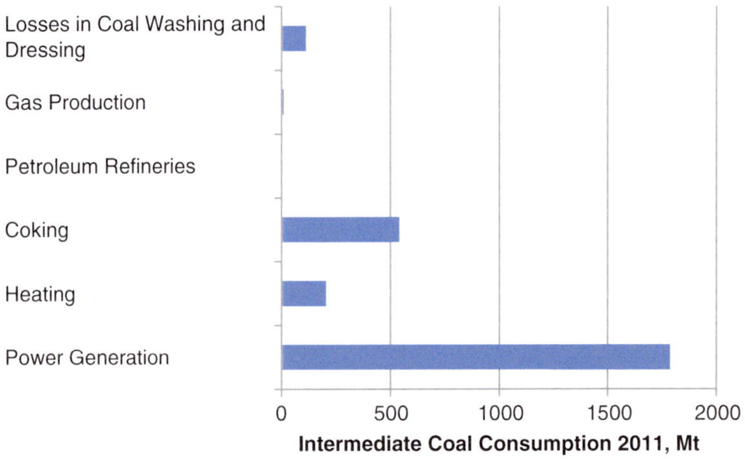

Fig. 2.17 Coal consumption in sectors in China 2011 (Data source: Department of Industry and Transport Statistics & National Energy Administration of China, China Energy Statistical Year Book 2013, China Statistics Press, Beijing, China)

We can see that the intermediate coal consumption was around 2566 Mt in 2011, while power generation used coal 1785 Mt, cooking 540.68 Mt, and heating 202.5 Mt as shown in Fig. 2.17. Table 2.1 exhibits the coal balance from 1980 to 2012 (note: the unit shown in Fig. 2.17 and Table 2.1 is in the unit of Mt which is different from Mtce). The coal import is increasing and export is decreasing due to the lower coal price and high quality in the international markets.

Table 2.1 Coal balance sheet (Mt)

Item	1980	1985	1990	1995	2000	2005	2010	2012
Total coal available for consumption	626.01	827.76	1022.21	1334.61	1367.94	2269.41	3297.72	3800.33
Output	620.15	872.28	1079.88	1360.73	1384.18	2349.51	3235	3645
Imports	1.99	2.307	2.003	1.635	2.1788	26.1711	163.095	288.411
Exports (−)	6.32	7.77	17.29	28.617	55.0647	71.7244	19.1037	9.2748
Stock changes in the year	10.19	−39.05	−42.385	0.868	36.6468	−34.554	−81.272	−123.80
Total coal consumption	610.09	816.03	1055.23	1376.76	1410.92	2318.51	3122.36	3526.47

Source: Department of Industry and Transport Statistics & National Energy Administration of China, China Energy Statistical Year Book 2013, China Statistics Press, Beijing, China

2.1.3 Oil Production and Consumption

There is a lack of oil resource in China, and the oil production in recent years is further limited comparing to the high demand after joining WTO in the early of the twenty-first century. Crude oil production had increased about double during 1978–2012. As shown in Fig. 2.18, it was 148.77 Mtce with the share of 23.7 % in 1978, 151.69 Mtce with the share of 23.8 % in 1980, 197.45 Mtce with the share of 19 % in 1990, 232.28 Mtce with the share of 17.2 % in 2000, and 295.34 Mtce with the share of 8.9 % in 2012. The share in the total energy production declined continually during 1978–2012 since the oil resource was not enough to meet the demand.

Oil consumption has grown fast in the past three and a half decades. It was 129.72 Mtce in 1978 and 680.06 Mtce in 2012 (Fig. 2.19). The average annual growth was 4.99 % for 34 years. The share of oil consumption in the total energy consumption waved around 20 % during 1978–2012 as shown in Fig. 2.19. The distribution of oil consumption in some sectors is shown in Fig. 2.20, in which the secondary industry use was 180.5 Mt, transport and storage used 160 Mt, and non-production consumption was 40 Mt in 2011.

As for the balance of oil production and international trade in 2011 (see Fig. 2.21), the total production of oil was 202.9 Mt, the import oil was 315.9 Mt, the net import was 274.76 Mt, and the share of import in the total oil consumption was 60.55 %, while it was 61.9 % in the USA in 2009. With the growth of living standard level, the oil consumption in transport will increase fast, and then, the share of import oil will grow fast too in the near future.

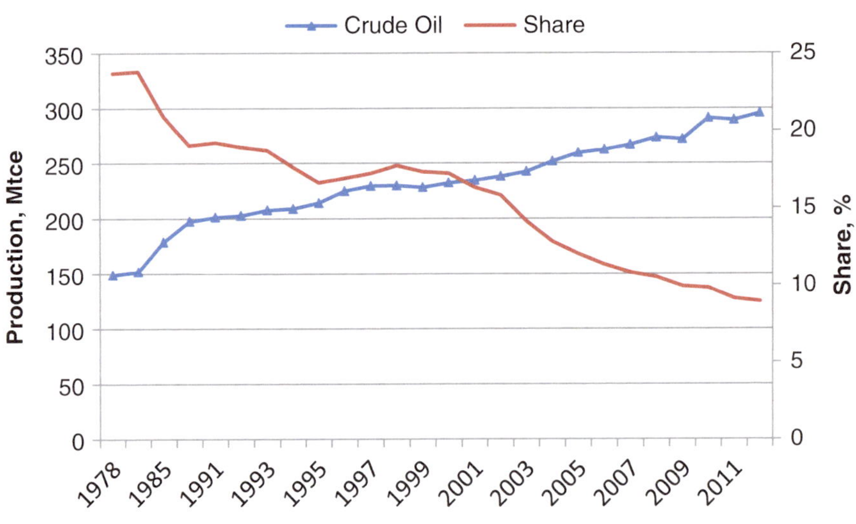

Fig. 2.18 Crude oil production and the share in total energy production in China (Data source: Department of Industry and Transport Statistics & National Energy Administration of China, China Energy Statistical Year Book 2013, China Statistics Press, Beijing, China)

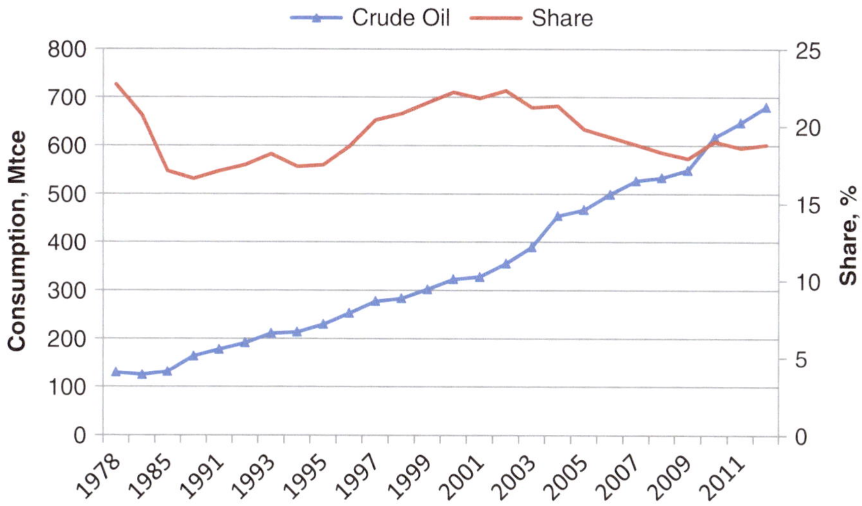

Fig. 2.19 Oil consumption and the share in total energy consumption in China (Data source: Department of Industry and Transport Statistics & National Energy Administration of China, China Energy Statistical Year Book 2013, China Statistics Press, Beijing, China)

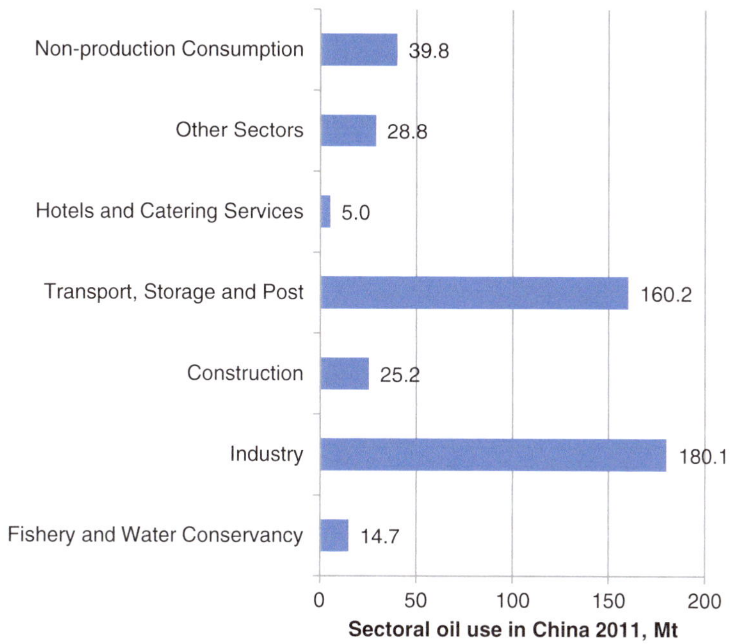

Fig. 2.20 Oil consumption in sectors in China 2011 (Source: Department of Industry and Transport Statistics & National Energy Administration of China, China Energy Statistical Year Book 2013, China Statistics Press, Beijing, China)

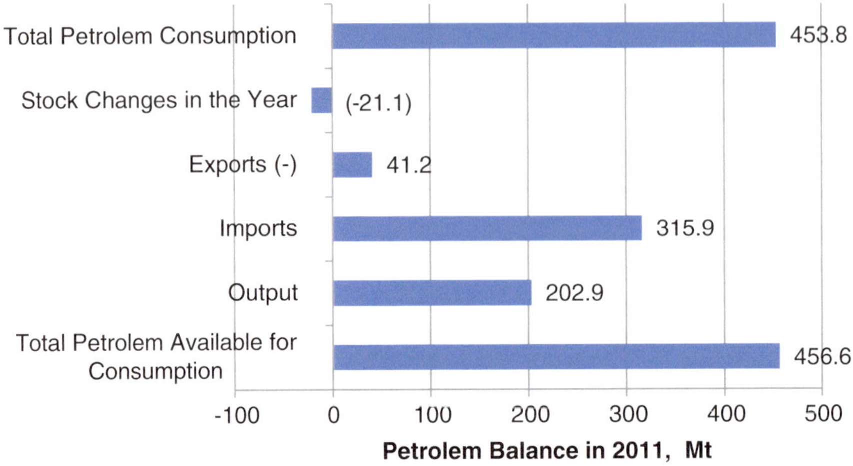

Fig. 2.21 Oil balance in China in 2011 (Source: Department of Industry and Transport Statistics & National Energy Administration of China, China Energy Statistical Year Book 2013, China Statistics Press, Beijing, China)

2.1.4 *Natural Gas Production and Consumption*

Natural gas resource is not enough to meet the high demand and the gas production is limited in China. Figure 2.22 illustrates the natural gas production and the share in the total energy production in 1978–2012. The production increased slowly from

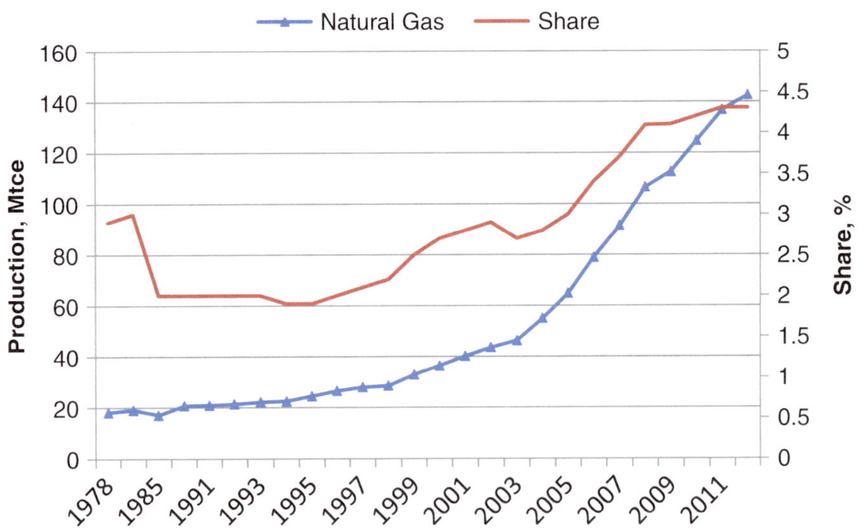

Fig. 2.22 Natural gas production and the share in total energy consumption in China (Data source: Department of Industry and Transport Statistics & National Energy Administration of China, China Energy Statistical Year Book 2013, China Statistics Press, Beijing, China)

18.2 Mtce in 1978 to 43.69 Mtce in 2002, while it was 142.7 Mtce in 2012. The share increased from 2 % in 1985 to 4.35 % in 2012.

Comparing Figs. 2.22 and 2.23, we can see that the curve of gas consumption is similar to the curve of gas production. It means there were very few gas imports before 2010. The share of gas in the total energy use appeared like a "U" form as shown in Fig. 2.23, from 3.1 % in 1980 down to 1.8 % in the 1990s and up to 5.2 % in 2012

Total gas consumption was 146.3 billion cubic meters in 2012 in China, while secondary industry used 94.7 billion cu.m, residential consumption consumed 28.8 billion cu.m, and transport and storage and post used 15.5 billion cu.m as depicted in Fig. 2.24.

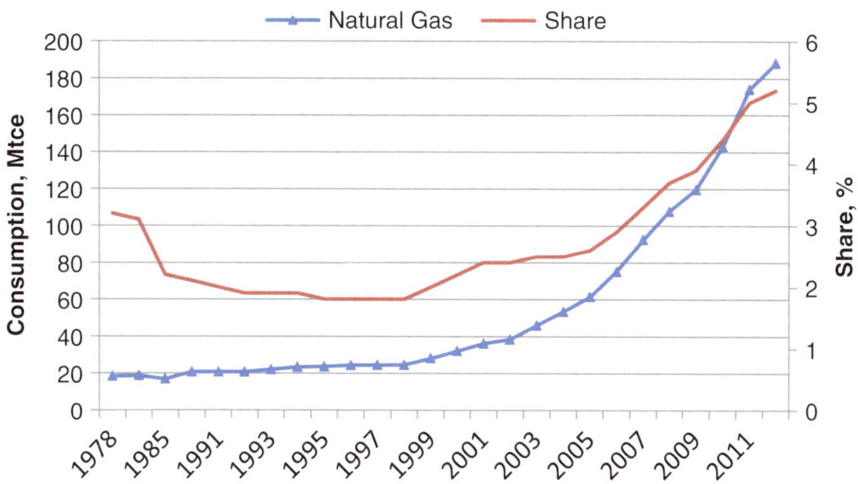

Fig. 2.23 Gas consumption and the share in total energy consumption in China (Data source: Department of Industry and Transport Statistics & National Energy Administration of China, China Energy Statistical Year Book 2013, China Statistics Press, Beijing, China)

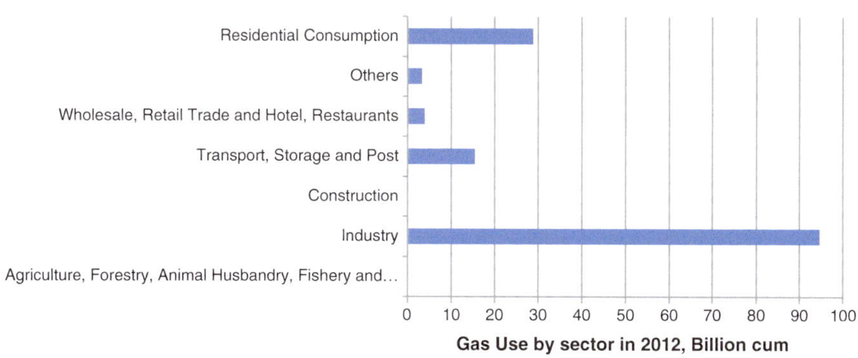

Fig. 2.24 Natural gas consumption by sector in 2012 in China (Data source: Department of Industry and Transport Statistics & National Energy Administration of China, China Energy Statistical Year Book 2013, China Statistics Press, Beijing, China)

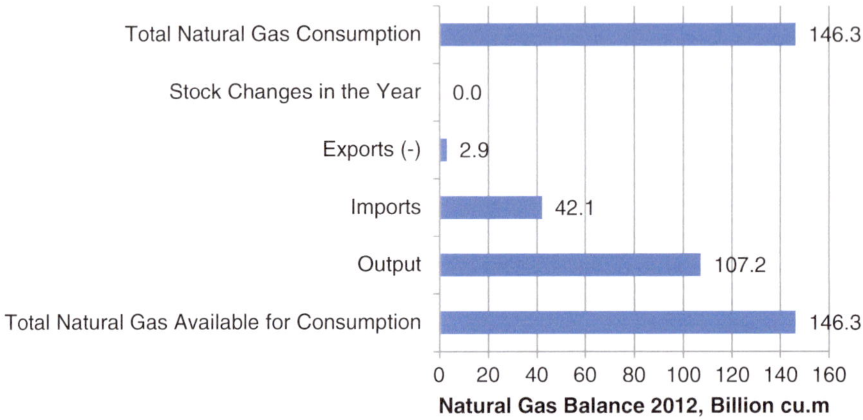

Fig. 2.25 Natural gas balance in China in 2012 (Data source: Department of Industry and Transport Statistics & National Energy Administration of China, China Energy Statistical Year Book 2013, China Statistics Press, Beijing, China)

As for natural gas balance in 2012 (see Fig. 2.25), the total gas production was 107.2 billion cu.m, the import was 42.1 billion cu.m, and the net import was 39.8 billion cu.m, which takes 26.77 % in the total gas consumption. It is expected that the share will increase fast in the future since the demand grows quickly and gas resource is limited in China.

2.1.5 Hydro-, Nuclear, and Wind Power Production and Consumption

Electricity production from hydropower, nuclear power, and wind power is shown in Fig. 2.26 (note: it is measured in Mtce). It was 19.46 Mtce in 1978 and 341.8 Mtce in 2012 and increased 17.57 times in 34 years. The share in the total energy production increased from 3.1 to 10.3 % in the period.

The electric power consumption from hydropower, nuclear power, and wind power is shown in Fig. 2.27. It is the production minus line loss. It was 340 Mtce with the share of 9.4 % in the total energy consumption in 2012.

2.1.6 Total Energy Production and Consumption

As displayed in Fig. 2.28, the total energy production was 627.7 Mtce in 1978, 1039.22 Mtce in 1990, 1350.48 Mtce in 2000, and 3318.48 Mtce in 2012. The growth rate was −2.72 % in the Asian crisis in 1998 and 5.37 % in global financial crisis in 2008.

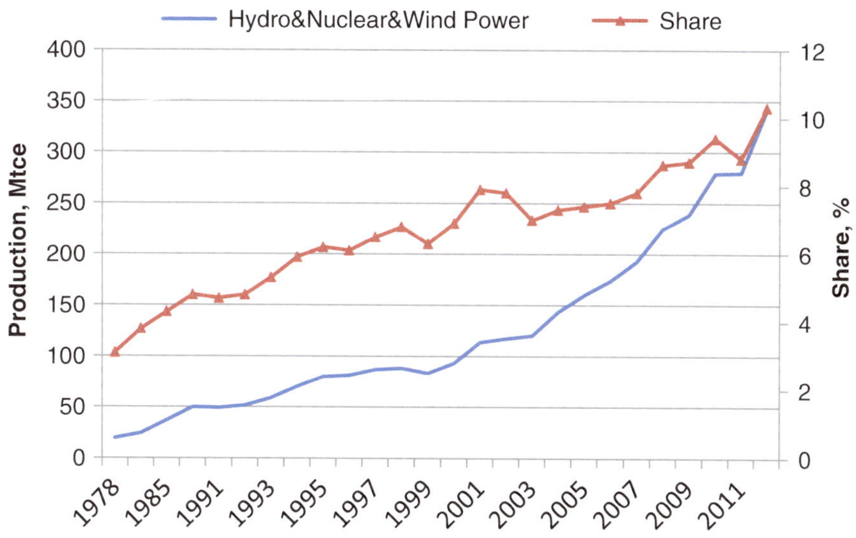

Fig. 2.26 Hydro-, nuclear, and wind power generation in China

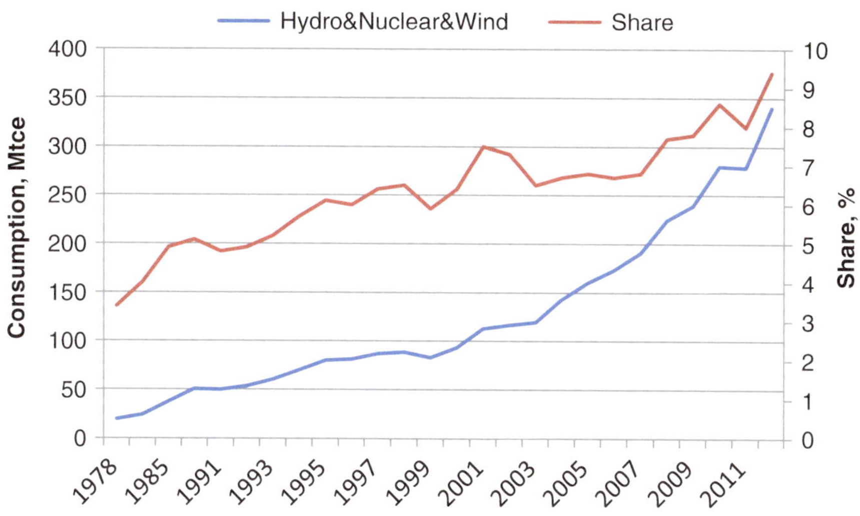

Fig. 2.27 Hydro-, nuclear, and wind power consumption and the share in total energy consumption in China

The composition of the energy production is illustrated in Fig. 2.29. The highest share was coal around 76.5 %; the second one was hydro-, nuclear, and wind power in 10.3 %; the third one was oil in 8.9 %; and the last one was gas in 4.3 % in 2012.

The total energy consumption is shown in Fig. 2.30. It was 571.44 Mtce in 1978, 1455.31 Mtce in 2000, and 3617.32 Mtce in 2012. The composition of the energy consumption is shown in Fig. 2.31. The highest share was coal around 66.6 %; the

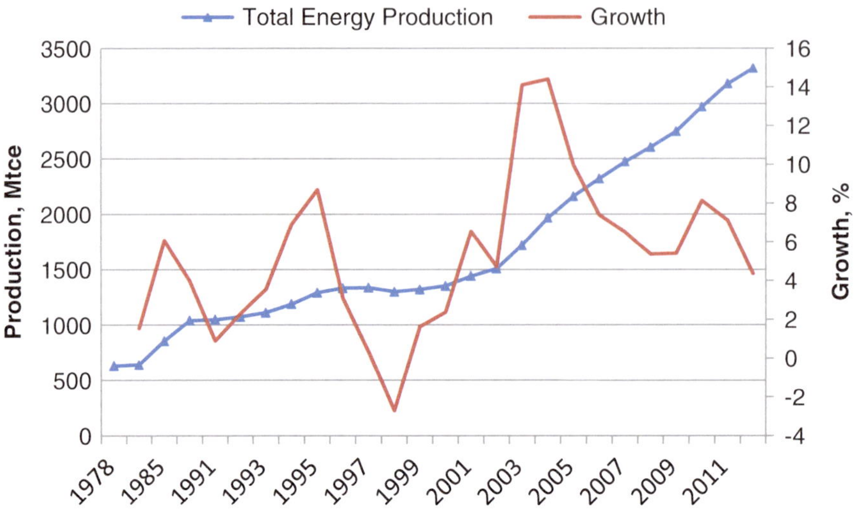

Fig. 2.28 Total energy production and growth in China (Source: Department of Industry and Transport Statistics & National Energy Administration of China, China Energy Statistical Year Book 2013, China Statistics Press, Beijing, China)

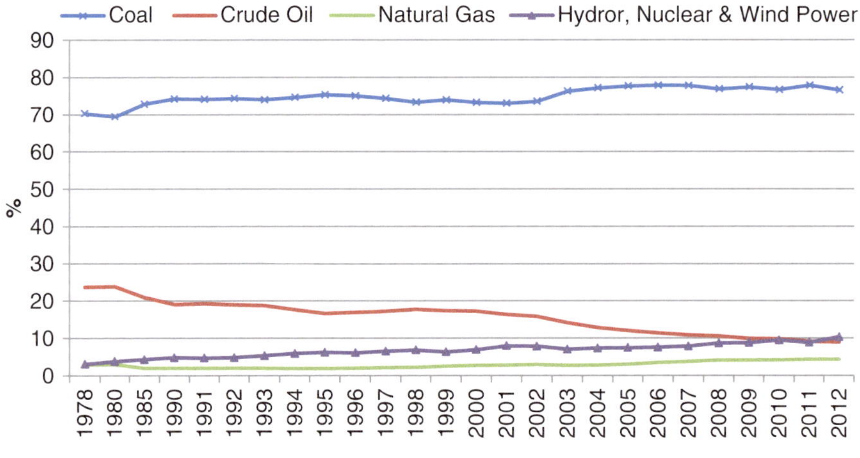

Fig. 2.29 Composition of energy production in China (Source: Department of Industry and Transport Statistics & National Energy Administration of China, China Energy Statistical Year Book 2013, China Statistics Press, Beijing, China)

second one was oil in 18.8 %; the third one was hydro-, nuclear, and wind power in 9.4 %; and the last one was gas in 5.2 % in 2012.

Energy consumption in manufacture in 2011 is exhibited in Fig. 2.32; iron and steel take the highest one at 588.97 Mtce, manufacture of nonmetallic mineral products takes the second highest one at 300 Mtce, and the third one is electric power, gas, and water supply at 260 Mtce.

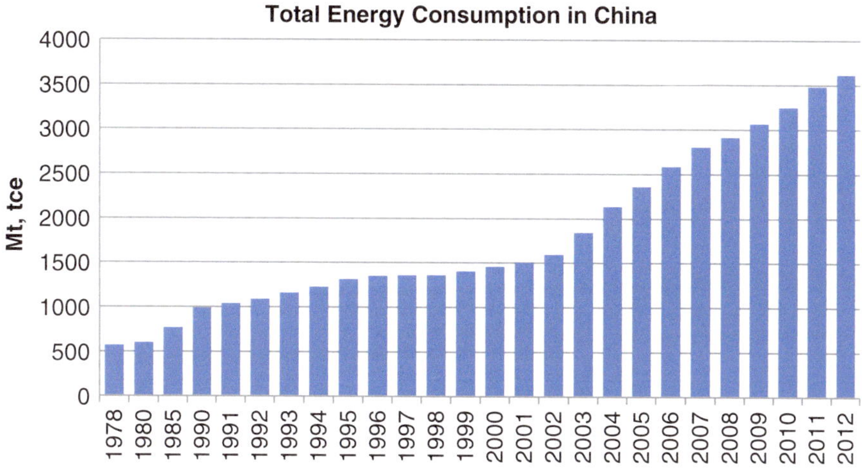

Fig. 2.30 Total energy consumption in China (Source: Department of Industry and Transport Statistics & National Energy Administration of China, China Energy Statistical Year Book 2013, China Statistics Press, Beijing, China)

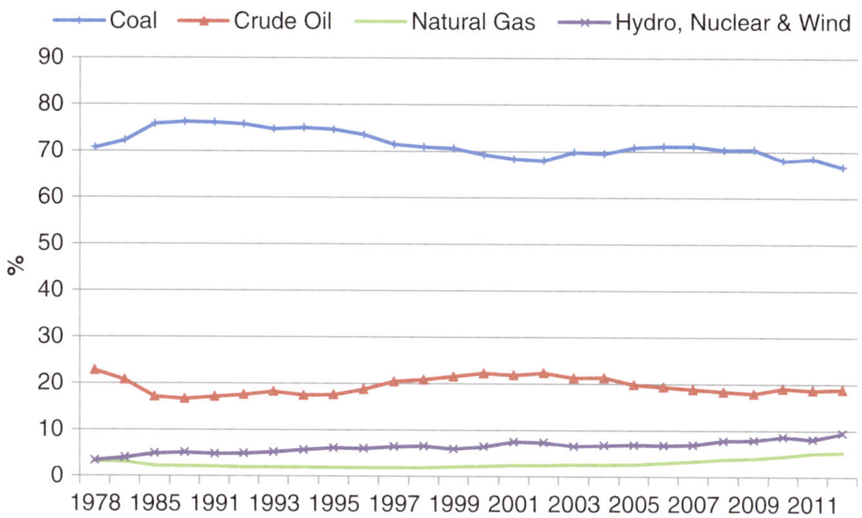

Fig. 2.31 Composition of energy consumption in China (Source: Department of Industry and Transport Statistics & National Energy Administration of China, China Energy Statistical Year Book 2013, China Statistics Press, Beijing, China)

The energy balance in 2011 is shown in Fig. 2.33. We can see that the total energy consumption was 3480 Mtce, primary energy production was 3179.86 Mtce, energy import was 622.6 Mtce, and export was 84.47 Mtce. The net energy import in the total primary energy production was 17 % in 2011, and it will enlarge fast in the future.

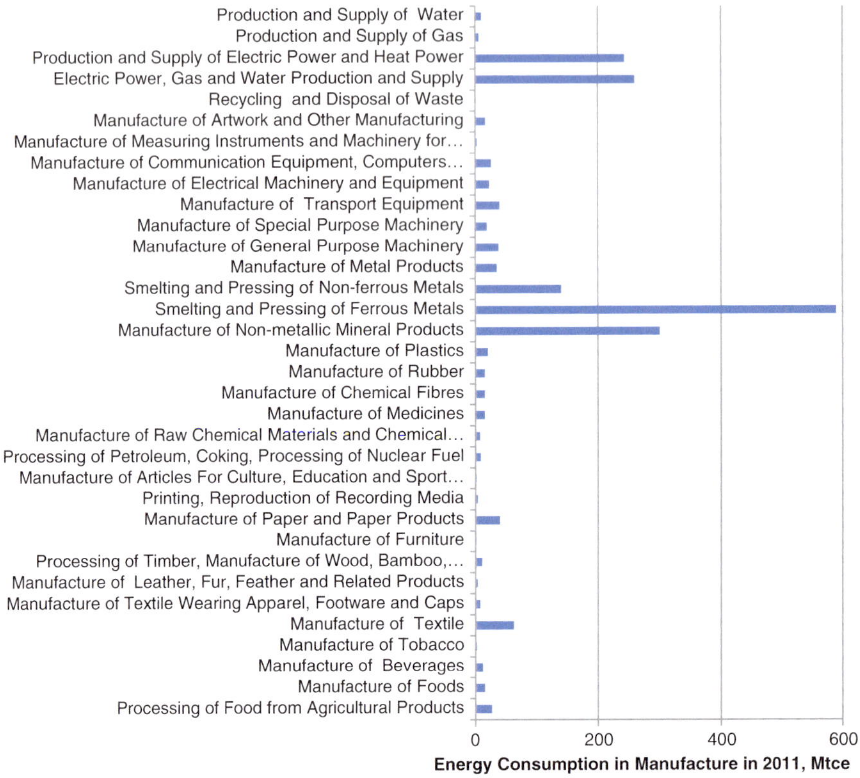

Fig. 2.32 Energy use in manufacture in China 2011 (Source: Department of Industry and Transport Statistics & National Energy Administration of China, China Energy Statistical Year Book 2013, China Statistics Press, Beijing, China)

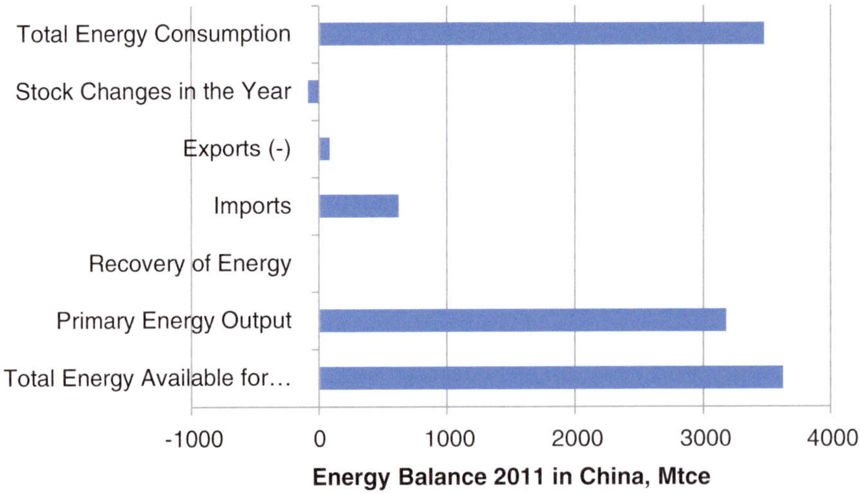

Fig. 2.33 Energy balance in China 2011 (Source: Department of Industry and Transport Statistics & National Energy Administration of China, China Energy Statistical Year Book 2013, China Statistics Press, Beijing, China)

Among the energy resources in China, coal resources account for more than 96 %, petroleum resources account for 2.4 %, and natural gas resources account for 1.2 %, indicating that petroleum and natural gas resources are very limited. Per capita possession of natural energy resources of China is less than half of the world average level, wherein per capita possession of petroleum resources only accounts for 17 % of the per capita value of the world and natural gas resources only account for 13.2 % of the per capita value of the world. The conditions of insufficiency of energy resources characterized by major shortage in coal resources and poor gas and oil will not change greatly in the next 20 years.

2.2 Emissions

Large proportion of coal will lead to more serious environmental issue and health problems. From the energy consumption mixture, it can be seen that the coal resources accounted for approximately 66.6 %, petroleum for approximately 18.8 %, natural gases for 5.2 %, and nonfossil energy for 9.4 % in 2012. Due to excessively large proportion of fossil energy, pollutant discharge will significantly worsen living environment of the people.

Large import volumes of petroleum impose high risks on the security of energy resources. The potential of petroleum production is only limited in around 200 million tons due to the resource, and China has to import great amount of petroleum to meet the demand; therefore, its dependence on import share was very high and reached America's record of 61.9 % in 2009 [4]. Such high dependence on the foreign supply will increase the risks to procurement and transportation of petroleum and the like.

With the fast growth of energy consumption, especially the coal that contributes a high share of energy use, air pollution from coal combustion was, is, and will be the main source of emissions. On the other hand, as a result of the rapid growth of China's economy, especially in the period of industrialization, lots of manufacturing firms have productions in their business. Therefore, big amount of industrial wastewater was, is, and will be a serious problem for people's health in China.

2.2.1 Air Pollutions

Since coal is the major energy resource in China, there would be serious problem on emissions such as carbon dioxide (CO_2), sulfur dioxide (SO_2), nitrogen oxides (NOx), and other pollutants. The emissions seriously affected China's air quality and people's health. It will be poorer and poorer in the next few years. It is shown in Fig. 2.34 that the CO_2 emissions were less than 1000 Mt before 1976; it was 1950 Mt in 1989, less than 2000 Mt; it was 2635 Mt in 2002; however, it reached 5818 Mt in 2009.

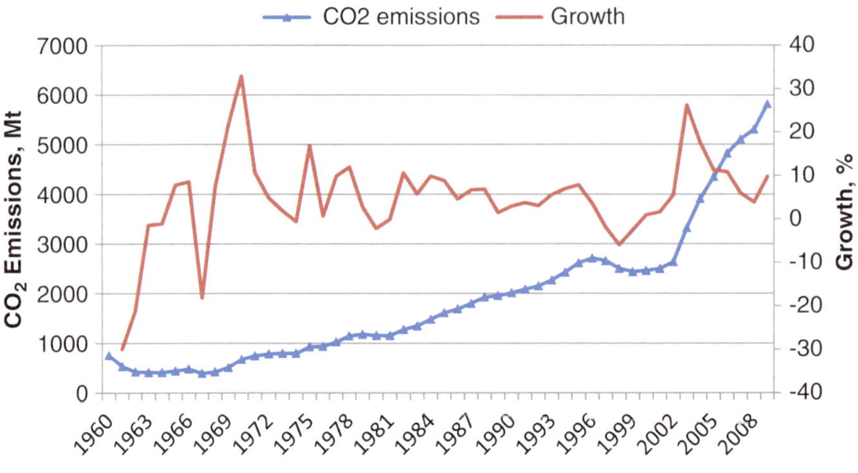

Fig. 2.34 CO_2 emissions in China (Source: The World Bank, http://data.Worldbank.org/country/china)

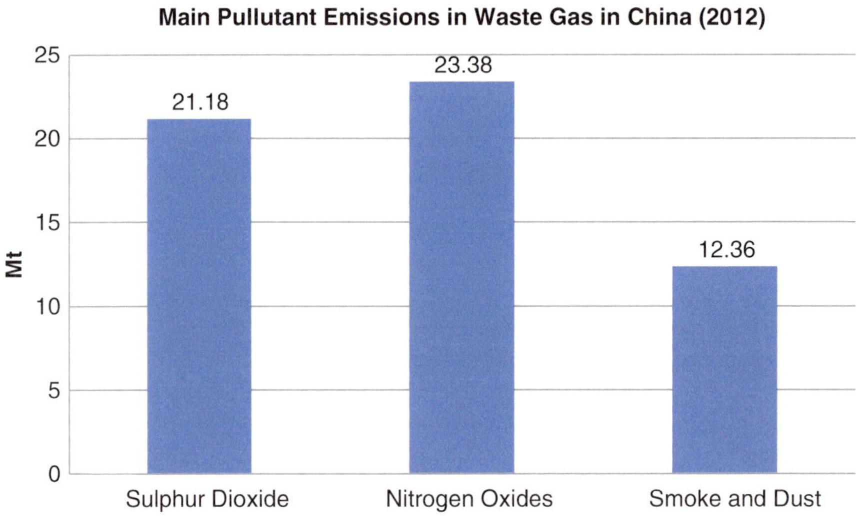

Fig. 2.35 Main pollutant emissions in waste gas in China (Source: China Statistic Press, China Statistical Year Book 2013)

The emissions of sulfur dioxide SO_2 is also very serious in China. According to China Statistic Yearbook, SO_2 emission was 13.25 Mt in 1985, 14.94 Mt in 1990, and 18.91 Mt in 1995. The average growth rate of SO_2 emission was 3.62 % annually during 1985–1995, and it was 1.14 % during 1995–2005. The main reason for the lower growth rate of SO_2 emission during 1995–2005 is that the filter equipment has been promoted in coal-fired power plant. Figure 2.35 shows that SO_2 emission was 21.18 Mt, nitrogen oxide emission was 23.38 Mt, and smoke and dust emission was 12.36 Mt in 2012. The emissions have resulted in chemical rains in large scale time and time again, especially in the middle and east region of China. The acid rain area was very little, and the range of annual average *pH* value of precipitation was in 5–5.6 in 1980; the area grew to about one-third of the country and the *ph* was in 5–5.6 in 1990; the area was about half of the country with *pH* of 4.5–5 in 2000; it was poor in 2005 wherein the area had reached to 75 % of the country, having a *pH* less than 4.5.

2.2.2 Water Pollutions

China has seven major river systems, and nearly one-half of them are polluted. The quality of 86 % of the urban river sections is below official standards [5]. Total wastewater discharge was around 36 billion tons in 1985 and was 37.2 billion tons including 22.19 billion tons for industrial discharge in 1995. However, after that, it increased quickly. The industrial wastewater discharge was 24.3 billion tons and residential wastewater discharge was 28.14, and the total was 52.44 billion tons in 2005 and 68.47 billion tons in 2012 [6]. The wastewater discharge in main cities is shown in Fig. 2.36; we can see that Shanghai has contributed the biggest at 2192.44 Mt; the second is Guangzhou, capital of Guangdong province, at 1528.33 Mt; and the third one is Beijing, capital of China, at 1402.74 Mt.

Chemical oxygen demand (COD) is an index to measure the water by reducing the pollution; the higher the value, the more polluted the water. Figure 2.37 shows the COD discharged in main cities in China. Clearly, the highest one is Chongqing, the city near Yangzi river, with the COD of 402.8 kt; the second one is Harbin, capital of Helongjiang province in the northeast of China, with the COD of 314.69 kt; and the third one is Shenyang, capital of Liaoning province, with the COD of 260.49 kt in 2012.

Mercury discharge in the wastewater is also serious as shown in Fig. 2.38. According to the amount of mercury discharge, the number one city was Urumqi, capital of Xinjiang municipality region, at 14.43 Mt; the second city was Fuzhou, capital of Fujian province, at 12.99 Mt; and the third one was Naning, capital of Guangxi Gui municipality region, at 5.22 Mt in 2012.

The organic water pollution emissions from 2003 to 2007 in China are shown in Fig. 2.39. It was 7.066 kt per day in 2003, 7.96 kt per day in 2004, and 9.43 kt per day in 2007.

2.2.3 People's Health Problems

The air pollution, water pollution, and other pollutions have resulted in serious problems on people's health in China. It makes people to pay more attention to health issues, such as taking exercise, Chinese medicine, medical insurance, and

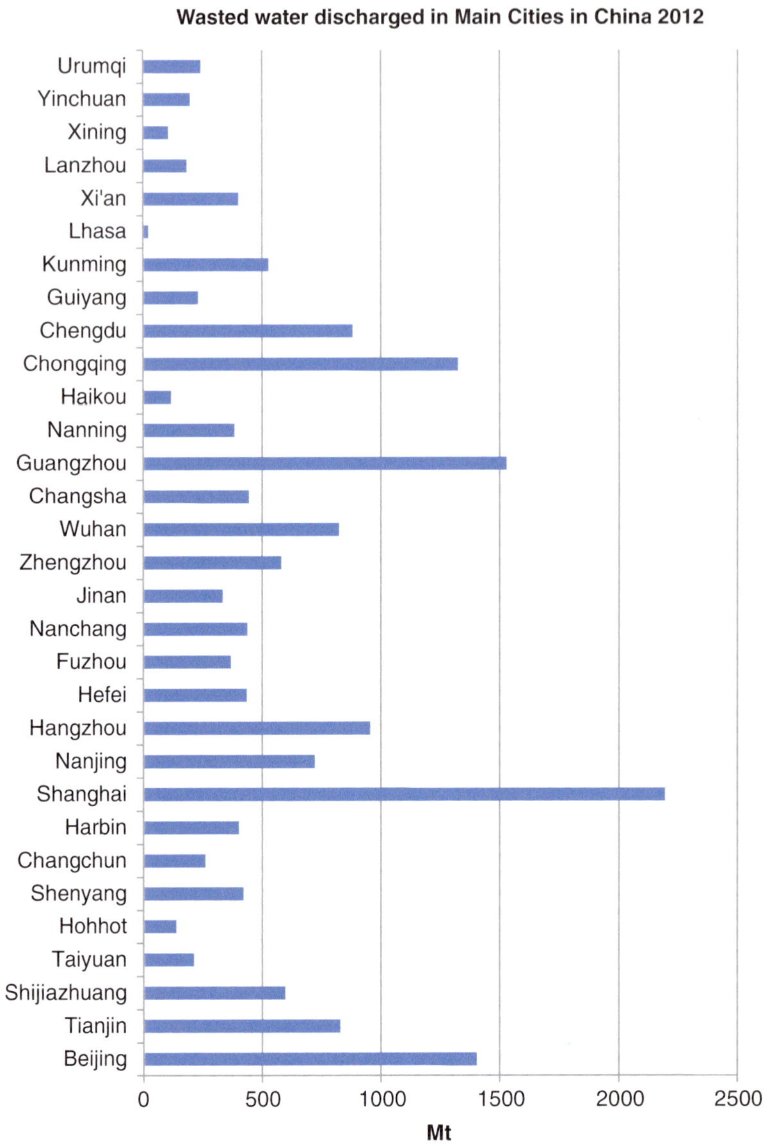

Fig. 2.36 Wastewater in main cities in China (2012) (Source: China Statistic Press, China Statistical Year Book 2013)

health expenditure. It is clear that the payment on health is growing fast, especially in the last 10 years. Although GDP growth is in high speed, the share of health expenditure in GDP grows quickly too. It is shown in Fig. 2.40 that the share of public health expenditure in GDP was almost unchanged around 1.7–1.8 %, while

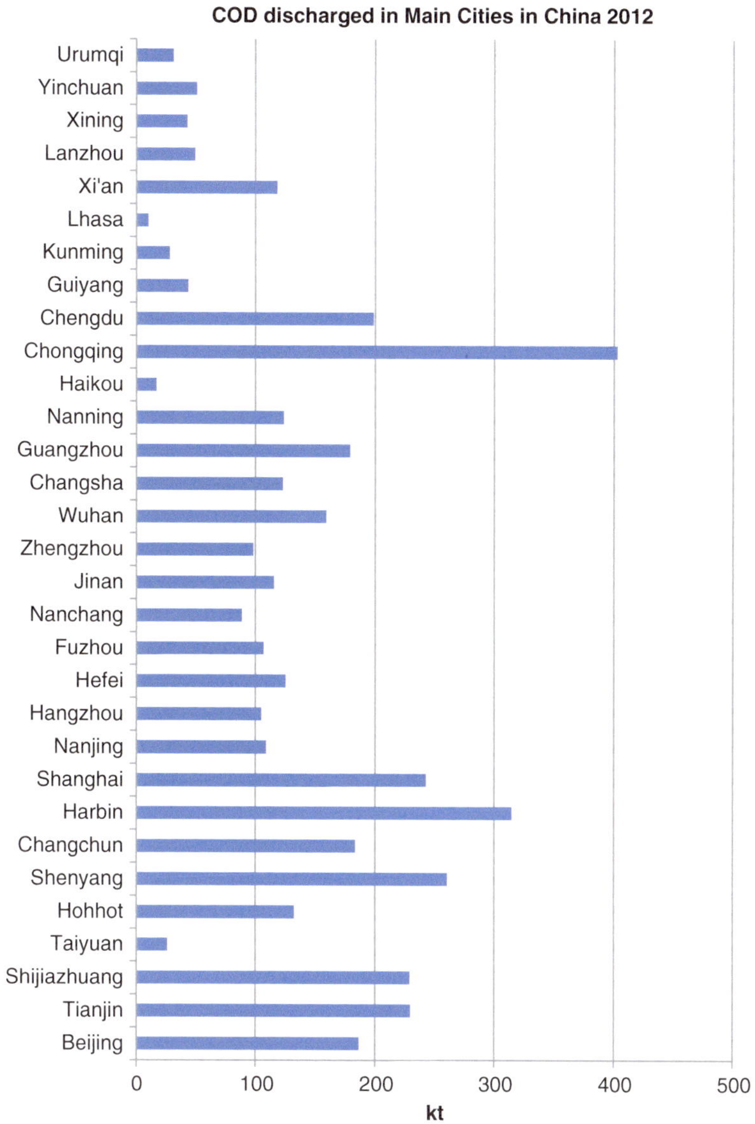

Fig. 2.37 COD in wastewater in main cities in China (2012) (Source: China Statistic Press, China Statistical Year Book 2013)

total health expenditure share in GDP had grown from 3.7 % in 1995 to 4.78 % in 2005. It reveals that the personal health expenditure is growing due to the reform from government pay the cost to personal pay it. The reform has led to distrustful and mistrustful state between patient and doctor; thus, the government tries to pay more on health cost. Therefore, the share of public health expenditure in GDP

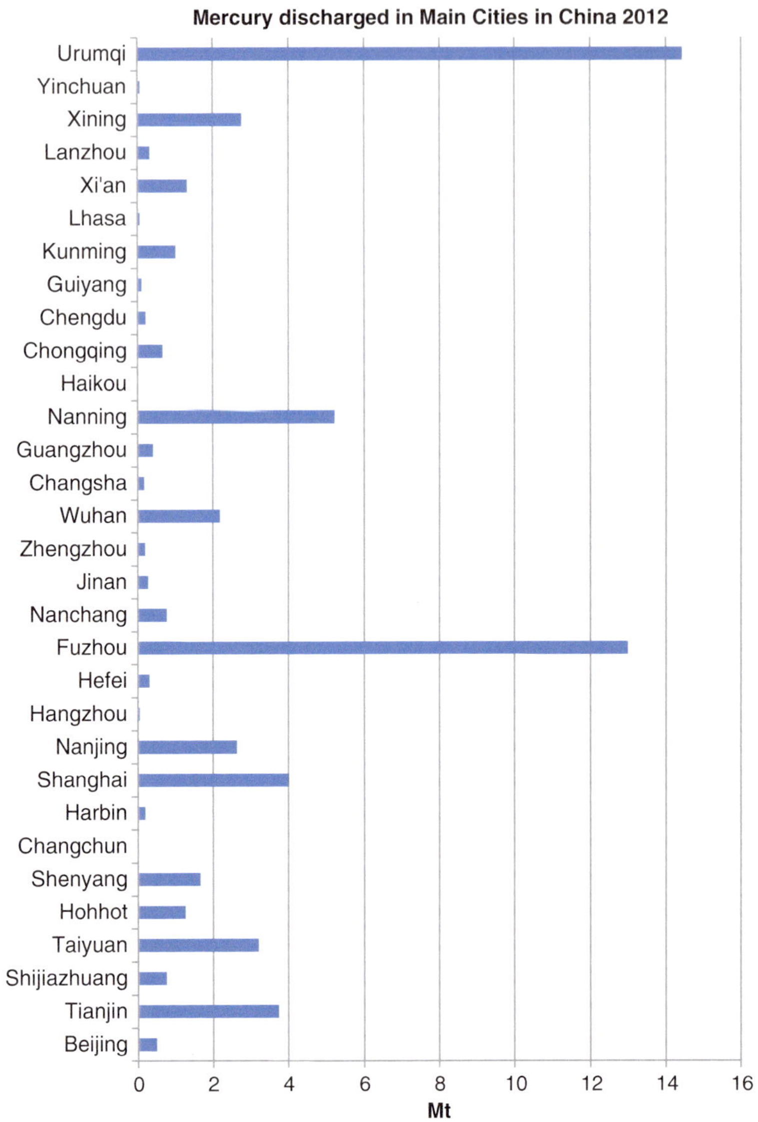

Fig. 2.38 Mercury in wastewater in main cities in China (2012) (Source: China Statistic Press, China Statistical Year Book 2013)

increased from 2006 as shown in Fig. 2.40. The total health expenditure has also grown considerably since 2006 because the people's health has become poorer and poorer. The health expenditure per capita varied from 62.3$ (constant 2005 international dollar and exchange rate by purchasing power parity) in 1995 to 214.88$ in 2006 and then to 378.9$ in 2010 (see Fig. 2.41). The average annual growth rate of the health expenditure per capita was 13.71 % during 1995–2006, and it was

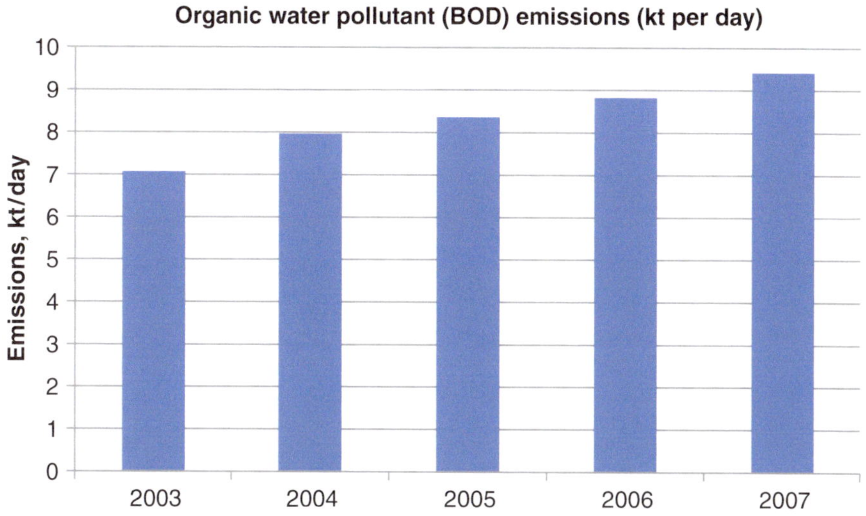

Fig. 2.39 Organic water pollutant emissions each day in China (Source: The World Bank, http://data.Worldbank.org/country/china)

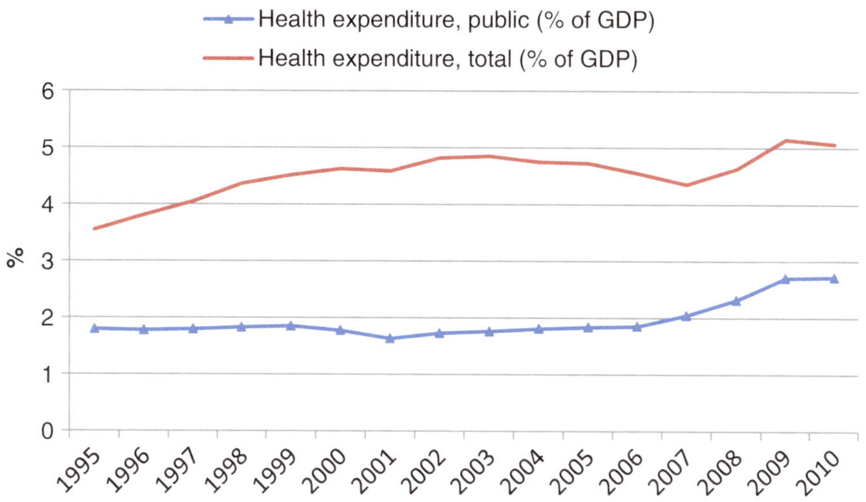

Fig. 2.40 Health expenditure in GDP in China (Source: World Bank, http://data.Worldbank.org/country/china)

16.37 % during 2007–2010. However, the average annual GDP growth was 9.96 % during 1995–2006, and it was 8.43 % during 2007–2010. The situation requires people to rethink: What is the goal of economic growth? Does it have more value to purse economic growth itself or focus on the ultimate goal of the economic development?

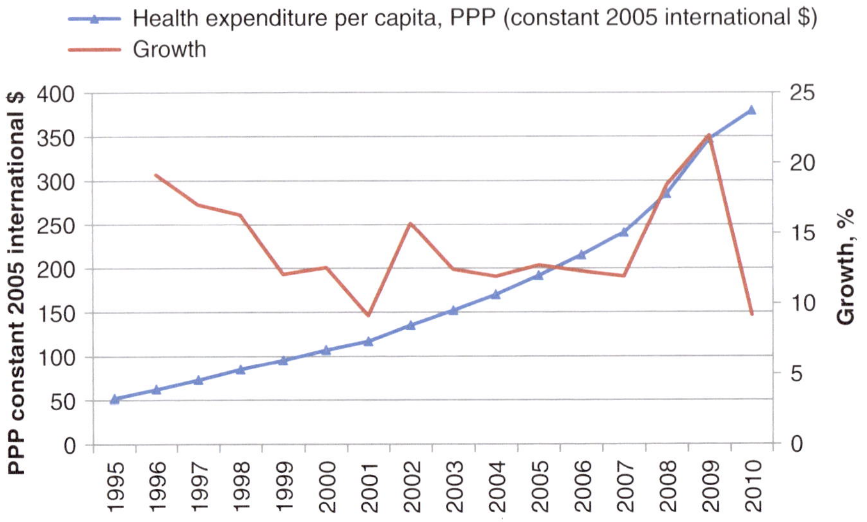

Fig. 2.41 Health expenditure per capita in China (Source: World Bank, http://data.Worldbank.org/country/china)

2.3 Economic Mutation

Pollutions are a serious problem in China because people are paying more attention to health. Is it worth having more pollution for economic growth? Is it possible to decrease the growth of energy demand to get blue sky and clean water to ensure comfortable living conditions? It has to be studied from economic system. If the gene of the economic system changes, perhaps it would be an opportunity for China' economy to solve pollution problems in the near future. As introduced in Chap. 1, *E-GDP* function has four characteristics of gene: replication, mutation, evolution, and uniqueness. A mutation of economy will happen if the intercept of the *E-GDP* function changes from negative to positive or from positive to negative, and then, the slope of the function would change greatly. Economic mutation will indicate a poor economic state, such as lower marginal representative factor productivity and stagflation of the economy; and it may also mean a good economic state, such as high marginal representative factor productivity and lower energy demand. It depends on the integration of technology improvement and innovation, interrelationships between sectors of the economy, and other factors of global economy. It would make the marginal representative factor productivity change greatly. We will compare the economic mutations in the USA, Japan, and China to find if there is a way to convert challenge to opportunity in China's economy development.

In order to discuss conveniently, average GDP (*AGDP*) of electricity can be calculated in two ways: *AGDP(S)* means the published GDP(S) divided by electricity use *E* and *AGDP(E)* is based on the calculated GDP(E) from *E-GDP* function divided by electricity use *E*. Both of them are called *AGDP* and the differences between them are very little unusually.

On the other hand, the slope of the *E-GDP* function signifies the *MGDP* in the period. A mutation implies that *MGDP* goes up and down alternatively. If the intercept of the *E-GDP* function is negative, the slope of the function will be large, i.e., the *MGDP* will be up. The mutation with *MGDP* going up is called negative mutation, and the function is called negative *E-GDP* function. If the intercept of the *E-GDP* function is positive, the slope of the function will be small relatively, i.e., the *MGDP* will be down, and the mutation is called positive mutation and the function is called positive *E-GDP* function. Thus, the negative mutation means the economy grows healthily, while the positive mutation indicates that the economy is in some problems or sick. The slope of the *E-GDP* function shows the *MGDP*, and it is also the marginal representative factor productivity of the economy.

2.3.1 Mutations of the Economy in the USA

For the US economy, three mutations happened in 1976, 1986, and 1991 since 1949. Thus, there were four periods of the US economy: 1949–1976, 1977–1986, 1987–1991, and 1992–2011. The *E-GDP* functions are shown as follows:

1949–1976:

$$\mathrm{GDP}(E) = 3.153E + 1362.2 \tag{2.1}$$

1977–1986:

$$\mathrm{GDP}(E) = 6.3484E - 2875.7 \tag{2.2}$$

1987–1991:

$$\mathrm{GDP}(E) = 2.098E + 4001.9 \tag{2.3}$$

1992–2013:

$$\mathrm{GDP}(E) = 10.703E - 13407 \tag{2.4}$$

The *E-GDP* function in formula (2.1) was a positive function; the *MGDP*, marginal representative factor productivity, was relatively small (as shown in Fig. 2.42); the *AGDP* was larger than *MGDP*; and *AGDP* decreased with an increasing *E* during 1949–1976. Since a negative mutation happened in 1977, the *MGDP* had increased from 3.153$/kWh to 6.3484$/kWh, and the intercept of the *E-GDP* function had changed from positive to negative as shown in formulas (2.1) and (2.2). Why can the mutation show that *MGDP* experiences a big growth? One reason is the growth of productivity caused by technology improvement and

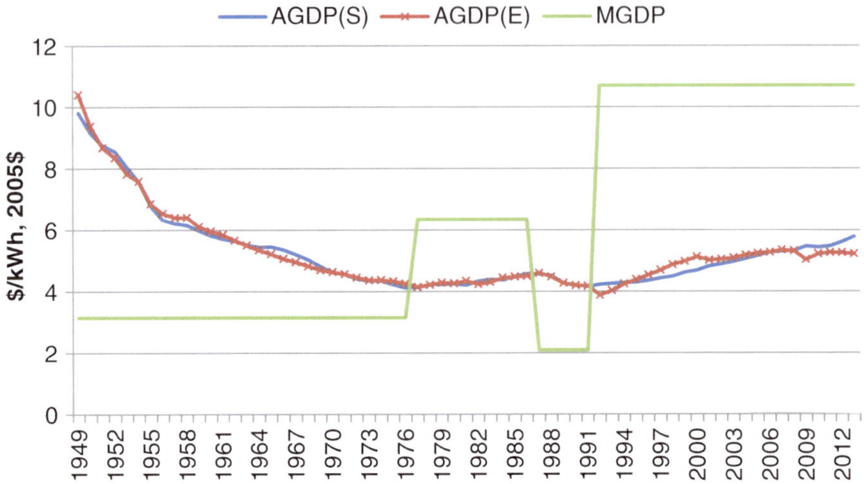

Fig. 2.42 *AGDP* and *MGDP* with mutations in the USA (Data source: Bureau of Economic Analysis, http://www.bea.gov/ Energy Information Administration USA, Annual Energy Review 2013 http://www.eia.gov/aer)

innovation. Another reason is the energy efficiency impacted by oil crisis in 1973, and also the high oil price makes the production of high energy intensity products transfer to other countries, i.e., the change of economic mix. It can be verified from the share change of the secondary industry value added in GDP. The share of secondary industry in GDP had decreased to 3.616 % during 1949–1976, while it had decreased to 4.191 % during 1977–1986. As for *AGDP*, it was decreasing from 1949 to 1976 because the intercept of formula (2.1) was positive. It is clear that the negative mutation renders the economy a growing productivity.

Another mutation happened in 1987; it was a positive mutation in which *MGDP* decreased to 2.9765$/kWh deeply during 1987–1991 (Fig. 2.42). What were the reasons for that? It is very difficult to know the details. Since *AGDP* was bigger than *MGDP*, it was decreasing with an increasing *E* during 1987–1991. It is a poor mutation for the economy. Fortunately, it takes only a short period, and another mutation, a negative mutation, happened in 1992.

The *E-GDP* function is negative as shown in formula (2.4). The *MGDP* has increased to 10.703$/kWh since 1992; the main reason is probably the Internet technology applications. The *AGDP* is less than *MGDP*, and *AGDP* keeps increasing since 1992. Figure 2.42 shows that there is a big space for the growth of *AGDP*. The negative mutation is beneficial to the economy.

The primary energy consumption in the USA was 31.98 quadrillion Btu in 1949. The electricity consumption in the three industries was 188 TWh in 1949. The annual energy and electricity growth rates are in decline trends as depicted in Fig. 2.43. However, it is very clear to reflect the feature of each period (Fig. 2.44) as follows:

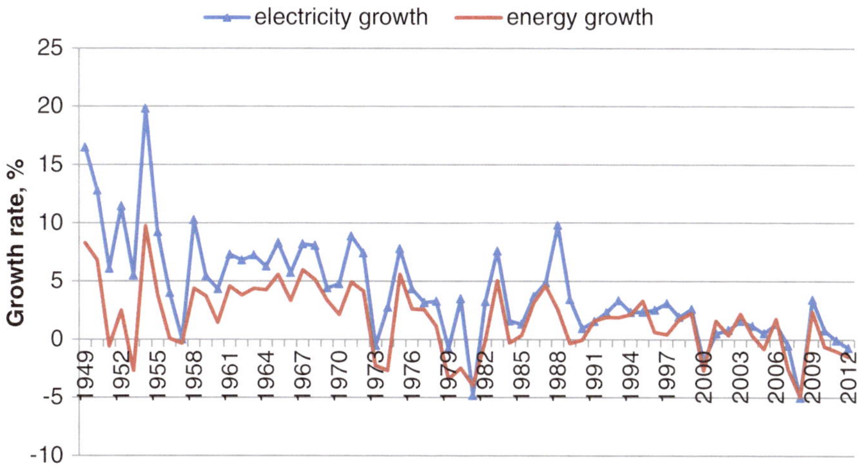

Fig. 2.43 Energy and electricity growth rate in the USA from 1950 to 2012 (Data source: Bureau of Economic Analysis, http://www.bea.gov/ Energy Information Administration USA, Annual Energy Review 2012 http://www.eia.gov/aer)

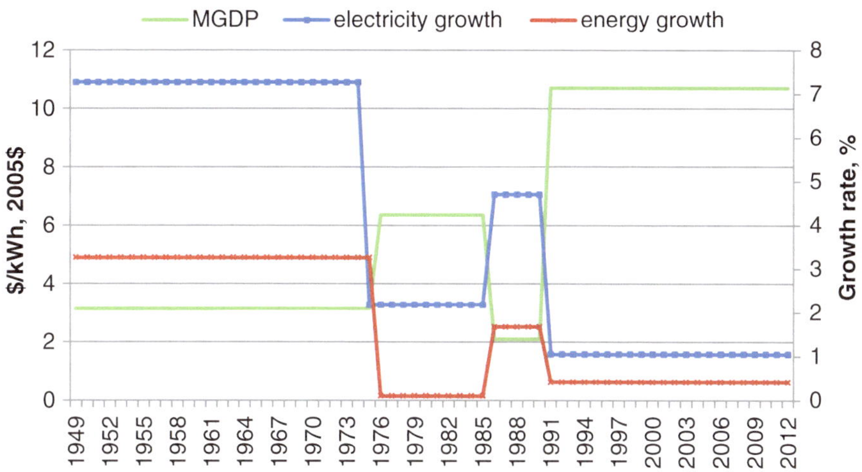

Fig. 2.44 *MGDP*, electricity growth rate, and energy growth rate with mutations in the USA (Data source: Bureau of Economic Analysis, http://www.bea.gov/ Energy Information Administration USA, Annual Energy Review 2013 http://www.eia.gov/aer)

1949–1976: a positive mutation happened in 1949, the *MGDP* was relatively lower (3.15$/kWh), and annual average growth rates of primary energy consumption and electricity consumption were relatively higher (3.26 and 7.27 %)..

1977–1986: a negative mutation happened in 1977, the *MGDP* was relatively higher (6.349$/kWh), and annual average growth rates of primary energy consumption and electricity consumption were relatively lower (2.18 and 0.098 %)..

1987–1991: a positive mutation happened in 1987, the *MGDP* was lower (2.098$/kWh), and annual average growth rates of primary energy consumption and electricity consumption were higher (1.68 and 4.7 %)..

1992–2013: a negative mutation happened in 1992, the *MGDP* was higher (10.703 $/kWh), and annual average growth rates of primary energy consumption and electricity consumption were higher (0.59 and 1.18 %).

We can see that when a negative mutation happened, the *MGDP* gets higher and the growth rates of energy and electricity consumption turn lower; and when positive mutation happened, the *MGDP* is lower and energy and electricity growth rates will be higher. It shows that mutation could indirectly affect the energy and electricity demand. It is also reflected in Figs. 2.43 and 2.44 that the energy growth rate is always lower than electricity growth, which means the growing up of electrification level is the direction of energy consumption. Shown in Fig. 2.44, we can find that energy efficiency had played an important role during 1977–1991 because the electricity growth was relatively lower; and electrification has played a significant role since 1992 because the electricity growth is relatively higher compared with primary energy consumption growth.

In summary of the mutations in the US economy, the positive mutation during 1949–1976 locked longer in lower marginal representative factor productivity; the negative mutation in 1977 showed a large increase in marginal representative factor productivity; it was a good opportunity to increase the productivity with lower energy consumption. The positive mutation in 1987 brought about a big decrease of *MGDP* with higher energy growth; it indicated a sick economy which is a challenge. However, the period took only 5 years. Fortunately, another negative mutation happened in 1992, and the *MGDP* had increased largely, which meant the opportunity of enhancing productivity with lower energy growth in the USA.

2.3.2 Mutations of the Economy in Japan

In order to show the mutation feature of an economy, we take a case study in Japan, for example. As mentioned in Sect. 1.1.2 in Chap. 1, the *E-GDP* function of Japan's economy during 1965–2010 is shown in formula (1.8). However, it can be divided into four periods, namely, 1965–1973, 1974–1988, 1989–2000, and 2001–2010, and three mutations happened in 1974, 1989, and 2001. The mutations mean big changes of marginal GDP of electricity. Thus, the four *E-GDP* functions for the four periods are shown as follows:

1965–1973:

$$\mathrm{GDP}(E) = 579.02E + 42613 \tag{2.5}$$

Formula (2.5) is a positive *E-GDP* function; the slope of function is 579.02, which indicates that the increase of one unit electricity consumption in the three industries can show an increase of GDP 579.02 Yen. The positive intercept of the

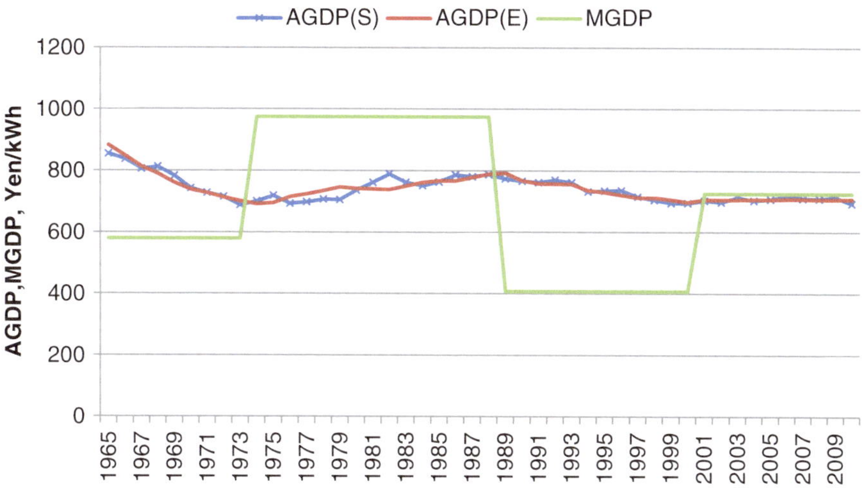

Fig. 2.45 Average GDP of electricity and marginal GDP of electricity in Japan

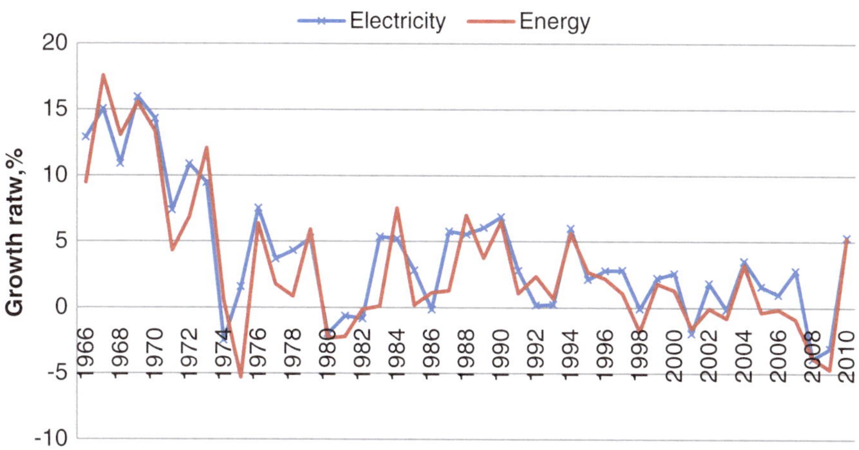

Fig. 2.46 Growth rates of primary energy and electricity in Japan (Data source: Handbook of Energy & Economic Statistics in Japan 2012)

function shows that the average GDP of one unit electricity use is in decrease with the growth of electricity consumption. As illustrated in Fig. 2.46, *AGDP* was higher than *MGDP* and it keeps declining in that period.

A negative mutation happened in 1974; thus, the *E-GDP* function becomes a negative function as shown in formula (2.6). The slope has increased to 973.62, i.e., the marginal GDP of electricity has a big increase from 579.02Yen/kWh to 937.62Yen/kWh. It shows the negative mutation makes the marginal representative factor productivity grow fast in Japan. And also *AGDP* is lower than *MGDP* and *AGDP* is increasing in this period (Fig. 2.45). The negative *E-GDP* function is shown as follows in this period:

1974–1988:

$$\mathrm{GDP}(E) = 973.62E - 96337 \tag{2.6}$$

There was another mutation, a positive mutation, which happened in 1989 in Japan. The marginal GDP of electricity decreased to 407.33Yen/kWh in the 1989–2000 period (formula 2.7). Since the mutation has shown a so big change on *MGDP*, it is poor for the economy. Unfortunately, the period lasted for 12 years until another mutation happened. Figure 2.45 shows that the *AGDP* was higher than *MGDP*, and it had decreased during 1989–2000. The positive *E-GDP* functions are shown as follows in this period:

1989–2000:

$$\mathrm{GDP}(E) = 407.33E + 211829 \tag{2.7}$$

There was a negative mutation in 2001, and it has no more data to learn if there was another mutation that happened after 2010. Thus, the period was from 2001 to 2010 in our discussion. The negative *E-GDP* function during 2001–2010 was as follows:

2001–2010:

$$\mathrm{GDP}(E) = 726.72E - 14501 \tag{2.8}$$

The marginal GDP of electricity had increased to 726.72Yen/kWh: however, it was still lower than 973.62Yen/kWh in the 1974–1988 period. The average GDP of electricity takes up trend since the intercept of formula (2.8) is negative (Fig. 2.45). However, the *AGDP* is very close to *MGDP*, and there is no enough space for *AGDP* to grow. Maybe a mutation will happen in the near future after 2010. Although the negative mutation in 2001 meant the growth of *MGDP*, it is too low to promote the *AGDP* increase. Thus, the negative mutation is too weak to drive the economic growth.

As for primary energy consumption and industrial electricity use, their growth rates are illustrated in Fig. 2.46. The difference between energy growth rate and electricity growth rate is very small, which demonstrated that the energy efficiency in electricity use has played an important role in Japan.

The feature of each period shown in Fig. 2.48 can be discussed as follows:

1965–1973: a positive mutation happened in 1965, the *MGDP* was relatively lower (579.02Yen/kWh), annual average growth rates of primary energy and electricity consumption were relatively higher (11.45 and 5.21 %), and energy growth was higher than electricity growth.

1974–1988: a negative mutation happened in 1974, the *MGDP* was relatively higher (973.62Yen/kWh), and annual average growth rates of primary energy and electricity consumption were 1.51 % and 2.61 %, respectively, in relatively lower, while the energy growth was lower than that of electricity..

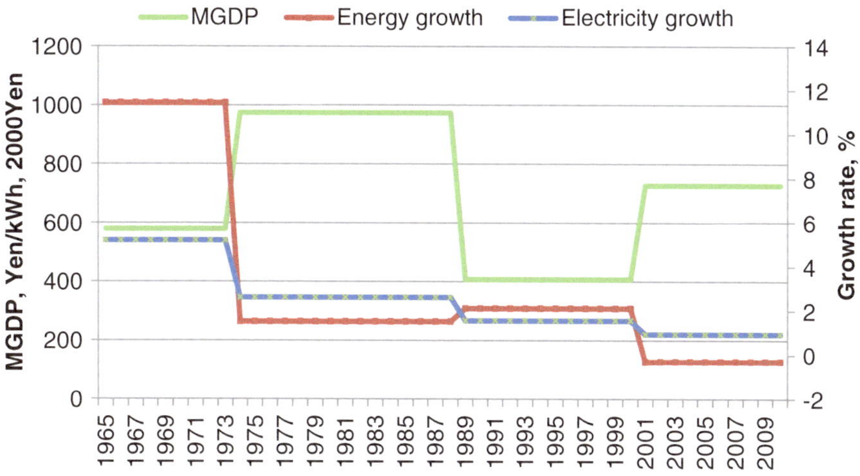

Fig. 2.47 *MGDP*, electricity growth rate, and energy growth rate with mutations in Japan (Data source: Handbook of Energy & Economic Statistics in Japan 2012)

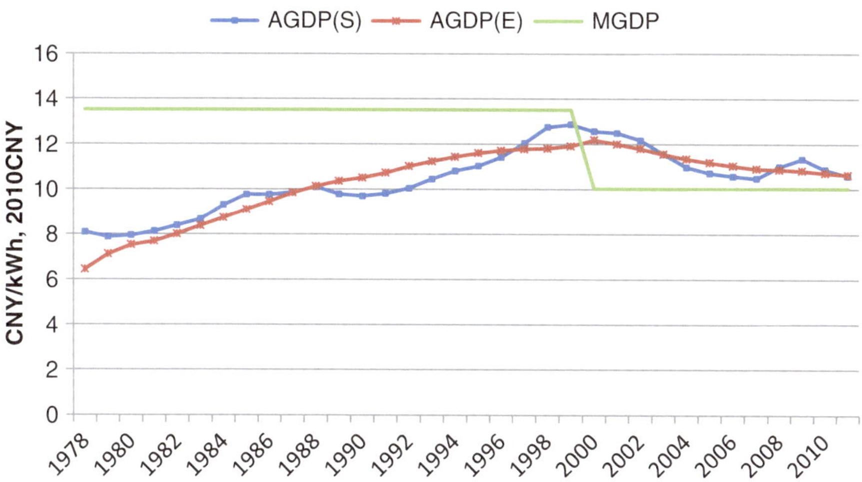

Fig. 2.48 *AGDP* and *MGDP* in China

1989–2000: a positive mutation happened in 1989, the *MGDP* was lower (407.33Yen/kWh), and annual average growth rates of primary energy consumption and electricity use were higher (2.11 and 1.56 %), and energy growth was higher than electricity growth.

2001–2010: a negative mutation happened in 2001, the *MGDP* was higher (726.72Yen/kWh), and annual average growth rates of primary energy consumption and electricity use were higher (−0.31 and 0.94 %), and energy growth was lower than electricity growth.

It is also verified that when a negative mutation happened, the *MGDP* is relatively higher, and the growth rates of energy and electricity consumption will be relatively lower; and when a positive mutation happened, *MGDP* is lower, and energy and electricity growths will be relatively higher. Thus, mutation could indirectly affect the energy and electricity demand. In other words, a positive mutation indicates the energy demand would be lower, and negative mutation means that the energy demand would be higher. It is also interesting that the energy growth was lower than electricity growth when a negative mutation happened in Japan, and vice versa when a positive mutation happened. (Why? If some readers can provide the answer, it will be valuable.)

In summary, the positive mutation period was only 9 years from 1965 to 1973; it looked good with shorter period of the low marginal representative factor productivity; the negative mutation in 1974 indicated that *MGDP* increased largely; it meant the increase of the marginal representative factor productivity with lower energy consumption; and it also looked good for the longer negative mutation period of 16 years. The positive mutation in 1989 had shown a big decrease of *MGDP* with higher energy growth, showing a sick economy. The period took 12 years. Fortunately, another negative mutation happened in 2001; the marginal representative factor productivity increased a little.

Why do the mutations happen in the national economy? What factors will affect or result in economic mutation? One factor is technology improvement. It is clear that advanced technology will improve energy efficiency, i.e., use less electricity and produce more products. Another factor is technology innovation. The new technology to produce new product will increase more electricity use in the economy, such as electric train and electric vehicle, both of them use electricity to replace gasoline. This is the growing process of electrification, which is to replace primary energy such as oil, natural gas, coal, etc. by using more electricity. Management is perhaps also a booster in energy efficiency, electrification, and production. Market demand will drive producers and consumers to balance supply and demand, for which managers have to do decision making. Market demand will reflect the economic stage, i.e., the economic mixture. In the middle stage of industrialization, the heavy industry grows fast, while in the late and post-industrialization stage, tertiary industry grows fast. Therefore, the economic mix is another factor of economic mutation.

In the case of Japan's economic mutations, in 1974, the factor of technology improvement played a crucial role in energy efficiency, which made the *MGDP* very high during 1974–1988. In that period, Japan's economy was in the late and post-industrialization stage, and economic structure transferred from heavy industry to tertiary industry. However, for the mutation in 1989, the factor of management played a role in lowering marginal representative factor productivity in the 1989–2000 period. The mutation rendered Japan's economy low marginal representative factor productivity, and the *MGDP* was the lowest one. For the mutation in 2001, although the negative mutation showed the *MGDP* was up to 726.72 Yen/kWh, it was still lower than 973.62 Yen/kWh, and the *AGDP* grew very slowly. It still demonstrated that the national economy was not strong. The study of mutations can be conducted more deeply in the three industries, many sectors, and firms or companies if more data are available. However, it can be concluded that the next mutation will be a positive one and the *MGDP* will be lower than 726.72 Yen/kWh. Since there was more space for the growth of *AGDP* in 2010, the positive mutation may happen in the next few years in Japan's economy. Then, the following mutation will be a negative one, and the *MGDP* will increase to a high level, which will bring about a bright future for the economy in Japan.

2.3.3 Mutations of the Economy in China

Reviewing the *E-GDP* function of China from 1978 to 2011, there was a mutation that happened in 2000 that divided China's economy into two periods with *E-GDP* functions as follows:

1. 1978–1999:

$$\mathrm{GDP}(E) = 13.52E - 1709.3 \tag{2.9}$$

   It is a negative *E-GDP* function. The *MGDP*, slope of the *E-GDP* function, is 13.52 CNY/kWh, and the intercept is negative, which means the *AGDP* will rise with the increase of *E*. As shown in Fig. 2.47, the *AGDP* changed from 6.43 CNY/kWh in 1978 to 11.9 CNY/kWh in 1999. Then, a positive mutation happened in 2000.
2. 2000–2011:

$$\mathrm{GDP}(E) = 10.037E + 2526.5 \tag{2.10}$$

   It is a positive function. The mutation shows *MGDP* decreases from 13.52 CNY/kWh to 10.037 CNY/kWh, and the intercept of the *E-GDP* function changes from −1709.3 to 2526.5. As illustrated in Fig. 2.47, *MGDP* is constant, while *AGDP* decreases from 12.55 to 10.59 CNY/kWh. *AGDP* is very close to 10.038 CNY/kWh of *MGDP*, which means a negative mutation will happen in a few years after 2011.

The period featured can be summarized as follows:

1978–1999: a negative mutation happened, the *MGDP* was relatively higher (13.52CNY/kWh), and annual average growth rates of primary energy consumption and electricity use were relatively lower (4.38 and 7.31 %).

2000–2011: a positive mutation happened in 1999, the *MGDP* was lower (10.037/kWh), and annual average growth rates of primary energy consumption and electricity consumption were higher (8.75 and 12.42 %).

We can see that when a negative mutation happens, the *MGDP* gets higher, the growth rates of energy and electricity consumption turn lower; on the contrary, when a positive mutation happens, the *MGDP* is lower, and energy and electricity

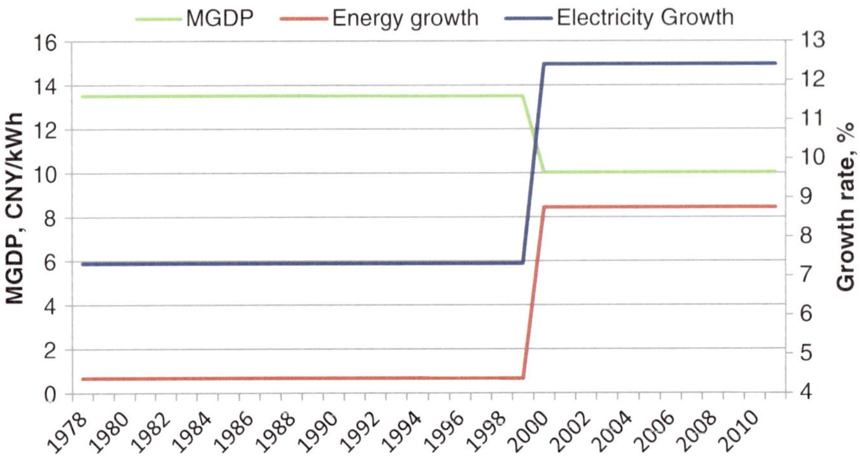

Fig. 2.49 *MGDP*, energy growth rate, and electricity growth rate in China

growth rates will be higher. It also shows that mutation could indirectly affect the energy and electricity demand, i.e., negative mutation means relatively low energy demand and positive mutation indicate high energy demand. It is also showed in Fig. 2.49 that the energy growth rate is always lower than electricity growth, which means the increase of electrification level is the trend of energy consumption.

In summary, the positive mutation in 1999 appeared with a decrease of *MGDP* with higher energy growth, which implies a challenge in the economy.

The next economic mutation will be a negative one with the *MGDP* increasing. It means an opportunity of growing productivity with lower energy growth in China. Then, the pollutions would be better as people expected. However, if the mutation were not powerful enough to increase the *MGDP* greatly, the economy would be in stagflation. As for the case in Japan, there was a negative mutation in 2001; the *MGDP* only increased 1.784 times. This is a warning. What is the power of the mutation to increase *MGDP* greatly? It must be the technology improvement and innovation in a suitable economic mechanism. For example, the negative mutation happened in 1992 in the US economy; the *MGDP* increased more than 4.67 times.

2.4 Findings

Based on the above discussions, the following findings can be summarized as:

1. Positive mutation and negative mutation in an economy will happen in turn alternatively.
2. Positive mutation will show a smaller marginal representative factor productivity and higher energy demand, while negative mutation will show a bigger marginal representative factor productivity and lower energy demand relatively. It is the cycle mutation feature.

3. If the next economic mutation in China will be a negative mutation, then marginal representative factor productivity would have a growth, the energy demand would be lower, and the situation of pollutions in China would be better. However, according to the experience in Japan since 2001, if the negative mutation were not powerful enough to increase marginal representative factor productivity greatly, the economy would be in stagflation in that period.
4. A health economy has two features: one is that the negative mutation can be powerful enough to increase marginal representative factor productivity greatly and last for a longer period without another mutation happening, and another feature is that the positive mutation shows that marginal representative factor productivity decreases only a little and another mutation will happen as soon as possible.

References

1. The Energy Conservation Center (2014) Handbook of energy & economic statistics [M], The Energy Data and Modelling Center, The Institute of Energy Economics, Japan, p 308
2. Maria van der Hopositive (2014) How we must build the electric future, energy [J]. J Int Energy Agency, Issue 6 – 2nd quarter, p 3, Paris
3. International Energy Agency (2012) Electricity information [R]. International Energy Agency Statistics, Paris
4. BP Statistical Review of World Energy [R], June 2009
5. Huijiong Wang (2003) Integrated study of China's development and reform [M]. Foreign Languages Press, Beijing, p 473
6. Source: China Statistical Year Book [R]. China Statistic Press, Beijing, 1996, 2006, and 2013

Chapter 3
Economic Gene Mapping of China

China's economy bids farewell to the period of high growth near 10 % in the past three decades. What will it be in the future? How do we avoid risk of stagflation and convert the challenges to opportunities of economic development by the fiscal and monetary policies? It will be tried in this chapter by studying the economic gene mapping of China. Simulations of economic development until 2025 have been executed with our multi-Agent Response Equilibrium (ARE) model which will be introduced in Part Two in this book. Policy selection is like doing experiments to test many policy simulations by ARE and then to find a better one. It has tested many scenarios of combining fiscal and monetary policies each year during 2015–2025. Two different policy simulations have been studied in this chapter: one is based on the fiscal and monetary policies in 2014 without any new policy during 2015–2025 as BAU scenario, and another one is to select some fiscal and monetary policies in 2015–2016 as policy scenario. Two economic gene mappings have been studied based on the two simulations. One finding is that the BAU would increase marginal representative factor productivity of the economy a little higher than policy scenario. Another finding is that a mutation occurred in primary industry, secondary industry, and national economy in 2012, respectively, a mutation in tertiary industry happened in 2014, and there would be no mutations in the three industries and national economy until 2025. The third finding is marginal representative factor productivity in primary industry would have a great increase during 2012–2025 based on the gene map, and perhaps it could be a productivity revolution. The fourth finding is there would be an economic cycle in China's economy in 2015–2025, the peak growth would be around 2019–2021, and the low growth periods would be in 2015–2017 and 2023–2025. Policy scenario would decline the peak growth and rise lower growth of the economy. As for the inherent rule of the economy in China, there would be an increment of the productivity in the next few years if there were no economic events and strategic mistakes on national economy. Therefore, the mutation of national economy in 2012 would be an era to turn up productivity from lower level to higher one. In this case, the challenges could be an opportunity in China's economic development.

Z. Hu et al., *China's Economic Gene Mutations*,
DOI 10.1007/978-3-662-47298-9_3

3.1 Gene Mapping of an Economy

As we have discussed in Chaps. 1 and 2 that economic mutation can show poor economic situation, such as lower marginal representative factor productivity and higher energy demand, it will also indicate good economic situation, such as high productivity and lower energy demand. It depends on the integration of technology improvement and innovation and interrelationships between sectors of the economy and other factors of global economy. We will discuss gene mapping of an economy and case studies for the USA and Japan in this section. In order to study the gene mapping of China for a long period, economic simulations on BAU and policy scenarios will be introduced in Sects. 3.2 and 3.3, respectively. And Sect. 3.4 will show the economic gene mapping of China and study the impacts of the two scenarios on the gene mapping. Comparing gene maps between Japan and China, economic diagnostic analysis has been conducted in Sect. 3.5, and some findings are presented in Sect. 3.6 in this chapter.

3.1.1 Economic Gene Mapping

As mentioned in Chap. 1, *electricity consumption can be called as representative production factor (RPF)* (see formula (1.3)) in the economic activities such as firm's manufacturing in secondary industry and company's servicing in tertiary industry. It is clear that the more inputs of production factors, the more electricity has to be consumed in the manufacturing and servicing. Production function with electricity is the map between RPF, electricity use e, and the output of production, such as value added v. Production function with electricity (e–v function) for a firm or a service company is a linear function. The slope of e–v function is marginal value added of the electricity (mv); it is also the marginal representative factor productivity (MRFP) of the firm or company, which has the characteristic of replication of gene, and the change of intercept of e–v function from positive to negative or from negative to positive, reflecting the characteristic of economic mutation of gene [1]. The relationship expresses the economic gene map, and the e–v function can be called economic gene function of the firm or company. It is clear that every firm (company) has its own economic gene map. This is the economic gene map on firm's level.

A sector contains many firms or companies that produce the same kind of product or service. Production function with electricity of a sector, $Es-Vs$ function, is a linear function, and it has the gene characteristics. It can be called economic gene function or economic gene map on sectoral level.

Many sectors can compose an industry, such as primary industry, secondary industry, or tertiary industry. Industrial production function with electricity, $Ei-Vi$ function $i = 1,2,3$, as a linear function, has also the gene characteristics, and it can be called economic gene function or economic gene map on industrial level.

National economy consists of three industries. Then, national production function with electricity (*E-GDP* function) is a linear function too, and it also has the gene characteristics. It can be called economic gene function or economic gene map on national level.

Gene mapping of an economy includes four level gene maps. The top level is national economic gene maps which are formed by a few lines of *E-GDP* functions in different periods, and each period starts with a mutation. The second level is the three industries' economic gene maps which are formed by $Ei-Vi$ ($i=1, 2, 3$) functions for primary, secondary, and tertiary industry, respectively. For each industry, it consists of a lot of sectors. Thus, the third level is sectoral economic gene maps which are formed by *Es–Vs* functions in the industry. The fourth level is firm or company economic gene maps which are formed by e_j–v_j functions for each firm j in the sector. It is the bottom level that shows the economic gene feature of a firm, company, shopping center, etc., which are the base of the economic gene mapping. Therefore, the economic gene mapping contains millions of production functions with electricity for firms, sectors, industries, and national economy. It covers all of the economic units in the economy.

As discussed in Chap. 2, a positive mutation will form a positive gene function with a positive period, i.e., a positive gene map, and a negative mutation will form a negative gene map. An ideal gene map must have two features:

1. For negative gene function, production function with electricity, the MRFP is big and the negative period is longer.
2. For positive gene function, the MRFP is as high as possible and the positive period is shorter.

For the national economic gene map, it means the MRFP must be higher; and the negative gene period ought to be as long as possible, while the positive gene period has to be short.

3.1.2 Economic Gene Maps of the USA

For the US economy as mentioned in Chap. 2, three mutations have happened in 1976, 1986, and 1991 since 1949. Thus, four *E-GDP* functions are shown as follows:

1949–1976:

$$\mathrm{GDP}(E) = 3.153E + 1362.2 \tag{3.1}$$

1977–1986:

$$\mathrm{GDP}(E) = 6.3484E - 2875.7 \tag{3.2}$$

1987–1991:

$$\mathrm{GDP}(E) = 2.098E + 4001.9 \tag{3.3}$$

1992–2013:

$$\mathrm{GDP}(E) = 10.703E - 13407 \tag{3.4}$$

It shows the positive mutation function lasts too long from 1949 to 1976. If it were short, it would be better for the US economy. Fortunately, the negative mutation in 1977 indicated that MRFP grew 101.34 % by comparing formula (3.2) with (3.1), the positive mutation in 1987 showed MRFP growth of −66.95 % by comparing formula (3.3) with (3.2), and the negative mutation in 1992 revealed that MRFP grew 410.15 % by comparing formula (3.4) with (3.3). The four national gene functions can be denoted as four lines in Fig. 3.1. It is the gene maps of the national economy in the USA. The red dots in the left of each line show the years when mutation happened. The national gene map is nearly ideal one based on the gene map features except the longer positive mutation line in the left of Fig. 3.1. It means that the US economy is in good health generally. However, there was a risk in 2009, when *AGDP (E)* was lower as shown in Fig. 2.43 in Chap. 2. It may be a bad news for the economy. On the review of gene in the US economy, would the financial crisis in 2009 show a mutation? It will depend on the new data for in-depth study.

3.1.3 Economic Gene Maps of Japan

As a case study for Japan, there were three mutations in 1974, 1989, and 2001, respectively. Thus, the four *E-GDP* functions for the four periods are shown as follows:

1965–1973:

$$\mathrm{GDP}(E) = 579.02E + 42613 \tag{3.5}$$

1974–1988:

$$\mathrm{GDP}(E) = 973.62E - 96337 \tag{3.6}$$

1989–2000:

$$\mathrm{GDP}(E) = 407.33E + 211829 \tag{3.7}$$

2001–2010:

$$\mathrm{GDP}(E) = 726.72E - 14501 \tag{3.8}$$

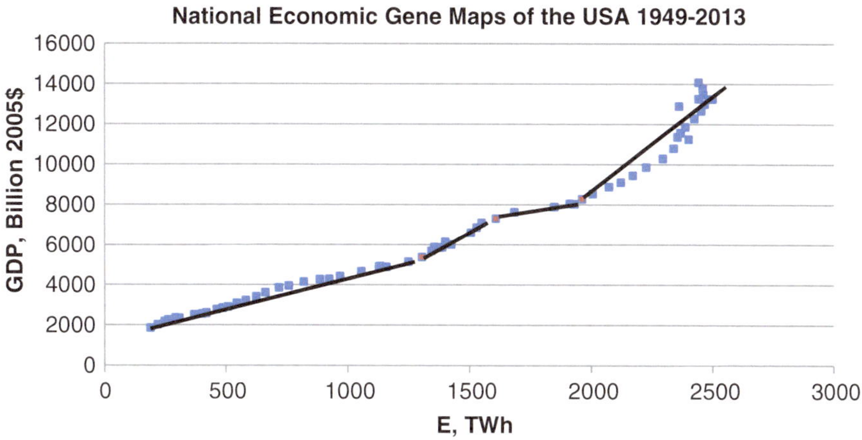

Fig. 3.1 National gene maps of the US economy from 1949 to 2013 (Data source: Bureau of Economic Analysis, http://www.bea.gov/ Energy Information Administration USA, Annual Energy Review 2014 http://www.eia.gov/aer)

A negative mutation in 1974 indicates great increase of slope in formula (3.6). The MRFP was 973.62, the biggest one during 1965–2010, with growth rate equaling 68.15 % by comparing formula (3.5) with (3.6). Unfortunately, the positive mutation in 1989 shows the MRFP decreases sharply and it takes in 12 years. The negative mutation in 2001 reveals that MRFP is increasing, but it is lower than that of 1974–1988. The national economic gene maps of Japan are shown in Fig. 3.2. It clearly shows the mutations happened in national economy and the gene feature in Japan. However, if the third line from the left in Fig. 3.2 were shorter and the fourth line were sharper, it would be better. It means Japan's economy would be in better condition if the 1989–2000 period of the positive function were short and the MRFP in formula (3.8) were greater than 726.72 by diagnostic gene map of the national economy.

3.2 Economic Simulation in China (BAU Scenario)

We select the version of input–output table with 42 sectors in this book. The 42 sectors are listed in Table 3.1. Sector 1, Agriculture, Forestry, Animal Husbandry, and Fishery, is regarded as the primary industry, Sectors 2–26 can be classified as secondary industry, and Sectors 27–42 can be put as tertiary industry in China.

Based on the official input–output table of 2010 and some fiscal and monetary policies, the economic simulations from 2011 to 2014 will be introduced in Chap. 9.

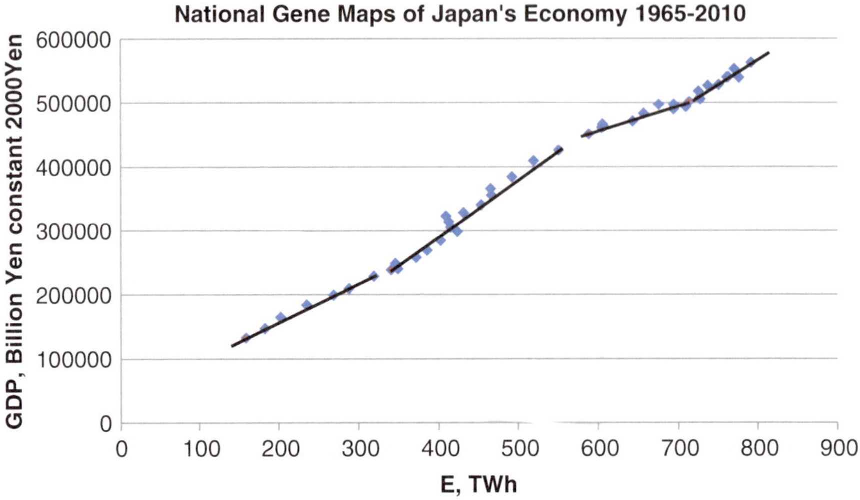

Fig. 3.2 Economic mutation lines for Japan (Data source: Handbook of Energy & Economic Statistics in Japan 2012)

The following monetary policies and fiscal policies are implemented in 2014, and there is no any policy changes in 2015–2025 in our ARE simulation. This is the business as usual (BAU) scenario in our economic simulation in this book.

The monetary policies are set as follows: Liabilities to nonfinancial institutions and households are 112,662 billion CNY.[1] Central bank bonds are 702.62 billion CNY. Claims on nonfinancial institutions are 67,311.56 billion CNY. Total assets are 175546.60 billion CNY. Deposit reserve ratio is 20.0 %. One-year benchmark lending rate is 6.00 %. One-year benchmark deposit rate is 3.00 %.

The fiscal policies are set as follows: domestic value-added tax rate is 13.03 %; domestic consumption tax rate is 1.86 %; import value-added tax and consumption tax rate is 5.18 %; sales tax rate is 5.54 %; corporate income tax rate is 7.39 %; resource tax rate is 0.22 %; urban maintenance and construction tax rate is 0.97 %; stamp rate is 1.90 %; urban land use tax rate is 0.32 %; personal income tax rate is 2.68 %; and tariff rate is 1.21 %.

It is worth noting that the tax rates in our ARE model don't have the same meanings as those in practice, since the tax bases of all the different taxes are diverse and difficult to be reflected in the ARE simulation. The rate of a certain kind of tax is redefined as the total amount of this kind of tax divided by the overall tax and surplus in the starting year of the simulation. Actually, the rate is treated as a certain portion of the overall tax and surplus instead of a proportion of its tax base in our model. For example, the total amount of domestic value-added tax is 1,547.02 billion CNY in 2007, and the overall tax and surplus is 11,874.03 billion CNY according to China's 2007 input–output table. Therefore, the "tax rate" in our

[1] All the CNY in this book is in constant 2010 CNY.

Table 3.1 42 sectors in China's economy

No.	Sectors	No.	Sectors
1	Agriculture, Forestry, Animal Husbandry, and Fishery	22	Scrap and Waste
2	Mining and Washing of Coal	23	Production and Supply of Electric Power and Heat Power
3	Extraction of Petroleum and Natural Gas	24	Production and Distribution of Gas
4	Mining of Metal Ores	25	Production and Distribution of Water
5	Mining and Processing of Nonmetal Ores and Other Ores	26	Construction
6	Manufacture of foods and tobacco	27	Traffic, Transport, and Storage
7	Manufacture of Textile	28	Post Service
8	Manufacture of Textile Wearing Apparel, Footwear, Caps, Leather, Fur, Feather(Down), and Its products	29	Information Transmission, Computer Services, and Software
9	Processing of Timbers and Manufacture of Furniture	30	Wholesale and Retail Trades
10	Papermaking, Printing, and Manufacture of Articles	31	Hotels and Catering Services
11	Processing of Petroleum, Coking, and Processing of Nuclear Fuel	32	Financial Intermediation
12	Chemical Industry	33	Real Estate
13	Manufacture of Nonmetallic Mineral Products	34	Leasing and Business Services
14	Smelting and Rolling of Metals	35	Research and Experimental Development
15	Manufacture of Metal Products	36	Comprehensive Technical Services
16	Manufacture of General Purpose and Special Purpose Machinery	37	Management of Water Conservancy, Environment, and Public Facilities
17	Manufacture of Transport Equipment	38	Services to Households and Other Services
18	Manufacture of Electrical Machinery and Equipment	39	Education
19	Manufacture of Communication Equipment, Computer, and Other Electronic Equipment	40	Health, Social Security, and Social Welfare
20	Manufacture of Measuring Instrument and Machinery for Cultural Activity and Office Work	41	Culture, Sports, and Entertainment
21	Manufacture of Artwork and Other Manufacture	42	Public Management and Social Organization

model is 13.03 % that is calculated based on the above two numbers, which is different from 17 %, the actual rate in real life.

3.2.1 Production Growths of 42 Sectors

The price index of a certain sector reflects the supply and demand balance of its product in the market. ARE model can provide product price index each week dynamically in the simulation period. It is very important information for us to know the market demand and analyze manufacturer's production situation. Although the simulation of the period is from 2015 to 2025, we will show the price index weekly from 2015 to 2020 since the space is limited. There are 52 weeks each year and 13 weeks each quarter. Thus, the 1st–13th weeks are in the first quarter, 14–26th weeks are in the second quarter, 27–39th weeks are in the third quarter, and 40th–52nd weeks are in the fourth quarter.

Figure 3.3 depicts weekly the price indexes of Sector 6, Sector 7, Sector 8, and Sector 9 during the 2015–2020 period. Sector 6 is Manufacture of Foods and Tobacco; Sector 7 is Manufacture of Textile; Sector 8 is Manufacture of Textile Wearing Apparel, Footwear, Caps, Leather, Fur, Feather (Down), and Its Products; and Sector 9 is Processing of Timbers and Manufacture of Furniture. Their initial values are 1.0556, 2.4313, 1.1240, and 0.4732 in the first week of 2015, and their last states in this period, the last week of 2020, are 1.5303, 1.7458, 2.4183, and 3.2234, respectively. These sectors all belong to the light industry. Their products contribute a lot to domestic household consumption and China's export. The overall trends of their price indexes are up during the 2015–2020 period, which means the market demand will be enhanced and the international economy situation will turn better. The product price index of Manufacture of Foods and Tobacco (No.6) looks stable in wavelets, and it is also in wavelets for Sector 8. While the waves of Sectors 7 and 9 are bigger relatively, especially for Sector 7, it is 4.78 in the 50th week of 2019 and then falls to 1.77 in the last week of 2020. It shows the demand would be higher from the third quarter of 2019 to the second quarter of 2020 in both international and domestic markets. However, the demand would fall deeply in the second half of year 2020.

Figure 3.4 shows the growth rates of the value added of Sector 6, Sector 7, Sector 8, and Sector 9 during the 2015–2025 period. The average growth rates of the four sectors in the 2015–2020 period are, respectively, 6.85 %, 6.25 %, 4.82 %, and 3.76 % annually. And the average growth rates of the four sectors in the 2021–2025 period are, respectively, 3.59 %, 2.98 %, 2.50 %, and 1.61 % annually. The sector of Manufacture of Foods and Tobacco (No. 6) has the best performance in these sectors in the whole 10-year period, while the sector of Processing of Timbers and Manufacture of Furniture (No. 9) has the worst performance. From the perspective of these two periods, each sector has higher growth rates in the 2015–2020 period than in the 2021–2025 period, which reflects the changes in Chinese economic trend in the future.

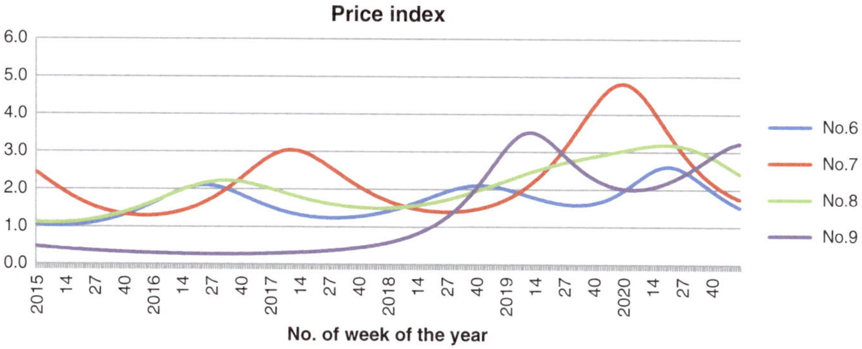

Fig. 3.3 Market price indexes of Sector 6, Sector 7, Sector 8, and Sector 9 during the period 2015–2020

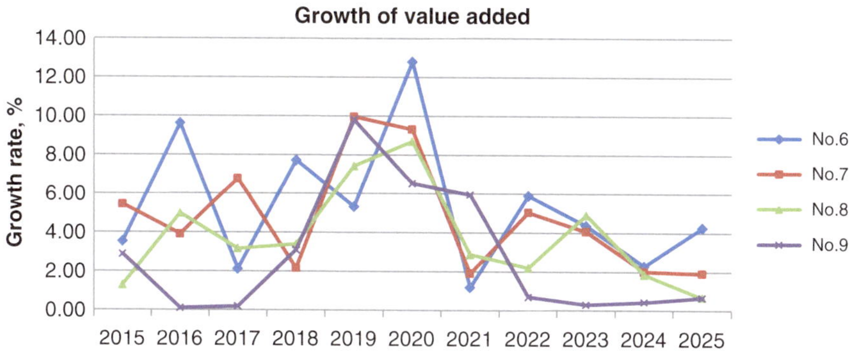

Fig. 3.4 Annual growth rates of the value added of Sector 6, Sector 7, Sector 8, and Sector 9 during the period 2015–2025

Figure 3.5 displays the price indexes of Sector 11, Sector 12, Sector 13, and Sector 14 during the 2015–2020 period. Sector 11 is Processing of Petroleum, Coking, and Processing of Nuclear Fuel; Sector 12 is Chemical Industry; Sector 13 is Manufacture of Nonmetallic Mineral Products; and Sector 14 is Smelting and Rolling of Metals. As shown in the figure, the overall trends of most price indexes are up owing to the increasing demand. The initial levels of these price indexes (the first week of 2015) are 0.8641, 1.4318, 1.1237, and 1.2535, while their final values are 1.7895, 1.4011, 2.8317, and 3.2845 (the last week of 2020). We can see that the price index of Sector 13, Manufacture of Nonmetallic Mineral Products, is going to grow fast for a long period of time from 2017, while the price indexes of other sectors encounter wave-like uplifts. Since these sectors provide various fundamental materials, such as cement, rubber, iron, and steel, for the development of China's economy, the increasing demand reflects the rapid growth of China's national economy in the period.

Fig. 3.5 Market price indexes of Sector 11, Sector 12, Sector 13, and Sector 14 during the period 2015–2020

Figure 3.6 illustrates the growth rates of the value added of Sector 11, Sector 12, Sector 13, and Sector 14 during the 2015–2025 period. The annual average growth rates of the four sectors in the 2015–2020 period are, respectively, 7.49 %, 5.79 %, 5.86 %, and 6.28 %. Their average growth rates in the 2021–2025 period are 2.41 %, 2.19 %, 2.39 %, and 2.57 % in proper order. Sector 11, Processing of Petroleum, Coking, and Processing of Nuclear Fuel, has the fastest growth in these sectors in the former 6-year period, while Sector 12, Chemical Industry, has the lowest growth rate. As for the latter 5-year period, Sector 14, Smelting and Rolling of Metals, has the highest growth, while Sector 12, Chemical Industry, is still the one with the lowest growth in the four sectors. It is clear that Chemical Industry will go through a tough time in the future in China. Besides, all sectors have greater growth rates in the 2015–2020 period. It also shows that the growth peak period is in 2019–2020, which is over 8 % for the four sectors.

Figure 3.7 displays the price indexes of Sector 16, Sector 17, Sector 18, and Sector 19 during the 2015–2020 period. Sector 16 is Manufacture of General Purpose and Special Purpose Machinery; Sector 17 is Manufacture of Transport Equipment; Sector 18 is Manufacture of Electrical Machinery and Equipment; and Sector 19 is Manufacture of Communication Equipment, Computer, and Other Electronic Equipment. These four sectors are all important manufacturing. Their price indexes increase from 1.2349, 0.5856, 1.7053, and 1.3641 to 2.9898, 3.4320, 2.7669, and 2.9752, respectively, which reflect the continuous prosperity of Chinese economy. As for Sector 16, Sector 18, and Sector 19, their price indexes grow fast from the fourth quarter in 2016 to the first quarter in 2019, which means their product is quite popular in this period, while for Sector 17, its price index experience obvious ascents both from the third quarter in 2015 to the first quarter in 2018 and from the first quarter in 2019 to the second quarter in 2020, when the demand for its product is in high level.

Figure 3.8 shows the growth rates of the value added of Sector 16, Sector 17, Sector 18, and Sector 19 during the 2015–2025 period. The average growth rates of the four sectors in the 2015–2020 period are, respectively, 7.46 %, 5.80 %,

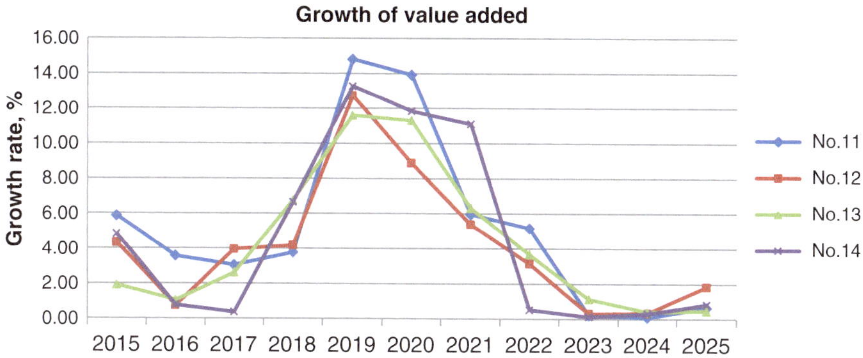

Fig. 3.6 Annual growth rates of the value added of Sector 11, Sector 12, Sector 13, and Sector 14 during the period 2015–2025

Fig. 3.7 Market price indexes of Sector 16, Sector 17, Sector 18, and Sector 19 during the period 2015–2020

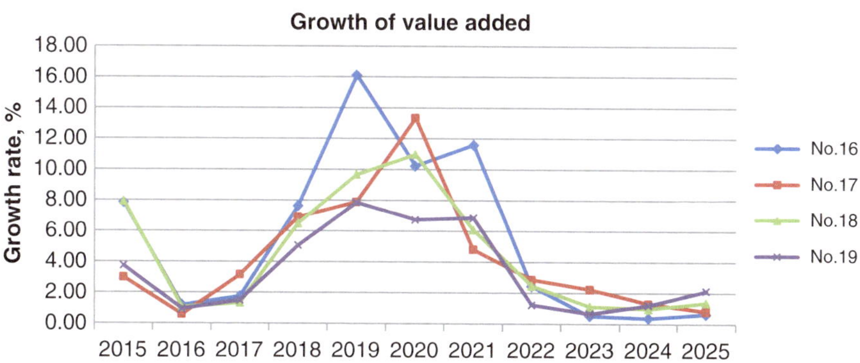

Fig. 3.8 Annual growth rates of the value added of Sector 16, Sector 17, Sector 18, and Sector 19 during the period 2015–2025

6.24 %, and 4.30 %, and their average annual growth rates in the 2021–2025 period are successively 3.12 %, 2.41 %, 2.41 %, and 2.42 %. Among these four sectors, Sector 16, Manufacture of General Purpose and Special Purpose Machinery, has the greatest growth rate in the two periods. Oppositely, Sector 19, Manufacture of Communication Equipment, Computer, and Other Electronic Equipment, has the lowest growth rate in the 2015–2020 period, while Sector 17, Manufacture of Transport Equipment, gains the worst performance in the 2021–2025 period. Evidently, Sector 16 will experience a good time for its value added. From the perspective of the two periods, all sectors have growth higher than those in the former one. We can see from Fig. 3.8 that the growth peak years are 2018, 2019, and 2020 for the sectors.

Figure 3.9 depicts the price indexes of Sector 30, Sector 31, Sector 32, and Sector 33 during the 2015–2020 period. Sector 30 is Wholesale and Retail Trades; Sector 31 is Hotels and Catering Services; Sector 32 is Financial Intermediation; and Sector 33 is Real Estate. Their price indexes are initially 1.2135, 1.4516, 0.1955, and 1.1123, respectively, in the 1st week in 2015, and they arrive at 1.7348, 4.0506, 1.1028, and 2.6965 in proper order in the last week in 2020. The price index of Sector 32 experiences high volatility, which reflects the special feature in the development of Financial Intermediation; however, it will fall deeply in 2020. It is also shown in Fig. 3.9 that Sector 31, Hotels and Catering Services, is going to have a big growth during 2017–2019, which demonstrates more expenditure will be spent on hotels and restaurants within this time .

Figure 3.10 illustrates the growth rates of the value added of Sector 30, Sector 31, Sector 32, and Sector 33 during the 2015–2025 period. The average growth rates of the four sectors in the 2015–2020 period are, respectively, 8.09 %, 6.31 %, 8.57 %, and 5.59 %, and their average growth rates in the 2021–2025 period are 3.17 %, 3.42 %, 3.23 %, and 3.09 % in proper order. The sector of Financial Intermediation (No. 32) has the fastest growth in these sectors in the 2015–2020 period, while the sector of Real Estate (No. 33) has the lowest growth rate. As for the situations in the 2021–2025 period, the sector of Hotels and Catering Services (No. 31) has the highest growth, while the sector of Real Estate (No. 33) still has the lowest growth among these four sectors. Obviously, the sector of Real Estate will go through depression in the future. In terms of time, all sectors have higher growth in the 2015–2020 period. Besides, what is worth noting is that the growths of these sectors are obviously higher than the above sectors in the real economy which have been shown before.

Figure 3.11 displays the price indexes of Sector 35, Sector 36, Sector 37, and Sector 38 during the 2015–2020 period. Sector 35 is Research and Experimental Development; Sector 36 is Comprehensive Technical Services; Sector 37 is Management of Water Conservancy, Environment, and Public Facilities; and Sector 38 is Services to Households and Other Services. The price indexes of these four sectors are initially 0.3094, 0.5995, 1.4270, and 0.9350 at the start of the period, while their final values are 2.1165, 1.3875, 1.9874, and 3.0144, respectively, at the end of 2020. Their overall trends are up from 2016, even though they go through fluctuations.

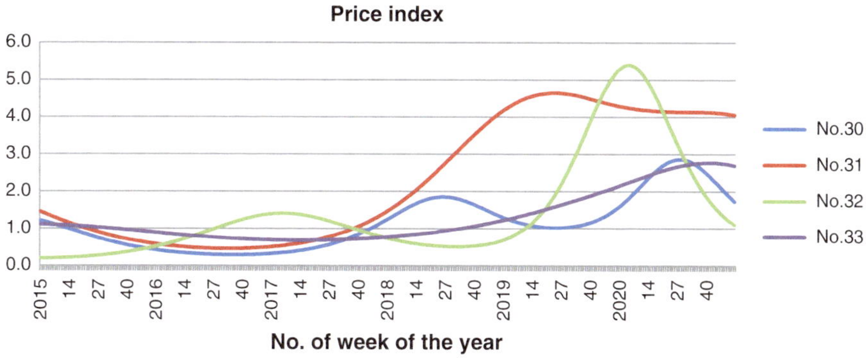

Fig. 3.9 Market price indexes of Sector 30, Sector 31, Sector 32, and Sector 33 during the period 2015–2020

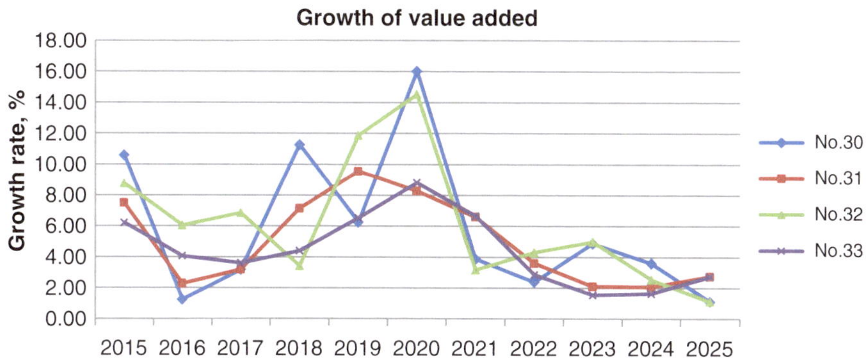

Fig. 3.10 Annual growth rates of the value added of Sector 30, Sector 31, Sector 32, and Sector 33 during the period 2015–2025

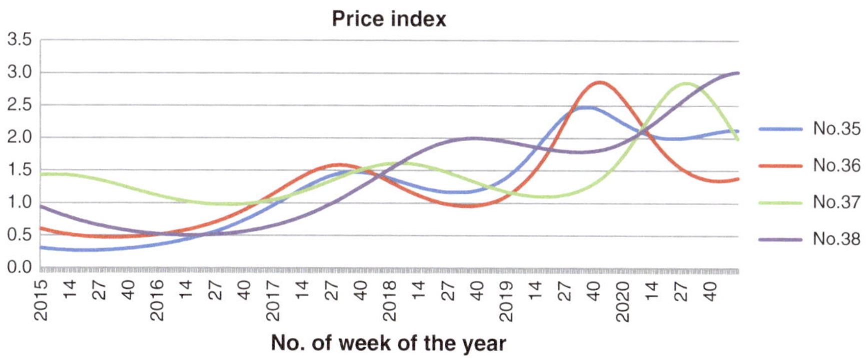

Fig. 3.11 Market price indexes of Sector 35, Sector 36, Sector 37, and Sector 38 during the period 2015–2020

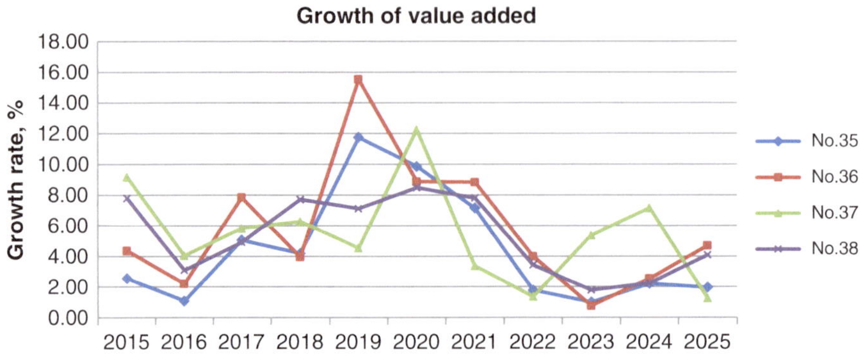

Fig. 3.12 Annual growth rates of the value added of Sector 35, Sector 36, Sector 37, and Sector 38 during the period 2015–2025

Figure 3.12 shows the growth rates of the value added of Sector 35, Sector 36, Sector 37, and Sector 38 during the 2015–2025 period. The average growth rates of the four sectors in the 2015–2020 period are, respectively, 5.74 %, 7.12 %, 7.01 %, and 6.51 % and become 2.81 %, 4.16 %, 3.70 %, and 3.87 % in the 2021–2025 period. Among these four sectors, Sector 36, Comprehensive Technical Services, has the highest growth, while Sector 35, Research and Experimental Development, has the lowest growth in the 2015–2020 period. As for the next period, the sector of Comprehensive Technical Services remains the highest one, while the sector of Research and Experimental Development is still the lowest one. Sector 36 will encounter an excellent era, while Sector 35 is not so lucky. However, the condition of Sector 35 is not that bad considering the fact that the growth rate of 2.81 % is still greater than the majority of sectors in manufacturing. Similar to the above sectors, these sectors have relative higher growth rates in the 2015–2020 period.

Figure 3.13 depicts the price indexes of Sector 39, Sector 40, Sector 41, and Sector 42 during the 2015–2020 period. Sector 39 is Education; Sector 40 is Health, Social Security, and Social Welfare; Sector 41 is Culture, Sports, and Entertainment; and Sector 42 is Public Management and Social Organization. These four sectors are very closely related to the daily life of residents in China. Compared with most sectors discussed before, the price indexes of these sectors are relatively stable except the price of Sector 41, which goes through a rapid increase from the third quarter in 2018 to the second quarter in 2020. The price indexes of the other three sectors fluctuate within the area of 1.0–2.3 in most time. Their initial statuses are, respectively, 1.1791, 1.5402, 0.8608, and 1.4707 in the 1st week in 2015, while their final values are 1.5787, 1.5076, 2.4200, and 1.3332 in the 52nd week in 2020.

Figure 3.14 illustrates the growth rates of the value added of Sector 39, Sector 40, Sector 41, and Sector 42 during the 2015–2025 period. The average growth rates of the four sectors in the 2015–2020 period are, respectively, 6.88 %, 6.64 %, 7.80 %, and 6.51 %, and their average annual growth rates in the 2021–2025 period are 5.37 %, 4.89 %, 4.17 %, and 5.09 % in proper order. In the former period, the

Fig. 3.13 Market price indexes of Sector 39, Sector 40, Sector 41, and Sector 42 during the period 2015–2020

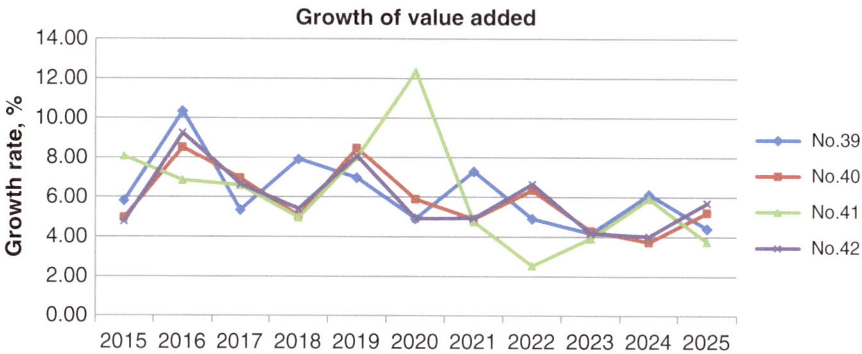

Fig. 3.14 Annual growth rates of the value added of Sector 39, Sector 40, Sector 41, and Sector 42 during the period 2015–2025

sector of Culture, Sports, and Entertainment has the fastest growth among these sectors, while the sector of Public Management and Social Organization has the lowest growth rate. In the latter period, the sector of Education owns the best performance, while the sector of Culture, Sports, and Entertainment gains the worst performance. Anyway, all of these sectors have great growth rates of their valued added in the whole national economy. To sum up, the sectors in the tertiary industry will grow faster than the sectors in the second industry, which is in line with the industrial structure adjustment conducted by Chinese government.

3.2.2 *Value-Added Growths of Three Industries*

The GDP would be 82.772 trillion CNY in 2020 and 96.628 trillion CNY in 2025. People used to pay more attention on economic growth. Figure 3.15 shows the

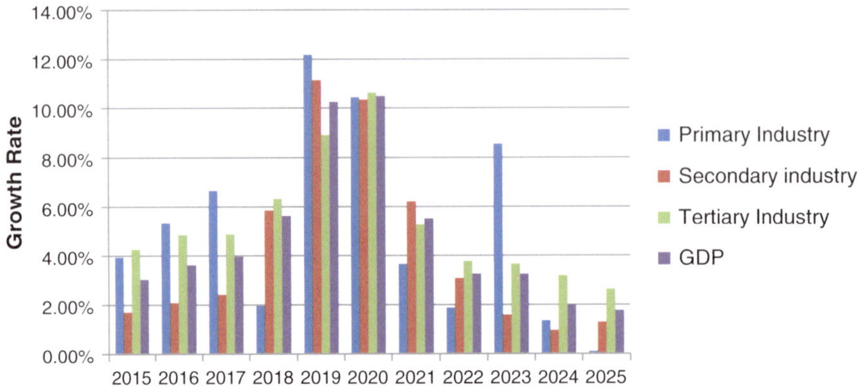

Fig. 3.15 Annual growth rates of the value added of primary industry, secondary industry, and tertiary industry and GDP during the period 2015–2025

growth rates of the value added of primary industry, secondary industry, tertiary industry, and GDP in each year during the 2015–2025 period. It seems that there is a peak growth in 2019–2020, and it would be in low growth in 2015–2016 and 2024–2025, which shows a cycle in about 10 years. On average, the annual growth rate of GDP during the 2015–2020 period is 6.16 %; the annual growth rate of value added of the primary industry, the secondary industry, and the tertiary industry during the 2015–2020 period is, respectively, 6.75 %, 5.59 %, and 6.64 %. Within the 2021–2025 period, the average annual growth rate of GDP is 3.15 %; the average annual growth rates of value added of the primary industry, the secondary industry, and the tertiary industry during the 5-year period are, respectively, 3.11 %, 2.63 %, and 3.71 %. It is clear that the growth rate of China's national economy will slow down in the future. Particularly, the secondary industry will go through a relatively low growth. This is because the government in China is trying to adjust the economic structure and pursue a sustainable development path. It is significant for China to search for a new position apart from the World Factory which is destined to become a historical memory sooner or later.

3.2.3 Energy Demand and Supply

The demand and supply of energy resource are also our concern. Besides, energy consumption is an important booster and indicator in China's national economy. In this subsection, we present some results of our ARE simulation on electricity and coal. Figure 3.16 displays the electricity demand of primary industry, secondary industry, tertiary industry, and the whole industry in each year during the 2015–2025 period. It is evident that the demand keeps growing steadily. But the growth

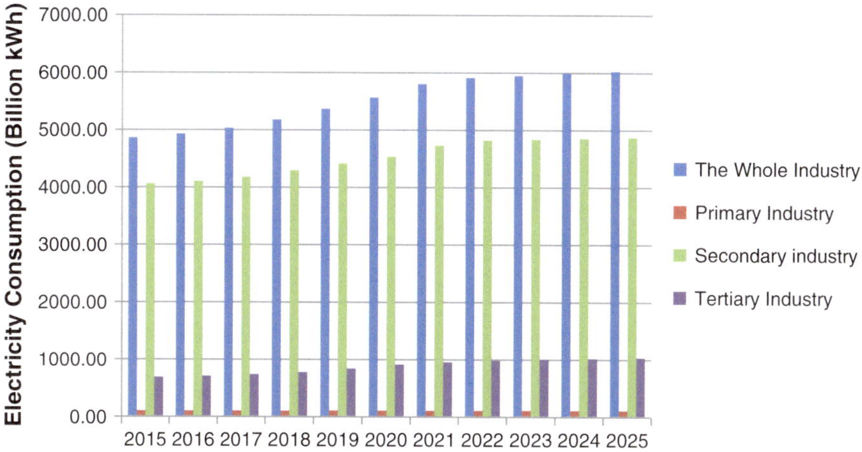

Fig. 3.16 Annual electricity demand by the whole industry, primary industry, secondary industry, and tertiary industry during the period 2015–2025

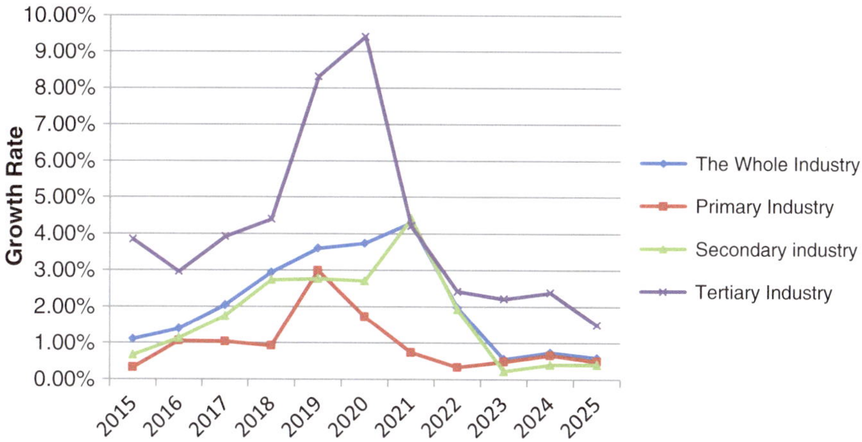

Fig. 3.17 Annual growth rates of electricity consumption of the whole industry, primary industry, secondary industry, and tertiary industry during the period 2015–2025

rate will turn quite slower from 2022 to 2025, which is a significant indicator for the research of saturated electricity demand.

Figure 3.17 shows the growth rate of electricity demand of primary industry, secondary industry, tertiary industry, and the whole industry in each year during the 2015–2025 period. It is evident that the growth rates are slightly growing to a peak within the 2015–2020 period. This is because the electricity will play an increasingly significant role in the national economy due to its features of clean, efficient,

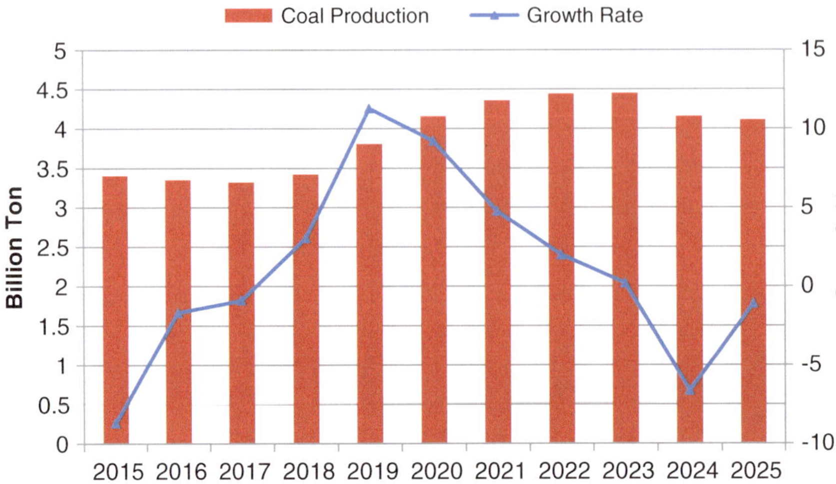

Fig. 3.18 Annual coal production and growth rate during the period 2015–2025

and convenient. Especially, the tertiary industry electricity consumption has the largest growth rate because of the relatively rapid development of tertiary industry in the next period of Chinese economy, which is in line with the goal of Chinese government to adjust the economic structure. As for the 2021–2025 period, the growths of electricity consumption are obviously slowing down. The reasons for this situation are mainly related to the slowdown of the economic growth.

Figure 3.18 displays the coal production and its growth rate in each year during the 2015–2025 period. Clearly, the production is in negative growth in 2015–2017, and then, it keeps growing steadily to 2019, but then goes down until the end of the period. The peak coal production is 4.45 billion tons in 2023, and it will be 4.11 billion tons in 2025. The coal is an indispensable energy resource in China. The simulation results verify that it will continue playing the crucial role, but its contribution will decline with the maturity of the technologies on the utilization of renewable energy resources.

3.3 Economic Simulation of Fiscal and Monetary Policy (Policy Scenario)

From BAU scenario, we can see that economic growth is in a wavelet in which the low growth period is in 2015–2017, and it gradually rise up to the peak in 2019–2020; then, the growth rate begins to fall in 2022–2025. It shows around a 10-year cycle of economic growth in China. It used to be around a 9-year cycle in the 1978–2008 period [2]. In order to stimulate the low growth in 2015–2017, some fiscal and monetary policies will be designed as a policy scenario. In this section, we will present the detailed results with policy adjustments to see the impacts of different fiscal policies and monetary policies on China's economy.

As mentioned in Sect. 3.2, the initial monetary policies are set as follows: Liabilities to nonfinancial institutions and households are 112,662 billion CNY. Central bank bonds are 702.62 billion CNY. Claims on nonfinancial institutions are 67,311.56 billion CNY. Total assets are 175,546.60 billion CNY. Deposit reserve ratio is 20.0 %. One-year benchmark lending rate is 6.00 %. One-year benchmark deposit rate is 3.00 %.

In 2015, the deposit reserve ratio is adjusted to 19 %; 1-year benchmark lending rate is amended to 5.0 %; and the 1-year benchmark deposit rate is set to 2.5 %. In 2016, the deposit reserve ratio is adjusted to 15 %; 1-year benchmark lending rate is amended to 3.0 %; and the 1-year benchmark deposit rate is set to 1.5 %.

The initial fiscal policies are set as follows: domestic value-added tax rate is 13.03 %; domestic consumption tax rate is 1.86 %; import value-added tax and consumption tax rate is 5.18 %; sales tax rate is 5.54 %; corporate income tax rate is 7.39 %; resource tax rate is 0.22 %; urban maintenance and construction tax rate is 0.97 %; stamp rate is 1.90 %; urban land use tax rate is 0.32 %; personal income tax rate is 2.68 %; tariff rate is 1.21 %.

In 2015, the rates of all sorts of taxes except tariff are adjusted. They are reset as follows: domestic value-added tax rate is 10.42 %; domestic consumption tax rate is 1.49 %; import value-added tax and consumption tax rate is 4.15 %; sales tax rate is 4.43 %; corporate income tax rate is 5.91 %; resource tax rate is 0.18 %; urban maintenance and construction tax rate is 0.78 %; stamp rate is 1.52 %; urban land use tax rate is 0.26 %; personal income tax rate is 2.14 %.

In 2016, the rates of all sorts of taxes except tariff are adjusted again. They are reset as follows: domestic value-added tax rate is 7.30 %; domestic consumption tax rate is 1.04 %; import value-added tax and consumption tax rate is 2.90 %; sales tax rate is 3.10 %; corporate income tax rate is 4.14 %; resource tax rate is 0.12 %; urban maintenance and construction tax rate is 0.55 %; stamp rate is 1.07 %; urban land use tax rate is 0.18 %; personal income tax rate is 1.50 %.

During 2017–2025, the fiscal and monetary policies will be the same as the abovementioned without any change. This is the design of policy scenario.

3.3.1 Production Growths of 42 Sectors

We can see that Fig. 3.19 depicts the price indexes of Sector 6, Sector 7, Sector 8, and Sector 9 during the 2015–2020 period. As mentioned before, Sector 6 is Manufacture of Foods and Tobacco; Sector 7 is Manufacture of Textile; Sector 8 is Manufacture of Textile Wearing Apparel, Footwear, Caps, Leather, Fur, Feather (Down), and Its Products; and Sector 9 is Processing of Timbers and Manufacture of Furniture. Their initial values are 1.0556, 2.4313, 1.1240, and 0.4732, and their last states in this period are 1.2814, 0.9107, 1.5239, and 1.4357, respectively.

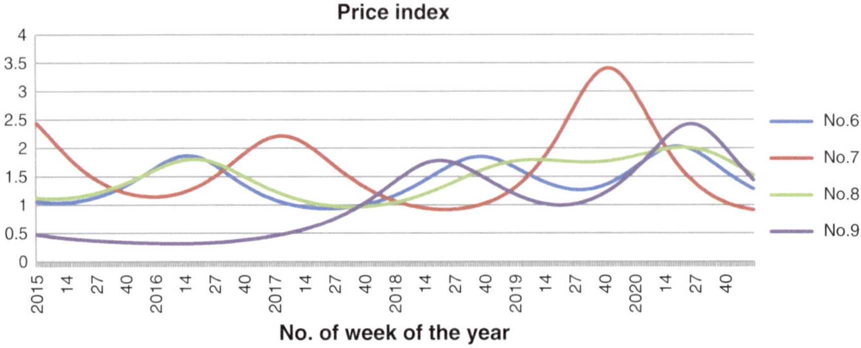

Fig. 3.19 Market price indexes of Sector 6, Sector 7, Sector 8, and Sector 9 during the period 2015–2020

Compared with the results in BAU scenario introduced in 3.2.1, the levels of their price trends are relatively lower. The highest value of these price indexes in this scenario is 3.4107 for Sector 7 that happens in the 41st week in 2019, while that in BAU scenario is 4.8035 for Sector 7 in the 52nd week in 2019. The reason is that the new policy settings implemented in this scenario provide better economic environments for companies to expand their production, which leads to the change in the supply–demand relationship of the products of these sectors.

Figure 3.20 shows the growth rates of the value added of Sector 6, Sector 7, Sector 8, and Sector 9 during the 2015–2025 period. The average growth rates of the four sectors in the 2015–2020 period are, respectively, 6.65 %, 6.18 %, 4.68 %, and 4.00 %, and the average growth rates in the 2021–2025 period are 3.78 %, 3.34 %, 2.48 %, and 1.62 % in proper order. In the former period, Sector 6, Manufacture of Foods and Tobacco, has the highest growth in these sectors, while Sector 9, Processing of Timbers and Manufacture of Furniture, has the lowest growth. In the latter period, Sector 6 remains the sector with the highest growth rate, while Sector 9 is still the sector with the lowest growth rate. The situation is the same like that in the BAU scenario.

Figure 3.21 displays the price indexes of Sector 11, Sector 12, Sector 13, and Sector 14 during the 2015–2020 period. As mentioned above, Sector 11 is Processing of Petroleum, Coking, and Processing of Nuclear Fuel; Sector 12 is Chemical Industry; Sector 13 is Manufacture of Nonmetallic Mineral Products; and Sector 14 is Smelting and Rolling of Metals. As shown in the Fig. 3.20, the overall trends of most price indexes are up owing to the increasing demand. The initial levels of these price indexes are 0.8641, 1.4318, 1.1237, and 1.2535 in the 1st week in 2015, while their final values are 1.3095, 0.9052, 1.7967, and 2.2763 in the 52nd week in 2020. Compared with the results of BUA introduced in Sect. 3.2.1, the final price levels are relatively lower. Especially for Sector 13, Manufacture of Nonmetallic Mineral Products, its final price index is only 63.45 % of its original level in the BAU scenario. The reason is that the new policies in this scenario promote the

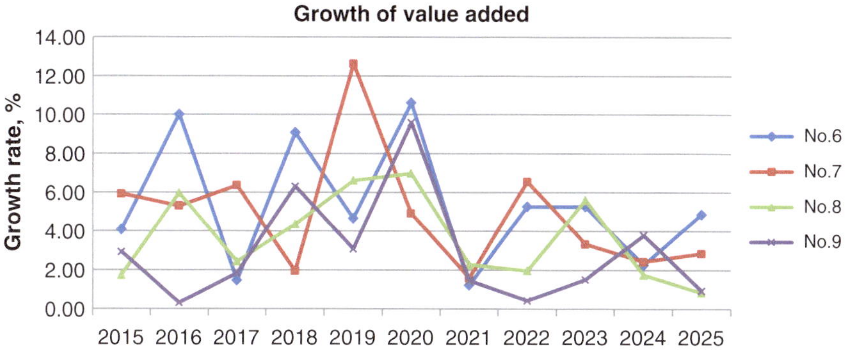

Fig. 3.20 Annual growth rates of the value added of Sector 6, Sector 7, Sector 8, and Sector 9 during the period 2015–2025

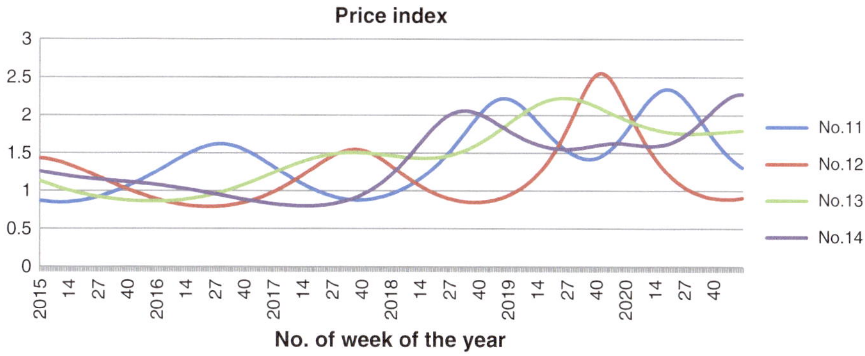

Fig. 3.21 Market price indexes of Sector 11, Sector 12, Sector 13, and Sector 14 during the period 2015–2020

expansion of production for the companies in these sectors, which results in the change in the supply–demand relationship of their products.

Figure 3.22 illustrates the growth rates of the value added of Sector 11, Sector 12, Sector 13, and Sector 14 during the 2015–2025 period. The average growth rates of the four sectors in the 2015–2020 period are, respectively, 6.96 %, 4.96 %, 5.49 %, and 5.84 %. And their average annual growth rates are, respectively, 3.34 %, 2.64 %, 3.14 %, and 2.29 %. In the former period, the sector of Processing of Petroleum, Coking, and Processing of Nuclear Fuel (No. 11) has the fastest growth in these sectors, while the sector of Chemical Industry (No. 12) has the lowest growth rate. In the latter period, the sector of Processing of Petroleum, Coking, and Processing of Nuclear Fuel (No. 11) remains the one with the highest, while the sector of Smelting and Rolling of Metals (No. 14) becomes the one that has the worst growth in the four sectors.

Figure 3.23 displays the price indexes of Sector 16, Sector 17, Sector 18, and Sector 19 during the 2015–2020 period. As mentioned before, Sector 16 is

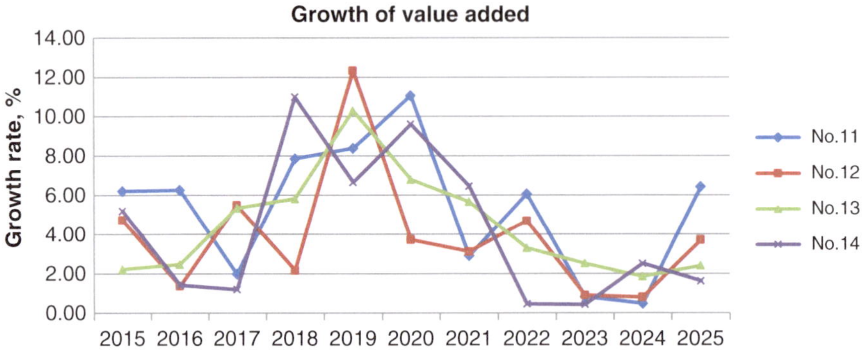

Fig. 3.22 Annual growth rates of the value added of Sector 11, Sector 12, Sector 13, and Sector 14 during the period 2015–2025

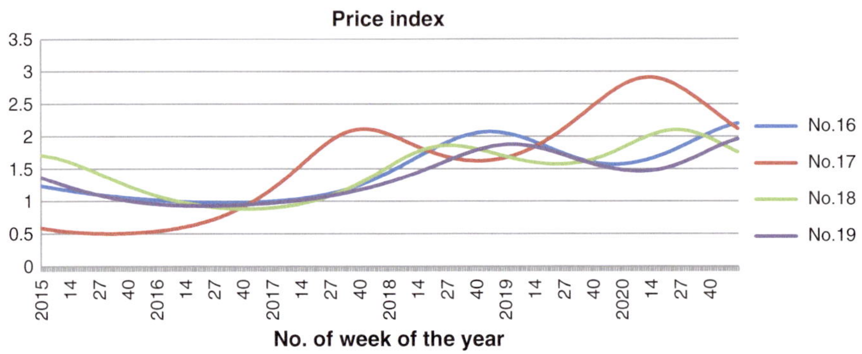

Fig. 3.23 Market price indexes of Sector 16, Sector 17, Sector 18, and Sector 19 during the period 2015–2020

Manufacture of General Purpose and Special Purpose Machinery; Sector 17 is Manufacture of Transport Equipment; Sector 18 is Manufacture of Electrical Machinery and Equipment; and Sector 19 is Manufacture of Communication Equipment, Computer, and Other Electronic Equipment. These four sectors are all important manufacturing. Their price indexes increase from 1.2349, 0.5856, 1.7053, and 1.3641 to 2.1982, 1.8466, 2.1982, and 2.1199, respectively, by the end of 2020, which are clearly lower than the levels in the BAU scenario due to the adjustments in fiscal policies and monetary policies. For instance, the highest index is 2.9070 in the 13rd week in 2020 for Sector 17, while the biggest value is 4.7599 in the BAU scenario. The decline rate is 38.93 %.

Figure 3.24 shows the growth rates of the value added of Sector 16, Sector 17, Sector 18, and Sector 19 during the 2015–2025 period. The average growth rates of the four sectors in the 2015–2020 period are, respectively, 7.06 %, 6.05 %, 5.90 %, and 4.24 %. Their average growth rates in the 2021–2025 period are 3.51, 3.03, 2.55, and 2.57 % in proper order. The sector of Manufacture of General

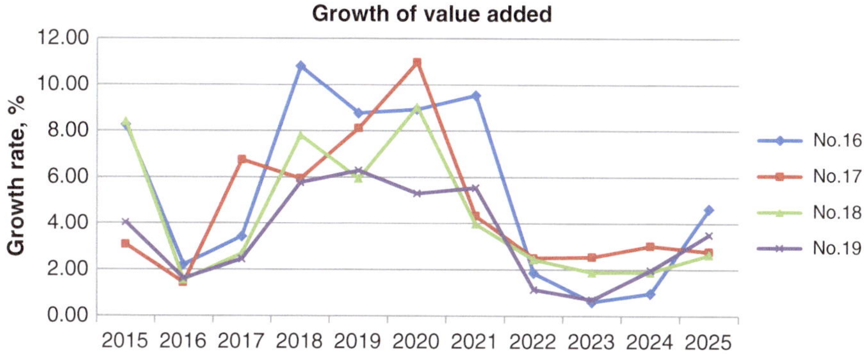

Fig. 3.24 Annual growth rates of the value added of Sector 16, Sector 17, Sector 18, and Sector 19 during the period 2015–2025

Purpose and Special Purpose Machinery (No.16) has the highest growth in these sectors, while the sector of Manufacture of Communication Equipment, Computer, and Other Electronic Equipment (No. 19) has the lowest growth in the 2015–2020 period. And the sector of Manufacture of General Purpose and Special Purpose Machinery (No. 16) has the greatest growth, while the sector of Manufacture of Electrical Machinery and Equipment (No. 18) has the lowest growth rate in the 2021–2025 period.

Figure 3.25 depicts the price indexes of Sector 30, Sector 31, Sector 32, and Sector 33 during the 2015–2020 period. As mentioned previously, Sector 30 is Wholesale and Retail Trades; Sector 31 is Hotels and Catering Services; Sector 32 is Financial Intermediation; and Sector 33 is Real Estate. Their price indexes are initially 1.2135, 1.4516, 0.1955, and 1.1123, respectively, in the 1st week in 2015, and they arrive at 0.7588, 1.1378, 0.3644, and 1.1348 in proper order at the end of the period. Compared with the results in BAU scenario introduced in 3.2.1, the levels of their price trends are relatively lower. The average price indexes of these four sectors in the period are, respectively, 0.7676, 1.0029, 0.7109, and 0.7173. On the contrary, their average values are 1.1486, 2.2811, 1.4808, and 1.2821 in the BAU scenario. This is because the new policy settings implemented in this scenario provide relaxed environments for companies to expand their production, which leads to the change in the supply–demand relationship of their products.

Figure 3.26 illustrates the growth rates of the value added of Sector 30, Sector 31, Sector 32, and Sector 33 during the 2015–2025 period. The average growth rates of the four sectors in the 2015–2020 period are, respectively, 8.43 %, 6.87 %, 8.77 %, and 6.36 %, and their average growth rates in the 2021–2025 period are 3.33 %, 3.95 %, 4.07 %, and 3.54 % successively. In the former period, Sector 32, Financial Intermediation, has the fastest growth, while Sector 33, Real Estate, has the slowest growth in these sectors. In the latter period, Sector 32, Financial Intermediation, keeps the fastest growth, while Sector 30, Wholesale and Retail Trades, obtains the worst performance.

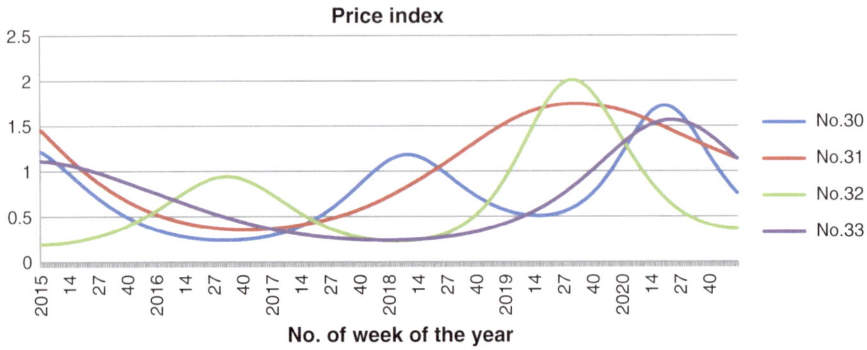

Fig. 3.25 Market price indexes of Sector 30, Sector 31, Sector 32, and Sector 33 during the period 2015 2020

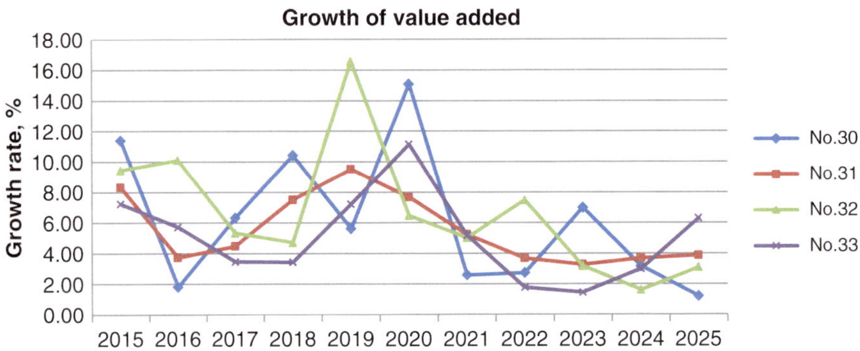

Fig. 3.26 Annual growth rates of the value added of Sector 30, Sector 31, Sector 32, and Sector 33 during the period 2015–2025

Figure 3.27 displays the price indexes of Sector 35, Sector 36, Sector 37, and Sector 38 during the 2015–2020 period. Sector 35 is Research and Experimental Development; Sector 36 is Comprehensive Technical Services; Sector 37 is Management of Water Conservancy, Environment, and Public Facilities; and Sector 38 is Services to Households and Other Services. The price indexes of these four sectors are initially 0.3094, 0.5995, 1.4270, and 0.9350, while their final values are 1.4160, 0.9498, 0.9622, and 0.8696, respectively. Contrary to the results introduced in Sect. 3.2.1, the final price levels are relatively lower. The most evident case is Sector 38, Services to Households and Other Services. Its maximum value and average value are 3.0144 and 1.4127 in the BAU scenario, while these two numbers turn to 1.1266 and 0.6983. These price declines reveal that the policies in this scenario have its unique advantages for the residents to enjoy the same services with obviously lower price from these sectors, which is good for the residents' livelihood and is in accordance with the intentions of Chinese government.

Fig. 3.27 Market price indexes of Sector 35, Sector 36, Sector 37, and Sector 38 during the period 2015–2020

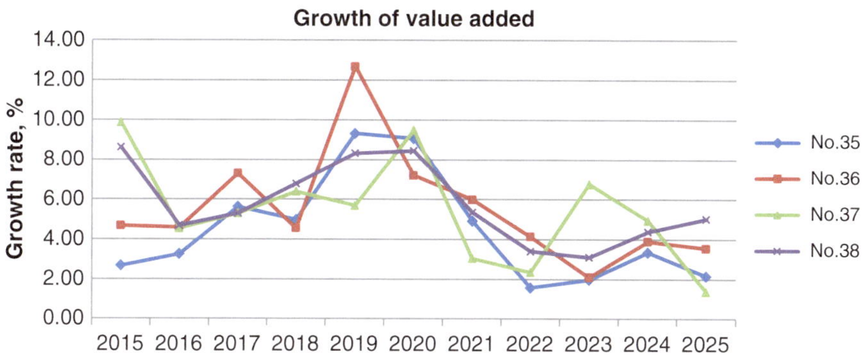

Fig. 3.28 Annual growth rates of the value added of Sector 35, Sector 36, Sector 37, and Sector 38 during the period 2015–2025

Figure 3.28 shows the growth rates of the value added of Sector 35, Sector 36, Sector 37, and Sector 38 during the 2015–2025 period. The average growth rates of the four sectors in the 2015–2020 period are, respectively, 5.82 %, 6.83 %, 6.88 %, and 7.02 %. The average growth rates become 2.79 %, 3.93 %, 3.69 %, and 4.25 % in the 2021–2025 period. In the former period, Sector 38, Management of Water Conservancy, Environment, and Public Facilities, has the best performance in these sectors, while Sector 35, Research and Experimental Development, has the worst performance. In the latter period, Sector 38 remains the best sector, while Sector 35 is still the worst sector.

Figure 3.29 depicts the price indexes of Sector 39, Sector 40, Sector 41, and Sector 42 during the 2015–2020 period. As mentioned above, Sector 39 is Education; Sector 40 is Health, Social Security, and Social Welfare; Sector 41 is Culture, Sports, and Entertainment; and Sector 42 is Public Management and Social Organization. These four sectors are very closely related to the daily life of residents in China. Their initial statuses are, respectively, 1.1791, 1.5402, 0.8608, and 1.4707 at

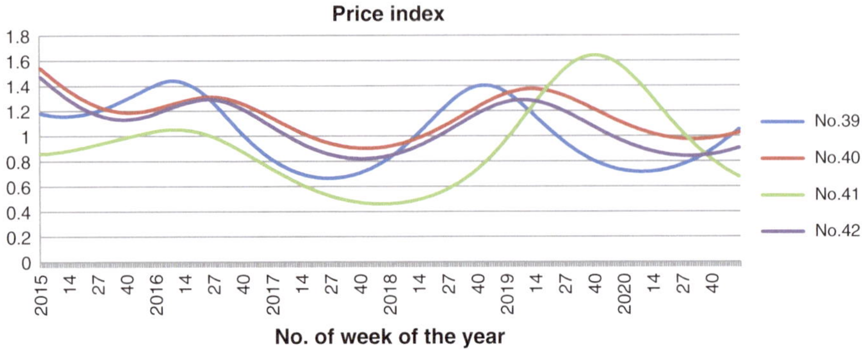

Fig. 3.29 Market price indexes of Sector 39, Sector 40, Sector 41, and Sector 42 during the period 2015–2020

the beginning of the period, while their final values were 1.0516, 1.0267, 0.6774, and 0.9055 at the end of 2020, which are obviously lower than the levels in the BAU scenario owing to the adjustments in fiscal policies and monetary policies. The most obvious case is Sector 41, Culture, Sports, and Entertainment. Its average price indexes within the period in this scenario and in the BAU scenario are, respectively, 0.9277 and 1.7472. The decline rate is 46.91 %. It is evident that the residents in China can enjoy the products of culture, sports, and entertainment with less money by virtue of the new policy settings, which is a good thing for both the policy makers and Chinese households.

Figure 3.30 illustrates the growth rates of the value added of Sector 39, Sector 40, Sector 41, and Sector 42 during the 2015–2025 period. The average growth rates of the four sectors in the 2015–2020 period are, respectively, 6.79 %, 6.53 %, 7.85 %, and 6.41 % and in the 2021–2025 period 5.57, 5.14, 4.19, and 5.25 % in proper order. Sector 41, Culture, Sports and Entertainment, has the fastest growth, while Sector 42, Public Management and Social Organization, has the lowest growth rate in these sectors in the former period. Sector 39, Education, performs best in the latter 5-year period, while Sector 41, Culture, Sports, and Entertainment, performs worst in the same period.

3.3.2 Value-Added Growths of Three Industries

The GDP would be 82.271 trillion CNY (13.059 trillion USD in constant 2010$) in 2020 and 97.813 trillion CNY (15.525 trillion USD in constant 2010$)[2] in 2025. In 2020, the GDP of BAU is a little higher than that of policy scenario, while in 2025, the GDP of policy is a little higher than BAU. Figure 3.31 depicts the growth rates

[2] Exchange rate: 1$ = 6.3CNY in 2010.

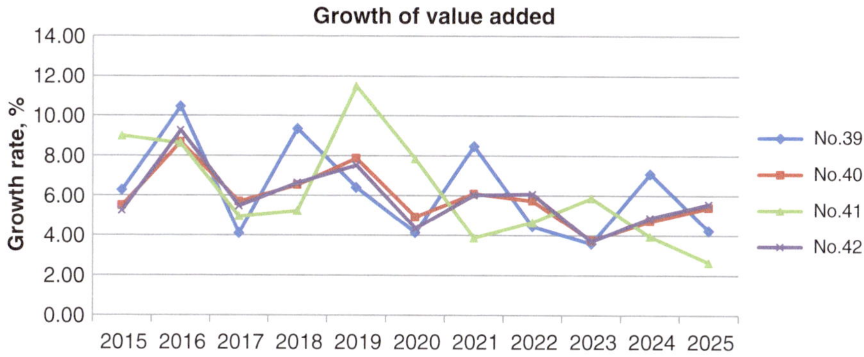

Fig. 3.30 Annual growth rates of the value added of Sector 39, Sector 40, Sector 41, and Sector 42 during the period 2015–2025

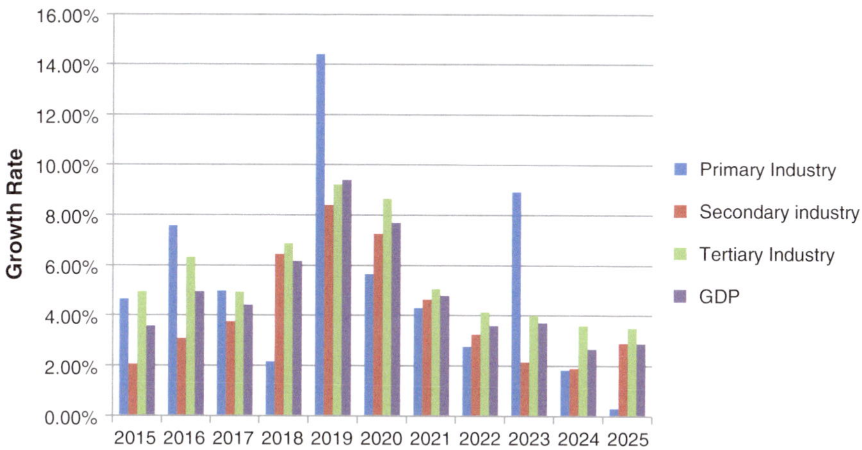

Fig. 3.31 Annual growth rates of the value added of primary industry, secondary industry, tertiary industry, and GDP during the period 2015–2025

of the value added of primary industry, secondary industry, and tertiary industry and GDP in each year during the 2015–2025 period in the scenario with policy adjustments. On average, the annual growth rate of GDP during the 2015–2020 period is 6.03 %; the annual growth rates of value added of the primary industry, the secondary industry, and the tertiary industry during the 2015–2020 period are, respectively, 6.57 %, 5.16 %, and 6.83 %. Within the 2021–2025 period, the average annual growth rate of GDP is 3.52 %; the average annual growth rates of value added of the primary industry, the secondary industry, and the tertiary industry during the 5-year period are, respectively, 3.62 %, 2.96 %, and 4.06 %.

Compared with the results in the BAU scenario shown in Sect. 3.2.2, it is clear that the policy scenario will make some differences on the growth rates. In the 2015–2020 period, the growth rate of GDP will be 0.13 percentage point lower than that of BAU; the growth rate of value added of the primary industry will be 0.18 percentage point lower; and the growth rate of value added of the secondary industry will be 0.43 percentage point lower, but the growth rate of value added of the tertiary industry will be 0.18 percentage point higher. As for the 2021–2025 period, the growth of GDP and three industrial growths are a little higher than those of BAU. The growth rate of GDP in the period is 0.37 percentage point higher; the growth rate of value added of the primary industry is 0.51 percentage point higher; the growth rate of value added of the secondary industry is 0.34 percentage point higher; and the growth rate of value added of the tertiary industry is 0.35 percentage point higher. Evidently, the new policy settings are beneficial to the steady and sustained growth of China's national economy.

If the two periods are put together, we can observe the impacts of the policy scenario in 2015–2025. On average, the growth rate of GDP in policy is 0.10 percentage point higher than that of BAU; the growth rate of value added of the primary industry is 0.13 percentage point higher than BAU; the growth rate of value added of the secondary industry is 0.08 percentage point lower than BAU; and the growth rate of value added of the tertiary industry is 0.26 percentage point higher than BAU.

On the other hand, in the peak growth period of 2019–2021, the yearly growth rates of policy are lower than that of BAU on GDP, while in the other years, the growth rates of policy are higher. It shows the policy scenario plays a role in lowering peak growth and rising valley growth of China's economy. Therefore, the policy scenario seems to make the economic growth flatter than BAU in 2015–2025 in China.[3]

3.3.3 Energy Demand and Supply

In this subsection, we present some results of our ARE simulation on electricity and coal in the scenario with policy adjustments. Figure 3.32 displays the electricity demand of primary industry, secondary industry, tertiary industry, and the whole industry in each year during the 2015–2025 period. Obviously, the demand keeps growing steadily. The electricity demands of policy scenario are mostly higher than those results in the BAU scenario each year in the period (see Figs. 3.16 and 3.32).

Figure 3.33 shows the growth rate of electricity demand of primary industry, secondary industry, tertiary industry, and the whole industry in each year during the 2015–2025 period in the policy scenario. It is clear that the tertiary industry electricity demand has the largest growth rate, which is similar to the situation in

[3] It must be noted that the economic simulations by ARE model in this book are based on a series of assumptions; BAU and policy are only two different scenarios.

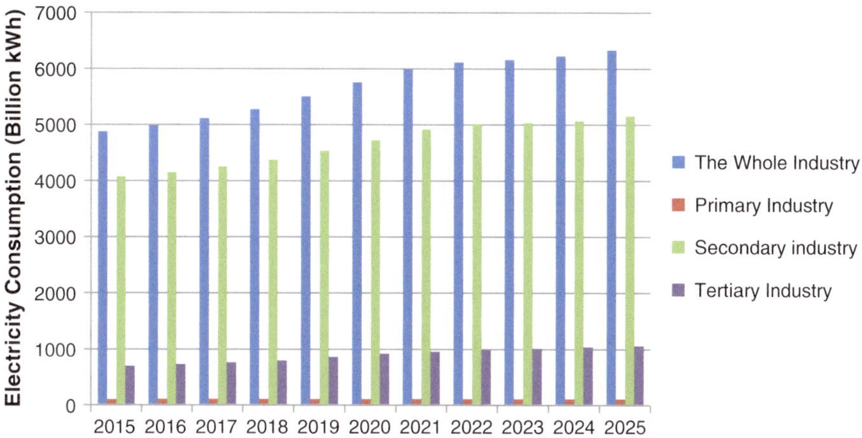

Fig. 3.32 Annual electricity demand by the whole industry, primary industry, secondary industry, and tertiary industry during the period 2015–2025

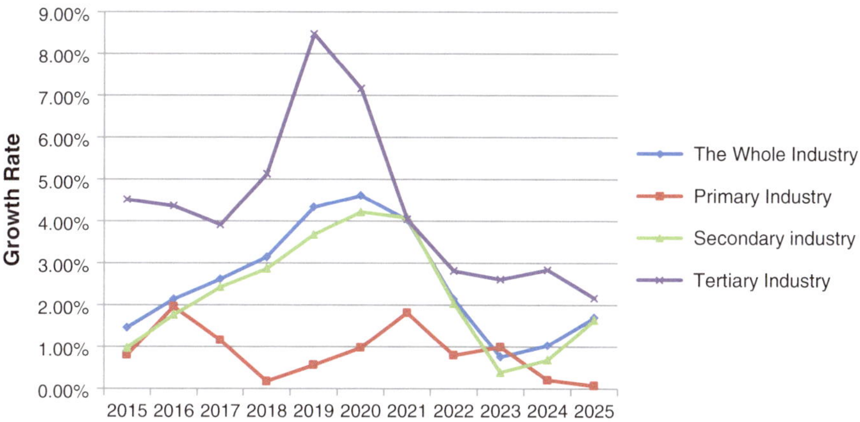

Fig. 3.33 Annual growth rates of electricity consumption of the whole industry, primary industry, secondary industry, and tertiary industry during the period 2015–2025

the BAU scenario (see Fig. 3.17). The average growth rates of the electricity demand of primary industry, secondary industry, tertiary industry, and the whole industry in the 2015–2020 period are, respectively, 0.94 %, 2.65 %, 5.59 %, and 3.05 %. The growth rates are higher than those in the BAU scenario except the growth rate of the primary industry. In the 2021–2025 period, the average growth rates of the electricity demand of primary industry, secondary industry, tertiary industry, and the whole industry are, respectively, 0.77 %, 1.76 %, 2.89 %, and 1.93 %. All the growth rates are greater than those in the BAU scenario.

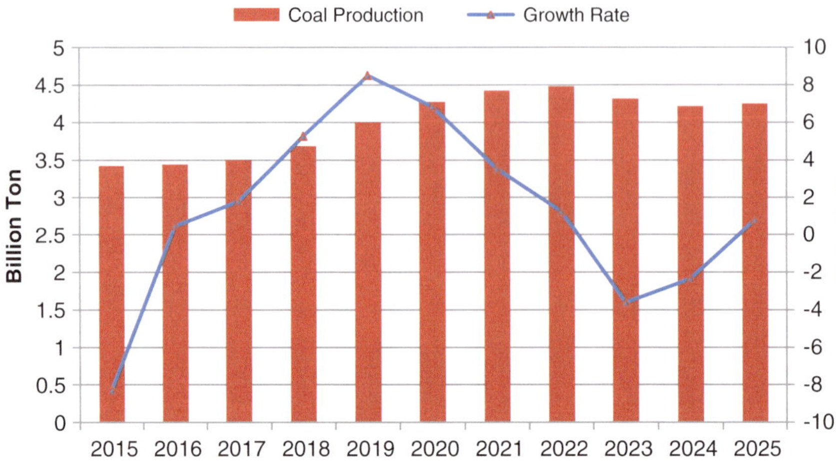

Fig. 3.34 Annual coal production and growth rate during the period 2015–2025

Figure 3.34 displays the coal production and its growth rate in each year during the 2015–2025 period in the policy scenario. Clearly, the production is in negative growth in 2015 as −8.3 % and 2016 as −0.48 %, and then, it keeps growing steadily from 2017 to 2022. The peak coal production is 4.48 billion tons in 2022 and then falls slightly in 2024 at 4.21 billion tons and 2025 at 4.25 billion tons. Since the peak value is 4.45 billion tons in 2023 in the BAU scenario, the value in the policy scenario is 0.03 billion tons higher.

3.4 Economic Gene Mapping of China

Production function with electricity is a linear function with stable slope and an intercept. However, a mutation will happen if the intercept changes from positive to negative or vise versa, and thus, the slope of the function will have a change too. Then, the function will be in different form when a mutation happens. It means the mutation can be regarded as a key year of the two economic periods since one production function with electricity can denote an economic period.

3.4.1 Economic Gene Maps of China's National Economy

As discussed in Chap. 2, there was a mutation in Chinese national economy in 2000, and the *E-GDP* function in the 1978–1999 period is shown in formula (3.9):

1978–1999:

$$\mathrm{GDP} = 13.52E - 1709.3 \tag{3.9}$$

The slope of the function is 13.52, i.e., the increase of a unit electricity use can result in GDP increase of 13.52 CNY in the period. It is a little higher since the national economy is in the primary commodity stage (1978–1982) and early stage of industrialization (1983–2000), and the share of heavy industry is relatively lower. However, the situation had been changed after the year 2000; the heavy industry had grown fast until 2012. The *E-GDP* function is shown in formula (3.10).

2000–2011:

$$\mathrm{GDP} = 10.037E + 2625.5 \tag{3.10}$$

The *MGDP* has fallen to 10.037 CNY/kWh, around 25.76 % decrease. It is a positive function which lasts for 12 years. Fortunately, another mutation happened in China's economy in 2012. It is a negative mutation.

There are two economic simulations as BAU discussed in Sect. 3.2 and policy discussed in Sect.3.3 by ARE. Based on BAU scenario, there will be no mutation until 2025; then, the *E-GDP* function is shown as follows:

2012–2025:

$$\mathrm{BAU:} \quad \mathrm{GDP} = 30.657E - 89658 \tag{3.11}$$

The negative mutation shows a *MGDP* increase of 205.44 % comparing formula (3.10) and (3.11), it is a great growth, and the period takes at least 14 years. As we have mentioned in Sect. 3.1, the ideal negative gene function has high *MGDP* and takes a long period. It can be regard as an ideal gene map of the national economy in China. The national gene maps of China's economy in BAU from 1978 to 2025 are illustrated in Fig. 3.35; the red dots indicate the two mutations and the black lines are the gene maps in BAU. It would be better if the middle line were shorter, for example, in 5–8 years rather than 12 years.

Based on the policy scenario, there will be no mutation until 2025, and the *E-GDP* function is shown as follows:

2012–2025:

$$\mathrm{Policy:} \quad \mathrm{GDP} = 25.905E - 67{,}232 \tag{3.12}$$

The negative mutation reveals an increase of 158.095 % in the marginal representative factor productivity by comparing formula (3.10) and (3.12). The national gene maps of China's economy in policy scenario from 1978 to 2025 can be depicted in Fig. 3.35, where the red line represents the 2012–2025 period.

Comparing BAU and policy scenario, the *MGDP* of BAU scenario is 1.1834 times higher than that of policy one, and *AGDP* of BAU is around 1.04 times higher than *AGDP* of policy scenario after 2019 (Fig. 3.36). It demonstrates that the policy

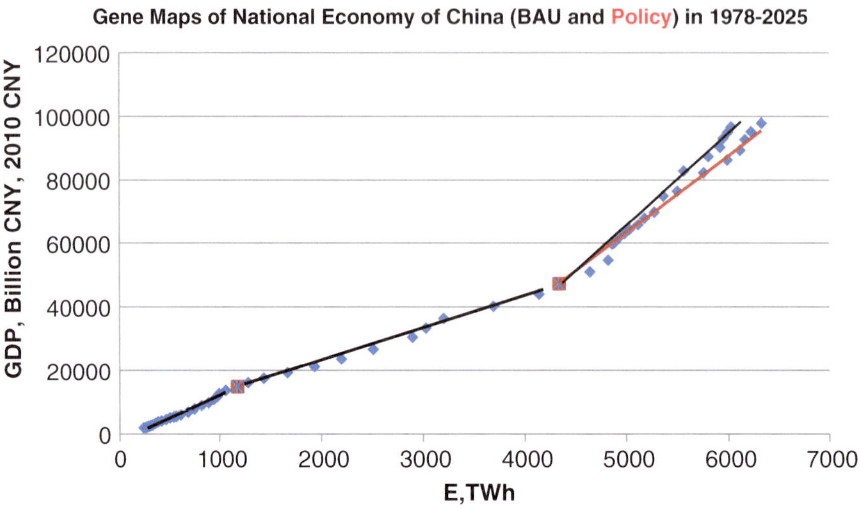

Fig. 3.35 Gene maps of national economy of China in BAU and policy during 1978–2025

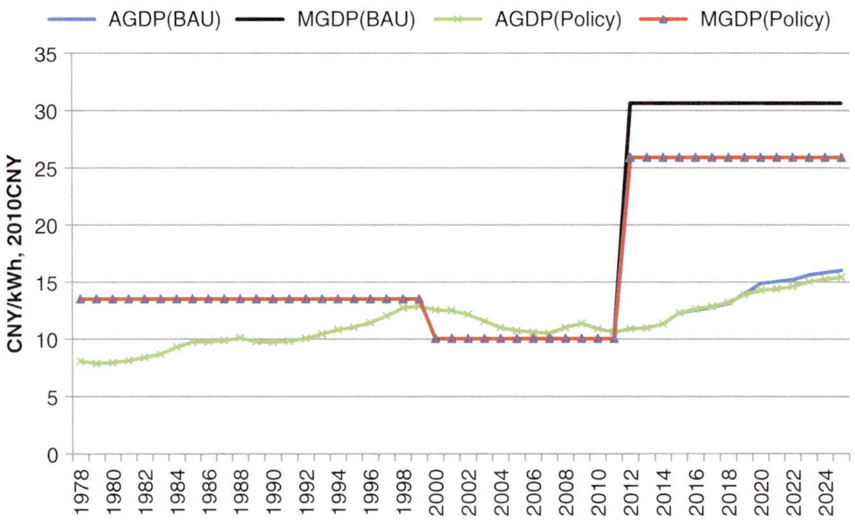

Fig. 3.36 *MGDP* and *AGDP* for BAU and policy in China during 1978–2025

scenario is not better than BAU on the productivity of the national economy. On the other hand, since the *MGDP*s are bigger than *AGDP*s as displayed in Fig. 3.36, there is enough space for *AGDP*s to increase in both BAU and policy scenarios.

As for the diagnostic analysis of the economic gene maps of China, there are two negative production functions with electricity in the 1979–1999 and 2012–2015 periods and one positive function in the 2000–2011 period. Based on the features of the ideal gene map discussed in Sect. 3.1.1, the *MGDP*s of production function with electricity in China are high for the two negative mutation periods, especially in

2012–2015. However, the 12-year positive period seems longer. In fact, there was an opportunity in 2008 to have a mutation because the *AGDP* in 2007 was 10.48 CNY/kWh, very close to *MGDP* at 10.037 CNY/kWh, and *AGDP* was up to 10.99 CNY/kWh in 2008 and 11.35 CNY/kWh in 2009. Unfortunately, it was down to 10.87 CNY/kWh in 2010 and 10.59 CNY/kWh in 2011; the mutation did not happen until 2012. If a mutation were to happen in 2008, the positive period would only be 8 years—what an ideal gene map it could be.

3.4.2 Economic Gene Maps for Three Industries

For primary industry, the E_1–V_1 function, production function with electricity, can be set up based on the electricity consumption E_1 and the value added V_1. Since there was a mutation happened in primary industry in 2012 in China, the E_1–V_1 function during 1978–2011 is shown as follows:

1978–2011:

$$V_1 = 37.099E_1 + 344.78 \tag{3.13}$$

It is a positive function which shows the marginal value added of electricity is near 37.1 CNY/kWh, and the positive intercept implies that the average value added of electricity is declining with the growth of electricity consumption E_1. It takes 34 years and the period is too long. The mutation that occurred in the primary industry in 2012 was a negative mutation which means a new economic period starts from 2012. Based on the economic simulations of BAU and policy scenarios by ARE, there would be no mutation in the primary industry until 2025. Then, the E_1–V_1 function in the period in both BAU and policy scenarios is shown as follows:

2012–2025:

$$\text{BAU}: \quad V_1 = 410.25E_1 - 36,576 \tag{3.14}$$

$$\text{Policy}: \quad V_1 = 481.72E_1 - 44,072 \tag{3.15}$$

The slopes of the functions have a big increase for both BAU and policy scenarios, which means the marginal value added of electricity is very high. It shows the productivity will increase greatly in the primary industry in China during 2012–2025. And the negative intercept indicates the average value added is rising with the growth of E_1. Such a higher slope of the E_1–V_1 function also shows that the primary industry would have a bright future in China, and perhaps there would be a revolution of production in the primary industry.

The primary gene maps are shown in Fig. 3.37; the two black lines are for BAU and one red line is for the policy scenario in the 2012–2025 period; thus, gene maps for policy scenario consist of the black line and the red line in Fig. 3.37. The slope of policy scenario is 1.174 times higher than that of BAU which verifies the policy

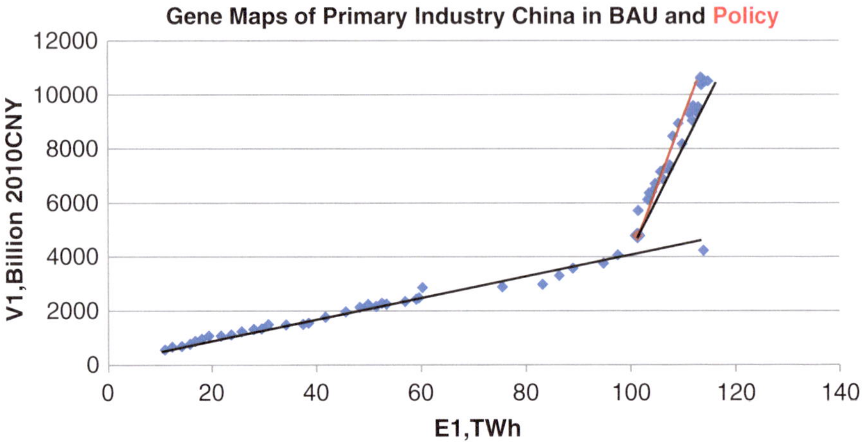

Fig. 3.37 Gene maps of primary industry of China in BAU and policy scenarios

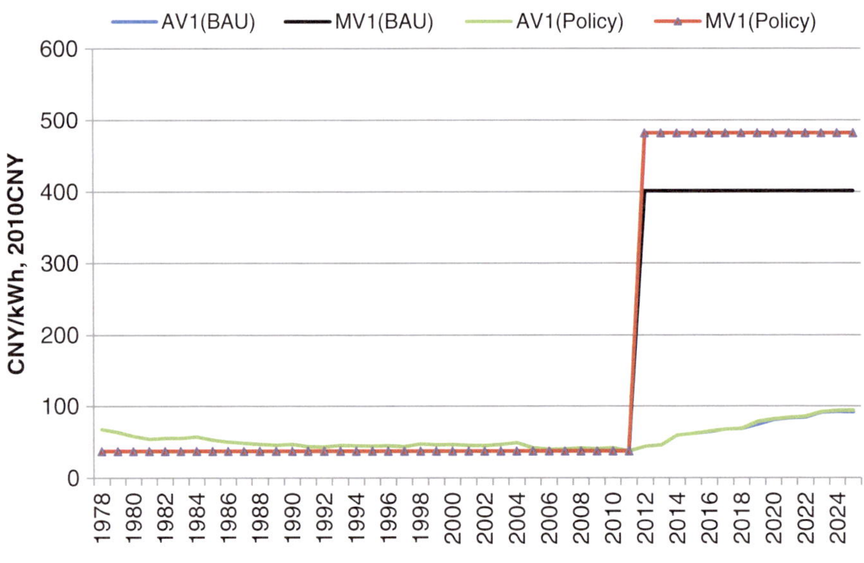

Fig. 3.38 AV_1 and MV_1 for BAU and policy scenarios in primary industry

scenario would promote growth of productivity more than BAU in primary industry. It also shows the gene of primary industry is relatively stable, which is reflected by the fact that only one mutation happened during 1978–2025. However, the slope of the gene line has great increase. Shall we conclude that the stable gene would experience a big change on slope?

The average value added of electricity and marginal value added of electricity for both BAU and policy can be depicted as shown in Fig. 3.38. The AV_1 for the BAU scenario is nearly equal to that of policy after 2012, and there is still very big space for them to grow in the future.

For the secondary industry, there was no mutation before 2000. The E_2–V_2 function in the 1978–1999 period is shown as follows:

1978–1999:

$$V_2 = 7.4768E_2 - 1430.9 \tag{3.16}$$

The marginal value added of electricity is 7.4768 CNY/kWh, and the average value added of electricity is rising since the interception of the function is negative. The period lasts for 23 years, which is relatively longer as the negative function.

A mutation happened in 2000, which was a positive mutation. Thus, the marginal value added of electricity falls to 5.6916 CNY/kWh in the 2000–2011 period as shown in formula (3.13). It shows a decreasing productivity in the period. The $E_2\_V_2$ function is shown as follows:

2000–2011:

$$V_2 = 5.6916E_2 + 469.24 \tag{3.17}$$

Another negative mutation happened in 2012, and there will be no mutation in the secondary industry until 2025 based on the simulation by ARE model for both BAU and policy scenarios. And the E_2–V_2 functions for both BAU and policy scenarios in the 2012–2025 period are shown as follows:

2012–2025:

$$\text{BAU}: \quad V_2 = 17.929E_2 - 45257 \tag{3.18}$$

$$\text{Policy}: \quad V_2 = 13.595E_2 - 28016 \tag{3.19}$$

The marginal value added of electricity increases largely in both BAU and policy scenarios due to the negative mutation in 2012–2025; it means the marginal representative factor productivity has been improved greatly in the period. However, the slope in policy scenario is lower than that in BAU, which demonstrates the stimulating policies are not better for increasing productivity of the secondary industry during 2012–2025. The economic gene maps of the secondary industry for both BAU and policy (red line for policy in 2012–2025) can be shown in Fig. 3.39. The AV_2 and MV_2 in both BAU and policy scenarios are shown in Fig. 3.40.

Based on the gene features, the negative gene line (the left line in Fig. 3.39) looks better since the MV_2 is relatively high and the period lasts for 22 years. The second line from the left in Fig. 3.39 is a positive gene line. It is better on MV_2 since it falls from 7.48 to 5.69, which is not too much. However, it lasts for 12 years; the period seems longer. The third gene line is better for anyone of BAU and policy since the *MGDP* has a great increase, and it takes at least 14 years until 2025. The gene map of the secondary industry looks good generally. Of course, if the second gene line were in shorter period, it would be very good.

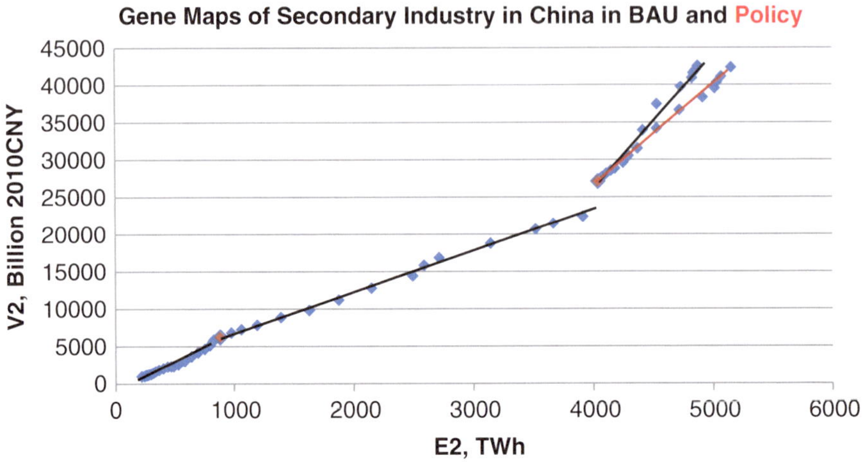

Fig. 3.39 Economic gene maps of secondary industry in BAU and policy

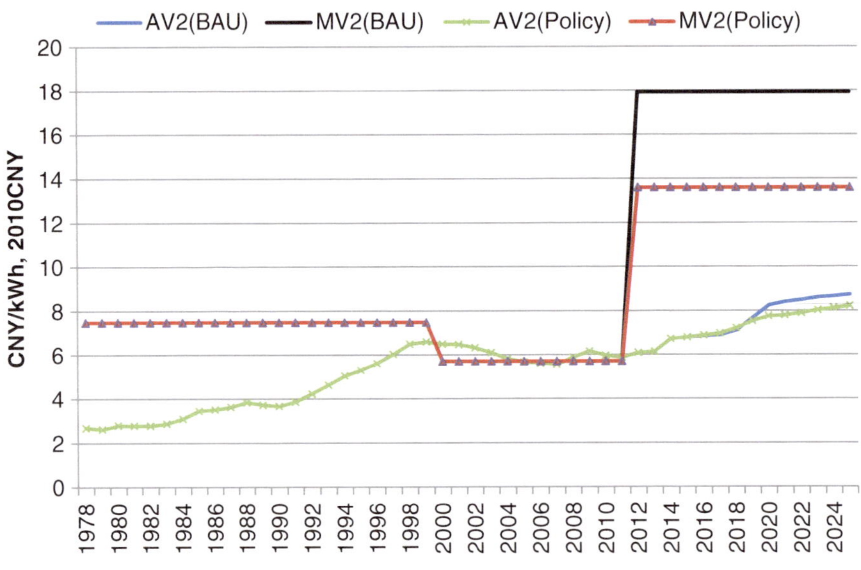

Fig. 3.40 AV_2 and MV_2 for BAU and policy in secondary industry

For the tertiary industry, there were two mutations which happened in 1984 and 2014, and there would be no mutation until 2025 by ARE simulations in BAU and policy scenarios. The E_3–V_3 functions in each period are as follows:

1978–1983:

$$V_3 = 81.044E_3 - 324.3 \tag{3.20}$$

1984–2013:

$$V_3 = 36.265E_3 + 615.64 \tag{3.21}$$

2014–2025:

$$\text{BAU}: \quad V_3 = 49.266E_3 - 8235.2 \tag{3.22}$$

$$\text{Policy}: \quad V_3 = 49.582E_3 - 8575.5 \tag{3.23}$$

We can see that the slope in formula (3.14) is the largest, and then, it is 36.265 in formula (3.15) which has a big decrease, and it last for 30 years due to the positive mutation in 1984. Fortunately, it increases to 49.266 in formula (3.22) in BAU or 49.582 in formula (3.23) in policy because the mutation in 2014 was a negative one. However, it is still only 61 % of the biggest one in the formula (3.14). And the slope of policy scenario is a little higher than that of BAU scenario which verifies that the stimulating policy scenario would be a little better than BAU on productivity in the tertiary industry in 14 years in China.

The economic gene maps of the tertiary industry can be depicted in Fig. 3.41; the three gene lines denote the three periods 1978–1983, 1984–2013, and 2014–2025, respectively; and the slope of the line indicates the marginal value added of electricity; the two red dots show the mutations which happened in 1984 and 2014, respectively; the red line in the right of the picture stands for the policy scenario and black one stands for BAU scenario. Unfortunately, comparing with Fig. 3.42, we can see that the action of positive mutation takes a longer period when marginal value added of electricity remains quite low, and the negative mutation does not have enough power to increase MV_3 largely. Based on the rule of positive and negative mutations alternatively, it would be impossible to expect the high MV_3 in the future. This is a bad news for the tertiary industry in China.

3.4.3 Economic Gene Maps for Sectors

Based on the input–output table of 2010 published by Chinese government, economic growths in 42 sectors from 2011 to 2025 have been simulated by suing ARE model as shown in Sect. 3.2 for BAU and Sect. 3.3 for policy scenario in this chapter. Therefore, the gene map of sectoral level can be studied.

For Sector 9 as Processing of Timbers and Manufacture of Furniture, the products of the sector mainly are exported to the international market. The share of value added in GDP is around 0.7 %, and it would fall to 0.48 % in 2025 by the ARE simulation. There would be a mutation in the sector in 2015; then, the production functions with electricity in two periods are shown as follows:

2010–2014:

$$V_{s9} = 6.307E_{s9} + 87.569 \tag{3.24}$$

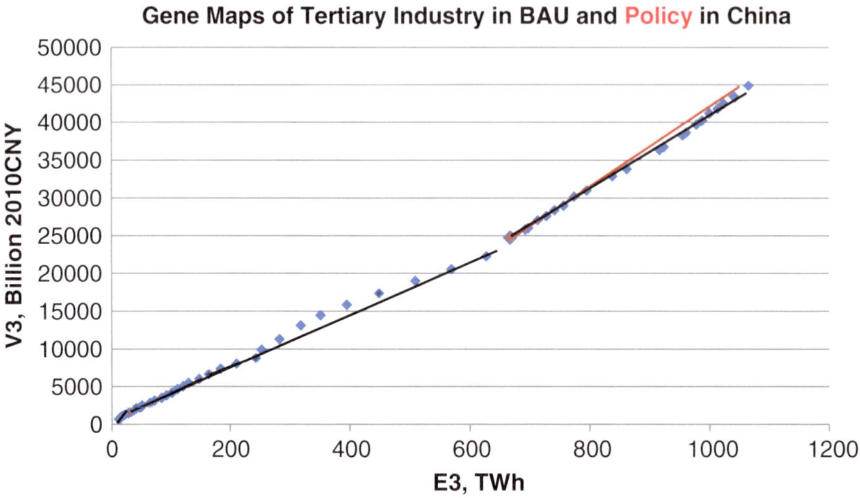

Fig. 3.41 Economic gene maps of tertiary industry in BAU and policy

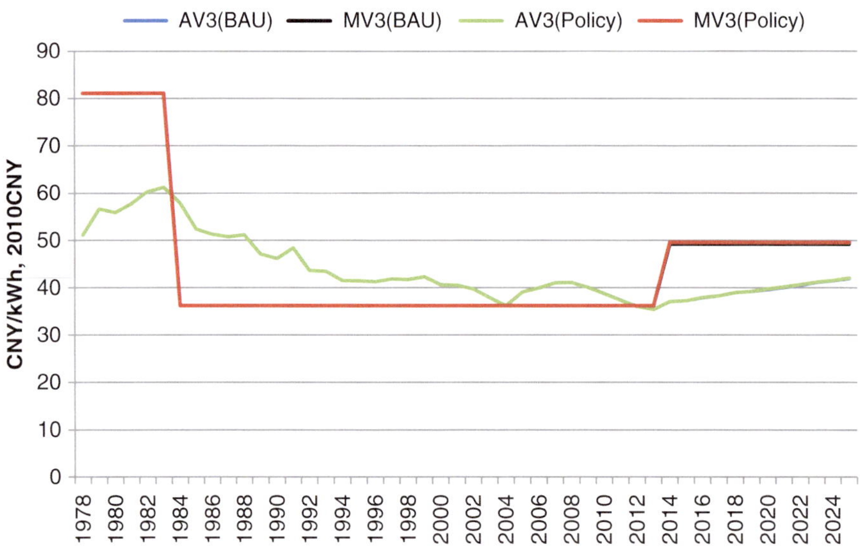

Fig. 3.42 AV_3 and MV_3 of BAU and policy for tertiary industry

2015–2025:

$$\text{BAU}: \quad V_{s9} = 31.727E_{s9} - 927.39 \tag{3.25}$$

$$\text{Policy}: \quad V_{s9} = 16.186E_{s9} - 297.58 \tag{3.26}$$

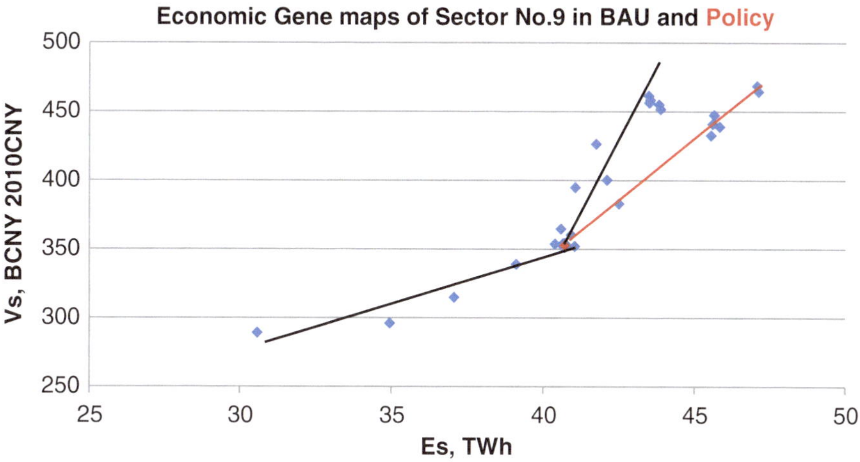

Fig. 3.43 Economic gene maps of processing of timbers and manufacture of furniture in BAU and policy

It is a negative mutation and the mutation shows MVs_9 rising greatly in BAU and slightly in policy scenario. It means the policy scenario with stimulating policies would not be better than BAU for increasing the productivity of the sector in China's economy. The gene maps of the sector are shown in Fig. 3.43. Average value added of electricity and marginal value added of electricity can be illustrated as in Fig. 3.44.

For Sector 13, Manufacture of Nonmetallic Mineral Products, the share of value added in GDP is around 2 %, and it would fall to 1.94 % in 2025 by our simulation. There will be no mutation in the sector in both BAU and policy scenarios until 2025. The *Es*–*Vs* functions for BAU and policy are shown as follows:

2010–2025:

$$\text{BAU}: \quad V_{s13} = 7.9831E_{s13} - 1283.1 \tag{3.27}$$

$$\text{Policy}: \quad V_{s13} = 7.4059E_{s13} - 1112.7 \tag{3.28}$$

Both of them are negative function, and the slope of BAU is a little higher than that of policy scenario. However, the slope 7.98 is relatively lower in the 42 sectors. The gene maps of the sector can be seen in Fig. 3.45; AVs_{13} and MVs_{13} of the sector are shown in Fig. 3.46.

As for Manufacture of Metal Products, Sector 15, the share of value added in GDP is 1.1 % and will decline to 0.83 % in 2025 by the simulation of ARE model. A negative mutation happened in the sector in 2014. The production functions with electricity for both BAU and policy can be expressed as follows:

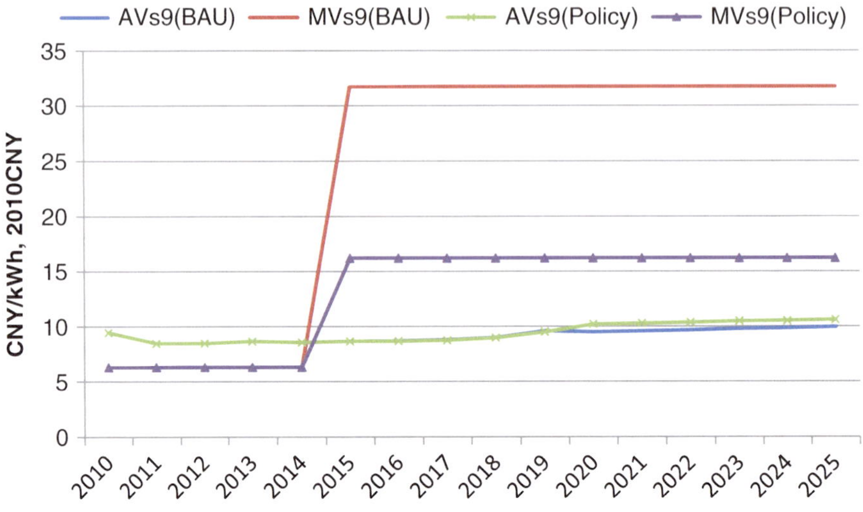

Fig. 3.44 AVs_9 and MVs_9 for BAU and policy scenarios

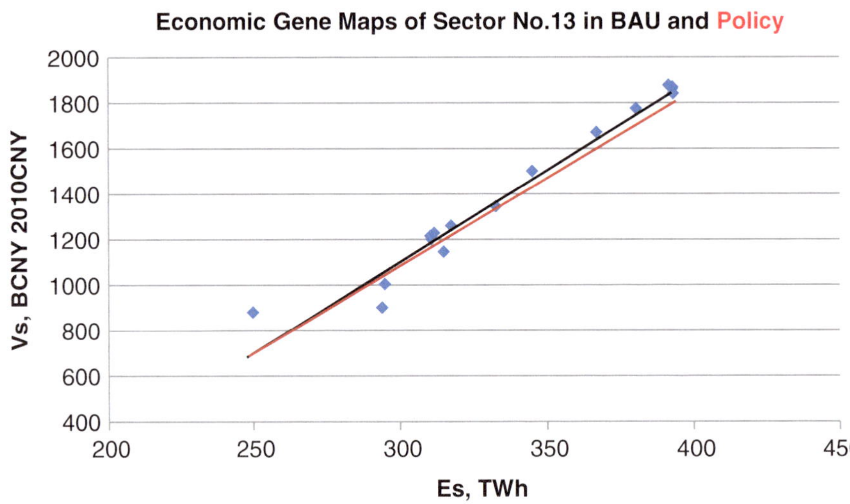

Fig. 3.45 Economic gene maps of manufacture of nonmetallic mineral products in BAU and policy

2010–2013:

$$V_{s15} = 1.8342E_{s15} + 254.68 \tag{3.29}$$

2014–2025:

$$\text{BAU}: \quad V_{s15} = 5.458E_{s15} - 299.41 \tag{3.30}$$

$$\text{Policy}: \quad V_{s15} = 5.0874E_{s15} - 236.44 \tag{3.31}$$

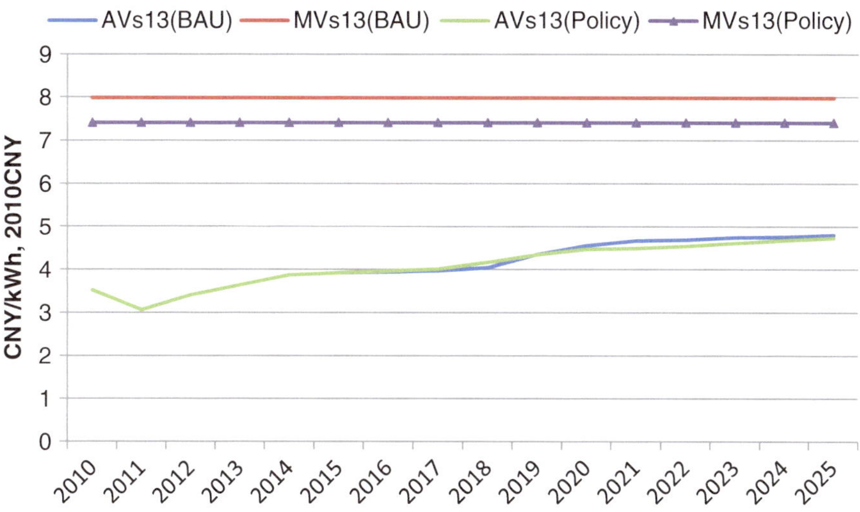

Fig. 3.46 AVs_{13} and MVs_{13} for BAU and policy scenarios

The slope of the function in formula (3.17) is very low. After the negative mutation, it increases to 5.458 in BAU and 5.0874 in policy, which demonstrates that the BAU is better than policy scenario. The gene maps of the sector, AVs_{15} and MVs_{15}, are shown in Figs. 3.47 and 3.48, respectively.

For Sector 34, Leasing and Business Services in the tertiary industry, there would be no mutation until 2025 by the simulation. The production functions with electricity for BAU and policy are the same as follows:

2011–2025:

$$V_{s34} = 41.144E_{s34} - 316.72 \tag{3.32}$$

The gene map of the Leasing and Business Services sector looks fine since it is a negative gene line (Fig. 3.49), and the AVs_{34} remains in the growing process until 2025 (Fig. 3.50).

3.5 Comparison of Gene Maps of China and Japan

Economic development in Japan has some similarity with China in the period of industrialization. Some economists like to compare the economic growth between the two countries. We can try to compare national economic gene maps of Japan and China in this section. First, the comparison ought to be conducted in the same

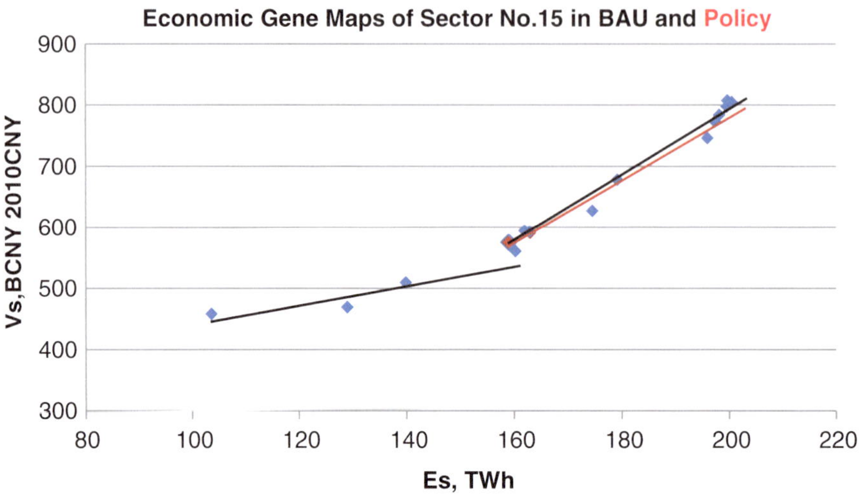

Fig. 3.47 Economic gene maps of manufacture of metal products in BAU and policy

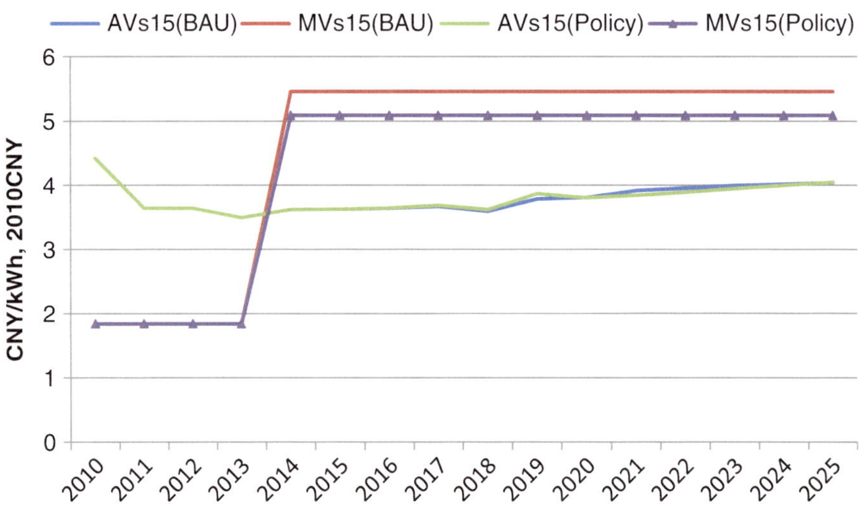

Fig. 3.48 AVs_{15} and MVs_{15} for BAU and policy scenarios

economic stage of the two counties. Second, it can be compared by the economic gene maps of Japan and China in the stage. And third, the analysis includes the similarity and difference between them to diagnose China's economic development.

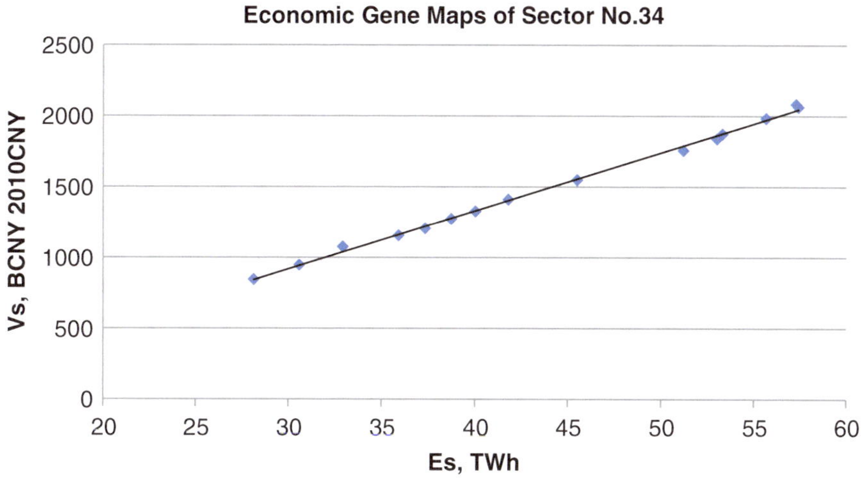

Fig. 3.49 Economic gene maps of leasing and business services

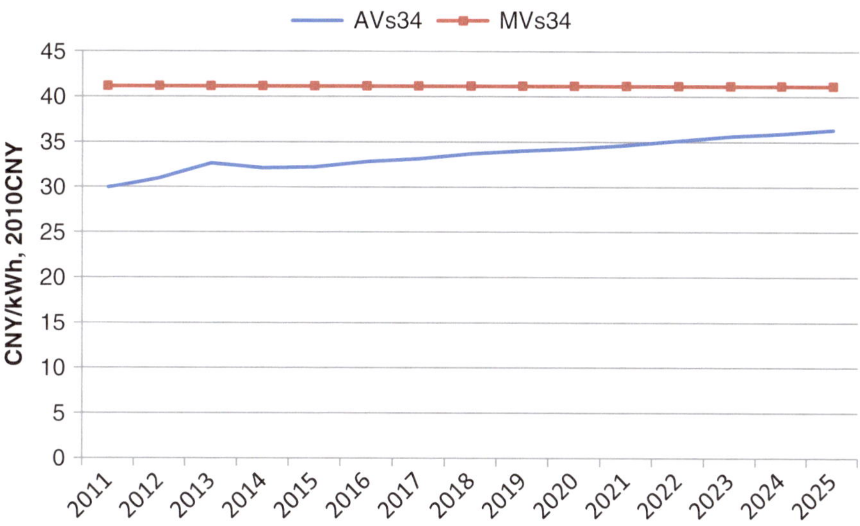

Fig. 3.50 AVs_{34} and MVs_{34} of the leasing and business services sector

3.5.1 *Economic Development Stage*

As mentioned in Chap. 1, per capita electricity consumption can be used to show the stages of industrialization of economic development for a country. In the primary commodity stage, total electricity consumption per capita TE/p is less than 300 kWh, and residential electricity consumption per capita RE/p is less than

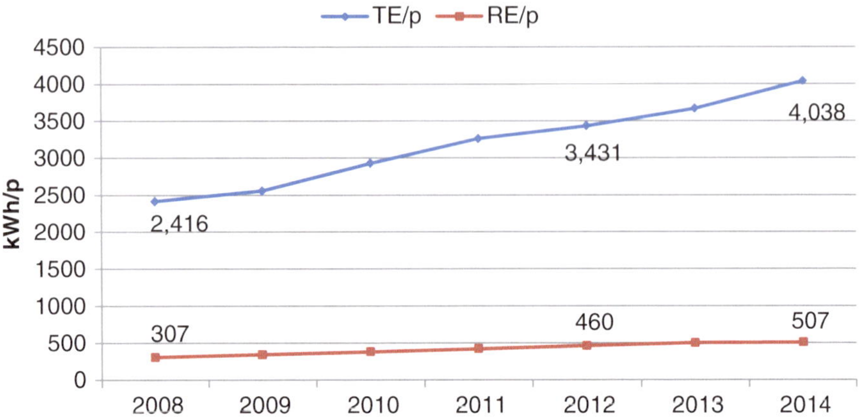

Fig. 3.51 Electricity consumption per capita in late industrialization in China (Data source: statistical material of China Electricity Council over years)

20 kWh. In the early period of industrialization, *TE*/*p* is in 300–1000 kWh/p and *RE*/*p* is in 20–80 kWh/p; in the middle period of industrialization, T*E*/*p* is in 1000–2400 kWh/p and *RE*/*p* is in 80–240 kWh/p; in the late period of industrialization, *TE*/*p* is in 2400–4500 kWh/p and *RE*/*p* is in 240–810 kWh/p; in the completion period of industrialization, *TE*/*p* is in 4500–5000 kWh/p and *RE*/*p* is in 810–900 kWh/p. China's economy has entered into the later period of industrialization since 2008. *TE*/*p* and *RE*/*p* are shown in Fig. 3.51, in which *TE*/*p* was 4038 kWh/p and *RE*/*p* was 507 kWh/p in 2014.

Comparing with Japan, China has entered into late period of industrialization from 1969, when *TE*/*p* was 2729 kwh/p and *RE*/*p* was 442 kWh/p (see Fig. 3.52). It means the economic level of 2008 in China is the same with the level of 1969 in Japan, and the level of 2014 in China is the same with that of 1975 in Japan. However, the *RE*/*p* was 307 kWh/p in 2008 in China, while it was 442 kWh/p in 1969 in Japan, and *RE*/*p* was 507 kWh/p in 2014 in China, while it was 552 kWh/p in 1971 in Japan. It reveals that the *RE*/*p* in China is much lower than that of Japan though the *TE*/*p* in China was higher than Japan. Anyway, the economic stage of China in 2014 is similar to the period of Japan in 1974–1975.

3.5.2 A Negative Mutation in the Economic Stage

A negative mutation happened in Japan's economy in 1974, the late period of industrialization of economic development. And a negative mutation occurred in China's economy in 2012, the late period of industrialization of economic

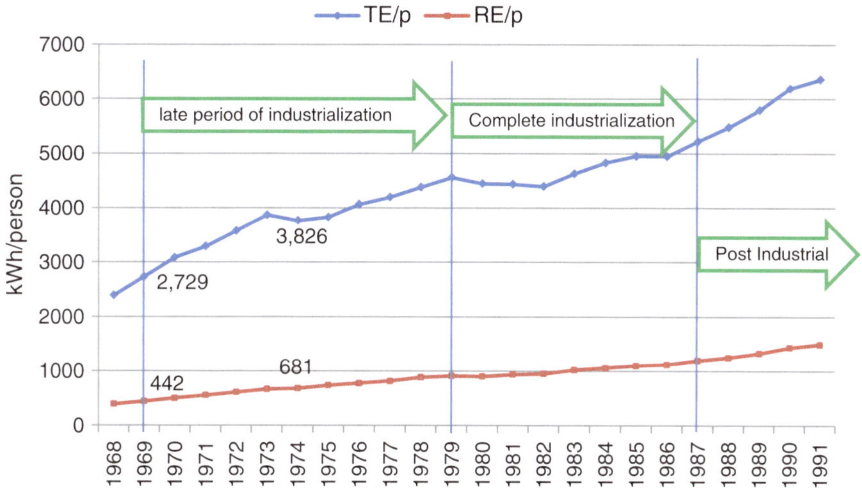

Fig. 3.52 Electricity consumption per capita in late industrialization in Japan (Data source: The Energy Data and Modelling Center, The Institute of Energy Economics Japan, Handbook of Energy & Economic Statistics in Japan 2013, The Energy Conservation Center, Japan)

development. Now, we can compare economic gene map of Japan in 1974–1988 and the gene map of China in 2012–2025. The production function with electricity for Japan, *E-GDP* function, is a negative gene function as shown in formula (3.6). The *E-GDP* function for China is a negative function too (formulas (3.11) and (3.12)). For Japan's economy, the negative period lasts for 15 years, and it would be at least 14 years for China's economy. The *MGDP* is highest at 973.62 which is 1.68 times higher than the early period in Japan; it is highest at 30.657 which is 3.05 times higher than the 2000–2011 period in China. It looks more similar comparing the gene map of Japan, the second line from the left in Fig. 3.3, with the gene map of China, the third line in Fig. 3.35.

As for shares of E_1, E_2, E_3, and *RE* in *TE*, we can see from Fig. 3.53 that E_2 shared 62.89 %, E_3 was 12.29 %, and *RE* shared 19.34 % in Japan in the negative mutation year 1974. For China as shown in Fig. 3.54, E_2 was 72.16 %, E_3 was 12.25 %, and *RE* was 13.39 % in the negative mutation year 2012. The share of *RE* shows the potential of final consumption; the higher share of *RE* in *TE* means people have enough income to do consumption in the market, i.e., they are rich to consume products produced by secondary industry. The big difference of *RE* shares for Japan in 1974 and China in 2012 shows China lacks potentials of final consumption. One reason is the house/apartment price has increased 5–8 times since 2000, the high house/apartment price makes the economic bobble, and people have to pay more money to the bank and leave no more extra money for consumption. Then, the final consumption has declined, while investment is very difficult to increase fast since the surplus of production capacity in the secondary industry.

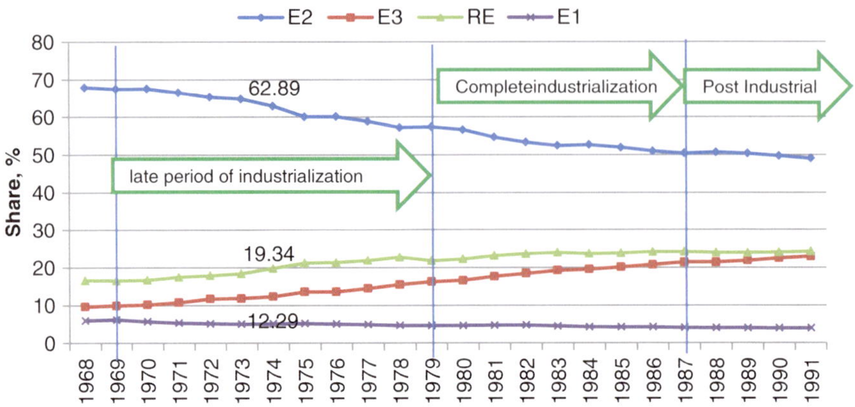

Fig. 3.53 Shares of E_1, E_2, E_3, and *RE* in *TE* in later industrialization in Japan (Data source: The Energy Data and Modelling Center, The Institute of Energy Economics Japan, Handbook of Energy & Economic Statistics in Japan 2013, The Energy Conservation Center, Japan)

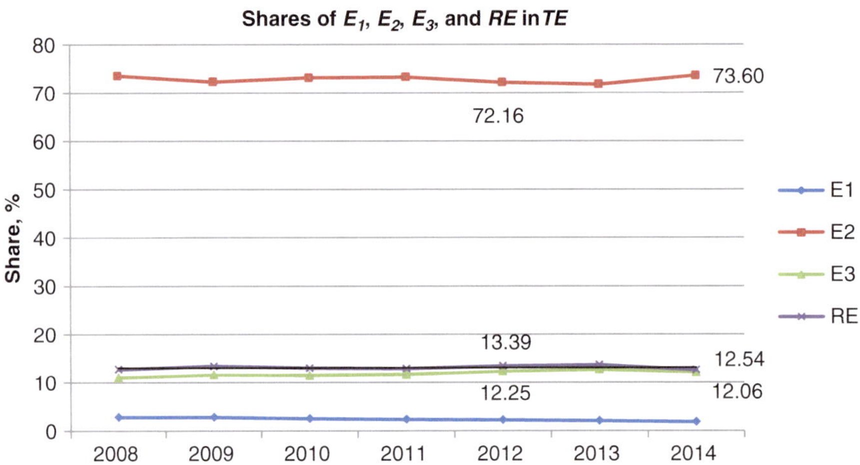

Fig. 3.54 Shares of E_1, E_2, E_3, and *RE* in *TE* in late industrialization in China (Data source: statistical material of China Electricity Council over years)

Moreover, China's export is losing its competitiveness in the world market due to the rising labor cost and lack of energy resources.

On the other hand, the real-estate bobble in Japan started from the middle of 1980s, and it was broken out in 1990s, while the price of real estate in China grew fast from 2007, and it is still in the peak in 2014. It seems that the real-estate bobble in China is early than Japan and the risk of the bobble is very high in China. Comparing with Japan, the household final consumption expenditure per capita in

China in the negative gene line period was 1043 USD (constant 2000 US$) in 2012, while it was 11,376 USD in 1974 and 17,343 in 1988 in Japan [3]. Electricity consumption and household electricity use per capita were 3,431 kWh and 460 kWh, respectively, in 2012 in China, while they were 3,761 kWh and 681 kWh, respectively, in 1974 in Japan. It is clear that the potential of final consumption per capita in China is lower than that in Japan in the same negative economic mutation year. This is the risk in China's economic growth in 2012–2025.

As for the heavy government debt, a lot of debt is the investment of infrastructure in central and local governments, and it is really a great risk in China's economic development. However, state has a great amount of property and financial capital in China. We can perhaps see an opportunity for lowering the risk from there.

3.5.3 Similarities and Differences of Gene Maps of China and Japan

Comparing China and Japan's economic gene map, there are some similarities summarized as follows:

1. A negative mutation happened in the late stage of the industrialization for both Japan and China. It was in 1974 in Japan and 2012 in China.
2. The negative gene function period lasts for around 15 years. It is 15 years during 1974–1988 in Japan and would be at least 14 years from 2012 to 2025 in China.
3. The electricity consumption per capita is around 3,600 kWh. It was 3,761 kWh/p in 1974 in Japan and 3,431 kWh/p in 2012 in China.

And there are some differences between Japan and China as follows:

1. The growth of marginal representative factor productivity, i.e., *MGDP*, in the transfer from the positive mutation to the negative mutation in China is greater than that in Japan. It would grow from 158 to 205 % in China and it has been 68.15 % in Japan.
2. The household final consumption expenditure per capita (in constant 2000US$) in China is less than 10 % of that in Japan in the negative mutation year of 1974. It shows the growth of people's income is lower in China in 2012.
3. The household electricity consumption per capita in China in 2012 is only two-thirds of that in Japan in the negative mutation year of 1974.
4. The real-estate bobble in China is earlier than the bobble in Japan in the negative gene function period. It shows the risk is high in China.
5. The great government debt would be a great risk if there were no further reform on economic system and mechanism in China. It is quite different on government debt in Japan.

Since 2012, China is being in a negative mutation period with relatively high marginal representative factor productivity until 2025. It reveals that China's economy has an opportunity to increase productivity though there are some risks and a series of problems in the economic growth. However, since economic system is very complex, it is very difficult to forecast some economic events in China and in the world. The powerful economic events and its long-term duration would stimulate economic mutation to happen earlier if the technology innovation and improvement were not strong enough to decline the impacts of the events. If a mutation were to happen before 2025 in China, it would be a positive mutation, and the marginal representative factor productivity would decline to a lower level. Therefore, it is a very good opportunity for economic development in China in the negative mutation period which started in 2012. It has to use the opportunity to lengthen the period by stimulating technology innovation and improvement, promoting primary industry revolution growth, and searching sustainable path of tertiary industry growth on economic system and mechanism in China.

It is a warning signal in differences (2) and (3) mentioned above for China's economy in the future. It shows that the economic events such as real-estate bobble and government debt would be very difficult to cope with due to the lower income in China. It also shows the difficulty of increasing household consumption expenditure to promote tertiary industry growth. What are the measures the government should take in facing the challenges? Maybe the following fiscal measures would be helpful to remove the barriers of household consumption:

1. Decreasing the income tax rate to allow people have enough money for consumption
2. Paying much share on health-care cost of Chinese residents by government to remove the barriers of consumption
3. Paying more on education, i.e., ensuring free education from primary school to high school and providing more payment to college and university students for their education
4. Paying more on endowment insurance for old-age people for living out one's life after retirement

3.6 Findings

Economic mutation in 2012 in China is a negative one. As for the inherent rule of the economy in China, there would be a great opportunity for China's economy to increase marginal representative factor productivity by promoting innovation and improvement of technology and adjusting the economic structures, i.e., increasing the share of tertiary value added and decreasing the share of value added of secondary industry in the GDP based on the interrelationship of the economy. It would be true if there were no other powerful economic events and strategic mistakes. Therefore, the mutation of national economy in 2012 would be an era to turn the marginal representative factor productivity from lower level to higher one.

On the other hand, the energy demand, especially the coal demand, would be lower to meet China's economic growth. The average annual coal demand growth during 2016–2025 would be only 2.06 % in BAU and 2.1 % in the policy scenario. And the peak of the coal demand would be 4.45 billion tons in 2023 in BAU and 4.47 billion tons in 2022 in the policy scenario; then, the coal demand would decline in China. Thus, the pollution situations would be better and better in China after 2023. In this case, the challenges discussed in Chap. 2 would be an opportunity in China's economic development.

Based on the study in this chapter, we can summarize the following findings:

1. Economic mutation is resulted by technology innovation and improvement, change of economic mix, and other factors. It is necessary for the evolution of the economic system, and a series of mutations is the process of the evolution of the economy. Economic events would stimulate a mutation to happen earlier or later.
2. Economic system has its special rule in development and mutation, while fiscal policy and monetary policy do not have enough power to change the rule. However, it could affect the marginal representative factor productivity of the economy slightly.
3. The BAU scenario would increase marginal representative factor productivity of the economy a little than that of policy one based on economic simulation of China's economy by ARE model.
4. There would be an economic cycle in China's economy in 2015–2025. The peak growth would be around 2019–2021. The low growth periods would be in 2015–2017 and 2023–2025. And policy scenario would decline the peak growth and raise the lower growth of the economy in China.
5. There was a mutation that occurred in primary industry and secondary industry in 2012, respectively, a mutation in tertiary industry in 2014, and a mutation in Chinese national economy in 2012, and there would be no mutations in the three industries and national economy until 2025 in China by ARE simulations if there were no big economic accidents in the period.
6. There is a great opportunity for primary industry to increase the marginal representative factor productivity based on the gene map of the primary industry in China. Technology innovation and improvement and mechanism reform will result in productivity revolution in the primary industry.
7. There is a great opportunity of applying new technologies to increase marginal representative factor productivity in tertiary industry in China. However, the economic gene maps of the tertiary industry in BAU and policy are quite different. It means that applications of the technologies would be limited by some unsuitable mechanism and management. A mechanism reform is essential to capture the potentials of rising marginal representative factor productivity of the tertiary industry.
8. Economic gene mapping can serve as a tool to analyze and diagnose an economy.

References

1. Zheng Hu, Zhaoguang Hu (2013) Production function with electricity consumption and its applications [J]. Energy Econ 39:171–192
2. Zhaoguang Hu, Zheng Hu (2013) Electricity economics: production functions with electricity [M]. Springer, Heidelberg, p 152
3. World Bank Data. http://data.worldbank.org/

Part II
Methodology of Agent Response Equilibrium

Chapter 4
Review of Economic Modeling

Economic system is a large and complex system. It consists of a great number of economic men that interlink and interact with each other, which renders the whole system particular functions and characteristics. To carry out research on economic issues, appropriate models are indispensable tools. Plenty of scholars in the world have done a wide variety of researches based on certain economic modeling techniques. In this chapter, we will provide a brief review of popular models in the area of national economy. To start with, the input–output table will be introduced, since it is the most classical means to express the national economy situations effectively and clearly, and it serves as the foundation of not only some existing economic models but also the Agent Response Equilibrium (ARE) model proposed in this book. Then, we focus on the Computable General Equilibrium (CGE) model and the Agent-Based Computational Economics (ACE) model, which are useful tools that appear frequently in both academic papers and practical applications. In fact, our model is proposed and developed based on the analysis of the strengths and weaknesses of these models. Therefore, we would like to present our knowledge of existing economic modeling methodologies before introducing our model in detail in the following chapters.

4.1 Economic Activities

There are diverse kinds of economic man that have decision-making ability, communication ability, and learning ability in the economic system. They make appropriate actions in order to enlarge their own benefits. Their activities interconnect with each other. The overall effects of all the activities determine the trends of various macroeconomic indicators. Typical economic man in the economic system includes diverse producers, the government, the central bank, commercial banks, consumers, etc.

Z. Hu et al., *China's Economic Gene Mutations*,
DOI 10.1007/978-3-662-47298-9_4

The activities of each producer include obtaining market information, making production plans, hiring employees, seeking financial support through applying loans from commercial banks, purchasing products from other sectors to meet the requirement of its own production, selling products in the market, paying all sorts of taxes and fees to the government, paying labor compensation, etc.

The activities of the government include consuming the final products of various producers, collecting taxes and fees, implementing fiscal policies to affect the national economy, etc. The rates of all kinds of taxes, such as business tax, value-added tax, and income tax, can be adjusted by the government if necessary. The adjustment may be valid either for all the producers or only for certain producers, which is called structural (sectoral) adjustments. The aims of the adjustments are to promote economic growth, adjust the economic mix, and increase employments.

The activities of the central bank include setting the deposit and lending rates, setting the deposit reserve ratio, and conducting open-market operations. These activities are known as monetary policy. Their aim is to impact the financial environment in which all the economic men are in better operating.

The activities of commercial banks include receiving deposits from residents and lending money to production and business entities that are in need of funding. The decisions of loaning money are made according to relevant monetary policies and the features of the borrowing entities.

The activities of residents include participating in the production as labor force, obtaining remunerations, consuming the final products of various producers, and depositing money at commercial banks.

It can be seen clearly that economic men are considerably diverse. They have different economic activities and decision-making behaviors, and they play different roles in the macroeconomic system. Moreover, even for one certain kind of economic man, there are hundreds and thousands of individuals with diverse characteristics. For example, varied producers will definitely have different requirements for the labor, loan, and raw materials, in terms of both type and quantity. Therefore, the economic system is considered as quite a complex system.

The complexity of economic system is further exacerbated by the intricate relationships among all sorts of economic man. The communications and interactions of them bring about appropriate markets. Accordingly, many complicated economic issues that are worthy of being researched come into beings. Three typical examples are given here. Firstly, the interactions between the producers and the commercial banks bring about the issue of loan. The demand quantity of loans of the producers, the supply potential of loans of the commercial banks, and the distribution of loans among various producers are determined in the financial market. Secondly, the interactions between the producers and the consumers including both the government and residents give rise to the issue of trading of goods and services. The supply quantity of goods and services of the producers, the consumption willingness of goods and services of the consumers, and the price of each kind of product are determined in the product market. Thirdly, the interactions between the producers and the employees lead to the issue of labor. The demand quantity of labor force of the producers, the supply potential of labor force of the residents, the supply–demand balance of the labor resources, and the remuneration levels are determined in the labor market.

Furthermore, apart from the conditions of each domestic economic man, the operation of a national economic system has something to do with the situations of other countries and regions. The influences of international environment can be immensely significant. A typical example is the damages of the economic crisis that started in 2008 to the macroeconomic system of a large number of countries. It even poses huge threats to some country's economy, such as the UK, Iceland, Greece, Italy, etc. The impacts of such external environment are mainly reflected by the fluctuations in the import and export market and the international financial market.

In summary, the national economic system is such a complicated system that consists of intricately interlinked economic man whose activities and features are totally different and is also evidently dependent on the international situations. As a result, it is very difficult to do some simulations of the national economy, which is a challenge in the researches and applications of economics. It is widely known that experiments are important means to promote scientific researches and facilitate practical applications. Most significant science discoveries are based on experimental foundations. However, it is almost impossible to do experiments related to economy in reality. To make matters worse, it is difficult to build a simulation platform for the economic system because of its complexity.

With the increasingly high standards of economic research and policy making, subjective judgment and qualitative analysis cannot meet the requirement of either academic community or governing parties. Effective quantitative analysis is essential under the background of new era. Considering the intricate relationship among various elements in the national economic system, the analysis approach needs to be powerful enough to reflect the interactions and operations of them. Such quantitative analysis and simulations have not only great academic values but also practical meanings, because they can provide the policy makers with the opportunities to know the effects of certain policies in advance so as to make sure their policies are appropriate and reliable.

To cope with the challenging task of simulating the national economic system, it is essential to find proper tools and models.

4.2 Input–Output Table

As described in Sect. 4.1, the national economic system is a large and complicated system. Diverse elements in the system interlink with each other. Furthermore, the interdependent relationships among them can be both direct and indirect. For example, the generation of electricity needs some raw materials such as coal; the production of coal needs corresponding mining machines; the production of these machines needs a lot of steels; the production of steels is done with many intermediate inputs including electricity. In order to depict such a complex system, input–output table is a very suitable tool. Input–output table and input–output analysis approach are proposed by W. Leontief [1, 2]. It has been quite popular to observe and research the national economic system by means of input–output table at present.

In terms of unit of measurement, input–output table is divided into two different types, physical input–output table and value-based input–output table. The content in physical input–output table is classified by product. The data in this kind of table will not be influenced by the price level. But the addition and subtraction processing of the data is limited, because different products are measured by different units. Therefore, the value of analysis and research is kind of limited. On contrary, the content in value-based input–output table is classified by sector. The data is entirely measured in unit of money. All sorts of intermediate inputs of a certain sector can be added together, which means the value formation process of the sector is able to be observed. Some key indicators of the national economy, such as the total production output, the social output value, and the national income, are contained in the value-based input–output table. Moreover, sector structure analysis can be done by means of the value-based input–output table. Therefore, this kind of input–output table is more likely to be used in practice nowadays. The schematic view of a typical value-based input–output table is shown as follows (Table 4.1).

As reflected in the schematic diagram, input–output table usually contains three valid quadrants. The fourth quadrant in the lower right corner tends to be blank and neglected in current studies.

Quadrant I is the intermediate use quadrant. It is the core part of the input–output table. The flow of products among various sectors is reflected in this quadrant. Each row and column is corresponding to a sector, and the number of row is equal to the number of column. The row represents the supply side of the intermediate products, while the column represents the demand side of the intermediate products. Therefore, each data in Quadrant I have dual meaning of both production and consumption. Let us take the element in the *i*th row and the *j*th column, noted as x_{ij}, for example. From the perspective of the sector corresponding to the *i*th row, x_{ij} means the value of the product sold to the sector corresponding to the *j*th column. From the perspective of the sector corresponding to the *j*th column, x_{ij} means the value of the product bought from the sector corresponding to the *i*th row as intermediate use. The technical and economic links of interdependence and mutual restraint among various sectors are reflected in this quadrant.

Quadrant II is the final use quadrant. The final consumption of the product of each sector is listed in this quadrant. Similar to the expenditure approach of GDP accounting, consumption, investment, and net exports are considered in Quadrant II. Sometimes, more detail information can be reflected. There could be several columns representing rural household consumption, urban household consumption, government consumption, gross fixed capital formation, changes in inventories, exports, and imports, respectively. For each row, the summation of the data in all the columns in Quadrant II is the overall final demand of the product of the sector corresponding to the row.

Quadrant III is the initial input quadrant, also noted as the value-added quadrant. Apart from intermediate inputs, the products from other sectors and the production

Table 4.1 The schematic view of the value-based input–output table

	Output								
		Intermediate use				Final use			
Input		Sector 1	Sector 2	...	Sector n	Final consumption	The sum of capital formation	Net export	Total Output
Intermediate input	Sector 1	Quadrant I				Quadrant II			
	Sector 2								
	...								
	Sector n								
Total value added	Labor remuneration	Quadrant III				...			
	Net taxes on production								
	Depreciation of fixed assets								
	Operating surplus								
Total input									

of a certain sector need initial inputs. Typical initial inputs include labor remuneration, net taxes on production, depreciation of fixed assets, and operating surplus. They are depicted in Quadrant III. The initial inputs are also known as the value added. Therefore, the formation of Quadrant III is nearly the same as the income approach of GDP accounting. For each column, the summation of the data in all the rows in Quadrant III is the overall initial inputs of the sector corresponding to the column. From another perspective, it is the total value added created by the production of the sector.

The input–output table can also be analyzed across multiple quadrants, which is the essence of input–output analysis [3]. If Quadrant I and Quadrant II are observed together, the data in each row represents the distribution of the product of the sector corresponding to the row. The sum of the intermediate use reflected in Quadrant I and the final use reflected in Quadrant II equals the total use of the sector's product. If Quadrant I and Quadrant III are considered together, the data in each column represents the input source of the production of the sector corresponding to the column. The sum of the intermediate input reflected in Quadrant I and the initial input reflected in Quadrant III equals the total input of the sector's production.

For each sector, the total input is theoretically equal to the total output. As a whole, the overall input of all the sectors should equal the overall output. Besides, the overall intermediate use of all the sectors must be equal to the overall intermediate input of all the sectors, which is shown in Quadrant I. Accordingly, the overall final use of the products of all the sectors is theoretically equal to the overall value added created by all the sectors. To put it another way, the sum of the data in Quadrant II should be equal to the sum of the data in Quadrant III.

The reason why the input–output table has been widely used to deal with national economy issues is that it has a solid foundation of scientific methodology, the systematic theory. The systematic theory is a way of thinking in the research on general systems. It has been regarded as sub-philosophy together with information theory and control theory. The input–output analysis approach borrows ideas from the systematic theory in many respects. The first aspect is the feature of wholeness. The national economy is considered as a complete economic system in this approach. The input–output table includes almost all of the national economic activities in a country. Some significant indicators, such as the total output and input of the products, are reflected in this table. The second aspect is the feature of structure. Apart from the overall indicators, the input–output table embodies the structure of the national economy. This is its unique advantage compared with other tools and models of economic issues. The third aspect is the feature of interdependence. Not only the linkages and interactions among different elements in the national economy in a country but also the relationship and interdependence between the national economic system and the international environment have been expressed in the input–output table. The fourth aspect is the feature of multilevel divisibility. The input–output table can be constructed on different levels, such as national version, regional version, and industrial version, according to actual needs.

In a word, the input–output table provides us with an effective and powerful tool to observe and research the issues on national economic system. The research in the book is mainly the agent-based simulation of the national economic system and its applications in policy analysis, which is on the basis of the input–output table. Besides, the Computable General Equilibrium (CGE) model, one of the most popular models in dealing with economic researches, is conducted based on the Social Accounting Matrix. The foundation of the Social Accounting Matrix is the input–output table.

In practice, the input–output table is compiled by the statistics department in the national or regional government. It is constructed based on a great deal of observed economic data. The time span of the data is a year under most circumstances. Compiling a new input–output table based on a great deal of statistical data needs lots of time. As a result, the input–output table tends to be published only once every several years instead of being published yearly. Chinese National Bureau of Statistics constructs the new national input–output table in the years end with numbers 2 and 7 [4]. Additionally, an updated version of the input–output table will be published in the years end with numbers 0 and 5. The researches in this book are mainly based on China's 2007 annual 42-sector input–output table and China's 2010 annual 42-sector input–output table.

4.3 Computable General Equilibrium Model

4.3.1 Social Accounting Matrix

The Computable General Equilibrium (CGE) model is the most popular tool of national economic simulation at present. It is widely used to conduct academic researches and form policy recommendations. The Social Accounting Matrix (SAM) is the data foundation of the CGE model [5]. It is the cornerstone of the calculation and analysis in economic research by means of the CGE model. To start with, let us have a brief view of the SAM.

Similar to the input–output table, the SAM is also a two-dimensional table that reflects the relatively dependent relationships among different elements in the national economy. Each figure in the SAM represents the value flow from the economic entity corresponding to the row to the economic entity corresponding to the column. Therefore, the SAM is also an effective tool in the analysis and research of national economy.

Precisely, the SAM is a development and evolution of the input–output table. As described in Sect. 4.2, the input–output table mainly focuses on the relationship between diverse producers and the relationship between producers and other elements. In general, it is based on the input flow and output flow of various producers. On contrast, the SAM pays attention to the flow relationship among various accounts in the national economy [6]. For example, the transfer payment of the government to residents is considered in the SAM, while it cannot be reflected in the input–output table. Accordingly, Quadrant IV, which is not considered in the input–output table, must exist in the SAM.

In addition, the SAM is a square matrix. Each row and each column is, respectively, corresponding to an accounting in the national economy, such as the government, residents, producers, other regions in the world, etc. Each row represents the income of the accounts from other accounts. Each column represents the expenditure of the accounts to other accounts. The SAM and the input–output table share the same feature: the sum of each row should be equal to the sum of corresponding column.

4.3.2 Introduction of the CGE Model

The general equilibrium theory is the theoretical basis of the CGE model. The theory is originally proposed by a French economist, Leon Walras, in his book *Elements of Pure Economics* [7]. According to the general equilibrium theory, the egoistic behaviors of consumers and producers will lead to the equilibrium of the supply and demand in each product market and each factor market. All the equilibrium relationships are depicted by a set of equations. The most typical feature of the general equilibrium theory is that it pays attention to all the markets corresponding to various products and factors. The supply of each product or factor must be equal to the demand of it [8]. The equilibrium point is corresponding to a certain equilibrium quantity and a certain equilibrium price.

The overall cyclic of the national economy is reflected in CGE model, i.e., production brings about income; the income raises demand; and the demand leads to the production. At the specific operational level, it is essential to work out the target state value by means of solving the set of equations, while all the equilibrium relationships are the convergence conditions. The step size is determined by the constraints or exogenous variables.

The first CGE model is constructed by Johansen in 1960 [8]. After a long period of development and improvement, the CGE model has become a standard and popular model that is used by the think tanks and governments in many countries. It is a powerful tool in the researches in fields of national economy. The circumstances and interactions of all sorts of economic man in various situations can be observed. The main issues concerned by economists and government officials, such as employment issues, tax issues, and economic growth issues, are reflected in CGE model. Moreover, it is able to serve as the policy simulation laboratory in which diverse policy experiments can be carried out. The influences of certain policies or changes in external environments can be reflected in the output of the model. Therefore, it is possible to evaluate the advantages and disadvantages of various policies in the laboratory platform, in order to help policy makers to avoid losses caused by policy mistakes.

However, the CGE model has some typical drawbacks that restrict its academic and practical value. Firstly, the SAM is obtained based on the input–output table that is published only once every several years because of the great amount of manpower and time to compile a national yearly input–output table. The input–output table may not be effective enough for neighboring years, considering the rapid development and variation of the economic system. Once valid SAM cannot be obtained, the data foundation of the CGE model is not solid. Relevant results of policy simulation and analysis may be affected and lack accuracy. Secondly, there are quite a number of parameters that need to be determined exogenously in the CGE model. As a result, nearly every equation can be impacted by the users of the model. This is a bad characteristic for a policy simulation model. The reason is not only that the builders and users of CGE model must be skilled enough to cope with the considerably high challenges, but also that the objectivity of the results tends to be questionable. Thirdly, the mechanism of CGE model is to find a solution to the target equilibrium state of an economic system. It is impossible to depict the details of the process of evolution of the economic system. Policy makers hope to be aware of the operation of the economic system during a certain period after the policy implementation. Apart from the final state, the time series of some chief variables are considerably valuable, but the CGE model is unable to get them. Last but not the least, significant technological advances cannot be reflected in the CGE model. Science and technology is an important force to promote economic development. In the modern world, technological development is so increasingly rapid that should not be ignored when we do researches in area of national economy and policy recommendations. As a result, the simulation of long-term economy by means of CGE may deviate from the reality.

4.3.3 Applications of the CGE Model

In the past few decades, the CGE model is widely utilized all over the world with the promotion of the World Bank and the International Monetary Fund. There are a broad range of applications.

The international trade is a typical aspect of the applications of the CGE model. Economic globalization renders the foreign trade an essential role in national economy. Whalley built a CGE model considering the trades among diverse countries [9]. As a result of the increasingly fierce competition, international trade conflicts are escalated and a number of countries had to deal with trade negotiations. The aim of his research was to evaluate the results of trade negotiations. Lee constructed a CGE model that quantitatively depicted the consequences of the agricultural protection policy in the USA and Japan [10]. He found that setting up a high tariff and nontariff barriers for the sake of the interests of domestic farmers was an improper decision. These protection policies tended to make bad financial effects and distort the resource allocation, which actually undermined the national economy and welfare. Zhi Wang, Huijiong Wang, and Shantong Li built their multi-country dynamic CGE model to analyze the impacts of China's accession to WTO on international market of labor-intensive products and agricultural exports of the USA [11]. The typical conclusion was that it would be a win-win decision for the USA to weaken the import restrictions of labor-intensive products from China in exchange for China's open attitude to the food and agricultural products from the USA. They also found that China should negotiate with other exporter of agricultural products apart from the USA to seek the best price and maximum benefits. Chinese Academy of Social Sciences worked with Policy Research Centre at Monash University in Australia to develop a CGE model specifically applicable to the Chinese economy. They studied the impacts of trade liberalization on China's national, regional, and industrial economy and discussed relevant policy implications [12]. They drew two chief conclusions. Firstly, the impacts were different for different sectors or different regions with diverse input–output characteristics and trade structures, so various measures should be carried out to cope with the trade liberalization. Secondly, the advantages of China's accession to WTO outweighed the disadvantages, but the price leverage should be fully utilized under current conditions to promote the restructuring of the national economy.

The tax policy is also a typical field of CGE's applications. Tax reform was implemented in 1985 in Australia. Meagher built a CGE model in 1991 to analyze the influences of the tax reform in detail [13]. The conclusion was that the key to carry out the reform is to ensure a reasonable redistribution of the national income among all levels of society. Tao Wang and Jianjun Zhou built a CGE model to deal with tax issues in China [14]. They focused on three sorts of turnover tax, value-added tax, consumption tax, and business tax, in Chinese tax system. Quantitative analysis on tax incidence and tax shifting was done thoroughly. Specifically, they used their CGE model to study the total amount of tax, the price changes of factors, factor mobility among various sectors, the changes in residents' income and their consumption behavior, and social welfare.

The economic reform and development are other application areas of CGE model. Garbaccio constructed a CGE model that consists of 30 sectors to research the impacts of the reform of price mechanism in China in the 1980s on the economic development and industrial structure [15]. The conclusion was that enhancing the market force instead of maintaining planned economy control would be conducive to reduce the government deficit and promote the economic development. Dianqing Xu built a CGE model considering Chinese dual-pricing system [16]. He assumed that the market economy and the planned economy coexisted in China, which might be the feature of Chinese economic reform. He also did a simulation of the shock therapy, a fast and absolute way to reform. He found that the unemployment rate would be higher than 20 % if the shock therapy was implemented. Obviously, China could not afford this situation. Therefore, the dual-pricing system was necessary in China. Fan Zhai and Shantong Li simulated the medium- and long-term trends of economic development, employment, and trade structure in China by virtue of DRCCGE, an improved version of CGE model [17]. They emphasized the important role of technological development. Jie Bai did some work to study the source of living allowance to workers laid off from state-owned enterprises with the help of the CGE model [18]. His team aimed to find the reasonable allocation scheme of the allowance among the government, state-owned enterprises, and non-state-owned enterprises and its influences on the national economy.

4.4 Intelligent Agent and Simulation

Artificial intelligence provides us with a brand-new way to construct national economic model and conduct relevant simulations. Particularly, intelligent agent and multi-agent system have attracted the attention of many academics. We would like to depict some background knowledge on them in this subsection and focus on the specific applications of agent-based model in economics, the so-called Agent-Based Computational Economics.

Artificial intelligence is a branch of computer science. Scientists in this area attempt to understand the essence of intelligence and develop intelligent machine that can act similarly to human beings. Human intelligence is summarized, imitated, and extended. Therefore, the artificial intelligence may approach or even surpass the human intelligence, although breakthroughs on scientific theory and engineering still need to be made. Artificial intelligence is an extremely challenging science because a comprehensive background is essential. Apart from computer science, artificial intelligence may also involve the ideas or the knowledge on information theory, cybernetics, automation, bionics, biology, psychology, mathematical logic, linguistics, philosophy, and other subjects. Owing to its high technical level, profound practical significances, and broad application prospects, artificial intelligence is considered to be one of three cutting-edge technologies in the twenty-first century with genetic engineering and nanotechnology.

As relevant theory and technology become more and more mature, the areas of research and application are continuously expanding. Knowledge representation, automated reasoning and search methods, machine learning and knowledge acquisition, knowledge processing systems, natural language understanding, computer vision, intelligent robots, soft computing, imprecise and uncertain management, artificial life, neural networks, and complex systems are all typical fields. The researches on these topics have achieved fruitful results. Consequently, artificial intelligence has exceeded the scope of computer science and has gradually become an independent subject. Researchers in a great number of fields are paying increasing attention to artificial intelligence to deal with some issues more efficiently with the brand-new approach. In the future, the products in this area are believed to be the container of human intelligence.

The agent is the most typical concept in modern research of artificial intelligence. An agent means an entity that can complete smart tasks autonomously in certain environment. It usually acts towards some goals and tend to use its knowledge base or learning ability to make decisions in order to achieve its goals successfully. Accordingly, it is kind of intelligent and has its unique behavior characteristics. In the computer platform, an agent corresponds to a special section of program that can realize its specific functions.

The agent has six fundamental properties [19]. The first one is the feature of autonomy. An agent is an autonomous entity that can do its work partly or wholly under its own willingness. It can control its behaviors and status to deal with practical issues under the continuous variable environment. The second one is the feature of proxy. An agent stands for a particular entity, which is a basic characteristic of the agent technology. When we pay attention to a certain agent, we tend to focus on the entity corresponding to the agent instead of the agent itself. The third one is the feature of reactivity. The agent is able to perceive the external environment and make actions or adjust its status accordingly. The fourth one is the feature of purpose. The behaviors of agent are goal oriented. No matter what happens, an agent always has its goals to achieve. All its operations are for its goals. The fifth one is the feature of society. Various agents can communicate mutually. They may not only work together to do something jointly but also compete or even withstand with each other, just like the human society. The last property is the feature of intelligence. The agent is intelligent to some extent. For example, it can make decisions by means of inference. And it also has the ability of learning from both other agents and its historical behaviors and results.

Generally speaking, an agent consists of communication manager, inference control unit, state set, knowledge base, etc. Typical internal behaviors of an agent include the following steps. To start with, communication manager obtains the external environment information that could be the messages from other agents. Then, inference control unit analyzes and understands these information and makes inferences on the basis of these information, the statues of the agent, and the knowledge base of it. Consequently, relevant actions are made and the status of the agent is updated. The inference function may work again if necessary. It is noteworthy that the activities of the agent are conducted actively. The inference control unit tends to check the information that the communication manager receives and the status of the agent, followed by an inference activity. Evidently, the communication manager is the interface of the agent with external environment. By virtue of it, the environment information is transferred into the agent, and the messages from the agent are transmitted outwards. The inference control unit acts like the brain of the agent. The behaviors of the agent are determined by the inference control unit after the inference. The state set of the agent reflects various status variables of the agent. The knowledge base of the agent is the important basis of the inference and decision making of the agent.

Seen from the above, the agent has some kinds of intelligence. Its successive actions have the basic characteristics of an intelligent element. It behaves somewhat like the human beings, observing, inferring, and taking actions, again and again. Therefore, the agent can be utilized to simulate the behaviors of a real intelligent entity, such as a person with certain role and a unit with certain functions. In this way, some social science issues can be researched by means of programming on the computer. It is very meaningful to do some simulation on the platform of computer program, because the bottleneck of some social science subjects is that the scientists do not have the chance to do experiments and the conclusions are drew only from observing the facts that have happened actually.

In current researches and relevant applications, the agent technique often appears in the form of multi-agent system. The multi-agent system is the important branch of distributed artificial intelligence. It arises at the end of the twentieth century and is considered to be the state-of-the-art artificial intelligence. As the name implies, the multi-agent system is a collection of a number of agents. The meaning of the multi-agent system is to construct a large and complex system by building models of many small agents that are on behalf of the elements in the system. These agents communicate and coordinate with each other. The multi-agent system has the characteristics of a system in terms of both the internal structure and the overall parameters. The multi-agent technique can resolve some complicated issues that cannot be handled by a single agent.

Scientists have realized that the nature of the human intelligence is a social intelligence. The vast majority of human activities involve community groups composed of more than one person. Solving complex problems and completing complicated tasks require the cooperation of large quantities of individuals and organizations. The operation of the society is dependent on individual behavior and group interaction. In order to research the social intelligence, we should focus on the simulation from both the perspective of individuals and that of groups. In the approach of multi-agent system, the agent corresponds to individual person, while the whole system corresponds to the society. On the one hand, the study of the system includes the social division, internal architecture, and communication mode. On the other hand, the study follows with interest the overall performance and the variation of the variables of the whole system. The multi-agent system is a qualitative leap in the research and applications of the agent technology.

Experts and scholars in diverse sectors summarize the typical advantages of the multi-agent system in coping with practical issues after long-term depth study. To sum up, the main points are as follows. Firstly, each agent in the multi-agent system is autonomous and independent to some extent. It can solve certain minor problems by observing, inferring, and making decisions. A single agent has its impacts on the environment in a particular manner. Secondly, the multi-agent system is ideal for distributed applications. The module is easy to design and expand, which overcomes the difficulties in the management and expansion when building a huge system. Thirdly, the building approach of a multi-agent system is to construct diversified agents on multiple levels, instead of seeking a single large and complex system. Accordingly, the complexity of the system and relevant issues is reduced. Fourthly, the multi-agent system is a coordinated system. The agents coordinate with each other to deal with complicated problems. From another perspective, the multi-agent system is also an integrated system. It uses information integration technology to integrate the information of each subsystem. Fifthly, each agent in the multi-agent system works in parallel, which effectively improves the efficiency of the system and the ability of solving problems. Sixthly, the multi-agent technology breaks the restriction that only one expert system is used in artificial intelligence. Different experts in various fields can collaborate to solve the problems that cannot be handled well by a particular expert, which enhances the ability of the system to solute problems. Seventhly, various agents are heterogeneous and distributed. The programs that correspond to different agents can be developed by different professionals or organizations with different design methods and computer languages. Last but not the least, the processing of the multi-agent system is asynchronous. Since each agent is autonomous, each agent has its own process, which enables it to operate asynchronously according to its own manner.

As a result of its advantages in technical performance and application scope, the multi-agent system has been widely used in a large number of fields. It is very suitable for the situations in which the resources and information are located in distributed environment. Especially, if the evolution of diverse agents and the interactions among them coexist and have mutual influences, the multi-agent technology is the most appropriate choice. Typical application fields include the simulation of the power system operation, the analysis of the transportation system, the flexible manufacturing, the research of the game theory in market environment, supply chain management, the training of military campaign, distributed monitoring and diagnosis machine, etc.

4.5 Agent-Based Computational Economics

4.5.1 Introduction of the ACE Model

Due to the complexity of economics, models based on traditional mathematical approaches are unable to reflect some features of national economy. Conventional methods of mathematical modeling are inseparable from lots of abstractions and simplifications, which means they may fail to deal with complicated economic issues. With the development of computer science and artificial intelligence, doing researches on economic systems by means of program simulations becomes possible and feasible. It is believed that the complex behaviors and intricate interactions of various entities could be considered and reflected.

Leigh Tesfatsion firstly proposed the concept of Agent-Based Computational Economics (ACE) [20]. The ACE is a brand-new approach to develop models for economic system by building an evolutionary multi-agent system. It is a calculation research way to cope with complicated economic issues. The economic system is considered to be a complex adaptive system. The bottom-up methods could be a novel subject that is worthy of long-term exploration and research to enhance the ability to do economic studies.

The particular methodology of ACE can be summarized as follows. To start with, construct an economic system with a number of agents on the platform of computers. Afterwards, define the initial states of the economic system according to existing theories and empirical knowledge. Typical contents include the characteristics of various agents and the rules of their interactions. Then, render the system the capability to operate and evolve with the spontaneous behaviors of these agents. The cornerstone of the ACE approach is that the economic system is an autonomous developing world, which can be regarded as a special dynamic model. Certain economic issues can be simulated by virtue of the multi-agent system. Accordingly, scholars are able to carry out some experiments, which eliminate a major flaw of the researches of economics. Existing theories of economics can be verified and novel discoveries can be achieved.

Although ACE is a novel approach, there are some relations between ACE and traditional methods. Firstly, this approach essentially depends on conventional economic theory and interpretation. It is necessary to have a solid foundation of economics to build the ACE model. Each step in the development of the ACE model should be done according to economic theories. For example, the structure of the multi-agent system is expected to be the same with a real economic system that contains various typical entities, and the characteristics and the relationship of them should also be correct. Secondly, the ACE approach cannot be separated from mathematical methods. Although the complexity of the economic system poses a great challenge to conventional mathematical methods, mathematical tools play an important role in the ACE approach. In fact, the setting of initial conditions and analysis of simulation results are on the basis of mathematical operations. Thirdly, the ACE approach is based on the idea of experimental economics. The research object of economics is the real world that we live in. As a result, it is not so easy to carry out experiments of economics in practice as the ones of natural sciences. However, the ACE is a feasible way to realize the spirit of experimental economics. With the help of computer program, we can do the experiments in the virtual world. The settings of the scenarios and relevant observations shall comply with the essence of experimental economics.

The chief goal of the ACE approach is not only the positivism but also the norm. On the one hand, the complicated phenomenon in economic system should be observed and interpreted. Existing economic methodologies can be validated on the virtual platform. On the other hand, the results of the ACE research should be of guiding significance. Through the comparison of the trends of certain economic indicators under different policy scenarios, qualitative and quantitative analysis can help policy makers choose appropriate policies.

4.5.2 Applications of the ACE Model

The idea of the ACE model was firstly put forward by Professor Leigh Tesfatsion in Iowa State University [20]. Her main research fields included the macroscopic and microscopic policies, the electric power market, and the financial market. Each agent in the ACE model had parallel communication ability. The agents complied with the market rules and behavior paradigms. The researchers built the autonomous multi-agent system by means of the computation laboratory. They focused on the learning ability of the agents, the bottom-up modeling process, the composition of the economic network, and the establishment of the ACE laboratory. The ACE underlined the self-directed learning and evolution, which was like the perfect competitive market without the role of the governments to make actions to intervene the market environment.

Sandia National Laboratory in the USA developed a multi-agent model for its national economy in the mid-1990s [21]. They called it ASPEN. By means of the multi-agent technology, the model was operated on the distributed system. The Monte-Carlo simulation was also considered. Diverse economic indicators were observed or predicted in the ASPEN model. All these indicators had solid microscopic foundation. They were the overall results of the behaviors of the economic actors that had self-determination functions. This model resembled the real national economic system in terms of micro and macro structure and operation mechanism. Therefore, a true representation of some complicated characteristics of economic phenomena could be obtained. Compared with traditional mathematical approach, the agent-based model had better performances in depicting the nonlinearity, volatility, and differentiation.

Herbert Dawid and his team in Bielefeld University in Germany built an agent-based macroeconomic model. They called it Eurace@Unibi model [22]. It was a macroeconomic simulation platform established for economic policy analysis. The model described an economic system containing labor, consumption goods, capital goods, and financial and credit markets in a regional context. The economy was inhabited by numerous instances of different types of agents: firms (consumption goods producers and capital goods producers), households, and banks in the economy. Each agent was located in one certain region. Additionally, there was a single central bank and a government that collected taxes and finances social benefits as well as potentially some economic policy measures, where policies might be different in different regions. By means of the model, they discussed the dynamics of main economic variables on the macro level and on different types of markets. They pointed out some business cycle properties. They also carried out some experiments under the situations in which substantial crashes of the economy followed by recoveries occurred. Then, they found that many of the qualitative properties of the generated simulation output match documented empirical stylized facts.

Kenneth Steiglitz and his team in Princeton University in the USA developed an Economics via Object-Oriented Simulation (EOS) model [23, 24]. It was an ACE model of a complete economy, and it sought to create a computational laboratory wherein economists could test out hypotheses in a controlled environment and explore potential effects of policies on the economy. Since its inception in 2007, several different baseline versions of EOS have been created. It contains four agent classes (laborer, firm owner, farm, and utility factory), three goods (labor, food, and utility), and three markets (labor market, food market, and utility market). Utility here is an abstract good that represents everything besides food that brings happiness. By way of the EOS model, they have done some economic experiments. They have paid close attention to interest rate shock, money supply shock, technological progress, etc. They found that the experiment results seemed to make sense and conformed to economic prediction.

Giorgio Fagiolo in Sant'Anna School of Advanced Studies in Italy and Andrea Roventini in Université Paris Ouest Nanterre La Défense in France have done some researches on the ACE model, and they have compared this novel approach with their Dynamic Stochastic General Equilibrium (DSGE) model [25]. As far as they are concerned, there are still some open issues that should be addressed in the ACE approach. The most important ones concern empirical validation, over-parameterization, estimation, and calibration. Nevertheless, the success of ACE models in delivering policy implications while simultaneously explaining the observed micron and macro stylized facts is encouraging for the development of a new way of doing macroeconomic theory. They have outlined the most prominent values added deriving from performing policy experiments within an ACE approach. These include ACE's extreme modeling flexibility, the friendly relation of agent-based models with empirical data, the easiness of carrying out empirical-validation exercises, the almost infinite possibility of experimentation, and the positive impacts that a more realistic and algorithmically structured model can have on political decision makers as compared to obscure and un-intuitive mathematical neoclassical models.

Domenico DelliGatti in Catholic University of Milan in Italy, Saul Desiderio and Mauro Gallegati in Università Politecnica delle Marche in Italy, Edoardo Gaffeo in the University of Trento in Italy, and Pasquale Cirillo in the University of Bern in Switzerland proposed the concept of macroeconomics from the bottom up [26]. They constructed a model called Bottom-up Adaptive Macroeconomics (BAM). They treated the whole economic system as a complex adaptive dynamic system. The complexity derived from the nonlinearity of the interactions among various agents. We could observe and measure the macroeconomic indicators, but we could not deduce the economic statues through observing isolated individual behaviors. Therefore, the multi-agent technology deserved to be utilized. In their BAM model, all the matching processes between supply and demand happened under a decentralized environment. The transactions occurred not in accordance with an equilibrium price. On the simulation platform, they mainly observed the dynamic trend of revenue in front of sudden crisis, the spontaneous macroeconomic law, and collaborative changes and sequence relationships among different variables.

Ketter Wolfgang in Erasmus University in the Netherlands and Collins John in the University of Minnesota in the USA made efforts to predict the trend of the economy on the basis of multi-agent system [27]. In their model, the agent could made tactical decisions such as pricing and strategic decisions such as determining product mix and production planning via observing the situations in the market. They proposed a novel learning strategy based on historical data with the help of Gaussian mixture model. The agents kept obtaining the market information and would not stop learning. The Markov correction and exponential smoothing were also added in the agent-based model. Under these conditions, the prediction of the market condition in the future could be made.

Shiwei Zhang and his team in Jilin University in China have been conducting the agent-based economic simulation researches for many years [28]. They put forward an agent-based macroeconomic model which was called ASMEC-C. The purpose of the model was to analyze the effects of macroeconomic policies. The multi-agent system consisted of a lot of households, enterprises, one bank, and one government. Microscopic individuals had quite different states and behaviors. All of them could adjust their behaviors according to the changes of the environment. Macroeconomic dynamic was the result of the accumulation of microscopic individuals' interactions. The effects of monetary policies and fiscal policies were simulated with the ASMEC-C model. The simulation results were basically consistent with the economic theories. Afterwards, they built an agent-based transition economy model to analyze the dynamic correlation of economic growth and income gap during the process of the transition economy. They found that with the continued economic growth, the income gap tends to narrow after experiencing an expansion process. A moderately progressive tax policy could promote the rapid economic growth while narrowing the income gap.

Gaoxiang Gu and his team in Chinese Academy of Sciences in China combined the agent-based simulation model with sequence innovation theory [29]. They developed an agent-based macroeconomic process simulation system and simulated the scenarios in which economic crisis happened. Through comparing different financial policies, they studied the reasons and solutions of the economic crisis. Their conclusion was that the overly liberal financial policies might be the chief reason of crisis because it tended to bring about severe inflations, even though it could make the market very prosperous, and the society develop rather rapidly. They also found that appropriately liberal financial policies could be applied after the economic crisis, in order to recover the employment and innovations and improve production technologies. Then, the financial policies should be prudent to guard against fluctuant inflations.

References

1. Leontief W (1936) Quantitative input–output relations in the economic system of the United States. Rev Econ Stat 18:105–125
2. Leontief W (1941) The structure of American economy, 1919–1929. An empirical application of equilibrium analysis. Harvard University Press, Cambridge, MA
3. Ronald EM, Peter DB (1987) Input–output analysis, 2nd edn. Cambridge University Press, Cambridge, UK
4. Duan W, Hu ZG, Zhou YH, Xiao X (2012) Input output table updating based on agent-responses equilibrium model. In: IEEE 2012 computational intelligence for financial engineering & economics conference, New York, pp 1–4
5. Wang Q, Li S, Gao Y (2008) The social accounting matrix: principles, approaches and applications. Tsinghua University Press, Beijing (In Chinese)
6. United Nations Statistics Division (1993) System of national accounts 1993. http://unstats.un.org/unsd/nationalaccount/docs/1993sna.pdf
7. Walras L (1969) Elements of pure economics. A.M. Kelley, New York

8. Tan X (2008) Study on the building and applications of electricity computable general equilibrium model. PhD thesis, North China Electric Power University, Beijing. (In Chinese)
9. Whalley J (1985) Trade liberalization among major world trading areas. MIT Press, Cambridge, MA
10. Lee H, Roland D (1991) Bilateral trade between the United States and Japan and the implications of agricultural protection. Working paper prepared for the second international CGE modeling conference, Waterloo, Canada
11. Wang Z, Wang H, Li S (1997) The influences of China's joining the World Trade Organization on the international labor-intensive product market and American agricultural exports – dynamic recursive computable general equilibrium analysis. Econ Res 4:54–65 (In Chinese)
12. Fan M, Zheng Y (2000) The influences of trade liberalization on China's economy based on the computable general equilibrium analysis. World Econ 4:16–26 (In Chinese)
13. Meagher G (1990) A general equilibrium analysis of fiscal incidence in Australia. Working paper prepared for the first international CGE modeling conference, Waterloo, Canada
14. Zhou J, Wang T (2004) CGE models and tax reform in China. J Syst Sci Inf 2:319–324
15. Garbaccio RF (1994) Price reform and structural change in the Chinese economy: policy simulation using a CGE model. China Econ Rev 6:1–34
16. Xu D (1996) The Chasm in the transition: a CGE analysis of Chinese economic reform. J Policy Model 18:117–139
17. Zhai F, Li S, Feng S (1999) Medium-term economic growth and structural change—recursive dynamic general equilibrium analysis. Syst Eng Theory Pract 2:88–95. (In Chinese)
18. Bai J (2001) Study on the living allowance source of laid-off workers from state-owned enterprises—computable general equilibrium analysis. J Peking Univ (Philos Soc Sci) 38:126–131. (In Chinese)
19. Hu Z, Shan B et al (2009) Simulation experiments of electric power supply and demand—Soft Science Laboratory based on the intelligent engineering. China Electric Power Press, Beijing (In Chinese)
20. Tesfatsion L (2001) Introduction to the special issue on agent-based computational economics. J Econ Dyn Control 25:281–293
21. Basu, Pryor, Quint, Arnold (1996) ASPEN: a microsimulation model of the economy [EB/OL]. Sandia report 1996, http://www-aspen.cs.sandia.gov
22. Herbert D, Simon G, Philipp H et al (2012) The Eurace@Unibi model: an agent-based macroeconomic model for economic policy analysis. Bielefeld working papers in economics and management No. 05-2012
23. Xu Z (2011) Developing a realistic baseline for the economics via object-oriented simulation framework. Thesis, Princeton University
24. DeLuise A (2010) Micro-validated baseline for the economics via object-oriented simulation framework. Thesis, Princeton University
25. Fagiolo G, Roventini A (2012) Macroeconomic policy in DSGE and agent-based models. Document de Travail Working Paper 2012–17.
26. Gatti D, Desiderio S, Gaffeo E, et al (2011) Macroeconomics from the bottom-up. Springer Press, New York
27. Ketter W, Collins J, Gini M et al (2009) Detecting and forecasting economic regimes in multi-agent automated exchanges. Decision Support Syst 47:307–318
28. Zhang S, Deng C (2004) Study on the simulation experiments of economic policies. China Soft Sci 3:26–30 (In Chinese)
29. Gu G (2012) Multi agent-based simulation on technology diffusion of China. In: 10th international conference on cellular automata for research and industry, Greece, pp 370–374

Chapter 5
Introduction of Intelligent Engineering

The system of national economy is extremely large and complex. As a result, it is an undoubtedly difficult task to build models for economic system and simulate its operation. In order to cope with this issue, we need to find some advanced methodologies from the perspective of complex system. The system engineering is an option, but its effect is comparatively limited to deal with the uncertainty problems and distributed intelligence in the system. With the development of innovative ideas and techniques of artificial intelligence, authors have improved the theory and proposed the methodology of intelligent engineering [1]. From our point of view, it is a reasonable and feasible choice to serve as the foundation for modeling complex economic systems. It plays an important role in designing our model and coding our programs. In this chapter, the basic theory of intelligent engineering will be introduced. Particularly, we focus on two aspects to express the typical contents of intelligent engineering, i.e., generalized model and intelligent space. We hope it will be useful and inspiring for researchers to solve problems related to large complex systems with distributed intelligence.

5.1 Introduction

As described in the previous chapter, it is very difficult to build model for the economic system and simulate its operation. Although the computable general equilibrium (CGE) model and the agent-based computational economics (ACE) model have their advantages to cope with the simulation of national economy, both of them have their limitations. The reason is that the economic system is considerably complicated. As a result, it should be treated as a complex system and analyzed from the perspective of the characteristics of complex system.

The complex systems have the following common features [2]. Firstly, they are large scale. There are a great number of elements that have different characteristics and statuses in the system. A complex system has not only large space but also long

Z. Hu et al., *China's Economic Gene Mutations*,
DOI 10.1007/978-3-662-47298-9_5

time to evolution. Secondly, they have complicated structure. Various elements play diverse roles and they have intricate relationships with each other. The relationships tend to be dynamic and nonlinear, which poses great challenges to construct a model for the system. Thirdly, they have comprehensive functions. Different elements have different goals. Some of the goals may be contradictory, which pull the evolution of the system towards diverse destinations.

Since complex systems exist in many areas of human life and production and even many fields of nature, a number of issues cannot be discussed without considering the complex systems. Therefore, experts and scholars all over the world attach importance to complex systems. Actually, systems can be divided into various sorts according to their specific properties. For example, they can be divided into natural systems and social systems, physical systems and conceptual systems, static systems and dynamic systems, open systems and closed systems, and open-loop systems and the closed-loop systems.

It is difficult to execute research on complex system due to their features such as nonlinear, time varying, high dimensional, and uncertainty. As a result, traditional methods seem to be powerless. Firstly, precise mathematical model is the cornerstone of traditional methods, but it is sometimes not feasible to construct conventional mathematical models for complex systems in practical. Secondly, many simplifications are necessary in traditional approaches to deal with large systems. However, these simplifications make it impossible to depict certain features of the systems. Thirdly, complex systems could be an open system in which human plays an essential role. Human learns from the system and takes actions on the system. Conventional mathematical models fail to reflect the participation of human beings.

With the development of artificial intelligence, intelligent computing technology, and uncertainty theory, the research of complex systems can be carried out with more approaches. Particularly, intelligent engineering is a feasible choice [1].

Intelligent engineering is the supplement and expansion of system engineering. The mathematical model in system engineering is expanded to the generalized model [3], which will be discussed in detail in Sect. 5.2. The state space in system engineering is expanded to intelligent space and the state transition matrix is expanded to intelligent path, which will be introduced in Sect. 5.3. The methodology of intelligent engineering is proposed by Minjie Xu [4], as shown in Fig. 5.1. It is obvious that the methodology of intelligent engineering is quite comprehensive. It inherits the essence of system engineering, and it develops to a more complex extent. Moreover, intelligent engineering has more foundations including complex adaptive system, artificial intelligence, and computer science, which can be treated as the state-of-the-art human technology. Intelligent engineering can serve as a novel powerful tool to study the operation of complex systems. The typical feature of intelligent engineering is that human's participation and impacts on the evolution of the system are considered [5].

Artificial intelligence and computer technology are the foundation of intelligent engineering. Moreover, the subjective initiative of human is emphasized in this brand-new theory. While most codes for multi-agent are conducted on the computer, a special open agent is needed for the role of human outside of the computer.

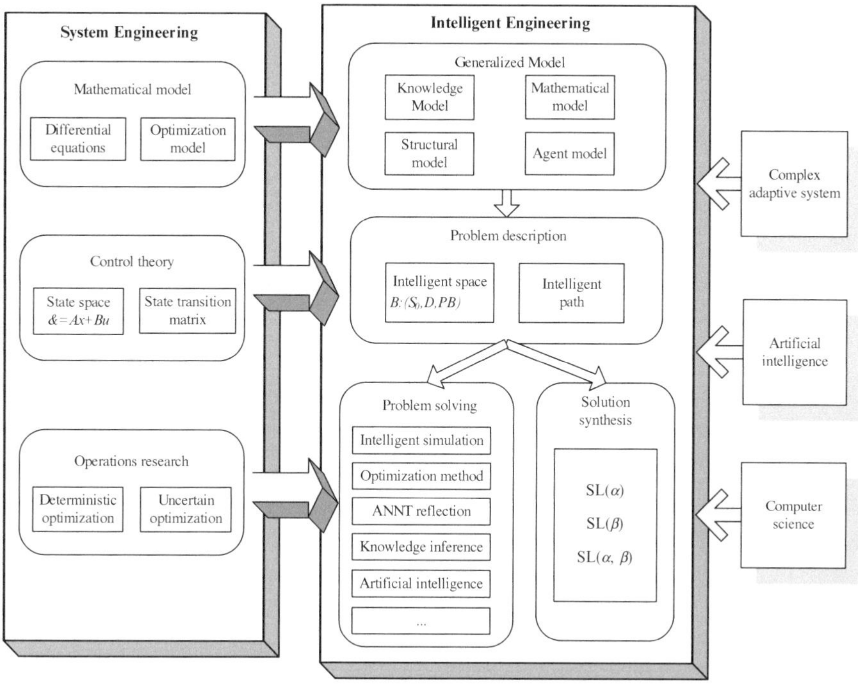

Fig. 5.1 Schematic diagram of methodology of intelligent engineering

After all, the experience and decision-making capacity cannot be replaced by computer, so the wisdom of human should be associated with the intelligence of computer. The participation of human is good for the solution of complex problems. Viewed from another perspective, this kind of mechanism design is also beneficial for the actual research of some issues. For example, if researchers focus on the public policies, the government can be set as an open agent, which means that the researchers can play the role of the government partly. In this way, they can find the optimal policies by virtue of experiments.

The contributions of the intelligent engineering can be summarized as the following three aspects [6]. Firstly, novel approaches are studied. Because of the complicated features of the complex systems, the analysis and simulation are quite difficult to conduct. Conventional system engineering based on control theory, operations research, and information theory cannot meet the requirements of effective researches. Intelligent engineering makes a breakthrough and provides us a brand-new approach to cope with these issues. Secondly, new methods are proposed. Traditional mathematical methods are suitable for relatively ascertain system. New methods are needed to study the increasingly complex and uncertain systems. Thirdly, new technologies are explored. The knowledge and experience of human are crucial for the operation and evolution of complex systems. As a result, new technologies such as human–computer interaction technology and intelligent decision support system should be developed on the basis of the idea of intelligent engineering.

To sum up, the intelligent engineering not only inherits the principles and means of system engineering to cope with complicated issues but also brings about some modern theories, such as artificial intelligence, intelligent computing technology, and uncertainty theory, to enrich and expand the system engineering. It can serve as a novel idea to deal with complex systems. Through the combination of system engineering and artificial intelligence, an advanced approach is generated to build models for large complex systems and analyze the evolution path of the complex systems [5].

5.2 Generalized Model

As discussed in the book *Simulation of Power Supply and Demand by Intelligent Engineering* (Chinese) [6], traditional mathematical model cannot meet the requirements of depicting and analyzing problems in complex systems. Especially, mathematical model tends to fail to describe some practical issues, in which there may be certain correspondence relationship. For example, a serious cold air in December in a city in southern China will lead to the increase of the power load demand. The underlying principle is that people tend to turn on the air condition more frequently, since there is no central heating service in southern China. As a result, the electric power demand will be raised up. The inherent logic relationship exists and is easy to understand. However, it is not easy to express this issue by means of conventional mathematical model.

In order to describe these objective existences of the relationship between various things that cannot be expressed by mathematical model, we put forward the concept of generalized model in intelligent engineering theory [6]. The generalized model is the spread of traditional mathematical model. The key idea is to treat certain relationship between any set X and Y as a kind of mapping. The generalized model can be noted as follows:

$$f : x \to y \quad x \in X, \quad y \in Y \tag{5.1}$$

There are five types in the concept of generalized model as follows:

- The first one is mathematical model. It means the model expressed by mathematical equations, such as $ax + by = 0$.
- The second one is the rule-based model. It indicates the model expressed in the form of rules, such as "if x..., then y...."
- The third one is the fuzzy inference model. The principle is to depict the elements that are difficult to quantify with fuzzy set and draw conclusions with fuzzy inference.

- The fourth one is the neural network model. A great deal of historical data is utilized to train the neural network to establish the relationship among different issues.
- The fifth one is the hybrid model. The concept includes the combination of two or more above kinds of models. In addition, the agent model and expert system model also belong to the hybrid model.

We will briefly introduce the above five models in the following subsections.

5.2.1 Mathematical Model

Mathematical model can quantitatively depict the static characteristics and dynamic process of certain things by virtue of mathematical equations. With the help of mathematical models, these things can be researched quantitatively. Typical examples include the equations of state, econometric models, and general equilibrium models.

The history of mathematical model can be traced back to the era when humans began to use number. As number was used more and more widely, humans started to build diverse mathematical models to solve a wide variety of practical problems. In fact, the mathematical model serves as an essential bridge that links the actual issues and mathematical methods.

There are some requirements to build a mathematical model. Firstly, it should be true and complete. The objective phenomenon ought to be described accurately and vividly with the help of mathematical model. The problems can be solved successfully in terms of all the requisites. Secondly, it should be concise and practical. The essence and natural relationship should be reflected in the model, while some negligible items that have very limited influences on certain issue can be abandoned. The aim is to make it relatively easy to collect data and solve the mathematical model under certain conditions of accuracy demand. Thirdly, it should be adaptable to changes. Relevant variables and parameters can be adjusted if certain conditions change or people's understandings develop, in order to make it suitable in more cases and in longer period.

There are a variety of classification methods for mathematical models. They can be divided into static models and dynamic models depending on the variables, and their interrelations are fixed or mutative. Algebraic equations are the typical forms in static models, while differential equations and difference equations tend to be used in dynamic models. They also can be divided into linear models and nonlinear models. As for the linear models, the relationships between various variables are linear, which means that superposition principle can be applied. It is the opposite case in nonlinear models. Nonlinear model can be transferred into linear model under certain circumstances. The approach is to conduct Taylor series expansion in the neighborhood of the operating point, retain an order term, and omit higher order terms. They can also be divided into continuous-time models and discrete-time models according to the time continuity of the variables in the model. Evidently, the variables in continuous-time models are continuous in time, while it is opposite for the discrete-time models.

5.2.2 Rule-Based Model

In order to understand the rule-based model, let us take one example. "If the price of coal decreases by 10 %, the stock market price of a power generation group will rise by 5 %." This is a simple rule on the basis of the experience and knowledge of the stock analyst. It can be broken down into two propositions shown as follows:

A: the price of coal decreases by 10 %.
B: the stock market price of a power generation group will rise by 5 %.

There are two typical characteristics as for a proposition. Firstly, it is a declarative sentence. Secondly, it can be true or false. In the two-valued logic, the value of the proposition is 1 if it is true and the value is 0 if it is false. So the rule mentioned above can be expressed by the value. If $A = 1, B = 1$ means the same as "if A is true, then B is true" [6].

There may be a question why the stock market price of a power generation group will rise by 5 % instead of 10 %. This is the judgment of the stock analyst based on his/her previous experiences. It is the summary of related knowledge. The rule builds the connection between specific event and the stock price. This connection cannot be expressed by mathematical equations. However, it is of usage and importance to depict the mapping relationship between certain events and stock price of a company. Models that are similar to this example is called rule-based model. It can be noted as follows:

$$A \to B \tag{5.2}$$

A is the prerequisite, which could be a fact. B is the conclusion.

Some complicated rules derive from the basic rule-based model. There is a so-called rule-based deduction system in artificial intelligence to cope with the issue. Forward chaining and backward chaining are the two typical ways. Forward chaining is to infer the result from the prerequisite. On the contrary, backward chaining is to infer the prerequisite from the result.

The rule-based model is usually used to present the experience and knowledge of human beings on certain things, so as to conduct analysis and inference. It may be either the understanding of a certain thing or the judgment of the relationships among various things [4]. Furthermore, it can also be reflected on the techniques of artificial intelligence.

5.2.3 Fuzzy Inference Model

There are many things that cannot be depicted by exact data or sentences. For example, when asked about the feeling of a certain thing, we would like to say "not bad"; when asked about the pressure of economic crisis on a certain company, its manager may say "a lot."

A valuable solution to deal with these issues effectively is the fuzzy sets and systems, which is also called as fuzzy mathematics. It is the collective concept that contains fuzzy topology, fuzzy measure theory, and other related subjects in fields of mathematics based on fuzzy sets and fuzzy logic. It acts as the tool to research the problems in which the boundaries are vague to some extent in the real world.

The set theory is the cornerstone of modern mathematics. All the actual theory systems have the possibility to be included in the mathematical framework described by the set. Classical set theory only focuses on the concept and relevant things that have clear extension. According to the statements in classical set, each set must be constituted by the determined element, and the affiliation of elements to the set should be clear. With the development of fuzzy mathematics, these statements may not be the only choice to cope with practical issues with mathematical methods. Each element can belong to a certain set to a certain extent and belong to another set to another extent. For instance, we have now two sets; one is the set in which people are tall, while the other one is the set in which people are short. If we take Ming Yao, the famous NBA player, for example, he belongs to the first set with no doubts. However, if a man who is 175 cm in height appears, we may find it difficult to link him with either set. In such a case, we could state that he belongs to not only the first set but also the second set. An essential variable should be defined here to reflect the extent to which he belongs to each set. It is called the degree of membership, which is the most important and representative concept in fuzzy mathematics. Now the idea of fuzzy set should be clear. Fuzzy mathematics develops on the basis of fuzzy set.

The fuzzy inference is the inference theory in fuzzy mathematics based on the idea of fuzzy set. Inexact conclusions can be inferred from inexact prerequisite set. It is a kind of inference with uncertainties. Sometimes, people would also like to call it approximate reasoning. It is closer to the characteristics of the thinking way of humans, compared with classical mathematical tools. Moreover, it is more suitable to a variety of practical issues. Therefore, it has broad application prospects and great value in scientific research and actual production.

5.2.4 Neural Network Model

The neural network model is also named artificial neural networks or connection model. It is a novel tool in which the neural and brain networks of human are imitated. The process of information is distributed in the system. The complex characteristic of the neural networks makes it possible to deal with complicated problems in reality by virtue of adjusting interconnected relationship among a large number of internal nodes.

It is generally believed that there are three basic ways of thinking in the human brain, the abstract logical thinking, the intuitive thinking, and the inspiration thinking. By means of the abstract logical thinking, inferences are conducted according to logic rules. It is similar to the operation mode of computer and can be depicted in mathematical expressions. On the contrast, the intuitive thinking is more obscure. People probably come up with ideas suddenly because of the comprehension of distributed information. There are two features of this kind of thinking. Firstly, all the information is distributed on the nodes in the network through the excitement mode of neurons. Secondly, the information processing is performed by a dynamic process of interaction between neurons at the same time.

The neural networks simulates the second way of thinking of humans. It is a nonlinear dynamical system. As for a single neuron, the structure is simple and the function is limited. However, the network system consists of huge amounts of neurons that have the ability to realize diverse complex behaviors. Accordingly, it provides us with a relatively simple and feasible solution to some complicated problems in reality.

To render the neural networks able to work, it is necessary to make it learn from historical data according to certain learning rules. For example, we set up a neural network and want to distinguish letter A and letter B. It is formulated that 1 should be output when letter A is the input, while 0 should be output when letter B is the input. The learning principle of the neural network is to reduce the possibility of making the same mistake next time via its learning. To start with, set the connection weights in the network as the random number between 0 and 1. We make A to be the input. The network will obtain its output after weighted sum and nonlinear operation. In this case, the output of 1 and 0 has the same probability, 50 %, because it is a wholly random process. If the output is 1, we enlarge the connection weights so that the network can make the correct judgment when A is input next time. After a process of learning, the neural networks have reliable functions and can be utilized by researchers effectively.

5.2.5 Hybrid Model

The hybrid model is a composite model that consists of two or more models mentioned in previous subsections. For example, the compound of mathematical model and rule-based model can form a hybrid model [6].

Typically, the agent-based model is a hybrid model. It is the state of the art with the development of artificial intelligence and has been paid close attention by a large number of researchers. In an agent-based model, conventional mathematical model is essential in many aspects, because there should be many equations for the environment and each agent. Then, the rule-based model is also necessary, with the aim to formulate the behavior rule of each agent. Besides, the fuzzy inference model may also be included in the agent-based model, especially when the agent system is an open system for the users. Furthermore, the neural network model can be considered to some extent, because the agent-based model also has the ability to learn, which may borrow the principle of the neural networks. All in all, the agent-based model is quite a representative hybrid model in the background of intelligent engineering.

We have offered a brief view about the agent-based model in the previous chapter. The communication manager, inference control unit, state set, and knowledge base of the agent have been introduced in Sect. 4.4.

The basic principle of an agent to interact with the environment can be described as follows [6]. It is assumed that the initial state of the environment is e_0. The agent chooses a certain action under the command of its inference control unit. We assume the action is called α_0. The result of the action is that the state of the environment is changed. The new state of the environment is assumed to be e_1. The agent makes further decision on the next action after observing the environment changes. The second action of the agent is called α_1. As a result, the state of the environment will be changed into e_2. So the cycle continues.

In order to depict the influences of the action of the agent, the state transition function needs to be introduced. The mapping between the action and the state of the environment is established by virtue of the state transition function.

In the reality, every entity that has independent behavioral objectives and the ability to take actions according to the information from the external environment can be abstracted as an agent. It may be as big as a country or a large economic system. It can also be as small as a company, a family, and even an individual person [7, 8]. With the help of the agent model, it is possible to simulate the behaviors of these active entities and the interactions among them. As a consequence, the complex large system can be modeled from the bottom up. It provides the scholars in various fields with a feasible approach to conduct the researches on these complex systems. At least, it is an effective way to study the system of national economy, which has been verified by our researches shown in this book.

5.3 Intelligent Space

5.3.1 Concepts

In this the basic concepts and theorems in the theory of intelligent space will be introduced [6, 7, 9].

Definition 5.1 *The intelligent path means all kinds of relationship or mapping from the initial state set* S_0 *to the target state set* D. *It can be denoted as follows:*

$$P : S_0 \rightarrow D \tag{5.3}$$

Definition 5.2 *It is assumed that* $s \in S_0$ *and* $d \in D$, *and* P *is the set of the intelligent paths from* S_0 *to* D. *If* $IP \subset P$, *the intelligent equation can be defined as follows:*

$$d = IP(s) \tag{5.4}$$

Definition 5.3 *The intelligent space can be expressed as follows:*

$$I =< P, S > \tag{5.4}$$

S is the state set, and *P* is the intelligent path set.

Definition 5.4 *The basic problem in the theory of intelligent space is noted as Problem* B. *It is defined as follows:*

$$B =< S_0, D, \mathrm{PB} > \tag{5.6}$$

S_0 is the initial state set, *D* is the target state set, and PB is the set of the intelligent paths from S_0 to *D*.

Definition 5.5 *If the initial state* S_0 *and the target state* D *are known, while the intelligent path set PB is unknown, the problem is called Problem* B_1. *It is one specific kind of Problem* B, *which can be expressed as follows:*

$$B_1 =< S_0, D, \mathrm{PB} > B_1 : (S_0, D) \Rightarrow \mathrm{PB} \tag{5.7}$$

A specific solution of Problem B_1 *is denoted as* ps, *which is one element of PB:*

$$ps \in \mathrm{PB} \tag{5.8}$$

Definition 5.6 *If the initial state* S_0 *and the intelligent path set PB are known, while the target state* D *is unknown, the problem is called Problem* B_2. *It is one specific kind of Problem* B, *which can be expressed as follows:*

$$B_2 =< S_0, D, \mathrm{PB} > B_2 : (S_0, \mathrm{PB}) \Rightarrow D \tag{5.9}$$

A specific solution of Problem B_2 *is denoted as* $d = IP(s)$, *which is the solution of the intelligent equation.*

Definition 5.7 *If target state* D *and the intelligent path set PB are known, while the initial state* S_0 *is unknown, the problem is called Problem* B_3. *It is one specific kind of Problem* B, *which can be expressed as follows:*

$$B_3 =< S_0, D, \mathrm{PB} > B_3 : (\mathrm{PB}, D) \Rightarrow S_0 \tag{5.10}$$

A specific solution of Problem B_3 *is the inverse of the solution of the intelligent equation:*

$$s = \mathrm{IP}^{-1}(d) \tag{5.11}$$

Definition 5.8 *As for Problem* B_1, *if the intelligent path set PB has a fuzzy set,*

$$fp : \mathrm{PB} \rightarrow [0, 1] \tag{5.12}$$

the *α-smart* solution of Problem B_1 is denoted as SL(α). It is defined as follows:

$$\mathrm{SL}(\alpha) = \{\mathrm{ps} | \mu_{fp}(\mathrm{ps}) \geq \alpha,\ \mathrm{ps} \in \mathrm{PB}\}, \quad \alpha \in [0, 1] \tag{5.13}$$

Definition 5.9 $<P_s, S_t>$ *is the subspace of* <P, S> *if and only if*

$$P_s \subset P \tag{5.14}$$

$$S_t \subset S \tag{5.15}$$

Definition 5.10 *The forward inference in the artificial intelligence is the approach to find the* α-smart *solution, SL(α), of Problem* B_1 *in the subspace* $<P_{AI}, S>$. P_{AI} *is the set of all the relationships in* S, *and* fp *in SL(α) is a fuzzy set in* P_{AI}.

$$P_{AI} \subset P \tag{5.16}$$

$$< P_{AI}, S > \quad \subset \quad < P, S > \tag{5.17}$$

Definition 5.11 *The main issue in the neural networks is the approach to find the* α-smart *solution, SL(α), of Problem* B_1 *in the subspace* $<P_{NN}, S>$. P_{NN} *is a mapping set of* S, *and* fp *in SL(α) is a fuzzy set in* P_{NN}.

$$P_{NN} \subset P \tag{5.18}$$

$$< P_{NN}, S > \quad \subset \quad < P, S > \tag{5.19}$$

Definition 5.12 *The aim of the fuzzy inference is to find the* α-smart *solution, SL(α), of Problem* B_1 *in the subspace* $<P_{FS}, S>$. P_{FS} *is the set of conversion relationships from a fuzzy set to another fuzzy set in* S, *and* fp *in SL(α) is a fuzzy subset in* P_{FS}.

$$P_{FS} \subset P \tag{5.20}$$

$$< P_{FS}, S > \quad \subset \quad < P, S > \tag{5.21}$$

Definition 5.13 *The intelligent engineering is a method to solve Problem* B *in the intelligent space* <P, S>.

Definition 5.14 *If a complex system can be divided into a serial of some subparts, the path problem of each subpart is called the* and *subproblem, which is illustrated in Fig. 5.2.*

Definition 5.15 *If a complex system can be divided into a parallel of some subparts, the path problem of each subpart is called the* or *subproblem, which is illustrated in Fig. 5.3.*

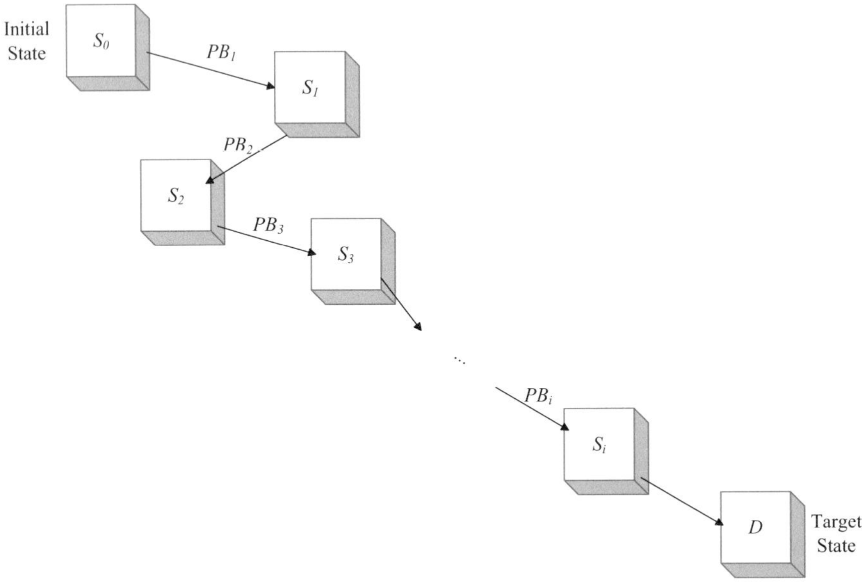

Fig. 5.2 The *and* subproblem of Problem B_1

Fig. 5.3 The *or* subproblem of Problem B_1

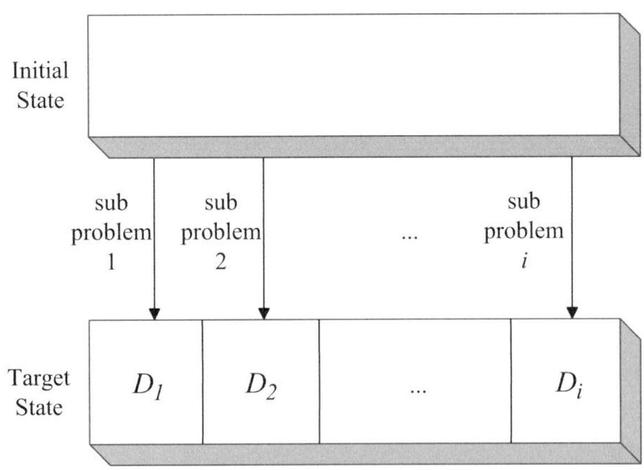

Theorem 5.1 *As for Problem* B_1*, if there is a* α*-smart solution in the intelligent subspace* $<P_{AI}, S>$*, there should be a* α*-smart solution in the intelligent space* $<P, S>$*.*

Theorem 5.2 *As for Problem* B_1*, if there is a* α*-smart solution in the intelligent subspace* $<P_{NN}, S>$*, there should be a* α*-smart solution in the intelligent space* $<P, S>$*.*

Theorem 5.3 *As for Problem* B_1*, if there is a* α*-smart solution in the intelligent subspace* $<P_{FS}, S>$*, there should be a* α*-smart solution in the intelligent space* $<P, S>$*.*

Theorem 5.4 *As for Problem* B_1*, if there is a solution* $ps \in PB$*,*

$$ps = (P_0, P_1, P_2, \ldots, P_n) \tag{5.22}$$

there is a α*-smart* solution SL(α) as follows:

$$\mathrm{SL}(\alpha) = \left\{\mathrm{ps} \middle| \mu_{fp}(\mathrm{ps}) \geq \alpha,\ \mathrm{ps} \in P_{AI}\right\}, \quad \alpha \in [0, 1] \tag{5.23}$$

Theorem 5.5 *As for Problem* B_1*, if there is a solution ps that is like*

$$\mathrm{ps} = \left(P_0, P_1, P_2, \ldots, P_i, \ldots, P_j, \ldots, P_k, \ldots, P_n\right) \in PB \tag{5.24}$$

$$\text{and} \quad (P_0, P_1, P_2, \ldots, P_i) \in P_{AI} \tag{5.25}$$

$$\left(P_{i+1}, P_{i+2}, \ldots, P_j\right) \in P_{NN} \tag{5.26}$$

$$\left(P_{j+1}, P_{j+2}, \ldots, P_k\right) \in P_{FS} \tag{5.27}$$

$$(P_{k+1}, P_{k+2}, \ldots, P_n) \in P_X \tag{5.28}$$

there is a α*-smart* solution SL(α) to the problem.

Theorem 5.6 *If there is a* α*-smart solution, SL(*α*), of Problem* B_1 *in the subspace* $<P_{AI}, S>$*, there is a* α'*-superior solution in the intelligent space* $<P, S>$*, and*

$$\alpha^{'} \geq \alpha, \alpha \in [0, 1], \alpha^{'} \in [0, 1] \tag{5.29}$$

Theorem 5.7 *If there is a* α*-smart solution, SL(*α*), of Problem* B_1 *in the subspace* $<P_{NN}, S>$*, there is a* α'*-superior solution in the intelligent space* $<P, S>$*, and*

$$\alpha^{'} \geq \alpha, \alpha \in [0, 1], \alpha^{'} \in [0, 1] \tag{5.30}$$

Theorem 5.8 *If there is a α-smart solution, SL(α), of Problem B_I in the subspace <P_{FS}, S>, there is a α′-superior solution in the intelligent space <P, S>, and*

$$\alpha^{'} \geq \alpha, \alpha \in [0,1], \alpha^{'} \in [0,1] \tag{5.31}$$

Theorem 5.9 *ps = (P_0, P_1, P_2,. . .,P_n) is one solution of Problem B_I, if and only if*

$$P_0 : s \rightarrow ls \in S_0 \tag{5.32}$$

$$P_1 : l \rightarrow ml, m \in S \tag{5.33}$$

$$P_0 : t \rightarrow dl, m, \ldots, t \in S \; d \in D \tag{5.34}$$

Theorem 5.10 *B_I: (S_0, D)⇒PB. If all the or subproblem B_{Ii} have α′-superior solution, $SL_i(\alpha_i)$ (i = 1, 2, . . ., m), the α-smart solution of Problem B_I can be denoted as follows:*

$$\mathrm{SL}(\alpha) = \cup \mathrm{SL}_i(\alpha_i) \quad (i = 1, 2, \ldots, m) \tag{5.35}$$

$$\alpha = \cap \alpha_i \quad (i = 1, 2, \ldots, m) \tag{5.36}$$

5.3.2 *Problem* B_I

As for Problem B_1, the *α-smart* solution, SL(α), reflects the closeness of the solutions to the global optimal solution [10]. The parameter α indicates the degree of closeness. Viewed from another perspective, it is the degree of membership of a certain solution to the optimal solution. Actually, SL(α) is a set in which there are all the solutions with degree of membership larger than α. Specifically, if α equals 1, the set of the *α-smart* solution only contains the optimal solution. On the other hand, if α equals 0, the set of the *α-smart* solution contains all the feasible solutions. Since a large number of issues are essential to strike an optimal solution, it is necessary to make a definition about the optimal solution here [6].

Definition 5.16 *If the* α-smart *solution of Problem* B_I *is denoted as SL(α), SL (1) represents the optimal solution of Problem* B_I. *The set of all the optimal solutions is defined as the core of SL(α), which is denoted as* K_{er}*(SL). Obviously, we have the following equation:*

$$K_{er}(\mathrm{SL}) = \{\mathrm{SL}(\alpha)\,|\alpha = 1\} \tag{5.37}$$

Theorem 5.11 *If* $\mathrm{B}_1 = \mathrm{B}_{11} \cup \mathrm{B}_{12} \cup \ldots \cup \mathrm{B}_{1n}$, *Problem* B_1 *has the optimal solution if and only if the core of* α_i*-superior solution of at least one subproblem* B_{1i} $(i = 1,2, \ldots, n)$ *is not empty.*

Theorem 5.12 *If* $B_1 = B_{11} \cap B_{12} \cap \ldots \cap B_{1n}$, *Problem* B_1 *has the optimal solution if and only if the core of* α_i*-superior solution of each subproblem* B_{1i} $(i = 1,2, \ldots, n)$ *is not empty.*

If the intelligent path of all the subspaces of Problem B_1 *can be found and combined to form the intelligent path set, the optimal solution can be obtained by means of comparison.*

Theorem 5.13 *The* α_1*-superior solution of Problem* B_1 *is denoted as* $SL(\alpha_1)$. *Similarly, the* α_2*-superior solution of Problem* B_1 *is denoted as* $SL(\alpha_2)$. *If* $\alpha_1 \geq \alpha_2$, $\mathrm{SL}(\alpha_1) \subseteq \mathrm{SL}(\alpha_2)$.

Theorem 5.14 *The* α_1*-superior solution of Problem* B_1 *is denoted as* $SL(\alpha_1)$. *Similarly, the* α_2*-superior solution of Problem* B_1 *is denoted as* $SL(\alpha_2)$. *If* $\alpha' > \alpha$, $\mathrm{SL}(\alpha') \subset \mathrm{SL}(\alpha)$; *if* $\alpha' < \alpha$, $\mathrm{SL}(\alpha') \equiv \mathrm{SL}(\alpha)$.

Theorem 5.15 *If the* α*-smart solution of Problem* B_1 *exists, and the* α_i*-superior solution of subproblem* B_{1i} *exists,* $\alpha = \min\{\alpha_i\}, \alpha \in [0,1]$, $\alpha_i \in [0,1]$.

Definition 5.17 *The absolute optimal level in the intelligent space is denoted as* α'. *If* $\alpha' > \alpha$, $SL(\alpha') \subset SL(\alpha)$; *if* $\alpha' < \alpha$, $SL(\alpha') \equiv SL(\alpha)$.

The essence of Problem B_1 is to research the optimization of development path. If the target has been determined, the development path needs to be searched according to the initial state and relevant constraints [10].

The definitions and theorems that are shown above may be kind of obscure. Now let us go through an example in real life. It is assumed that Tom is a student who wants to improve his performances at school. The current state of Tom is indicated by S_0, which is actually a set $\{s_{01}, s_{02}, s_{03}, \ldots, s_{0m}\}$. The elements in set S_0 express Tom's initial status in various aspects. Some elements are deterministic variables. They describe the math scores, physical scores, chemical scores, etc. Other elements are fuzzy variables. They depict the health condition, popularity, morality, etc. The development target of Tom is denoted by D, which is a set corresponding to S_0. How to get to the target state from the initial state? It is essential for Tom to seek the development path. There is at least one path from S_0 to D, or the target is impossible to realize and has no practical significance. All the possible actions and measures can be integrated into a strategy set that is also called the path set in this book. It is denoted by $\mathrm{PB} = \{\mathrm{PB}_1, \mathrm{PB}_2, \ldots, \mathrm{PB}_n\}$. The issue can be treated as one kind of Problem B_1 that is shown as follows:

$$B_1 = < S_0, D, \mathrm{PB} > B_1 : (S_0, D) \Rightarrow \mathrm{PB} \tag{5.38}$$

The main point is to solve the path set. The global optimal solution may be difficult to be found. Accordingly, people tend to get an *α-smart* solution in most instances. Sometimes, it is such an arduous work to solve the whole problem entirely at once that we can convert the problem to a series of subproblems. The ultimate solution can be obtained on the basis of the solutions of all the subproblems.

5.3.3 *Problem* B_2

As introduced in previous paragraphs, Problem B_2 is another basic problem model in the intelligent engineering [4, 6].

Definition 5.18 *As for Problem* B_2, $B_2 = <S_0, D, PB>$, B_2*:* $(S_0, PB) \Rightarrow D$. *If there is a fuzzy set* fd *on the target state* D, $DL(\beta) = \{d / \mu_{fd}(d) \geq \beta, d \in D\}$, $\beta \in [0,1]$ *is regarded as the β-superior target set of the problem. If* $\beta = 1$, *DL(1) is called the optimal target set.*

Definition 5.19 *As for Problem* B_2, *if there is a fuzzy set* fd *on the target state* D, $PB(\beta) = \{x / PB_\beta: S_0 \rightarrow DL(\beta), x \in PB_\beta\}$ *is regarded as the β-superior target path set of the problem.*

Definition 5.20 *If neither the β-superior target set nor the β-superior target path set of Problem* B_2 *is empty, the β-superior target of the problem is accessible. Otherwise, it is inaccessible.*

Definition 5.21 *The* (α, β)-satisfied *solution of Problem* B_2 *is defined as* $SL(\alpha, \beta) = \{x / \mu_{fp}(x) \geq \alpha, x \in PB_\beta\}$, *in which* $\alpha \in [0,1]$, $\beta \in [0,1]$.

Definition 5.22 *As for the fuzzy target set of Problem* B_2, $K_{er}(fd) = \{d / \mu_{fd}(d) = 1, d \in D\}$ *is called the core of the target of Problem* B_2.

Theorem 5.16 The partial superior relationships between the (α, β)-satisfied solution of Problem B_2*,* $SL(\alpha, \beta)$*, and the α-smart solution of Problem* B_1*,* $SL(\alpha)$*, exist as follows.* $\mathrm{SL}(\alpha) \succ \mathrm{SL}(\alpha, \beta)$, *if* $D = K_{er}(\widetilde{D})$, *in which D is the target of SL(α).*

Theorem 5.17 If $\alpha = 1$ *and* $\beta = 1$, $SL(\alpha) = SL(\alpha, \beta)$ *is the global optimal solution of Problem* B_1.

Theorem 5.18 The (α_i, β_i)*-satisfied solution of m and subproblem* B_i *is denoted as* $SL_i(\alpha_i, \beta_i)$, $(i = 1, 2, \ldots, m)$. *The (α, β)-satisfied solution of Problem* B_2 *is* $SL(\alpha, \beta) = \cup SL_i(\alpha_i, \beta_i)$, $\alpha = \wedge \alpha_i$, $\beta = \wedge \beta_i$, $(i = 1, 2, \ldots, m)$.

The essence of Problem B_2 is a problem of forward reasoning. Given the initial state and possible development paths of a certain system or entity, the aim of Problem B_2 is to determine the target state along each development path [4]. The meaning is that Problem B_2 is studying all the possible target states of the system or entity. Typically, scenario analysis belongs to Problem B_2.

Now let us have a brief view of a practical example. It is assumed that there is a project to forecast the energy demand in China in 2050 on the basis of the data in 2014. Since the energy demand is highly correlated with the economic situation, different development situations and growth speeds of the national economy between 2014 and 2050 will have evidently influences on the energy demand. The forecast result should be considered separately in different scenarios of the national economy. We depict the development mode and growth speed of Chinese national economy with variable M and variable S, respectively. Then the variable corresponding to energy demand, D, is impacted by M and S. It is assumed that economic development mode and growth speed in Scenario 1 are, respectively, m_1 and s_1, while the energy demand is d_1. Similarly, the economic development mode and growth speed in Scenario 2 are, respectively, m_2 and s_2, while the energy demand is d_2. We can set as many scenarios as we want and get different forecast results in line with different scenarios. It is an instance of Problem B_2 shown as follows:

$$B_2 =< S_0, D, \mathrm{PB} > B_2 : (S_0, \mathrm{PB}) \Rightarrow D \tag{5.39}$$

S_0 depicts the initial states in 2014. PB represents the economic development mode and growth speed here. D is the energy demand that needs to be forecasted.

References

1. Zhaoguang Hu (1995) Intelligent engineering-its application [C]. In: Proceedings of the IEEE international conference on system, man and cybernetics. IEEE, Vancouver, Canada, pp 609–614
2. Tu X, Wang C, Guo Y (2005) Cybernetics of large systems [M]. Beijing University of Posts and Telecommunications Press, Beijing
3. Zhaoguang Hu (2007) Intelligent modeling [C]. In: Asia energy environment modeling forum: how to model low carbon economy in Asia. University, Beijing
4. Minjie Xu (2008) Intelligent engineering and its applications in the analysis and forewarning of the supply and demand of electric power. PhD thesis, Beijing Jiaotong University, Beijing
5. Zhaoguang Hu (2005) Study on electronic laboratory of power economy by intelligent engineering [J]. Electric Power (Chinese) 38(1):7–11.
6. Zhaoguang Hu, Baoguo Shan et al (2009) Simulation of power supply and demand by intelligent engineering [M] (Chinese). China Electric Power Press, Beijinh, pp 19–23, 27–38, 39–54
7. Zhaoguang Hu (1999) Intelligent space. In: IEEE international conference on fuzzy systems, pp 1621–1625
8. Xiaoyou Jiao, Zhaoguang Hu (2008) Distributed intelligent control and its application in simulation of macro-policy to electric power supply and demand [J]. Proc Chine Soc Electric Eng (in Chinese) 23(5):121–127.
9. Xiaoyou Jiao, Zhaoguang Hu (2008) Intelligent engineering and its application in policy simulation [C]. In: Proceedings 9th international conference IDEAL2008, Korea, Nov 2008, pp 404–411
10. Wei Ding (2007) Transmission planning approaches and investment analysis in the market mechanism. PhD thesis, North China Electric Power University, Beijing

Chapter 6
Agent Response Equilibrium Model

This chapter is regarded to be the core of the study tool, because it covers the introduction of our Agent Response Equilibrium (ARE) model. To deal with the modeling of national economy which is quite a large and complex system, authors propose ARE model on basis of intelligent engineering theory and multi-agent system technique. Principal elements in the national economy and their interactions are all considered in the model, i.e., production sectors, banks, residents, the government, and various kinds of markets. With the help of ARE model, it is possible to simulate the operation of the economic system and conduct some policy experiments. All the simulation results in other chapters in this book come from the ARE model. In this chapter, we will provide readers with the foundation and overview of the model. Firstly, the concept of agent will be described, and the multi-agent system techniques will be outlined. Then, the framework, components, and features of our ARE model will be introduced. Last but not least, we will present some details of the model, such as the assumptions, database, rule base, communication mechanism, etc.

6.1 Agent

6.1.1 Definition

Professor John H. Holland referred to the element in a complex adaptive system as adaptive agent, abbreviated as agent [1]. It is regarded as an entity that can observe and impact the environment and make decisions autonomously. The agent is able to interact with the environment in order to make the system evolve according to its benefits. As a consequence, agent can be widely applied in the area of economy, society, and ecology in which other approaches have limited effects.

Z. Hu et al., *China's Economic Gene Mutations*,
DOI 10.1007/978-3-662-47298-9_6

The concept of agent can be understood from different perspectives. Wooldridge provided a weak concept and a strong concept of agent, respectively [2]. According to the weak concept, an agent has the following abilities: The first one is autonomy, which means the agent has a certain degree of control over its own behaviors and internal states, and its actions are taken without outside interventions. The second one is sociability, which indicates the agent can exchange information with other agents and have influences on other agents. The third ability is reactivity, which means the agent is able to perceive the environment in which it lives and make certain reactions to impact the environment. The fourth one is pre-activeness, which suggests the agent takes measures to pursue its goal on the basis of the information it has. On the other hand, the agent should have some characteristics of human apart from the above features according to the strong concept. Typical examples are knowledge, beliefs, intentions, commitments, and other mental states. Shoham considered the agent as the entity composed of faiths, capabilities, choices, and commitments [3]. Nwana proposed the three-level concept of agent, including definition level, organization level, and cooperation level [4]. On the definition level, the agent is described from the perspectives of learning mechanism, goals, resources, and abilities. On the organization level, the relationships of the agent with other agents are defined. On the cooperation level, the social ability of agent is explained. Muller offered the agent a concept that contains knowledge, reasoning, and decision-making skills, communication skills, and awareness [5].

6.1.2 Classification

As for the majority of agents, it is essential to interact with the environment to grasp the outside information and to process and explain the information so as to realize their own goals. In terms of the processing mechanism of the agent for the information, agents can be divided into diverse sorts:

1. Cognitive Agent
   Cognitive Agent is a symbolic model, including logical reasoning capacities of the environment and its intelligent behaviors. Usually, the environment model is achieved previously. The Cognitive Agent has knowledge representation, problem-solving representation, environment representation, and specific communication protocols. The agent is regarded as a kind of intentional system.
2. Reactive Agent
   Reactive Agent does not consist of the world model expressed by the symbols and does not use complicated symbols for inferring. The intelligence of agent depends on perception and action. The model of its intelligent behavior is a kind of perception–action model. The outside impacts are in correspondence with the reactions of agents. Reactive Agent has the capability of evolving similar to humans without knowledge and inference. Its behaviors can be reflected only in the interactions between the real world and the surrounding environment.

3. Hybrid Agent
   Hybrid Agent is a mixture of Cognitive Agent and Reactive Agent. Reactive Agent can quickly respond to changes in external information and the environment, but it has low degree of intelligence and is lack of sufficient flexibility. Cognitive Agent has relatively higher intelligence level, but cannot quickly respond to changes in the environment, and the efficiency is relatively low. Hybrid Agent combines the advantages of them and has strong flexibility and fast response speed.

6.1.3 Features

As independent intelligent entity, an agent must have a wide range of intelligent qualities [6]. Agents in specific environments should have the following several characteristics:

1. Autonomy
   An agent should have independent knowledge and knowledge processing methods. With limited computing resources and behavior control mechanisms, the agent can continue to operate without direct intervention and guidance from human and other agents and respond to environmental changes in a particular manner. Besides, the agent should be able to make their own decisions and control their own states and behaviors according to its internal states and the perception of environmental information. Autonomy is a key characteristic to distinguish agents from other similar concepts.
2. Reactivity
   An agent takes action according to its goal when it observes and responds to the outside environment. In some cases, the agent can take proactive behaviors to change the surrounding environment in order to achieve its goals.
3. Sociability
   The agent usually does not exist alone. Similar to the biological communities in the real world, a number of agents tend to live together to form a social group. The agent cannot only operate autonomously but also collaborate with other agents in the environment. Moreover, it is necessary to conduct conflict resolution by virtue of the mutual coordination.
4. Evolution
   An agent is an open entity that gradually adapts to the surrounding environment and evolves. Furthermore, it is capable of constantly expanding their knowledge and ability with the changes of the environment to improve intelligence and reliability of the whole system.

6.1.4 Self-Learning

Most decision-making rules of agents are not static except for some simple reactive agents. When it comes to the modeling of these intelligent units, it is essential to take into account the self-learning function of them. They tend to improve their knowledge and skills through the learning mechanism based on the living environment and their own historical behaviors. Consequently, their knowledge bases are expanded and their decision-making rules are adjusted. In this way, these agents dynamically adapt to the environment.

Self-learning is a crucial characteristic of intelligent living creature. An organism is believed to have the self-learning ability if it is able to improve its performance through the implementation of certain process. As for the agent, the function of self-learning means it modifies its behavior strategies and enhances its performance in the process of continuous interactions according to its perception and observation. Common self-learning methods include reinforcement learning, genetic algorithms, artificial neural networks, etc.

6.2 Multi-agent Model

6.2.1 Multi-agent System

Even though the agent is defined to have specific functions and can make independent decisions, it is usually impossible for single agent to describe and solve complicated and large-scale problems in reality. Therefore, an applied system should consist of large quantities of agents. They have their own goals and the ability to solve problems, and they have the skill of mutual cooperation to achieve common overall goal. Multi-agent system (MAS) refers to the set that contains many agents. Diverse agents can interact with each other to realize the evolution of the system and the emergence of the macroscopic phenomenon [7].

Multi-agent system has outstanding robustness and reliability and high efficiency to solve practical problems. The operation results of the whole system depend on the cooperation of various agents. The relationship of different agents is not only unidirectional causality but also calling and collaboration in their respective time axis.

6.2.2 Multi-agent Model

Multi-agent-based modeling refers to the approach in which multi-agent system is utilized to build dynamic models for complex systems, such as ecological systems, social systems, and economic systems. Corresponding system model is regarded as the multi-agent model. The social and economic system is a typical complex complicated system that consists of a huge number of diverse types of economic agents. The structural changes and evolution of the system is the results of the unconscious and selfish behaviors of these individual agents. Here lies the distinctive feature of the multi-agent systems, i.e., the complicated interplay of various agents renders the system some characteristics which single agent does not have, so that the overall performance of the system is better than the simple sum of the individual actions. As a novel research method, multi-agent-based modeling is playing an increasingly important role in the model research of social science.

Multi-agent-based modeling has the following features [8, 9]:

1. Each agent is an active and alive entity. This is the key difference between multi-agent model and other kinds of models. This feature makes the approach of multi-agent-based modeling able to be utilized to deal with economic systems, social systems, ecological systems, and other sorts of complex systems.
2. There are interactions among different agents and the environment in the multi-agent model, which acts as the main driving force of the evolution of the system. With traditional modeling approach, the individual's own internal properties tend to be put in a prime location, while the interplay of individuals and environment has not drawn enough attention.
3. Multi-agent-based modeling organically combines macroscopic and microscopic together instead of completely separating them as many conventional methods. The microscopic phenomena are the foundation of the microscopic phenomena, while the microscopic phenomena have influences on the microscopic phenomena.
4. Random factors have been introduced in the multi-agent-based modeling, which makes it have stronger properties in description and presentation of practical issues.

6.3 Introduction to ARE Model

6.3.1 ARE Model

Based on the methodology of intelligent engineering as introduced in Chap.5, Professor Zhaoguang Hu combined the agent technique and generalized model [10, 11] and researched the ideas and methods on simulation of economic system operation by means of man–machine integration system that gave full play to both people's initiative and computer's advantages. He proposed the Agent Response Equilibrium (ARE) model [12, 13]. In fact, the economic simulation by ARE is a B_2 problem in the intelligent space as introduced in Chap.5. The initial state S_0 in the state space is the economic state in the starting year, in our study is the initial input-output table; the intelligent path *PB* is the policy scenario settings and international commodity prices; the target state *D* is the last simulated economic state, in our study is the input-output table of the last simulated year; the economic simulation is to search a (α, β)-*satisfied* solution of Problem B_2 in the intelligent space. When a set of policy and price of international market is selected, i.e., a set of *PB* is selected, the economic system will be simulated from the starting year dynamically to reach the target state *D* in the intelligent space. This is the way to get a (α, β)-*satisfied* solution of the economic simulation, and it can be executed many times to select a better (α, β)-*satisfied* solution. Therefore, by means of ARE model, economic research can be conducted like the physical experiments after the intelligent level of computers in simulating the operation of complex systems has been enhanced. It becomes possible to explore the performances of the system under different conditions, which provides the scholars a valuable way to cope with complicated economic issues and develop research ideas.

It has been used to renew input–output table. China's national input–output table in 2010 has updated by virtue of the ARE model, offering us the latest detailed data on the relationship among production, consumption, and income in China's national economy [13]. Xiao Xiao, Jianwei Tian, and Minjie Xu improved and refined the structure of the ARE model and simulated the specific influences of different fiscal policies on the indicators of economic output, prices, etc. [14]. Specifically, they focused on the effects of stimulating policy (such as reducing taxes, increasing government spending) and tightening policy (such as increasing taxes) on economic growth [15].

6.3.2 ARE Model Framework

The economic system is simplified according to related economic theory. The model mainly contains four parts, i.e., residents, sectors, government, and external environment, which acts as the foundation of cycle operation of the national economy. Sectors have technologies and capitals to carry out production to maximize their profits. They are both the demander of production factors and the supplier of productions. Residents provide labor force and consume production. The government is the economic regulator with some policy tools. Besides, it also acts as a consumer that has revenue sources through taxation. The external environment is either the demander of domestic production and production factors or the supplier of production and production factors. All these four kinds of entities have their own goal and behaviors and the ability to respond to the changes in the surrounding environment to adjust their behaviors to realize their objectives.

The agents in the model are classified into the following categories: The first category is four sorts of market agents, including commodity market agent, financial market agent, labor market agent, and international commodity market agent. The second category is three sorts of special role agents, including sector agent, resident agent, and bank agent. The third category is two sorts of regulator agents, including government agent and central bank agent. The interrelationships of these

Fig. 6.1 Structure diagram of ARE model

agents constitute the operation diagram of the economic system, as shown in Fig. 6.1.

6.3.3 Functions of Agents

Commodity market agent determines the supply–demand relationship of the products of each sector considering the supply from the sector agents and the demand from the residents, government, and sector agents. The price of certain product will decrease if its supply is greater than its demand, while the price will increase if the supply is lower than the demand. Then, the price information is sent to other agents.

Labor market agent determines the supply–demand relationship of labor force and the salary level according to the amount of labor from the residents and the demand from all the sector agents. The salary level will decrease if the supply of certain sort of labor force is greater than the demand; otherwise, it will go up.

Financial market agent is in charge of the deposit and loan. Its main function is the determination of the loan to each sector agent in the light of the deposit from residents, the benchmark deposit and lending rates, deposit reserve ratio, and the discount rate published by the central bank agent.

International commodity market agent determines the import and export of each kind of product according to the international price level and domestic price level. Conversely, the domestic price level will also be impacted by the import and export decided by the international commodity market agent.

Sector agents make decisions to increase or decrease the production to maximize their own profit, considering the market price of their product and the labor force and the cost of financing. Each sector agent is the producer of its product, the demander of production factors, and the demander of the products of other sectors as its intermediate input. The conditions in commodity market, international commodity market, labor market, and financial market will affect its decisions. Unlike the several sorts of agents mentioned above, we usually have a number of sector agents that represent the actual sectors in the national economy to provide better platform for researchers and scholars to carry out simulation and analysis.

Resident agent acquires salary through providing labor force as production factor. The income is either deposited in the commercial bank or consumed in the commodity market. Resident agent decides its labor time, deposit, and consumption in accordance with the price of commodities, the salary of labor force, and its consumption tendency.

The government agent is typically the regulator of economic system. Its target is to ensure the economic growth and social fairness. The principal behavior of the government agent is to adjust the operation of the economic system by means of fiscal policies. Its income is from taxation, while its expenditure consists of transfer payments, government consumption, and public infrastructure investment. Fiscal policies include the adjustments of various tax rates and the changes in government expenditure. The actions of government agent in the model are to determine the consumption for each kind of product and the expenditure and conduct some fiscal policies if necessary according to its observation on the economic situation. On the other hand, the government agent can also be set as an open agent, which means its decisions may be made by the researchers who are running the model as exogenous variables. In this way, the researchers are able to carry out experiments through the settings of tax rates, government expenditure, etc.

The central bank agent determines the monetary policies including benchmark deposit and lending rates, bank discount rate, reserve ratio, and open market operations to affect the economic operation condition. It can also be treated as an open agent like the government agent. The researchers have the competence to decide the monetary policies and try different scenarios.

Now we have reviewed the components in ARE model and the interrelationships between them. The main functions and actions of each sort of agent have been briefly introduced. More detail on various sorts of agent will be covered in Chap. 7.

6.3.4 Features of ARE Model

The ARE model has the following features:

1. Multi-objective
   The foundation of the model is various agents whose objectives are quite diverse. They conduct certain behaviors according to their own target. There is no form of equations to reflect the endogenous relationship in the system. The system has nonlinear dependence relationship. There is no central control over all the agents.
2. Multi-mission
   Different agents are on behalf of different economic entities with different missions. Consequently, the behaviors of diverse agents in the system are considerably complicated, which increases the complexity of calculations and enhances the practical value of the system.
3. Autonomous optimization
   Through continuous learning and updating the knowledge base, each agent autonomously pursues the maximum benefits.
4. Dynamic
   All the functions of the model are on the basis of dynamic simulation, through which the reasons and prospects of economic fluctuations are analyzed. Therefore, the economic system can be regarded as an evolution system.
5. Exogenous intervention
   We pay attention to the intervention of experts, especially in the area of policies. Although the agents in the economic system are independently and mutually operated as stated before, the overall result may not be a good state for a country. Accordingly, the intervention of researchers is essential for a perfect agent-based simulation system. This is the unique feature of ARE model compared with other modeling approaches.
6. Complexity
   In order to solve the multi-objective and multi-mission model and realize dynamic simulation, the complexity of the multi-agent system is quite high. The whole system is in the continuous cycle of "complexity–adaptation–new complexity–the new adaptation."

6.4 Design of ARE Model

6.4.1 Model Structure

The agents in ARE model are classified into three sorts, i.e., control agent, coordination agent, and function agent. Their roles are listed in the following: Tables 6.1, 6.2, and 6.3.

Table 6.1 The role of policy maker

Name	Policy maker (control agent)
Member	Government agent, central bank agent
Activity	Observe the operation status of the national economy and offer exogenous variables to intervene in the economy activity
Competence	Get: GDP and energy consumption
	Change: Policy
Action	Rate change, taxation, subsidize
Target	Maintain the economic growth, stabilize prices, guarantee employment, ensure the balance of payments between domestic and international environment

Table 6.2 The role of market

Name	Market (coordination agent)
Member	Commodity market agent, financial market agent, labor force agent, international commodity market agent
Activity	Gather all the supply and demand information; calculate and send price signal to related agents
Competence	Get: Supply and demand, price
	Send: Price signal
Action	Calculate price, deliver price
Target	Provide the supply and demand information and relevant price signal and guide the next interaction among various agents

Table 6.3 The role of producer and consumer

Name	Producer and consumer (function agent)
Member	Sector agent, resident agent, commercial bank agent
Activity	Maximize their own benefit
Competence	Production, consumption, loan
Action	Produce goods, consume, loan
Target	Sector agent pursues the maximum profit; commercial bank agent pursues the revenue through interest rate differential; resident agent pursues the maximum utility of the residents

6.4.2 Interaction Mechanism Assumptions

The interactions among all the different sorts of agents mentioned above consist of three kinds of flow, i.e., material flow, capital flow, and information flow.

1. Material flow
   Material flow mainly refers to the flow of product, labor, and intermediate input factors. Material flow provides various agents with necessary products for the production and consumption of each agent. As for the function agents, the resident agent receives the products from the commodity market and offers labor force in the labor market; the sector agents get labor from the labor market, send

commodities into the commodity market, and consume other products in the commodity market as intermediate inputs. As for the coordination agent, the commodity market agent acts as the bridge on which the products flow from the supplier to the demander; similarly, the labor market agent serves as the bridge on which the labor force flow from the supplier to the demander; the international commodity market agent exchanges products with the commodity market agent by way of international trade. As a control agent, the government agent is also a consumer in the commodity market, so it receives commodities like resident agent.

2. Capital flow
   This kind of flow means the capital flow among different agents, especially the loan between the commercial bank agent and various sector agents and the payment between the consumer and the producer of certain products. Typical examples include the payment of residents for their consumptions and their salary for the labor force they provide, the deposit from the residents to the commercial bank agent, and the loan from the bank agent to the financial market agent.
3. Information flow
   Information flow exists when communication occurs in the interaction between the various agents. Typical examples include the price information sent by the market agents to other agents and the loan information sent by the financial market agent to other agents.

6.4.3 Assumptions

In the ARE model, there are some assumptions as follows:

Assumption 1: The target of each agent is to maximize its own benefit.

Assumption 2: There is no externality, which means private gain or loss will not affect the overall benefits of other agent or the whole society. Pareto optimal state does not exist in the model.

Assumption 3: The competition is sufficient. The price is determined by the relationship between supply and demand in the market.

Assumption 4: Financial market only refers to the credit market. Other financial elements, such as bonds and futures, associate with many other issues and theories apart from the characteristics of capital supply and demand, so they are not considered in ARE model.

Assumption 5: The international commodity market is considered in the model. The influence of international financial market is also considered, but the exchange rate movements are not covered. The main reason is that the indicators and results of the model will be too complicated and the conductivity of price will not be obvious enough if the exchange rate movements are added into the ARE model. Therefore, the model deals with the semi-open economic system.

Assumption 6: Since too many variables may lead to the curse of dimensionality, the direct consumption coefficients, final consumptions, fixed capital formation, and the amount of import and export in the national input–output table with 42 sectors are utilized in the model to express the interrelationship among different agents.

6.4.4 Database Design

Database is important in the ARE model, and it is designed as follows:

1. Based on the 42 sector input–output table, 42 corresponding sector agents are designed. Meanwhile, only one resident agent is set in the model to represent the overall behaviors of all the residents. Therefore, the database of the sector agents and resident agent is chiefly the input–output table.
2. Only one commercial bank agent is designed in the model. As a result, it only serves as the intermediary of money flow. There is no competence between different commercial banks. Moreover, the interbank rates and differences in the interest rate are not considered in the system either. The database of the commercial bank agent is balance sheet of other depository corporations that is on behalf of the overall assets and liabilities of depository corporations.
3. The database of central bank agent is primarily from the central bank balance sheets published by the People's Bank of China.
4. The database of the government agent mainly comes from the data on taxes and government expenditures from the China Financial Yearbook.
5. The database of international commodity market is the data on the amount of import and export of goods and services from the China Statistical Yearbook and China Customs Statistics Yearbook.

6.4.5 Rule Base Design

An agent organization consists of a number of heterogeneous agents. Since it is in an open environment with constant changes, each agent does not have the perfect knowledge. Furthermore, these are behavioral dependencies and resource consumption constraints among different agents. So the conflicts are unavoidable if all the agents make decisions according to their own goal perfectly and absolutely. To deal with the issue, it is necessary to set rules for the agent organization. The rules are actually the limits on the behaviors of these agents.

The design of rule base ought to meet the following tips:

1. No conflict between agents will happen under the rules.
2. The benefit of the agent organization should be maximized under the premise of (1).

3. The benefit of individual agent should be maximized under the premise of (1) and (2).

Each agent takes the surrounding information as input, finds related rule in the rule base, and makes appropriate response in the light of the rule.

In ARE model, sector agents make decisions on the production according to the price signals in commodity market, labor market, and financial market; resident agent determines its consumption according to the price signal in commodity market and labor market; the government agent and central bank agent take relevant measures according to certain indicators. Therefore, the general form of the rules in the model can be abstracted as follows:

$$\text{IF} < \text{condition} > \quad \text{THEN} \quad < \text{action} > \tag{6.1}$$

Specifically, the rule can be divided into response rule and conversion rule.

The response rule guides the stress reaction according to the changes in surrounding environment. For example, the production scale will be revised if the price of product or labor changes, as shown in the following equation:

$$\text{IF} < \text{products' price rise} > \quad \text{THEN} \quad < \text{produce more} > \tag{6.2}$$

The conversion rule is designed mainly for the coordination agents. For example, the commodity market agent will convert the signal of supply and demand imbalance into the price changes and send to relevant agents, as shown in the following equation:

$$\begin{aligned} &\text{IF} < \text{products' supply is lower than the demand} > \quad \text{THEN} \\ &\quad < \text{price increases} > \end{aligned} \tag{6.3}$$

6.4.6 Self-Learning Rules of Agents

An agent will receive a reward after its each action. The self-learning function of agents renders them the capability to find better actions by pursuing larger reward. The result is that the agent becomes aware of the optimal strategies, with which it can have the best performance.

The product of output and price is set as the reward in the model. It should be noted that there is no substitutability between different sectors. Each sector is the only producer of a certain kind of product. The profit function of each sector can be expressed as follows:

$$\pi = P \times \left(AK^{\alpha}L^{\beta}\right) - \gamma K - \omega L - T \tag{6.4}$$

where π is the profit of the sector, P is the price of the product of the sector, A is the technology productivity, K is the capital input, L is the labor input, α and β are the replacement rates of these two production factors, γ is the interest rate, ω is the labor price, and T is the taxes.

The cost function can be expressed as follows:

$$c_j = c_j(Q) = \sum_{i=1}^{n} \left(p_{ij} \times q_{ij}\right) \tag{6.5}$$

where Q is the output, p_{ij} is the price of the product of sector i consumed by sector j, and q_{ij} is the quantity of the product of sector i consumed by sector j.

Sector agent adjusts the output according to the market price in each loop. The result has something to do with the previous behaviors and results, which is consistent with the original purpose of self-learning characteristic. After each loop, the agent has the judgment of its behaviors. As a result, it will become smarter if the similar situations appear the next time.

6.4.7 Communication Mechanism

Since the agent has the capability of interaction with other agents, the communication mechanism is a crucial issue in the multi-agent system. Typically, the agents communicate with each other with agent communication language (ACL), as shown in Fig. 6.2. Some ACLs have been developed, such as KQML and COOL. However, it is the job of the designer of the multi-agent system to decide the time to use interaction protocol, the message to be sent, and the way to describe the session structure. For a close multi-agent system, direct coding is feasible because all the interaction protocol can be forecasted and designed in advance. On the contrary, it is not the case for an open system, because some unexpected situation such as the addition of new agents and new functions will result in new interaction protocol, which poses great challenges for the design of a multi-agent system.

FIPA ACL, an internationally popular agent communication language is utilized in ARE model. Specifically, we have designed the following three ways to deal with the communication in the multi-agent system:

1. Bulletin board
   It is also known as the blackboard communication. Each agent sends information into a storage room, the bulletin. Then, all the information is organized and coordinated by the bulletin (Fig. 6.3).
2. Direct communication
   The agents send letters to each other. The expected receiver and its address are listed on the envelope, so that the letter can be sent to designated recipient. In turn, the recipient will reply to the sender after receiving and processing (Fig. 6.4).

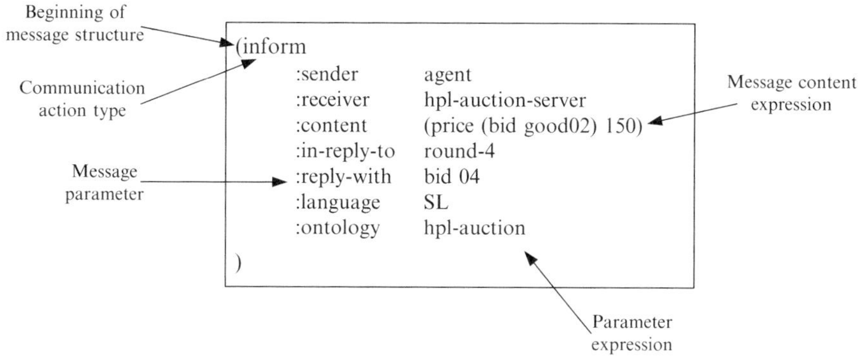

Fig. 6.2 The structure of ACL

Fig. 6.3 The schematic diagram of blackboard communication

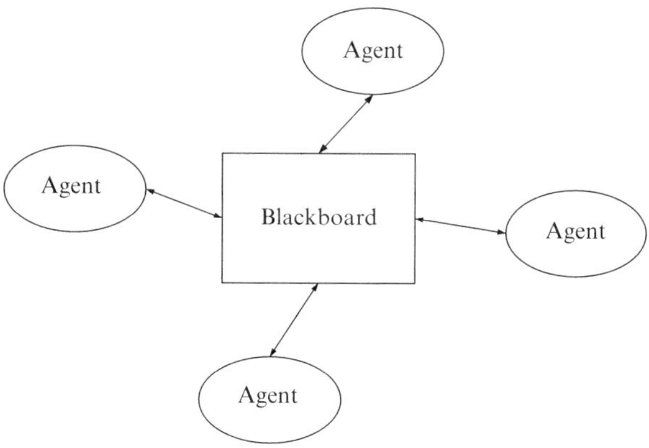

Fig. 6.4 The schematic diagram of direct communication

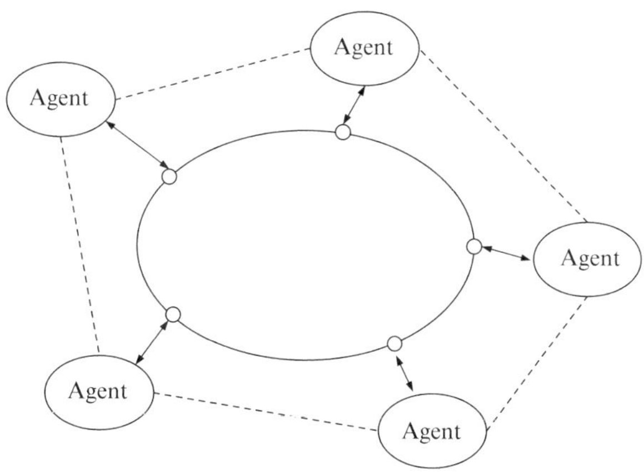

The dotted line means the loosely coupled states among various agents in the system. The circle drawn in solid line represents the channel through which the letters are sent to the designated agents.

3. Calling relations in the program
   This way refers to the input–output relationship of the functions in the program when the communication relationship is considered in the process of coding. It is suitable for the situation in which the communication relationship is relatively simple and the communication object is relatively fixed. The difficulty of designing the communication part will be reduced in this way.

6.4.8 Energy Consumption Calculation

The value added of various sectors and the price of their products can be obtained in the model. Then, the production and consumption of each sort of energy source can be calculated. Diverse kinds of data on energy consumption are very important for us to carry out the analysis and research by means of ARE model, because energy acts as the lifeblood of the national economy and energy data is able to indicate the situation of the economy from certain perspectives. Moreover, our background is related to electricity and energy, so we are particularly concerned about the energy consumption data and make the function of energy consumption calculation a representative feature of our ARE model.

Let us take coal sector, for example, to introduce the approach of energy consumption calculation in our program in detail. As mentioned in previous chapters, the row in the input–output table reflects the destination of the product of the sector that corresponds to the row, either as the intermediate input of other sectors or as the final consumption of residents, government, export, etc. The sum of the row represents the output of the sector. The unit of output is money. Therefore, it equals its yield quantity multiplied by the price of its product, as shown in the following equation:

$$Y = \sum_i q_i p_i \tag{6.6}$$

where Y is the overall output of a certain sector, i represents the sort of product of the sector, q_i is the yield of product i, and p_i is the price of product i.

Since it is assumed that each sector only produces one kind of product in our model, we can calculate the yield of the coal sector based on the output and the price of its product:

$$Q = \frac{Y}{P} \tag{6.7}$$

where Q is the total yield of the sector and P is the price of the only kind of product of the sector.

The price can be calculated according to the relevant price level in the simulation and the real price benchmark of the coal sector as shown in the following equation:

$$P = P_A \times P_L \tag{6.8}$$

where P_A is actual price benchmark of the sector and P_L is the price level in the simulation of the model.

The price benchmark of a sector is determined on the basis of historical output and yield data as shown below. Then, it is stored in the knowledge base of the agent.

$$P_A = \frac{Y_H}{Q_H} \tag{6.9}$$

where Y_H is output of the sector in a certain year and Q_H is the yield of the sector in the year.

In order to decide on the price level, the overall situation of the national economy should be taken into consideration. The prices of different sectors have different impact factors on the price level.

$$P_L = \sum_{i=1}^{42} p_l^i \times \alpha_i \tag{6.10}$$

where p_l^i represents the price level of sector i and α_i is the price impact factor of sector i.

Thus far we can calculate the price of the coal sector. Then the yield of the coal sector is determined according to Eq. (6.7), since the output of the sector is included in the input–output table that acts as the result of the simulation. The coal consumption in the national economy is just the yield of the coal sector minus net export.

Similarly, we are able to calculate the consumption of other kinds of energy source, such as electricity, in the operation of the economy by virtue of the approach introduced above.

Moreover, it is possible to obtain the energy consumption by a certain sector as shown in the following equation:

$$Q_j^i = \frac{Y_i \times a_{ij}}{P_i} \tag{6.11}$$

where Q_j^i is the consumption of the product of sector i by sector j, Y_i is the total output of sector i, a_{ij} is the direct consumption coefficient of sector i by sector j, and P_i is the price of the product of sector i.

If sector j is the coal sector, the coal consumption of each sector can be determined. If sector j corresponds to another energy sector, the consumption can be calculated in the same way.

References

1. Holland JH (1996) Hidden order: how adaptation builds complexity [M]. Basic Books, pp 1–3
2. Wooldridge M, Jennings NR (1995) Intelligent agents: theory and practice [J]. Knowl Eng Rev 10(2):115–152
3. Shoham Y (1993) Agent-oriented programming [J]. Artif Intell 60:51–92
4. Nwana HS (1996) Software agents: an overview [J]. Knowl Eng Rev 11(3):1–40
5. Muller JP, Wiley J (1999) Architectures and applications of intelligent agents: a survey [J]. Knowl Eng Rev 13(4):353–380
6. Yuan J, Li M, Hu Z (2005) Research on large customer response under time-of-use electricity pricing policy based on multi-agent system simulation[J]. Power Syst Technol (in Chinese) 29 (23)):44–49
7. Chen S, Huang Y (2008) Risk preference, forecasting accuracy and survival dynamics: simulations based on a multi-asset agent-based artificial stock market [J]. J Econ Behav Organ 67(3–4):702–717
8. Shi Zhongzhi, Zhang Haijun, Dong Mingkai (2003) MAGE:multi-agent environment [C]. In: Proceedings of international conference on computer and mobile computing, Shanghai, China
9. Jiahai Yuan, Zhaoguang Hu (2005) A multi-agent based negotiation simulation system for electricity contract market [J]. Power Syst Technol (Chinese) 29(11):49–53
10. Zhaoguang Hu, Xiantong Tan, Zhaoyuan Xu (2013) An exploring economic development and electricity demand by the year 2050 [M]. Elsevier Insights and China Electric Power Press, p 46
11. Zhaoguang H (2005) Study on electronic laboratory of power economy by intelligent engineering. Electric Power (Chinese 38(1):7–11
12. Zhaoguang Hu (2011) Study on agents response equilibrium model [J]. Energy Technol Econ (Chinese) 23:9–15
13. Zhaoguang Hu, Wei Duan, Xiao Xiao, Jianwei Tian (2011) Derivation of China's 2010 input–output table based on Agent Response Equilibrium (ARE) model [J]. Energy Technol Econ (Chinese) 23(11):8–14
14. Xiao Xiao, Zhaoguang Hu, JianweiTian, Minjie Xu (2011) Agent-based system for hybrid policy analysis on economy and power [C]. In: International conference on system science, engineering design and manufacturing informatization
15. Jianwei Tain, Zhaoguang Hu, Junyong Wu, Xiao Xiao, Minjie Xu (2010) Dynamic economy and power simulation system based on multi-agent modelling [J. Proc Chine Soc Elect Eng (Chinese) 30(7):85–91

Chapter 7
Individual Agent Functions and Computer Programming

In this chapter, we will go on with the introduction of our original ARE model, which is a tool to simulate the operation of national economic system. The foundation and overview of the model has been introduced in Chap. 6. More details will be focused on in this chapter. First and foremost, we will offer the detailed presentation of various sorts of agents in ARE model one by one, i.e., the sector agent, market agent, international commodity market agent, government agent, central bank agent, commercial bank agent, and resident agent. Then, we will provide some information of our computer program for readers especially who are interested in computer programming and want to know more about our computer program which is the basis of building ARE model. The flow chart and computer programming environment are described, respectively. Furthermore, some typical statements and the input data statements of our computer program will be shown. Lastly, we will exhibit the parameter settings. We hope that readers can come to understand our ARE model after reading Chap. 6 and this chapter, and we even look forward to the improvement and criticism of the model from readers.

7.1 Individual Agent

There are seven sorts of agents in our model in total. In this section, the functions and structures of each sort of agent will be introduced in detail.

7.1.1 Sector Agent

Production sectors play an important role in the economic system. In our model, 42 different sector agents are set according to China's 2007 national input–output table. They are shown in detail in Table 7.1 [1].

Z. Hu et al., *China's Economic Gene Mutations*,
DOI 10.1007/978-3-662-47298-9_7

Table 7.1 42 sector agents in our model

No.	Sector agent	No.	Sector agent
1	Agriculture, Forestry, Animal Husbandry, and Fishery Agent	22	Scrap and Waste Agent
2	Mining and Washing of Coal Agent	23	Production and Supply of Electric Power and Heat Power Agent
3	Extraction of Petroleum and Natural Gas Agent	24	Production and Distribution of Gas Agent
4	Mining of Metal Ores Agent	25	Production and Distribution of Water Agent
5	Mining and Processing of Nonmetal Ores and Other Ores Agent	26	Construction Agent
6	Manufacture of Foods and Tobacco Agent	27	Traffic, Transport, and Storage Agent
7	Manufacture of Textile Agent	28	Post Service Agent
8	Manufacture of Textile Wearing Apparel, Footwear, Caps, Leather, Fur, Feather (Down), and Its Products Agent	29	Information Transmission, Computer Services, and Software Agent
9	Processing of Timbers and Manufacture of Furniture Agent	30	Wholesale and Retail Trades Agent
10	Papermaking, Printing, and Manufacture of Articles Agent	31	Hotels and Catering Services Agent
11	Processing of Petroleum, Coking, Processing of Nuclear Fuel Agent	32	Financial Intermediation Agent
12	Chemical Industry Agent	33	Real Estate Agent
13	Manufacture of Nonmetallic Mineral Products Agent	34	Leasing and Business Services Agent
14	Smelting and Rolling of Metals Agent	35	Research and Experimental Development Agent
15	Manufacture of Metal Products Agent	36	Comprehensive Technical Services Agent
16	Manufacture of General Purpose and Special Purpose Machinery Agent	37	Management of Water Conservancy, Environment, and Public Facilities Agent
17	Manufacture of Transport Equipment Agent	38	Services to Households and Other Services Agent
18	Manufacture of Electrical Machinery and Equipment Agent	39	Education Agent
19	Manufacture of Communication Equipment, Computer, and Other Electronic Equipment Agent	40	Health, Social Security, and Social Welfare Agent
20	Manufacture of Measuring Instrument and Machinery for Cultural Activity and Office Work Agent	41	Culture, Sports, and Entertainment Agent
21	Manufacture of Artwork and Other Manufacture Agent	42	Public Management and Social Organization Agent

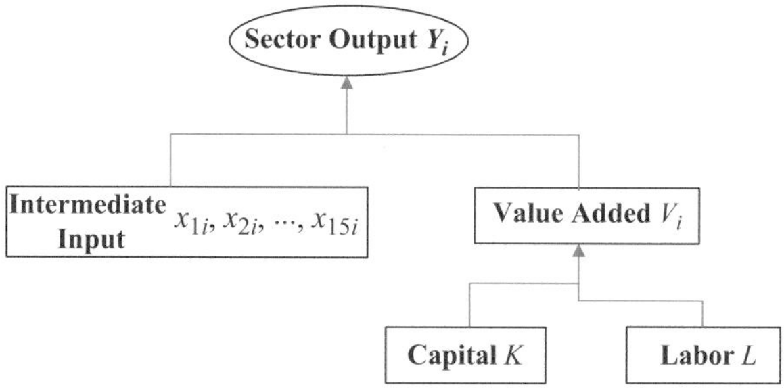

Fig. 7.1 Input and output of each sector

Sector agents serve as both the producer of their own products and the demander of production factors and the products of other sectors as intermediate input. They made decisions on production to maximize their benefits in the light of conditions in production factor market and commodity market. This process in ARE model is designed to simulate the situation in practice. To start with, each sector agent makes the production plan for the next period on the basis of accumulated knowledge and experience and expectations of the market and determines the demand of fixed assets input, labor, and intermediate inputs. Then, each sector agent obtains capital from the commercial banks via financial market and acquires labor from residents via labor market. Afterwards, these sector agents send the information on its input and output to market agents and get the prices decided by the market agents. Lastly, the sector agents will adjust their production plan in each loop until the supply–demand balances of all the products and production factors are achieved.

The input–output structure of the sector agents is illustrated in Fig. 7.1. It is assumed that the product of a certain sector cannot be replaced by that of another sector. The production function is shown in the following equations:

$$Y_i = \min\left\{\frac{x_{1i}}{a_{1i}}, \frac{x_{2i}}{a_{2i}}, \ldots, \frac{x_{15i}}{a_{15i}}, \frac{V_i}{1 - a_{1i} - a_{2i} - \cdots - a_{15i}}\right\} \tag{7.1}$$

$$V_i = A_i K_i^{\alpha} L_i^{1-\alpha} \tag{7.2}$$

where Y_i is the total output of sector i, x_{ji} is the consumption of the product of sector j by sector i as intermediate input, a_{ji} is the intermediate input coefficient which is regarded as the direct consumption coefficient in the input–output table, K_i is the total fixed assets consumed by sector i, L_i is the total labor input of sector i, and V_i is the value added of sector i.

The target of each sector agent is to maximize its profit in each production loop. Its actions include determining the production plan, sending its product to the commodity market to gain revenue, paying for labors and capital, and paying taxes to the government. Its decision-making behavior can be denoted as follows:

$$\max \Pi_i = \left(p_i Y_i - \sum_{j=1}^{15} p_j x_{ji} - wL_i - rK_i\right)(1 - t) \tag{7.3}$$

S.t.

$$Y_i = \min\left\{\frac{x_{1i}}{a_{1i}}, \frac{x_{2i}}{a_{2i}}, \ldots, \frac{x_{15i}}{a_{15i}}, \frac{V_i}{1 - a_{1i} - a_{2i} - \cdots - a_{15i}}\right\}$$

$$V_i = AK_i^{\alpha} L_i^{1-\alpha}$$

$$I_{i,in} \leq I_{i,\max}$$

where Π_i is the profit of sector i, w is the wage rate, r is the interest rate, t is the tax rate, p_i is the price factor of the product of sector i, $I_{i,in}$ is the capital input of sector i, and $I_{i,\max}$ is the maximum available capital of sector i. The meanings of other variables are the same with those in Equation (7.1).

Since each sector agent is in charge of the whole production and decision-making process, its structure is quite complex. Its method base is the model shown in Eq. (7.3). Its internal states include the status and results of last action and the supply–demand situation in the commodity market. Its knowledge base consists of two parts. Part 1 contains its internal states. Part 2 is the module in which the production adjustments are made according to information in Part 1. The rule of sector agents to make adjustments is listed below:

① If (Benefit= =0 and Sign= =0 and Balance= =0), then Action = $-aS$
② If (Benefit= =0 and Sign= =0 and Balance= =1), then Action = bS
③ If (Benefit= =0 and Sign= =1 and Balance= =0), then Action = $-aS$
④ If (Benefit= =0 and Sign= =1 and Balance= =1), then Action = bS
⑤ If (Benefit= =1 and Sign= =0 and Balance= =0), then Action = $-aS$
⑥ If (Benefit= =1 and Sign= =0 and Balance= =1), then Action = $-aS$
⑦ If (Benefit= =1 and Sign= =1 and Balance= =0), then Action = bS
⑧ If (Benefit= =1 and Sign= =1 and Balance= =1), then Action = bS

where Benefit represents the changes in last profit and 0 indicates the profit decreases, while 1 means the profit increases; Sign represents the status of last action and 0 verifies the production scale decreases, while 1 means the production scale increases; and Balance represents the status of supply–demand balance in last loop and 0 means that the supply is greater than the demand, while 1 indicates that the supply is lower than the demand. a and b are two positive constants that are used to express the production adjustment. $S = (|B_{i-1} - B_{i-2}|/B_{i-1}) \times (|\Delta SD_{i-1}|/Y_{i-1})Y_0$, B_i is the profit in the ith loop, $|\Delta SD_i|$ is the difference between supply and demand in the ith loop, Y_0 is the initial output, and Y_i is the sectoral output in the ith loop.

Sector agents can adjust the production scale according to their own goals and the situations in the market. Let us take Rule ①, for example. If both the profit and the production scale were decreased while the supply was still greater than the demand in last loop, the agent will keep curtailing the production scale.

In our computer program, each sector agent has four methods, which are in accordance with the four chief steps in the production process, production factor

Table 7.2 Components and functions of a sector agent

Name	Function
public void InvNewDemand()	Calculate the capital demand
public double ProdFunction()	Expand reproduction based on the allocation of funds
public double FixAssetsFormation()	Fix asset formation after production
public void OutputTable()	Update input–output table

preparation, production, fixed capital formation, and result output, as shown in Table 7.2.

Typical statements in our computer program for the sector agent are shown below for your reference:

```
public void InvNewDemand(){
/**
  * This section is for the calculation of capital demand of a sector
agent to enlarge its production scale.
*/
/**
* In the first loop, the capital demand is calculated with the perpetual
inventory method.
*/
Capital=CapitalForamtion(t+1)- depreciationRate* CapitalForamtion
(t);
/**
* In the following loops, the capital demand is calculated based on the
price signals from the market and the relevant price elasticity.
*/
Production=Production*(1+(price-preprice)*Elasticity;
/**
* The capital distribution is determined.
*/
CapitalDistribution
}
-------------------
public double ProdFunction(){
%Production Function of the Sector Agent
/**
* Through the comparison of the capital demand and available loan from
the commercial bank, the production adjustments are determined.
*/
RealProduction=Production*CapitalDistribution/TaxRate;
/**
* Read the data of labor payment from the input-output table.
*/
```

```
preL=data[43][0];
/**
 *The sector agent conducts its production plan. Related results are
written into the input-output table.
*/
data=RealProduction*aijAgr[i]* Price(0)/Price()[i];
}
___________________
public double FixAssetsFormation(){
/*
Calculate the fixed capital formation after the production and write
relevant data into the input-output table.
*/
}
___________________
public void OutputTable(){
/*
Update the input-output table and output.
*/
}
```

7.1.2 Market Agent

The main function of the market agent is to determine the price according to the supply–demand balance situations according to the market mechanisms. Its structure is relatively simple as shown in Fig. 7.2. The market agent receives the demand

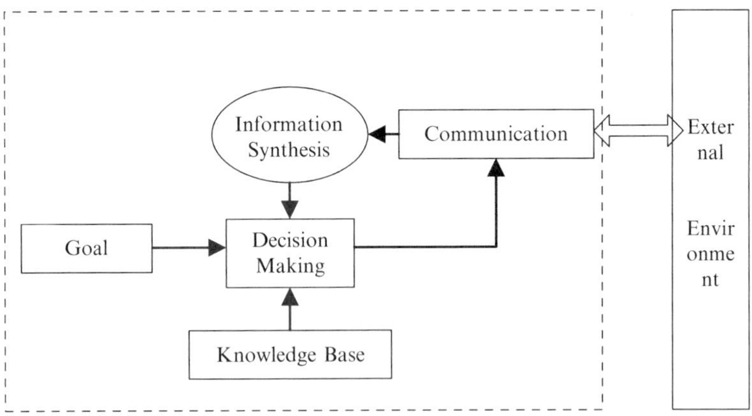

Fig. 7.2 The structure of market agent

of various sector agents, the resident agent and the government agent, and the supply of various sector agents. Then, the price of each kind of product is adjusted according to the current supply–demand relationship.

The target of the market agent is to provide other agents with the market environment. Chief actions include determining the market prices and sending these prices to related sector agents, for which the signals are crucial to make the production plans for the next loop. The supply and demand information is gathered and various prices are decided according to the following equations:

$$S_i = q_i \quad (i = 1, 2, \ldots, N) \tag{7.4}$$

$$D_i = d_R^i + d_G^i + d_O^i + \sum_{j=1}^{N} d_j^i \quad (i = 1, 2, \ldots, N; \quad j = 1, 2, \ldots, N) \tag{7.5}$$

$$P_i^{t+1} = P_i^t \times \left(1 - \frac{S_i^t - D_i^t}{S_i^t \vee D_i^t}\right) \quad (i = 1, 2, \ldots, n) \tag{7.6}$$

where S_i is the supply of sector i, which is equal to the product output q_i; D_i is the overall demand from the whole economic system for the product of sector i, which is the sum of the residential demand d_R, the government demand d_G, the export demand d_O, and the demand from other sectors; P_i^t is the price of the product of sector i in time t; S_i^t and D_i^t are, respectively, the supply and the total demand of the product of sector i in time t; and operator $\vee$ is to take a larger value.

The market continually publishes the price information in time to help each sector agent to adjust its production scale. Equation (7.6) indicates that the price in the next loop is related to the price in current loop and the difference between supply and demand. Moreover, the relatively greater value of the supply and the demand is chosen to serve as the baseline of changing rate, in order to avoid sharp fluctuations of the price.

Take the product of textile sector, for example. The market agents gather the information on the demand for the product of textile sector and compare the total amount with the output of textile sector. If the supply cannot meet the demand, the price will increase; otherwise, the price will decrease. The specific rate of increase or decrease is related to the difference between the supply and the demand of this product.

In our computer program, the market agents are coded on the basis of public void init() in JAVA, as shown in Table 7.3.

Table 7.3 Components and functions of a market agent

Name	Function
public void init().InputSum	Count the total input of each sector
public void init().OutputSum	Count the total demand of each sector
public void init().Price	Determine the product price
public void init().setprice	Send the price to each sector

Related statements are shown below for your reference:

```
public void init(){
/**
* Set the initial price.
*/
IndustryAgent.setPreprice;
//Calculate the total input and total output.
Calculate TotalInputSum;
Calculate TotalOutputSum;
//Determine the price according to the supply-demand situation.
Price=abs(TotalOutputSum-TotalInputSum)/Max(TotalInputSum,
TotalOutputSum);
//Send the price signal to other agents.
IndustryAgent.setprice;
//Update the input-output table.
}
```

7.1.3 International Commodity Market Agent

International trade consists of export and import. Export means that the country sells its product to other countries or provides labor force in the international environment. Oppositely, import refers to the product and labor from other countries. In our model, export and import are considered as a whole, regarded as net export.

Net export equals the export minus import. From the perspective of international trade, exchange rate, domestic income level, and some other factors may have influences on the net export. If the exchange rate is low, domestic commodity is relatively cheap in the international market. As a result, foreigners would like to buy more products from the country, while domestic households tend to curtail their budget on imported products. Consequently, the net export will increase in this situation. On the contrary, the net export will decrease if the exchange rate is relatively high. In a word, the net export of a country has negative correlation with the exchange rate level.

Apart from the exchange rate, the real domestic income is also a crucial issue. If the income level is enhanced, the consumption of domestic household will increase. They will spend more money on the imported commodities. Considering that the export may have little to do with the domestic income, the net export of a certain country has negative correlation with the domestic income level.

In our model, the international commodity market agent is designed to predict the international product price based on historical trends. The net export depends on the ratio between domestic price and international price instead of exchange rate and real income level for simplicity.

According to the content in China Statistical Yearbook and China Customs Statistics Yearbook, those commodities with high import amount or export amount, such as oil and steel, are selected. The price ratios of these products are calculated. We assume that the price levels in the international market are mainly determined by these important products, and the price levels of those products with low import or export amount are impacted by the main products. Then it is possible to get the international prices of all the different products. Furthermore, the price ratio of each kind of product can be obtained, which determines the net export as shown in the following equation:

$$nx_{\mathrm{i}} = \alpha - \gamma p_i \quad (i = 1, 2, \ldots N) \tag{7.7}$$

where nx_i is the net export of Product i and p_i is the ratio between domestic price and international price of the product. It is evident that the amount of net export of a certain product is inversely related to the ratio. The larger the ratio is, the greater the amount will be.

7.1.4 Government Agent

The actions of the government agent are related to its income and expenditure. It gets income through taxation and expends money via consumptions and transfer payments.

(1) The Income of the Government Agent

The tax is the main approach to gain income for the government agent in our model. The taxes can be divided into the following six sorts in China:

1. Turnover taxes, including value-added tax, consumption tax, business tax, and tariff. They are taxed in the areas of production, distribution, or service. Value-added tax consists of domestic value-added tax and import value-added tax.
2. Income taxes, including corporate income tax and personal income tax. They are taxed according to the corporate profit or personal income.
3. Resource taxes, including resource tax and urban land use tax. Those people who are exploiting natural resources or utilizing urban land need to pay the tax.
4. Property taxes, mainly including real estate tax.
5. Specific purpose taxes, including urban maintenance and construction tax, farmland occupation tax, land value number, and vehicle purchase tax. They are set for various specific purposes.
6. Behavior taxes, including travel tax, tonnage tax, stamp duty, and the deed tax.

To sum up, the income function of the government can be expressed by Equation (7.8):

$$INC_G = \sum_{i=1}^{N}\sum_{j=1}^{M} T_{ij} \quad (i = 1, 2, \ldots, N; j = 1, 2, \ldots, M) \tag{7.8}$$

where T_{ij} is the amount of the jth tax paid by Agent i.

(2) The Expenditure of the Government Agent

The expenditure of government consists of purchase expenditure and transfer expenditure. The purchase expenditure refers to the consumption of goods and services for daily government affairs or for the sake of national investment. The transfer expenditure means the gratuitous and unilateral transfer of the government's funds. Typical examples include subsidies, donations, and interest on debt. The overall expenditure of government in the model can be expressed by the following equation:

$$EXP_G = \sum_{i=1}^{N} \left(Q_G^i + T_G^i\right) \quad (i = 1, 2, \ldots, N) \tag{7.9}$$

where Q_G^i and T_G^i are, respectively, the purchase expenditure and transfer expenditure of the government on sector i.

We assume that the purchase expenditure has fixed structure and its overall amount is proportional to the income, as shown in Eq. (7.10). The transfer expenditure can be determined according to actual conditions of the economic system:

$$Q_G = \lambda INC_G \tag{7.10}$$

7.1.5 Central Bank Agent

The central bank agent serves as both the actor and the regulator in the financial market [2]. On the one hand, it acts as the lender of last resort. On the other hand, it publishes and conducts monetary policies. Its behaviors will impact the supply and demand of capital in the financial market and the behaviors of other agents, although they are not for profit.

It is assumed the central bank agent has a high decision-making authority in designing and conducting monetary policies. Possible policy tools are listed as follows:

(1) The statutory reserve ratio

The central bank agent owns the right to decide the statutory reserve ratio for commercial banks. If it is necessary to increase the money supply from the point of view of the central bank, the statutory reserve ratio may be reduced, which will lead to the increment of bank reserves. Changing the statutory reserve ratio is the easiest way to adjust the money supply in theory. However, it is not the favorite method of the central bank because the frequent changes in statutory reserve ratio tend to disturb the normal credit business of financial institutions.

(2) Open market operations

Open market operations refer to the public transactions of government bonds to control the money supply and interest rates. It is the most commonly used method of the central bank. The reserve of commercial banks will increase if the central bank buys the bonds in the open market, while it will decrease if the central bank sells the bonds. The changes in the reserve will lead to the changes in money supply in terms of multiplier effects.

(3) The discount rate

The discount rate is the lending rate of the central bank to commercial banks and other financial institutions. It usually serves as the supplementary means to be carried out with the open market operations. If the discount rate increases, the loan of commercial banks from the central bank will decrease and then the money supply will reduce. Otherwise, the money supply will increase. The adjustment of discount rate is kind of passive because it works only when commercial banks borrow money from the central bank.

In our model, the central bank agent focuses on the money supply and the health and stability of the financial market. It is in charge of setting deposit and lending rates and deposit reserve ratio. Relevant functions are shown in the following equations:

$$S_L = S + \Delta M^S \tag{7.11}$$

$$\Delta M^S = f(S, \theta_1, \theta_2, \theta_3) \tag{7.12}$$

$$\theta_1 = (1 - \tau DR) \tag{7.13}$$

$$\theta_2 = (1 - \upsilon CR) \tag{7.14}$$

$$\theta_3 = (1 - \omega BR) \tag{7.15}$$

where S_L is current money supply; ΔM^S represents the changes in the money supply, which is related to benchmark deposit and lending interest rates, changes in the rate of the deposit reserve ratio, and the money supply in last loop; $\theta_1, \theta_2, \theta_3$ are, respectively, the elasticity of benchmark deposit and lending interest rates and changes in the rate of the deposit reserve ratio to the money supply; and *DR*, *CR*, and *BR* are, respectively, benchmark deposit rate, the benchmark lending rate, and the deposit reserve ratio.

7.1.6 Commercial Bank Agent

The commercial bank agent is the main participator in the financial market. Different from the central bank, the commercial bank is a profit-oriented financial institution. Only one commercial bank is set in ARE model for the sake of simplicity. Its functions include receiving deposit from residents and lending moneys to sector agents.

The total amount of deposit depends on the historical deposit amount and the new deposits. The total lendable fund is calculated on the basis of the total deposit amount, as shown below:

$$D_L = \mathrm{D} + \Delta \mathrm{M}^D \tag{7.16}$$

$$\Delta \mathrm{M}^D = f(RI, DR, DE) \tag{7.17}$$

$$C_T = D_L{}^*(1 - \mathrm{BR})^*\mathrm{LR} \tag{7.18}$$

where D_L is the current total amount of deposit; $\Delta \mathrm{M}^D$ is the new deposit, which is related to the residential income, the deposit interest rate, and the residential saving rate elasticity; *RI* is the residential income; *DR* is the benchmark deposit rate; *DE* is the residential saving rate elasticity; C_T is the current total lendable fund; BR is the deposit reserve ratio; and LR is the maximum deposit ratio, which will also impact the lendable fund and is assumed to be 75 % in the program.

7.1.7 Resident Agent

The resident agent acquires salary through providing labor force in the production of various sectors. The salary may be utilized to consume commodities or deposited in the commercial banks.

The saving rate of the resident agent is dependent on both the changes in deposit interest rate and residential saving rate elasticity, as shown in the following equation:

$$\mathrm{DC}^R = \Delta DR^*DE \tag{7.19}$$

where DC^R is the residential saving rate, ΔDR is the change in deposit interest rate, and *DE* is the residential saving rate elasticity.

The change in residential consumption is determined by the changes in disposable income and the residential consumption elasticity, as shown below:

$$\Delta \mathrm{RC} = \Delta \mathrm{RI}^*\mathrm{CE} \tag{7.20}$$

where ΔRC is the change in residential consumption, ΔRI is the change in disposable income, and CE is the residential consumption elasticity.

7.2 Computer Program of ARE Model

In this section, we provide the flowchart, development environment, and some typical statements of our computer program for those readers who are interested in the program.

7.2.1 The Computer Program Flowchart

The main flow of the computer program is illustrated in Fig. 7.3.

7.2.2 Computer Programming Environment

The simulation of the multi-agent system should be carried out on the basis of the functions of distributed control, the design of individual agent, and the interaction of different agents, statistic, and display. As a result, an appropriate program development environment is essential to build a multi-agent system model. In our research, Swarm is chosen to serve as the platform to develop the program.

Swarm is designed by the Santa Fe Institute (SFI) in the USA [3]. It is a software experiment platform with relatively high efficiency and reliable characteristics. It provides the professional researchers and scholars in certain areas with both a standard software toolset and a programming environment. The core of it is an object-oriented framework, in which the individual behaviors and interactions of various agents are developed and researched.

The modeling ideology of Swarm is to make various independent agents interact with each other standing on their own feet, so that the complex adaptive behaviors of the system that consists of many individuals can be studied. Developers are able to build the simulation system by means of the class library of Swarm to realize the interactions of all the agents. Besides, functions of analysis, control, and display are offered by Swarm to have a deeper review on the simulation process of the complex multi-agent system.

In the development environment provided by Swarm, there are active agents that all have independent thinking and their surroundings. Some rules are necessary for the interactions among different agents and the interactions between the agents and their surrounding environment. These rules tend to be in the form of "if-then," which is typical conditional statement in the programming by the system developer. The sequence of the behaviors of all the agents is possibly controlled by the schedule of model users. When the system developers need to build the multi-agent system model, they are required to deal with various agents and their

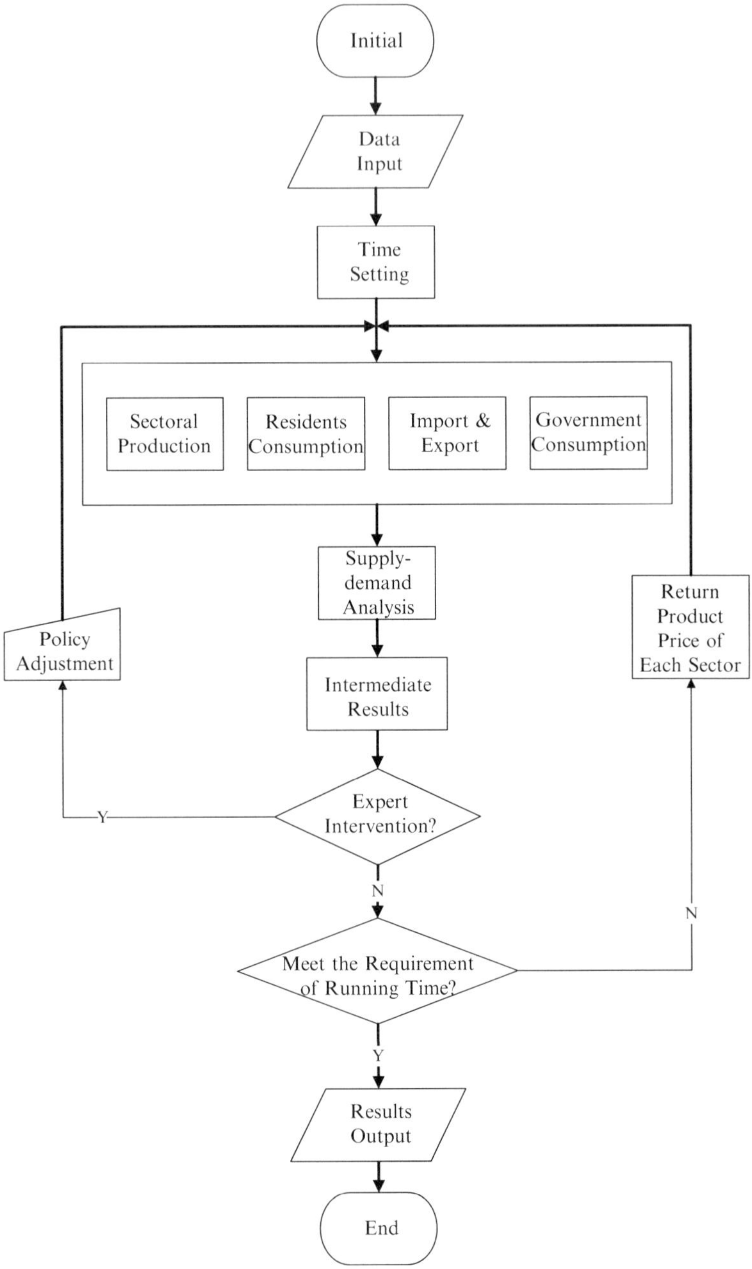

Fig. 7.3 The flowchart of the computer program

activities and the living environment of the agents, which have been integrated in the module of Model Swarm in the framework of the Swarm platform.

Apart from the settings of behavior rules of agents and their surrounding environment, the observation of the simulation is also of crucial importance. Accordingly, another module called Observer Swarm is contained in the framework of Swarm. Principal elements in this module include Probe that is responsible for monitoring the operation of the agents, Control Panel that is in charge of the interactions of the simulation system, Zoom Raster that is designed to display the operation environment of the agents, and all sorts of figures which are meaningful for the statistics of the simulation results. Let us put it in another way; the module of Model Swarm serves as the environment in which diverse models can be run, while the module of Observer Swarm acts as the window from which the model users can observe the results of operations.

Since Swarm is designed for object-oriented programming, it contains a great number of function packages. In our program, several packages have played significant role for the realization of our ideas. We select some chief function packages in Swarm here, as listed in Table 7.4.

The hardware environment of our computer program is introduced briefly as follows. CPU higher than Intel Core Duo T6600 and RAM larger than 2G are recommended. There is no specific requirement on the hard drive capacity. The computer program can be developed and run on both notebooks and desktop computers.

Table 7.4 Chief function packages in Swarm

Name	Function
swarm.activity	Control the execution of Swarm on the whole; provide the support of timetable
swarm.analysis	Serve as a set of tools for analysis
swarm.defobj	Provide standard objects for the expansion of GNU and Objective-C
swarm.collections	Provide a series of lists of set objects, such as list, array, set, etc.
swarm.gui	Offer interfaces for GUI in swarm
swarm.objectbase	Support objects and probes in Swarm
swarm.random	Provide the users with a random number generator
swarm.simtools	Provide tools of data analysis and display; have the functions of collecting data, drawing timing diagram, etc.
swarm.simtoolgui	Contain tools that are related to GUI
swarm.space	Serve as the space library to help users observe objects in different spaces

7.2.3 Typical Statements of the Computer Program

In this subsection, some representative statements in our program will be provided for your reference. Firstly, it is necessary to initialize the program as follows:

```
begin

inputrunningyears
input data
input policy

buildObjects
buildActions
```

Sector agents are very significant role in the model. The following statements are typical actions of each sector agent, including determining investment, seeking loan, conducting production, etc.:

```
//Sector i
{
Production=Production*(1+(Price-prePrice)*Elasticity/prePrice);
NewInvestment=FixAssets*(1+Production/TotalProduction)-
(1-DepreciationRate/52)* FixAssets;
Credit=CommercialBankCredit*DebtDistribution;
Capital=min(NewInvestment,Credit);
RealProduction=Production*Capital*parameter;
data[j]=RealProduction*aij;
TotalProduction=TotalProduction+Realproduction;
FixAssets=FixAssets*(1+RealProduction/TotalProduction);
}
```

The government agent is mainly in charge of collecting various sorts of taxes and expending in forms of consumption, subsidy, and investment. Typical statements are shown as follows:

```
//The Government
{
setTaxRate
GovernmentConsumption=NewTaxSum*parameter;
Cal DomesticVAT;
Cal DomesticConsumpionTax;
Cal VATandConsumptionTaxfromImport;
Cal BusinessTax
Cal CorporateIncomeTax;
Cal ResourceTax;
```

```
Cal CityMaintenanceAndConstructonTax;
Cal StampTax;
Cal UrbanLandUseTax;
Cal IndividualIncomeTax;
Cal ImportTax;
Cal ExportTax;
Cal GovernmentSubsidy;
Cal GovernmentInvestment;
}
```

Markets are essential in our program. Their function is chiefly to decide the price levels of every kind of commodity and production factor, according to the relationship between supply and demand:

```
//Markets
{
InterInput[i]=sum(InputOutputData[j][i]);
InterOutput[i]=sum(InputOutputData[i][j]);
AddedValue[i]=sum(InputOutputData[m][i]);
Input[i]=InterInput[i]+AddedValue[i];
Output[i]=InterOutput[i]+sum(InputOutputData[i][n]);
sdd[i]=Input[i]-Output[i];
if (abs(sdd[i])<parameter)
{Price[i]=Price[i];}
elseif(sdd[i]>0)
{Price[i]=Price[i]-Price[i]*abs(sdd[i])/Input[i]*parameter;}
else
{Price[i]=Price[i]+Price[i]*abs(sdd[i])/Input[i]*parameter;}
output Price[i]
}
```

Residents serve as labor force to take part in the production of each sector. Meanwhile, they act as consumers who spend money on commodities and services in the market. Relevant statements are shown as follows:

```
//Residents
{
TotalLaborIncome=sum(data[k][i]);
IncomeChangeRate[i]=(Income[i]+GovernmentSubsidy[i])/(preIncome
[i]+preGovenmentSubsidy[i])-1;
ResidentsDepositsChange=DepositsChange*DepositsElasticity;
RuralResidentsConsumption=RuralResidentsConsumption*(1
+IncomeChangeRate)*(1-       ResidentsConsumpChangeRate)*(1-(Price
[i]-prePrice[i])/prePrice[i])*parameter;
```

```
UrbanResidentsConsumption=UrbanResidentsConsumption*(1
+IncomeChangeRate)*(1-      ResidentsConsumpChangeRate)*(1-(Price
[i]-prePrice[i])/prePrice[i])*parameter;
ResidentsDepositsChange=
ResidentsDeposits*DepositsIncomeElasticity*parameter;
}
```

The function of the central bank is to publish the fiscal policies, including reserve rate, deposit interest rate, and credit interest rate, as shown below:

```
//The Central Bank
{
setReserveRate
setDepositsInterestRate
setCreditInterestRate
}
```

Commercial banks in the program are responsible for managing deposits and determining the credit scale. Typical statements are listed as follows:

```
//Commercial Banks
{
Deposits=Deposits+ResidentsDeposits*(1+ResidentsDepositsChange)
+ResidentsDepositsChangeIncome;
Deposits=Deposits*(1-ReserveRate);
Deposits=Deposits+OpenMarketAmmount;
CommercialBankCredit=Deposits/parameter;
}
```

International commodity market is set to reflect the influences of price fluctuations in the international market on the operation of China's economy. Its statements are shown below:

```
//International Commodity Market
{
getexpPrice
getimpPrice
Export=ExportConsumption*(1-(Price-expPrice)/expPrice);
Import=ImportConsumption*(1+(Price-impPrice)/impPrice);
}
```

Last but not least, it is essential to judge the year of simulation to determine whether the program should be stopped:

```
if count==the end of one year
```

```
pause
outputInputOutputTable
return to
input policy
continue

if count==the end of runningyears
stop

end
```

7.3 Input Data of ARE Model

This section focuses on the input of our model. There are totally five kinds of input data: (1) data in input–output tables; (2) data related to financial market and monetary issues; (3) labor data, such as employment and salary; (4) coefficients in the production function of each sector agent; and (5) data related to various sorts of tax. We provide some computer program statements in detail on these inputs in the following subsections.

7.3.1 IO Data Input

Since IO tables serve as the data foundation of the simulation, IO data is essential input. In this subsection, the program statement of reading IO table is provided for readers:

```
//Read input-output tables to the program    for(inti=0;i<52;i++){
        for(int j=0;j<42;j++){
            cellvalue=rs2.getCell(31, j+1);
            numc10=(NumberCell) cellvalue;
            WeekAccount[j][i]=numc10.getValue();
        }
    }

    for(inti=0;i<52;i++){
        for(int j=0;j<4;j++){
            cellvalue=rs4.getCell(i+1,j+7);
            numc10=(NumberCell) cellvalue;
            PowerAccount[j][i]=numc10.getValue();
        }
    }
```

```
for(inti=0;i<52;i++){
      for(int j=0;j<4;j++){
            cellvalue=rs4.getCell(i+1,j+1);
            numc10=(NumberCell) cellvalue;
            GDPAccount[j][i]=numc10.getValue();
      }
}

for(inti=0;i<42;i++){
      for(int j=0;j<42;j++){
            cellvalue=rs.getCell(i+5,j+10);
            numc10=(NumberCell) cellvalue;
            InputOutputData[j][i]=numc10.getValue();
      }
}

for(inti=0;i<42;i++){
      for(int j=43;j<47;j++){
            cellvalue=rs.getCell(i+5,j+10);
            numc10=(NumberCell) cellvalue;
            InputOutputData[j][i]=numc10.getValue();
      }
}
//Read the intermediate use of each sector
for (int j=0;j<42;j++){
cellvalue = rs.getCell(j+5, 10);
numc10 = (NumberCell)cellvalue;
   Variables.AgricultureSupply[j] = numc10.getValue();
}

for (int j=0;j<42;j++){
cellvalue = rs.getCell(j+5, 11);
numc10 = (NumberCell)cellvalue;
Variables.MiningSupply[j] = numc10.getValue();
}

for (int j=0;j<42;j++){
cellvalue = rs.getCell(j+5, 12);
numc10 = (NumberCell)cellvalue;
Variables.PetroleumSupply[j] = numc10.getValue();
}
......

//Read rural household consumption, urban household consumption,
total household consumption and total household income
for (int j=0;j<42;j++){
```

```
cellvalue = rs2.getCell(3, j+1);
numc10 = (NumberCell)cellvalue;
Variables.RuralResidentsConsump[j] = numc10.getValue();
}

for (int j=0;j<42;j++){
cellvalue = rs2.getCell(4, j+1);
numc10 = (NumberCell)cellvalue;
Variables.UrbanResidentsConsump[j] = numc10.getValue();
}

   cellvalue = rs.getCell(50,52);
numc10 = (NumberCell)cellvalue;
Variables.ResidentsConsumpSum= numc10.getValue();

cellvalue = rs2.getCell(1,1);
numc10 = (NumberCell)cellvalue;
Variables.ResidentsSave= numc10.getValue();

//Read government consumption and net export data
for (int j=0;j<42;j++){
cellvalue = rs2.getCell(5, j+1);
numc10 = (NumberCell)cellvalue;
Variables.GovernmentConsump[j] = numc10.getValue();
    }

for (int j=0;j<42;j++){
cellvalue = rs2.getCell(6, j+1);
numc10 = (NumberCell)cellvalue;
Variables.ExportConsump[j] = numc10.getValue();
    }

//Read the operation surplus of each sector
for (int j=0;j<42;j++){
cellvalue = rs.getCell(j+5, 56);
numc10 = (NumberCell)cellvalue;
Variables.OperatingSurplus[j] = numc10.getValue();
    }

    cellvalue = rs.getCell(47, 56);
    numc10 = (NumberCell)cellvalue;
    Variables.OperatingSurplusSum = numc10.getValue();

//Read gross fixed capital formation of each sector
```

```
for (int j=0;j<42;j++){
cellvalue = rs.getCell(55, j+10);
numc10 = (NumberCell)cellvalue;
Variables.CapitalFormation[j] = numc10.getValue();
    }

//Read labor remuneration
cellvalue = rs.getCell(47, 53);
numc10 = (NumberCell)cellvalue;
Variables.ResidentsIncome = numc10.getValue();

cellvalue = rs.getCell(27, 53);
numc10 = (NumberCell)cellvalue;
Variables.ElecIncome = numc10.getValue();

//Read the ratio of value added to intermediate input
for (int j=0;j<42;j++){
cellvalue = rs.getCell(j+5, 69);
numc10 = (NumberCell)cellvalue;
Variables.ValueAddedRatio[j] = numc10.getValue();
    }

//Read the intermediate output of each sector
    cellvalue = rs.getCell(60, 10);
    numc10 = (NumberCell)cellvalue;
    Variables.AgricultureSupplySum = numc10.getValue();

    cellvalue = rs.getCell(60, 11);
    numc10 = (NumberCell)cellvalue;
    Variables.MiningSupplySum = numc10.getValue();

    cellvalue = rs.getCell(60, 12);
    numc10 = (NumberCell)cellvalue;
    Variables.PetroleumSupplySum = numc10.getValue();

    ......

//Read the direct consumption coefficient of each input factor

    for (int j=0;j<47;j++){
cellvalue = rs1.getCell(2, j+4);
numc10 = (NumberCell)cellvalue;
Variables.aijAgr[j] = numc10.getValue();
        }
```

```
    for (int j=0;j<47;j++){
cellvalue = rs1.getCell(3, j+4);
numc10 = (NumberCell)cellvalue;
Variables.aijMin[j] = numc10.getValue();
        }

    for (int j=0;j<47;j++){
cellvalue = rs1.getCell(4, j+4);
numc10 = (NumberCell)cellvalue;
Variables.aijPet[j] = numc10.getValue();
        }

        ......

//Read the intermediate input of each sector
    for (int j=0;j<42;j++){
cellvalue = rs.getCell(5, j+10);
numc10 = (NumberCell)cellvalue;
Variables.qijAgr[j] = numc10.getValue();
        }

    for (int j=0;j<42;j++){
cellvalue = rs.getCell(6, j+10);
numc10 = (NumberCell)cellvalue;
Variables.qijMin[j] = numc10.getValue();
        }

    for (int j=0;j<42;j++){
cellvalue = rs.getCell(7, j+10);
numc10 = (NumberCell)cellvalue;
Variables.qijPet[j] = numc10.getValue();
        }

    ......


    //Read the initial production scale of each sector
    for(int j=0;j<42;j++){
         //cellvalue=rs3.getCell(1, j+1);
         cellvalue=rs3.getCell(16, j+1);
         numc10=(NumberCell) cellvalue;
         Variables.initial[j]=numc10.getValue();
    }
```

7.3.2 Financial Data Input

Financial data determines the environment in which all the sector agents and resident agent live and make decisions on loan and deposit. Relevant program statements are shown below:

```
Cell cellvalue = null;
NumberCell numc10 =null;

//Read the overall deposit
cellvalue = rs.getCell(1, 21);
numc10 = (NumberCell)cellvalue;
Variables.TotDeposits = numc10.getValue();
// Read the central bank bonds
cellvalue = rs.getCell(1, 14);
numc10 = (NumberCell)cellvalue;
Variables.CentralBankBonds = numc10.getValue();

//Total loan of non-financial institutions
cellvalue = rs.getCell(1, 17);
numc10 = (NumberCell)cellvalue;
Variables.ClaimsOnNonFinancial = numc10.getValue();

//Read overall assets
cellvalue = rs.getCell(1, 38);
numc10 = (NumberCell)cellvalue;
Variables.TotAssets = numc10.getValue();

//Read the initial monetary policy
cellvalue = rs.getCell(4, 7);
numc10 = (NumberCell)cellvalue;
Variables.BaseDepositsInterestRate = numc10.getValue();

cellvalue = rs.getCell(5, 7);
numc10 = (NumberCell)cellvalue;
Variables.BaseCreditInterestRate = numc10.getValue();

cellvalue = rs.getCell(6, 7);
numc10 = (NumberCell)cellvalue;
Variables.BaseReserveRate = numc10.getValue();

//Read the distribution function
for(int j=0;j<42;j++){
      cellvalue=rs1.getCell(2, j+2);
```

```
        numc10=(NumberCell) cellvalue;
        Variables.DebtDistribution[j]=numc10.getValue();
    }
```

7.3.3 Labor Data Input

As another necessary input, labor data is of vital significance for the relationship between production sectors and residents in the labor market. Data that is related to the initial employment situations is read by the following statements:

```
    Cellcellvalue = null;
    NumberCell num10 = null;

    //Read the average weekly wage of each sector
    for (int j=0;j<42;j++){
    cellvalue = rs.getCell(2, j+1);
    num10 = (NumberCell)cellvalue;
    Variables.WeeklyIncome[j] = num10.getValue();
    }

    //Read the total amount of the average weekly wage of each sector
    for (int j=0;j<42;j++){
    cellvalue = rs.getCell(3, j+1);
    num10 = (NumberCell)cellvalue;
    Variables.BaseWeeklyIncome[j] = num10.getValue();
    }

    //Read the initial employment of the primary industry
    cellvalue = rs.getCell(5,1);
  num10 = (NumberCell)cellvalue;
    Variables.InitialEmployed[1] = num10.getValue();

    // Read the initial employment of the secondary industry
    cellvalue = rs.getCell(5,2);
    num10 = (NumberCell)cellvalue;
    Variables.InitialEmployed[2] = num10.getValue();

   //Read the initial employment of the tertiary industry    cellvalue
= rs.getCell(5,3);
    num10 = (NumberCell)cellvalue;
    Variables.InitialEmployed[3] = num10.getValue();
```

```
// Read the overall initial employment
cellvalue = rs.getCell(5,4);
num10 = (NumberCell)cellvalue;
Variables.InitialTotalEmployed[4] = num10.getValue();
```

7.3.4 Production Function Data Input

As mentioned previously, production function plays a crucial role in making decisions on production scale for each sector agent. Therefore, related data has to been read into the program. Statements for this module are shown as follows:

```
//Read the coefficients of the production function of each sector
for(inti=0; i<6; i++){
    cellvalue = rs.getCell(2+i, 1);
    numc10 = (NumberCell)cellvalue;
    Variables.AgricultureFunction[i] = numc10.getValue();
    }

    for(inti=0; i<5; i++){
      cellvalue = rs.getCell(2+i, 2);
      numc10 = (NumberCell)cellvalue;
      Variables.MiningFunction[i] = numc10.getValue();
      }

    for(inti=0; i<5; i++){
      cellvalue = rs.getCell(2+i, 3);
      numc10 = (NumberCell)cellvalue;
      Variables.PetroleumFunction[i] = numc10.getValue();
      }

    ......

    //Read the inflation rate
    cellvalue = rs.getCell(9, 7);
    numc10 = (NumberCell)cellvalue;
    Variables.BaseRate = numc10.getValue();
```

7.3.5 Tax Data Input

Tax data is also a necessary input of the program for ARE model. Data related to all sorts of taxes that have been mentioned in Sect. 7.1.4 is read by means of the statements shown below:

```
    Cell cellvalue = null;
    NumberCell numc10 =null;

    //Read the domestic value added tax

    for (int j=0;j<42;j++){
    cellvalue = rs1.getCell(2, 45);
    numc10 = (NumberCell)cellvalue;
        Variables.DomesticValueAddedTax[j] = numc10.getValue();
   GovernmentAgent.Instance().setDomesticVATChangeRate(Variables.
DomesticValueAddedTax[j], j);
}

    //Read the domestic consumption tax
    for (int j=0;j<42;j++){
    cellvalue = rs1.getCell(2, 46);
    numc10 = (NumberCell)cellvalue;
        Variables.DomesticConsumptionTax[j] = numc10.getValue();
    GovernmentAgent.Instance().setDomesticCTChangeRate(Variables.
DomesticConsumptionTax[j], j);
    }

    //Read the value added and consumption tax from imports
    for (int j=0;j<42;j++){
    cellvalue = rs1.getCell(2, 47);
    numc10 = (NumberCell)cellvalue;
            Variables.VATandConsumptionTaxFromImports[j] = numc10.
getValue();
   GovernmentAgent.Instance().setVATAndCTFromImportsChangeRate(Var-
   iables.VATandConsumptionTaxFromImports[j], j);
    }

    //Read the business tax
    for (int j=0;j<42;j++){
    cellvalue = rs1.getCell(2, 49);
    numc10 = (NumberCell)cellvalue;
        Variables.BusinessTax[j] = numc10.getValue();
```

```
        GovernmentAgent.Instance().setBusinessTaxChangeRate(Vari-
ables.BusinessTax[j], j);
    }

    //Read the corporate income tax
    for (int j=0;j<42;j++){
    cellvalue = rs1.getCell(2, 50);
    numc10 = (NumberCell)cellvalue;
        Variables.CorporateIncomeTax[j] = numc10.getValue();
        GovernmentAgent.Instance().setCorporateIncomeTaxFromImports-
ChangeRate(Variables.CorporateIncomeTax[j], j);
    }

    //Read the resource tax
    for (int j=0;j<42;j++){
    cellvalue = rs1.getCell(2, 52);
    numc10 = (NumberCell)cellvalue;
        Variables.ResourceTax[j] = numc10.getValue();
        GovernmentAgent.Instance().setResourceTaxChangeRate(Var-
iables.ResourceTax[j], j);
    }

    //Read the city maintenance and construction tax
    for (int j=0;j<42;j++){
    cellvalue = rs1.getCell(2, 53);
    numc10 = (NumberCell)cellvalue;
        Variables.CityMaintenanceAndConstructionTax[j] = numc10.
        getValue();
        GovernmentAgent.Instance().setCityMaintenanceAndConstruc-
tionTaxChangeRate(Variables.CityMaintenanceAndConstructionTax
[j], j);
    }

    //Read the stamp tax
    for (int j=0;j<42;j++){
    cellvalue = rs1.getCell(2, 55);
    numc10 = (NumberCell)cellvalue;
        Variables.StampTax[j] = numc10.getValue();
        GovernmentAgent.Instance().setStampTaxChangeRate(Variables.
StampTax[j], j);
    }


    //Read the urban land use tax
    for (int j=0;j<42;j++){
```

```
    cellvalue = rs1.getCell(2, 57);
    numc10 = (NumberCell)cellvalue;
        Variables.UrbanLandUseTax[j] = numc10.getValue();
      GovernmentAgent.Instance().setUrbanLandUseTaxChangeRate(Vari-
ables.UrbanLandUseTax[j], j);
    }


    //Read the production tax subsidy
    for (int j=0;j<42;j++){
    cellvalue = rs.getCell(j+3,7);
    numc10 = (NumberCell)cellvalue;
      Variables.ProductionNetTax[j] = numc10.getValue()/50;
    }


    //Read the net production tax
    cellvalue = rs.getCell(45, 7);
    numc10 = (NumberCell)cellvalue;
       Variables.ProductionTaxNet = (double) (numc10.getValue()/50);

    //Read the personal income tax
        cellvalue = rs1.getCell(1, 51);
        numc10 = (NumberCell)cellvalue;
        Variables.IndividualIncomeTax = numc10.getValue();

    //Read the tariff
        cellvalue = rs1.getCell(1, 62);
        numc10 = (NumberCell)cellvalue;
        Variables.Tariffs = numc10.getValue();

        //Read the proportion of government investment
        for (int j=0;j<42;j++){
        cellvalue = rs2.getCell(2, j + 1);
        numc10 = (NumberCell)cellvalue;
           GovernmentAgent.Instance().setGovInv(numc10.getValue()
  * 100, j);
           Variables.govInvRate0708[j] = GovernmentAgent.Instance().
getGovInv(j);
       }

        //Read the proportion of government consumption
        for (int j=0;j<42;j++){
        cellvalue = rs2.getCell(3, j + 1);
        numc10 = (NumberCell)cellvalue;
```

```
            GovernmentAgent.Instance().setGovConsump(numc10.getValue
() * 100, j);
            Variables.govConsumpRate0708[j] = GovernmentAgent.Instance
().getGovConsump(j);
        }

        // Read the proportion of government transfer payments
        for (int j=0;j<42;j++){
        cellvalue = rs2.getCell(10, j + 1);
        numc10 = (NumberCell)cellvalue;
          GovernmentAgent.Instance().setGovSubsidy(numc10.getValue
() *
  100, j);
            Variables.govSubsidyRate0708[j] = GovernmentAgent.Instance
().getGovSubsidy(j);
        }

        cellvalue=rs2.getCell(12,41);
        numc10=(NumberCell) cellvalue;
        GovernmentAgent.setTotalSubsidy(numc10.getValue());



        //Read the actual amount of government investment
        for (int j=0;j<42;j++){
        cellvalue = rs2.getCell(6, j + 1);
        numc10 = (NumberCell)cellvalue;
        }   Variables.govInv07[j] = numc10.getValue() * 10000;
```

7.4 Parameters

In this section, some important parameters that appear in the equations of each agent introduced in the Sect. 7.1 are listed, for those readers who may have interests in rewriting the code.

7.4.1 *Elasticity*

Elasticity is the key data that determines the production scale of each sector according to the fluctuations of the price. The price elasticity of each sector is shown in Table 7.5 based on our knowledge and experiments of China's economy.

Table 7.5 Elasticity of each sector in our program

Sector agent	Elasticity
1. Agriculture, Forestry, Animal Husbandry, and Fishery Agent	2.0
2. Mining and Washing of Coal Agent	3.0
3. Extraction of Petroleum and Natural Gas Agent	3.0
4. Mining of Metal Ores Agent	4.0
5. Mining and Processing of Nonmetal Ores and Other Ores Agent	2.0
6. Manufacture of Foods and Tobacco Agent	3.8
7. Manufacture of Textile Agent	2.0
8. Manufacture of Textile Wearing Apparel, Footwear, Caps, Leather, Fur, Feather (Down), and Its Products Agent	2.0
9. Processing of Timbers and Manufacture of Furniture Agent	2.0
10. Papermaking, Printing, and Manufacture of Articles Agent	1.2
11. Processing of Petroleum, Coking, Processing of Nuclear Fuel Agent	3.0
12. Chemical Industry Agent	2.8
13. Manufacture of Nonmetallic Mineral Products Agent	2.0
14. Smelting and Rolling of Metals Agent	3.2
15. Manufacture of Metal Products Agent	1.8
16. Manufacture of General Purpose and Special Purpose Machinery Agent	2.5
17. Manufacture of Transport Equipment Agent	1.8
18. Manufacture of Electrical Machinery and Equipment Agent	2.5
19. Manufacture of Communication Equipment, Computer, and Other Electronic Equipment Agent	2.3
20. Manufacture of Measuring Instrument and Machinery for Cultural Activity and Office Work Agent	2.0
21. Manufacture of Artwork and Other Manufacture Agent	2.0
22. Scrap and Waste Agent	2.5
23. Production and Supply of Electric Power and Heat Power Agent	1.5
24. Production and Distribution of Gas Agent	2.0
25. Production and Distribution of Water Agent	1.5
26. Construction Agent	3.9
27. Traffic, Transport, and Storage Agent	1.4
28. Post Service Agent	1.5
28. Information Transmission, Computer Services, and Software Agent	1.5
30. Wholesale and Retail Trades Agent	1.4
31. Hotels and Catering Services Agent	0.7
32. Financial Intermediation Agent	0.9
33. Real Estate Agent	0.8
34. Leasing and Business Services Agent	1.8
35. Research and Experimental Development Agent	2.0
36. Comprehensive Technical Services Agent	2.0
37. Management of Water Conservancy, Environment, and Public Facilities Agent	1.5
38. Services to Households and Other Services Agent	0.8
39. Education Agent	1.7
40. Health, Social Security, and Social Welfare Agent	1.6
41. Culture, Sports, and Entertainment Agent	1.0
42. Public Management and Social Organization Agent	1.6

7.4.2 *Monetary Parameters*

Some other important parameters apart from production elasticity of each sector are introduced in this subsection.

The first one is the rate of salvage value, which is set as 5 % in our program. The second is the depreciation period that is regarded as 8.4 years on average.

Besides, there are some parameters that have something to do with the elasticity of interest rate. They determine the changes in certain variables, such as residential deposit and consumption, with the change in the deposit interest rate. New investment elasticity on interest rate is set as −0.800. The rural household deposit elasticity on interest rate is set as 0.079, while the urban household deposit elasticity on interest rate is set as 0.021. The consumption elasticity is set similarly as the deposit elasticity, which means the rural household consumption clasticity on interest rate is 0.079 and the urban household consumption elasticity on interest rate is 0.021.

Last but not least, the program also includes several parameters that are related to elasticity on income as follows. They are crucial to decide the changes in residential consumption and deposit with certain change range of their income level. The rural household consumption elasticity on income is set as 0.823, while the urban household consumption elasticity on income is set as 0.438. As for the deposit elasticity, the rural household deposit elasticity on income is set as 0.177, and the urban household deposit elasticity on income is set as 0.562.

References

1. Wei Duan, Zhaoguang Hu, Yuhui Zhou, Xiao Xiao (2012). Input output table updating based on agent-responses equilibrium model[C]. In: IEEE 2012 computational intelligence for financial engineering & economics conference
2. Xiao Xiao, Zhaoguang Hu, JianweiTian, Minjie Xu (2011) Agent-based system for hybrid policy analysis on economy and power [C]. In: International conference on system science, engineering design and manufacturing information
3. Eric Bonabeau, Marco Dorigo, Guy Theraulaz (1999). Swarm intelligence: from natural to artificial systems [M]. Oxford University Press, Oxford, pp 1–2

Part III
China's Economic Simulations by Agents Response Equilibrium

Chapter 8
China's Economic Simulation for the Period of Global Crisis

There is a cycle of around 9 years in economic growth in China, which is like a sine curve. It can be roughly seen from the curve that each of the four periods of valley, growth, peak, and fall will take about 2.5 years. As we discussed in Chap. 1, the economic growth valley years were 1981, 1989, and 1998; thus, on the basis of the economic cycle, the next growth valley would be in around 2007–2008. Reviewing electricity data, the growth of electricity demand was falling monthly to enter the valley period in 2007, which means the economic cycle is active in China's economic operation. However, the global economic crisis happened in 2008, and China's economy was affected sharply by both China's economic cycle and the global crisis because imports and exports in the Chinese economy are closely related to the global economy. Realizing the economic recession was serious, the Chinese government decided to make a huge investment to stimulate the economy in early 2009, and then the economy grew fast for 2 years, especially the growth in electricity demand, which was in double digits. However, there were some problems such as bubbles in real estate, excess capacity in many sectors of manufacture, etc., and some economists criticized the action of the investment. It will be simulated by using our Agent Response Equilibrium (ARE) model as discussed in Chaps. 5 and 6. In this chapter, the evaluation rules of economic simulation will be introduced to show the test and evaluation of the ARE model, and the economic growth in 2007 will be discussed as the basis for the simulation based on the input–output (I/O) table in 2007. Then the economic simulations for 2008, 2009, and 2010 by the ARE model will be tested by the rules, which show very small errors. Experiments on different policies for economic operation can be simulated on a computer by the ARE model.

Z. Hu et al., *China's Economic Gene Mutations*,
DOI 10.1007/978-3-662-47298-9_8

8.1 Evaluation Rules of Economic Simulation by the ARE Model

As we introduced in Chaps. 4 and 5, the ARE model could simulate the national economy on the basis of the I/O table dynamically. What is the best way to test the simulation and evaluate the ARE model? Generally, it can be tested by historical data such as the gross domestic product (GDP) and growth, the value added in some sectors, investment, consumption, import and export, etc.; if the simulation results are close to the historical data, the simulation is accepted and the model is applicable. In our case, the economic simulation using the ARE model can provide an I/O table for the national economy in each year; then the model and simulation results will be accepted if the errors between the published I/O table and the I/O table produced by the ARE model are small. Thus, it can be used as a rule for evaluating the ARE model and the simulation results. However, the I/O table is published every 5 years, in China, such as in 1997, 2002, and 2007, and then it is further updated in 2000, 2005, and 2010. As a model for economic simulation over a few years, there are not enough historical data for the I/O table to be used to test the errors. On the other hand, some data from statistical methods also include some big errors in China, and it is very difficult to test the simulated results of the ARE model. We select 42 sectors for the I/O table in this book. The 42 sectors are shown in Table 3.1 in Chap. 3 of this book. For the three areas of industry, the No. 1 sector (Agriculture, Forestry, Animal Husbandry & Fishery) is primary industry, sectors Nos. 2–26 can be classified as secondary industry, and sectors Nos. 27–42 can be classified as tertiary industry in China.

As was explained in our introduction of the concept of electricity economics in Chap. 1, electricity data have characteristics such as very high accuracy and are widely used in economic activities, and updates of detailed electricity data are made available each month. For electricity consumption in different sectors of the economy, it will be easy to calculate the errors between the published data and the simulated data. Thus, smaller errors mean that the ARE model is applicable for economic simulation. This is another rule for evaluating the ARE model and the simulation results.

To summarize the above discussion, we have set the following rules:

Rule 1: Some data errors are compared between the published I/O table and the simulated I/O table, such as GDP and growth, and value added and growth in primary industry, secondary industry, and tertiary industry. In this chapter, the ARE simulation is based on the published I/O table and some data on fiscal policy such as tax rates for different sectors and government expenditures, and on monetary policy such as the reserve requirement ratio, discount rate, and open-market operations; then the resulting I/O table for 2008, 2009, and 2010 will be produced. A comparison will be done between the published I/O table and the I/O table resulting from the ARE model in 2010.

Rule 2: Errors are compared between published electricity consumption and simulated electricity consumption according to the ARE model for the 42 sectors, and also for the three areas of industry, and the total industrial electricity consumption in China. In this chapter, a comparison will be done between the published electricity use and the resulting electricity use for the 42 sectors in the three areas of industry, and the total industrial electricity consumption in 2008, 2009, and 2010 according to the ARE model.

8.2 Review of China's Economy in 2007

An I/O table has enough data to show the national economy and the relationships between sectors. In order to review the economy in China, an I/O table for 2007 published by the National Bureau of Statistics of China was used to analyze sectoral economic activities. Figure 8.1 shows the value added in the 42 sectors in China. As a primary industry, the value added in Sector No. 1 was 2865.9 billion CNY (in constant 2007 CNY). For secondary industry in Sector Nos. 2–26, the value added was 13,449.53 billion CNY, while the biggest value was 1451.35 billion CNY in Sector No. 26 (Construction), the second biggest one was 1259.28 billion CNY in Sector No. 12 (Chemical Industry), the third one was 1192.87 billion CNY in Sector No. 14 (Smelting and Rolling of Metals), and the value added in Sector No. 6 (Manufacture of Foods and Tobacco) was 1017.85 billion CNY as the fourth one. For tertiary industry Sector Nos. 27–42, the total value added was 10,288.94 billion CNY, while the biggest value was 1733.24 billion CNY in Sector No. 30 (Wholesale and Retail Trades), the next was 1462.44 billion CNY in Sector No. 27 (Traffic, Transport and Storage), the third one was 1343.13 billion CNY in Sector No. 32 (Financial Intermediation), and the fourth one was 1231.92 billion CNY in Sector No. 33 (Real Estate). We can see that the construction sector in secondary industry and real estate in tertiary industry take high shares in the value added, and this is a special feature of China's economy. Thus, GDP is 26,604.38 billion CNY, while primary industry shared 10.77 %, secondary industry shared 50.55 % and tertiary industry shared 38.67 %. The shares of the 42 sectors are shown in Fig. 8.2.

Intermediate use of the products of each sector can show the production process. It can be shown as Fig. 8.3, where the total intermediate use was 55,281.52 billion CNY, while the biggest value was 6,158.79 billion CNY in Sector No. 14 (Smelting and Rolling of Metals), the next was 6154.69 billion CNY in Sector No. 12 (Chemical Industry), and other values were 3434.4 billion CNY in primary industry Sector No. 1 (Agriculture), 3030.7 billion CNY in Sector No. 23 (Production and Supply of Electric Power and Heat Power), 2924.3 billion CNY in Sector No. 19 (Manufacture of Communication Equipment, Computer and Other Electronic Equipment), and 2447.5 billion CNY in Sector No. 27 (Traffic, Transport and Storage) in tertiary industry. The shares of intermediate use in GDP are shown in Fig. 8.4.

Fig. 8.1 Value added in the 42 sectors in 2007 (data source: National Bureau of Statistics of China, 2014; input–output table for China, 2007)

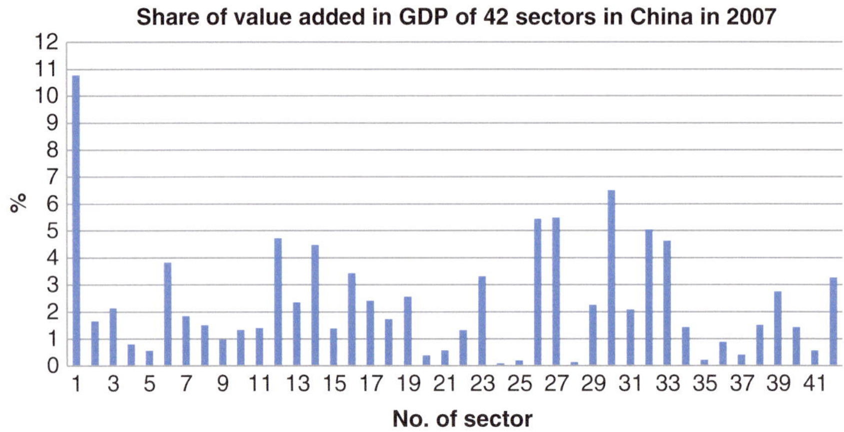

Fig. 8.2 Shares of value added in gross domestic product (GDP) for the 42 sectors in 2007

Labor payments in the 42 sectors can be seen in Fig. 8.5. The biggest value was in primary industry at 2718.16 billion CNY; however, there were 307.31 million people working in primary industry, and the average payment was 8845 CNY/person-year. The payment in Sector No. 42 (Public Management and Social Organization) in tertiary industry was 753.3 billion CNY, and in Sector No. 26 (Construction) in secondary industry it was 740.53 billion CNY. The total payment in secondary industry was 4599.42 billion CNY, with an average labor payment of 22,785.19 CNY/person-year, and it was 15,108.79 CNY/person-year for tertiary industry.

Total production taxes differ in the 42 sectors, as shown in Fig. 8.6. The biggest figure was 420.45 billion CNY in Sector No. 30 (Wholesale and Retail Trades), and

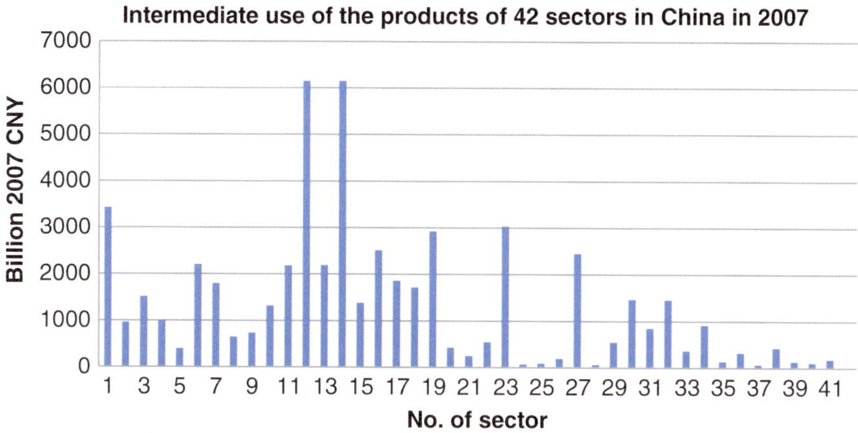

Fig. 8.3 Total intermediate use in the 42 sectors in 2007 (data source: National Bureau of Statistics of China; input–output table for China, 2007)

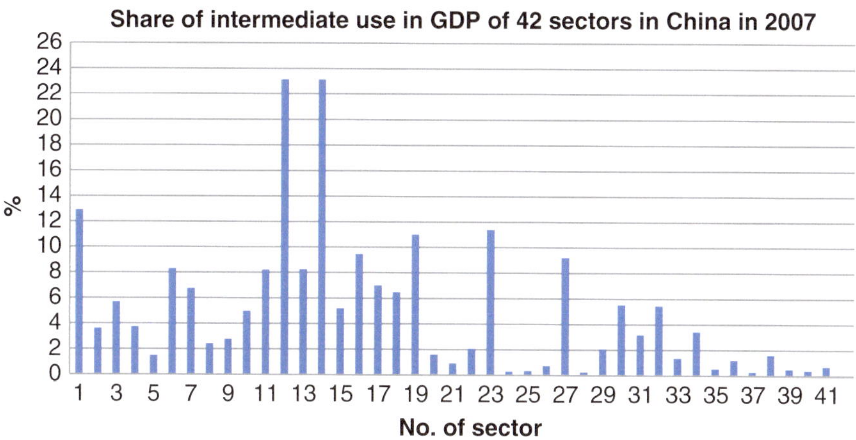

Fig. 8.4 Shares of intermediate use in gross domestic product (GDP) in the 42 sectors in 2007

the next one was 324.19 billion CNY in Sector No. 6 (Manufacture of Foods and Tobacco), followed by 313.27 billion CNY in Sector No. 14 (Smelting and Rolling of Metals) and 241.93 billion CNY in Sector No. 12 (Chemical Industry). The other values were less than 200 billion CNY.

Production profits in the 42 sectors are shown in Fig. 8.7. The three biggest ones were in Sector No. 32 (Financial Intermediation), Sector No. 30 (Wholesale and Retail Trades), and Sector No. 27 (Traffic, Transport and Storage). All of them were in tertiary industry, and their profit was over 600 billion CNY. However, there was a negative profit of −4.84 billion CNY in Sector No. 28 (Post Service).

The total population in 2007 was 1.32 billion, of whom 0.606 billion residents lived in urban areas and 0.715 billion farmers lived in rural areas. Values for final

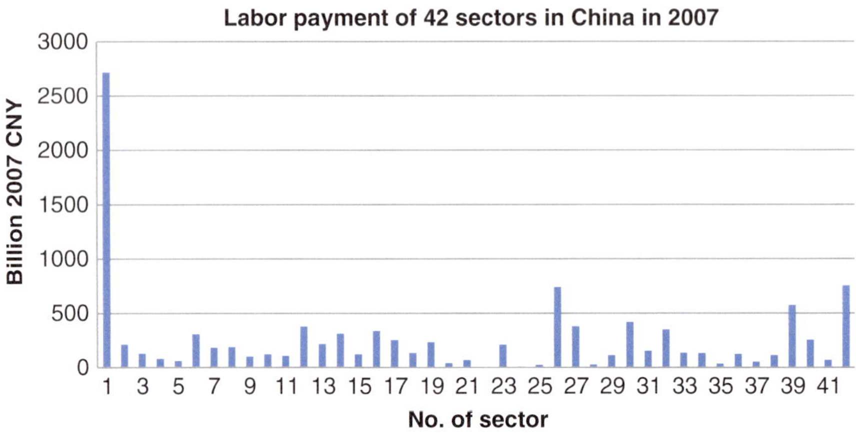

Fig. 8.5 Labor payments in the 42 sectors in 2007 (data source: National Bureau of Statistics of China; input–output table for China, 2007)

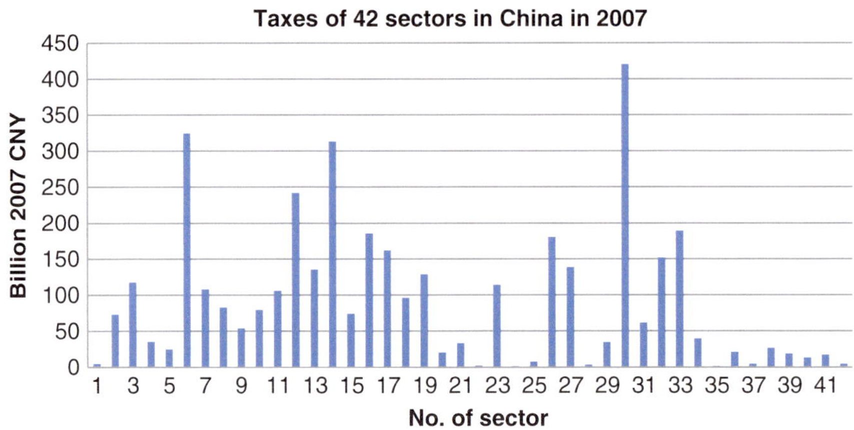

Fig. 8.6 Production taxes in the 42 sectors in 2007 (data source: National Bureau of Statistics of China; input–output table for China, 2007)

consumption of products from the 42 sectors by farmers and urban residents are shown in Fig. 8.8. The values of products from Sector No. 6 (Manufacture of Foods and Tobacco) were 1228.62 billion CNY for purchases by urban residents and 440.17 billion CNY for purchases by farmers. The second biggest values for products from primary industry were 599.67 billion CNY for purchases by urban residents and 515.94 billion CNY for purchases by farmers, and the third biggest values were in the tertiary industry of Wholesale and Retail Trade, with purchases of 568.74 billion CNY by urban residents and 206.03 billion CNY by farmers. The values for Real Estate were 529.77 billion CNY for purchases by urban residents and 226.67 billion CNY for purchases by farmers. For the Manufacture of Textile

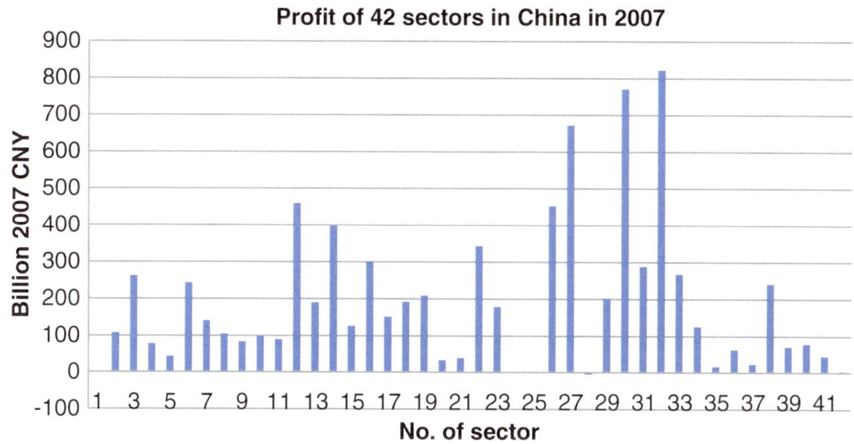

Fig. 8.7 Profits in the 42 sectors in 2007 (data source: National Bureau of Statistics of China; input–output table for China, 2007)

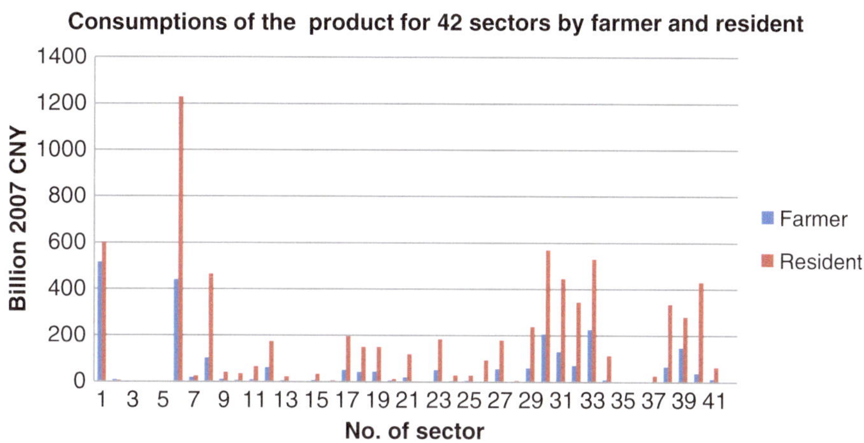

Fig. 8.8 Final consumption of products of the 42 sectors by farmers and urban residents in 2007 (data source: National Bureau of Statistics of China; input–output table for China, 2007)

Wearing Apparel, Footwear, Caps, Leather, Fur, Feather (Down) and Its Products sector, the values were 465.56 billion CNY for purchases by urban residents and 101.57 billion CNY for purchases by farmers. For the costs of Health, Social Security and Social Welfare, the values were 431.03 for urban residents and 36.83 billion for farmers. The costs of Hotels and Catering Services were 445.07 billion for urban residents and 129.73 billion for farmers. Others product costs included in the final consumption were less than 400 billion CNY.

For exports and imports, the biggest export figure was 2137.75 billion CNY for the products of Manufacture of Communication Equipment, Computer and Other

Electronic Equipment (Sector No. 19), and its import value was 1629.87 billion CNY. For Sector No. 18 (Manufacture of Electrical Machinery and Equipment), the export value was 682.57 billion CNY and the import value was 343.52 billion CNY. Sectors No. 18 and 19 have the feature of being a worldwide center for manufacturing, as China imports intermediate products from abroad and produces them as final products for export to other countries. The Manufacture of Textile sector (No. 7), Manufacture of Textile Wearing Apparel, Footwear, Caps, Leather, Fur, Feather (Down) and Its Products sector (No. 8), Processing of Timbers and Manufacture of Furniture sector (No. 9), and Papermaking, Printing and Manufacture of Articles sector (No. 10) in secondary industry, and the Traffic Transport and Storage sector (No. 27) and Wholesale and Retail Trades sector (No. 30) in tertiary industry are export sectors because their exports are much higher than imports.

As for the import sectors, Agriculture, Forestry, Animal Husbandry & Fishery (Sector No. 1), Extraction of Petroleum and Natural Gas (Sector No. 3), Mining of Metal Ores (Sector No. 4), Import of the Product of Chemical Industry (Sector No. 12), Manufacture of General Purpose and Special Purpose Machinery (Sector No. 16), Manufacture of Measuring Instrument and Machinery for Cultural Activity & Office Work (Sector No. 20), and Scrap and Waste (Sector No. 22), their imports are much higher than exports, as shown in Fig. 8.9.

Electricity consumption data are important for review of the sectoral economy. The biggest sector of electricity consumption was Smelting and Rolling of Metals (Sector No. 14), which used 612.62 billion kWh of electricity, the second one was Production and Supply of Electric Power and Heat Power (Sector No. 23), which used 464.92 billion kWh, the third one was Chemical Industry (Sector No. 12), which used 389.94 billion kWh, the fourth was Manufacture of Nonmetallic Mineral Products (Sector No. 13), which used 186.88 TWh, and another one was Manufacture of Textile (Sector No. 7), which used 113.93 TWh (see Fig. 8.10). We can see that all of them are in secondary industry, which is a feature of the industrialization of China's economy.

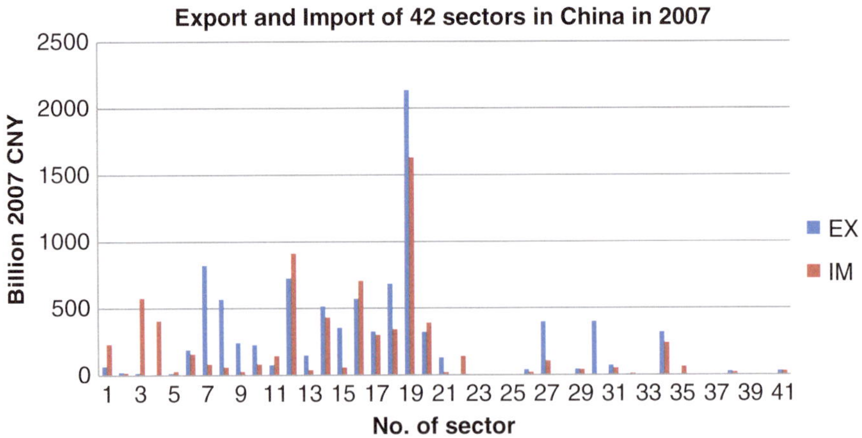

Fig. 8.9 Exports and imports in the 42 sectors in 2007 (data source: National Bureau of Statistics of China; input–output table for China, 2007)

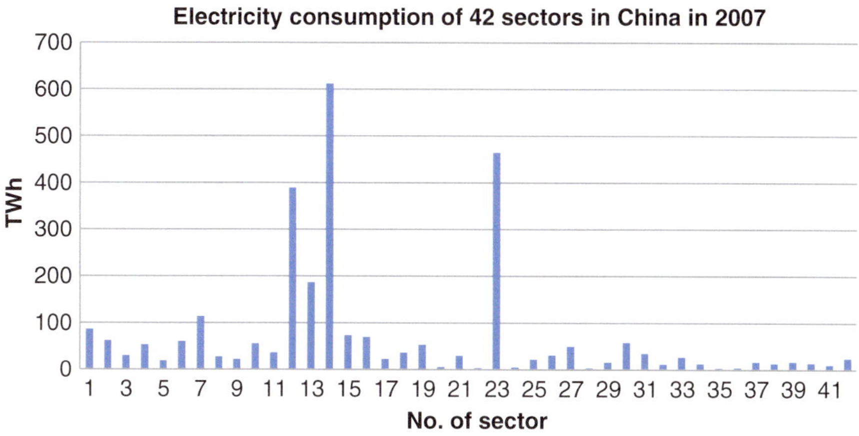

Fig. 8.10 Electricity consumption in the 42 sectors in 2007 (data source: calculated from statistical material from China Electricity Council, 2008)

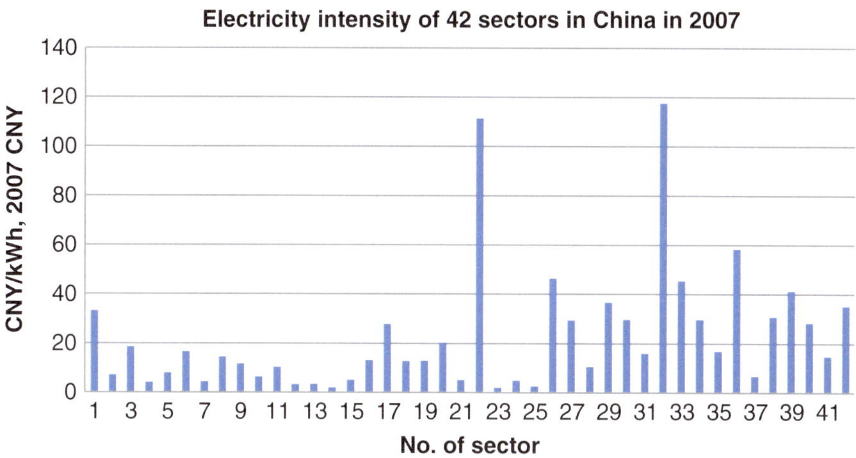

Fig. 8.11 Electricity intensity in the 42 sectors in 2007

Electricity intensity is the value added divided by the electricity consumption. It is clear that we expect to produce more value added with lower electricity consumption, i.e. lower electricity intensity. However, the electricity intensity values are quite different in the 42 sectors because of their different technologies and production processes. For a sector, electricity intensity can be used to show technology improvements in series of years. For the national economy, it can show the mixture of the economy or the shares of the sectors in the national economy. Usually, the electricity intensity is higher for the tertiary industry than for the secondary industry. Figure 8.11 shows that it was 117.5 CNY/kWh for the

Financial Intermediation sector (No. 32) and 111.26 CNY/kWh for the Scrap and Waste sector (No. 22). Electricity intensity for primary industry was 33.19 CNY/kWh, 5.4 CNY/kWh for secondary industry, and 32.31 CNY/kWh for tertiary industry. For the national economy, it was 9.19 CNY/kWh.

In order to simulate the Chinese economy during 2008–2010 by the ARE model, 2007 was used as the start year and 2010 was used as the goal year. Then the I/O table for 2007 was used as the initial state of the economy as one input of the ARE model, and another input was an assumption of policy study, i.e. some data on fiscal and monetary policies must be inputted into the ARE model to do the simulations.

8.3 Economic Simulation for 2008

8.3.1 Assumptions

In order to simulate China's economy in 2008, the I/O table for 2007 and the data on monetary policies and fiscal policies in 2008 were inputted into the ARE model.

The data are as follows. Liabilities of non-financial institutions and households are 37,290.512 billion CNY. Central Bank bonds are 3863.340 billion CNY. Claims on non-financial institutions are 23,491.288 billion CNY. Total assets are 54,120.301 billion CNY. The deposit reserve ratio is 14.5 %. The 1-year benchmark lending rate is 7.47 %. The 1-year benchmark deposit rate is 4.41 %. Changes of these rates in 2008 are also considered. In the fourth week, the deposit reserve ratio is raised by 0.5 percentage point. In the 13th week, the deposit reserve ratio is raised by 0.5 percentage point. In the 17th week, the deposit reserve ratio is raised by 0.5 percentage point. In the 21st week, the deposit reserve ratio is raised by 0.5 percentage point. In the 23rd week, the deposit reserve ratio is raised by 0.5 percentage point. In the 38th week, the benchmark lending rate is lowered by 0.27 percentage point. In the 41st week, both the benchmark lending rate and the benchmark deposit rate are lowered by 0.27 percentage point. In the 42nd week, the deposit reserve ratio is lowered by 0.5 percentage point. In the 44th week, both the benchmark lending rate and the benchmark deposit rate are lowered by 0.27 percentage point. In the 48th week, both the benchmark lending rate and the benchmark deposit rate are lowered by 1.08 percentage points. In the 49th week, the deposit reserve ratio is lowered by 1 percentage point. In the 52nd week, the deposit reserve ratio is lowered by 0.5 percentage point, while both the benchmark lending rate and the benchmark deposit rate are lowered by 0.27 percentage point (see Table 8.1).

The fiscal policies are set as follows. The domestic value-added tax rate is 13.03 %, the domestic consumption tax rate is 1.86 %, the import value-added tax and consumption tax rate is 5.18 %, the sales tax rate is 5.54 %, the corporate income tax rate is 7.39 %, the resource tax rate is 0.22 %, the urban maintenance and construction tax rate is 0.97 %, the stamp duty rate is 1.90 %, the urban land use

Table 8.1 Monetary policies in 2008

Date (year/month/day)	Changed rate: Deposit reserve ratio	Deposit rate	Lending rate	Final rate: Deposit reserve ratio (%)	Deposit rate (%)	Lending rate (%)
2008/01/25	Up 0.5 %	–	–	15	4.41	7.47
2008/03/25	Up 0.5 %	–	–	15.50	4.41	7.47
2008/04/25	Up 0.5 %	–	–	16.00	4.41	7.47
2008/05/20	Up 0.5 %	–	–	16.50	4.41	7.47
2008/06/07	Up 0.5 %	–	–	17.50	4.41	7.47
2008/09/16	–	–	Down 0.27 %	17.50	4.41	7.20
2008/10/09	–	Down 0.27 %	Down 0.27 %	17.50	3.87	6.93
2008/10/15	Down 0.5 %	–	–	17.00	3.87	6.93
2008/10/30	–	Down 0.27 %	Down 0.27 %	17.00	3.60	6.66
2008/11/27	–	Down 1.08 %	Down 1.08 %	17.00	2.52	5.58
2008/12/05	Down 1 %	–	–	16.00	2.52	5.58
2008/12/23	–	Down 0.27 %	Down 0.27 %	16.00	2.25	5.31
2008/12/25	Down 0.5 %	–	–	15.50	2.25	5.31

tax rate is 0.32 %, the personal income tax rate is 2.68 %, the tariff rate is 1.21 %, and net taxes on production are overall 3851.872 billion CNY.

8.3.2 Results

The ARE model can provide results such as the 2008 I/O table, price trends of the products of the 42 sectors, etc. The product price index for the 42 sectors is based on the start price at the start of 2008, i.e. the price on January 1, 2008. All of the prices of the products of the 42 sectors are set at 1, thus, they are varied on that basis in the market. If the supply of the product for a sector is higher than the demand in the market, the price will decrease, otherwise the price will increase.

Sector No. 2 is Mining and Washing of Coal, and the main product of the sector is coal; Sector No. 3 is Extraction of Petroleum and Natural Gas, which produces oil and natural gas; Sector No. 4 is Mining and Dressing of Metal Ores, which produces metal ores; Sector No. 5 is Mining and Dressing of Nonmetal and Other Ores, of which the main product is nonmetal ores. The price indexes for coal, oil and gas, metal ores, and nonmetal ores in each week of 2008 are shown in Fig. 8.12; we can see that all of them experienced falling trends. The oil and gas price index fell before the price indexes for other products, and it reached a lower point at week

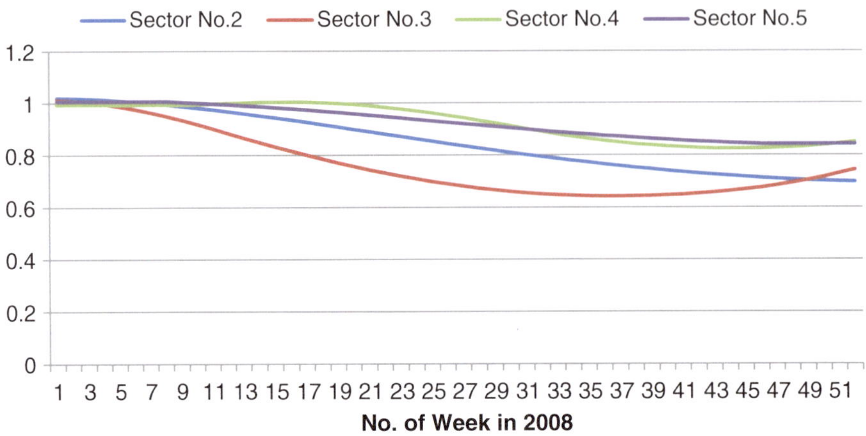

Fig. 8.12 Product price indexes for coal, oil and gas, metal ores, and nonmetal ores in each week of 2008

No. 36–37 (0.64156) and then started to rise to 0.74364 by the end of 2008. The coal price index fell during the year to 0.6969 in the last week of 2008. The price indexes for metal ores and nonmetal ores followed the same trend and were around 0.8412 in week No. 52. These figures show that the market demands for coal, oil and gas, metal ores, and nonmetal ores were weak in 2008.

Sector No. 29 is Information Transmission, Computer Services and Software; it can also be referred to as IT (information technology) services. Sector No. 30 is Wholesale and Retail Trades, Sector No. 31 is Hotels and Catering Services, Sector No. 32 is Financial Intermediation, and Sector No. 33 is Real Estate. Figure 8.13 shows a small decline to 0.9539 in the price index for Hotels and Catering Services in week No. 52 in 2008, while the price indexes for Sector Nos. 33, 29, 30, and 32 declined sharply to 0.6815, 0.494, 0.4645, and 0.2814, respectively, in the last week of 2008. This means that consumption of hotel and restaurant services was little affected by the global crisis in China in 2008. However, the price index trend for real estate shows that house/apartment prices fell by around 32 % in 2008, and the Real Estate, IT, Wholesale and Retail Trades, and Financial sectors were sensitive to the global economy in 2008.

Resident Services and Other Services (Sector No. 38) and Culture, Sports and Entertainment (Sector No. 41) experienced falling trends in their price indexes in 2008, as shown in Fig. 8.14. This shows that people tightened their spending on entertainment during the crisis. However, people's payments on Education (Sector No. 39) and Health, Social Security and Social Welfare (Sector No. 40) still grew during the crisis because Chinese people place a high priority on the education of their children and their health.

The market price indexes in 2008 formed the basis on which firms and other business companies made their production and business plans as economic

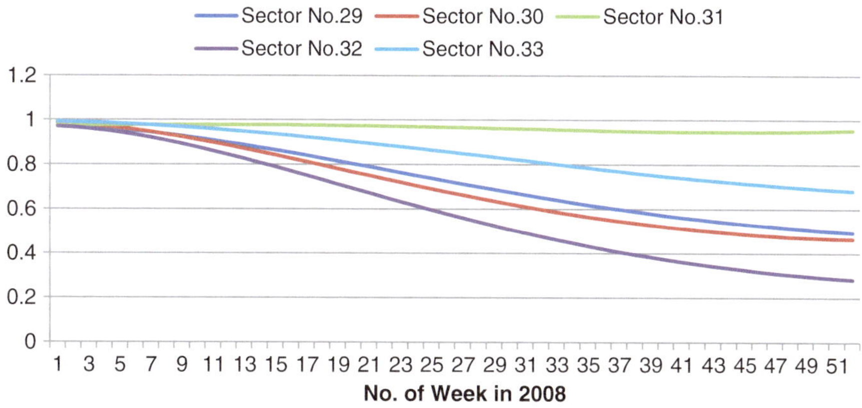

Fig. 8.13 Price indexes for information technology (IT), wholesale and retail trades, hotel and catering services, financial and real estate services in each week of 2008

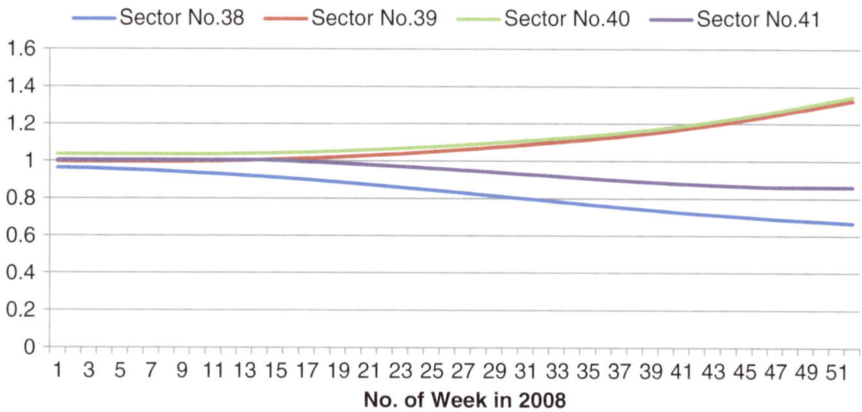

Fig. 8.14 Price indexes for resident services, education, health, and entertainment weekly in 2008

activities in the Chinese economy. This is simulated by our ARE model, resulting in the I/O table for 2008. It will be analyzed as follows.

Value added and growth (year over year [YoY]) for the 42 sectors in 2008 is shown in Fig. 8.15. Value added for Agriculture was 3059.24 billion CNY, reflecting growth of 6.75 %. The highest growth was 10.05 % in the Real Estate sector (Sector No. 33), though the price index for real estate fell by 32 % in 2008. This shows the bubble in the sector. The second highest growth was 10 % in Financial Intermediation (Sector No. 32). There may be some misunderstanding about why this sector would show such high growth in the global financial crisis, because the economic performance of this sector was so poor in other parts of the world, such as the USA and Europe. This is a special feature of the Chinese Financial Intermediation sector. The main reason is the large difference between

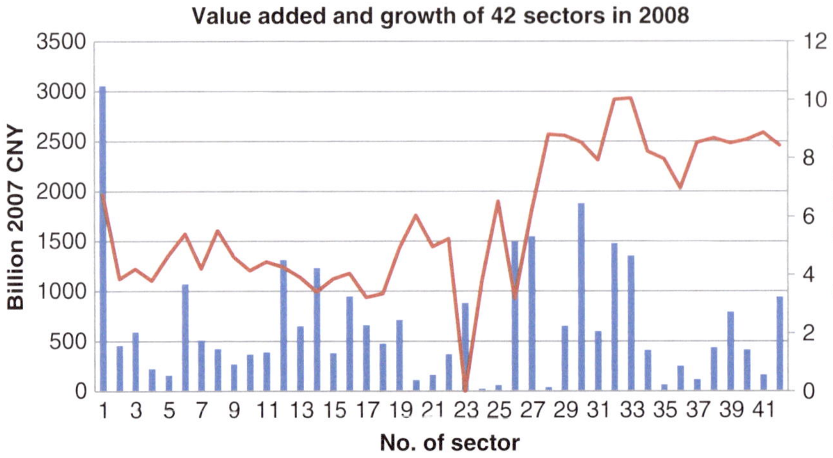

Fig. 8.15 Value added and growth in the 42 sectors in 2008

the loan rate and the deposit rate, which is more than 3 %. It is clear to see from Fig. 8.15 that the growth rates for Sector Nos. 27–42 (tertiary industry) were higher than those for Sector Nos. 2–26 (secondary industry), meaning that manufacturing was affected by the global crisis more deeply than services in China in 2008. The GDP was 28,192.755 billion CNY, reflecting growth of 5.97 %; the value added in primary industry was 3059.24 billion CNY, reflecting growth of 6.75 %. The value added in secondary industry reached 13,968.78 billion CNY, reflecting growth of 3.86 %, and the value added in tertiary industry reached 13,968.78 billion CNY, reflecting growth of 8.51 % in 2008. The shares of the value added for the 42 sectors of GDP in 2008 are shown in Fig. 8.16, which is little different than Fig. 8.2. The shares of the three areas of industry (primary, secondary, and tertiary) in GDP were 10.85 %, 49.55 %, and 39.6 %, respectively, in 2008 in China; secondary industry declined by about 1 %, and tertiary industry increased by 1 %, in comparison with 2007.

Intermediate use in the 42 sectors and their shares of total intermediate use in 2008 are shown in Figs. 8.17 and 8.18. In comparison with 2007 (Fig. 8.3), Financial Intermediation (Sector No. 32) grew by 8.15 %, the largest growth in the 42 sectors in 2008; the second highest growth was in Real Estate (Sector No. 33), which grew by 8.1 %. In third place was Post Service (Sector No. 28), which grew by 7.69 %, while Smelting and Rolling of Metals (Sector No. 14) grew by only 3.78 %. Total intermediate use grew by 6.34 %. This shows that the low growth of heavy industry products in intermediate use was affected by the weak market demand in the crisis in 2008.

What about employment salaries in the crisis in 2008? As shown in Fig. 8.19, the growth of labor payments in tertiary industry (Sector No. 27–42) was higher than that in secondary industry (Sector Nos. 2–26) generally. The highest growth was in Real Estate (Sector No. 33), which grew by 10.05 %, and another growth sector was

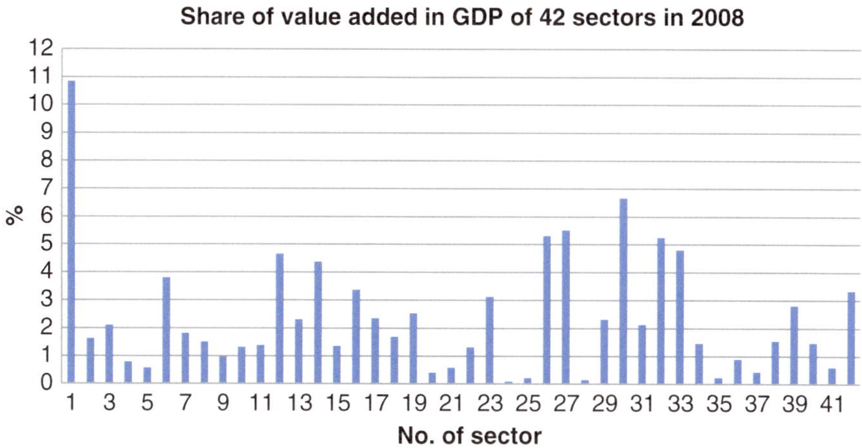

Fig. 8.16 Shares of value added in gross domestic product (GDP) for the 42 sectors in 2008

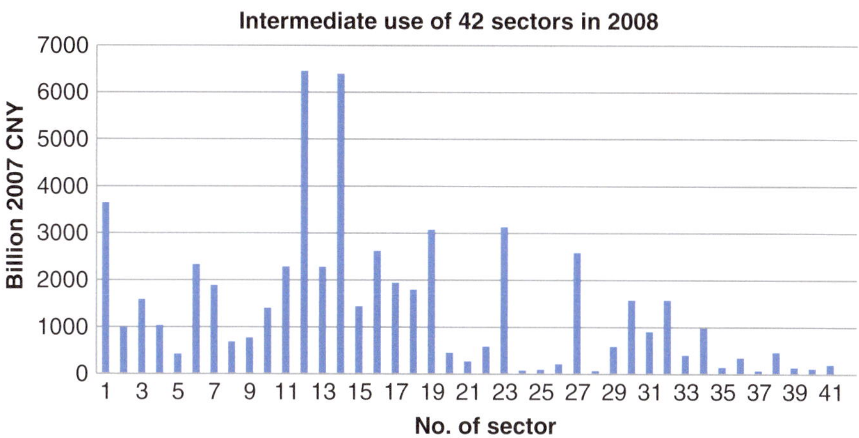

Fig. 8.17 Total intermediate use in the 42 sectors in 2008

Financial Intermediation (Sector No. 32), which grew by 10 %. The lowest value in tertiary industry was in Traffic, Transport and Storage (Sector No. 27), which grew by 6.2 %. In secondary industry, the highest growth was in Production and Distribution of Water (Sector No. 25), which grew by 6.51 %, and the lowest growth was in Production and Supply of Electric Power and Heat Power (Sector No. 23), with zero growth. The growth of primary industry was 6.75 %, which was better than that of secondary industry. Total labor payments grew by 6.1 % in 2008.

As the financial income of the Chinese government, Fig. 8.20 shows production taxes and the growth of the 42 sectors in 2008. We can see that the growth of production taxes in the 42 sectors in Fig. 8.20 is similar to that of labor payments in

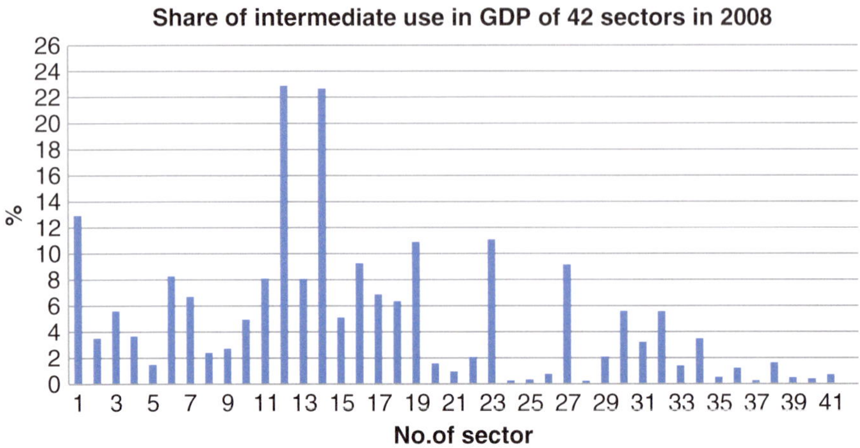

Fig. 8.18 Shares of intermediate use in gross domestic product (GDP) for the 42 sectors in 2008

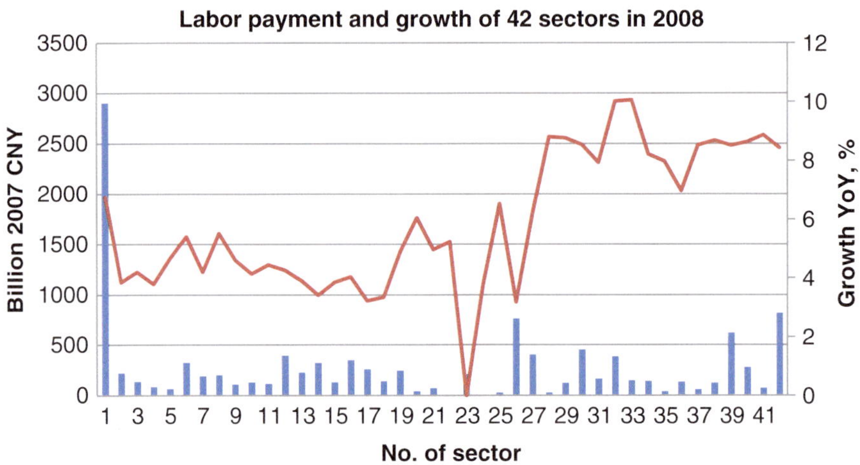

Fig. 8.19 Labor payments and growth in the 42 sectors in 2008

Fig. 8.19. However, total production taxes grew by 5.38 %, which was lower than the growth of total labor payments.

Profit can show the business benefits for firms and even sectors. Total profits are shown in Fig. 8.21, and the biggest profit is in Sector No. 32, Financial Intermediation, which was 905.67 billion CNY and grew by 10.01 % in 2008; it shared 10.65 % of total profits. The second biggest figure is in Wholesale and Retail Trades (Sector No. 30), which was 836.57 billion CNY and grew by 8.53 %; it shared 9.83 % of total profits. The third one is the Information Transmission, Computer Services and Software sector (No. 27), which profited by 713.43 billion CNY and

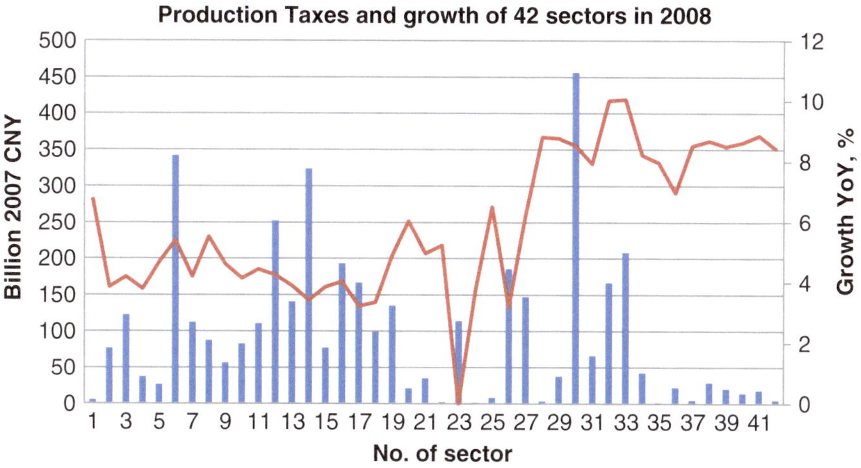

Fig. 8.20 Production taxes and growth in the 42 sectors in 2008

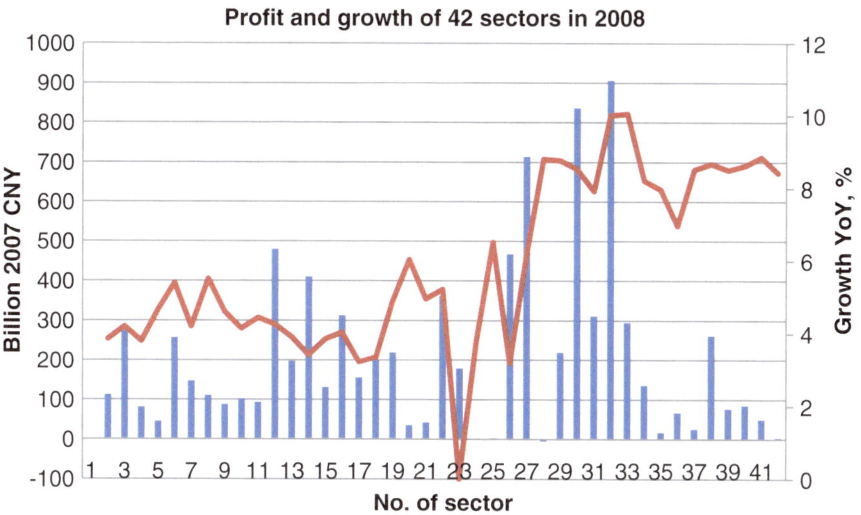

Fig. 8.21 Profits and growth in the 42 sectors in 2008

shared 8.39 %. All of these are in tertiary industry. For the three areas of industry, primary industry had zero profit, the profit in secondary industry was 4505.51 billion CNY, with a share of 52.96 %, and the profit in tertiary industry was 4002.48 billion CNY, with a share of 47.04 %.

Final consumption of the products from the 42 sectors by urban residents and rural farmers in 2008 can be seen in Fig. 8.22; this is similar to Fig. 8.8 in 2007.

With regard to exports and imports in the global market for the 42 sectors in 2008 (see Fig. 8.23), Sector No. 19 (Manufacture of Communication Equipment,

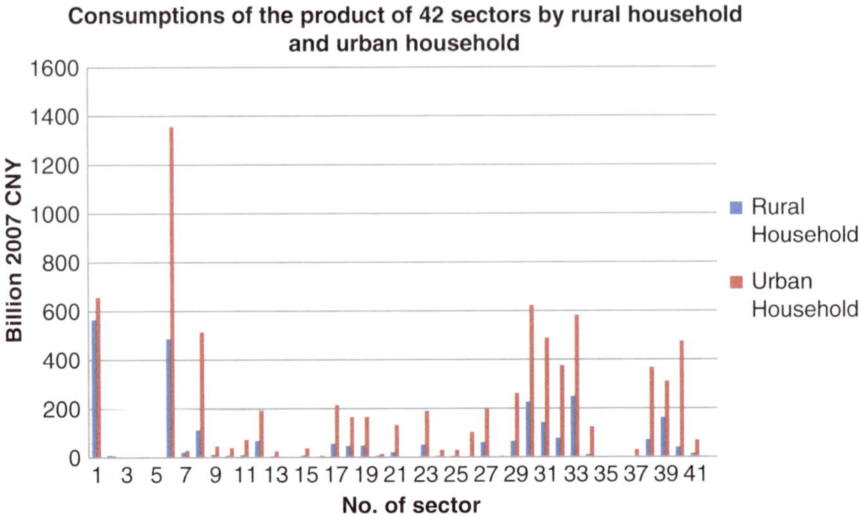

Fig. 8.22 Final consumption of products of the 42 sectors by rural households and urban households in 2008

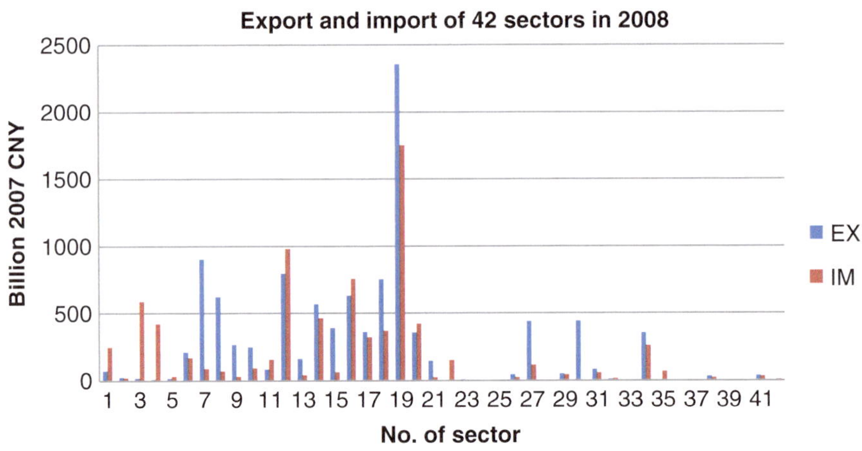

Fig. 8.23 Exports and imports in the 42 sectors in 2008

Computer and Other Electronic Equipment) is still the biggest sector in China; its exports totaled 2355.66 billion CNY and grew by 10.19 %, and its imports were 1752.42 billion CNY and grew by 7.52 %. The average export growth for the 42 sectors was around 10 %, and import growth was around 7.5 %, showing that competition in the global market is still strong in China.

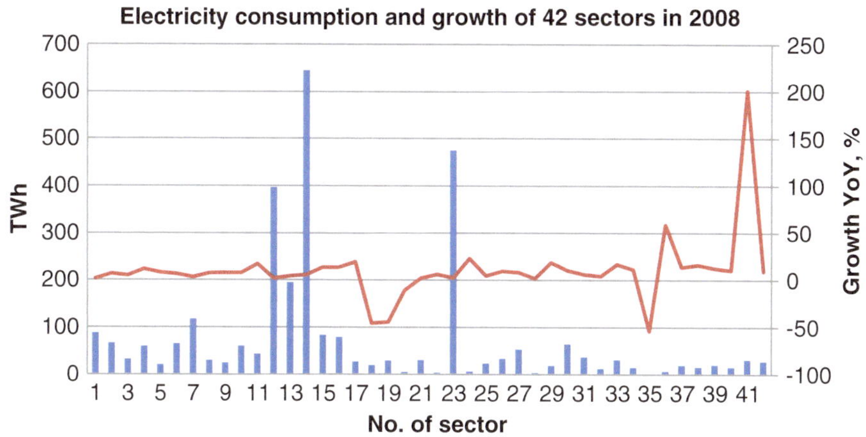

Fig. 8.24 Simulated electricity consumption and growth in the 42 sectors in 2008

8.3.3 *Calibrations*

In order to verify the accuracy and reliability of our agent-based simulation, we observe some key indicators in the results of our model and compare them with the official published data. In particular, we focus on electricity consumption and value added. Considering that GDP and value added in each industry are often utilized to reflect the situation of the national economy, we use them to test the results of our simulation. However, some scholars have cast doubts on the accuracy of them because they are based on artificial statistics. In contrast, electricity consumption data measured by electric meters is far more accurate and can also be used to depict economic conditions, according to the theory of electricity economics, as mentioned in Chap. 1. Therefore, we pay more attention to the comparison of the electricity consumption data.

Electricity consumption and growth rates in the 42 sectors simulated by the ARE model are shown in Fig. 8.24. The errors between the official published electricity consumption and our simulated electricity consumption in each sector are shown in Fig. 8.25 and Table 8.2. It can be seen that the relative errors in terms of electricity consumption are quite small, except in Sector No. 41. (Fig. 8.25). The biggest error is 190 % in Culture, Sports and Entertainment (Sector No. 41), for which simulated electricity use is 32.26 TWh and the published figure is 10.78 TWh; the second biggest error is −56.19 % in Sector No. 35 (Research and Experimental Development), for which simulated electricity use is 1.66 TWh and the published figure is 3.78 TWh; the third biggest error is 50.86 % in Sector No. 36 (Comprehensive Technical Service), for which simulated electricity use is 6.44 TWh and the published figure is 4.27 TWh; the fourth one is −49.49 % in Sector No. 18 (Manufacture of Electrical Machinery and Equipment), for which simulated electricity use is 19.8 TWh and the published figure is 39.22 TWh; the fifth one is −48.26 % in Sector No. 19 (Manufacture of Communication Equipment, Computer

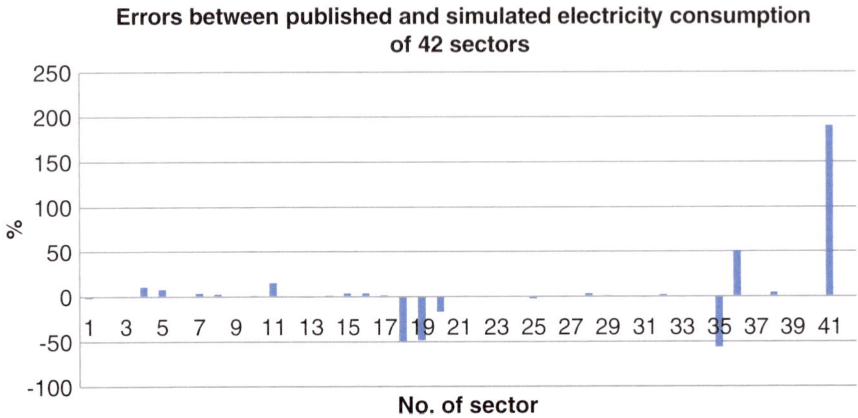

Fig. 8.25 Errors between published and simulated electricity consumption in the 42 sectors in 2008

Table 8.2 Errors between published and simulated electricity consumption in the 42 sectors in 2008

Sector no.	Relative error (%)	Sector no.	Relative error (%)	Sector no.	Relative error (%)
1	−1.67	15	4.02	29	1.54
2	0.52	16	4.02	30	−0.39
3	−0.11	17	1.72	31	−1.30
4	10.95	18	−49.49	32	2.36
5	8.21	19	−48.26	33	−1.26
6	0.66	20	−17.16	34	0.35
7	3.88	21	0.24	35	−56.19
8	2.86	22	0.70	36	50.86
9	1.10	23	0.19	37	−0.52
10	1.42	24	0.47	38	4.56
11	15.46	25	−2.35	39	−0.15
12	−0.40	26	−0.43	40	−0.64
13	−1.03	27	−0.35	41	190.03
14	1.33	28	3.66	42	−0.95

and Other Electronic Equipment), for which simulated electricity use is 29.57 TWh and the published figure is 57.16 TWh; the other errors are very small. The error in total electricity use is 9.17 %, with errors of 6.17 % in primary industry, 0.61 % in secondary industry, and 5.72 % in tertiary industry. Most sectors have errors with absolute values of less than 10 %. In 25 sectors, the errors are less than 2 %.

Errors in GDP and value added in the three areas of industry are shown in Table 8.3. The growth rates obtained from the simulation results of our model tend to be lower than the published ones. For example, on the basis of the *China*

Table 8.3 Errors between published and simulated value added data in 2008

	GDP growth (%)	Primary industry growth (%)	Secondary industry growth (%)	Tertiary industry growth (%)
Published	9.60	5.40	9.90	10.40
ARE simulated	5.97	6.75	3.86	8.51
Error	3.63	−1.35	6.04	1.89

ARE Agent Response Equilibrium, *GDP* gross domestic product

Statistical Yearbook 2007, published by the National Bureau of Statistics of China, the GDP growth rate was 9.60 % in 2008, while our simulation result was 5.97 %. The absolute error is 3.63 %. In our opinion, the officially published GDP and value added in each industry in China may not be as accurate as some people expect. The official data are probably higher than the actual values. The reasons are numerous and complicated, and are beyond the scope of our discussion in this book. According to the theory of electricity economics, the electricity consumption data are more accurate and reliable to indicate the situation in the national economy. Therefore, we think that the results are reasonable if the errors in terms of electricity consumption data are relatively small even though the simulated GDP and value added data may be lower than the official data, more or less.

8.4 Economic Simulation for 2009

8.4.1 Assumptions

The input parameter settings are the same as in the initial situation mentioned in Sect. 8.3, since it is a continuous simulation started from 2008. Besides, there is no evident change in the deposit reserve ratio, the benchmark lending rate, the benchmark deposit rate, and other monetary policies and fiscal policies in 2009.

8.4.2 Results

Figure 8.26 illustrates the price indexes for Sector Nos. 2, 3, 4, and 5 during the year 2009. As mentioned previously, Sector No. 2 is Mining and Washing of Coal, Sector No. 3 is Extraction of Petroleum and Natural Gas, Sector No. 4 is Mining and Dressing of Metal Ores, and Sector No. 5 is Mining and Dressing of Nonmetal and Other Ores. As shown in the figure, the overall trends in these four price indexes were rising, because demand for these energy and raw material products was raised. In particular, the price indexes for Sector Nos. 2 and 5 increased sharply to 1.4092 and 1.6440, respectively, reflecting the fact that the demand for coal and nonmetal

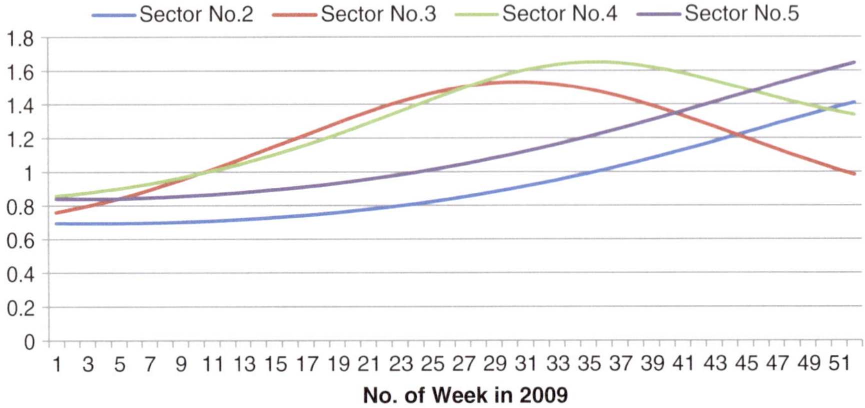

Fig. 8.26 Price indexes for Sector Nos. 2, 3, 4, and 5 in 2009

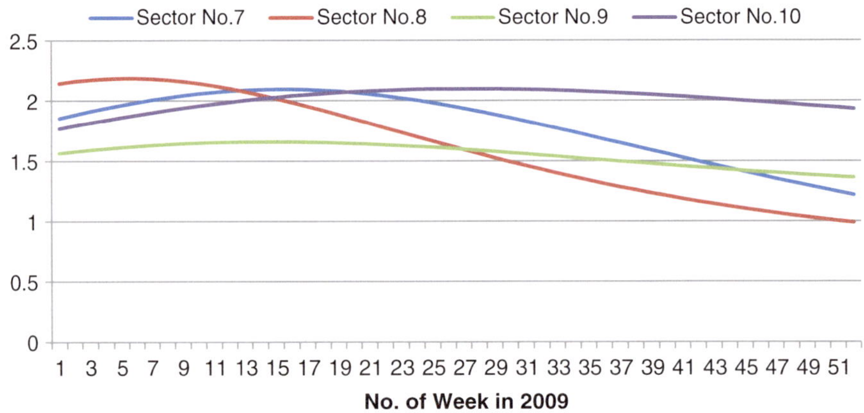

Fig. 8.27 Price indexes for Sector Nos. 7, 8, 9, and 10 in 2009

ores was particularly large. This is due to the huge investments made by the Chinese government, because the construction of various infrastructures, i.e. the main area of the government's huge investment plan, requires a great deal of energy and material.

Figure 8.27 illustrates the price indexes for Sector Nos. 7, 8, 9, and 10 during the year 2009. Sector No. 7 is Textile Industry; Sector No. 8 is Manufacture of Textile Garments, Footwear and Headgear & Manufacture of Leather, Down and Their Products; Sector No. 9 is Processing of Timber and Manufacture of Furniture; and Sector No. 10 is Papermaking, Printing and Manufacture of Articles for Culture, Education and Sport Activities. As shown in the figure, the price indexes for these four sectors started to decline in the second half of the year. There were two reasons for this situation. On the one hand, as typical representatives of light industry, they

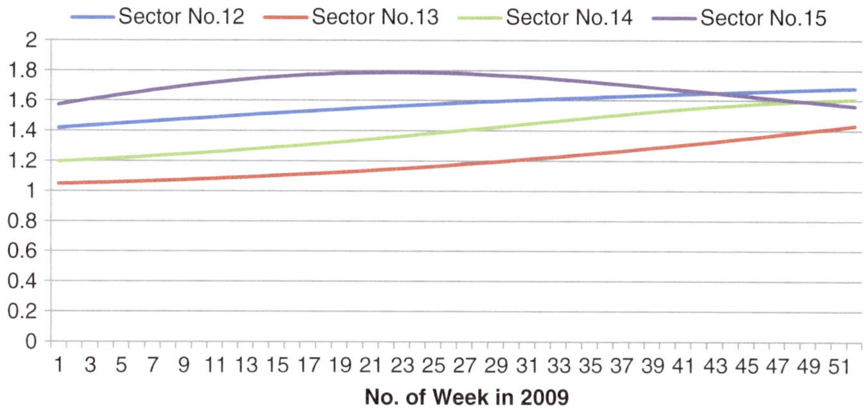

Fig. 8.28 Price indexes for Sector Nos. 12, 13, 14, and 15 in 2009

were not evidently promoted by the huge investment conducted by the Chinese government. On the other hand, the poor international economic environment led to a decline in exports, which posed serious threats to these sectors, especially Sector Nos. 7 and 8. Consequently, the price indexes for Sector Nos. 7 and 8 declined to 1.2171 and 0.9893, respectively.

Figure 8.28 depicts the price indexes for Sector Nos. 12, 13, 14, and 15 during the year 2009. Sector No. 12 is Chemical Industry; Sector No. 13 is Manufacture of Nonmetallic Mineral Products; Sector No. 14 is Smelting and Pressing of Metals; and Sector No. 15 is Manufacture of Metal Products. As illustrated in the figure, the overall trends of the price indexes for the four sectors were up. They reached 1.6801, 1.4305, 1.6053, and 1.5623, respectively, at the end of 2009. The main reason was the increasing demand for these basic industrial products as a result of the huge investment. Sector No. 13 was influenced, evidently, because cement, the product of Sector No. 13, is an essential material in construction. Sector No. 14 was also impacted, clearly, since steel and iron, the products of Sector No. 14, play nearly the same role in construction. The price index for Sector No. 12 kept increasing too.

Figure 8.29 shows the price indexes for Sector Nos. 30, 31, 32, and 33 during the year 2009. Sector No. 30 is Wholesale and Retail Trades; Sector No. 31 is Hotels and Catering Services; Sector No. 32 is Financial Intermediation; and Sector No. 33 is Real Estate. As depicted in the figure, the overall trends of the price indexes for these four sectors were up from the low level. Specifically, the price index for Sector No. 30 increased from 0.4635 to 1.3022, the price index for Sector No. 31 increased from 0.9559 to 1.5557, the price index for Sector No. 32 increased from 0.2773 to 1.1113, and the price index for Sector No. 33 increased from 0.6775 to 0.8029. This reflects the fact that residential daily life was getting better. The trough brought about by the crisis appeared in the first half of 2009.

The value added and its growth rate in each sector in 2009 are shown in Fig. 8.30. Among the 42 sectors, Sector No. 1 (Agriculture, Forestry, Animal Husbandry & Fishery) was No. 1 in terms of value added. It was as high as

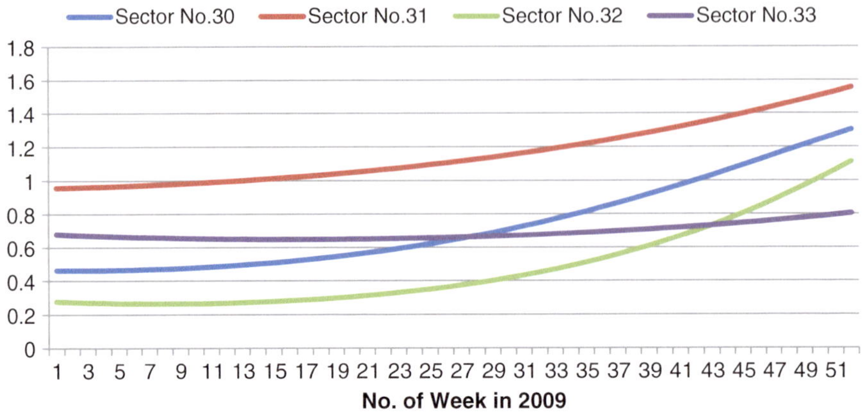

Fig. 8.29 Price indexes for Sector Nos. 30, 31, 32, and 33 in 2009

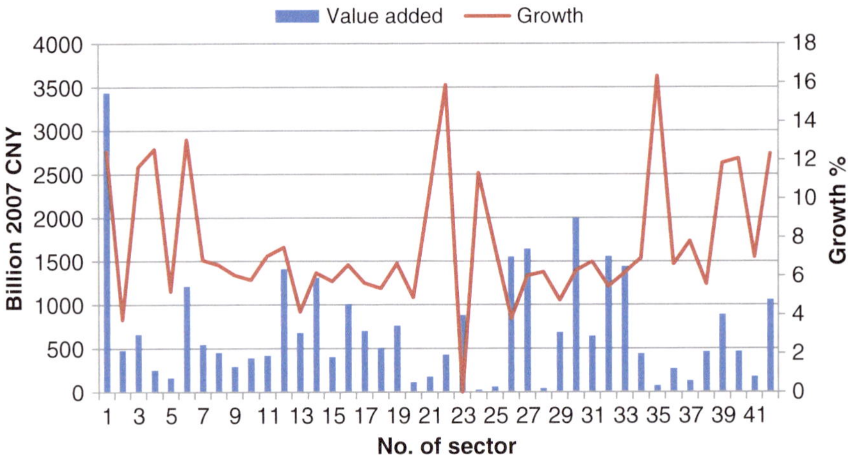

Fig. 8.30 Value added and growth in the 42 sectors in 2009

3438.43 billion CNY, which reflects the foundation status of agriculture in China's national economy. Sector No. 30 (Wholesale and Retail Trades) ranked second, with the value added equaling 1999.04 billion CNY. Then came Sector No. 27 (Traffic, Transport and Storage), Sector No. 32 (Financial Intermediation), and Sector No. 26 (Construction). Their value added figures were 1646.42 billion CNY, 1558.10 billion CNY, and 1554.30 billion CNY, respectively. In terms of the growth rate of the value added, the relative ranking of the 42 sectors was obviously different from that of the amount of value added. The first sector was Sector No. 35 (Research and Experimental Development), whose growth rate was 16.32 %, which reflects China's emphasis on science and technology. Sector

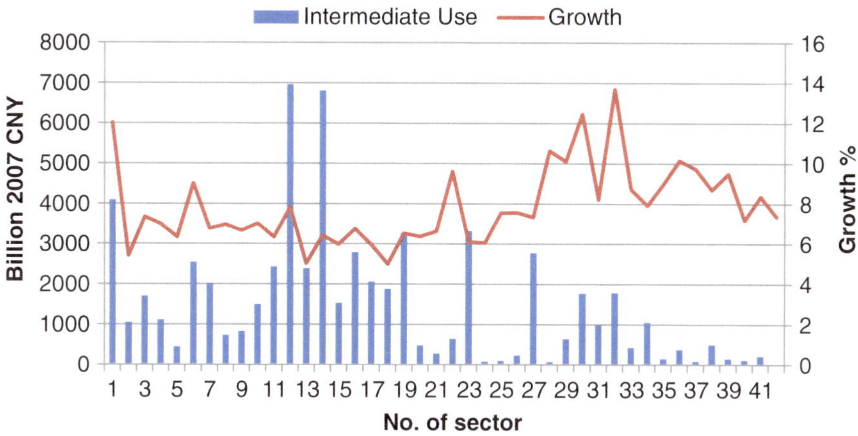

Fig. 8.31 Total intermediate use and growth in the 42 sectors in 2009

No. 22 (Scrap and Waste) came second, with a growth rate of 15.87 %, which reflects the development of the circular economy in China. The next three sectors were Sector No. 6 (Manufacture of Foods and Tobacco), Sector No. 4 (Mining of Metal Ores), and Sector No. 1 (Agriculture, Forestry, Animal Husbandry & Fishery). The growth rates in these sectors were 13.02 %, 12.53 %, and 12.39 %, respectively. They were all basic sectors in China's national economy.

The total intermediate use of products of each sector and its growth rate in 2009 are depicted in Fig. 8.31. In terms of the amount of intermediate use, the top five sectors were Sector No. 12 (Chemical Industry), Sector No. 14 (Smelting and Rolling of Metals), Sector No. 1 (Agriculture, Forestry, Animal Husbandry & Fishery), Sector No. 23 (Production and Supply of Electric Power and Heat Power), and Sector No. 19 (Manufacture of Communication Equipment, Computer and Other Electronic Equipment) in proper order, which demonstrates the fundamental role of the products of these sectors, such as cement, steel and electricity, in the infrastructure construction driven by the huge investment. According to the growth rate of intermediate use, the top five sectors were successively Sector No. 32 (Financial Intermediation), Sector No. 30 (Wholesale and Retail Trades), Sector No. 1 (Agriculture, Forestry, Animal Husbandry & Fishery), Sector No. 28 (Post Service), and Sector No. 36 (Comprehensive Technical Services). Four of them belong to tertiary industry, which reflects the economic restructure in China.

Labor payments and their growth in each sector in 2009 are shown in Fig. 8.32. According to the growth rate, the top five sectors were Sector No. 35 (Research and Experimental Development), Sector No. 22 (Scrap and Waste), Sector No. 6 (Manufacture of Foods and Tobacco), Sector No. 4 (Mining of Metal Ores), and Sector No. 1 (Agriculture, Forestry, Animal Husbandry & Fishery). The growth rates in these sectors were 16.32 %, 15.87 %, 13.02 %, 12.53 %, and 12.39 %, respectively.

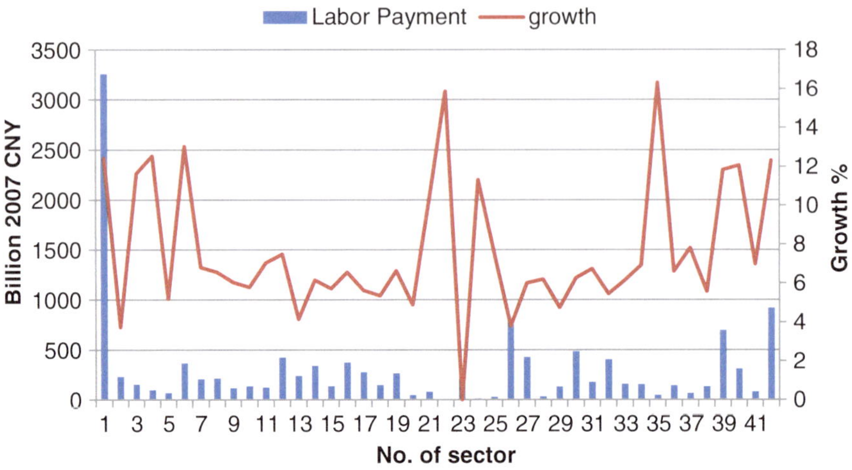

Fig. 8.32 Labor payments and growth in the 42 sectors in 2009

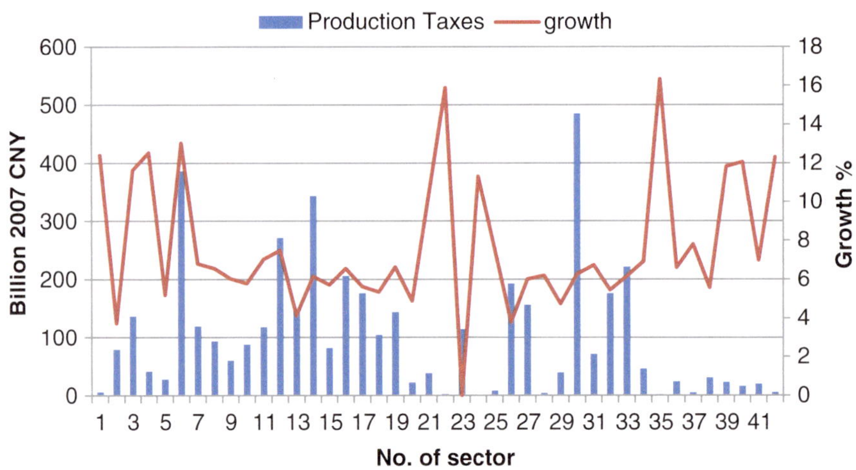

Fig. 8.33 Production taxes and growth in the 42 sectors in 2009

Production taxes and the growth rate in each sector in 2009 are depicted in Fig. 8.33. In terms of the growth rate, the top five sectors were Sector No. 35 (Research and Experimental Development), Sector No. 22 (Scrap and Waste), Sector No. 6 (Manufacture of Foods and Tobacco), Sector No. 4 (Mining of Metal Ores) and Sector No. 1 (Agriculture, Forestry, Animal Husbandry & Fishery). The growth rates in these sectors were 16.32 %, 15.87 %, 13.02 %, 12.53 %, and 12.39 %, respectively.

The profit of each sector and its growth rate in 2009 are illustrated in Fig. 8.34. The top five sectors were Sector No. 35 (Research and Experimental Development),

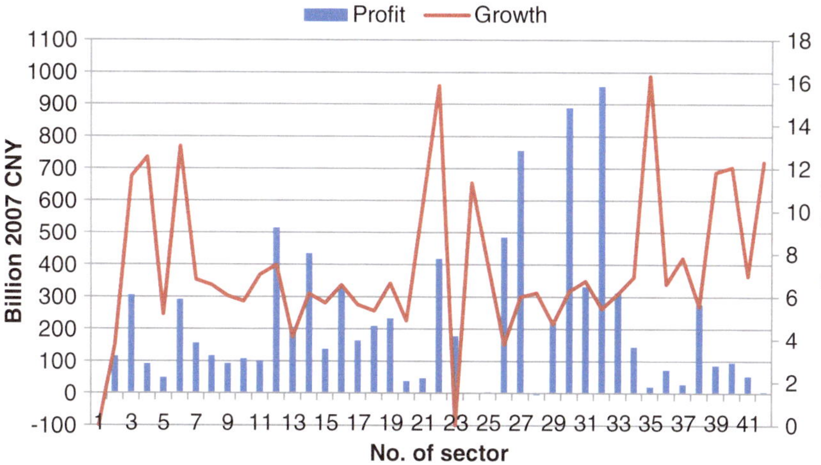

Fig. 8.34 Profits and growth in the 42 sectors in 2009

Sector No. 22 (Scrap and Waste), Sector No. 6 (Manufacture of Foods and Tobacco), Sector No. 4 (Mining of Metal Ores), and Sector No. 3 (Extraction of Petroleum and Natural Gas), according to the growth rates of their profits, which were 16.32 %, 15.87 %, 13.02 %, 12.53 %, and 11.64 %, respectively. This reflects the fact that companies in areas of high-tech and the recycling economy tended to gain more profit, which was consistent with the policy guidance for national development in China.

Rural household consumption and urban household consumption of the products of each sector in 2009 are shown in Fig. 8.35. The consumption structure was nearly the same as that in 2008, as depicted in Fig. 8.22. However, the overall consumption amount increased by 9.22 %, which reflects the fact that residential daily life was improved with the increment in GDP.

Exports and imports of the products of each sector in 2009 are shown in Fig. 8.36. The structure was similar to that in 2008, as depicted in Fig. 8.23. The total amounts of exports and imports had a slight increment. The growth rate of overall exports was almost 0.00 %, and the growth rate of overall imports was 1.15 %, which verifies the negative influences of the recession on the economic situation around the world.

8.4.3 *Calibrations*

To verify the accuracy and reliability of our model in terms of the simulation results in 2009, we proceed to observe the electricity consumption of each sector and value added in each industry. The simulated data is compared with the officially published data. As mentioned before, we pay more attention to electricity

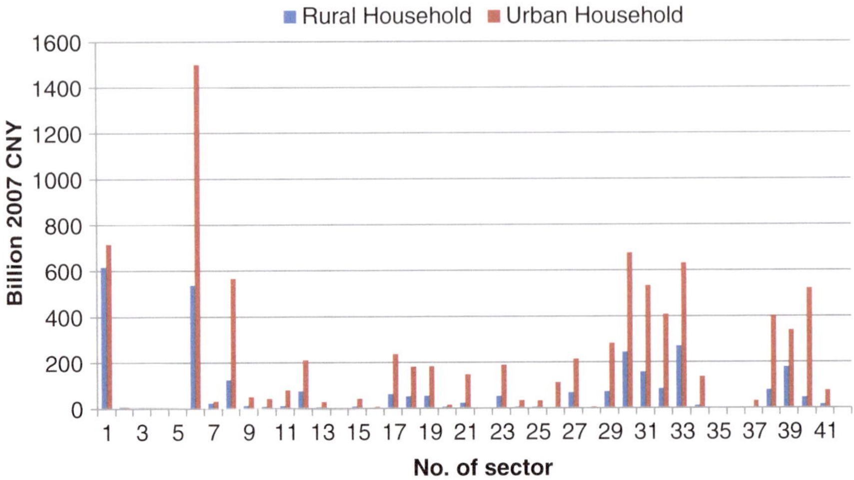

Fig. 8.35 Final consumption of products of the 42 sectors by rural households and urban households in 2009

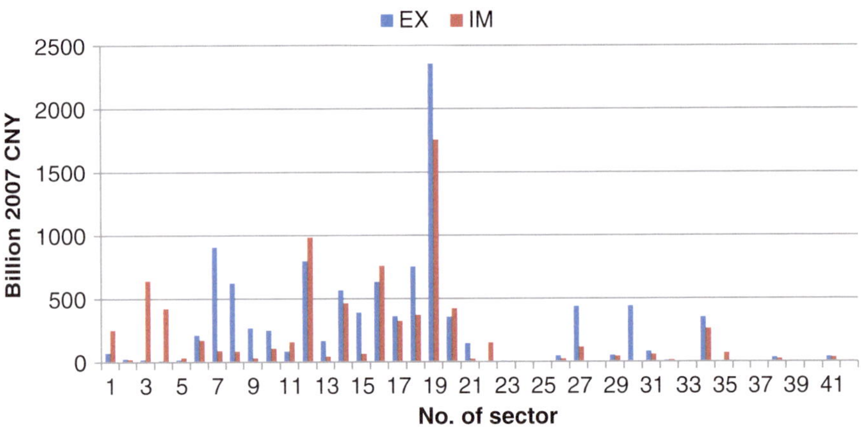

Fig. 8.36 Exports and imports in the 42 sectors in 2009

consumption data, considering its reliable source. Electricity consumption and its growth rate in each sector are shown in Fig. 8.37. The relative errors between the official published electricity consumption and our simulated electricity consumption in each sector are shown in Fig. 8.38 and Table 8.4. It is clear that the relative errors in 2009 are very small, except in Sector No. 41. The five sectors whose errors are larger than 10 % are Manufacture of Electrical Machinery and Equipment (Sector No. 18), Manufacture of Communication Equipment, Computer and Other Electronic Equipment (Sector No. 19), Research and Experimental Development (Sector No. 35), Comprehensive Technical Services (Sector No. 36), and

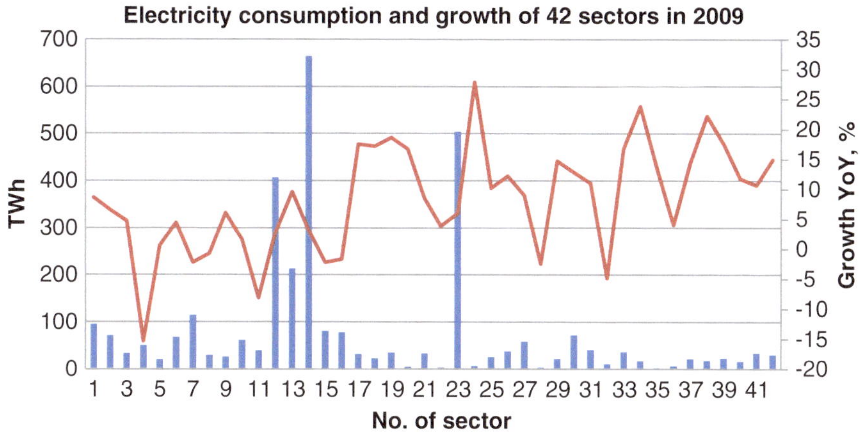

Fig. 8.37 Simulated electricity consumption and growth in the 42 sectors in 2009

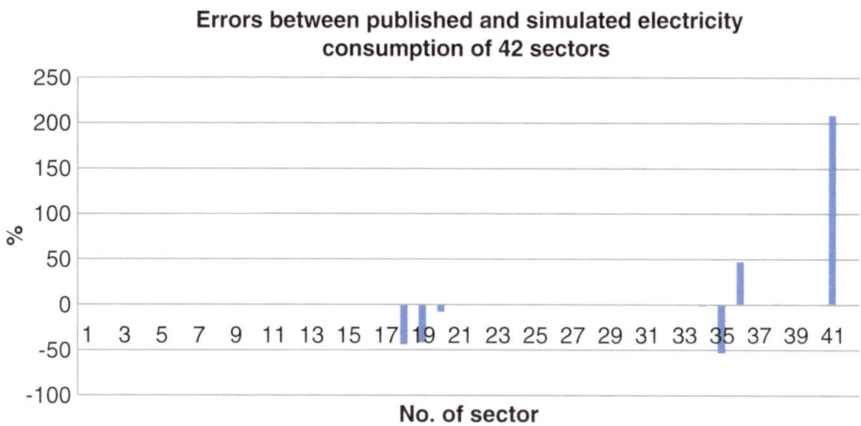

Fig. 8.38 Errors between published and simulated electricity consumption in the 42 sectors in 2009

Culture, Sports and Entertainment (Sector No. 41). Most sectors have errors that are very close to 0 %. If we ignore the error in Sector No. 41, the average error is 4.81 %. All in all, the results of our simulation should be reliable and reasonable through the comparison of electricity consumption data.

The errors between the published GDP and value added in each industry and the simulated ones obtained from our model are listed in Table 8.5. Evidently, the errors are small, even though they are not as ideal as those of the electricity consumption data—which is similar to the situation in 2008. Moreover, it shares the same condition with 2008 that the simulated GDP and value added tend to be lower than the officially published ones. For example, the published growth rate of

Table 8.4 Errors between published and simulated electricity consumption in the 42 sectors in 2009

Sector number	Relative error	Sector number	Relative error	Sector number	Relative error
1	0.00	15	0.00	29	0.00
2	0.00	16	0.00	30	0.00
3	0.00	17	0.00	31	0.00
4	0.00	18	−43.75	32	−0.01
5	0.00	19	−41.68	33	0.00
6	0.00	20	−8.15	34	−1.43
7	0.00	21	0.00	35	−53.36
8	0.00	22	0.03	36	47.20
9	0.00	23	0.00	37	0.00
10	0.00	24	0.01	38	1.40
11	0.00	25	0.00	39	0.00
12	0.00	26	0.00	40	0.00
13	0.00	27	0.00	41	208.61
14	0.00	28	0.03	42	0.00

Table 8.5 Errors between published and simulated value added data in 2009

	GDP growth	Primary industry growth	Secondary industry growth	Tertiary industry growth
Published	9.21	4.18	9.94	9.56
ARE simulated	7.52	12.39	6.66	7.25
Error	1.69	8.21	3.28	2.31

ARE Agent Response Equilibrium, *GDP* gross domestic product

GDP is 9.21 %, while our simulated growth rate is 7.52 %. But the absolute error is 1.69 %, which is a small number.

8.5 Economic Simulation for 2010

8.5.1 Assumptions

The input of the ARE model mostly remains the same as the original level. Only some parameters are reset to reflect the changes that happened in 2010. In the fourth week, the deposit reserve ratio is raised by 0.5 percentage point. In the 9th week, the deposit reserve ratio is raised by 0.5 percentage point. In the 20th week, the deposit reserve ratio is raised by 0.5 percentage point. In the 43rd week, both the

Table 8.6 Monetary policies in 2010

Date (year/month/day)	Changed rate: Deposit reserve ratio	Changed rate: Deposit rate	Changed rate: Lending rate	Final rate: Deposit reserve ratio	Final rate: Deposit rate	Final rate: Lending rate
2010/01/18	Up 0.5 %	–	–	16.00	2.25	5.31
2010/02/25	Up 0.5 %	–	–	16.50	2.25	5.31
2010/05/10	Up 0.5 %	–	–	17.00	2.25	5.31
2010/10/20	–	Up 0.25 %	Up 0.25 %	17.00	2.50	5.56
2010/11/16	Up 0.5 %	–	–	17.50	2.50	5.56
2010/11/29	Up 0.5 %	–	–	18.00	2.50	5.56
2010/12/20	Up 0.5 %	–	–	18.50	2.50	5.56
2010/12/26	–	Up 0.25 %	Up 0.25 %	18.50	2.75	5.81

benchmark lending rate and the benchmark deposit rate are raised by 0.25 percentage point. In the 47th week, the deposit reserve ratio is raised by 0.5 percentage point. In the 49th week, the deposit reserve ratio is raised by 0.5 percentage point. In the 52nd week, the deposit reserve ratio is raised by 0.5 percentage point, while both the benchmark lending rate and the benchmark deposit rate are raised by 0.25 percentage point (see Table 8.6).

8.5.2 Results

Figure 8.39 illustrates the price indexes for Sector Nos. 7, 8, 9, and 10 during the year 2010. Sector No. 7 is Textile Industry; Sector No. 8 is Manufacture of Textile Garments, Footwear and Headgear & Manufacture of Leather, Down and Their Products; Sector No. 9 is Processing of Timber and Manufacture of Furniture; and Sector No. 10 is Papermaking, Printing and Manufacture of Articles for Culture, Education and Sport Activities. As shown in the figure, the overall trends of the price indexes for the four sectors were down. Compared with the situation in 2009, the decreasing amplitude was less. The price indexes for Sector Nos. 8 and 9 remained approximately stable. The price indexes for Sector Nos. 7 and 10 declined slightly to 0.4971 and 1.6639, respectively. The negative influences of the crisis were faint, because the most terrible period had passed by. The economic situations of some foreign countries were even recovering. Consequently, exports of the products of these sectors did not worsen as sharply as in 2009.

Figure 8.40 shows the price indexes for Sector Nos. 12, 13, 14, and 15 during the year 2010. Sector No. 12 is Chemical Industry; Sector No. 13 is Manufacture of Nonmetallic Mineral Products; Sector No. 14 is Smelting and Pressing of Metals; and Sector No. 15 is Manufacture of Metal Products. As illustrated in the figure, the price indexes for these four sectors remained relatively high at 1.3–1.7, which

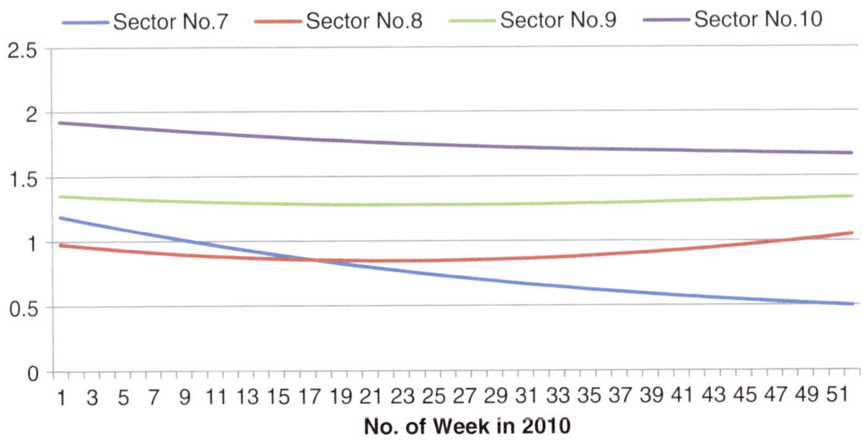

Fig. 8.39 Price indexes for Sector Nos. 7, 8, 9, and 10 in 2010

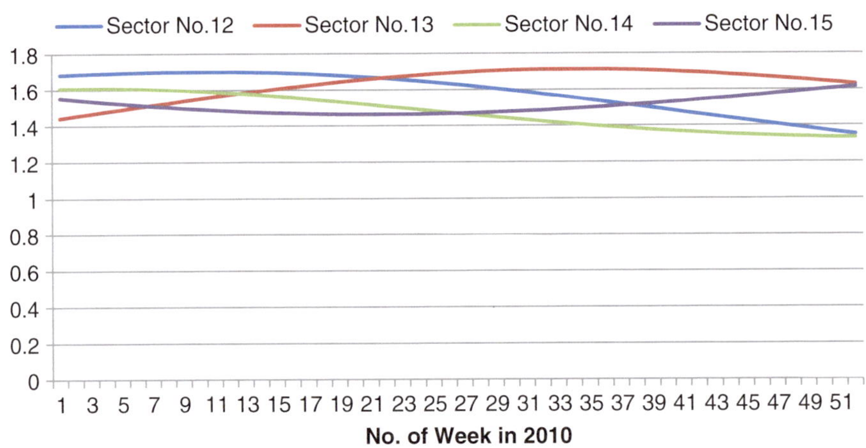

Fig. 8.40 Price indexes for Sector Nos. 12, 13, 14, and 15 in 2010

reflects the effects of the huge investment that still existed. The demand for cement, steel, and other products of these sectors were promoted. However, the trends were different from the situation in 2009, in which the price indexes for the four sectors kept growing. The reason is that the effects of the huge investment had arrived at a peak and a saturated level.

Figure 8.41 depicts the price indexes for Sector Nos. 38, 39, 40, and 41 during the year 2010. Sector No. 38 is Resident Services and Other Services; Sector No. 39 is Education; Sector No. 40 is Health, Social Security and Social Welfare; and Sector No. 41 is Culture, Sports and Entertainment. As illustrated in the figure, the price indexes for Sector Nos. 38 and 41 were up, while the price indexes for

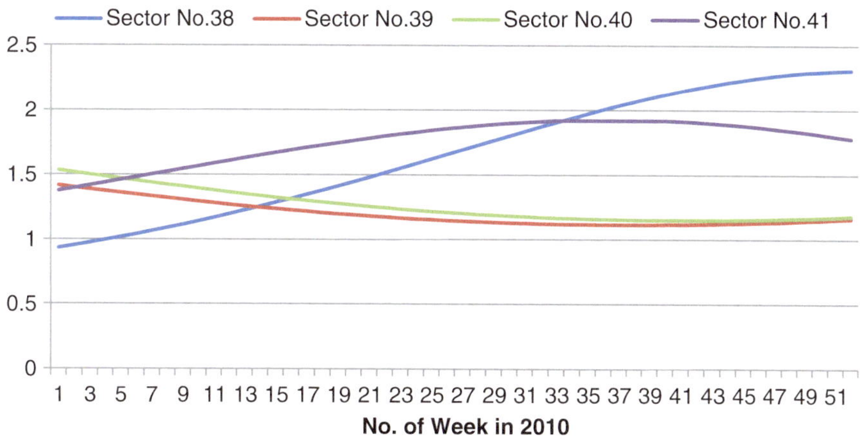

Fig. 8.41 Price indexes for Sector Nos. 38, 39, 40, and 41 in 2010

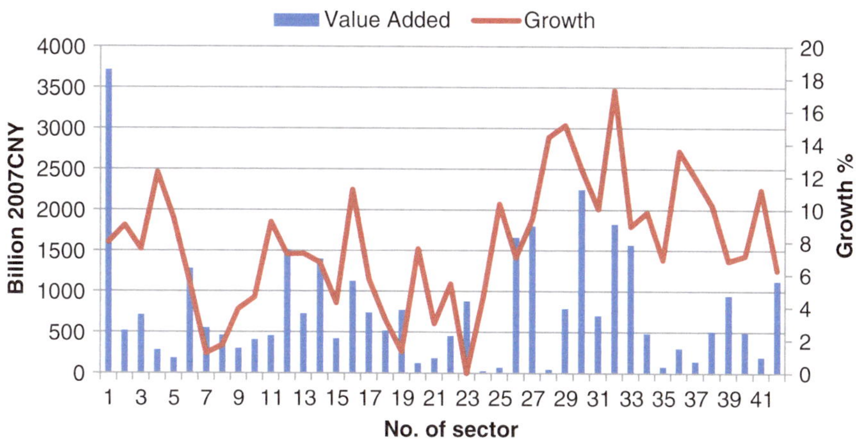

Fig. 8.42 Value added and growth in the 42 sectors in 2010

Sector Nos. 39 and 40 were relatively stable in 2010. The price index for Sector No. 38 increased sharply from 0.9335 to 2.3114, which reflects the fact that the Chinese government had taken measures to adjust the economic structure by attaching importance to services. This is a meaningful attempt because a key point of the Chinese national economy lies in economic restructuring instead of huge investment. Meanwhile, the increase in the price index for Sector No. 41 from 1.3750 to 1.7794 reflects the fact that the Chinese government paid more attention to the Culture sector to create more possible ways to develop the economy and enhance the quality of life of the residents of China.

The value added and its growth rate in each sector in 2010 are shown in Fig. 8.42. It is clear that Sector No. 1 (Agriculture, Forestry, Animal Husbandry

& Fishery) retained the position of No. 1 in terms of value added. It is as high as 3714.40 billion CNY, which reflects the fact that promotion of the development of agriculture has always been the focus of the Chinese government. Similar to the previous situation, Sector No. 30 (Wholesale and Retail Trades) ranked second, with the value added equaling 1999.04 billion CNY. Then came Sector No. 32 (Financial Intermediation), Sector No. 27 (Traffic, Transport and Storage), and Sector No. 26 (Construction). The value added figures in these sectors were 1827.87 billion CNY, 1802.13 billion CNY, and 1664.19 billion CNY, respectively. According to the growth rate of the value added, the first sector was Sector No. 32 (Financial Intermediation), with a growth rate of 17.31 %, which illustrates the development of finance under the policies of the Chinese government in the crisis. Sector No. 29 (Information Transmission, Computer Services and Software) came second, with a growth rate equaling 15.17 %, which reflects the increasingly important role of information technology in modern society. The next three sectors were Sector No. 28 (Post Service), Sector No. 36 (Comprehensive Technical Services), and Sector No. 30 (Wholesale and Retail Trades). The growth rates in these sectors were 14.44 %, 13.58 %, and 12.49 %, respectively. It should be noted that the top five sectors in terms of the growth rate of value added all belonged to tertiary industry, resulting from the decision on economic structural adjustment made by the Chinese government.

Total intermediate use of the products of each sector and its growth rate in 2010 are shown in Fig. 8.31. According to the amount of intermediate use, the top five sectors were successively Sector No. 12 (Chemical Industry), Sector No. 14 (Smelting and Rolling of Metals), Sector No. 1 (Agriculture, Forestry, Animal Husbandry & Fishery), Sector No. 23 (Production and Supply of Electric Power and Heat Power), and Sector No. 19 (Manufacture of Communication Equipment, Computer and Other Electronic Equipment). It was the same as the situation in 2009. When it comes to the growth rate of intermediate use, the top five sectors were successively Sector No. 33 (Real Estate), Sector No. 22 (Scrap and Waste), Sector No. 3 (Extraction of Petroleum and Natural Gas), Sector No. 34 (Leasing and Business Services), and Sector No. 42 (Public Management and Social Organization). The growth rates in these sectors were 10.24 %, 9.51 %, 9.21 %, 9.18 %, and 8.82 %, respectively (Fig. 8.43).

Labor payments and their growth in each sector in 2010 are depicted in Fig. 8.44. According to the growth rate, the top five sectors were Sector No. 32 (Financial Intermediation), Sector No. 29 (Information Transmission, Computer Services and Software), Sector No. 28 (Post Service), Sector No. 36 (Comprehensive Technical Services) and Sector No. 4 (Mining of Metal Ores). The growth rates in these sectors were 17.31 %, 15.17 %, 14.44 %, 13.58 %, and 12.27 %, respectively. This reflects the fact that people who work in the fields of finance and information technology tended to have larger salary increments, which is consistent with our knowledge in real life.

The production taxes and related growth of each sector in 2010 are shown in Fig. 8.45. In terms of the growth rate, the top five sectors were Sector No. 32 (Financial Intermediation), Sector No. 29 (Information Transmission,

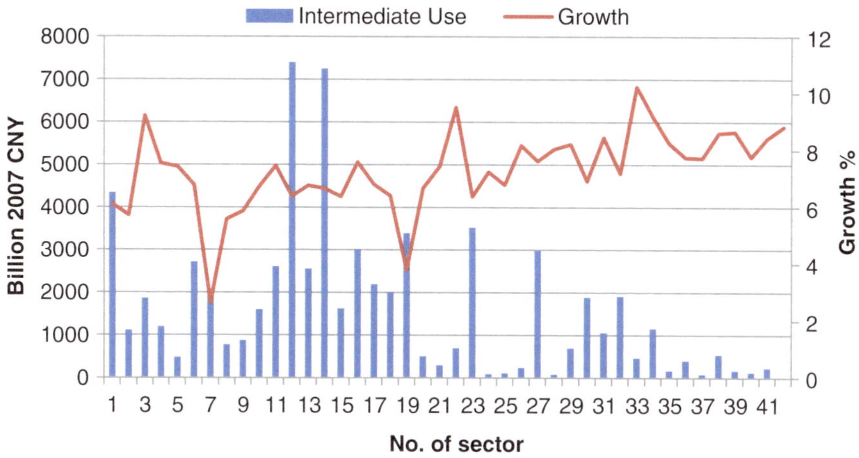

Fig. 8.43 Total intermediate use in the 42 sectors in 2010

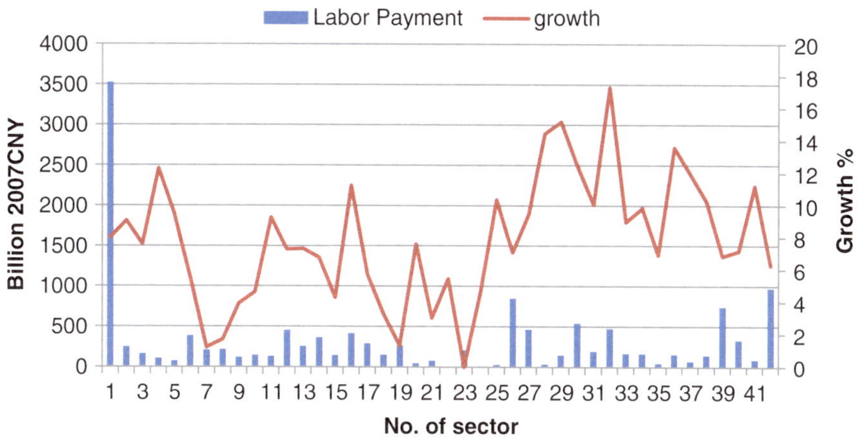

Fig. 8.44 Labor payments and growth in the 42 sectors in 2010

Computer Services and Software), Sector No. 28 (Post Service), Sector No. 36 (Comprehensive Technical Services), and Sector No. 4 (Mining of Metal Ores). The growth rates in these sectors were 17.31 %, 15.17 %, 14.44 %, 13.58 %, and 12.27 %, respectively.

The profit of each sector and its growth rate in 2010 are illustrated in Fig. 8.46. The top five sectors were Sector No. 32 (Financial Intermediation), Sector No. 29 (Information Transmission, Computer Services and Software), Sector No. 28 (Post Service), Sector No. 36 (Comprehensive Technical Services), and Sector No. 4 (Mining of Metal Ores), according to the growth rates of their profits, which were 17.31 %, 15.17 %, 14.44 %, 13.58 %, and 12.27 %, respectively.

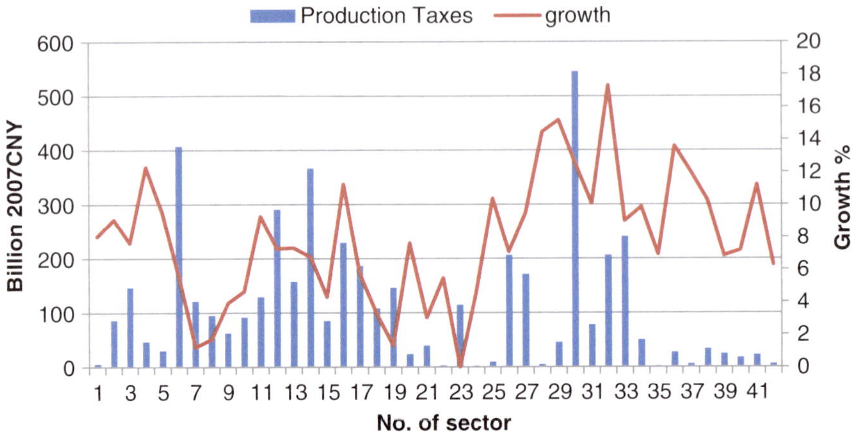

Fig. 8.45 Production taxes and growth in the 42 sectors in 2010

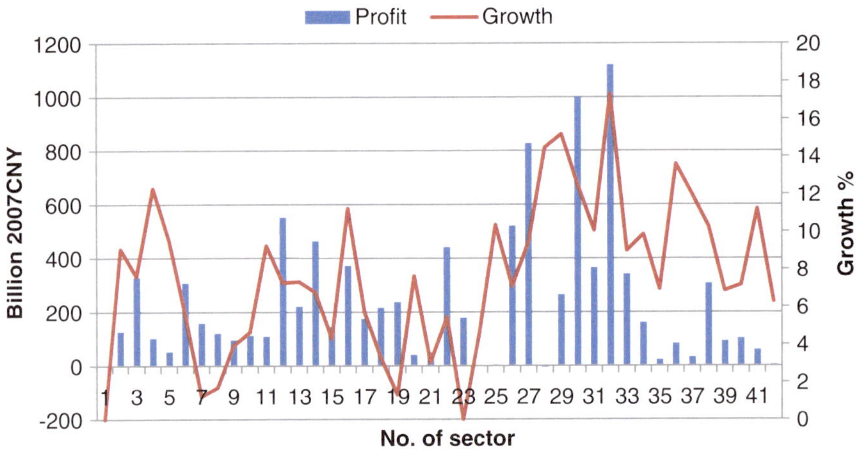

Fig. 8.46 Profits and growth in the 42 sectors in 2010

Consumption of the products of each sector by rural households and urban households in 2010 is depicted in Fig. 8.47. The consumption structure was similar to that in 2009, as shown in Fig. 8.35. However, the overall consumption amount increased by 18.37 %, which demonstrates that domestic demand was stimulated by the Chinese government.

Exports and imports of the products of each sector in 2010 are shown in Fig. 8.48. The structure was similar to that in 2009, as illustrated in Fig. 8.36. The total amount of exports and imports maintained a very slight increment in 2010. The growth rate of overall exports was nearly 0.00 %, and the growth rate of overall imports was 1.83 %, which reflects the fact that the negative influences of

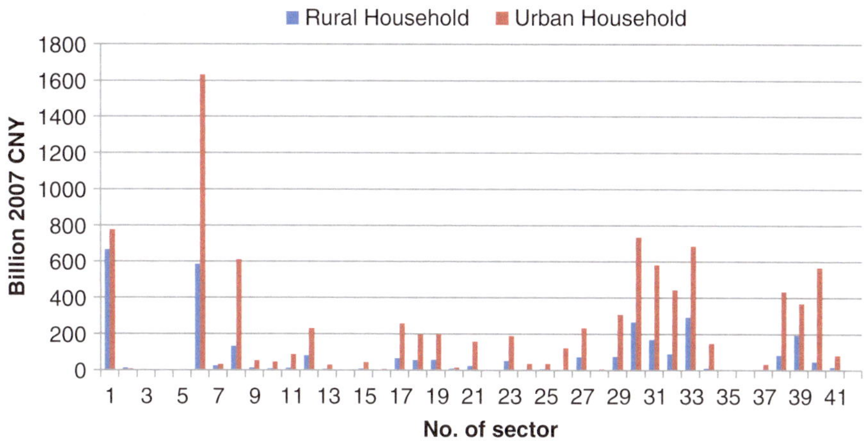

Fig. 8.47 Final consumption of products of the 42 sectors by rural households and urban households in 2010

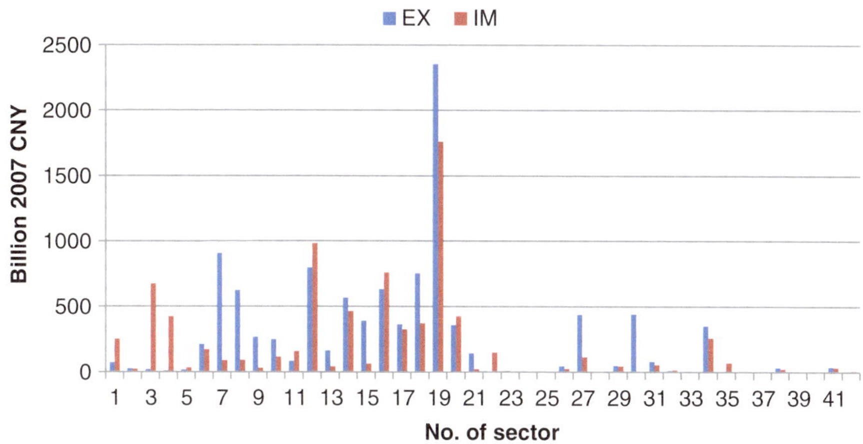

Fig. 8.48 Exports and imports in the 42 sectors in 2010

the recession on the world economy continued, but the situation recovered a little bit, because the growth rate in 2010 was higher than that in 2009, although it was still at a low level.

8.5.3 Calibrations

In order to test the accuracy and reliability of our simulation results in terms of the situation in 2010, the electricity consumption in each sector and value added in each

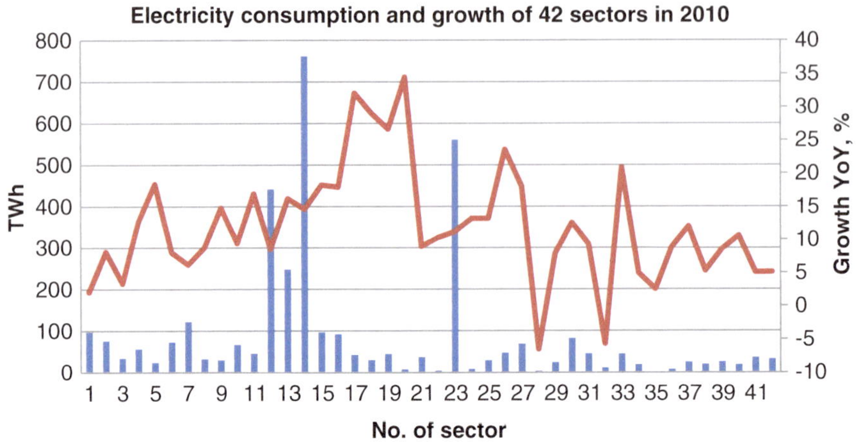

Fig. 8.49 Simulated electricity consumption and growth in the 42 sectors in 2010

industry are chosen as the key indicators again. As in Sects. 8.3.3 and 8.4.3, we compare the simulated data with the officially published data, and we pay more attention to the electricity consumption data, according to our theory of electricity economics. The electricity consumption and its growth rate in each sector are illustrated in Fig. 8.49. The relative errors between the official published electricity consumption and our simulated electricity consumption in each sector are calculated and shown in Fig. 8.50 and Table 8.7. Clearly, the relative errors in 2010 are small, except in Sector No. 41. Most sectors have errors that are less than 5 %. Only several sectors have errors that are beyond the acceptable range. They are Manufacture of Electrical Machinery and Equipment (Sector No. 18), Manufacture of Communication Equipment, Computer and Other Electronic Equipment (Sector No. 19), Research and Experimental Development (Sector No. 35), Comprehensive Technical Services (Sector No. 36), and Culture, Sports and Entertainment (Sector No. 41). All of these sectors have errors that are larger than 30 %. Therefore, we may draw the conclusion that the simulation results of our model are accurate and reliable through the comparison of electricity consumption data.

The errors in terms of GDP and value added in each industry are listed in Table 8.8. It is evident that the situation in 2010 has two features that are similar to those in 2008 and 2009. On the one hand, these errors are greater than those in terms of the electricity consumption data. On the other hand, the growth rate in the GDP and value added in each industry are lower than the officially published data. We think there may be some problems in the statistical processes used to generate the official GDP and value added data. Since the errors are within the acceptable range, we think our simulation is reliable through the comparison.

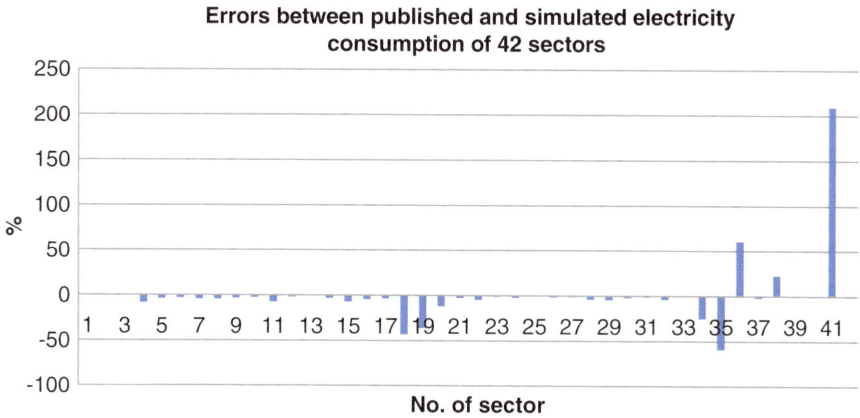

Fig. 8.50 Errors between published and simulated electricity consumption in the 42 sectors in 2010

Table 8.7 Errors between published and simulated electricity consumption in the 42 sectors in 2010

Sector number	Relative error (%)	Sector number	Relative error (%)	Sector number	Relative error (%)
1	0.69	15	7.27	29	4.74
2	1.08	16	4.74	30	2.03
3	0.61	17	3.77	31	1.55
4	8.48	18	43.25	32	3.95
5	3.90	19	36.26	33	0.02
6	3.46	20	12.06	34	25.19
7	4.59	21	3.10	35	59.19
8	4.72	22	4.98	36	60.48
9	3.50	23	1.48	37	2.05
10	2.88	24	2.59	38	22.87
11	7.57	25	0.00	39	1.18
12	1.96	26	1.94	40	1.11
13	0.90	27	1.55	41	209.25
14	3.29	28	3.89	42	1.11

Table 8.8 Errors between published and simulated value added data in 2010

	GDP growth (%)	Primary industry growth (%)	Secondary industry growth (%)	Tertiary industry growth (%)
Published	10.45	4.27	12.25	9.75
ARE simulated	8.13	8.03	5.92	10.91
Error	2.32	3.76	6.33	1.16

ARE Agent Response Equilibrium, *GDP* gross domestic product

8.6 Update of Input–Output Table

The foundation of the ARE model is the I/O table introduced in detail in previous chapters. Our program is designed on the basis of the I/O table and I/O analysis. Each iteration of the code actually means a change for all national economic elements, which can be reflected by updating the I/O table. At the end of every year, it is feasible to obtain the I/O table for the year. Since the compilation of an I/O table is based on a huge number of materials, it requires a lot of time and energy. The official I/O table tends to be published every several years, and there tends to be a 3-year delay in publication of an I/O table for a certain year. Therefore, the novel ARE model proposed by us has its advantage in updating the I/O table easily and in a timely fashion, in addition to its meaningful functions in policy simulation and analysis.

In the simulation of China's national economy from 2008 to 2010, the starting point is the official 2007 I/O table published by the National Bureau of Statistics of China. By virtue of our model, we have obtained yearly I/O tables for 2008, 2009, and 2010. Since the National Bureau of Statistics of China did not publish I/O tables for 2008 and 2009, we focus on the update of the 2010 I/O table. By means of comparison of the updated I/O table through our program simulation and the officially published I/O table, the feasibility and reliability of updating the I/O table by our model can be verified.

The updated 2010 national I/O table for China is shown in the appendix of this book. The official 2010 national I/O table for China can be found in the publicly available materials from the National Bureau of Statistics of China. To save space, we only present the comparison between the two versions of the I/O table.

It should be noted that the prices in our simulation are the same as those at the starting point, the level of 2007, because inflation and price changes are not embodied in our model. The updated 2010 I/O table created by our program is at the price level of 2007, while the officially published 2010 I/O table is at the price level of 2010. As a result, our updated I/O table cannot be compared directly with the official one. To cope with this issue, we focus on the comparison of the direct consumption coefficients of each sector, which have nothing to do with inflation and the price level.

Another fact that should not be ignored is that the official 2007 I/O table—the starting point of our simulation—has 42 sectors, which means the results of our simulation are all expressed in terms of the 42 sectors. However, two sectors are combined in the official 2010 I/O table, making the number of sectors change from 42 to 41. The three relevant sectors are Sector No. 20 (Manufacture of Measuring Instrument and Machinary for Cultural Activity & Office work), Sector No. 21 (Manufacture of Artwork and 720 Other Manufacture) and Sector No. 22 (Scrap and Waste). They are such small sectors that they have very low status and limited impacts on the national economy of China. Consequently, they can be ignored in our analysis. By comparing the direct consumption coefficients of our simulated I/O table and the officially published I/O table for 2010, the relative

Table 8.9 Average relative error of each sector between published input–output table and updated input–output table

Sector no.	Relative error	Sector no.	Relative error
1	0.1510	23	0.1946
2	0.2581	24	0.2159
3	0.2376	25	0.1496
4	0.1717	26	0.1840
5	0.2014	27	0.2296
6	0.1425	28	1.9170
7	0.1616	29	0.2164
8	0.1506	30	0.4387
9	0.1479	31	0.3862
10	0.1506	32	0.2148
11	0.3180	33	0.2818
12	0.1397	34	0.2824
13	0.2192	35	0.2141
14	0.1890	36	0.1702
15	0.1388	37	0.3011
16	0.1407	38	0.2261
17	0.1302	39	0.8791
18	0.1508	40	0.4950
19	0.2037	41	0.3604

errors of all of the elements in the I/O table can be calculated. The averages of the relative errors in the 42 sectors, except for Sector Nos. 20, 21 and 22, are listed in Table 8.9.

As shown in the table, the average relative errors of most sectors are small. The overall average relative error of all of the sectors is 0.2806, which is within the acceptable scope. Unlike other sectors, Sector No. 28 (Post Service) has a relatively high error. But it is an unimportant sector with a very limited influence on the national economy and is beyond the scope of the analysis and research in this book. If the relative error in Sector No. 28 is disregarded, the overall average relative error is 0.2386. It may be safe to draw the conclusion that this novel approach to update the national I/O table by ARE simulation is feasible and reliable.

Chapter 9
China's Economic Simulations in 2011–2014 by Agent Response Equilibrium Model

Economic system is very complex with a lot of uncertainties, and it is very difficult to be simulated and modeled mathematically. However, it can be simplified to a few sectors of the productions and economic activities. Business decision making in the economic activities can be simulated by an intelligent agent based on some rules such as maximum profits of the productions and economic activities. This is the methodology of ARE model. Based on the economic simulations of China's economy from 2011 to 2014, the simulation results in each year are shown and discussed, and the errors between simulated results by ARE model and published data by National Bureau of Statistics of China have been compared in this chapter.

ARE model can be used to study China's economy from sector level to national economy level dynamically. As one of the simulation results by ARE model, the price indexes of the products of 42 sectors weekly are useful information to analyze the trends of the productions. Another useful result by ARE model is the dynamical input–output table year by year in the simulation period; it can show the national economy completely and also reveal the production process of the sectors in the economy. Through the analysis of the simulation results and errors in this chapter, it is safe to draw the conclusion that ARE model has the ability to simulate China's national economy feasibly and accurately.

9.1 Summary of Evaluation Rules on Economic Simulation of ARE Model

There are 42 sectors selected in input–output table in Chinese economy. As shown in Table 3.1 in Chap. 3, Sector 1 as Agriculture, Forestry, Animal Husbandry, & Fishery is the primary industry, Sectors 2–26 can be classified as secondary industry, and Sectors 27–42 can be put as tertiary industry in China.

Z. Hu et al., *China's Economic Gene Mutations*,
DOI 10.1007/978-3-662-47298-9_9

As mentioned in Chap. 8, the evaluated rules can be listed as follows:

Rule 1: Comparing some data errors between published input–output table and simulated input–output table such as GDP and the growth and value added and the growths for primary industry, secondary industry, and tertiary industry. In this chapter, the simulation by ARE is based on the simulated input–output table 2010 provided in Chap. 7. And some data on fiscal policy such as tax rates for different sectors and government expenditures and on monetary policy such as reserve requirement ratio, discount rate, and open-market operations are collected on the published data. Then, we can get the input–output table for 2011, 2012, 2013, and 2014 resulting from ARE model.

Rule 2: Comparing the errors between published electricity consumptions and simulated electricity consumptions by ARE model for 42 sectors, and also for three industries, and the total industrial electricity consumption in China. In this chapter, the published electricity use and resulted electricity use for 42 sectors in three industries and the total industrial electricity consumption in 2011, 2012, 2013, and 2014 by ARE model will be compared.

9.2 Economic Simulation for 2011 by ARE Model

9.2.1 Fiscal and Monetary Policies

As the input data of ARE model, input–output table of 2010 and some data on fiscal policy such as tax rates of productions of 42 sectors, value-added tax rate, consumption tax rate, income tax rate, etc., and monitory policy data such as reserve requirement ratio, deposit and lending rate, and open-market operations are necessary information for the simulation. The published input–output table of 2010 as shown in the appendix has been used as the basic data for ARE model. The data of fiscal and monetary policies are set according to the actual situation in 2011.

Fiscal policy data has been shown in Fig. 9.1. It is calculated in average rates. For instance, a custom duty of 1.21 % is the average rate of thousands of imported goods and an income tax of 2.68 % is the total income tax over the total income in China. It looks like the tax rates are very low in China. The reason is that the tax rates in our program don't have the same meanings as those in practice, since the tax bases of all the different taxes are diverse and difficult to be reflected in the ARE simulation. As illustrated in Sect. 7.3.1, the rate of a certain kind of tax is redefined as the total amount of this kind of tax divided by the overall tax and surplus in the starting year of the simulation. Besides, there are some other fees and payments which firms and business people have to pay in China, and they are not put into the taxes in our program. For instance, the land use payment is the part of money that people have to pay to local governments for 50 years of usage of the land for manufacturing and business activities, and it may even be 70 years in case of real perspective. The cost of the land is quite different between the eastern area and western area in China. Maybe the fees and payments take around 80 % of the total taxes and fees.

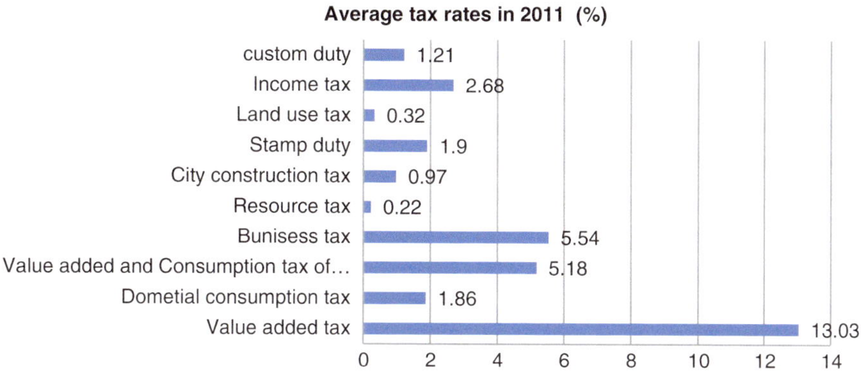

Fig. 9.1 Average tax rates based on real taxes of 42 sectors in 2011

Monetary policy has been shown in Table 9.1. Reserve requirement rate was 18.5 %, deposit rate was 2.75 %, and lending rate was 5.81 % at the end of 2010. It shows that the reserve requirement rate was up 0.5 % reaching 19 % on Jan. 20, 2011, and deposit rate and lending rate were up 0.25 % reaching 3 % and 6.06 %, respectively, on Feb. 9, 2011. There were ten times when monetary policy was adjusted in 2011 as shown in Table 9.1. The central bank took a little more tighten monetary policies in order to avoid serious inflation in the national economy.

9.2.2 Simulation Results

Product price is the key element in the production. The business people will adjust the production plan based on the market price of their product. In the simulation of Chinese economy in 2011 by ARE model, it is supposed that the product prices of the 42 sectors in the last week of 2010 are 1, which is the initial price of each product of the 42 sectors, and the prices vary every week based on the initial price of 1. The market prices of the products from Mining and Washing of Coal (Sector 2), Extraction of Petroleum and Natural Gas (Sector 3), Mining of Metal Ores (Sector 4), and Mining and Processing of Nonmetal Ores and Other Ores (Sector 5) varied weekly in 2011 as shown in Fig. 9.2. It can be seen that the product price of Mining of Metal Ores declined fast after the 19th week to 0.4242 at the end of the year, the prices of products from Sectors 2 and 5 were declined slightly, while for the product price of Sector 3, it was down to 0.5276 in the 30th week and then up to 0.7656 in the last week of 2011.

Figure 9.3 depicts the price indexes of Sector 11, Sector 12, Sector 13, Sector 14, Sector 15, and Sector 16 during the year of 2011. As mentioned previously, Sector 11 is Processing of Petroleum, Coking, and Processing of Nuclear Fuel; Sector 12 is Chemical Industry; Sector 13 is Manufacture of Nonmetallic Mineral

Table 9.1 Changes of reserve requirement ratio and interest rates in 2011

Date	Adjusted rate			Final rate		
	Reserve requirement ratio	Deposit rate	Lending rate	Reserve requirement ratio (%)	Deposit rate (%)	Lending rate (%)
By the end of 2010	–	–	–	18.50	2.75	5.81
20/01/2011	Up 0.5 %	–	–	19.00	2.75	5.81
9/09/2011	–	Up 0.25 %	Up 0.25 %	19.00	3.00	6.06
24/02/2011	Up 0.5 %	–	–	19.50	3.00	6.06
25/03/2011	Up 0.5 %	–	–	20.00	3.00	6.06
6/04/2011	–	Up 0.25 %	Up 0.25 %	20.00	3.25	6.31
21/04/2011	Up 0.5 %	–	–	20.50	3.25	6.31
18/05/2011	Up 0.5 %	–	–	21.00	3.25	6.31
20/06/2011	Up 0.5 %	–	–	21.50	3.25	6.31
7/07/2011	–	Up 0.25 %	Up 0.25 %	21.50	3.50	6.56
5/12/2011	Down 0.5 %	–	–	21.00	3.50	6.56

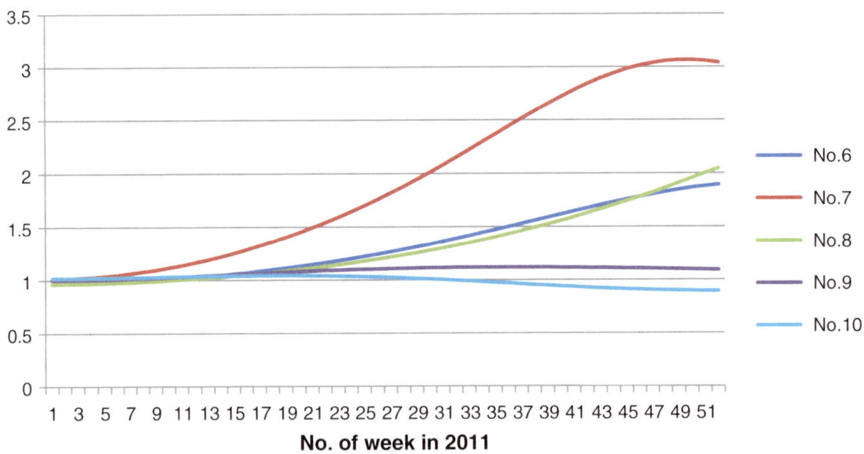

Fig. 9.2 Market price indexes of Sector 6, Sector 7, Sector 8, Sector 9, and Sector 10 in 2011

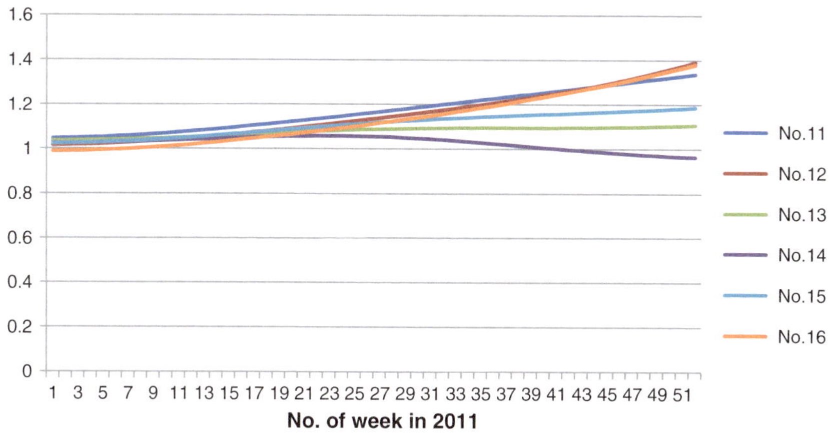

Fig. 9.3 Market price indexes of Sector 11, Sector 12, Sector 13, Sector 14, Sector 15, and Sector 16 in 2011

Products; Sector 14 is Smelting and Rolling of Metals; Sector 15 is Manufacture of Metal Products; and Sector 16 is Manufacture of General-Purpose and Special Purpose Machinery. As shown in the figure, the overall trends of most price indexes were up. The initial levels of these price indexes were all around 1. Their final statuses arrived at 1.3364, 1.3899, 1.1065, 0.9636, and 1.1853, respectively. This was due to the increasing demand for the products of these sectors, such as cement and steel, which verifies the real economic rebound in China.

Figure 9.4 displays the price indexes of Sector 17, Sector 18, Sector 19, and Sector 20 during the year of 2011. Sector 17 is Manufacture of Transport Equipment; Sector 18 is Manufacture of Electrical Machinery and Equipment; Sector 19 is Manufacture of Communication Equipment, Computer, and Other Electronic Equipments; and Sector 20 is Manufacture of Measuring Instrument and Machinery for Cultural Activity & Office Work. Sector 17, Sector 18, and Sector 19 are all important manufacturing. Their price indexes reached 2.0078, 1.4979, and 1.8675, respectively, after continuous increase in 2011, which reflects the recovery of Chinese economy. The price index of Sector 17 had the largest increase, due to the development of transportation such as urban rail transit in China.

Figure 9.5 shows the price indexes of Sector 38, Sector 39, Sector 40, and Sector 41 during the year of 2011. As mentioned in previous sections, Sector 38 is Services to Households and Other Services; Sector 39 is Education; Sector 40 is Health, Social Security, and Social Welfare; and Sector 41 is Culture, Sports, and Entertainment. It is clear that overall trends of their price indexes were up except the price of Sector 38. Their initial statuses were nearly the same around 1, while their final values were 0.8967, 2.1361, 2.3934, and 1.2398, respectively. This verifies the residents in China have increasingly high demand for qualified education, health, and culture services.

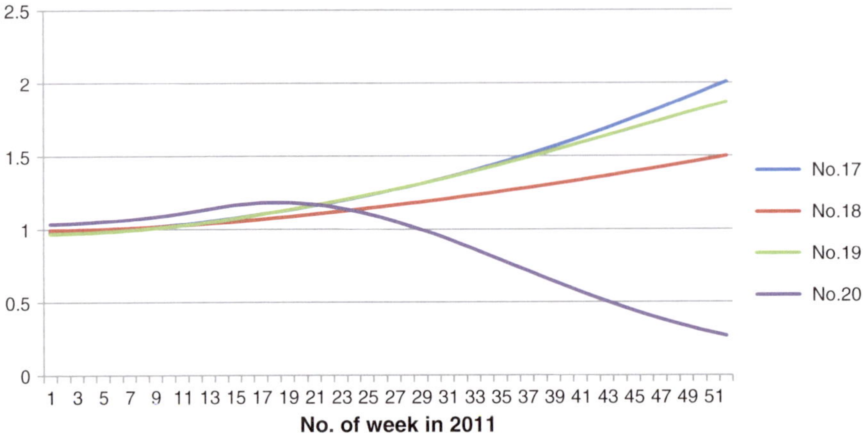

Fig. 9.4 Market price indexes of Sector 17, Sector 18, Sector 19, and Sector 20 in 2011

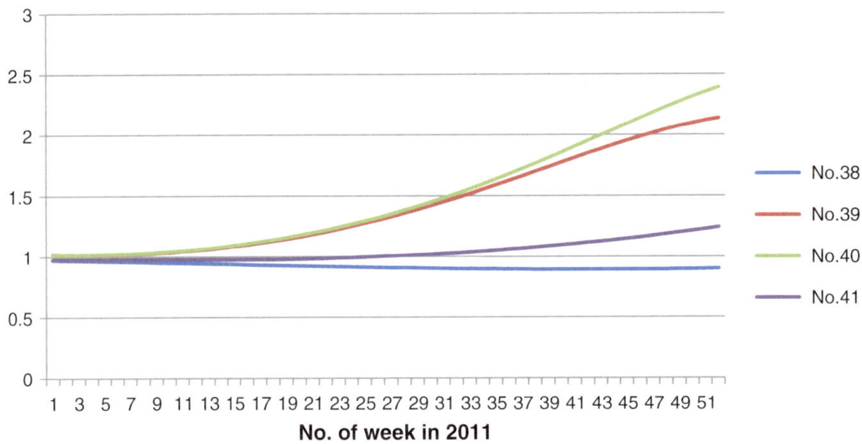

Fig. 9.5 Market price indexes of Sector 38, Sector 39, Sector 40, and Sector 41 in 2011

The value added and its growth rate of each sector in 2011 are shown in Fig. 9.6. Among the 42 sectors, Sector 1 (Agriculture, Forestry, Animal Husbandry, & Fishery) was no. 1 according to value added. It is as high as 4336.28 billion CNY, which reflects the foundation status of the agriculture in China's national economy. Sector 30 (Wholesale and Retail Trades) ranked second with the value added equaling 3269.79 billion CNY. Then came Sector 26 (Construction), Sector 33 (Real Estate), and Sector 32 (Financial Intermediation). Their value added was 2738.21 billion CNY, 2428.06 billion CNY, and 2254.05 billion CNY, respectively. In terms of the growth rate of the value added, the relative ranking of the 42 sectors was obviously different from that of the amount of value added. It seems that Sector 21 had the largest growth rate. This is because this officially published

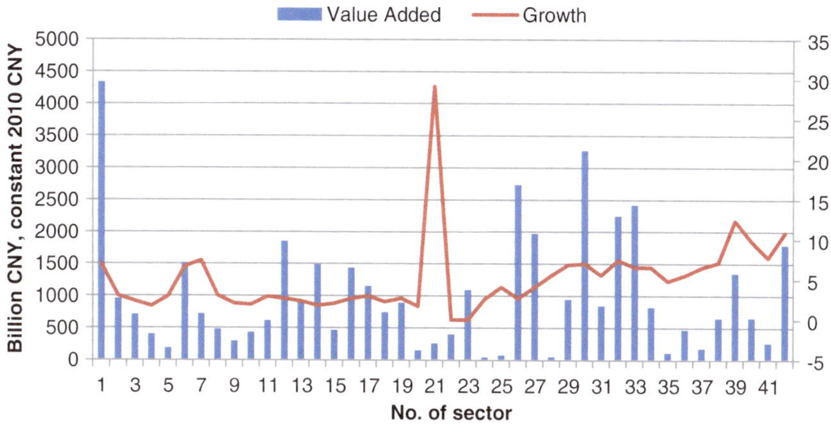

Fig. 9.6 Value added and growth of 42 sectors in 2011

2010 input–output table combined Sector 21 and Sector 22 together, but our program was still on the basis of the 42 sectors similar to that during the period of 2008–2010. Therefore, the simulated result of 2011 input–output table and the officially published 2010 input–output table cannot be compared reasonably for Sector 21 and Sector 22, which means we can just ignore Sector 21 and Sector 22 in Section 8.2.2. Consequently, the first sector in terms of growth rate was Sector 39 (Education) whose growth rate was 12.41 %, which reflects the China's emphasis on education. Sector 42 (Public Management and Social Organization) came second with the growth rate equaling 10.84 %. The next three sectors were Sector 40 (Health, Social Security, and Social Welfare), Sector 41 (Culture, Sports, and Entertainment), and Sector 32 (Financial Intermediation). Their growth rate was 9.82 %, 7.77 %, and 7.43 %, respectively. They all belong to tertiary industry, which reflects the structural readjustment in China's economy.

The overall taxes of each sector and the growth rate in 2011 are depicted in Fig. 9.7. In terms of the amount of taxes, the top 5 sectors were Sector 30 (Wholesale and Retail Trades), Sector 6 (Manufacture of Foods and Tobacco), Sector 26 (Construction), Sector 33 (Real Estate), and Sector 12 (Chemical Industry) in proper order. The taxes of these sectors were 890.84 billion CNY, 497.81 billion CNY, 433.09 billion CNY, 382.44 billion CNY, and 364.41 billion CNY, respectively. These sectors were the main contributor to the revenue in China.

The profit of each sector and its growth rate in 2011 are illustrated in Fig. 9.8. The top 5 sectors were Sector 30 (Wholesale and Retail Trades), Sector 32 (Financial Intermediation), Sector 27 (Traffic, Transport, and Storage), Sector 33 (Real Estate), and Sector 26 (Construction) according to the amount of profit. Their profit was 1272.12 billion CNY, 1211.07 billion CNY, 640.82 billion CNY, 608.94 billion CNY, and 594.09 billion CNY, respectively.

The total intermediate use of the product of each sector and the growth rate in 2011 are depicted in Fig. 9.9. According to the amount of intermediate use, the top

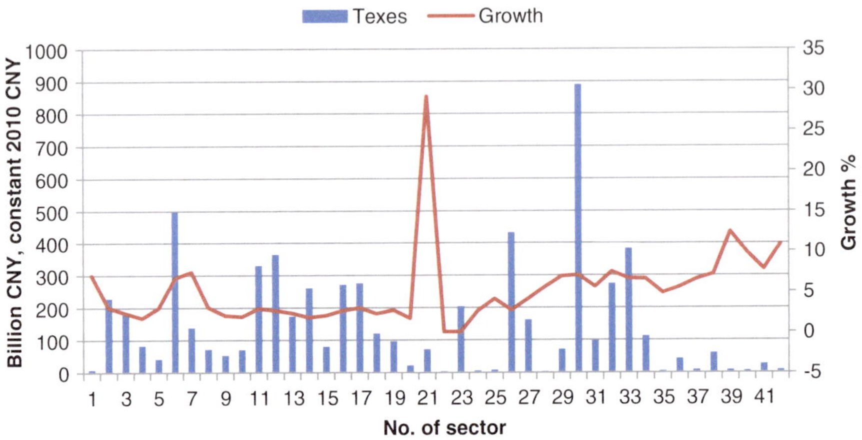

Fig. 9.7 Taxes and growth of 42 sectors in 2011

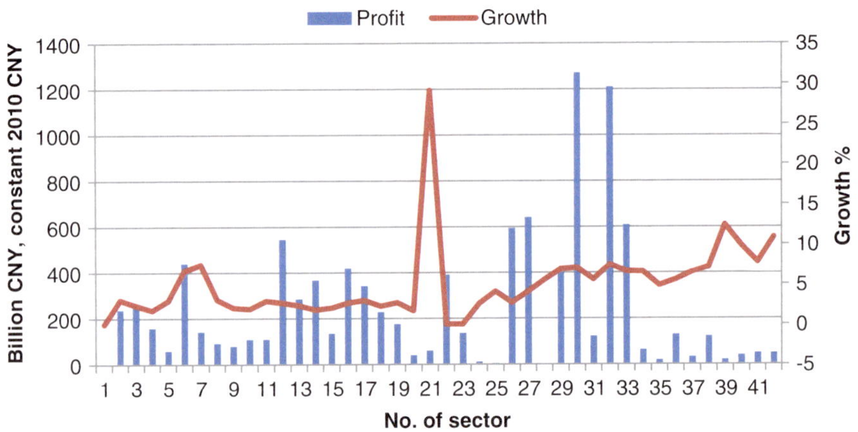

Fig. 9.8 Profit and growth of 42 sectors in 2011

5 sectors were Sector 12 (Chemical Industry), Sector 14 (Smelting and Rolling of Metals), Sector 1 (Agriculture, Forestry, Animal Husbandry, & Fishery), Sector 27 (Traffic, Transport, and Storage), and Sector 23 (Production and Supply of Electric Power and Heat Power) in proper order. The intermediate use of the product of these sectors was 9648.50 billion CNY, 8684.08 billion CNY, 5798.32 billion CNY, 4496.44 billion CNY, and 4359.70 billion CNY, respectively. These sectors were of fundamental significance in the national economy in China.

The total output of each sector and its growth rate in 2011 are shown in Fig. 9.10. The top 5 sectors were Sector 26 (Construction), Sector 12 (Chemical Industry), Sector 14 (Smelting and Rolling of Metals), Sector 1 (Agriculture, Forestry, Animal Husbandry, & Fishery), and Sector 6 (Manufacture of Foods and Tobacco) according to the amount of output. The total output of them was 10552.76 billion

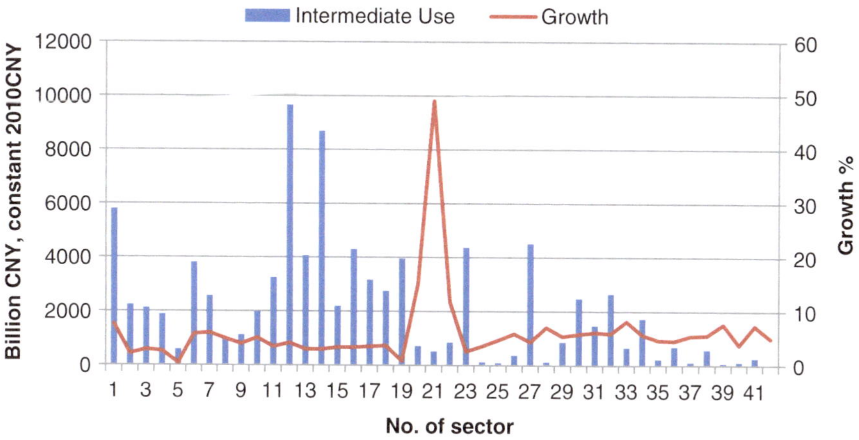

Fig. 9.9 Intermediate use and growth of 42 sectors in 2011

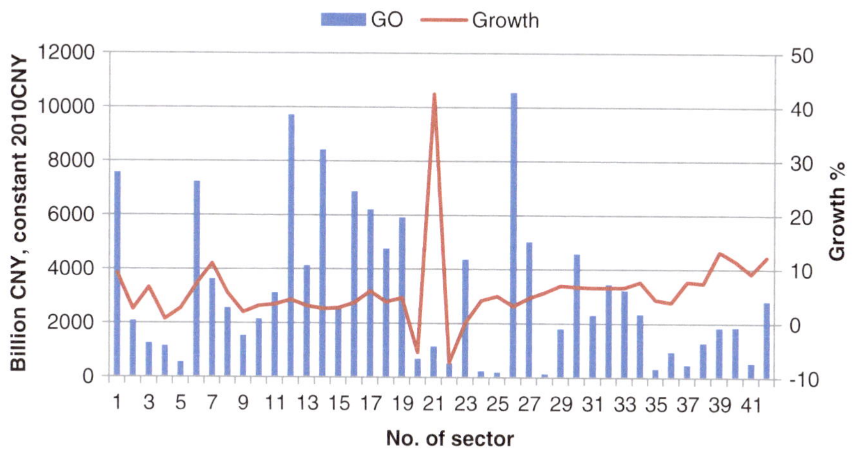

Fig. 9.10 Total output and growth of 42 sectors in 2011

CNY, 9722.07 billion CNY, 8422.43 billion CNY, 7569.67 billion CNY, and 7228.12 billion CNY, respectively. It is evident that the construction played a central role in China's national economy.

The export of the product of each sector and its growth rate in 2011 are depicted in Fig. 9.11. According to the amount of export, the top 5 sectors were Sector 19 (Manufacture of Communication Equipment, Computer, and Other Electronic Equipments), Sector 7 (Manufacture of Textile), Sector 18 (Manufacture of Electrical Machinery and Equipment), Sector 12 (Chemical Industry), and Sector 16 (Manufacture of General-Purpose and Special Purpose Machinery) in proper order. The export of the product of these sectors was 2635.96 billion CNY, 1163.34 billion CNY, 1000.36 billion CNY, 944.20 billion CNY, and 823.44 billion CNY,

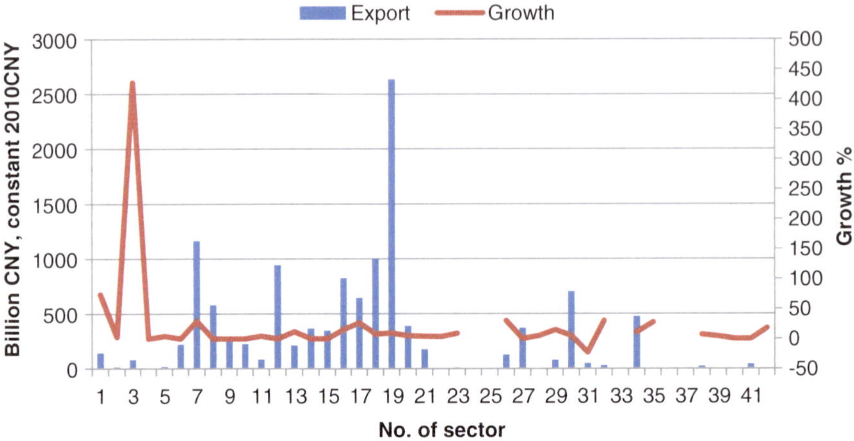

Fig. 9.11 Export and growth of 42 sectors in 2011

respectively. These sectors all belonged to manufacturing, which verified the role of China as the world factory.

The electricity consumption and its growth rate of each sector in 2011 simulated by the ARE model are shown in Fig. 9.12. The top 5 sectors were Sector 14 (Smelting and Rolling of Metals), Sector 23 (Production and Supply of Electric Power and Heat Power), Sector 12 (Chemical Industry), Sector 13 (Manufacture of Nonmetallic Mineral Products), and Sector 7 (Manufacture of Textile) in terms of the amount of consumption. When it comes to the growth rate, the top 5 sectors were Sector 34 (Leasing and Business Services), Sector 15 (Manufacture of Metal Products), Sector 4 (Mining of Metal Ores), Sector 22 (Scrap and Waste), and Sector 5 (Mining and Processing of Nonmetal Ores and Other Ores).

9.2.3 *Error Analysis*

In order to analyze the error of electricity use in 2011, the published one and simulated one by ARE model are compared. The published electricity consumption of the 42 sectors is 4140.58 TWh and the simulated one is 4067.88 TWh; the relative error is −1.76 %. The error is very low. For primary industry, the published electricity consumption is 101.29 TWh while the simulated one is 97.08 TWh, indicating that the error is −4.15 %; for secondary industry, the published electricity consumption is 3528.84 TWh while the simulated one is 3430.54 TWh, indicating that the error is −2.79 %; and for tertiary industry, the published electricity consumption is 510.45 TWh while the simulated one is 533.49 TWh, indicating that the error is 4.51 %. The errors of 42 sectors are shown in Fig. 9.13 and listed in Table 9.2. The highest error is 214.96 % for sector 41, of which the published electricity use is 11.23 TWh and the simulated one is 40.42TWh. The errors of

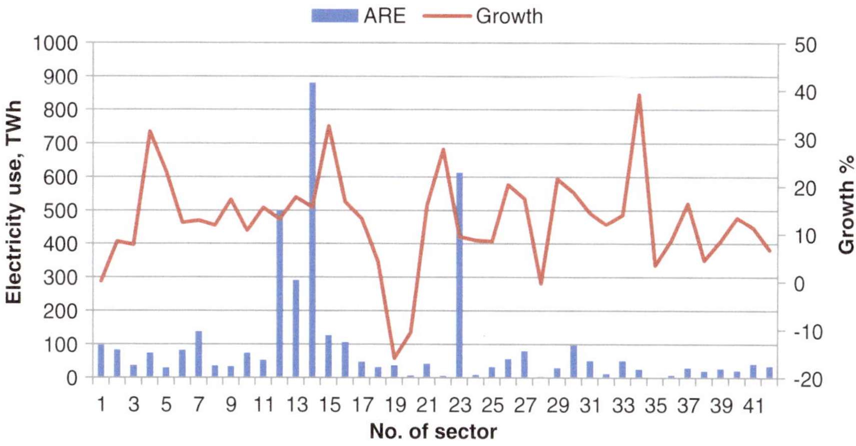

Fig. 9.12 Electricity consumption and growth rate of 42 sectors simulated by ARE in 2011

Sectors 35 and 36 are −60.8 % and 61.9 %, respectively. Their electricity consumptions are 5.085 TWh and 4.9 TWh; it shows that big errors tend to appear in the sector with lower electricity use. The errors of most other sectors are very small. More than half of them have the errors whose absolute values are less than 1 %, which verifies that the simulation of the ARE model is accurate and reliable.

9.3 Economic Simulation for 2012 by ARE Model

9.3.1 Fiscal and Monetary Policies

For fiscal policy data in 2012, it is the same with that in 2011 which has been shown in Fig. 9.1. The monetary policy data is listed in Table 9.3. As shown in the table, the reserve requirement rate declined by 0.5 % twice on Feb. 24 and May 18, and it was down from 21 to 20 %. The deposit rate and lending rate also decreased twice on June 8 and July 6; the deposit rate declined from 3.5 to 3 % and lending rate from 6.56 to 6 %. It shows that the monetary policy is slightly relaxed to provide more money for investment to stimulate the economy.

9.3.2 Simulation Results

Figure 9.14 illustrates the price indexes of Sector 2, Sector 3, Sector 4, and Sector 5 during the year of 2012. As mentioned previously, Sector 2 is Mining and Washing of Coal, Sector 3 is Extraction of Petroleum and Natural Gas, Sector 4 is Mining and Dressing of Metal Ores, and Sector 5 is Mining and Dressing of

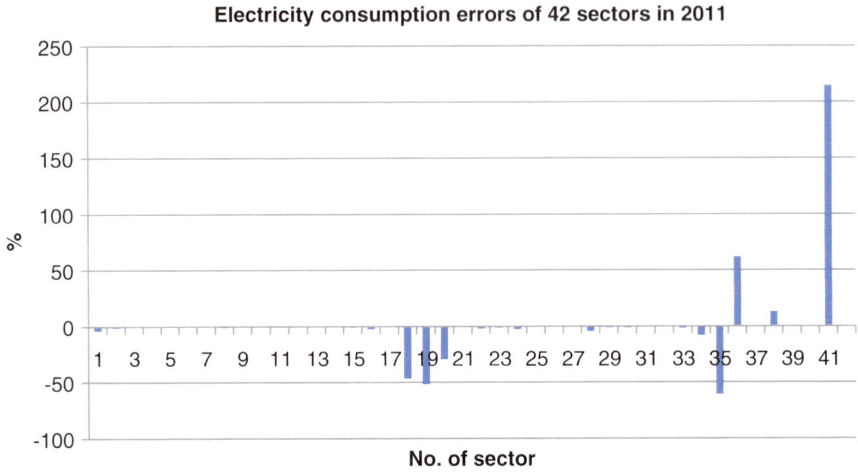

Fig. 9.13 Errors between published and simulated electricity consumption of 42 sectors in 2011

Table 9.2 Errors between published and simulated electricity consumption of 42 sectors in 2011

Sector number	Relative error (%)	Sector number	Relative error (%)	Sector number	Relative error (%)
1	−4.15	15	−1.19	29	−1.54
2	−1.47	16	−2.62	30	−1.51
3	−0.61	17	−0.04	31	−0.38
4	−0.37	18	−46.88	32	−0.02
5	−0.79	19	−51.87	33	−1.89
6	−0.88	20	−29.39	34	−8.42
7	−0.30	21	−0.24	35	−60.80
8	−1.29	22	−2.22	36	61.91
9	−0.98	23	−1.36	37	−0.45
10	−0.33	24	−2.77	38	12.91
11	−0.55	25	−0.50	39	−0.74
12	−0.28	26	−0.27	40	−0.63
13	−0.78	27	−0.51	41	214.97
14	−0.68	28	−4.56	42	−0.81

Nonmetal and Other Ores. As shown in the figure, the overall trends of these four price indexes were up. The initial values of the price indexes were 0.7475, 0.7970, 0.4200, and 0.8174, while their final values were 0.9661, 1.0358, 1.4183, and 1.8924, respectively. The reason was that the demand for these energy and raw material products was raised, which verifies the recovery of China's national economy.

Figure 9.15 depicts the price indexes of Sector 6, Sector 7, Sector 8, Sector 9, and Sector 10 during the year of 2012. As mentioned previously, Sector 6 is Manufacture of Foods and Tobacco; Sector 7 is Manufacture of Textile; Sector 8 is

Table 9.3 Changes of reserve requirement ratio and interest rates in 2012

Date of adjusted rate in 2012	Adjusted rate: Reserve requirement ratio	Deposit rate	Lending rate	Final rate: Reserve requirement ratio (%)	Deposit rate (%)	Lending rate (%)
By the end of 2011	–	–	–	21.00	3.50	6.56
24/02/2012	Down 0.5 %	–	–	20.50	3.50	6.56
18/5/2012	Down 0.5 %	–	–	20.00	3.50	6.56
8/06/2012	–	Down 0.25 %	Down 0.25 %	20.00	3.25	6.31
6/07/2012	–	Down 0.25	Down 0.31	20.00	3.00	6.00

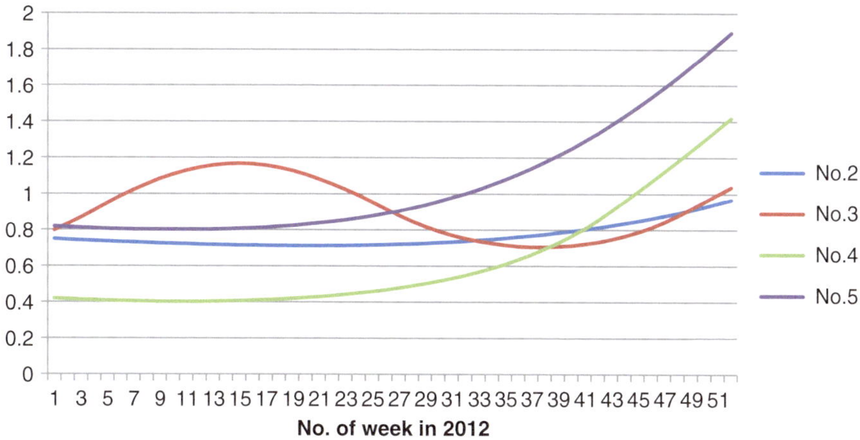

Fig. 9.14 Market price indexes of Sector 2, Sector 3, Sector 4, and Sector 5 in 2012

Manufacture of Textile Wearing Apparel, Footwear, Caps, Leather, Fur, Feather (Down), and Its Products; Sector 9 is Processing of Timbers and Manufacture of Furniture; and Sector 10 is Papermaking, Printing, and Manufacture of Articles. Their initial values were 1.9019, 3.0156, 2.0932, 1.0867, and 0.8895, and their last states in this year were 1.3865, 1.6459, 1.5223, 1.1700, and 2.2349, respectively.

Figure 9.16 displays the price indexes of Sector 11, Sector 12, Sector 13, Sector 14, Sector 15, and Sector 16 during the year of 2012. As mentioned before, Sector 11 is Processing of Petroleum, Coking, and Processing of Nuclear Fuel; Sector 12 is Chemical Industry; Sector 13 is Manufacture of Nonmetallic Mineral Products; Sector 14 is Smelting and Rolling of Metals; Sector 15 is Manufacture of Metal Products; and Sector 16 is Manufacture of General-Purpose and Special Purpose Machinery. As shown in the figure, the overall trends of most price indexes were up owing to the increasing demand. The growth rates were more obvious than those in 2011. The initial levels of these price indexes were 1.3434, 1.4062, 1.1087, 0.9619,

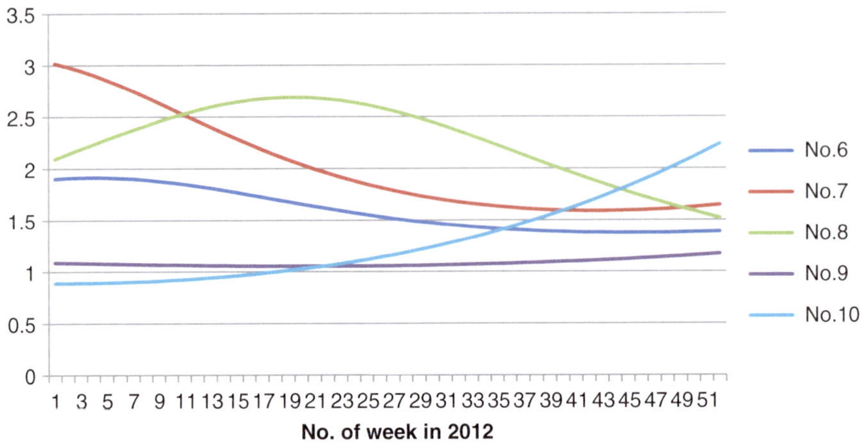

Fig. 9.15 Market price indexes of Sector 6, Sector 7, Sector 8, Sector 9, and Sector 10 in 2012

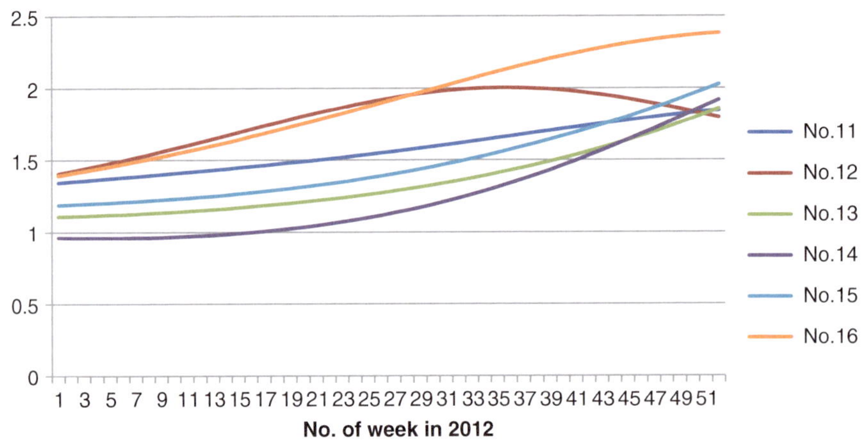

Fig. 9.16 Market price indexes of Sector 11, Sector 12, Sector 13, Sector 14, Sector 15, and Sector 16 in 2012

1.1886, and 1.3931, while their final values were 1.8452, 1.7953, 1.8563, 1.9169, 2.0256, and 2.3809, respectively. Since these sectors provided various fundamental materials, such as cement, rubber, iron, and steel, for the development of China's economy, the increasing demand verified the rapid improvement of China's national economy.

Figure 9.17 illustrates the price indexes of Sector 30, Sector 31, Sector 32, Sector 33, and Sector 34 during the year of 2012. Sector 30 is Wholesale and Retail Trades, Sector 31 is Hotels and Catering Services, Sector 32 is Financial Intermediation, Sector 33 is Real Estate, and Sector 34 is Leasing and Business Services. Evidently, the overall trend of these price indexes was up. They were initially 0.4886, 1.1633,

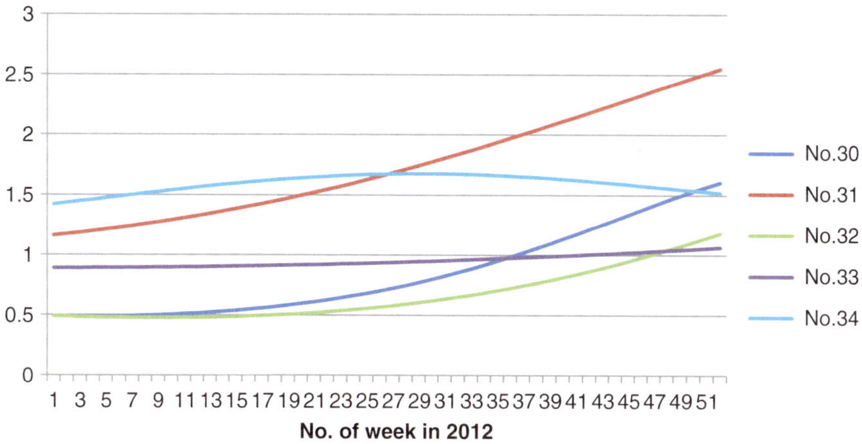

Fig. 9.17 Market price indexes of Sector 30, Sector 31, Sector 32, Sector 33, and Sector 34 in 2012

0.4899, 0.8886, and 1.4186 respectively, and they arrived at 1.6031, 2.5475, 1.1786, 1.0631, and 1.5176 in proper order. The price of Sector 30, Sector 31, and Sector 32 increased obviously due to the growing demand, which exhibited the potential for the development of the tertiary industry in China.

The total output of each sector and its growth rate in 2012 are shown in Fig. 9.18. The top 5 sectors were Sector 26 (Construction), Sector 12 (Chemical Industry), Sector 14 (Smelting and Rolling of Metals), Sector 1 (Agriculture, Forestry, Animal Husbandry, & Fishery), and Sector 6 (Manufacture of Foods and Tobacco) according to the amount of output. Their total output was 11189.42 billion CNY, 10525.14 billion CNY, 8980.39 billion CNY, 8433.10 billion CNY, and 7933.71 billion CNY, respectively. In terms of the growth rate, the top 5 sectors were Sector 21 (Manufacture of Artwork, Other Manufacture), Sector 20 (Manufacture of Measuring Instrument and Machinery for Cultural Activity & Office Work), Sector 39 (Education), Sector 1 (Agriculture, Forestry, Animal Husbandry, & Fishery), and Sector 42 (Public Management and Social Organization). Their growth rates were 17.42 %, 18.51 %, 11.76 %, 11.41 %, and 10.84 %, respectively.

The total intermediate use of the product of each sector and the growth rate in 2012 are depicted in Fig. 9.19. According to the amount of intermediate use, the top 5 sectors were Sector 12 (Chemical Industry), Sector 14 (Smelting and Rolling of Metals), Sector 1 (Agriculture, Forestry, Animal Husbandry, & Fishery), Sector 27 (Traffic, Transport, and Storage), and Sector 23 (Production and Supply of Electric Power and Heat Power) in proper order. The intermediate use of the product of these sectors was 10454.82 billion CNY, 9277.25 billion CNY, 6389.57 billion CNY, 4822.23 billion CNY, and 4607.92 billion CNY, respectively. These sectors played fundamental roles in the national economy in China.

The capital formulation and the growth rate of each sector in 2012 are shown in Fig. 9.20. The top 5 sectors were Sector 26 (Construction), Sector 16 (Manufacture

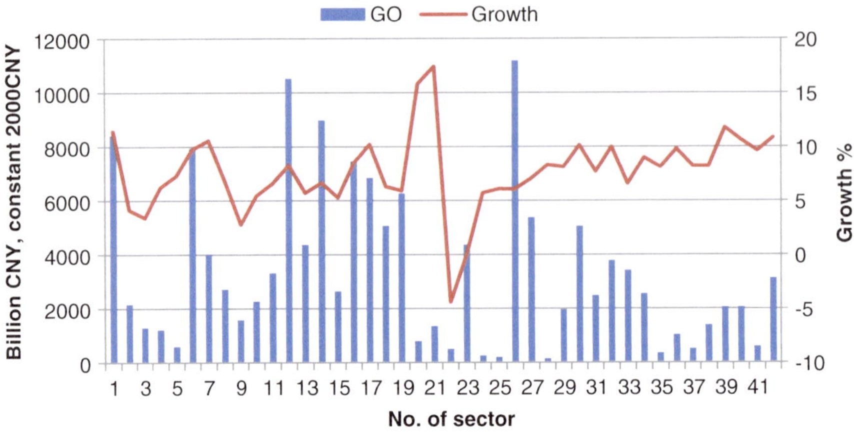

Fig. 9.18 Total output and growth of 42 sectors in 2012

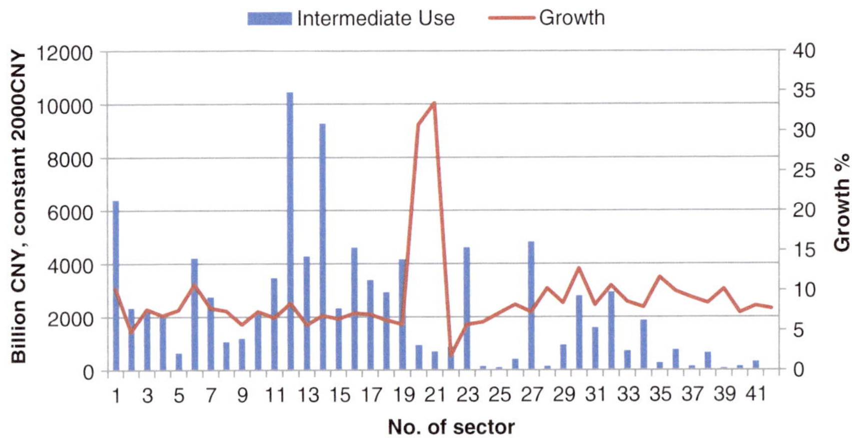

Fig. 9.19 Intermediate use and growth of 42 sectors in 2012

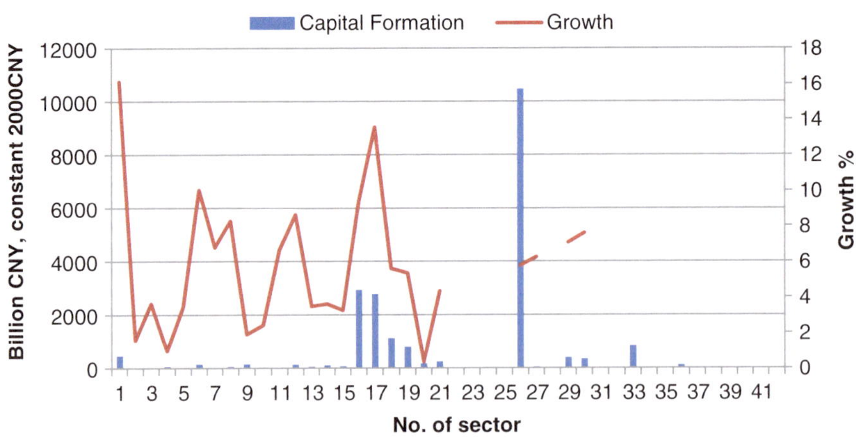

Fig. 9.20 Capital formation and growth of 42 sectors in 2012

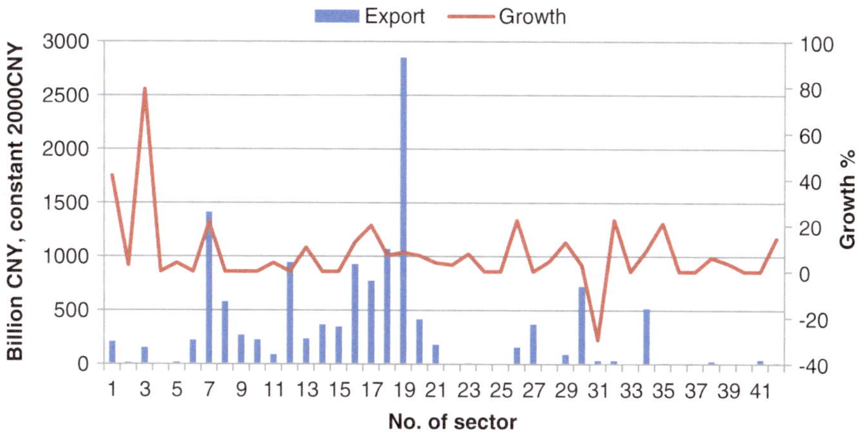

Fig. 9.21 Export and growth of 42 sectors in 2012

of General-Purpose and Special Purpose Machinery), Sector 17 (Manufacture of Transport Equipment), Sector 18 (Manufacture of Electrical Machinery and Equipment), and Sector 33 (Real Estate) according to the amount of output. Their capital formulation was 10457.57 billion CNY, 2933.10 billion CNY, 2722.42 billion CNY, 1134.16 billion CNY, and 842.33 billion CNY, respectively. These sectors were sensitive to the investment promoted by Chinese government.

The export of the product of each sector and its growth rate in 2012 are depicted in Fig. 9.21. According to the amount of export, the top 5 sectors were Sector 19 (Manufacture of Communication Equipment, Computer, and Other Electronic Equipments), Sector 7 (Manufacture of Textile), Sector 18 (Manufacture of Electrical Machinery and Equipment), Sector 12 (Chemical Industry), and Sector 16 (Manufacture of General-Purpose and Special Purpose Machinery) in proper order. The export of the product of these sectors was 2854.68 billion CNY, 1415.31 billion CNY, 1069.88 billion CNY, 944.20 billion CNY, and 927.45 billion CNY, respectively.

The value added and its growth rate of each sector in 2012 are shown in Fig. 9.22. Among the 42 sectors, Sector 1 (Agriculture, Forestry, Animal Husbandry, & Fishery) was no. 1 according to value added. It is as high as 5035.04 billion CNY, which reflects the foundation status of agriculture in China's national economy. Sector 30 (Wholesale and Retail Trades) ranked second with the value added equaling 3517.92 billion CNY. Then came Sector 26 (Construction), Sector 33 (Real Estate), and Sector 32 (Financial Intermediation). Their value added was 2896.17 billion CNY, 2571.14 billion CNY, and 2384.86 billion CNY, respectively. In terms of the growth rate of the value added, the relative ranking of the 42 sectors was obviously different from that of the amount of value added. The first sector was Sector 24 (Production and Distribution of Gas), which reflects China's emphasis on the development of natural gas. Sector 4 (Mining of Metal Ores) came second. The next three sectors were Sector 2 (Mining and Washing of Coal),

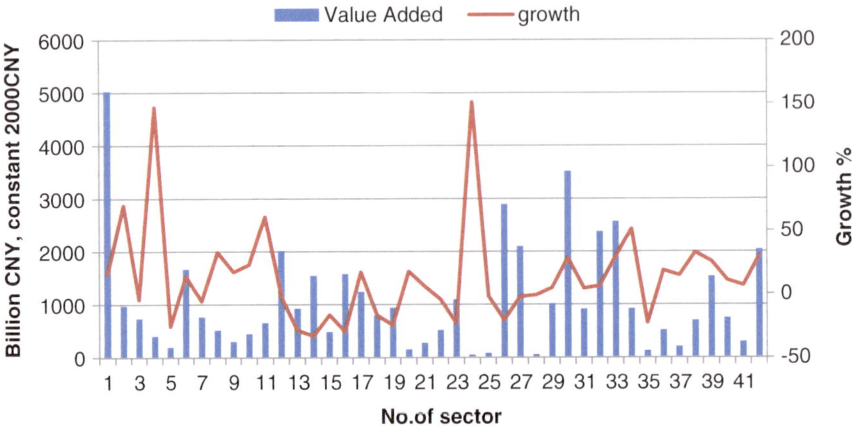

Fig. 9.22 Value added and growth of 42 sectors in 2012

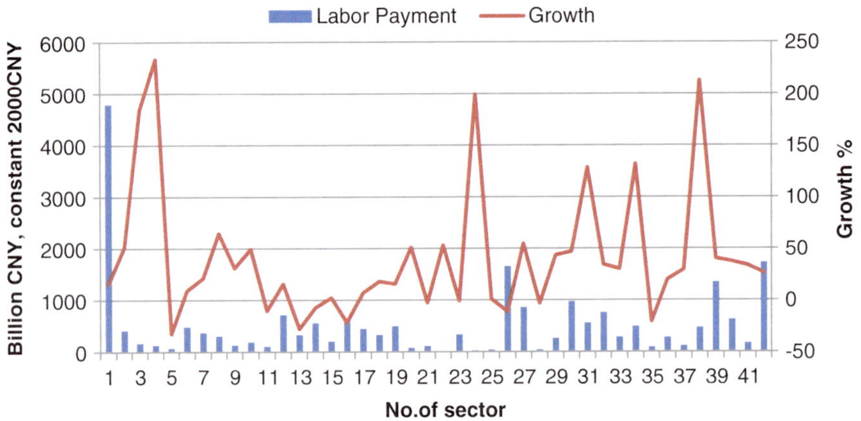

Fig. 9.23 Labor payment and growth of 42 sectors in 2012

Sector 11 (Processing of Petroleum, Coking, and Processing of Nuclear Fuel), and Sector 34 (Leasing and Business Services).

The labor payment and growth rate of each sector in 2012 are depicted in Fig. 9.23. According to the amount of labor payment, the top 5 sectors were Sector 1 (Agriculture, Forestry, Animal Husbandry, & Fishery), Sector 42 (Public Management and Social Organization), Sector 26 (Construction), Sector 39 (Education), and Sector 30 (Wholesale and Retail Trades) in proper order. The labor payment of these sectors was 4790.24 billion CNY, 1724.97 billion CNY, 1649.48 billion CNY, 1344.77 billion CNY, and 966.22 billion CNY, respectively.

The overall taxes of each sector and the growth rate in 2012 are illustrated in Fig. 9.24. In terms of the amount of taxes, the top 5 sectors were Sector 30 (Wholesale

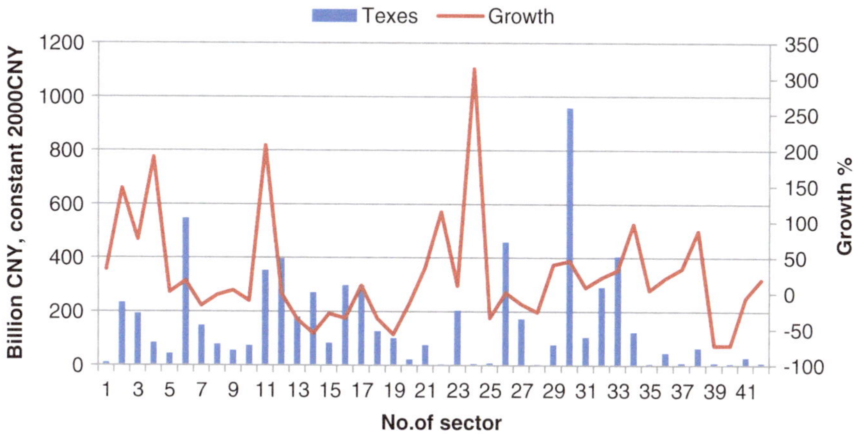

Fig. 9.24 Taxes and growth of 42 sectors in 2012

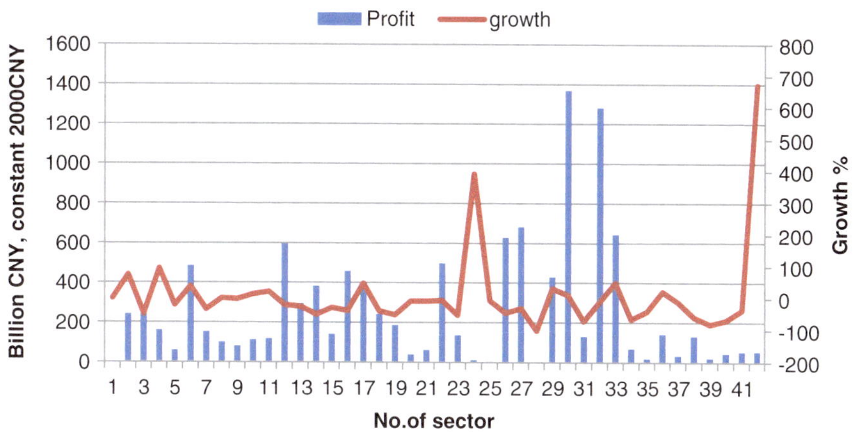

Fig. 9.25 Profit and growth of 42 sectors in 2012

and Retail Trades), Sector 6 (Manufacture of Foods and Tobacco), Sector 26 (Construction), Sector 33 (Real Estate), and Sector 12 (Chemical Industry) in proper order. The taxes of these sectors were 958.44 billion CNY, 547.75 billion CNY, 458.07 billion CNY, 404.98 billion CNY, and 395.80 billion CNY, respectively. These sectors were the main contributor to the revenue in China.

The profit of each sector and its growth rate in 2012 are shown in Fig. 9.25. The top 5 sectors were Sector 30 (Wholesale and Retail Trades), Sector 32 (Financial Intermediation), Sector 27 (Traffic, Transport, and Storage), Sector 33 (Real Estate), and Sector 26 (Construction) according to the amount of profit. Their profit was 1368.66 billion CNY, 1281.35 billion CNY, 680.75 billion CNY, 644.82 billion CNY, and 628.36 billion CNY, respectively.

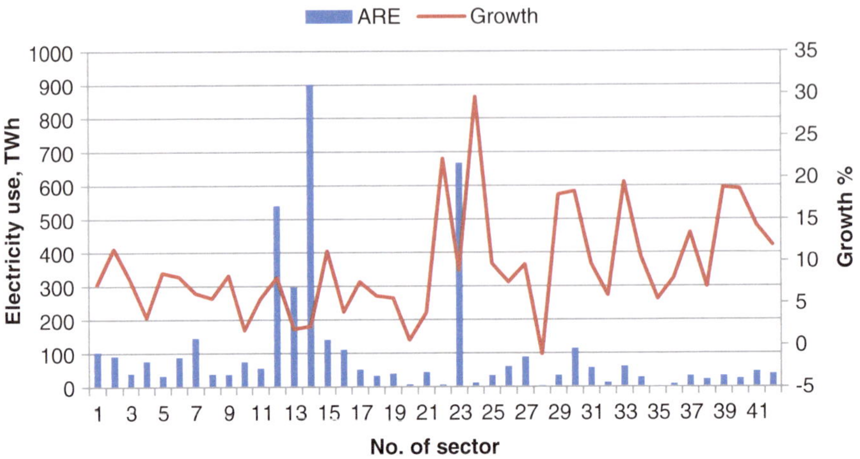

Fig. 9.26 Electricity consumption and growth rate of 42 sectors simulated by ARE in 2012

The electricity consumption and the growth rate of each sector in 2012 simulated by the ARE model are shown in Fig. 9.26. The top 5 sectors were Sector 14 (Smelting and Rolling of Metals), Sector 23 (Production and Supply of Electric Power and Heat Power), Sector 12 (Chemical Industry), Sector 13 (Manufacture of Nonmetallic Mineral Products), and Sector 7 (Manufacture of Textile) in terms of the amount of consumption. When it comes to the growth rate, the top 5 sectors were Sector 24 (Production and Distribution of Gas), Sector 22 (Scrap and Waste), Sector 33 (Real Estate), Sector 39 (Education), and Sector 40 (Health, Social Security, and Social Welfare).

9.3.3 Error Analysis

The published electricity consumption of the 42 sectors is 4342.88TWh and the simulated one is 4333.65TWh and the relative error is −0.21 %. The simulation is very accurate. For primary industry, the published electricity consumption is 100.25 TWh while the simulated one is 104.09 TWh, indicating that the error is 3.83 %; for secondary industry, the published electricity consumption is 3673.33 TWh while the simulated one is 3430.54 TWh, indicating that the error is −6.61 %; and for tertiary industry, the published electricity consumption is 569.3 TWh while the simulated one is 607.47 TWh, indicating that the error is 6.71 %. The errors of 42 sectors are shown in Fig. 9.27 and listed in Table 9.4. The highest error is 214.4 % for sector 41, of which the published electricity use is 13.53 TWh and the simulated one is 46.18 TWh. The errors of sector 35 and 36 are −61.08 % and 64.54 %, respectively. Their electricity consumptions are 5.4 and 5.2 TWh. On the contrast, the errors of most other sectors are so small that half of their absolute

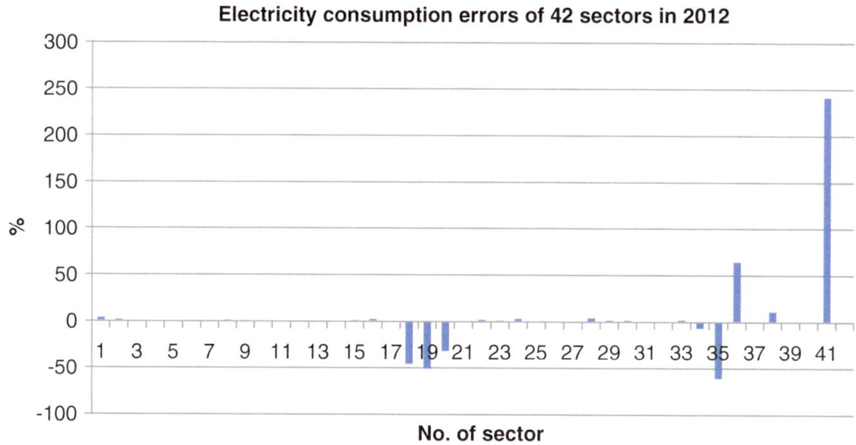

Fig. 9.27 Errors between published and simulated electricity consumption of 42 sectors in 2012

Table 9.4 Errors between published and simulated electricity consumption of 42 sectors in 2012

Sector number	Relative error (%)	Sector number	Relative error (%)	Sector number	Relative error (%)
1	3.83	15	1.28	29	1.69
2	1.61	16	2.49	30	1.66
3	0.64	17	0.04	31	0.39
4	0.38	18	−46.11	32	0.02
5	0.83	19	−51.28	33	2.13
6	0.87	20	−31.91	34	−6.91
7	0.30	21	0.24	35	−61.08
8	1.25	22	2.12	36	64.54
9	1.04	23	1.48	37	0.47
10	0.33	24	3.33	38	11.06
11	0.55	25	0.51	39	0.77
12	0.28	26	0.27	40	0.66
13	0.76	27	0.52	41	241.40
14	0.67	28	4.18	42	0.80

values are less than 1 %, which demonstrate the simulation based on ARE model is accurate.

On the other hand, the growth rates of 42 sectors are quite similar between published electricity growths and simulated ones as shown in Fig. 9.28.

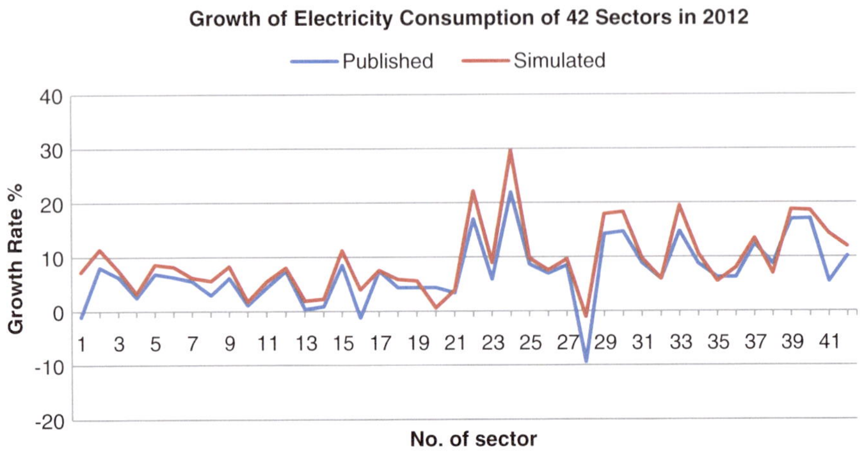

Fig. 9.28 Comparison between published and simulated electricity consumption growth of 42 sectors in 2012

9.4 Economic Simulation for 2013 by ARE Model

9.4.1 Fiscal and Monetary Policies

As we know, the input data to ARE model is the input–output table of the last year, fiscal policy data, and monetary data. The input data of input–output table for economic simulation of 2013 is the simulated input–output table of 2012. Since there were no changes on fiscal policy, we use the same data as that in 2012 as mentioned in Sect. 9.2.1 in this chapter (see Fig. 9.1). Monetary policy was not changed in 2013 either, i.e., the reserve requirement rate is input as 20 %, the deposit rate as 3 %, and lending rate as 6 %.

9.4.2 Simulation Results

Figure 9.29 illustrates the price indexes of Sector 2, Sector 3, Sector 4, and Sector 5 during the year of 2013. As mentioned before, Sector 2 is Mining and Washing of Coal; Sector 3 is Extraction of Petroleum and Natural Gas; Sector 4 is Mining and Dressing of Metal Ores; and Sector 5 is Mining and Dressing of Nonmetal and Other Ores. As shown in the figure, the overall trends of these four price indexes were down after a small period of going up, as a result of the changes of the supply–demand situation in domestic and international energy market. For instance, the price index of Petroleum and Natural Gas increased from 1.0757 to 1.4333 and then declined to 0.7921 in the 44th week.

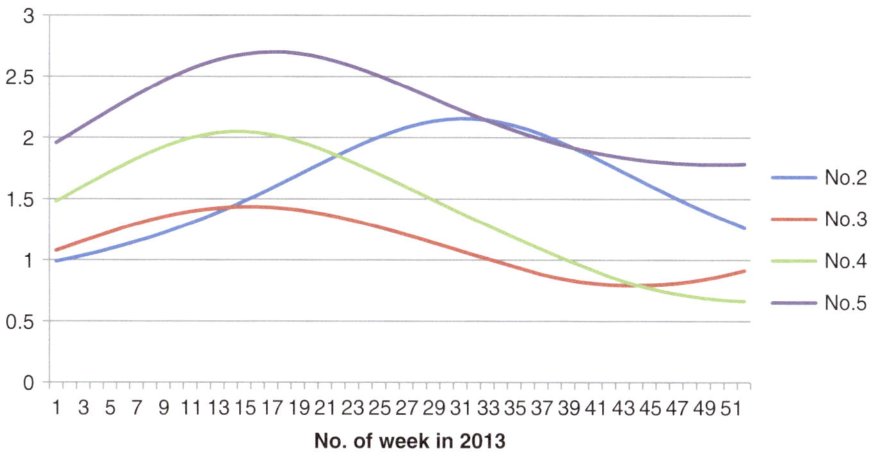

Fig. 9.29 Market price indexes of Sector 2, Sector 3, Sector 4, and Sector 5 in 2013

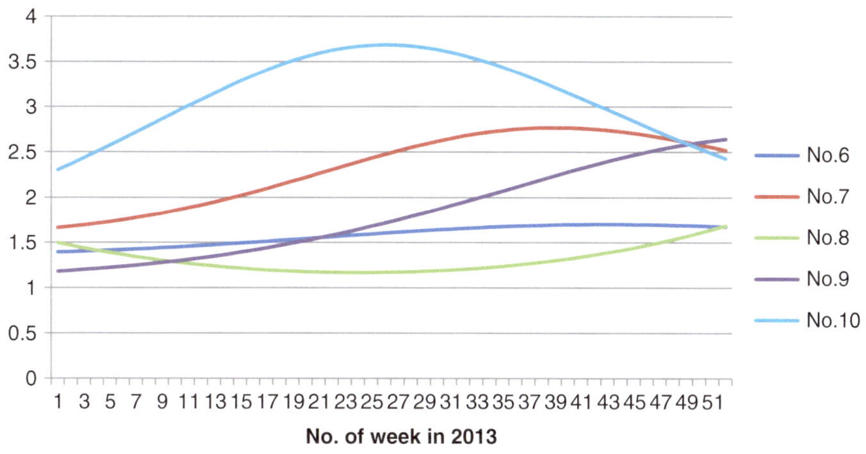

Fig. 9.30 Market price indexes of Sector 6, Sector 7, Sector 8, Sector 9, and Sector 10 in 2013

Figure 9.30 depicts the price indexes of Sector 6, Sector 7, Sector 8, Sector 9, and Sector 10 during the year of 2013. As mentioned previously, Sector 6 is Manufacture of Foods and Tobacco; Sector 7 is Manufacture of Textile; Sector 8 is Manufacture of Textile Wearing Apparel, Footwear, Caps, Leather, Fur, Feather (Down), and Its Products; Sector 9 is Processing of Timbers and Manufacture of Furniture; and Sector 10 is Papermaking, Printing, and Manufacture of Articles. Clearly, the overall trends of these price indexes were up. Their initial values were 1.3903, 1.6591, 1.4921, 1.1789, and 2.2997 respectively, while their last values were 1.6732, 2.5202, 1.6830, 2.6428, and 2.4272, respectively. Considering the

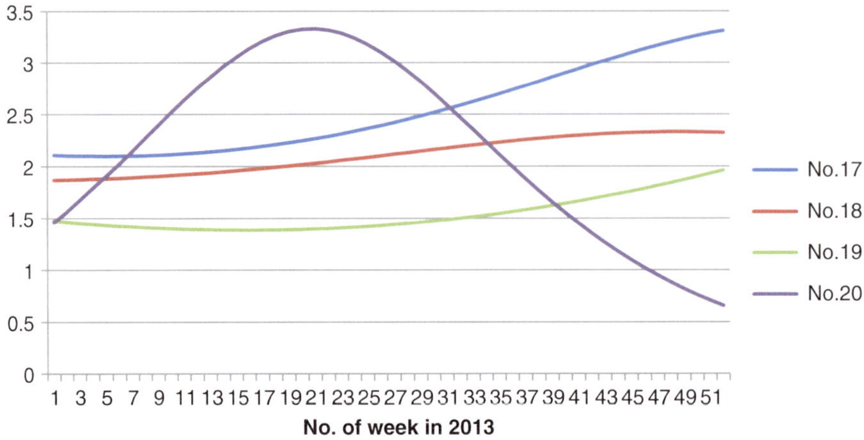

Fig. 9.31 Market price indexes of Sector 17, Sector 18, Sector 19, and Sector 20 in 2013

products of these sectors are all typical export goods, the increase of their price level verifies the recovery of the world economy.

Figure 9.31 shows the price indexes of Sector 17, Sector 18, Sector 19, and Sector 20 during the year of 2013. Sector 17 is Manufacture of Transport Equipment; Sector 18 is Manufacture of Electrical Machinery and Equipment; Sector 19 is Manufacture of Communication Equipment, Computer, and Other Electronic Equipments; and Sector 20 is Manufacture of Measuring Instrument and Machinery for Cultural Activity & Office Work. It is evident that the price indexes of Sector 17, Sector 18, and Sector 19 kept increasing. Their initial values were 2.1088, 1.8667, and 1.4711, respectively, while their last values were 3.3121, 2.3200, and 1.9616, respectively. This reflects the increasing demand for the products of equipment manufacturing, which is the feature of industrialization. Meanwhile, the price index of Sector 20 went up firstly from 1.4591 to 3.3284 but then started to decline in the 22th week reaching 0.6558 at the end of the year.

Figure 9.32 provides us the price indexes of Sector 38, Sector 39, Sector 40, and Sector 41 during the year of 2013. Sector 38 is Services to Households and Other Services; Sector 39 is Education; Sector 40 is Health, Social Security, and Social Welfare; and Sector 41 is Culture, Sports, and Entertainment. It can be seen that overall trends of their price indexes were up except the price of Sector 41. Their initial statuses were 1.4583, 1.0456, 1.1885, and 2.0337, while their final values were 2.0514, 1.8732, 1.7419, and 1.4284, respectively. This verifies the residents in China have increasingly high requests for the quality of life.

The total output of each sector and its growth rate in 2013 are shown in Fig. 9.33. The top 5 sectors were Sector 26 (Construction), Sector 12 (Chemical Industry), Sector 14 (Smelting and Rolling of Metals), Sector 1 (Agriculture, Forestry, Animal Husbandry, & Fishery), and Sector 6 (Manufacture of Foods and Tobacco) according to the amount of output. The total output of them was 13164.67 billion CNY, 11410.89 billion CNY, 10166.73 billion CNY, 9311.90 billion CNY, and

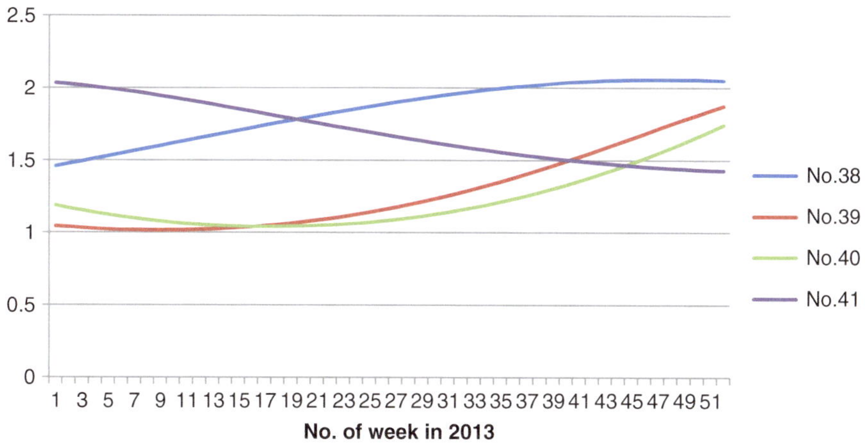

Fig. 9.32 Market price indexes of Sector 38, Sector 39, Sector 40, and Sector 41 in 2013

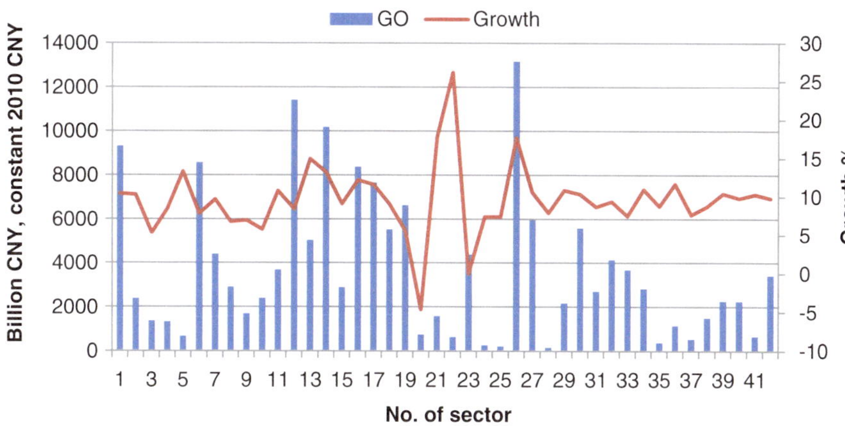

Fig. 9.33 Total output and growth of 42 sectors in 2013

8556.51 billion CNY, respectively. In terms of the growth rate, the top 5 sectors were Sector 22 (Scrap and Waste), Sector 21 (Manufacture of Artwork, Other Manufacture), Sector 26 (Construction), Sector 13 (Manufacture of Nonmetallic Mineral Products), and Sector 5 (Mining and Processing of Nonmetal Ores and Other Ores). Their growth rates were 26.17 %, 17.72 %, 17.65 %, 14.92 %, and 13.27 %, respectively.

The total intermediate use of the product and the growth rate of each sector in 2013 are depicted in Fig. 9.34. According to the amount of intermediate use, the top 5 sectors were Sector 12 (Chemical Industry), Sector 14 (Smelting and Rolling of Metals), Sector 1 (Agriculture, Forestry, Animal Husbandry, & Fishery), Sector 27 (Traffic, Transport, and Storage), and Sector 23 (Production and Supply of

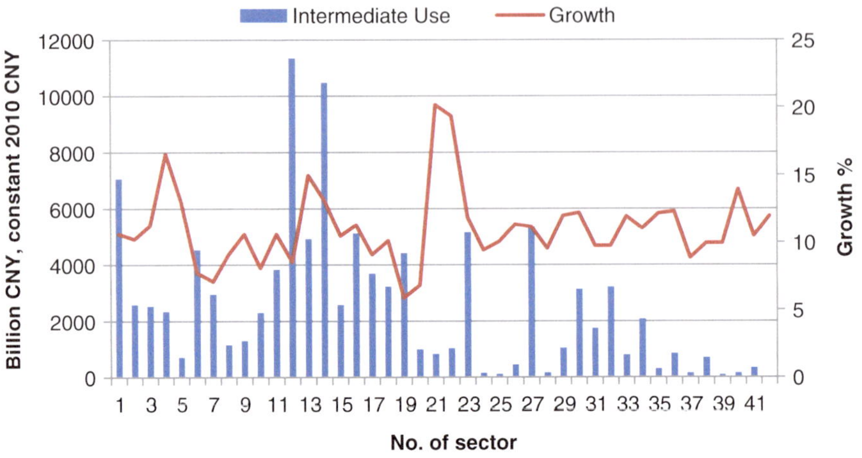

Fig. 9.34 Intermediate use and growth of 42 sectors in 2013

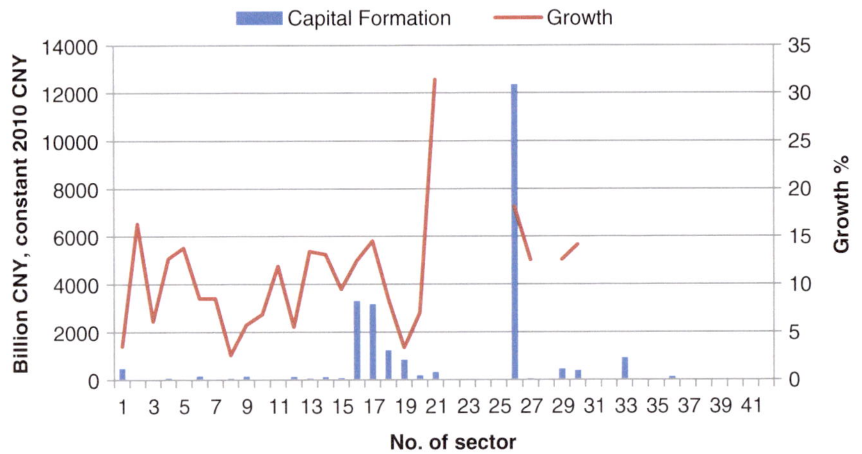

Fig. 9.35 Capital formation and growth of 42 sectors in 2013

Electric Power and Heat Power) in proper order. The intermediate use of the product of these sectors was 11347.13 billion CNY, 10488.70 billion CNY, 7067.48 billion CNY, 5359.35 billion CNY, and 5152.93 billion CNY, respectively. These sectors played fundamental roles in the national economy in China. Their products were widely utilized by other sectors.

The capital formulation and the growth rate of each sector in 2013 are shown in Fig. 9.35. The top 5 sectors were Sector 26 (Construction), Sector 16 (Manufacture of General-Purpose and Special Purpose Machinery), Sector 17 (Manufacture of Transport Equipment), Sector 18 (Manufacture of Electrical Machinery and Equipment), and Sector 33 (Real Estate) according to the amount of output. Their capital

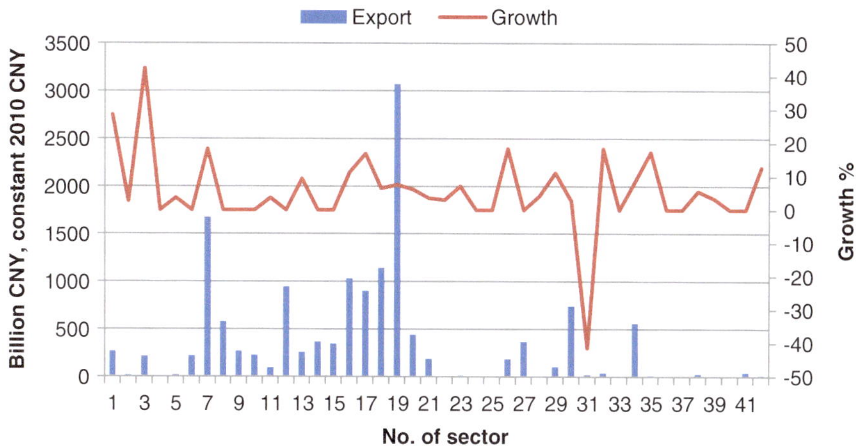

Fig. 9.36 Export and growth of 42 sectors in 2013

formulation was 12349.64 billion CNY, 3297.95 billion CNY, 3174.29 billion CNY, 1229.98 billion CNY, and 905.20 billion CNY, respectively.

The export of the product of each sector and its growth rate in 2013 are depicted in Fig. 9.36. According to the amount of export, the top 5 sectors were Sector 19 (Manufacture of Communication Equipment, Computer, and Other Electronic Equipments), Sector 7 (Manufacture of Textile), Sector 18 (Manufacture of Electrical Machinery and Equipment), Sector 16 (Manufacture of General-Purpose and Special Purpose Machinery), and Sector 12 (Chemical Industry) in proper order. The export of the product of these sectors was 3071.88 billion CNY, 1673.55 billion CNY, 1139.82 billion CNY, 1031.66 billion CNY, and 944.20 billion CNY, respectively. These manufacturing sectors were still the main contributors of the export of China's national economy.

The value added and its growth rate of each sector in 2013 are shown in Fig. 9.37. Among the 42 sectors, Sector 1 (Agriculture, Forestry, Animal Husbandry, & Fishery) was no. 1 according to value added. It is as high as 5210.95 billion CNY, which reflects the foundation status of the agriculture in China's national economy. Sector 30 (Wholesale and Retail Trades) ranked second with the value added equaling 4015.04 billion CNY. Then came Sector 26 (Construction), Sector 33 (Real Estate), and Sector 32 (Financial Intermediation). Their value added was 3420.17 billion CNY, 2763.04 billion CNY, and 2742.90 billion CNY, respectively. In terms of the growth rate of the value added, the relative ranking of the 42 sectors was obviously different from that of the amount of value added. The first sector was Sector 21 (Manufacture of Artwork, Other Manufacture). Sector 26 (Construction) came second. The next three sectors were Sector 2 (Mining and Washing of Coal), Sector 32 (Financial Intermediation), and Sector 30 (Wholesale and Retail Trades).

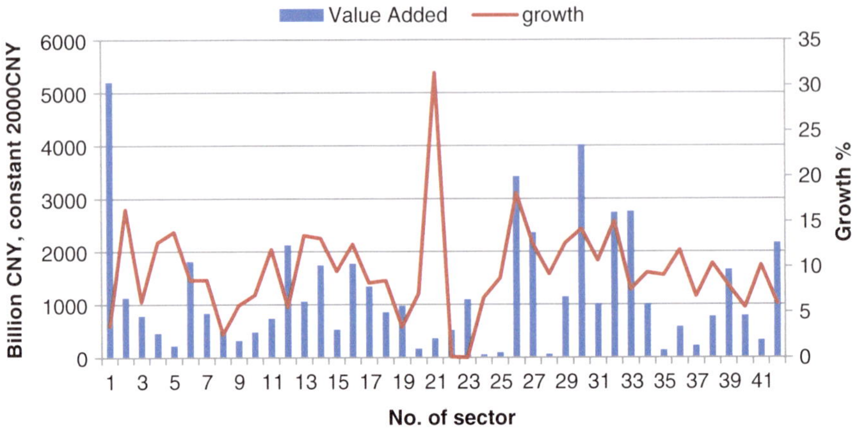

Fig. 9.37 Value added and growth of 42 sectors in 2013

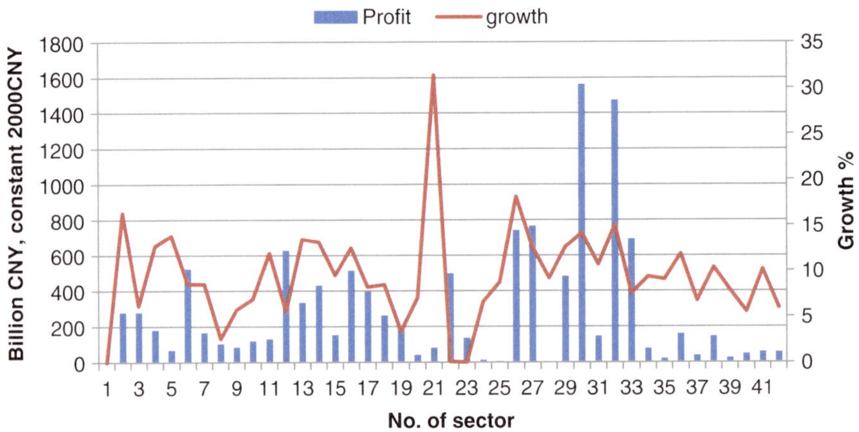

Fig. 9.38 Profit and growth of 42 sectors in 2013

The profit of each sector and its growth rate in 2013 are shown in Fig. 9.38. The top 5 sectors were Sector 30 (Wholesale and Retail Trades), Sector 32 (Financial Intermediation), Sector 27 (Traffic, Transport, and Storage), Sector 26 (Construction), and Sector 33 (Real Estate) according to the amount of profit. Their profit was 1562.06 billion CNY, 1473.72 billion CNY, 766.02 billion CNY, 742.05 billion CNY, and 692.95 billion CNY, respectively.

The overall taxes of each sector and the growth rate in 2013 are illustrated in Fig. 9.39. In terms of the amount of taxes, the top 5 sectors were Sector 30 (Wholesale and Retail Trades), Sector 6 (Manufacture of Foods and Tobacco), Sector 26 (Construction), Sector 33 (Real Estate), and Sector 12 (Chemical Industry) in proper order. The taxes of these sectors were 1093.88 billion CNY, 594.35 billion

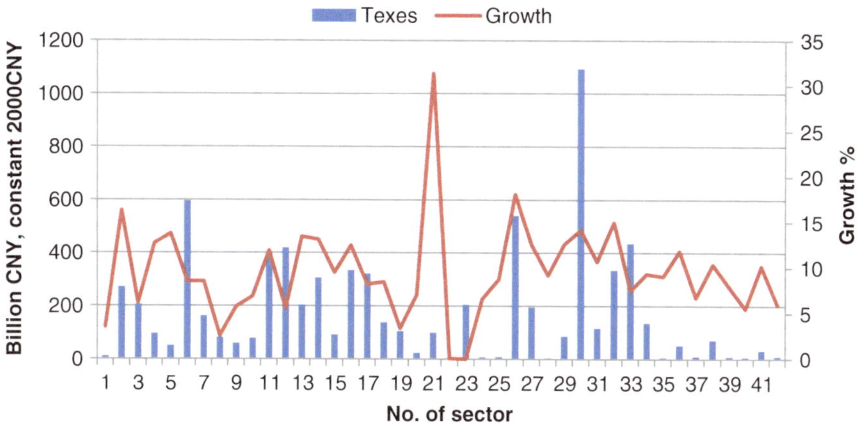

Fig. 9.39 Taxes and growth of 42 sectors in 2013

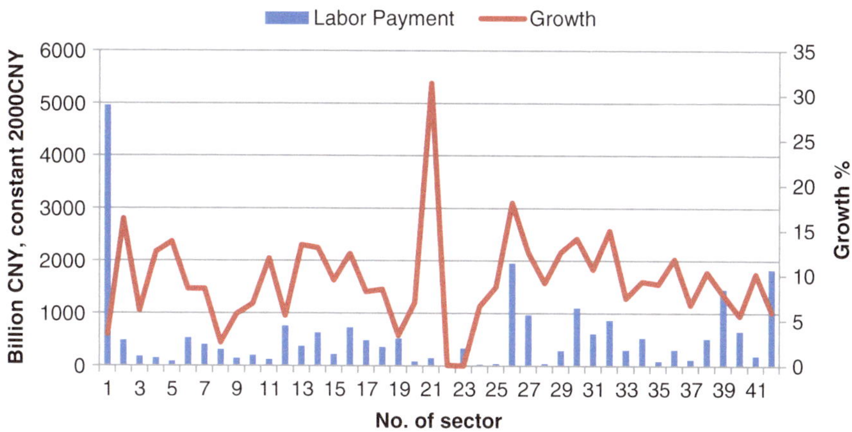

Fig. 9.40 Labor payment and growth of 42 sectors in 2013

CNY, 540.95 billion CNY, 435.20 billion CNY, and 417.70 billion CNY, respectively. These sectors were still the main contributors to the revenue in China.

The labor payment and growth rate of each sector in 2013 are depicted in Fig. 9.40. According to the amount of labor payment, the top 5 sectors were Sector 1 (Agriculture, Forestry, Animal Husbandry, & Fishery), Sector 26 (Construction), Sector 42 (Public Management and Social Organization), Sector 39 (Education), and Sector 30 (Wholesale and Retail Trades) in proper order. The labor payment of these sectors was 4957.59 billion CNY, 1947.92 billion CNY, 1827.39 billion CNY, 1450.46 billion CNY, and 1102.76 billion CNY, respectively.

The electricity consumption and its growth rate of each sector in 2013 simulated by the ARE model are shown in Fig. 9.41. The top 5 sectors were Sector 14 (Smelting and

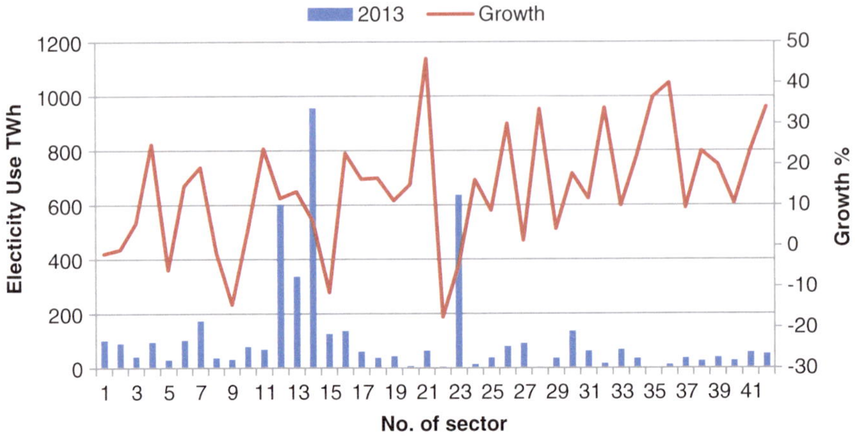

Fig. 9.41 Electricity consumption and growth rate of 42 sectors simulated by ARE in 2013

Rolling of Metals), Sector 23 (Production and Supply of Electric Power and Heat Power), Sector 12 (Chemical Industry), Sector 13 (Manufacture of Nonmetallic Mineral Products), and Sector 7 (Manufacture of Textile) in terms of the amount of consumption. When it comes to the growth rate in 2013, the top 5 sectors were Sector 21 (Manufacture of Artwork, Other Manufacture), Sector 36 (Comprehensive Technical Services), Sector 35 (Research and Experimental Development), Sector 41 (Culture, Sports, and Entertainment), and Sector 32 (Financial Intermediation). Most of them belong to the tertiary industry, which is in accordance with the spirit of the adjustment of economic structure in China.

9.4.3 *Error Analysis*

Error analysis of the published electricity use and simulated one is useful to evaluate the ARE model. The published electricity consumption of the 42 sectors in 2013 is 4663.43 TWh and the simulated one is 4698.94 TWh and the relative error is 0.76 %. The simulation is very accurate. For primary industry, the published electricity consumption is 102.69 TWh while the simulated one is 102.01 TWh, indicating the error is −0.66 %; for secondary industry, the published electricity consumption is 3933.19 TWh while the simulated one is 3897.21 TWh, indicating that the error is −0.92 %; and for tertiary industry, the published electricity consumption is 627.54 TWh while the simulated one is 699.72 TWh, indicating the error is 11.5 %. The errors of 42 sectors are shown in Fig. 9.42. The highest error is 306.23 % still for Sector 41. The errors of Sectors 35 and 36 are −50.9 % and 112.87 %, respectively. Other sectors have quite low errors in terms of electricity consumption as shown in Table 9.5. The accuracy and reliability of the simulation of ARE model have been verified through the comparison.

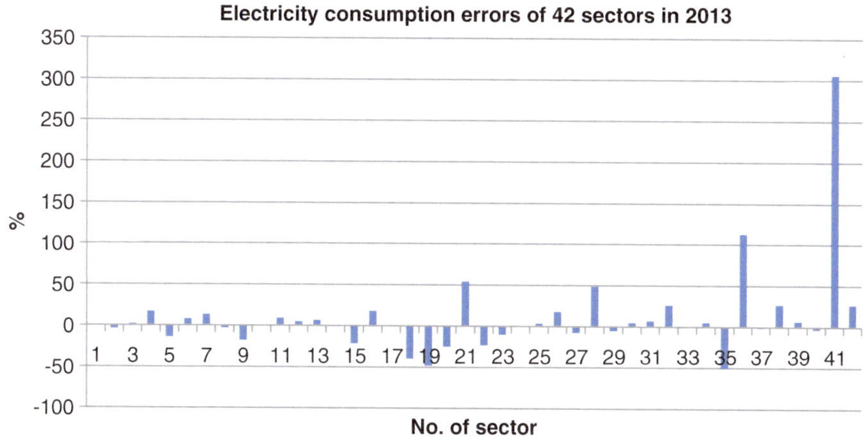

Fig. 9.42 Errors between published and simulated electricity consumption of 42 sectors in 2013

Table 9.5 Errors between published and simulated electricity consumption of 42 sectors in 2013

Sector number	Relative error	Sector number	Relative error	Sector number	Relative error
1	−0.66	15	−21.76	29	−5.05
2	−3.97	16	18.29	30	4.94
3	1.99	17	0.48	31	7.08
4	17.02	18	−39.76	32	26.55
5	−14.01	19	−48.11	33	−0.21
6	7.81	20	−24.91	34	5.55
7	13.48	21	54.01	35	−50.90
8	−3.06	22	−23.38	36	112.87
9	−18.11	23	−10.23	37	−1.27
10	0.54	24	0.61	38	27.11
11	9.09	25	3.88	39	6.76
12	5.08	26	17.97	40	−3.14
13	6.88	27	−7.52	41	306.32
14	0.12	28	48.60	42	27.11

9.5 Economic Simulation for 2014 by ARE Model

9.5.1 Fiscal and Monetary Policies

The input data to ARE model is the input–output table of the last year, fiscal policy data, and monetary data. The input data of input–output table for economic simulation of 2014 is the simulated input–output table of 2013. Since there were no changes on fiscal data, the same policy data is used as mentioned in Sect. 9.2.1 in this chapter (see Fig. 9.1). Monetary policy was still not changed in 2014, i.e., the reserve requirement rate is input as 20 %, the deposit rate as 3 %, and lending rate as 6 %.

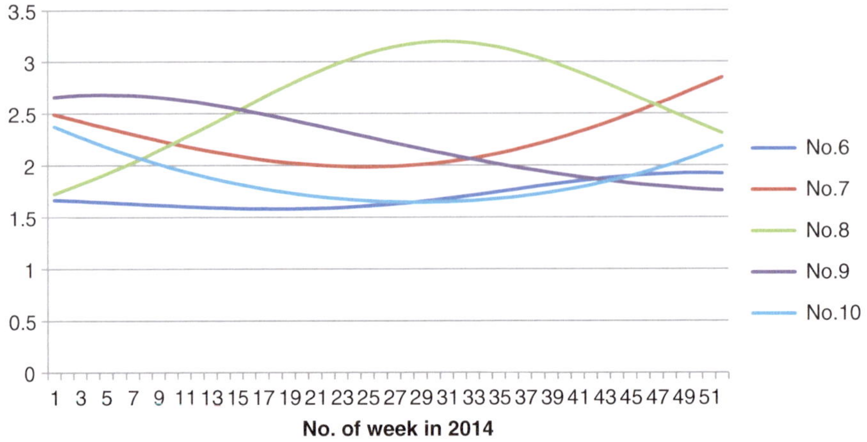

Fig. 9.43 Market price indexes of Sector 6, Sector 7, Sector 8, Sector 9, and Sector 10 in 2014

9.5.2 Simulation Results

Figure 9.43 depicts the price indexes of Sector 6, Sector 7, Sector 8, Sector 9, and Sector 10 during the year of 2014. As mentioned previously, Sector 6 is Manufacture of Foods and Tobacco; Sector 7 is Manufacture of Textile; Sector 8 is Manufacture of Textile Wearing Apparel, Footwear, Caps, Leather, Fur, Feather (Down), and Its Products; Sector 9 is Processing of Timbers and Manufacture of Furniture; and Sector 10 is Papermaking, Printing, and Manufacture of Articles. It is evident that the price indexes of the products of these sectors remained in high level. Their minimum values in the whole year were 1.5803, 1.9846, 1.7256, 1.7556, and 1.6440, respectively, which were all far greater than 1. The price index of Sector 8 even arrived at 3.1984 in the 31th week. This reflects the world economy is kind of prosperous compared with the situations in the previous years, leading to the increasing export demand for Chinese light industry products. Moreover, this is also due to the increase of domestic demand, which is the goal of China's economic transformation.

Figure 9.44 illustrates the price indexes of Sector 11, Sector 12, Sector 13, Sector 14, Sector 15, and Sector 16 during the year of 2014. As mentioned before, Sector 11 is Processing of Petroleum, Coking, and Processing of Nuclear Fuel; Sector 12 is Chemical Industry; Sector 13 is Manufacture of Nonmetallic Mineral Products; Sector 14 is Smelting and Rolling of Metals; Sector 15 is Manufacture of Metal Products; and Sector 16 is Manufacture of General-Purpose and Special Purpose Machinery. As shown in the figure, the price indexes were in high level and the overall trends of most price indexes were still up. The initial values of them were 1.8831, 1.8724, 2.5887, 1.5946, 2.2828, and 2.0876, while their final states were 2.4919, 1.9752, 3.4115, 2.2353, 2.9749, and 2.6371, respectively. This reflects the increasing demand for the basic materials.

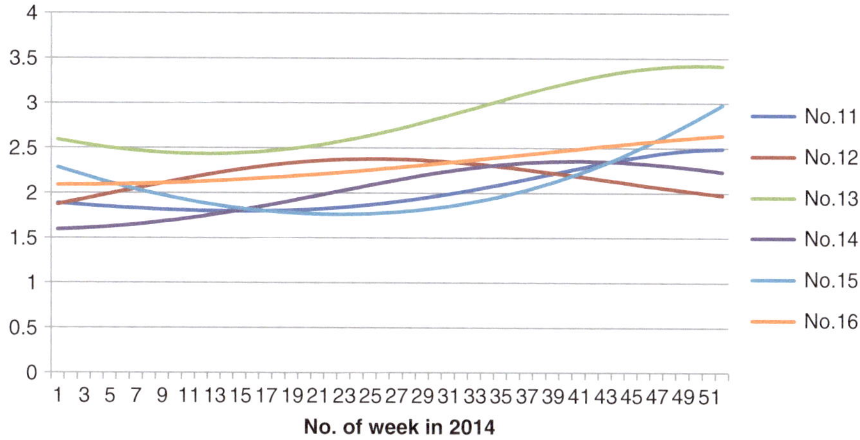

Fig. 9.44 Market price indexes of Sector 11, Sector 12, Sector 13, Sector 14, Sector 15, and Sector 16 in 2014

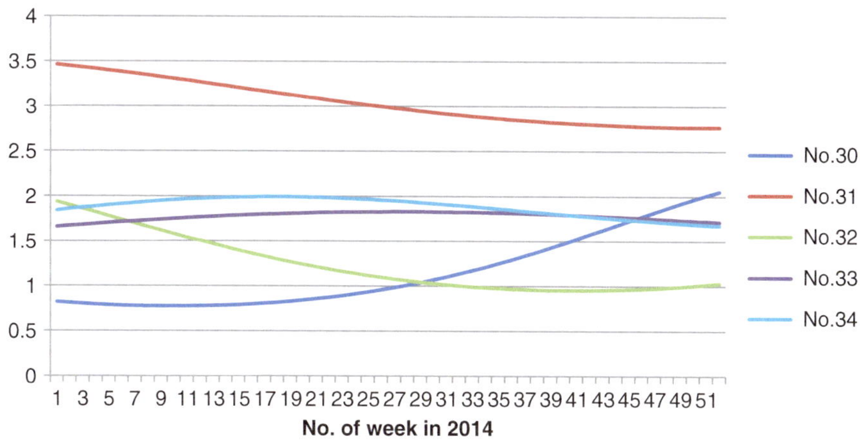

Fig. 9.45 Market price indexes of Sector 30, Sector 31, Sector 32, Sector 33, and Sector 34 in 2014

Figure 9.45 depicts the price indexes of Sector 30, Sector 31, Sector 32, Sector 33, and Sector 34 during the year of 2014. Sector 30 is Wholesale and Retail Trades; Sector 31 is Hotels and Catering Services; Sector 32 is Financial Intermediation; Sector 33 is Real Estate; and Sector 34 is Leasing and Business Services. The price index of Sector 30 increased from 0.8185 to 2.0517, which reflects the improvement of domestic demand. The price of Sector 31 and Sector 32 went down from 3.4592 and 1.9346 to 2.7657 and 1.0311, respectively. The price indexes of Sector 33 and Sector 34 remained at the same level between 1.7 and 2.0 through the whole year.

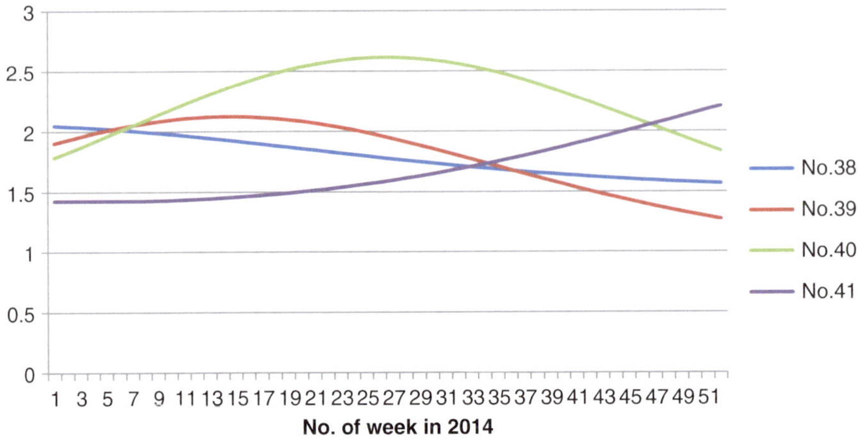

Fig. 9.46 Market price indexes of Sector 38, Sector 39, Sector 40, and Sector 41 in 2014

Figure 9.46 provides us the price indexes of Sector 38, Sector 39, Sector 40, and Sector 41 during the year of 2014. As mentioned in previous sections, Sector 38 is Services to Households and Other Services; Sector 39 is Education; Sector 40 is Health, Social Security, and Social Welfare; and Sector 41 is Culture, Sports, and Entertainment. It is clear that the overall trend of the price indexes of Sector 38, Sector 39, and Sector 40 was down. The price index of Sector 38 kept declining from 2.0471 to 1.5672. The price indexes of Sector 39 and Sector 40 went up firstly from 1.9026 and 1.7837 to 2.1224 and 2.6130 and then went down to 1.2731 and 1.8352, respectively, at the end of the year. The price index of Sector 41 kept increasing from 1.4258 to 1.8352, which reflects the increasing demand for qualified cultural products.

The value added and its growth rate of each sector in 2014 are shown in Fig. 9.47. Among the 42 sectors, Sector 1 (Agriculture, Forestry, Animal Husbandry, & Fishery) was no.1 according to value added. It is as high as 6143.45 billion CNY, which demonstrates that the foundation status of agriculture in China's national economy could not be changed. Sector 30 (Wholesale and Retail Trades) ranked second with the value added equaling 4393.10 billion CNY. The no. 3 to no. 5 was Sector 26 (Construction), Sector 33 (Real Estate), and Sector 32 (Financial Intermediation). Their value added in 2014 was, respectively, 4256.58 billion CNY, 3005.18 billion CNY, and 2976.37 billion CNY. In terms of the growth rate of the value added, the relative ranking of the 42 sectors was obviously different from that of the amount of value added. The first sector was Sector 26 (Construction). Sector 1 (Agriculture, Forestry, Animal Husbandry, & Fishery) came second. The next three sectors were Sector 36 (Comprehensive Technical Services), Sector 22 (Scrap and Waste), and Sector 40 (Health, Social Security, and Social Welfare). Their growth rate was 24.46 %, 17.90 %, 16.98 %, 16.37 %, and 14.21 %, respectively.

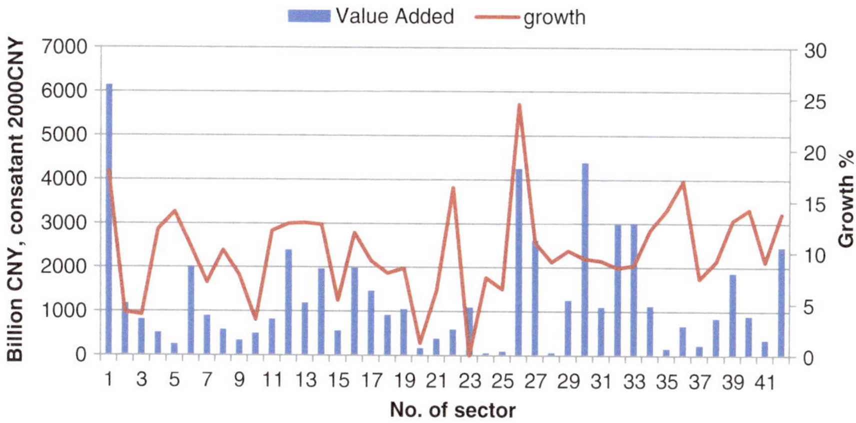

Fig. 9.47 Value added and growth of 42 sectors in 2014

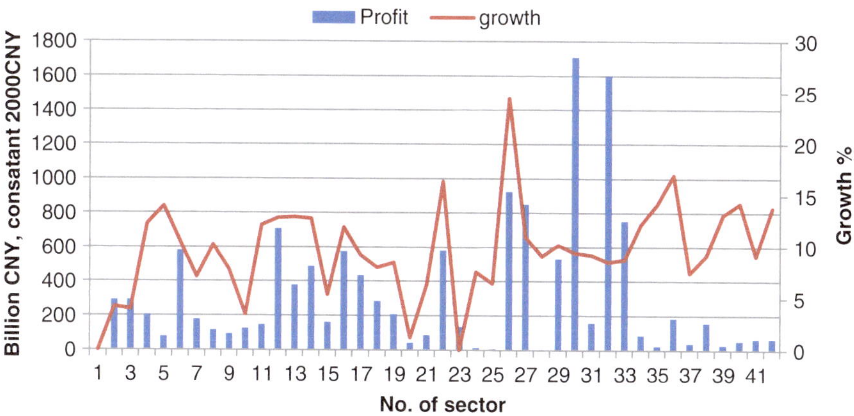

Fig. 9.48 Profit and growth of 42 sectors in 2014

The profit of each sector and its growth rate in 2014 are shown in Fig. 9.48. The top 5 sectors were Sector 30 (Wholesale and Retail Trades), Sector 32 (Financial Intermediation), Sector 26 (Construction), Sector 27 (Traffic, Transport, and Storage), and Sector 33 (Real Estate) according to the amount of profit. Their profit was 1709.15 billion CNY, 1599.16 billion CNY, 923.52 billion CNY, 850.05 billion CNY, and 753.68 billion CNY, respectively.

The total taxes of each sector and its growth rate in 2014 are illustrated in Fig. 9.49. In terms of the amount of taxes, the top 5 sectors were Sector 30 (Wholesale and Retail Trades), Sector 26 (Construction), Sector 6 (Manufacture of Foods and Tobacco), Sector 33 (Real Estate), and Sector 12 (Chemical Industry) in proper order. The taxes of these sectors were 1196.88 billion CNY, 673.25 billion CNY, 656.96 billion CNY, 473.34 billion CNY, and 471.28 billion CNY, respectively.

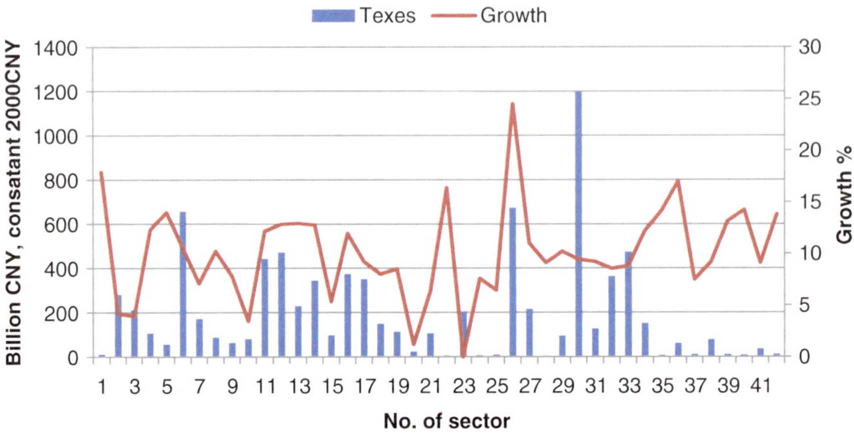

Fig. 9.49 Taxes and growth of 42 sectors in 2014

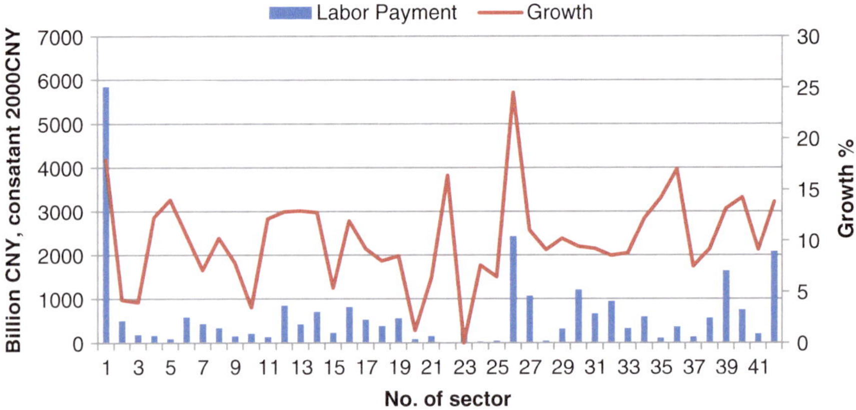

Fig. 9.50 Labor payment and growth of 42 sectors in 2014

These sectors were still the main contributors to the revenue of Chinese government.

The labor payment and growth rate of each sector in 2014 are depicted in Fig. 9.50. According to the amount of labor payment, the top 5 sectors were Sector 1 (Agriculture, Forestry, Animal Husbandry, & Fishery), Sector 26 (Construction), Sector 42 (Public Management and Social Organization), Sector 39 (Education), and Sector 30 (Wholesale and Retail Trades) in proper order. The labor payment of these sectors was 5844.76 billion CNY, 2424.29 billion CNY, 2079.23 billion CNY, 1640.85 billion CNY, and 1206.60 billion CNY, respectively.

The total output of each sector and its growth rate in 2014 are shown in Fig. 9.51. The top 5 sectors were Sector 26 (Construction), Sector 12 (Chemical Industry),

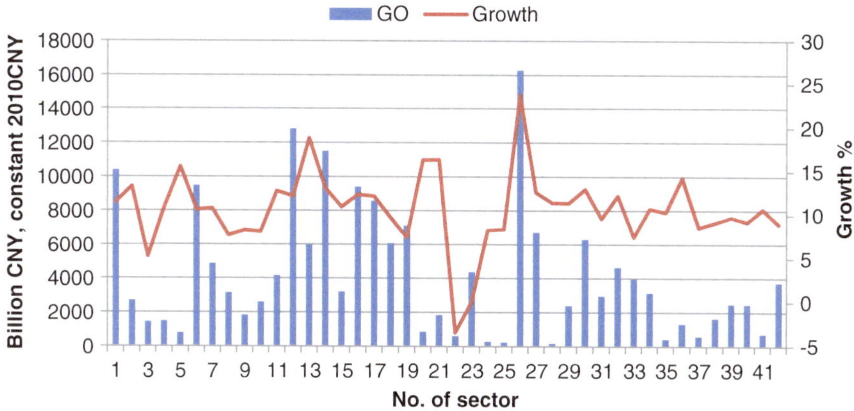

Fig. 9.51 Total output and growth of 42 sectors in 2014

Sector 14 (Smelting and Rolling of Metals), Sector 1 (Agriculture, Forestry, Animal Husbandry, & Fishery), and Sector 6 (Manufacture of Foods and Tobacco) according to the amount of output. Their total output was 16291.26 billion CNY, 12800.94 billion CNY, 11500.29 billion CNY, 10382.20 billion CNY, and 9467.38 billion CNY, respectively. In terms of the growth rate, the top 5 sectors were Sector 26 (Construction), Sector 13 (Manufacture of Nonmetallic Mineral Products), Sector 21 (Manufacture of Artwork, Other Manufacture), Sector 20 (Manufacture of Measuring Instrument and Machinery for Cultural Activity & Office Work), and Sector 5 (Mining and Processing of Nonmetal Ores and Other Ores). Their growth rates were 23.75 %, 18.83 %, 16.38 %, 16.32 %, and 15.58 %, respectively.

The total intermediate use of the product and the growth rate of each sector in 2014 are depicted in Fig. 9.52. According to the amount of intermediate use, the top 5 sectors were Sector 12 (Chemical Industry), Sector 14 (Smelting and Rolling of Metals), Sector 1 (Agriculture, Forestry, Animal Husbandry, & Fishery), Sector 27 (Traffic, Transport, and Storage), and Sector 23 (Production and Supply of Electric Power and Heat Power) in proper order. The intermediate use of the product of these sectors was 12730.06 billion CNY, 11845.84 billion CNY, 7876.18 billion CNY, 6078.48 billion CNY, and 5863.05 billion CNY, respectively. These sectors always play fundamental roles in China's national economy. Their products, such as cement and electricity, were widely utilized in the production of other sectors.

The capital formulation and the growth rate of each sector in 2014 are shown in Fig. 9.53. The top 5 sectors were Sector 26 (Construction), Sector 16 (Manufacture of General-Purpose and Special Purpose Machinery), Sector 17 (Manufacture of Transport Equipment), Sector 18 (Manufacture of Electrical Machinery and Equipment), and Sector 33 (Real Estate) according to the amount of output. Their capital formulation was 15369.80 billion CNY, 3691.50 billion CNY, 3660.74 billion CNY, 1328.32 billion CNY, and 984.52 billion CNY, respectively.

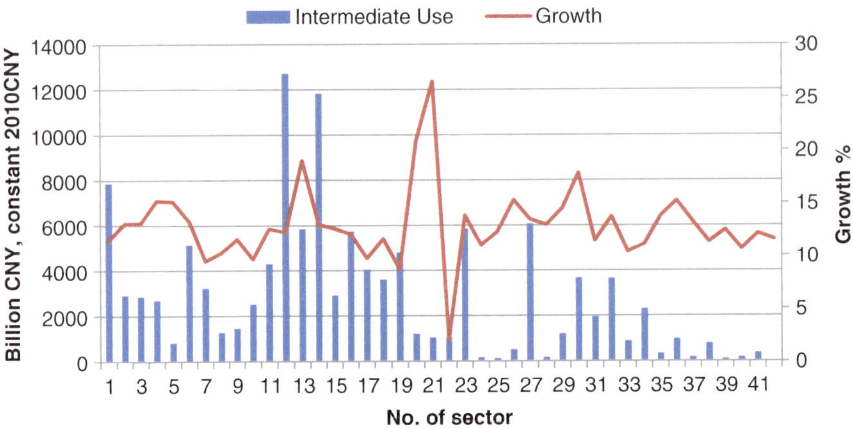

Fig. 9.52 Intermediate use and growth of 42 sectors in 2014

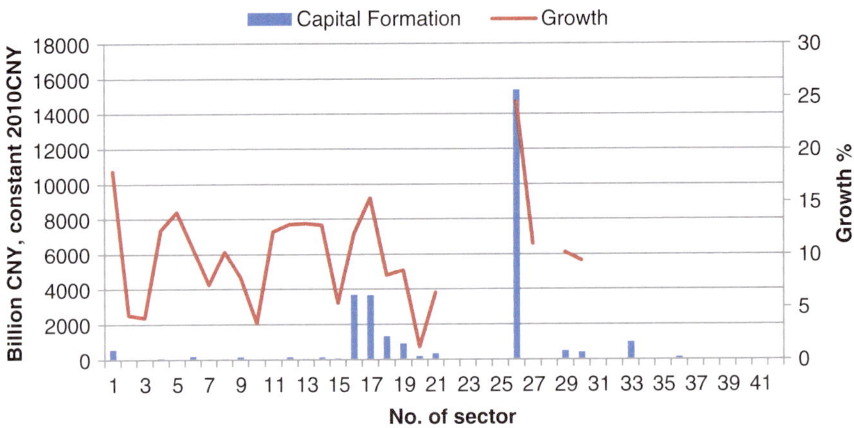

Fig. 9.53 Capital formation and growth of 42 sectors in 2014

The export of each sector and its growth rate in 2014 are depicted in Fig. 9.54. According to the amount of export, the top 5 sectors were Sector 19 (Manufacture of Communication Equipment, Computer, and Other Electronic Equipments), Sector 7 (Manufacture of Textile), Sector 18 (Manufacture of Electrical Machinery and Equipment), Sector 16 (Manufacture of General-Purpose and Special Purpose Machinery), and Sector 17 (Manufacture of Transport Equipment) in proper order. The export of the product of these sectors was 3253.31 billion CNY, 1932.74 billion CNY, 1209.81 billion CNY, 1136.37 billion CNY, and 1031.43 billion CNY, respectively. These sectors were still the main contributors of the export of China's national economy. The difference is that the Sector 17 was in top no. 5 in 2014 while Sector 12 (Chemical Industry) was no. 5 in 2013, which verifies the

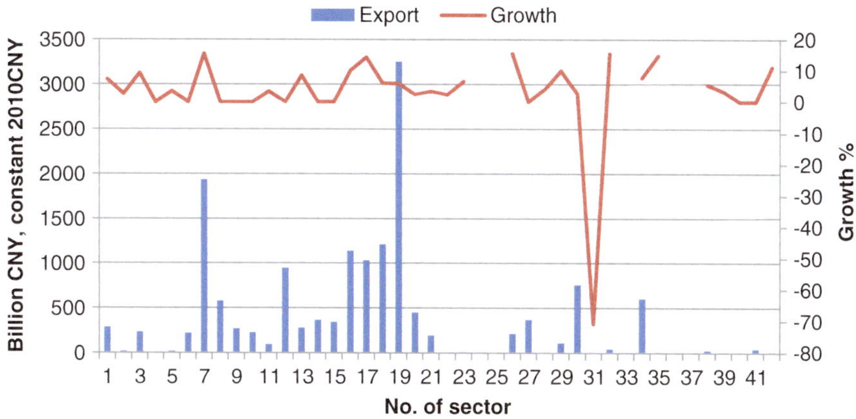

Fig. 9.54 Export and growth of 42 sectors in 2014

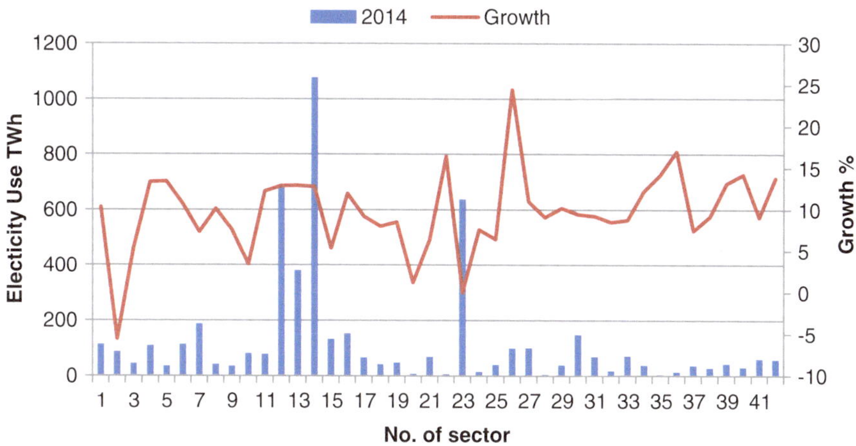

Fig. 9.55 Electricity consumption and growth rate of 42 sectors simulated by ARE in 2014

development of international business of transport equipment in China and the upgrade of the export of China's national economy.

The electricity consumption and its growth rate of each sector in 2014 simulated by the ARE model are shown in Fig. 9.55. The top 5 sectors were Sector 14 (Smelting and Rolling of Metals), Sector 12 (Chemical Industry), Sector 23 (Production and Supply of Electric Power and Heat Power), Sector 13 (Manufacture of Nonmetallic Mineral Products), and Sector 7 (Manufacture of Textile) in terms of the amount of consumption. When it comes to the growth rate in 2013, the top 5 sectors were Sector 26 (Construction), Sector 36 (Comprehensive Technical

Services), Sector 22 (Scrap and Waste), Sector 40 (Health, Social Security, and Social Welfare), and Sector 35 (Research and Experimental Development).

9.5.3 *Error Analysis*

The published electricity consumption and simulated electricity consumption are compared, in order to analyze the error of the simulation in 2014 of ARE model. The published whole industry electricity consumption is 4830.44 TWh while the simulated one is 4809.52 TWh, which means that the relative error is only −0.43 %. The error is considerably low. As for the primary industry, the published electricity consumption is 99.42 TWh while the simulated one is 103.24 TWh, indicating that the error is 3.84 %; as for the secondary industry, the published electricity consumption is 4065.02 TWh while the simulated one is 4039.29 TWh, indicating that the error is −0.63 %; and as for the tertiary industry, the published electricity consumption is 666.00 TWh while the simulated one is 666.99 TWh, indicating that the error is 0.15 %. Evidently, the errors are very low, which verifies that the simulation of the ARE model is accurate and reliable.

The errors of 42 sectors are shown in Fig. 9.56 and listed in Table 9.6. The highest error still appears in Sector 41, Culture, Sports, and Entertainment. But its electricity consumption is extremely low in the national economy, only accounting for 0.30 % in the whole industry electricity consumption. Meanwhile, its value added is quite low, too. Therefore, this error has very limited influences on the simulation results. Similarly, other several sectors with high errors are also small sectors. For example, Sector 36, Comprehensive Technical Services, has the second highest error, but its electricity consumption is less than 0.3 % of all the 42 sectors. On the whole, the errors of most other sectors are much lower.

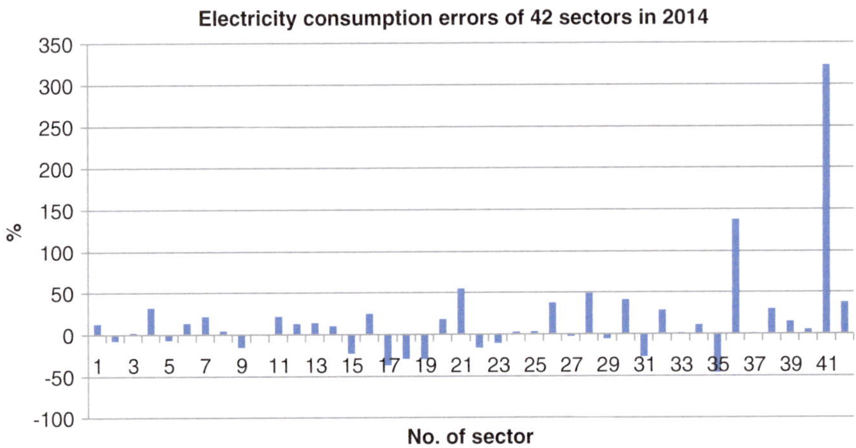

Fig. 9.56 Errors between published and simulated electricity consumption of 42 sectors in 2014

Table 9.6 Errors between published and simulated electricity consumption of 42 sectors in 2014

Sector number	Relative error	Sector number	Relative error	Sector number	Relative error
1	13.15	15	−23.01	29	−5.90
2	−7.59	16	24.93	30	41.42
3	2.39	17	−37.18	31	−27.31
4	32.24	18	−29.31	32	28.64
5	−6.89	19	−29.93	33	1.67
6	13.74	20	18.31	34	10.93
7	21.87	21	54.76	35	−46.49
8	4.42	22	−16.43	36	137.70
9	−15.49	23	−10.63	37	1.25
10	−0.53	24	3.16	38	29.98
11	21.61	25	3.58	39	15.28
12	12.94	26	37.67	40	5.59
13	14.25	27	−2.74	41	323.07
14	10.49	28	48.65	42	38.05

9.6 Summary and Conclusion

By comparing the published economic data and electricity data with simulated one, the errors are acceptable in a limited area. It shows that ARE model can be used to study China's economy from sector level to national economy level dynamically. As one of the simulation results by ARE model, the weekly price indexes of the products of 42 sectors are useful information to analyze the trends of the productions. Another useful result by ARE model is the input–output table; it can show the national economy completely and also reveal the production process of the sectors in the economy year by year. Therefore, it can be used to do policy studies, for example, some fiscal and monetary policies as input data to the ARE model, and review the simulated results. If the results are not satisfying, it can try to change some policies and simulate again, until optimal policies have been selected. ARE model is a policy study tool to do experiments on computer just like physical experiments on laboratory.

It will be studied by ARE model to simulate Chinese economy from 2015 to 2025 to select some optimal fiscal and monetary policies and try to find a sustainable way of the economy. The simulated results and input–output tables in 2015–2025 will be shown in the Chap. 10.

Chapter 10
China's Input–Output Tables of 2011–2025 Simulated by ARE

Input–output tables are considerably meaningful data foundation because they contain nearly all kinds of chief information on the national economy. From an input–output table, we cannot only find the intermediate commodity flow from each sector to another sector that reveals the positions and interrelationships of all the sectors but also figure out the compensation of employees, net taxes on production, depreciation of fixed assets, operating surplus, total value added and total input of each sector, residential consumption, government consumption, fixed capital formation, increase in inventories, export, import, and total output of each sector. A great number of researches can be carried out on the basis of the input–output table. Therefore, it is treated as an important tool by researchers and scholars in areas of national economy, international trade, public issues, etc. However, it is impossible to obtain the input–output table of each year since the input–output table will only be published once every few years. The reason is that compiling an input–output table requires a lot of human and material resources. To make matters worse, the time when the input–output table of a certain year is published tends to be several years later because it is very time-consuming to compile it. For example, only two input–output tables will be published every 5 years in China, and the delay can be as long as three years. In this chapter, we offer all the annual input–output tables (in the constant 2010 CNY) within the 2011–2025 period in China obtained by our simulation. Besides, we conduct some basic analysis of these tables. We hope that the information will meet the demand of some scholars who want to carry out various researches based on the input–output tables.

10.1 Input–Output Tables of 2011–2014 in China

In this section, we focus on the period of 2011–2014 that has just passed by in real life. However, no input–output tables of these years exist now. The input–output table of 2012 will be published in the future, while those tables of 2011, 2013, and

Z. Hu et al., *China's Economic Gene Mutations*,
DOI 10.1007/978-3-662-47298-9_10

2014 will not be published forever. We provide readers with these tables that are obtained by means of our program simulation in Sect. 10.1.1, since it may be useful for scholars and researchers as far as we are concerned. Then, we offer some brief analysis of these tables from the perspectives of influence coefficient and reaction coefficient of each sector in Sect. 10.1.2.

All the input–output tables in this book and in our model are the version of 42-sector tables. The correspondence between the number and the sectors is shown in Table 10.1.

Table 10.1 42 sectors in China's economy

No.	Sectors	No.	Sectors
1	Agriculture, Forestry, Animal Husbandry, and Fishery	22	Scrap and Waste
2	Mining and Washing of Coal	23	Production and Supply of Electric Power and Heat Power
3	Extraction of Petroleum and Natural Gas	24	Production and Distribution of Gas
4	Mining of Metal Ores	25	Production and Distribution of Water
5	Mining and Processing of Nonmetal Ores and Other Ores	26	Construction
6	Manufacture of Foods and Tobacco	27	Traffic, Transport, and Storage
7	Manufacture of Textile	28	Post Service
8	Manufacture of Textile Wearing Apparel, Footwear, Caps, Leather, Fur, Feather (Down), and Its Products	29	Information Transmission, Computer Services, and Software
9	Processing of Timbers and Manufacture of Furniture	30	Wholesale and Retail Trades
10	Papermaking, Printing, and Manufacture of Articles	31	Hotels and Catering Services
11	Processing of Petroleum, Coking, Processing of Nuclear Fuel	32	Financial Intermediation
12	Chemical Industry	33	Real Estate
13	Manufacture of Nonmetallic Mineral Products	34	Leasing and Business Services
14	Smelting and Rolling of Metals	35	Research and Experimental Development
15	Manufacture of Metal Products	36	Comprehensive Technical Services
16	Manufacture of General Purpose and Special Purpose Machinery	37	Management of Water Conservancy, Environment, and Public Facilities
17	Manufacture of Transport Equipment	38	Services to Households and Other Services
18	Manufacture of Electrical Machinery and Equipment	39	Education
19	Manufacture of Communication Equipment, Computers, and Other Electronic Equipment	40	Health, Social Security, and Social Welfare
20	Manufacture of Measuring Instrument and Machinery for Cultural Activity and Office Work	41	Culture, Sports, and Entertainment
21	Manufacture of Artwork and Other Manufacture	42	Public Management and Social Organization

10.1.1 Input–Output Tables

In this subsection, the input–output tables in the years of 2011, 2012, 2013, and 2014 are displayed. The unit of the numbers in all the tables is million CNY in constant 2010 CNY. In the following tables, we will refer to each sector by its number in order to save space. China's national input–output table in 2011, 2012, 2013, and 2014 obtained by the simulation of our model is shown in Tables 10.2, 10.3, 10.4, and 10.5 respectively.

10.1.2 Analysis

The input–output tables contain lots of useful information as stated before. Therefore, a large number of analyses can be conducted on the basis of diverse kinds of data in the input–output table. In this section, we will calculate the influence coefficient and the reaction coefficient of each sector, which is a very typical and popular application of the input–output table in 2011–2014.

Influence coefficient refers to the influence extent of the increment of a unit of final product of a certain sector on the demand of all the other sectors. The influence coefficient of a sector reflects the pulling effect of the sector on other sectors.

Reaction coefficient refers to the impact extent of the increment of a unit of final demand of all the sectors in the national economy on the demand of a certain sector. The reaction coefficient of a sector indicates its pushing effect on other sectors.

Table 10.2 China's national input–output table in 2011

	Sector 1	Sector 2	Sector 3	Sector 4	Sector 5	Sector 6	Sector 7	Sector 8
Sector 1	1,029,662	15,137	21	1439	165	2,928,912	608,337	114,195
Sector 2	5785	366,410	9658	4840	1382	20,201	16,160	9090
Sector 3	0	0	18,629	0	0	0	0	0
Sector 4	0	0	0	138,024	0	0	0	0
Sector 5	897	777	592	1182	39,495	2659	0	0
Sector 6	791,983	3389	2022	2608	1468	1,525,540	25,097	131,633
Sector 7	1051	554	895	666	312	3816	1,320,874	867,811
Sector 8	2898	6852	3987	4572	2231	9694	55,419	403,409
Sector 9	7643	15,675	4288	3552	965	7927	2975	4772
Sector 10	6920	2313	2395	2202	1162	138,006	35,385	39,086
Sector 11	55,114	23,072	42,442	67,144	26,048	20,776	8873	16,393
Sector 12	605,787	44,925	41,985	56,959	54,136	204,117	404,934	152,305
Sector 13	11,459	20,720	10,436	10,208	21,856	51,104	8954	3665
Sector 14	1872	86,518	55,031	15,370	3781	8342	2814	3635
Sector 15	25,090	39,521	8660	25,399	8824	34,237	6608	11,577
Sector 16	45,979	114,568	93,350	96,030	49,802	25,291	44,362	17,430
Sector 17	23,513	15,034	7679	12,589	12,135	9387	7036	4068
Sector 18	2163	31,340	10,440	15,631	5378	10,778	8195	4980
Sector 19	735	3260	1453	1594	607	2726	2627	2881
Sector 20	3071	11,400	10,650	4136	1331	5214	2974	1592
Sector 21	20,101	8479	1552	5056	3385	11,325	30,343	22,790
Sector 22	654	279	18	1164	6185	5761	7168	1451
Sector 23	82,396	68,868	82,380	155,640	48,316	76,735	74,889	18,284
Sector 24	298	822	1392	6180	2918	4381	1907	952
Sector 25	896	1385	885	1015	516	2998	2703	1149
Sector 26	1996	2501	1274	424	185	2088	810	782
Sector 27	145,516	98,887	21,869	54,328	35,459	212,246	66,687	63,307
Sector 28	5116	525	181	203	154	1670	2331	4559
Sector 29	26,643	5319	1912	1633	1150	13,145	6391	11,943
Sector 30	122,684	21,075	10,339	12,757	7334	178,777	64,282	46,112
Sector 31	21,103	23,354	10,012	13,905	8185	43,854	14,849	15,045
Sector 32	71,928	39,214	9247	12,426	7338	76,153	68,158	31,736
Sector 33	1801	2537	504	630	649	12,728	10,399	24,878
Sector 34	11,444	10,036	2050	4307	2129	122,263	12,653	54,823
Sector 35	12,086	2664	2062	369	123	6007	1796	1244
Sector 36	64,461	17,290	12,709	14,898	5430	25,472	7946	8463
Sector 37	19,321	2262	919	729	1254	2270	4704	320
Sector 38	25,329	13,868	4214	2473	1891	8924	3446	4185
Sector 39	2630	1562	208	379	572	1018	322	260
Sector 40	4452	4240	404	1954	989	3687	1549	1709
Sector 41	1753	2681	1080	2831	2076	7800	3620	4271
Sector 42	2303	432	341	208	90	1035	373	328
Total intermediate input	3,266,535	1,129,746	490,168	757,650	367,405	5,829,065	2,948,952	2,107,112
Compensation of employees	4,306,467	412,402	165,326	129,220	69,243	453,347	350,380	285,380
Net taxes on production	8750	230,045	187,731	83,379	42,134	511,334	139,443	74,054
Depreciation of fixed assets	211,334	78,816	110,365	30,498	17,951	140,494	92,027	33,613
Operating surplus	0	238,694	258,642	158,515	57,496	451,546	143,731	95,662
Total value added	4,526,551	959,958	722,064	401,612	186,824	1,556,722	725,581	488,709
Total input	7,793,085	2,089,704	1,212,232	1,159,262	554,229	7,385,787	3,674,533	2,595,821

Sector 9	Sector 10	Sector 11	Sector 12	Sector 13	Sector 14	Sector 15	Sector 16	Sector 17
180,176	94,797	70	312,025	1925	712	562	743	384
11,881	18,082	194,426	143,032	289,132	290,706	10,500	23,484	6233
0	0	1,761,267	188,413	7785	22,282	0	0	0
0	0	0	44,092	9641	1,641,200	24,751	20,294	2806
0	0	316	156,161	240,829	21,402	9295	4827	1447
8307	8422	14,561	156,995	14,395	27,406	7913	22,552	19,562
15,653	41,155	159	72,191	9244	962	3951	9675	11,053
27,745	15,321	1299	22,796	10,236	18,518	7358	14,779	62,462
473,675	45,977	959	15,633	27,405	6020	39,604	22,194	62,419
24,567	608,336	1697	139,570	90,953	8679	23,928	30,113	21,307
11,553	10,529	189,770	598,407	171,506	360,624	25,680	70,417	28,278
146,279	288,224	58,558	3,998,574	309,458	117,259	94,301	248,669	344,060
14,123	6146	10,600	79,890	783,309	180,039	28,875	49,502	49,072
36,171	25,010	2000	100,377	112,384	2,380,380	859,552	1,447,406	709,863
39,834	32,413	10,060	87,454	115,289	79,668	335,369	288,986	105,491
23,958	29,956	40,178	154,453	119,774	280,228	124,300	1,549,936	541,445
7630	17,841	7049	31,210	23,285	47,152	23,080	103,882	2,058,226
5371	14,523	3539	26,976	16,837	28,336	21,123	376,853	237,688
668	18,493	426	6073	2594	2179	2951	143,594	67,040
1188	9903	4440	45,140	6712	17,648	4044	36,024	76,954
7127	10,524	1379	43,016	19,094	18,830	10,672	47,620	48,721
5854	141,471	616	55,239	78,038	349,405	27,617	81,717	5653
44,477	53,597	51,606	435,066	246,905	406,542	145,788	153,909	73,958
745	656	4578	25,459	4866	11,563	2497	6139	4065
1020	4550	941	10,455	3934	5470	2006	2299	2871
411	516	1508	3733	1375	2634	456	2380	2012
53,830	55,508	59,595	283,654	184,512	172,485	58,949	197,033	125,258
1007	969	93	7016	989	2334	445	6107	1043
5572	5089	9624	32,158	8201	66,703	6819	21,741	13,260
34,843	49,643	36,732	183,476	94,448	136,582	53,598	176,214	227,771
9009	11,605	5476	59,947	32,863	32,546	30,643	53,195	26,976
24,752	32,932	18,330	162,157	107,956	120,249	22,792	73,228	60,514
6842	9264	487	21,249	9166	2090	10,184	19,215	8770
14,191	17,243	14,433	138,637	25,732	13,289	9528	60,765	77,326
894	1064	834	21,163	2788	12,241	1743	23,920	44,879
2738	9698	5137	55,713	14,766	26,455	11,966	41,191	46,391
211	5724	605	8943	1049	2270	311	1423	1478
1671	3644	2661	19,285	11,351	17,889	3705	8100	7413
233	251	121	1305	541	566	280	2186	957
1351	2328	782	5880	6587	6807	5102	16,346	10,150
1878	2236	691	11,112	6039	4843	2372	9650	5549
195	281	139	1121	604	744	286	934	704
1,247,628	1,703,920	2,517,744	7,965,246	3,224,497	6,943,936	2,054,894	5,469,240	5,201,510
131,118	184,565	101,479	671,956	318,134	537,312	194,627	602,983	423,633
54,442	71,084	335,031	371,103	173,233	260,130	81,026	276,691	281,442
31,316	70,162	75,205	287,606	123,588	320,465	56,833	160,523	122,788
79,103	109,446	111,538	556,633	284,989	367,452	136,317	426,614	348,760
295,980	435,257	623,252	1,887,298	899,943	1,485,359	468,804	1,466,811	1,176,623
1,543,607	2,139,177	3,140,996	9,852,543	4,124,440	8,429,295	2,523,697	6,936,051	6,378,133

(continued)

Table 10.2 (continued)

	Sector 18	Sector 19	Sector 20	Sector 21	Sector 22	Sector 23	Sector 24	Sector 25
Sector 1	75	0	27	138,412	175	78	39	0
Sector 2	3650	1779	1036	7567	807	769,486	11,122	241
Sector 3	0	0	0	11	0	27,359	108,972	0
Sector 4	5482	89	820	105	532	129	0	0
Sector 5	1980	41	76	3178	67	1176	9	18
Sector 6	17,612	22,130	2279	15,627	692	13,324	863	474
Sector 7	9148	1665	1069	103,392	76	225	27	54
Sector 8	10,098	10,963	2162	17,045	119	14,515	2598	895
Sector 9	12,086	9606	2025	35,003	89	2703	271	267
Sector 10	52,801	46,970	7204	20,507	121	8326	584	473
Sector 11	20,706	16,729	3404	8384	1190	128,623	4815	1879
Sector 12	389,393	394,872	62,792	87,350	11,627	13,583	1487	8959
Sector 13	102,703	115,035	25,919	23,996	1432	9312	535	530
Sector 14	1,265,737	193,483	28,705	109,083	5749	4221	571	210
Sector 15	214,695	138,572	28,926	48,278	603	25,941	2206	4890
Sector 16	248,052	66,474	20,628	4584	356	43,932	1870	1929
Sector 17	13,367	21,854	4614	11,706	123	48,927	5074	2252
Sector 18	712,531	180,336	37,090	7171	904	242,202	537	610
Sector 19	281,373	2,864,416	167,212	2852	46	2688	121	146
Sector 20	29,053	189,749	91,273	2476	54	58,872	1452	984
Sector 21	34,933	42,045	7603	125,653	107	2921	173	187
Sector 22	48,990	7743	88	227	5262	479	0	0
Sector 23	51,391	70,151	7641	15,415	2447	1,564,541	4629	42,524
Sector 24	2962	3311	503	1094	119	5088	12,900	591
Sector 25	1766	2394	518	778	114	5450	124	3803
Sector 26	1212	1144	198	179	56	1501	172	221
Sector 27	118,194	105,319	16,417	27,205	4174	53,364	6123	1720
Sector 28	3150	2069	805	324	12	517	30	56
Sector 29	13,948	54,147	2798	7728	167	26,815	330	671
Sector 30	128,499	210,923	16,742	22,803	2556	34,684	3888	1101
Sector 31	34,095	19,394	6914	8202	510	12,550	1062	1446
Sector 32	56,067	180,863	5742	12,521	2351	181,791	7677	10,450
Sector 33	13,498	15,138	3198	7778	178	1918	243	96
Sector 34	104,368	67,827	9232	9641	433	19,518	1020	893
Sector 35	16,694	41,348	3602	1122	45	2914	61	37
Sector 36	24,121	30,331	4240	2457	219	35,121	534	276
Sector 37	861	689	68	183	31	11,697	101	10,394
Sector 38	3735	3998	600	486	473	33,544	667	1432
Sector 39	613	543	131	191	5	677	54	139
Sector 40	7899	3030	1214	179	86	5229	650	40
Sector 41	5031	5091	782	1266	79	9876	437	545
Sector 42	478	589	91	163	0	625	20	36
Total intermediate input	4,063,047	5,142,850	576,389	892,323	44,189	3,426,441	184,048	101,473
Compensation of employees	315,505	481,228	76,369	103,418	6410	350,518	24,773	36,972
Net taxes on production	122,088	97,752	21,537	69,686	3839	213,148	5748	7286
Depreciation of fixed assets	86,095	151,090	15,578	23,341	2883	439,304	7692	31,841
Operating surplus	232,184	180,062	38,988	56,962	391,136	142,610	11,247	4269
Total value added	755,871	910,132	152,472	253,407	404,267	1,145,581	49,460	80,369
Total input	4,818,918	6,052,982	728,861	1,145,730	448,456	4,572,022	233,508	181,841

Sector 26	Sector 27	Sector 28	Sector 29	Sector 30	Sector 31	Sector 32	Sector 33	Sector 34
46,357	119,970	0	0	1041	263,625	0	264	5532
8371	17,311	188	0	1312	1522	0	683	3287
0	6245	0	0	0	0	0	0	0
0	0	0	0	0	0	0	0	0
106,589	1303	0	0	0	0	0	0	0
42,123	39,542	335	4396	22,382	793,439	9619	9053	27,584
8907	6495	203	221	5189	11,229	491	3583	30,136
40,157	35,900	969	3412	38,009	8616	9051	14,533	29,074
184,720	6124	669	1728	7103	3157	2580	3898	5889
17,632	18,447	3563	56,210	39,854	7767	66,706	18,895	208,909
136,016	936,753	3965	3476	14,401	11,234	21,704	42,105	74,223
325,486	88,421	446	9553	39,032	17,411	11,648	53,813	83,523
2,390,806	10,646	465	489	1172	4681	551	954	1238
1,315,344	25,873	0	18	2108	103	0	2890	1781
343,074	27,962	105	3625	1731	2668	3105	19,557	46,811
312,333	120,991	1408	14,585	9299	2679	14,620	11,419	15,305
43,611	353,653	10,625	12,712	37,715	7711	9993	12,303	70,945
399,108	14,127	238	124,138	29,361	16,365	7058	35,745	65,381
25,435	12,662	1179	107,738	16,825	931	2568	9657	134,695
23,041	8001	159	16,165	1405	72	13,182	6297	6942
34,346	4598	340	1910	3295	1423	3576	4164	14,131
0	0	0	0	0	0	0	0	0
78,803	60,215	1835	44,959	50,488	48,806	32,378	19,291	22,234
1248	4720	128	1026	883	8335	171	340	2251
2533	3026	216	2203	1652	3965	1624	588	1522
111,916	21,614	5681	3056	11,140	5226	4970	62,021	2276
982,957	424,075	19,154	55,283	302,664	27,777	85,201	24,322	114,730
2275	1578	5635	2356	5828	1312	14,789	2057	10,965
152,764	36,088	2730	78,126	35,119	7019	97,982	12,558	8717
259,590	64,988	10,074	56,641	21,913	48,295	14,570	13,750	54,077
105,507	80,054	966	35,414	82,910	15,456	126,734	52,499	153,841
99,822	296,358	1574	37,323	147,389	37,801	276,229	138,005	105,797
6148	24,515	2045	47,149	95,937	24,211	110,181	42,397	38,900
38,025	27,898	5215	90,391	210,490	27,287	209,583	108,768	102,083
7903	1545	73	3017	6734	20	4821	678	1164
122,036	5913	24	8814	16,647	421	2384	6176	1739
519	1516	0	505	998	749	2209	1035	20,027
14,775	114,312	214	11,544	39,922	21,313	13,359	23,710	36,108
1313	3751	135	2112	2581	595	6689	760	1122
3132	2710	26	537	694	385	2379	1113	381
14,657	10,341	1005	10,089	16,823	5901	20,584	17,028	10,006
1346	1417	44	687	1074	352	1962	2132	444
7,810,724	3,041,654	81,629	851,608	1,323,121	1,439,858	1,205,250	779,042	1,513,767
1,567,101	813,459	38,525	242,464	890,670	517,053	710,082	264,218	447,452
435,196	163,443	3473	71,321	883,501	98,189	272,388	380,871	113,478
152,245	361,764	9079	234,110	207,032	115,908	52,439	1,166,561	217,509
596,979	644,347	111	399,423	1,261,637	122,809	1,201,693	606,438	64,049
2,751,522	1,983,012	51,188	947,320	3,242,840	853,960	2,236,602	2,418,088	842,487
10,562,246	5,024,666	132,817	1,798,927	4,565,961	2,293,817	3,441,852	3,197,130	2,356,254

(continued)

Table 10.2 (continued)

	Sector 35	Sector 36	Sector 37	Sector 38	Sector 39	Sector 40	Sector 41	Sector 42
Sector 1	12,564	3774	11,703	14,802	3276	8753	2608	0
Sector 2	577	1270	2508	5157	3480	7381	433	7685
Sector 3	0	0	0	0	0	0	0	0
Sector 4	3144	589	844	0	0	0	0	0
Sector 5	51	29	315	67	632	934	262	1084
Sector 6	5934	2323	5092	28,752	13,803	15,240	29,962	9958
Sector 7	6753	626	2073	19,055	3000	10,169	1569	11,167
Sector 8	2579	1586	6257	8921	1921	25,067	8906	36,511
Sector 9	1132	1207	6173	11,938	20,881	13,636	6401	44,483
Sector 10	3720	13,368	3284	21,132	59,822	22,754	60,139	71,300
Sector 11	3050	19,392	17,778	23,244	13,493	7459	3474	39,765
Sector 12	36,267	20,897	32,564	110,341	15,384	778,856	21,946	27,711
Sector 13	2627	1839	6738	1346	8391	10,185	3425	13,352
Sector 14	1673	6966	3649	3249	1279	87	246	0
Sector 15	8505	20,267	3860	6288	3669	3814	2125	3241
Sector 16	5018	7623	10,130	3074	3412	81,597	1496	6363
Sector 17	2964	13,841	17,036	48,317	11,379	2026	2501	35,570
Sector 18	14,721	39,515	12,627	20,779	5041	5812	2951	6233
Sector 19	13,156	35,665	6961	31,224	5526	1344	3192	21,060
Sector 20	11,772	21,509	6664	12,333	16,642	4258	2004	4814
Sector 21	1522	2097	2538	10,070	2912	2292	1079	6198
Sector 22	0	0	0	0	0	0	0	0
Sector 23	6797	6560	13,791	14,049	31,184	21,196	4805	37,187
Sector 24	110	241	443	817	1100	848	358	823
Sector 25	518	1120	902	2079	2313	1216	423	1566
Sector 26	1188	2863	12,085	9713	8218	35,325	4355	38,302
Sector 27	13,844	56,303	11,180	26,612	43,806	9137	15,730	113,761
Sector 28	561	1131	463	1764	3575	585	1218	29,174
Sector 29	1899	4690	4298	4330	13,862	16,602	4315	50,880
Sector 30	6960	11,148	6911	22,651	12,587	54,439	10,980	27,955
Sector 31	12,938	38,945	11,556	19,957	41,017	17,562	16,653	173,755
Sector 32	1438	30,033	15,439	26,867	42,968	10,346	7998	43,434
Sector 33	1522	9720	1461	35,646	12,216	3481	6882	14,765
Sector 34	3160	14,308	7147	18,068	29,649	12,966	11,729	24,393
Sector 35	8012	2186	466	34	3421	827	538	37
Sector 36	4846	59,299	2496	976	3483	2548	364	3803
Sector 37	209	201	11,514	1839	2160	460	477	6326
Sector 38	994	8529	12,109	49,338	12,013	9948	5032	39,019
Sector 39	1123	940	800	389	7374	1734	510	20,804
Sector 40	881	884	1804	410	2423	4490	765	7449
Sector 41	5194	4831	3525	5093	9878	2614	16,342	33,938
Sector 42	86	303	191	293	319	262	110	730
Total intermediate input	210,010	468,617	277,377	621,018	477,509	1,208,250	264,307	1,014,596
Compensation of employees	81,569	253,575	105,497	434,794	1,207,545	560,989	154,449	1,554,545
Net taxes on production	5209	42,409	8660	60,064	7770	5853	26,628	8440
Depreciation of fixed assets	15,306	48,236	42,286	35,350	143,522	68,100	41,348	228,860
Operating surplus	17,771	130,297	31,739	122,421	20,259	40,171	48,551	49,197
Total value added	119,855	474,517	188,183	652,630	1,379,095	675,114	270,976	1,841,042
Total input	329,865	943,134	465,560	1,273,647	1,856,604	1,883,364	535,283	2,855,638

Total intermediate use	Rural household consumption expenditure	Urban household consumption expenditure	Government consumption expenditure	Gross fixed capital formation	Increase in inventories	Export	Import	Total output
5,922,335	643,828	735,849	56,284	422,229	221,672	145,637	433,664	7,714,168
2,277,855	9516	6165	0	21,772	−87,034	14,517	131,890	2,110,901
2,140,962	0	0	0	15,146	−55,601	82,385	996,660	1,186,233
1,892,543	0	0	0	62,027	−19,071	7635	767,394	1,175,740
597,690	0	0	0	5424	−22,672	16,451	34,646	562,247
3,896,390	839,495	2,225,673	0	138,026	255,511	220,825	242,567	7,333,354
2,596,545	23,420	34,984	0	19,755	−27,421	1,158,339	150,712	3,654,909
999,437	171,864	747,458	0	53,960	90,604	579,653	75,977	2,567,000
1,125,455	15,942	65,254	0	138,443	−1685	269,856	69,107	1,544,157
2,007,107	9704	73,138	0	27,467	−34,103	226,932	125,699	2,184,547
3,283,886	13,926	135,315	0	−4083	−134,219	84,934	225,523	3,154,235
9,813,895	87,148	284,205	0	121,808	−131,943	944,203	1,228,053	9,891,261
4,078,326	3595	31,490	0	44,371	−135,481	212,221	72,872	4,161,650
8,823,531	0	0	0	103,011	−188,641	365,674	541,125	8,562,449
2,218,994	8814	46,017	0	64,802	−54,600	345,348	83,920	2,545,455
4,360,218	625	9496	0	2,733,983	89,393	821,220	1,025,237	6,989,698
3,211,615	81,104	427,110	0	2,528,571	119,509	639,787	659,492	6,348,205
2,800,733	74,827	279,480	0	1,087,603	34,952	998,814	451,801	4,824,607
4,008,574	55,839	226,067	0	776,927	187,744	2,630,974	1,889,322	5,996,802
771,594	5925	29,521	0	170,142	−24,607	388,236	602,544	738,266
620,126	47,540	207,858	0	227,692	9937	174,355	52,121	1,235,387
831,081	0	0	0	0	6321	4061	369,448	471,991
4,472,672	60,878	224,532	0	0	−192,765	8571	1866	4,572,022
129,830	11,420	91,810	0	1532	358	0	0	234,950
87,505	6855	88,324	0	0	1226	0	0	183,909
369,715	0	139,114	0	9,935,282	72,498	126,753	39,789	10,603,572
4,548,174	68,832	280,042	151,222	35,830	−179,449	369,632	199,234	5,075,051
126,973	1039	6383	0	0	2855	5349	7489	135,110
875,886	95,669	384,598	0	365,939	43,230	80,038	31,138	1,814,222
2,575,392	266,732	788,309	0	307,405	47,359	701,057	0	4,686,255
1,482,506	161,191	619,631	0	0	98,640	48,745	85,796	2,324,917
2,715,390	108,838	542,978	31,425	0	108,023	26,783	25,300	3,508,137
660,619	310,422	1,327,537	0	792,188	120,162	0	0	3,210,928
1,744,942	15,750	229,495	84,618	0	59,141	472,897	249,350	2,357,495
243,175	0	0	180,535	0	2978	7132	99,662	334,158
709,989	1776	22,731	130,130	100,514	−1537	0	0	963,603
128,559	4463	57,532	280,526	0	−2160	0	0	468,919
597,228	98,101	561,976	0	0	35,106	23,137	29,469	1,286,079
68,476	172,317	392,103	1,235,010	0	−27,196	3628	10,758	1,833,580
123,105	86,268	896,461	775,773	0	−23,344	3828	2677	1,859,414
281,439	18,886	100,170	138,543	0	12,312	42,790	53,539	540,601
23,874	0	0	2,803,994	0	−12,349	7558	8288	2,814,789

Table 10.3 China's national input–output table in 2012

	Sector 1	Sector 2	Sector 3	Sector 4	Sector 5	Sector 6	Sector 7	Sector 8
Sector 1	1,120,258	20,653	23	2109	222	3,190,815	735,517	121,712
Sector 2	6332	427,659	10,416	5737	1596	22,007	17,778	9629
Sector 3	0	0	20,091	0	0	0	0	0
Sector 4	0	0	0	151,867	0	0	0	0
Sector 5	974	882	638	1337	44,257	2897	0	0
Sector 6	858,514	4043	2180	3247	1735	1,661,953	28,020	138,851
Sector 7	1130	625	965	774	350	4158	1,427,639	908,188
Sector 8	3090	8330	4300	5824	2686	10,560	62,619	422,258
Sector 9	8218	18,463	4625	4441	1128	8636	3266	5007
Sector 10	7447	2546	2583	2414	1269	150,347	37,588	40,874
Sector 11	59,221	26,421	45,773	78,812	29,582	22,634	9661	17,166
Sector 12	645,574	53,002	45,280	69,358	63,295	222,369	448,672	159,047
Sector 13	12,249	23,124	11,256	11,649	24,193	55,673	9560	3823
Sector 14	2023	99,475	59,350	17,649	4297	9088	3075	3822
Sector 15	26,837	43,905	9339	28,691	9733	37,298	7056	12,072
Sector 16	48,646	129,674	100,676	110,394	55,925	27,552	47,996	18,088
Sector 17	24,656	16,890	8281	14,664	13,553	10,226	7549	4199
Sector 18	2301	35,690	11,259	18,356	6080	11,742	8877	5184
Sector 19	792	3990	1567	2029	735	2970	2989	3034
Sector 20	3606	14,672	11,486	5793	1676	5680	3383	1772
Sector 21	24,787	13,184	1674	5811	5122	12,338	36,287	26,341
Sector 22	665	340	19	1221	7415	6276	8156	1488
Sector 23	92,962	90,293	88,846	212,681	62,415	83,597	88,606	19,839
Sector 24	316	933	1501	7159	3287	4773	2065	988
Sector 25	960	1591	954	1189	588	3266	2942	1201
Sector 26	2155	2963	1374	521	217	2274	896	823
Sector 27	153,773	109,582	23,586	61,623	39,028	231,225	70,903	65,565
Sector 28	5591	620	195	237	179	1819	2602	4830
Sector 29	29,036	6296	2062	1995	1348	14,321	7062	12,631
Sector 30	137,551	25,636	11,151	15,056	8785	194,763	72,552	49,665
Sector 31	22,025	25,254	10,798	15,326	8803	47,776	15,540	15,456
Sector 32	78,221	44,690	9972	14,549	8284	82,963	73,046	33,454
Sector 33	1963	3079	543	809	780	13,866	11,589	26,349
Sector 34	12,273	11,916	2211	5313	2505	133,196	14,060	57,471
Sector 35	13,221	3248	2224	450	147	6544	2040	1321
Sector 36	70,401	20,992	13,706	17,910	6498	27,750	8976	8979
Sector 37	21,210	2753	992	890	1507	2473	5352	340
Sector 38	27,114	15,839	4545	2926	2141	9722	3715	4373
Sector 39	2855	1948	225	504	706	1109	369	275
Sector 40	4798	5263	436	2559	1215	4017	1775	1801
Sector 41	1881	3086	1165	3283	2368	8498	3962	4470
Sector 42	2497	545	368	279	112	1128	431	347
Total intermediate input	3,538,125	1,320,094	528,638	907,435	425,760	6,350,299	3,294,172	2,212,733
Compensation of employees	4,685,375	481,341	178,301	142,180	77,593	493,886	378,701	298,714
Net taxes on production	9520	268,500	202,464	91,742	47,215	557,058	150,714	77,514
Depreciation of fixed assets	229,928	91,991	119,027	33,557	20,116	153,057	99,466	35,184
Operating surplus	0	278,595	278,941	174,413	64,428	491,923	155,349	100,132
Total value added	4,924,823	1,120,426	778,734	441,891	209,352	1,695,924	784,229	511,544
Total input	8,462,948	2,440,521	1,307,372	1,349,326	635,111	8,046,222	4,078,400	2,724,276

Sector 9	Sector 10	Sector 11	Sector 12	Sector 13	Sector 14	Sector 15	Sector 16	Sector 17
202,742	115,339	79	359,225	2621	929	717	852	502
12,954	19,856	220,840	159,969	339,620	331,921	11,865	26,947	7503
0	0	2,000,543	224,365	9622	27,767	0	0	0
0	0	0	53,887	12,504	1,946,364	30,084	23,286	3838
0	0	359	172,632	274,895	23,821	10,283	5539	1686
8994	9388	16,540	174,792	17,149	31,997	9070	25,877	23,425
16,637	44,304	181	78,829	10,424	1073	4337	11,102	12,640
30,000	17,280	1475	25,128	12,430	21,883	8559	16,958	74,325
503,741	50,944	1090	17,251	31,952	6998	44,805	25,467	71,955
26,237	643,732	1927	151,778	101,032	9408	25,852	34,553	24,252
12,378	11,429	215,551	655,165	196,910	405,441	28,542	80,801	32,765
157,201	318,785	66,513	4,373,704	365,713	134,880	107,247	285,337	401,341
14,922	6572	12,040	86,709	873,414	198,619	31,432	56,801	55,072
39,197	27,178	2272	110,740	130,503	2,669,145	961,558	1,660,840	838,194
42,234	34,533	11,426	94,764	128,414	87,372	364,069	331,600	118,970
25,450	32,267	45,637	166,388	136,154	310,756	136,992	1,778,490	611,554
7988	19,159	8007	33,316	26,064	52,266	25,212	119,200	2,255,959
5701	15,719	4020	29,235	19,171	31,739	23,351	432,423	269,505
729	20,936	484	6765	3180	2587	3456	164,768	81,576
1301	11,701	5043	55,676	8338	22,756	4887	41,337	97,811
11,343	13,767	1567	60,163	34,221	25,647	15,866	54,642	81,209
6340	162,469	700	58,625	100,725	405,819	33,551	93,767	6564
50,863	63,428	58,617	512,760	325,749	512,804	180,657	176,605	99,601
792	708	5200	27,432	5544	12,859	2757	7044	4598
1093	4948	1069	11,420	4537	6154	2236	2638	3314
442	572	1712	4138	1626	3055	519	2731	2374
56,426	59,089	67,691	303,991	204,183	188,898	63,788	226,087	137,591
1116	1070	105	7853	1188	2665	510	7008	1290
6050	5646	10,932	35,947	9721	77,325	7778	24,947	15,868
39,906	55,762	41,722	212,017	118,063	159,327	63,057	202,199	297,205
9321	12,172	6220	63,226	35,461	34,905	32,530	61,039	28,617
26,406	35,722	20,820	179,739	122,681	135,634	25,211	84,026	69,481
7392	10,462	554	23,909	11,033	2490	11,809	22,049	10,477
15,297	19,159	16,394	152,743	30,560	15,409	10,886	69,725	91,125
994	1199	947	23,887	3445	14,372	2045	27,447	56,198
3052	10,893	5835	62,743	18,240	30,853	14,009	47,265	57,937
235	6439	688	10,104	1297	2660	365	1632	1871
1769	3957	3022	21,058	12,909	20,169	4101	9294	8385
255	288	138	1470	670	687	332	2509	1176
1475	2661	889	6570	8159	8200	6031	18,757	12,401
2028	2432	785	12,158	7015	5435	2653	11,073	6507
214	325	158	1263	759	910	343	1072	873
1,351,216	1,872,291	2,859,790	8,823,533	3,757,896	7,983,995	2,313,351	6,275,734	5,977,540
139,441	195,304	115,265	734,996	354,729	602,494	211,283	691,899	464,331
57,898	75,220	380,546	405,919	193,160	291,686	87,960	317,492	308,481
33,304	74,244	85,422	314,588	137,805	359,341	61,697	184,194	134,584
84,124	115,814	126,691	608,854	317,771	412,027	147,983	489,523	382,265
314,767	460,582	707,924	2,064,356	1,003,465	1,665,549	508,923	1,683,107	1,289,661
1,665,983	2,332,873	3,567,714	10,887,889	4,761,361	9,649,544	2,822,274	7,958,842	7,267,201

(continued)

Table 10.3 (continued)

	Sector 18	Sector 19	Sector 20	Sector 21	Sector 22	Sector 23	Sector 24	Sector 25
Sector 1	91	0	29	177,041	177	89	42	0
Sector 2	4124	1844	1075	9679	815	881,582	12,190	266
Sector 3	0	0	0	14	1	31,344	119,439	0
Sector 4	6770	95	884	134	538	148	0	0
Sector 5	2200	43	78	4065	68	1347	10	20
Sector 6	19,954	22,897	2357	19,988	700	15,264	946	524
Sector 7	10,041	1712	1095	132,247	77	258	30	60
Sector 8	11,472	11,296	2241	21,802	120	16,630	2848	988
Sector 9	13,491	9912	2077	44,772	90	3097	297	295
Sector 10	57,516	48,254	7396	26,230	123	9539	640	523
Sector 11	22,926	17,217	3504	10,723	1203	147,360	5278	2076
Sector 12	435,798	406,019	64,851	111,729	11,751	15,562	1630	9896
Sector 13	111,859	118,099	26,499	30,693	1447	10,669	586	585
Sector 14	1,415,259	199,608	29,729	139,526	5810	4836	626	231
Sector 15	233,586	142,182	29,613	61,751	609	29,720	2418	5401
Sector 16	271,347	68,037	21,181	5864	359	50,332	2049	2130
Sector 17	14,446	22,304	4704	14,973	124	56,055	5561	2487
Sector 18	781,759	184,955	38,037	9172	914	277,485	589	674
Sector 19	322,861	2,962,105	173,993	3649	46	3079	133	162
Sector 20	35,068	204,264	94,089	3166	54	67,448	1592	1087
Sector 21	51,940	46,200	9818	160,721	108	3347	190	207
Sector 22	56,369	7887	92	291	5318	549	0	0
Sector 23	62,526	73,926	8112	19,717	2473	1,792,459	5073	46,967
Sector 24	3243	3389	516	1399	121	5829	14,139	653
Sector 25	1955	2462	534	995	115	6244	136	4201
Sector 26	1366	1182	204	230	57	1719	188	244
Sector 27	127,448	107,667	16,737	34,797	4218	61,138	6711	1900
Sector 28	3601	2144	843	414	13	592	33	62
Sector 29	15,797	56,103	2896	9884	168	30,722	362	741
Sector 30	151,730	220,776	17,785	29,167	2583	39,737	4262	1216
Sector 31	36,097	19,726	7010	10,491	516	14,379	1164	1597
Sector 32	62,086	187,012	5887	16,015	2376	208,274	8414	11,542
Sector 33	15,413	15,718	3304	9949	180	2198	267	106
Sector 34	117,453	69,912	9542	12,332	437	22,361	1118	986
Sector 35	19,374	42,953	3781	1435	45	3338	67	41
Sector 36	27,956	31,479	4458	3143	221	40,237	585	305
Sector 37	1001	715	72	234	32	13,401	111	11,481
Sector 38	4108	4112	614	621	478	38,431	732	1581
Sector 39	710	563	136	244	5	776	60	154
Sector 40	9110	3136	1263	229	87	5991	712	44
Sector 41	5599	5237	809	1619	80	11,314	479	602
Sector 42	557	612	94	209	0	716	22	40
Total intermediate input	4,546,004	5,323,754	597,938	1,141,356	44,658	3,925,595	201,726	112,076
Compensation of employees	346,159	497,640	78,725	132,281	6478	401,580	27,153	40,836
Net taxes on production	133,950	101,085	22,201	89,134	3880	244,199	6300	8047
Depreciation of fixed assets	94,459	156,243	16,059	29,855	2913	503,301	8431	35,168
Operating surplus	254,742	186,203	40,191	72,859	395,289	163,385	12,327	4715
Total value added	829,309	941,171	157,175	324,129	408,560	1,312,466	54,211	88,767
Total input	5,375,313	6,264,926	755,113	1,465,485	453,217	5,238,061	255,936	200,843

Sector 26	Sector 27	Sector 28	Sector 29	Sector 30	Sector 31	Sector 32	Sector 33	Sector 34
54,722	136,104	0	0	1193	295,556	0	287	6221
9881	19,639	210	0	1502	1706	0	741	3697
0	7085	0	0	0	0	0	0	0
0	0	0	0	0	0	0	0	0
125,821	1478	0	0	0	0	0	0	0
49,723	44,859	372	5001	25,629	889,541	11,145	9825	31,019
10,514	7369	226	251	5941	12,589	569	3888	33,889
47,403	40,728	1077	3882	43,522	9659	10,486	15,773	32,695
218,049	6948	744	1966	8133	3539	2989	4230	6622
20,814	20,928	3963	63,946	45,635	8708	77,286	20,507	234,926
160,557	1,062,732	4410	3954	16,490	12,595	25,146	45,697	83,466
384,214	100,312	496	10,868	44,694	19,519	13,496	58,404	93,924
2,822,178	12,077	517	556	1342	5248	639	1036	1392
1,552,671	29,352	0	21	2413	115	0	3137	2003
404,974	31,723	116	4124	1982	2991	3597	21,225	52,640
368,687	137,263	1566	16,592	10,648	3003	16,939	12,393	17,211
51,480	401,214	11,818	14,461	43,185	8645	11,577	13,352	79,780
471,119	16,027	264	141,224	33,620	18,347	8178	38,794	73,524
30,024	14,365	1311	122,566	19,266	1044	2976	10,481	151,469
27,198	9077	177	18,389	1608	80	15,273	6834	7807
40,543	5216	378	2173	3773	1595	4143	4519	15,891
0	0	0	0	0	0	0	0	0
93,021	68,313	2041	51,147	57,811	54,718	37,513	20,937	25,002
1473	5354	143	1167	1011	9344	198	369	2532
2990	3433	240	2506	1892	4445	1882	638	1712
132,109	24,520	6319	3476	12,755	5859	5758	67,313	2560
1,160,312	481,107	21,304	62,891	346,565	31,142	98,714	26,397	129,018
2686	1790	6268	2680	6673	1471	17,135	2233	12,330
180,327	40,941	3036	88,879	40,213	7869	113,522	13,629	9802
306,428	73,728	11,205	64,437	25,091	54,145	16,880	14,923	60,811
124,543	90,820	1075	40,289	94,936	17,329	146,834	56,978	173,000
117,833	336,213	1751	42,460	168,767	42,380	320,040	149,778	118,972
7257	27,812	2275	53,638	109,853	27,144	127,656	46,014	43,744
44,886	31,650	5800	102,832	241,022	30,592	242,824	118,047	114,796
9329	1753	81	3432	7711	23	5586	736	1309
144,054	6708	26	10,027	19,061	472	2762	6703	1956
612	1719	0	574	1143	839	2559	1123	22,521
17,441	129,686	238	13,133	45,713	23,895	15,477	25,733	40,605
1550	4256	150	2402	2956	667	7750	825	1261
3697	3074	29	611	795	431	2756	1207	428
17,302	11,732	1118	11,477	19,263	6616	23,849	18,481	11,252
1589	1607	49	781	1230	394	2274	2314	499
9,220,011	3,450,711	90,793	968,816	1,515,038	1,614,255	1,396,407	845,505	1,702,287
1,849,853	922,857	42,850	275,835	1,019,861	579,679	822,703	286,759	503,176
513,719	185,423	3863	81,138	1,011,652	110,082	315,590	413,364	127,610
179,715	410,416	10,098	266,331	237,061	129,947	60,756	1,266,084	244,597
704,691	731,002	124	454,397	1,444,636	137,684	1,392,285	658,175	72,026
3,247,978	2,249,697	56,934	1,077,701	3,713,210	957,392	2,591,334	2,624,383	947,408
12,467,989	5,700,409	147,727	2,046,516	5,228,248	2,571,647	3,987,741	3,469,888	2,649,696

(continued)

Table 10.3 (continued)

	Sector 35	Sector 36	Sector 37	Sector 38	Sector 39	Sector 40	Sector 41	Sector 42
Sector 1	13,799	4327	12,819	16,596	3609	9496	2949	0
Sector 2	634	1456	2747	5782	3834	8007	489	8327
Sector 3	0	0	0	0	0	0	0	0
Sector 4	3453	676	925	0	0	0	0	0
Sector 5	56	33	345	75	697	1013	296	1175
Sector 6	6518	2664	5578	32,238	15,207	16,532	33,881	10,790
Sector 7	7417	718	2270	21,365	3306	11,031	1775	12,100
Sector 8	2832	1818	6854	10,003	2117	27,193	10,071	39,561
Sector 9	1243	1384	6762	13,385	23,005	14,792	7238	48,198
Sector 10	4086	15,329	3597	23,694	65,908	24,684	68,007	77,255
Sector 11	3350	22,236	19,473	26,062	14,865	8092	3928	43,086
Sector 12	39,834	23,961	35,670	123,717	16,949	844,911	24,817	30,026
Sector 13	2886	2109	7381	1509	9245	11,049	3873	14,467
Sector 14	1838	7987	3996	3643	1409	95	278	0
Sector 15	9341	23,238	4229	7050	4043	4138	2403	3512
Sector 16	5512	8740	11,097	3447	3759	88,517	1692	6895
Sector 17	3256	15,870	18,661	54,174	12,537	2198	2828	38,541
Sector 18	16,169	45,310	13,832	23,298	5553	6305	3337	6753
Sector 19	14,450	40,895	7625	35,009	6088	1458	3609	22,819
Sector 20	12,930	24,663	7300	13,828	18,335	4619	2266	5216
Sector 21	1672	2404	2780	11,291	3209	2486	1220	6716
Sector 22	0	0	0	0	0	0	0	0
Sector 23	7466	7521	15,107	15,752	34,357	22,993	5434	40,293
Sector 24	121	277	485	916	1212	920	405	892
Sector 25	569	1285	988	2331	2548	1319	479	1697
Sector 26	1304	3283	13,238	10,891	9054	38,321	4925	41,502
Sector 27	15,205	64,559	12,246	29,838	48,263	9912	17,788	123,262
Sector 28	616	1297	507	1978	3939	634	1377	31,610
Sector 29	2086	5377	4708	4855	15,273	18,010	4880	55,130
Sector 30	7645	12,782	7570	25,397	13,868	59,056	12,417	30,289
Sector 31	14,211	44,655	12,658	22,376	45,190	19,052	18,832	188,267
Sector 32	1579	34,436	16,912	30,124	47,340	11,223	9045	47,061
Sector 33	1672	11,146	1600	39,967	13,459	3777	7782	15,998
Sector 34	3471	16,406	7828	20,259	32,666	14,066	13,264	26,430
Sector 35	8800	2506	511	39	3769	897	609	40
Sector 36	5323	67,994	2734	1095	3838	2764	412	4121
Sector 37	230	230	12,612	2061	2380	499	539	6854
Sector 38	1092	9780	13,264	55,319	13,235	10,792	5690	42,278
Sector 39	1233	1077	876	436	8124	1881	577	22,541
Sector 40	968	1014	1976	460	2669	4871	865	8071
Sector 41	5704	5539	3861	5711	10,884	2836	18,480	36,773
Sector 42	94	347	209	329	351	284	125	791
Total intermediate input	230,663	537,332	303,830	696,300	526,093	1,310,723	298,883	1,099,338
Compensation of employees	89,591	290,757	115,558	487,502	1,330,404	608,567	174,653	1,684,385
Net taxes on production	5722	48,628	9486	67,346	8560	6349	30,111	9145
Depreciation of fixed assets	16,811	55,309	46,319	39,635	158,125	73,876	46,758	247,975
Operating surplus	19,518	149,403	34,766	137,261	22,320	43,578	54,902	53,306
Total value added	131,642	544,096	206,129	731,745	1,519,409	732,370	306,424	1,994,811
Total input	362,306	1,081,428	509,959	1,428,045	2,045,501	2,043,093	605,307	3,094,149

Total intermediate use	Rural household consumption expenditure	Urban household consumption expenditure	Government consumption expenditure	Gross fixed capital formation	Increase in inventories	Export	Import	Total output
6,609,460	712,497	827,708	62,908	459,379	221,672	207,065	465,396	8,635,293
2,612,858	10,833	7214	0	25,412	−87,034	14,916	155,851	2,428,347
2,440,269	0	0	0	16,335	−55,601	148,591	1,267,229	1,282,364
2,235,453	0	0	0	68,248	−19,071	7635	906,938	1,385,326
679,019	0	0	0	6078	−22,672	17,066	39,559	639,932
4,288,927	907,798	2,445,095	0	150,369	255,511	220,825	242,567	8,025,958
2,804,093	26,240	39,466	0	21,352	−27,421	1,407,359	227,977	4,043,110
1,100,778	186,494	824,635	0	56,481	90,604	579,653	75,977	2,762,669
1,251,246	17,434	72,183	0	147,230	−1685	269,856	100,212	1,656,051
2,193,331	10,697	79,026	0	29,065	−34,103	226,932	202,041	2,302,907
3,694,678	15,366	147,339	0	−4638	−134,219	88,086	249,167	3,557,445
10,840,365	96,596	315,000	0	133,235	−131,943	944,203	1,281,577	10,915,880
4,685,049	4367	35,548	0	49,475	−135,481	234,056	94,048	4,778,967
10,042,991	0	0	0	115,507	−188,641	365,674	580,288	9,755,244
2,474,921	9883	51,967	0	70,348	−54,600	345,348	100,319	2,797,548
4,917,901	704	10,602	0	3,137,136	89,393	924,785	1,106,096	7,974,425
3,551,425	88,937	461,710	0	2,975,668	119,509	768,798	746,819	7,219,228
3,145,589	81,517	306,597	0	1,193,271	34,952	1,067,974	487,516	5,342,385
4,254,072	62,079	250,758	0	803,423	187,744	2,850,204	2,056,782	6,351,499
879,287	6746	31,350	0	175,390	−24,607	414,873	744,757	738,281
806,508	50,027	226,146	0	291,238	9937	180,615	57,417	1,507,053
964,647	0	0	0	0	6321	3910	478,513	496,365
5,291,004	60,899	224,621	0	0	−345,812	9227	1878	5,238,061
144,064	12,137	96,671	0	1679	358	0	0	254,909
97,690	7595	92,789	0	0	1226	0	0	199,300
416,769	0	156,020	0	11,727,901	72,498	155,171	47,303	12,481,056
5,102,168	77,128	310,626	182,663	40,649	−179,449	369,632	242,644	5,660,773
141,806	1152	7211	0	0	2855	5584	10,364	148,244
994,175	103,880	422,587	0	416,304	43,230	90,103	36,801	2,033,478
2,968,344	294,557	876,711	0	351,994	47,359	720,551	0	5,259,516
1,646,493	179,506	691,950	0	0	98,640	34,645	100,011	2,551,224
3,042,423	118,535	597,213	36,700	0	108,023	32,689	30,703	3,904,881
747,080	340,571	1,408,626	0	859,772	120,162	0	0	3,476,211
1,961,223	17,078	248,051	95,652	0	59,141	516,656	249,350	2,648,452
277,894	0	0	193,794	0	2978	8631	115,372	367,925
814,470	1776	22,731	145,122	115,253	−1537	0	0	1,097,814
144,349	4828	61,785	302,680	0	−2160	0	0	511,481
669,090	107,081	614,853	0	0	35,106	24,513	34,211	1,416,432
76,726	192,507	436,449	1,379,759	0	−27,196	3754	12,602	2,049,396
141,501	91,371	965,201	887,384	0	−23,344	3828	2677	2,063,263
315,417	20,736	110,265	156,167	0	12,312	42,790	53,539	604,148
27,141	0	0	3,107,594	0	−12,349	8655	8998	3,122,043

Table 10.4 China's national input–output table in 2013

	Sector 1	Sector 2	Sector 3	Sector 4	Sector 5	Sector 6	Sector 7	Sector 8
Sector 1	1,263,469	27,572	25	2459	278	3,497,911	896,463	134,050
Sector 2	7490	501,915	11,447	6169	1908	24,125	20,261	10,733
Sector 3	0	0	22,080	0	0	0	0	0
Sector 4	0	0	0	158,277	0	0	0	0
Sector 5	1083	1010	701	1430	50,516	3176	0	0
Sector 6	990,483	4950	2396	3593	2090	1,821,906	32,497	153,064
Sector 7	1292	720	1061	829	405	4558	1,565,482	990,018
Sector 8	3406	10,031	4725	6408	3135	11,577	71,624	452,928
Sector 9	9121	22,510	5083	4921	1339	9467	3673	5403
Sector 10	8779	3097	2839	2676	1560	164,817	40,939	45,599
Sector 11	67,546	30,774	50,305	84,869	34,465	24,812	10,776	18,697
Sector 12	722,034	63,869	49,763	76,460	74,633	243,771	503,303	171,986
Sector 13	13,424	26,068	12,370	12,350	27,100	61,031	10,453	4055
Sector 14	2259	116,454	65,225	19,176	4962	9962	3393	4109
Sector 15	30,686	49,821	10,264	30,466	11,212	40,888	7628	13,137
Sector 16	53,934	150,795	110,642	119,342	64,257	30,204	52,713	19,382
Sector 17	26,939	19,989	9101	16,013	15,667	11,210	8351	4474
Sector 18	2591	41,656	12,374	19,840	7065	12,872	9741	5610
Sector 19	877	4576	1722	2171	832	3256	3320	3246
Sector 20	4129	20,297	12,623	6904	2276	6227	3926	1988
Sector 21	31,433	14,782	1840	6133	6351	13,525	38,422	30,433
Sector 22	683	342	21	1223	7537	6881	8359	1514
Sector 23	115,132	123,698	97,641	251,107	83,406	91,643	109,255	23,122
Sector 24	353	1106	1650	7826	3839	5233	2280	1067
Sector 25	1105	1984	1049	1338	721	3580	3313	1325
Sector 26	2434	3595	1510	575	258	2493	1019	895
Sector 27	168,968	125,209	25,921	65,963	44,345	253,479	76,333	69,832
Sector 28	6580	721	215	255	211	1994	2882	5349
Sector 29	33,878	7914	2266	2248	1673	15,699	8029	14,067
Sector 30	166,719	30,409	12,255	16,389	10,658	213,508	81,093	56,108
Sector 31	23,816	29,658	11,867	16,688	10,104	52,374	17,106	16,356
Sector 32	92,371	60,304	10,959	17,169	11,024	90,948	82,981	37,870
Sector 33	2206	3791	597	899	937	15,201	13,403	28,664
Sector 34	13,919	14,808	2430	5957	3032	146,016	16,041	62,857
Sector 35	15,089	3974	2444	500	177	7174	2375	1446
Sector 36	80,446	25,269	15,063	19,721	7708	30,420	10,312	9813
Sector 37	25,252	3324	1090	979	1822	2711	6049	381
Sector 38	30,588	20,363	4994	3348	2658	10,658	4322	4785
Sector 39	3289	2527	247	576	882	1215	445	306
Sector 40	5413	6439	479	2831	1445	4404	2077	1961
Sector 41	2169	3720	1281	3621	2834	9316	4403	4916
Sector 42	2844	680	405	312	135	1236	513	381
Total intermediate input	4,034,228	1,580,721	580,968	1,000,012	505,458	6,961,476	3,735,554	2,411,927
Compensation of employees	5,284,341	564,917	195,951	148,181	88,566	541,419	415,265	320,410
Net taxes on production	10,736	315,120	222,506	95,614	53,892	610,671	165,266	83,144
Depreciation of fixed assets	259,322	107,964	130,809	34,973	22,961	167,788	109,069	37,739
Operating surplus	0	326,968	306,554	181,775	73,540	539,268	170,348	107,405
Total value added	5,554,400	1,314,969	855,821	460,543	238,958	1,859,146	859,949	548,698
Total input	9,588,627	2,895,689	1,436,789	1,460,554	744,416	8,820,622	4,595,503	2,960,625

Sector 9	Sector 10	Sector 11	Sector 12	Sector 13	Sector 14	Sector 15	Sector 16	Sector 17
226,161	140,462	91	428,917	3370	1203	935	998	604
14,524	24,058	254,073	189,544	417,953	418,144	14,050	31,557	9171
0	0	2,301,591	292,214	12,649	37,771	0	0	0
0	0	0	61,047	14,351	2,326,656	33,665	27,270	4376
0	0	413	194,764	320,091	28,269	11,823	6486	1928
9969	11,182	19,029	204,814	21,222	39,997	10,942	30,304	28,202
17,937	49,106	208	88,596	12,367	1265	4843	13,001	14,407
32,501	20,038	1697	28,560	14,805	26,328	10,193	19,860	85,322
542,147	57,045	1254	19,393	38,818	8405	52,450	29,824	80,979
28,326	686,586	2217	171,902	127,268	11,477	29,073	40,464	27,361
13,466	13,084	247,988	746,286	234,799	486,800	32,807	94,623	37,930
169,494	358,928	76,522	4,936,656	440,244	161,794	124,819	334,151	455,368
15,960	7404	13,851	95,992	996,324	229,190	35,417	66,518	61,428
42,152	30,467	2614	123,935	153,480	3,147,826	1,101,315	1,944,967	941,575
45,534	38,227	13,146	106,259	151,289	102,606	400,699	388,328	135,472
27,168	35,712	52,504	184,758	159,385	363,065	154,935	2,082,744	680,524
8470	21,004	9212	36,734	30,703	60,983	28,800	139,593	2,472,396
6119	17,390	4625	32,697	22,758	37,428	26,291	506,400	303,647
785	23,797	557	7582	3669	3022	3946	192,956	92,303
1448	13,086	5802	64,563	11,721	30,174	6017	48,408	112,511
13,400	17,254	1802	73,926	42,549	34,574	17,730	63,990	97,144
6645	181,495	806	62,311	103,135	432,788	34,493	109,808	7150
59,174	80,388	67,438	645,382	449,749	712,888	237,738	206,817	128,858
847	781	5983	30,628	6606	15,195	3141	8249	5141
1187	5529	1230	13,067	5695	7596	2629	3089	3807
484	667	1970	4758	1984	3748	616	3198	2776
59,635	63,734	77,877	332,051	236,176	216,349	70,327	264,765	149,531
1227	1229	121	9073	1430	3238	581	8206	1520
6659	6446	12,577	41,773	12,396	97,332	9265	29,215	18,621
45,053	66,796	48,000	252,801	146,821	202,299	73,527	236,790	362,910
9888	13,479	7156	69,565	41,453	40,668	37,152	71,481	31,384
29,117	39,545	23,953	209,982	168,431	178,179	30,485	98,400	80,841
8204	12,651	637	27,903	13,615	3118	14,394	25,821	12,531
16,638	21,806	18,861	174,935	37,860	18,945	12,969	81,653	105,086
1104	1457	1090	28,073	4218	18,057	2510	32,143	67,258
3395	13,270	6713	73,652	22,114	38,571	16,971	55,351	69,426
261	7449	791	11,830	1606	3310	427	1912	2231
1948	4659	3477	24,484	16,460	25,604	5066	10,885	9882
286	353	158	1760	861	890	422	2938	1447
1632	3212	1022	7686	9946	10,189	7358	21,965	14,892
2195	2699	903	13,813	8587	6582	3052	12,967	7424
240	401	182	1502	943	1155	427	1255	1072
1,471,382	2,092,879	3,290,141	10,126,167	4,519,903	9,593,675	2,664,302	7,349,351	6,726,438
150,072	208,305	132,611	829,599	404,648	710,544	232,541	810,265	508,879
62,312	80,228	437,812	458,166	220,342	343,997	96,809	371,807	338,076
35,843	79,187	98,276	355,079	157,197	423,784	67,904	215,704	147,496
90,538	123,524	145,756	687,221	362,490	485,920	162,872	573,267	418,940
338,765	491,243	814,454	2,330,066	1,144,677	1,964,246	560,126	1,971,044	1,413,391
1,810,147	2,584,122	4,104,595	12,456,232	5,664,580	11,557,921	3,224,428	9,320,395	8,139,829

(continued)

Table 10.4 (continued)

	Sector 18	Sector 19	Sector 20	Sector 21	Sector 22	Sector 23	Sector 24	Sector 25
Sector 1	115	0	31	221,131	177	107	47	0
Sector 2	4938	2105	1160	12,090	815	1,058,230	13,436	292
Sector 3	0	0	0	17	1	37,625	131,652	0
Sector 4	7681	104	949	168	538	177	0	0
Sector 5	2528	47	83	5078	68	1617	11	22
Sector 6	23,997	26,191	2522	24,966	700	18,323	1043	575
Sector 7	11,277	1905	1141	165,182	77	310	33	66
Sector 8	13,515	12,591	2364	27,231	120	19,962	3139	1085
Sector 9	15,564	11,116	2161	55,922	90	3717	328	324
Sector 10	64,458	55,325	7565	32,762	123	11,450	705	574
Sector 11	26,421	19,203	3689	13,394	1203	176,887	5817	2279
Sector 12	502,485	454,671	67,691	139,553	11,751	18,680	1797	10,862
Sector 13	125,902	127,595	27,714	38,337	1447	12,807	646	642
Sector 14	1,608,688	219,932	31,008	174,274	5810	5805	689	254
Sector 15	259,932	157,250	30,893	77,130	609	35,675	2665	5929
Sector 16	305,479	74,780	21,975	7324	359	60,417	2259	2339
Sector 17	16,297	24,534	4849	18,702	124	67,287	6130	2730
Sector 18	880,088	205,310	39,530	11,456	914	333,086	649	740
Sector 19	368,584	3,248,352	182,825	4557	46	3697	147	177
Sector 20	41,722	246,098	97,825	3955	54	80,963	1755	1193
Sector 21	59,909	53,178	11,032	200,747	108	4017	209	227
Sector 22	59,624	8038	96	363	5318	659	0	0
Sector 23	81,315	91,769	8911	24,627	2473	2,151,625	5592	51,555
Sector 24	3670	3765	535	1748	121	6997	15,585	717
Sector 25	2271	2821	555	1242	115	7495	150	4611
Sector 26	1610	1333	217	287	57	2064	207	268
Sector 27	140,084	117,178	17,207	43,463	4218	73,389	7397	2085
Sector 28	4147	2420	892	517	13	711	36	68
Sector 29	18,636	65,068	3041	12,346	168	36,877	399	813
Sector 30	179,387	254,478	19,217	36,431	2583	47,699	4698	1335
Sector 31	40,725	21,498	7254	13,104	516	17,260	1283	1753
Sector 32	73,422	223,586	6082	20,004	2376	250,007	9274	12,669
Sector 33	18,637	17,826	3560	12,427	180	2638	294	116
Sector 34	138,053	79,706	9987	15,403	437	26,842	1232	1083
Sector 35	23,580	48,805	4068	1792	45	4007	74	45
Sector 36	33,752	35,529	4813	3925	221	48,300	645	335
Sector 37	1175	824	76	293	32	16,086	122	12,602
Sector 38	4971	4725	650	776	478	46,132	806	1736
Sector 39	890	660	147	305	5	931	66	169
Sector 40	11,056	3560	1358	286	87	7191	785	49
Sector 41	6409	5933	840	2022	80	13,581	528	661
Sector 42	690	704	103	261	0	860	24	44
Total intermediate input	5,183,686	5,930,514	626,616	1,425,598	44,658	4,712,190	222,354	123,023
Compensation of employees	389,698	545,730	81,851	165,224	6478	482,047	29,929	44,825
Net taxes on production	150,798	110,854	23,083	111,332	3880	293,130	6944	8834
Depreciation of fixed assets	106,340	171,342	16,696	37,290	2913	604,150	9293	38,603
Operating surplus	286,783	204,197	41,787	91,004	395,289	196,124	13,587	5176
Total value added	933,619	1,032,122	163,417	404,850	408,560	1,575,452	59,754	97,437
Total input	6,117,305	6,962,636	790,032	1,830,449	453,217	6,287,642	282,108	220,461

Sector 26	Sector 27	Sector 28	Sector 29	Sector 30	Sector 31	Sector 32	Sector 33	Sector 34
66,923	156,814	0	0	1385	323,515	0	312	7053
12,085	22,627	238	0	1745	1868	0	808	4191
0	8162	0	0	0	0	0	0	0
0	0	0	0	0	0	0	0	0
153,875	1703	0	0	0	0	0	0	0
60,809	51,685	423	5688	29,773	973,691	12,659	10,707	35,169
12,859	8490	257	286	6902	13,779	646	4237	38,423
57,972	46,925	1223	4415	50,559	10,573	11,910	17,188	37,069
266,668	8005	845	2236	9448	3874	3395	4610	7508
25,454	24,112	4500	72,727	53,014	9532	87,782	22,347	266,355
196,357	1,224,439	5007	4497	19,157	13,786	28,561	49,797	94,633
469,882	115,576	563	12,360	51,920	21,366	15,328	63,645	106,490
3,451,443	13,915	587	633	1559	5745	726	1129	1578
1,898,872	33,819	0	24	2803	126	0	3418	2271
495,272	36,550	132	4690	2302	3274	4086	23,130	59,683
450,894	158,149	1778	18,871	12,370	3287	19,239	13,505	19,513
62,959	462,263	13,420	16,447	50,168	9462	13,150	14,550	90,453
576,164	18,466	300	160,617	39,056	20,083	9288	42,275	83,360
36,719	16,550	1489	139,398	22,381	1143	3380	11,421	171,733
33,262	10,458	201	20,915	1868	88	17,347	7447	8851
49,583	6010	429	2472	4383	1746	4706	4924	18,017
0	0	0	0	0	0	0	0	0
113,762	78,707	2318	58,171	67,158	59,894	42,608	22,816	28,347
1802	6169	162	1327	1175	10,228	225	402	2870
3657	3955	273	2850	2198	4865	2138	695	1941
161,565	28,251	7175	3954	14,818	6413	6540	73,352	2902
1,419,028	554,313	24,191	71,528	402,602	34,088	112,121	28,765	146,278
3285	2062	7117	3048	7752	1610	19,462	2433	13,980
220,534	47,171	3448	101,084	46,715	8614	128,940	14,852	11,113
374,752	84,946	12,723	73,285	29,148	59,267	19,173	16,262	68,947
152,313	104,639	1221	45,821	110,286	18,968	166,776	62,090	196,144
144,106	387,372	1988	48,291	196,055	46,389	363,506	163,217	134,889
8875	32,044	2583	61,004	127,615	29,711	144,993	50,143	49,596
54,894	36,466	6586	116,953	279,993	33,486	275,803	128,639	130,154
11,409	2019	92	3903	8958	25	6345	802	1484
176,174	7728	30	11,404	22,143	516	3137	7304	2217
749	1981	0	653	1328	919	2906	1224	25,534
21,330	149,419	271	14,937	53,104	26,155	17,579	28,042	46,037
1896	4903	170	2732	3434	730	8803	899	1430
4521	3542	33	695	923	472	3130	1316	486
21,160	13,517	1269	13,054	22,378	7242	27,088	20,139	12,757
1943	1852	56	888	1429	431	2582	2522	566
11,275,808	3,975,776	103,096	1,101,857	1,760,006	1,766,962	1,586,058	921,365	1,930,024
2,262,317	1,063,280	48,656	313,714	1,184,764	634,516	934,437	312,488	570,493
628,263	213,638	4386	92,280	1,175,227	120,495	358,451	450,452	144,682
219,786	472,865	11,467	302,905	275,392	142,240	69,008	1,379,680	277,319
861,817	842,232	140	516,796	1,678,221	150,709	1,581,376	717,228	81,661
3,972,184	2,592,014	64,649	1,225,694	4,313,604	1,047,960	2,943,273	2,859,849	1,074,156
15,247,991	6,567,790	167,745	2,327,551	6,073,610	2,814,922	4,529,331	3,781,214	3,004,180

(continued)

Table 10.4 (continued)

	Sector 35	Sector 36	Sector 37	Sector 38	Sector 39	Sector 40	Sector 41	Sector 42
Sector 1	15,542	5072	14,227	18,142	3965	10,516	3325	0
Sector 2	714	1706	3049	6321	4213	8868	552	9175
Sector 3	0	0	0	0	0	0	0	0
Sector 4	3889	792	1027	0	0	0	0	0
Sector 5	63	39	383	82	765	1122	333	1294
Sector 6	7341	3122	6191	35,241	16,708	18,308	38,200	11,890
Sector 7	8353	842	2520	23,356	3632	12,216	2001	13,333
Sector 8	3190	2131	7607	10,935	2326	30,114	11,355	43,593
Sector 9	1400	1622	7505	14,632	25,276	16,382	8161	53,110
Sector 10	4602	17,966	3992	25,902	72,415	27,335	76,674	85,128
Sector 11	3773	26,062	21,613	28,490	16,333	8961	4429	47,477
Sector 12	44,863	28,085	39,590	135,244	18,622	935,679	27,980	33,086
Sector 13	3250	2472	8192	1650	10,157	12,236	4367	15,941
Sector 14	2070	9362	4436	3983	1549	105	314	0
Sector 15	10,521	27,237	4693	7707	4442	4582	2709	3870
Sector 16	6208	10,244	12,316	3768	4130	98,026	1907	7598
Sector 17	3667	18,601	20,712	59,222	13,775	2434	3189	42,469
Sector 18	18,211	53,107	15,352	25,469	6102	6982	3762	7442
Sector 19	16,274	47,932	8463	38,271	6689	1615	4069	25,145
Sector 20	14,562	28,906	8102	15,116	20,145	5115	2555	5748
Sector 21	1883	2818	3086	12,343	3525	2753	1376	7400
Sector 22	0	0	0	0	0	0	0	0
Sector 23	8408	8816	16,767	17,220	37,748	25,463	6126	44,399
Sector 24	136	324	539	1002	1332	1019	456	983
Sector 25	641	1506	1096	2548	2800	1460	540	1870
Sector 26	1469	3848	14,693	11,905	9948	42,438	5552	45,731
Sector 27	17,125	75,668	13,592	32,618	53,027	10,977	20,055	135,824
Sector 28	694	1520	563	2162	4328	703	1553	34,832
Sector 29	2349	6303	5226	5308	16,780	19,945	5502	60,748
Sector 30	8610	14,982	8403	27,763	15,237	65,400	13,999	33,376
Sector 31	16,005	52,340	14,049	24,460	49,651	21,098	21,232	207,453
Sector 32	1779	40,362	18,770	32,930	52,013	12,429	10,197	51,857
Sector 33	1883	13,064	1776	43,690	14,788	4182	8774	17,628
Sector 34	3909	19,229	8689	22,146	35,890	15,577	14,954	29,123
Sector 35	9911	2937	567	42	4142	993	686	44
Sector 36	5995	79,694	3035	1197	4216	3061	464	4541
Sector 37	259	270	13,998	2253	2615	552	608	7553
Sector 38	1229	11,463	14,722	60,473	14,541	11,951	6415	46,586
Sector 39	1389	1263	972	476	8926	2083	651	24,838
Sector 40	1090	1188	2193	502	2933	5394	975	8894
Sector 41	6425	6493	4286	6243	11,958	3141	20,836	40,520
Sector 42	106	407	232	359	386	315	141	871
Total intermediate input	259,786	629,795	337,222	761,172	578,027	1,451,532	336,977	1,211,372
Compensation of employees	100,903	340,790	128,259	532,921	1,461,739	673,945	196,914	1,856,042
Net taxes on production	6444	56,996	10,528	73,620	9405	7032	33,949	10,077
Depreciation of fixed assets	18,934	64,827	51,410	43,328	173,734	81,813	52,717	273,247
Operating surplus	21,983	175,112	38,587	150,049	24,523	48,259	61,900	58,738
Total value added	148,263	637,724	228,784	799,919	1,669,401	811,048	345,479	2,198,103
Total input	408,050	1,267,519	566,006	1,561,091	2,247,428	2,262,580	682,456	3,409,475

Total intermediate use	Rural household consumption expenditure	Urban household consumption expenditure	Government consumption expenditure	Gross fixed capital formation	Increase in inventories	Export	Import	Total output
7,473,369	785,555	925,436	69,532	518,105	221,672	266,762	497,149	9,763,283
3,128,349	12,169	8278	0	29,824	−87,034	15,316	180,202	2,926,700
2,843,761	0	0	0	17,952	−55,601	213,020	1,570,078	1,449,054
2,640,968	0	0	0	71,128	−19,071	7635	1,208,068	1,492,592
790,803	0	0	0	6937	−22,672	17,687	44,434	748,322
4,802,563	977,415	2,668,740	0	164,841	255,511	220,825	242,567	8,847,329
3,098,218	29,152	44,094	0	23,413	−27,421	1,665,095	302,757	4,529,795
1,232,231	201,930	906,070	0	60,584	90,604	579,653	75,977	2,995,095
1,415,727	18,993	79,423	0	158,455	−1685	269,856	131,509	1,809,260
2,447,780	11,699	84,969	0	31,000	−34,103	226,932	261,682	2,506,596
4,272,291	16,842	159,673	0	−5336	−134,219	91,262	272,637	4,127,877
12,291,565	106,487	347,238	0	150,384	−131,943	944,203	1,333,940	12,373,994
5,559,605	5174	39,787	0	56,438	−135,481	255,952	115,168	5,666,306
11,723,483	0	0	0	136,222	−188,641	365,674	619,516	11,417,222
2,840,623	10,979	58,061	0	77,426	−54,600	345,348	117,034	3,160,803
5,658,802	785	11,740	0	3,673,819	89,393	1,028,677	1,186,707	9,276,509
3,957,232	96,874	496,768	0	3,465,056	119,509	898,331	833,801	8,199,969
3,626,911	88,435	334,639	0	1,343,360	34,952	1,137,737	522,928	6,043,106
4,709,673	69,055	278,363	0	881,063	187,744	3,066,448	2,225,085	6,967,261
1,026,354	7630	33,320	0	182,355	−24,607	440,896	888,154	777,793
962,167	53,040	248,304	0	363,767	9937	187,049	62,575	1,761,689
1,039,289	0	0	0	0	6321	3750	590,867	458,492
6,540,033	60,900	224,622	0	0	−545,905	9882	1890	6,287,642
163,211	12,863	101,594	0	1851	358	0	0	279,877
112,534	8335	97,251	0	0	1226	0	0	219,345
479,579	0	173,769	0	14,342,886	72,498	183,749	54,775	15,197,707
5,857,317	85,654	342,056	214,104	46,834	−179,449	369,632	285,159	6,450,990
160,742	1268	8064	0	0	2855	5820	13,242	165,506
1,162,039	112,199	461,074	0	473,472	43,230	100,173	42,460	2,309,727
3,494,238	323,211	967,750	0	408,909	47,359	740,074	0	5,981,541
1,868,136	197,618	763,463	0	0	98,640	20,457	114,141	2,834,173
3,566,132	128,204	651,286	41,974	0	108,023	38,731	35,988	4,498,362
852,601	371,178	1,490,948	0	936,913	120,162	0	0	3,771,802
2,239,048	18,432	266,968	106,687	0	59,141	560,273	249,350	3,001,199
325,862	0	0	207,052	0	2978	10,124	131,141	414,876
958,593	1776	22,731	160,115	135,085	−1537	0	0	1,276,763
166,036	5205	66,169	324,834	0	−2160	0	0	560,084
768,712	115,959	667,129	0	0	35,106	25,898	38,923	1,573,880
87,370	212,881	481,201	1,524,509	0	−27,196	3880	14,444	2,268,200
165,613	96,654	1,036,371	998,994	0	−23,344	3828	2677	2,275,439
359,039	22,605	120,469	173,790	0	12,312	42,790	53,539	677,468
31,455	0	0	3,411,194	0	−12,349	9758	9704	3,430,354
								170,773,981
								0
								0
								0
								0
								0
								0

Table 10.5 China's national input–output table in 2014

	Sector 1	Sector 2	Sector 3	Sector 4	Sector 5	Sector 6	Sector 7	Sector 8
Sector 1	1,392,585	29,226	26	2460	292	3,825,711	985,158	143,912
Sector 2	8286	537,590	11,600	6170	2021	26,386	22,926	11,501
Sector 3	0	0	22,375	0	0	0	0	0
Sector 4	0	0	0	158,305	0	0	0	0
Sector 5	1187	1082	710	1430	53,431	3474	0	0
Sector 6	1,100,304	5277	2428	3593	2202	1,992,642	35,624	164,166
Sector 7	1438	781	1075	829	429	4985	1,663,421	1,066,521
Sector 8	3701	10,550	4789	6409	3269	12,662	78,192	479,203
Sector 9	10,102	24,287	5151	4921	1426	10,354	4312	5833
Sector 10	10,039	3574	2877	2677	1739	180,262	51,607	51,266
Sector 11	74,228	32,993	50,978	84,880	36,429	27,137	12,068	19,998
Sector 12	802,448	69,057	50,428	76,472	79,162	266,616	551,373	185,676
Sector 13	14,683	27,926	12,535	12,352	28,590	66,751	11,420	4318
Sector 14	2515	127,046	66,098	19,180	5299	10,896	3722	4446
Sector 15	35,094	55,727	10,401	30,474	12,137	44,719	8297	14,469
Sector 16	59,019	161,397	112,122	119,358	67,694	33,035	57,757	20,671
Sector 17	29,241	21,258	9223	16,014	16,493	12,261	9664	4758
Sector 18	2883	45,238	12,539	19,843	7507	14,078	10,604	6057
Sector 19	966	4877	1745	2171	874	3561	3589	3467
Sector 20	4819	23,719	12,792	6906	2611	6811	5906	2293
Sector 21	39,996	21,268	1864	6143	8211	14,793	42,407	37,381
Sector 22	735	353	22	1223	7746	7525	8464	1587
Sector 23	133,329	136,577	98,947	251,166	90,055	100,231	122,009	25,775
Sector 24	391	1200	1672	7827	4084	5723	2550	1152
Sector 25	1232	2169	1063	1338	774	3915	3879	1442
Sector 26	2755	3941	1530	576	277	2727	1133	979
Sector 27	190,151	139,385	26,267	65,974	48,035	277,233	87,958	76,567
Sector 28	7554	806	217	255	229	2181	3158	5916
Sector 29	38,535	8806	2297	2249	1818	17,170	9463	15,566
Sector 30	196,051	34,725	12,419	16,395	11,749	233,516	88,880	63,135
Sector 31	25,215	30,854	12,026	16,690	10,430	57,282	18,149	16,995
Sector 32	105,706	67,736	11,106	17,171	12,254	99,471	116,798	42,436
Sector 33	2546	4205	605	899	1015	16,625	14,909	31,770
Sector 34	15,488	16,053	2462	5958	3229	159,699	18,122	68,137
Sector 35	16,788	4242	2476	500	188	7846	2849	1558
Sector 36	89,706	27,093	15,264	19,725	8174	33,271	11,503	10,564
Sector 37	28,989	3700	1104	980	1971	2965	6701	422
Sector 38	33,195	21,503	5061	3349	2780	11,656	4697	5063
Sector 39	3652	2684	250	576	926	1329	484	328
Sector 40	6033	6846	486	2832	1517	4816	2221	2104
Sector 41	2409	4059	1298	3622	3031	10,189	5027	5332
Sector 42	3197	725	410	312	142	1352	550	411
Total intermediate input	4,497,192	1,720,534	588,740	1,000,200	540,239	7,613,858	4,087,554	2,603,174
Compensation of employees	5,824,355	605,070	198,573	148,207	93,676	592,157	441,245	338,998
Net taxes on production	11,834	337,518	225,483	95,631	57,001	667,899	175,605	87,967
Depreciation of fixed assets	285,822	115,638	132,559	34,979	24,286	183,512	115,893	39,929
Operating surplus	0	350,208	310,655	181,807	77,783	589,804	181,005	113,635
Total value added	6,122,011	1,408,435	867,270	460,624	252,747	2,033,372	913,748	580,529
Total input	10,619,203	3,128,969	1,456,010	1,460,825	792,986	9,647,230	5,001,302	3,183,703

Sector 9	Sector 10	Sector 11	Sector 12	Sector 13	Sector 14	Sector 15	Sector 16	Sector 17
234,732	142,686	97	449,233	3516	1236	960	1064	659
15,224	24,654	269,794	200,406	440,666	433,551	14,623	33,659	10,320
0	0	2,444,009	324,458	13,708	40,292	0	0	0
0	0	0	65,580	15,608	2,464,737	36,168	29,086	4792
0	0	438	204,521	336,652	29,216	12,217	6918	2142
10,454	11,447	20,206	215,393	22,240	41,249	11,246	32,322	31,997
18,844	50,259	221	92,800	12,984	1302	4934	13,867	16,550
33,598	20,356	1802	29,755	15,382	26,994	10,454	21,182	92,701
563,098	57,980	1331	20,553	41,146	8722	55,074	31,810	88,724
29,780	699,820	2355	186,714	139,975	12,166	31,308	43,159	30,783
14,013	13,338	263,333	784,264	246,847	502,725	33,948	100,926	41,991
177,032	365,866	81,257	5,186,474	463,209	166,869	128,209	356,407	510,578
16,601	7545	14,708	100,404	1,045,048	236,119	36,414	70,949	68,020
44,194	31,133	2775	130,553	162,333	3,258,049	1,133,836	2,074,507	1,063,852
48,474	39,357	13,959	112,914	161,729	106,746	411,492	414,191	161,492
28,182	36,319	55,753	193,125	166,867	373,567	159,233	2,221,460	748,539
8701	21,246	9782	38,537	32,223	62,899	30,065	148,890	2,625,973
6407	17,757	4911	34,353	23,976	38,622	26,969	540,128	344,182
818	24,260	591	7927	3832	3106	4038	205,807	102,838
1545	13,398	6161	72,716	13,334	32,809	6900	51,632	126,944
16,022	19,212	1914	85,007	50,873	39,408	18,823	68,252	129,300
6943	186,313	855	64,639	105,876	442,672	34,713	117,121	8111
63,423	82,956	71,611	691,301	479,974	742,862	245,690	220,592	152,534
881	795	6353	32,220	6973	15,702	3246	8799	5697
1237	5627	1306	13,884	6066	7903	2757	3295	4209
512	684	2092	5048	2109	3891	637	3411	3203
62,438	64,948	82,696	352,993	253,273	225,439	73,398	282,399	168,771
1310	1268	129	9676	1530	3377	599	8753	1805
7026	6596	13,355	44,801	13,336	101,862	9748	31,161	21,251
49,085	69,716	50,971	273,205	159,086	213,019	75,995	252,560	447,697
10,128	13,640	7599	71,598	42,667	41,429	37,844	76,242	33,159
30,441	40,262	25,435	232,866	186,487	191,162	34,369	104,954	88,516
8771	13,034	676	29,839	14,592	3252	14,879	27,540	14,880
17,352	22,204	20,028	184,663	40,018	19,603	13,436	87,092	116,811
1150	1487	1157	29,960	4472	18,825	2660	34,284	74,116
3567	13,607	7129	77,944	23,312	39,995	17,592	59,038	78,554
277	7665	840	12,615	1716	3448	440	2039	2624
2020	4741	3692	25,526	17,136	26,281	5200	11,609	10,806
299	361	168	1847	900	916	432	3133	1629
1713	3287	1085	8058	10,383	10,477	7507	23,428	16,964
2288	2748	959	14,606	9106	6827	3176	13,831	8234
253	411	193	1580	986	1190	436	1339	1236
1,538,831	2,138,984	3,493,729	10,714,553	4,792,147	10,000,512	2,751,664	7,838,836	7,463,179
155,872	212,320	140,816	871,581	424,437	735,424	238,804	864,230	540,489
64,720	81,774	464,903	481,351	231,118	356,042	99,417	396,570	359,076
37,228	80,713	104,357	373,048	164,885	438,624	69,733	230,071	156,658
94,037	125,904	154,775	721,997	380,216	502,934	167,259	611,448	444,963
351,857	500,712	864,851	2,447,978	1,200,656	2,033,025	575,213	2,102,320	1,501,186
1,890,688	2,639,696	4,358,581	13,162,530	5,992,803	12,033,537	3,326,877	9,941,156	8,964,365

(continued)

Table 10.5 (continued)

	Sector 18	Sector 19	Sector 20	Sector 21	Sector 22	Sector 23	Sector 24	Sector 25
Sector 1	119	0	31	222,647	177	113	48	0
Sector 2	5177	2222	1183	12,173	815	1,114,863	13,982	308
Sector 3	0	0	0	17	1	39,638	137,004	0
Sector 4	8257	113	965	169	538	187	0	0
Sector 5	2633	49	84	5112	68	1703	11	24
Sector 6	24,938	27,485	2570	25,137	700	19,304	1085	605
Sector 7	11,656	1991	1161	166,315	77	326	34	69
Sector 8	13,969	13,134	2394	27,418	120	21,030	3266	1141
Sector 9	16,438	11,863	2185	56,305	90	3916	341	341
Sector 10	70,048	61,243	7644	32,987	123	12,063	734	604
Sector 11	27,565	20,223	3742	13,486	1203	186,354	6054	2397
Sector 12	522,386	479,221	68,658	140,511	11,751	19,680	1870	11,426
Sector 13	130,611	133,512	28,103	38,600	1447	13,492	672	676
Sector 14	1,676,447	232,239	31,499	175,469	5810	6115	718	267
Sector 15	271,911	167,006	31,556	77,659	609	37,584	2773	6237
Sector 16	316,845	78,333	22,242	7374	359	63,651	2350	2460
Sector 17	17,076	25,908	4885	18,831	124	70,888	6379	2872
Sector 18	914,558	216,028	40,137	11,535	914	350,912	676	778
Sector 19	381,251	3,397,346	185,608	4588	46	3894	153	187
Sector 20	47,643	285,511	99,134	3982	54	85,296	1826	1255
Sector 21	66,247	61,099	11,973	202,124	108	4232	218	239
Sector 22	60,867	8236	98	365	5318	694	0	0
Sector 23	85,542	98,589	9154	24,796	2473	2,266,774	5819	54,232
Sector 24	3832	3984	541	1760	121	7372	16,218	754
Sector 25	2402	3024	561	1251	115	7896	156	4850
Sector 26	1687	1419	221	289	57	2174	216	281
Sector 27	147,916	125,702	17,419	43,761	4218	77,316	7698	2194
Sector 28	4349	2582	914	521	13	749	38	72
Sector 29	19,834	70,431	3091	12,431	168	38,851	415	855
Sector 30	189,623	274,030	19,878	36,681	2583	50,252	4889	1405
Sector 31	41,705	22,104	7317	13,194	516	18,183	1336	1844
Sector 32	82,371	252,168	6137	20,141	2376	263,387	9651	13,327
Sector 33	19,599	19,096	3648	12,512	180	2779	306	122
Sector 34	144,547	84,702	10,119	15,508	437	28,278	1282	1139
Sector 35	25,051	52,286	4133	1805	45	4221	77	48
Sector 36	35,312	37,534	4911	3952	221	50,884	671	352
Sector 37	1234	882	78	295	32	16,947	127	13,256
Sector 38	5142	4922	659	781	478	48,600	839	1826
Sector 39	922	692	150	307	5	981	68	178
Sector 40	11,424	3720	1385	288	87	7576	817	51
Sector 41	6734	6319	850	2036	80	14,308	550	695
Sector 42	714	737	105	263	0	906	25	46
Total intermediate input	5,416,581	6,287,687	637,123	1,435,376	44,658	4,964,374	231,392	129,411
Compensation of employees	404,961	570,761	82,946	166,357	6478	507,845	31,146	47,152
Net taxes on production	156,704	115,938	23,391	112,096	3880	308,818	7226	9292
Depreciation of fixed assets	110,505	179,201	16,920	37,546	2913	636,483	9671	40,608
Operating surplus	298,016	213,563	42,346	91,628	395,289	206,620	14,140	5444
Total value added	970,186	1,079,463	165,603	407,627	408,560	1,659,766	62,183	102,497
Total input	6,386,767	7,367,150	802,726	1,843,003	453,217	6,624,140	293,575	231,908

Sector 26	Sector 27	Sector 28	Sector 29	Sector 30	Sector 31	Sector 32	Sector 33	Sector 34
69,904	163,136	0	0	1490	352,782	0	331	7577
12,623	23,540	252	0	1877	2037	0	856	4502
0	8492	0	0	0	0	0	0	0
0	0	0	0	0	0	0	0	0
160,729	1771	0	0	0	0	0	0	0
63,518	53,769	447	6035	32,021	1,061,776	13,634	11,345	37,777
13,431	8832	272	303	7423	15,026	696	4490	41,273
60,554	48,817	1296	4684	54,377	11,529	12,828	18,213	39,819
278,544	8328	895	2372	10,162	4224	3656	4885	8065
26,588	25,085	4765	77,161	57,018	10,394	94,545	23,680	286,110
205,102	1,273,804	5303	4771	20,604	15,033	30,761	52,767	101,651
490,810	120,236	596	13,114	55,842	23,299	16,509	67,440	114,388
3,605,162	14,476	622	671	1676	6264	782	1196	1695
1,983,443	35,182	0	25	3015	138	0	3622	2439
517,330	38,023	140	4976	2476	3570	4401	24,509	64,109
470,976	164,525	1883	20,021	13,304	3585	20,721	14,310	20,961
65,763	480,900	14,212	17,450	53,957	10,318	14,163	15,418	97,162
601,825	19,210	318	170,409	42,005	21,900	10,004	44,796	89,543
38,354	17,218	1577	147,896	24,072	1246	3640	12,103	184,471
34,744	10,879	213	22,190	2009	96	18,684	7891	9507
51,792	6252	454	2623	4714	1904	5069	5218	19,353
0	0	0	0	0	0	0	0	0
118,829	81,881	2455	61,717	72,231	65,312	45,891	24,176	30,450
1882	6418	171	1408	1264	11,153	242	426	3083
3820	4115	289	3024	2364	5306	2302	736	2085
168,761	29,390	7599	4195	15,937	6993	7044	77,726	3117
1,482,228	576,661	25,620	75,888	433,010	37,172	120,759	30,480	157,127
3431	2146	7538	3234	8338	1756	20,961	2578	15,017
230,356	49,072	3651	107,246	50,244	9393	138,874	15,737	11,938
391,443	88,371	13,474	77,753	31,350	64,628	20,650	17,231	74,060
159,096	108,858	1293	48,615	118,616	20,684	179,625	65,793	210,692
150,524	402,989	2105	51,235	210,863	50,585	391,511	172,949	144,894
9271	33,336	2736	64,723	137,254	32,399	156,163	53,133	53,275
57,339	37,937	6975	124,083	301,141	36,515	297,051	136,309	139,807
11,917	2101	97	4141	9634	27	6833	850	1594
184,021	8040	32	12,099	23,816	563	3379	7740	2382
782	2061	0	693	1428	1002	3130	1297	27,428
22,279	155,443	287	15,847	57,115	28,522	18,934	29,714	49,452
1980	5101	180	2899	3693	796	9481	953	1536
4723	3685	35	738	993	515	3371	1394	522
22,102	14,062	1344	13,849	24,068	7897	29,175	21,340	13,703
2029	1926	59	942	1537	471	2781	2672	608
11,778,005	4,136,067	109,186	1,169,032	1,892,938	1,926,811	1,708,251	976,305	2,073,172
2,363,075	1,106,148	51,530	332,840	1,274,248	691,918	1,006,428	331,121	612,806
656,245	222,251	4645	97,906	1,263,991	131,396	386,067	477,312	155,413
229,575	491,929	12,144	321,371	296,192	155,108	74,324	1,461,948	297,888
900,201	876,188	149	548,303	1,804,977	164,343	1,703,209	759,995	87,718
4,149,095	2,696,516	68,468	1,300,419	4,639,408	1,142,765	3,170,029	3,030,376	1,153,824
15,927,100	6,832,583	177,654	2,469,451	6,532,347	3,069,575	4,878,280	4,006,681	3,226,996

(continued)

Table 10.5 (continued)

	Sector 35	Sector 36	Sector 37	Sector 38	Sector 39	Sector 40	Sector 41	Sector 42
Sector 1	16,985	5573	15,052	19,847	4404	11,624	3570	0
Sector 2	781	1875	3226	6914	4679	9802	592	10,166
Sector 3	0	0	0	0	0	0	0	0
Sector 4	4250	871	1086	0	0	0	0	0
Sector 5	69	43	405	89	850	1241	358	1434
Sector 6	8023	3431	6549	38,553	18,558	20,237	41,010	13,173
Sector 7	9129	925	2666	25,550	4034	13,503	2148	14,773
Sector 8	3486	2342	8048	11,962	2584	33,287	12,190	48,298
Sector 9	1530	1782	7940	16,007	28,075	18,107	8761	58,843
Sector 10	5030	19,742	4223	28,335	80,433	30,215	82,316	94,317
Sector 11	4124	28,638	22,866	31,167	18,141	9905	4754	52,602
Sector 12	49,030	30,861	41,885	147,951	20,684	1,034,244	30,039	36,657
Sector 13	3552	2716	8667	1805	11,282	13,525	4688	17,662
Sector 14	2262	10,287	4693	4357	1720	116	337	0
Sector 15	11,498	29,930	4965	8431	4934	5065	2909	4287
Sector 16	6784	11,257	13,030	4122	4587	108,352	2048	8418
Sector 17	4008	20,440	21,912	64,786	15,300	2691	3423	47,053
Sector 18	19,902	58,356	16,242	27,862	6777	7717	4039	8245
Sector 19	17,786	52,670	8953	41,867	7430	1785	4369	27,859
Sector 20	15,915	31,764	8572	16,537	22,376	5654	2743	6368
Sector 21	2058	3097	3264	13,503	3916	3043	1477	8199
Sector 22	0	0	0	0	0	0	0	0
Sector 23	9189	9687	17,739	18,838	41,928	28,146	6577	49,192
Sector 24	149	356	570	1096	1479	1126	490	1089
Sector 25	700	1654	1160	2787	3110	1614	580	2072
Sector 26	1605	4229	15,544	13,024	11,049	46,908	5961	50,667
Sector 27	18,716	83,147	14,380	35,683	58,899	12,133	21,531	150,486
Sector 28	758	1670	595	2365	4807	777	1667	38,592
Sector 29	2568	6926	5529	5806	18,639	22,046	5907	67,305
Sector 30	9409	16,463	8890	30,372	16,924	72,290	15,029	36,979
Sector 31	17,492	57,513	14,864	26,759	55,148	23,321	22,794	229,847
Sector 32	1944	44,352	19,858	36,024	57,773	13,738	10,948	57,455
Sector 33	2058	14,355	1879	47,796	16,425	4623	9420	19,531
Sector 34	4272	21,130	9192	24,227	39,864	17,218	16,055	32,267
Sector 35	10,832	3228	600	46	4600	1098	737	49
Sector 36	6552	87,571	3210	1309	4683	3383	499	5031
Sector 37	283	296	14,809	2465	2905	610	653	8368
Sector 38	1344	12,596	15,576	66,155	16,151	13,210	6887	51,615
Sector 39	1518	1388	1028	521	9914	2303	699	27,520
Sector 40	1191	1305	2320	550	3257	5963	1047	9854
Sector 41	7021	7134	4534	6829	13,282	3472	22,369	44,894
Sector 42	116	447	246	393	429	348	151	965
Total intermediate input	283,918	692,047	356,771	832,692	642,032	1,604,436	361,771	1,342,133
Compensation of employees	110,276	374,475	135,694	582,994	1,623,596	744,938	211,402	2,056,391
Net taxes on production	7043	62,629	11,139	80,537	10,447	7772	36,447	11,164
Depreciation of fixed assets	20,693	71,234	54,390	47,399	192,972	90,431	56,596	302,742
Operating surplus	24,025	192,421	40,824	164,148	27,239	53,343	66,454	65,079
Total value added	162,036	700,759	242,046	875,079	1,854,253	896,484	370,899	2,435,376
Total input	445,954	1,392,806	598,817	1,707,770	2,496,285	2,500,920	732,670	3,777,509

Total intermediate use	Rural household consumption expenditure	Urban household consumption expenditure	Government consumption expenditure	Gross fixed capital formation	Increase in inventories	Export	Import	Total output
8,108,963	859,641	1,024,541	76,156	571,051	221,672	286,123	525,081	10,623,066
3,303,822	13,452	9300	0	31,944	−87,034	15,716	204,493	3,082,708
3,029,994	0	0	0	18,192	−55,601	232,923	1,835,772	1,389,735
2,790,713	0	0	0	71,141	−19,071	7635	1,352,041	1,498,376
830,092	0	0	0	7338	−22,672	18,312	49,276	783,794
5,234,472	1,046,885	2,891,910	0	180,288	255,511	220,825	242,567	9,587,325
3,297,344	31,990	48,604	0	24,878	−27,421	1,924,255	377,130	4,922,520
1,307,786	217,155	986,388	0	64,098	90,604	579,653	75,977	3,169,708
1,488,482	20,405	85,985	0	164,579	−1685	269,856	162,619	1,865,002
2,615,503	12,592	90,260	0	31,598	−34,103	226,932	276,965	2,665,817
4,483,116	18,196	170,980	0	−5666	−134,219	94,441	295,190	4,331,658
13,066,217	115,582	376,881	0	157,994	−131,943	944,203	1,386,324	13,142,609
5,817,939	5914	43,676	0	59,198	−135,481	277,903	136,235	5,932,913
12,320,634	0	0	0	140,991	−188,641	365,674	658,400	11,980,259
3,008,561	11,976	63,606	0	79,511	−54,600	345,348	133,314	3,321,088
5,996,597	859	12,781	0	3,918,505	89,393	1,133,669	1,266,484	9,885,319
4,193,147	104,211	529,181	0	3,812,311	119,509	1,028,293	920,502	8,866,150
3,840,743	94,743	360,208	0	1,395,975	34,952	1,207,905	558,137	6,376,389
4,940,836	75,634	304,398	0	921,475	187,744	3,248,399	2,378,346	7,300,141
1,132,138	8367	34,964	0	184,795	−24,607	451,610	989,301	797,965
1,083,045	55,487	266,301	0	366,262	9937	193,544	67,683	1,906,893
1,070,476	0	0	0	0	6321	3875	627,740	452,932
6,940,647	60,900	224,622	0	0	−610,666	10,539	1902	6,624,140
172,247	13,524	106,081	0	1926	358	0	0	294,137
120,068	9025	101,414	0	0	1226	0	0	231,734
511,596	0	189,184	0	14,981,684	72,498	212,465	62,211	15,905,216
6,238,070	93,183	369,808	245,545	48,722	−179,449	369,632	327,570	6,857,941
174,232	1378	8868	0	0	2855	6057	16,102	177,288
1,242,354	119,895	496,682	0	502,338	43,230	110,297	48,090	2,466,705
3,812,861	350,524	1,054,525	0	439,793	47,359	759,672	0	6,464,734
1,989,152	215,797	835,245	0	0	98,640	6346	128,345	3,016,835
3,876,477	137,278	702,031	47,249	0	108,023	44,783	41,264	4,874,577
916,656	400,860	1,570,780	0	992,779	120,162	0	0	4,001,238
2,397,749	19,710	284,821	117,721	0	59,141	604,365	249,350	3,234,158
350,608	0	0	220,310	0	2978	11,633	146,747	438,782
1,024,187	1776	22,731	175,107	148,438	−1537	0	0	1,370,701
179,556	5566	70,369	346,988	0	−2160	0	0	600,318
818,691	124,926	719,927	0	0	35,106	27,285	43,628	1,682,308
94,727	233,791	527,128	1,669,258	0	−27,196	4005	16,302	2,485,411
176,569	102,037	1,108,883	1,110,604	0	−23,344	3828	2677	2,475,901
385,388	24,396	130,240	191,414	0	12,312	42,790	53,539	733,002
33,640	0	0	3,714,794	0	−12,349	10,854	10,414	3,736,525

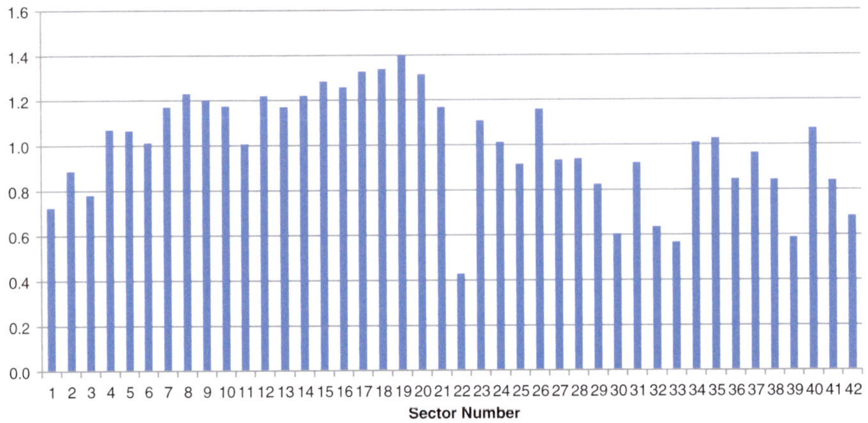

Fig. 10.1 The average influence coefficient of each sector during the 2011–2014 period

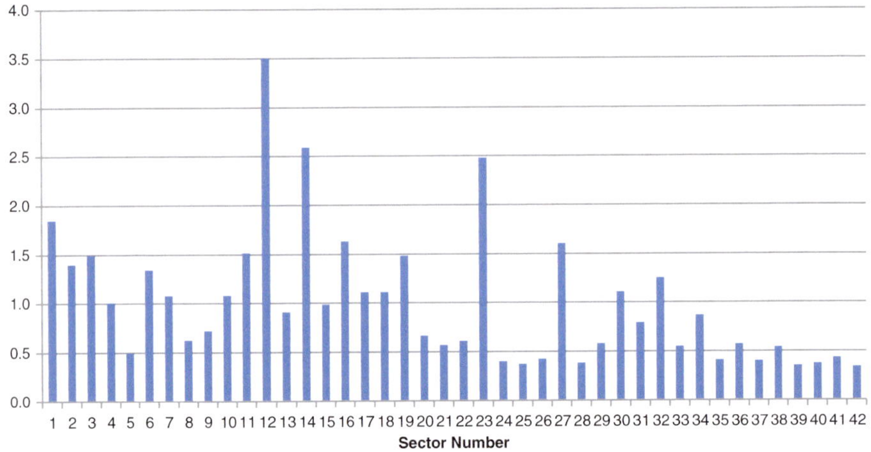

Fig. 10.2 The average reaction coefficient of each sector during the 2011–2014 period

The average annual influence coefficient of each sector during the 2011–2014 period is shown in Fig. 10.1. The first five sectors in terms of the top five influence coefficients are Manufacture of Communication Equipment, Computers, and Other Electronic Equipment (No. 19); Manufacture of Electrical Machinery and Equipment (No. 18); Manufacture of Transport Equipment (No. 17); Manufacture of Instruments, Meters, and Machinery for Cultural and Office Use (No. 20); and Manufacture of Metal Products (No. 15) in proper order. Their influence coefficients are, respectively, 1.4009, 1.3401, 1.3287, 1.3162, and 1.2861. These sectors have the greatest pulling effect on other sectors. It is worth noting that these five sectors all belong to manufacturing, which reflects the crucial role of manufacturing on the national economy in China.

The average annual reaction coefficient of each sector during the 2011–2014 period is illustrated in Fig. 10.2. The first five sectors in terms of the top five reaction coefficients are Chemical Industry (No. 12), Smelting and Rolling of Metals (No. 14), and Production and Supply of Electric Power and Heat Power (No. 23) in secondary industry; Agriculture, Forestry, Animal Husbandry, and Fishery (No. 1) in primary industry; and Manufacture of General and Special Purpose Machinery (No. 16) in tertiary industry in proper order. Their reaction coefficients are, respectively, 3.5043, 2.5904, 2.4795, 1.8503, and 1.6317. These sectors have the largest pushing effects on other sectors. They tend to be fundamental sectors, which provide basic materials for the production of other sectors, such as plastic, metals, and electricity.

10.2 Input–Output Tables of 2015–2025 (BAU Scenario)

10.2.1 Input–Output Tables

In this section, we focus on the period of 2015–2025 in the business as usual (BAU) scenario in which the policies have not been adjusted by us. The input–output table in each year within the period is going to be displayed. The unit of the numbers in all the tables is million CNY in constant 2010 CNY.

China's national input–output table in 2015 in the BAU scenario is shown in Table 10.6.

China's national input–output table in 2016 in the BAU scenario is displayed in Table 10.7.

China's national input–output table in 2017 in the BAU scenario is shown in Table 10.8.

China's national input–output table in 2018 in the BAU scenario is shown in Table 10.9.

China's national input–output table in 2019 in the BAU scenario is displayed in Table 10.10.

China's national input–output table in 2020 in the BAU scenario is shown in Table 10.11.

China's national input–output table in 2021 in the BAU scenario is shown in Table 10.12.

China's national input–output table in 2022 in the BAU scenario is displayed in Table 10.13.

China's national input–output table in 2023 in the BAU scenario is shown in Table 10.14.

China's national input–output table in 2024 in the BAU scenario is shown in Table 10.15.

China's national input–output table in 2025 in the BAU scenario is exhibited in Table 10.16.

Table 10.6 China's national input–output table in 2015

	Sector 1	Sector 2	Sector 3	Sector 4	Sector 5	Sector 6	Sector 7	Sector 8
Sector 1	1,447,366	29,238	26	2460	293	3,895,231	1,025,717	147,042
Sector 2	8968	538,234	11,600	6171	2047	26,866	25,203	12,079
Sector 3	0	0	22,375	0	0	0	0	0
Sector 4	0	0	0	158,306	0	0	0	0
Sector 5	1253	1083	710	1430	53,922	3537	0	0
Sector 6	1,139,104	5279	2428	3593	2213	2,028,852	37,321	167,312
Sector 7	1476	781	1075	829	431	5076	1,718,342	1,083,510
Sector 8	3837	10,555	4789	6409	3285	12,892	81,704	488,415
Sector 9	11,265	24,321	5151	4921	1449	10,542	4848	6230
Sector 10	10,488	3576	2877	2677	1752	183,538	55,027	52,441
Sector 11	77,612	33,015	50,978	84,881	36,686	27,630	12,799	20,497
Sector 12	831,201	69,087	50,428	76,473	79,557	271,461	574,087	189,587
Sector 13	15,351	27,944	12,535	12,352	28,795	67,964	12,050	4432
Sector 14	2602	127,107	66,098	19,180	5328	11,094	3884	4535
Sector 15	36,068	55,747	10,401	30,474	12,190	45,532	8575	14,730
Sector 16	61,373	161,482	112,122	119,359	68,090	33,635	60,549	21,138
Sector 17	31,893	21,282	9223	16,014	16,701	12,483	10,650	4995
Sector 18	2967	45,256	12,539	19,843	7539	14,334	10,987	6161
Sector 19	1004	4879	1745	2171	879	3626	3752	3545
Sector 20	5862	23,792	12,792	6906	2708	6934	7331	2629
Sector 21	40,161	21,270	1864	6143	8221	15,061	42,704	37,488
Sector 22	737	353	22	1223	7750	7662	8484	1592
Sector 23	137,820	136,635	98,947	251,167	90,491	102,052	127,258	26,295
Sector 24	404	1201	1672	7827	4104	5827	2661	1172
Sector 25	1297	2170	1063	1338	780	3986	4145	1483
Sector 26	2853	3943	1530	576	279	2776	1184	999
Sector 27	197,504	139,463	26,267	65,975	48,326	282,271	92,477	78,171
Sector 28	7813	806	217	255	230	2221	3285	6051
Sector 29	40,654	8813	2297	2249	1834	17,482	10,142	16,044
Sector 30	204,354	34,746	12,419	16,395	11,833	237,760	93,294	65,000
Sector 31	26,518	30,873	12,026	16,690	10,507	58,323	19,129	17,524
Sector 32	118,786	67,870	11,106	17,171	12,508	101,278	134,837	45,187
Sector 33	2638	4207	605	899	1021	16,927	15,567	32,455
Sector 34	16,064	16,061	2462	5958	3246	162,601	18,987	69,505
Sector 35	19,032	4250	2476	500	191	7988	3278	1675
Sector 36	96,431	27,121	15,264	19,725	8268	33,876	12,496	11,029
Sector 37	29,799	3702	1104	980	1979	3019	6942	429
Sector 38	35,198	21,519	5061	3349	2804	11,868	5007	5247
Sector 39	3767	2685	250	576	930	1353	504	333
Sector 40	6218	6848	486	2832	1524	4904	2303	2140
Sector 41	2510	4061	1298	3622	3050	10,374	5304	5453
Sector 42	3296	725	410	312	143	1377	570	418
Total intermediate input	4,683,544	1,721,981	588,740	1,000,209	543,883	7,752,214	4,263,387	2,654,970
Compensation of employees	6,053,471	605,795	198,573	148,209	94,537	602,918	455,814	345,515
Net taxes on production	12,299	337,922	225,483	95,632	57,525	680,036	181,403	89,658
Depreciation of fixed assets	297,066	115,777	132,559	34,980	24,509	186,846	119,719	40,696
Operating surplus	0	350,628	310,655	181,808	78,497	600,522	186,982	115,820
Total value added	6,362,836	1,410,121	867,270	460,628	255,067	2,070,322	943,918	591,689
Total input	11,046,380	3,132,102	1,456,010	1,460,837	798,950	9,822,536	5,207,305	3,246,659

Sector 9	Sector 10	Sector 11	Sector 12	Sector 13	Sector 14	Sector 15	Sector 16	Sector 17
234,870	144,106	97	459,518	3543	1250	990	1081	659
15,244	25,327	271,959	210,766	449,951	447,201	15,854	34,195	10,351
0	0	2,463,623	416,169	15,290	46,807	0	0	0
0	0	0	67,261	15,785	2,508,070	37,898	29,550	4797
0	0	442	211,828	341,658	29,865	12,934	7029	2146
10,460	11,552	20,368	220,405	22,432	41,772	11,666	32,838	32,012
18,851	50,714	223	94,250	13,065	1315	5057	14,088	16,549
33,617	20,537	1817	30,439	15,510	27,325	10,818	21,520	92,738
564,101	60,189	1342	21,965	42,243	9075	60,774	32,318	89,158
29,807	709,322	2373	193,350	141,783	12,410	33,307	43,848	30,826
14,023	13,509	265,446	808,011	249,724	511,404	35,638	102,535	42,035
177,126	370,054	81,909	5,297,471	466,976	168,952	132,361	362,092	510,795
16,612	7655	14,826	103,247	1,057,312	240,230	38,074	72,080	68,087
44,219	31,487	2798	133,489	163,775	3,302,447	1,175,066	2,107,597	1,064,362
48,494	39,793	14,071	114,783	162,855	107,896	422,520	420,798	161,506
28,199	36,762	56,200	197,901	168,464	378,988	165,414	2,256,894	749,007
8713	21,808	9860	40,648	32,896	64,872	32,566	151,265	2,635,246
6410	17,928	4951	34,972	24,145	39,035	27,740	548,743	344,211
818	24,538	596	8110	3866	3147	4179	209,090	102,893
1553	14,925	6211	83,416	14,133	35,921	8572	52,456	128,476
16,023	19,268	1929	85,357	50,966	39,548	19,015	69,340	129,283
6943	186,567	862	64,744	105,948	443,185	34,787	118,990	8103
63,459	83,822	72,186	706,485	483,728	751,807	254,106	224,110	152,624
882	802	6404	32,934	7031	15,896	3362	8939	5698
1239	5714	1316	14,391	6146	8064	2921	3347	4215
512	691	2109	5163	2127	3942	660	3465	3204
62,479	65,709	83,360	362,302	255,848	228,889	76,620	286,903	168,887
1311	1285	130	9875	1543	3422	618	8893	1806
7033	6710	13,462	46,512	13,524	104,056	10,360	31,658	21,288
49,119	71,036	51,380	280,434	160,859	216,855	79,322	256,589	448,167
10,136	13,881	7660	73,651	43,176	42,176	39,460	77,458	33,198
30,523	41,850	25,639	255,248	192,895	201,322	39,983	106,628	89,088
8776	13,184	682	30,527	14,724	3296	15,421	27,980	14,889
17,363	22,432	20,189	189,206	40,374	19,865	13,956	88,481	116,880
1153	1543	1167	32,505	4616	19,741	3007	34,831	74,525
3571	13,918	7186	81,626	23,742	41,119	18,889	59,980	78,749
277	7737	847	12,843	1727	3484	453	2072	2624
2022	4836	3722	26,433	17,376	26,837	5477	11,795	10,826
299	364	169	1884	906	926	446	3183	1629
1714	3314	1094	8206	10,453	10,582	7714	23,802	16,966
2289	2782	966	15,023	9201	6937	3325	14,051	8241
253	415	195	1609	993	1202	448	1360	1236
1,540,492	2,168,066	3,521,767	11,084,959	4,853,309	10,171,132	2,861,778	7,963,872	7,477,980
156,149	215,203	141,946	890,234	429,418	745,446	245,205	878,016	542,397
64,836	82,884	468,634	491,653	233,830	360,894	102,082	402,896	360,344
37,294	81,809	105,195	381,032	166,820	444,601	71,602	233,741	157,212
94,204	127,614	156,017	737,449	384,679	509,788	171,742	621,202	446,534
352,483	507,511	871,792	2,500,368	1,214,746	2,060,729	590,630	2,135,854	1,506,488
1,892,975	2,675,577	4,393,559	13,585,327	6,068,055	12,231,861	3,452,409	10,099,726	8,984,467

(continued)

Table 10.6 (continued)

	Sector 18	Sector 19	Sector 20	Sector 21	Sector 22	Sector 23	Sector 24	Sector 25
Sector 1	123	0	31	251,732	177	115	50	0
Sector 2	5579	2290	1183	13,763	815	1,133,610	14,331	314
Sector 3	0	0	0	20	1	40,305	140,419	0
Sector 4	8577	115	965	191	538	190	0	0
Sector 5	2774	50	84	5780	68	1732	12	24
Sector 6	25,774	27,852	2570	28,421	700	19,628	1112	617
Sector 7	11,937	2009	1161	188,040	77	332	35	70
Sector 8	14,418	13,305	2394	31,000	120	21,384	3348	1164
Sector 9	18,101	12,326	2185	63,660	90	3982	349	347
Sector 10	73,738	62,365	7644	37,296	123	12,266	752	616
Sector 11	28,790	20,568	3743	15,248	1203	189,488	6205	2445
Sector 12	538,775	485,090	68,659	158,865	11,751	20,010	1916	11,653
Sector 13	136,153	135,674	28,104	43,643	1447	13,719	689	689
Sector 14	1,732,387	235,201	31,500	198,390	5810	6218	735	273
Sector 15	278,990	168,627	31,557	87,803	609	38,216	2842	6360
Sector 16	328,393	79,422	22,242	8338	359	64,721	2409	2509
Sector 17	18,435	26,692	4885	21,290	124	72,080	6538	2929
Sector 18	939,303	218,241	40,137	13,041	914	356,813	692	793
Sector 19	394,029	3,443,381	185,610	5188	46	3960	157	190
Sector 20	58,533	309,357	99,147	4502	54	86,731	1872	1280
Sector 21	66,692	61,244	11,973	228,528	108	4303	223	244
Sector 22	61,025	8246	98	413	5318	706	0	0
Sector 23	88,244	99,869	9154	28,035	2473	2,304,890	5964	55,310
Sector 24	3955	4032	541	1990	121	7496	16,623	769
Sector 25	2530	3084	561	1414	115	8029	160	4947
Sector 26	1743	1437	221	326	57	2211	221	287
Sector 27	153,699	127,498	17,420	49,478	4218	78,616	7890	2237
Sector 28	4486	2614	914	589	13	762	39	73
Sector 29	20,944	71,915	3091	14,054	168	39,504	426	872
Sector 30	197,522	278,473	19,878	41,473	2583	51,097	5010	1432
Sector 31	43,502	22,465	7317	14,917	516	18,489	1369	1881
Sector 32	93,859	264,565	6138	22,772	2376	267,816	9892	13,592
Sector 33	20,262	19,353	3648	14,146	180	2826	314	124
Sector 34	149,632	85,857	10,119	17,534	437	28,754	1314	1161
Sector 35	28,065	54,861	4133	2040	45	4292	79	49
Sector 36	37,727	38,576	4911	4469	221	51,740	688	359
Sector 37	1267	891	78	333	32	17,232	130	13,520
Sector 38	5411	5022	659	883	478	49,418	860	1862
Sector 39	947	700	150	348	5	997	70	181
Sector 40	11,722	3760	1385	326	87	7703	837	52
Sector 41	7015	6418	850	2302	80	14,549	564	709
Sector 42	733	745	105	297	0	921	25	47
Total intermediate input	5,615,788	6,404,193	637,144	1,622,879	44,658	5,047,850	237,160	131,983
Compensation of employees	415,918	578,495	82,957	188,088	6478	516,385	31,922	48,089
Net taxes on production	160,944	117,509	23,394	126,739	3880	314,011	7407	9477
Depreciation of fixed assets	113,495	181,629	16,922	42,450	2913	647,185	9912	41,415
Operating surplus	306,079	216,457	42,351	103,598	395,289	210,094	14,492	5553
Total value added	996,436	1,094,090	165,624	460,875	408,560	1,687,675	63,733	104,533
Total input	6,612,224	7,498,283	802,769	2,083,754	453,217	6,735,525	300,893	236,516

Sector 26	Sector 27	Sector 28	Sector 29	Sector 30	Sector 31	Sector 32	Sector 33	Sector 34
71,436	167,763	0	0	1560	366,034	0	349	7900
12,900	24,207	264	0	1965	2113	0	902	4695
0	8732	0	0	0	0	0	0	0
0	0	0	0	0	0	0	0	0
164,253	1821	0	0	0	0	0	0	0
64,911	55,294	469	6206	33,529	1,101,662	13,921	11,948	39,390
13,726	9082	285	312	7773	15,590	710	4728	43,035
61,882	50,201	1358	4817	56,938	11,962	13,097	19,181	41,519
284,652	8564	938	2440	10,640	4383	3733	5144	8409
27,171	25,796	4996	79,353	59,703	10,784	96,531	24,938	298,327
209,600	1,309,930	5560	4907	21,574	15,598	31,408	55,570	105,992
501,572	123,646	625	13,486	58,471	24,174	16,856	71,022	119,272
3,684,211	14,887	652	690	1755	6500	798	1260	1767
2,026,933	36,180	0	26	3157	143	0	3815	2543
528,673	39,101	147	5118	2592	3704	4493	25,811	66,847
481,303	169,191	1974	20,590	13,931	3719	21,157	15,070	21,856
67,205	494,539	14,899	17,946	56,498	10,706	14,460	16,237	101,311
615,022	19,755	333	175,251	43,984	22,723	10,214	47,175	93,366
39,195	17,706	1653	152,098	25,205	1293	3717	12,745	192,348
35,505	11,188	223	22,820	2104	99	19,076	8311	9913
52,927	6430	476	2697	4936	1975	5175	5495	20,179
0	0	0	0	0	0	0	0	0
121,434	84,203	2574	63,471	75,632	67,765	46,855	25,460	31,750
1923	6600	180	1448	1323	11,572	247	448	3215
3903	4232	303	3110	2475	5505	2351	775	2174
172,461	30,224	7966	4314	16,688	7256	7192	81,855	3250
1,514,729	593,015	26,858	78,044	453,402	38,568	123,296	32,100	163,837
3506	2206	7902	3325	8731	1822	21,402	2715	15,658
235,407	50,464	3828	110,294	52,610	9746	141,792	16,573	12,447
400,026	90,877	14,126	79,962	32,826	67,056	21,084	18,147	77,223
162,585	111,945	1355	49,996	124,202	21,461	183,399	69,288	219,688
153,825	414,418	2207	52,691	220,793	52,486	399,737	182,136	151,080
9474	34,281	2868	66,562	143,717	33,616	159,444	55,955	55,550
58,596	39,012	7312	127,609	315,322	37,887	303,292	143,550	145,777
12,178	2160	102	4259	10,088	28	6977	895	1663
188,056	8268	33	12,443	24,937	584	3450	8151	2483
799	2119	0	712	1495	1040	3196	1366	28,599
22,768	159,852	301	16,298	59,805	29,593	19,332	31,293	51,563
2024	5245	189	2981	3867	826	9680	1003	1602
4826	3790	36	759	1040	534	3442	1468	544
22,587	14,461	1409	14,243	25,201	8194	29,788	22,474	14,288
2074	1981	62	969	1610	488	2840	2814	634
12,036,259	4,253,369	114,462	1,202,246	1,982,081	1,999,192	1,744,142	1,028,166	2,161,696
2,414,890	1,137,519	54,020	342,296	1,334,255	717,910	1,027,574	348,710	638,972
670,634	228,554	4869	100,687	1,323,515	136,332	394,179	502,667	162,049
234,609	505,881	12,731	330,502	310,141	160,934	75,886	1,539,606	310,608
919,939	901,037	156	563,881	1,889,977	170,517	1,738,994	800,366	91,464
4,240,072	2,772,991	71,776	1,337,366	4,857,887	1,185,693	3,236,632	3,191,349	1,203,093
16,276,331	7,026,359	186,238	2,539,612	6,839,968	3,184,884	4,980,773	4,219,515	3,364,788

(continued)

Table 10.6 (continued)

	Sector 35	Sector 36	Sector 37	Sector 38	Sector 39	Sector 40	Sector 41	Sector 42
Sector 1	17,029	5629	16,022	20,622	4709	12,286	3744	0
Sector 2	783	1894	3433	7185	5003	10,360	621	10,769
Sector 3	0	0	0	0	0	0	0	0
Sector 4	4261	879	1156	0	0	0	0	0
Sector 5	69	43	431	93	909	1311	376	1519
Sector 6	8043	3466	6971	40,059	19,844	21,390	43,015	13,955
Sector 7	9152	934	2838	26,548	4314	14,273	2253	15,649
Sector 8	3495	2366	8567	12,430	2763	35,183	12,786	51,165
Sector 9	1534	1800	8452	16,632	30,021	19,139	9190	62,336
Sector 10	5043	19,941	4496	29,442	86,007	31,937	86,340	99,915
Sector 11	4134	28,926	24,340	32,384	19,399	10,470	4987	55,724
Sector 12	49,156	31,171	44,584	153,730	22,117	1,093,177	31,507	38,833
Sector 13	3561	2743	9226	1876	12,064	14,295	4918	18,710
Sector 14	2268	10,391	4995	4527	1839	123	353	0
Sector 15	11,528	30,230	5285	8760	5275	5354	3051	4542
Sector 16	6801	11,370	13,870	4283	4905	114,526	2148	8917
Sector 17	4018	20,645	23,324	67,317	16,360	2844	3590	49,846
Sector 18	19,953	58,943	17,289	28,950	7247	8157	4236	8734
Sector 19	17,831	53,199	9530	43,503	7945	1886	4582	29,513
Sector 20	15,955	32,083	9124	17,183	23,926	5976	2877	6746
Sector 21	2063	3128	3475	14,030	4187	3217	1549	8685
Sector 22	0	0	0	0	0	0	0	0
Sector 23	9213	9784	18,882	19,574	44,834	29,750	6899	52,111
Sector 24	149	360	607	1138	1582	1190	514	1153
Sector 25	702	1671	1235	2896	3325	1706	608	2195
Sector 26	1610	4271	16,546	13,533	11,815	49,581	6252	53,674
Sector 27	18,763	83,983	15,307	37,077	62,981	12,825	22,583	159,417
Sector 28	760	1687	634	2458	5141	821	1749	40,882
Sector 29	2574	6995	5885	6033	19,930	23,302	6196	71,300
Sector 30	9434	16,629	9463	31,558	18,097	76,409	15,764	39,174
Sector 31	17,536	58,091	15,822	27,804	58,970	24,650	23,909	243,488
Sector 32	1949	44,798	21,138	37,432	61,776	14,521	11,483	60,865
Sector 33	2063	14,499	2000	49,663	17,563	4886	9880	20,690
Sector 34	4283	21,342	9785	25,173	42,627	18,199	16,840	34,182
Sector 35	10,860	3260	638	48	4919	1160	773	52
Sector 36	6569	88,452	3417	1360	5008	3576	523	5329
Sector 37	284	299	15,764	2561	3106	645	685	8864
Sector 38	1347	12,722	16,579	68,740	17,271	13,963	7224	54,678
Sector 39	1521	1402	1095	542	10,601	2434	733	29,153
Sector 40	1194	1319	2470	571	3483	6302	1098	10,439
Sector 41	7039	7206	4826	7096	14,202	3669	23,462	47,559
Sector 42	116	452	262	408	459	368	158	1022
Total intermediate input	284,643	699,006	379,764	865,220	686,525	1,695,861	379,458	1,421,785
Compensation of employees	110,557	378,241	144,439	605,768	1,736,113	787,387	221,738	2,178,433
Net taxes on production	7060	63,259	11,857	83,684	11,171	8215	38,229	11,827
Depreciation of fixed assets	20,746	71,951	57,895	49,251	206,345	95,584	59,363	320,709
Operating surplus	24,086	194,356	43,455	170,560	29,126	56,382	69,703	68,941
Total value added	162,449	707,806	257,646	909,263	1,982,755	947,568	389,033	2,579,910
Total input	447,092	1,406,812	637,409	1,774,483	2,669,280	2,643,429	768,491	4,001,695

Total intermediate use	Rural household consumption expenditure	Urban household consumption expenditure	Government consumption expenditure	Gross fixed capital formation	Increase in inventories	Export	Import	Total output
8,340,799	928,952	1,117,258	82,780	593,514	221,672	305,585	548,526	11,042,035
3,370,994	14,369	10,031	0	31,982	−87,034	16,120	228,617	3,127,846
3,153,740	0	0	0	18,192	−55,601	253,011	1,835,772	1,533,569
2,838,541	0	0	0	71,142	−19,071	7635	1,483,953	1,414,293
849,186	0	0	0	7405	−22,672	18,939	54,098	798,760
5,380,354	1,108,036	3,088,358	0	183,565	255,511	220,825	242,567	9,994,082
3,400,499	34,694	52,902	0	25,700	−27,421	2,181,570	452,029	5,215,914
1,343,050	231,085	1,059,874	0	65,330	90,604	579,653	75,977	3,293,621
1,528,940	21,503	91,082	0	164,872	−1685	269,856	193,388	1,881,180
2,694,920	13,485	95,559	0	32,027	−34,103	226,932	294,469	2,734,352
4,594,613	19,348	180,600	0	−5711	−134,219	97,645	305,480	4,446,795
13,399,737	123,987	404,278	0	161,376	−131,943	944,203	1,438,944	13,462,694
5,939,579	6562	47,087	0	59,892	−135,481	300,062	157,104	6,060,598
12,572,876	0	0	0	142,913	−188,641	365,674	697,170	12,195,652
3,071,950	12,973	69,153	0	81,642	−54,600	345,348	149,629	3,376,838
6,099,582	924	13,702	0	3,981,008	89,393	1,238,891	1,346,086	10,077,414
4,255,835	110,119	555,277	0	3,833,278	119,509	1,160,088	1,005,997	9,028,110
3,919,027	100,837	384,908	0	1,433,745	34,952	1,277,986	593,390	6,558,065
5,025,348	81,497	327,596	0	933,962	187,744	3,431,293	2,509,252	7,478,187
1,199,205	8915	36,187	0	184,818	−24,607	462,375	1,086,536	780,357
1,117,582	59,174	293,412	0	414,107	9937	199,862	72,933	2,021,141
1,073,757	0	0	0	0	6321	4000	664,580	419,497
7,067,122	60,900	224,622	0	0	−626,401	11,195	1914	6,735,525
175,992	14,167	110,445	0	1974	358	0	0	302,936
123,620	9671	105,315	0	0	1226	0	0	239,833
531,005	0	203,711	0	15,310,185	72,498	241,164	69,652	16,288,911
6,399,292	100,588	397,107	276,986	50,104	−179,449	369,632	369,698	7,044,563
180,649	1485	9657	0	0	2855	6295	18,957	181,984
1,280,467	127,182	530,395	0	516,610	43,230	120,454	53,702	2,564,636
3,904,885	376,724	1,137,765	0	460,504	47,359	779,315	0	6,706,552
2,057,030	232,604	901,606	0	0	98,640	−8107	142,219	3,139,556
4,056,265	145,615	748,657	52,524	0	108,023	50,881	46,500	5,115,464
947,363	430,369	1,650,149	0	1,045,516	120,162	0	0	4,193,559
2,469,227	20,948	302,114	128,755	0	59,141	648,496	249,350	3,379,330
366,103	0	0	233,568	0	2978	13,150	162,268	453,532
1,054,320	1776	22,731	190,100	149,930	−1537	0	0	1,417,320
185,037	5935	74,666	369,142	0	−2160	0	0	632,621
847,298	133,241	768,891	0	0	35,106	28,682	48,298	1,764,920
98,500	254,011	571,540	1,814,008	0	−27,196	4132	18,137	2,696,856
180,287	107,190	1,178,288	1,222,215	0	−23,344	3828	2677	2,665,787
398,680	26,147	139,794	209,038	0	12,312	42,790	53,539	775,223
34,552	0	0	4,018,394	0	−12,349	11,961	11,118	4,041,441

Table 10.7 China's national input–output table in 2016

	Sector 1	Sector 2	Sector 3	Sector 4	Sector 5	Sector 6	Sector 7	Sector 8
Sector 1	1,524,379	29,245	26	2460	294	4,269,855	1,069,502	158,336
Sector 2	10,185	538,743	11,607	6171	2072	29,450	28,618	14,580
Sector 3	0	0	22,387	0	0	0	0	0
Sector 4	0	0	0	158,306	0	0	0	0
Sector 5	1372	1084	711	1430	54,360	3877	0	0
Sector 6	1,181,583	5280	2430	3593	2219	2,223,977	38,743	175,888
Sector 7	1530	782	1076	829	433	5564	1,785,304	1,143,619
Sector 8	3977	10,557	4791	6409	3293	14,131	84,408	512,703
Sector 9	13,961	24,356	5154	4921	1476	11,556	5938	8370
Sector 10	10,805	3577	2878	2677	1757	201,190	56,656	54,665
Sector 11	81,006	33,024	51,004	84,881	36,816	30,288	13,345	21,804
Sector 12	901,579	69,119	50,455	76,473	80,036	297,568	622,030	211,968
Sector 13	16,533	27,957	12,542	12,352	28,979	74,500	12,980	4923
Sector 14	2790	127,163	66,133	19,180	5357	12,161	4178	4988
Sector 15	41,314	55,797	10,407	30,474	12,320	49,911	10,072	17,589
Sector 16	65,906	161,551	112,182	119,359	68,463	36,870	65,005	23,297
Sector 17	35,230	21,296	9228	16,014	16,840	13,684	11,626	5758
Sector 18	3180	45,274	12,546	19,843	7579	15,713	11,826	6769
Sector 19	1078	4881	1746	2171	884	3975	4024	3912
Sector 20	11,725	23,972	12,799	6906	2961	7601	14,755	6750
Sector 21	41,280	21,276	1865	6143	8248	16,510	44,347	39,034
Sector 22	1024	354	22	1223	7910	8399	11,965	2405
Sector 23	145,465	136,677	98,999	251,167	90,853	111,867	134,415	28,464
Sector 24	421	1201	1673	7827	4117	6387	2772	1240
Sector 25	1355	2171	1064	1338	783	4370	4315	1580
Sector 26	3013	3944	1531	576	280	3044	1250	1080
Sector 27	208,557	139,510	26,281	65,975	48,535	309,419	97,660	84,220
Sector 28	8937	807	218	255	232	2435	3813	7315
Sector 29	42,895	8816	2298	2249	1842	19,163	10,671	17,369
Sector 30	237,848	34,786	12,425	16,395	12,003	260,626	110,828	81,163
Sector 31	30,745	30,902	12,032	16,690	10,632	63,932	22,157	21,324
Sector 32	124,820	67,895	11,112	17,171	12,569	111,019	140,136	48,901
Sector 33	2856	4209	606	899	1027	18,555	16,879	36,237
Sector 34	16,895	16,065	2464	5958	3258	178,239	19,931	74,468
Sector 35	20,877	4253	2478	500	193	8757	3537	1937
Sector 36	103,988	27,135	15,273	19,725	8326	37,134	13,369	12,387
Sector 37	31,642	3703	1105	980	1989	3309	7398	467
Sector 38	39,424	21,534	5064	3349	2830	13,010	5581	6161
Sector 39	3920	2685	250	576	932	1483	523	352
Sector 40	6450	6850	486	2832	1528	5376	2384	2254
Sector 41	2625	4062	1298	3622	3061	11,371	5538	5811
Sector 42	3418	725	410	312	143	1509	590	440
Total intermediate input	4,986,587	1,723,219	589,051	1,000,209	547,429	8,497,785	4,499,071	2,850,527
Compensation of employees	6,375,573	606,368	198,678	148,209	95,306	660,903	473,576	362,696
Net taxes on production	12,954	338,242	225,602	95,632	57,993	745,438	188,472	94,117
Depreciation of fixed assets	312,872	115,886	132,629	34,980	24,708	204,816	124,385	42,720
Operating surplus	0	350,960	310,819	181,808	79,136	658,277	194,268	121,579
Total value added	6,701,399	1,411,456	867,728	460,628	257,143	2,269,435	980,701	621,112
Total input	11,687,987	3,134,675	1,456,779	1,460,837	804,572	10,767,220	5,479,772	3,471,639

Sector 9	Sector 10	Sector 11	Sector 12	Sector 13	Sector 14	Sector 15	Sector 16	Sector 17
234,915	150,583	101	461,639	3562	1257	992	1094	660
15,253	28,888	281,632	213,481	458,305	454,368	15,931	34,602	10,433
0	0	2,551,243	419,133	15,437	47,310	0	0	0
0	0	0	68,230	16,161	2,556,879	38,090	29,902	4844
0	0	457	213,680	346,104	30,193	12,976	7112	2156
10,461	11,920	21,093	221,104	22,529	41,921	11,679	33,228	32,024
18,854	52,671	231	94,601	13,129	1321	5065	14,255	16,552
33,621	21,154	1881	30,533	15,572	27,420	10,830	21,776	92,744
564,632	74,519	1390	22,434	43,510	9303	61,210	32,702	90,689
29,811	734,068	2458	193,922	142,349	12,455	33,345	44,369	30,831
14,026	14,090	274,887	811,360	251,128	513,995	35,696	103,754	42,071
177,179	398,916	84,823	5,336,219	471,556	170,373	132,692	366,396	512,711
16,617	8236	15,354	103,981	1,068,201	242,361	38,176	72,937	68,317
44,230	33,681	2897	134,333	165,260	3,327,457	1,177,646	2,132,650	1,067,602
48,517	45,210	14,572	116,161	165,811	109,386	424,179	425,800	162,899
28,207	39,280	58,199	199,176	169,964	381,873	165,790	2,283,722	751,263
8717	24,083	10,211	41,052	33,329	65,658	32,686	153,063	2,650,478
6411	19,127	5127	35,188	24,353	39,315	27,800	555,266	345,208
818	26,122	617	8161	3899	3170	4188	211,575	103,182
1562	36,731	6431	91,396	16,301	40,199	8823	53,079	140,765
16,025	20,247	1998	85,610	51,230	39,732	19,037	70,165	129,426
6950	233,644	893	66,410	110,179	453,644	34,941	120,404	8317
63,473	88,318	74,753	710,013	486,809	756,013	254,550	226,774	152,915
882	834	6632	33,057	7066	15,966	3366	9045	5702
1239	5985	1363	14,454	6181	8107	2926	3387	4220
512	728	2184	5190	2142	3965	661	3507	3210
62,492	69,370	86,324	364,136	257,667	230,290	76,761	290,314	169,196
1311	1473	135	10,001	1571	3474	621	8998	1821
7034	7115	13,941	46,766	13,620	104,745	10,382	32,034	21,330
49,151	84,150	53,207	284,802	164,526	221,081	79,779	259,639	452,389
10,141	15,899	7932	74,645	44,021	42,850	39,641	78,379	33,531
30,531	44,848	26,551	256,732	194,330	202,858	40,079	107,896	89,297
8778	14,187	706	30,745	14,873	3323	15,459	28,312	14,944
17,366	23,494	20,907	190,049	40,612	19,965	13,977	89,533	117,047
1153	1704	1208	32,832	4678	19,995	3019	35,245	74,904
3572	15,061	7441	82,276	23,993	41,535	18,948	60,693	79,018
277	8212	877	12,915	1740	3507	454	2096	2630
2023	5405	3855	26,722	17,635	27,184	5499	11,935	10,898
299	376	175	1890	910	929	446	3221	1630
1714	3418	1133	8233	10,497	10,621	7723	24,085	16,970
2290	2912	1001	15,087	9253	6973	3330	14,218	8250
253	427	201	1614	997	1206	449	1376	1237
1,541,299	2,367,086	3,647,021	11,169,961	4,910,990	10,294,176	2,869,843	8,058,540	7,524,314
156,296	222,711	146,995	896,745	433,840	751,092	246,167	888,453	545,532
64,897	85,776	485,301	495,249	236,238	363,627	102,482	407,685	362,427
37,329	84,663	108,936	383,819	168,538	447,968	71,883	236,519	158,120
94,293	132,066	161,566	742,843	388,640	513,649	172,416	628,586	449,115
352,815	525,217	902,798	2,518,656	1,227,257	2,076,336	592,949	2,161,243	1,515,195
1,894,114	2,892,303	4,549,819	13,688,617	6,138,247	12,370,511	3,462,791	10,219,783	9,039,509

(continued)

Table 10.7 (continued)

	Sector 18	Sector 19	Sector 20	Sector 21	Sector 22	Sector 23	Sector 24	Sector 25
Sector 1	124	0	31	256,373	177	117	52	0
Sector 2	5701	2334	1183	14,017	815	1,151,659	14,912	329
Sector 3	0	0	0	20	1	40,946	146,109	0
Sector 4	8787	118	965	195	538	193	0	0
Sector 5	2815	51	84	5887	68	1759	12	25
Sector 6	25,909	27,981	2570	28,945	700	19,941	1157	647
Sector 7	12,017	2020	1161	191,508	77	337	36	74
Sector 8	14,491	13,365	2394	31,571	120	21,724	3484	1220
Sector 9	18,732	12,707	2185	64,834	90	4045	364	364
Sector 10	74,116	62,610	7644	37,984	123	12,461	783	645
Sector 11	28,995	20,689	3743	15,529	1203	192,505	6456	2563
Sector 12	545,331	490,170	68,659	161,795	11,751	20,329	1994	12,217
Sector 13	137,816	137,052	28,104	44,447	1447	13,937	717	722
Sector 14	1,750,698	237,339	31,500	202,049	5810	6317	765	286
Sector 15	284,639	171,581	31,557	89,423	609	38,824	2957	6668
Sector 16	331,950	80,147	22,242	8491	359	65,751	2507	2630
Sector 17	18,744	27,060	4885	21,683	124	73,228	6803	3071
Sector 18	949,132	220,167	40,138	13,282	914	362,494	720	832
Sector 19	398,181	3,475,547	185,611	5284	46	4023	163	200
Sector 20	67,519	354,378	99,148	4585	54	88,112	1947	1342
Sector 21	67,039	61,519	11,973	232,742	108	4372	232	255
Sector 22	63,333	8605	98	421	5318	717	0	0
Sector 23	88,973	100,614	9154	28,552	2473	2,341,588	6206	57,983
Sector 24	3980	4053	541	2026	121	7615	17,296	806
Sector 25	2549	3102	561	1440	115	8157	166	5186
Sector 26	1758	1448	221	332	57	2246	230	301
Sector 27	155,027	128,393	17,420	50,390	4218	79,868	8210	2345
Sector 28	4584	2664	914	600	13	774	40	77
Sector 29	21,139	72,462	3091	14,314	168	40,133	443	915
Sector 30	202,831	285,020	19,878	42,238	2583	51,911	5213	1502
Sector 31	44,480	22,892	7317	15,192	516	18,784	1424	1972
Sector 32	94,799	266,593	6138	23,192	2376	272,080	10,293	14,249
Sector 33	20,501	19,557	3648	14,407	180	2871	326	130
Sector 34	150,731	86,401	10,119	17,858	437	29,212	1367	1218
Sector 35	28,533	55,667	4133	2078	45	4361	82	51
Sector 36	38,235	39,023	4911	4551	221	52,564	716	377
Sector 37	1279	899	78	339	32	17,506	135	14,173
Sector 38	5509	5100	659	900	478	50,205	895	1952
Sector 39	952	704	150	354	5	1013	73	190
Sector 40	11,786	3778	1385	332	87	7826	871	55
Sector 41	7066	6457	850	2344	80	14,780	587	743
Sector 42	737	749	105	302	0	936	26	49
Total intermediate input	5,691,519	6,511,017	637,146	1,652,804	44,658	5,128,221	246,770	138,363
Compensation of employees	420,271	583,899	82,958	191,556	6478	524,607	33,216	50,414
Net taxes on production	162,628	118,607	23,395	129,076	3880	319,010	7707	9935
Depreciation of fixed assets	114,683	183,326	16,922	43,233	2913	657,490	10,314	43,417
Operating surplus	309,282	218,479	42,352	105,508	395,289	213,439	15,079	5821
Total value added	1,006,863	1,104,311	165,627	469,373	408,560	1,714,546	66,316	109,587
Total input	6,698,383	7,615,328	802,773	2,122,177	453,217	6,842,768	313,086	247,950

Sector 26	Sector 27	Sector 28	Sector 29	Sector 30	Sector 31	Sector 32	Sector 33	Sector 34
73,241	173,346	0	0	1580	374,375	0	363	8345
13,226	25,013	267	0	1990	2162	0	938	4959
0	9023	0	0	0	0	0	0	0
0	0	0	0	0	0	0	0	0
168,404	1882	0	0	0	0	0	0	0
66,551	57,134	474	6540	33,948	1,126,764	14,760	12,432	41,607
14,073	9385	288	329	7870	15,946	753	4920	45,457
63,446	51,872	1373	5076	57,649	12,235	13,887	19,958	43,856
291,845	8849	949	2571	10,773	4483	3958	5353	8883
27,858	26,655	5050	83,618	60,448	11,030	102,354	25,948	315,118
214,896	1,353,527	5620	5171	21,843	15,953	33,302	57,822	111,957
514,247	127,761	632	14,211	59,201	24,725	17,873	73,901	125,986
3,777,312	15,382	659	727	1777	6648	846	1311	1867
2,078,154	37,384	0	28	3197	146	0	3969	2686
542,033	40,403	148	5393	2625	3789	4764	26,857	70,609
493,466	174,822	1995	21,696	14,104	3804	22,433	15,681	23,086
68,903	510,998	15,061	18,910	57,203	10,950	15,333	16,895	107,013
630,563	20,412	337	184,669	44,532	23,240	10,830	49,087	98,621
40,186	18,295	1671	160,272	25,520	1323	3941	13,262	203,174
36,403	11,560	226	24,046	2130	102	20,227	8647	10,471
54,265	6644	481	2842	4997	2020	5487	5718	21,315
0	0	0	0	0	0	0	0	0
124,503	87,005	2602	66,882	76,576	69,310	49,681	26,492	33,537
1972	6819	182	1526	1340	11,836	262	467	3396
4002	4372	306	3277	2506	5630	2492	807	2296
176,819	31,230	8053	4546	16,896	7422	7626	85,173	3433
1,553,006	612,752	27,150	82,239	459,060	39,447	130,734	33,401	173,058
3595	2280	7988	3504	8839	1863	22,693	2825	16,539
241,356	52,144	3870	116,221	53,266	9968	150,345	17,245	13,148
410,135	93,902	14,279	84,259	33,236	68,584	22,356	18,882	81,569
166,693	115,671	1370	52,683	125,752	21,950	194,461	72,096	232,053
157,712	428,211	2231	55,522	223,549	53,681	423,849	189,519	159,584
9713	35,422	2899	70,139	145,511	34,382	169,062	58,223	58,676
60,077	40,311	7392	134,467	319,257	38,750	321,587	149,368	153,982
12,486	2232	103	4487	10,214	29	7398	932	1756
192,808	8543	34	13,111	25,249	597	3658	8481	2623
819	2190	0	751	1514	1063	3389	1421	30,209
23,343	165,172	304	17,173	60,551	30,267	20,498	32,561	54,466
2075	5420	191	3142	3915	845	10,264	1044	1692
4948	3916	37	800	1053	546	3650	1528	575
23,157	14,942	1425	15,008	25,516	8381	31,584	23,385	15,093
2126	2047	63	1021	1630	499	3011	2928	670
12,340,418	4,394,929	115,708	1,266,856	2,006,816	2,044,746	1,849,349	1,069,843	2,283,365
2,475,914	1,175,378	54,609	360,691	1,350,906	734,269	1,089,557	362,845	674,936
687,581	236,161	4922	106,098	1,340,032	139,438	417,956	523,042	171,170
240,537	522,718	12,869	348,264	314,011	164,601	80,464	1,602,014	328,090
943,186	931,025	158	594,184	1,913,562	174,402	1,843,890	832,809	96,612
4,347,219	2,865,282	72,558	1,409,238	4,918,511	1,212,710	3,431,867	3,320,711	1,270,807
16,687,636	7,260,211	188,266	2,676,094	6,925,327	3,257,456	5,281,215	4,390,553	3,554,172

(continued)

Table 10.7 (continued)

	Sector 35	Sector 36	Sector 37	Sector 38	Sector 39	Sector 40	Sector 41	Sector 42
Sector 1	17,212	5753	16,671	21,258	5196	13,332	4000	0
Sector 2	791	1935	3572	7406	5521	11,242	664	11,764
Sector 3	0	0	0	0	0	0	0	0
Sector 4	4307	899	1203	0	0	0	0	0
Sector 5	70	44	449	96	1003	1423	401	1660
Sector 6	8130	3542	7254	41,295	21,896	23,211	45,956	15,245
Sector 7	9251	955	2953	27,367	4760	15,488	2407	17,096
Sector 8	3533	2417	8913	12,813	3048	38,179	13,660	55,895
Sector 9	1550	1840	8794	17,145	33,124	20,769	9818	68,099
Sector 10	5097	20,378	4678	30,351	94,897	34,656	92,243	109,152
Sector 11	4179	29,560	25,325	33,383	21,404	11,361	5328	60,875
Sector 12	49,684	31,854	46,388	158,473	24,403	1,186,257	33,661	42,423
Sector 13	3599	2804	9599	1933	13,311	15,512	5254	20,440
Sector 14	2292	10,618	5197	4667	2029	133	378	0
Sector 15	11,651	30,893	5499	9031	5821	5809	3259	4962
Sector 16	6875	11,619	14,431	4416	5412	124,278	2295	9742
Sector 17	4061	21,098	24,268	69,394	18,052	3086	3836	54,454
Sector 18	20,167	60,234	17,988	29,843	7996	8852	4526	9542
Sector 19	18,023	54,365	9916	44,845	8766	2047	4895	32,241
Sector 20	16,127	32,786	9494	17,713	26,399	6485	3074	7370
Sector 21	2085	3196	3615	14,463	4620	3491	1655	9488
Sector 22	0	0	0	0	0	0	0	0
Sector 23	9312	9999	19,646	20,178	49,468	32,283	7370	56,929
Sector 24	151	368	631	1174	1745	1291	549	1260
Sector 25	710	1708	1285	2985	3669	1851	649	2398
Sector 26	1627	4365	17,216	13,950	13,036	53,803	6680	58,637
Sector 27	18,965	85,824	15,926	38,221	69,491	13,917	24,127	174,155
Sector 28	768	1724	659	2534	5672	891	1868	44,662
Sector 29	2602	7149	6123	6219	21,990	25,286	6619	77,891
Sector 30	9535	16,993	9845	32,532	19,967	82,915	16,842	42,795
Sector 31	17,725	59,364	16,462	28,662	65,066	26,749	25,543	265,998
Sector 32	1970	45,779	21,994	38,587	68,162	15,758	12,268	66,492
Sector 33	2085	14,817	2081	51,195	19,379	5302	10,556	22,603
Sector 34	4329	21,810	10,181	25,950	47,033	19,749	17,991	37,342
Sector 35	10,976	3332	664	49	5427	1259	825	56
Sector 36	6639	90,390	3556	1402	5526	3880	559	5822
Sector 37	287	306	16,402	2640	3427	700	731	9684
Sector 38	1362	13,001	17,250	70,860	19,056	15,151	7718	59,733
Sector 39	1538	1432	1139	558	11,697	2641	783	31,848
Sector 40	1207	1347	2570	589	3843	6839	1173	11,404
Sector 41	7115	7364	5022	7315	15,670	3982	25,066	51,956
Sector 42	117	462	272	421	506	399	169	1117
Total intermediate input	287,701	714,324	395,130	891,912	757,486	1,840,257	405,398	1,553,231
Compensation of employees	111,745	386,529	150,283	624,456	1,915,562	854,429	236,896	2,379,832
Net taxes on production	7136	64,645	12,336	86,265	12,325	8915	40,842	12,920
Depreciation of fixed assets	20,969	73,527	60,238	50,770	227,673	103,722	63,421	350,359
Operating surplus	24,345	198,615	45,213	175,822	32,137	61,183	74,468	75,315
Total value added	164,195	723,316	268,071	937,314	2,187,697	1,028,249	415,627	2,818,426
Total input	451,896	1,437,640	663,201	1,829,226	2,945,183	2,868,506	821,024	4,371,657

Total intermediate use	Rural household consumption expenditure	Urban household consumption expenditure	Government consumption expenditure	Gross fixed capital formation	Increase in inventories	Export	Import	Total output
8,880,444	1,000,401	1,212,834	89,404	625,095	221,672	325,021	572,564	11,782,307
3,436,718	15,269	10,747	0	32,012	−87,034	16,524	252,727	3,171,509
3,251,608	0	0	0	18,202	−55,601	272,971	2,036,279	1,450,901
2,889,616	0	0	0	71,142	−19,071	7635	1,483,953	1,465,368
861,646	0	0	0	7465	−22,672	19,568	58,912	807,095
5,674,260	1,181,351	3,323,880	0	201,219	255,511	220,825	242,567	10,614,480
3,542,340	37,447	57,278	0	26,701	−27,421	2,441,938	526,058	5,552,226
1,391,982	246,471	1,141,048	0	68,579	90,604	579,653	75,977	3,442,362
1,578,243	22,537	95,886	0	165,027	−1685	269,856	224,155	1,905,708
2,803,708	14,476	101,431	0	33,144	−34,103	226,932	330,144	2,815,443
4,696,365	20,684	191,762	0	−5915	−134,219	100,835	328,847	4,540,666
13,804,017	131,602	429,097	0	162,556	−131,943	944,203	1,490,986	13,848,546
6,068,173	7204	50,460	0	60,509	−135,481	322,258	177,939	6,195,184
12,713,349	0	0	0	143,995	−188,641	365,674	735,896	12,298,481
3,140,225	13,802	73,765	0	81,963	−54,600	345,348	165,810	3,434,693
6,194,338	988	14,598	0	4,028,331	89,393	1,344,433	1,425,446	10,246,636
4,324,065	116,301	582,585	0	3,867,719	119,509	1,291,716	1,091,601	9,210,292
3,989,645	106,409	407,497	0	1,448,749	34,952	1,348,388	628,482	6,707,157
5,096,198	87,168	350,035	0	942,686	187,744	3,614,542	2,635,721	7,642,651
1,333,103	9358	37,175	0	184,821	−24,607	473,146	1,185,163	827,833
1,136,797	62,163	315,396	0	421,743	9937	206,304	78,084	2,074,256
1,147,174	0	0	0	0	6321	4132	699,369	458,259
7,188,842	60,900	224,622	0	0	−641,522	11,852	1926	6,842,768
179,594	14,846	115,046	0	2054	358	0	0	311,898
127,055	10,380	109,588	0	0	1226	0	0	248,248
553,918	0	218,471	0	15,697,075	72,498	269,905	77,082	16,734,786
6,584,047	108,138	424,941	308,427	51,772	−179,449	369,632	411,763	7,255,745
191,987	1579	10,345	0	0	2855	6534	21,799	191,502
1,326,379	134,967	566,411	0	544,373	43,230	130,590	59,325	2,686,626
4,084,599	399,977	1,211,641	0	466,251	47,359	799,027	0	7,008,855
2,156,328	248,677	965,070	0	0	98,640	−22,577	156,075	3,290,064
4,181,332	154,907	800,624	57,799	0	108,023	56,946	51,765	5,307,864
986,262	459,239	1,727,800	0	1,087,896	120,162	0	0	4,381,360
2,557,146	22,228	319,994	139,789	0	59,141	692,520	249,350	3,541,469
374,446	0	0	246,826	0	2978	14,665	177,817	461,098
1,083,350	1776	22,731	205,093	153,216	−1537	0	0	1,464,628
193,277	6293	78,829	391,296	0	−2160	0	0	667,534
882,272	141,420	817,053	0	0	35,106	30,081	52,962	1,852,972
104,264	275,242	618,175	1,958,758	0	−27,196	4257	19,985	2,913,515
184,697	112,598	1,251,138	1,333,825	0	−23,344	3828	2677	2,860,066
415,589	27,974	149,762	226,662	0	12,312	42,790	53,539	821,551
35,673	0	0	4,321,994	0	−12,349	13,061	11,826	4,346,553

Table 10.8 China's national input–output table in 2017

	Sector 1	Sector 2	Sector 3	Sector 4	Sector 5	Sector 6	Sector 7	Sector 8
Sector 1	1,625,772	29,403	26	2460	297	4,360,334	1,160,952	163,634
Sector 2	11,401	543,024	11,611	6171	2101	30,074	33,893	15,419
Sector 3	0	0	22,396	0	0	0	0	0
Sector 4	0	0	0	158,306	0	0	0	0
Sector 5	1528	1092	711	1430	55,140	3959	0	0
Sector 6	1,270,574	5311	2431	3593	2240	2,271,104	43,214	182,555
Sector 7	1589	785	1076	829	435	5682	1,906,115	1,166,733
Sector 8	4232	10,610	4793	6409	3320	14,431	92,019	528,873
Sector 9	17,279	24,773	5156	4921	1521	11,801	7971	9284
Sector 10	11,861	3625	2880	2677	1791	205,453	64,929	57,594
Sector 11	86,337	33,238	51,026	84,881	37,182	30,930	14,692	22,537
Sector 12	975,213	69,417	50,476	76,473	80,674	303,874	698,096	220,011
Sector 13	17,833	28,109	12,547	12,352	29,252	76,079	14,561	5108
Sector 14	3132	128,511	66,160	19,180	5451	12,419	4975	5292
Sector 15	51,536	56,891	10,411	30,474	12,730	50,969	14,277	19,675
Sector 16	72,127	162,620	112,228	119,359	69,229	37,651	74,294	24,341
Sector 17	37,849	21,380	9232	16,014	16,964	13,974	12,845	5954
Sector 18	3488	45,570	12,551	19,843	7663	16,046	13,637	7081
Sector 19	1176	4911	1746	2171	893	4059	4575	4083
Sector 20	15,845	24,469	12,804	6906	3113	7762	22,314	7745
Sector 21	44,935	21,586	1866	6143	8449	16,860	51,808	41,137
Sector 22	1429	358	22	1223	8170	8577	18,323	2897
Sector 23	156,607	137,420	99,040	251,167	91,692	114,237	150,309	29,589
Sector 24	450	1208	1673	7827	4155	6523	3056	1282
Sector 25	1438	2184	1064	1338	790	4462	4701	1630
Sector 26	3233	3965	1531	576	282	3108	1389	1118
Sector 27	225,474	140,427	26,292	65,975	49,058	315,975	109,714	87,543
Sector 28	9980	812	218	255	235	2486	4541	7715
Sector 29	45,859	8870	2299	2249	1860	19,570	11,773	17,986
Sector 30	272,778	35,089	12,430	16,395	12,206	266,149	138,098	86,948
Sector 31	35,409	31,168	12,037	16,690	10,803	65,287	27,181	22,766
Sector 32	132,205	68,391	11,117	17,171	12,706	113,371	150,848	50,494
Sector 33	3212	4250	606	899	1044	18,948	20,082	38,489
Sector 34	18,295	16,167	2465	5958	3291	182,016	22,338	77,508
Sector 35	22,271	4275	2479	500	195	8942	3858	2002
Sector 36	110,372	27,274	15,279	19,725	8396	37,921	14,563	12,771
Sector 37	33,949	3721	1105	980	2005	3379	8262	484
Sector 38	43,614	21,668	5066	3349	2863	13,285	6459	6467
Sector 39	4228	2700	250	576	941	1515	583	365
Sector 40	6828	6883	486	2832	1539	5490	2591	2320
Sector 41	2794	4086	1299	3622	3089	11,612	6074	6000
Sector 42	3631	729	410	312	144	1541	644	454
Total intermediate input	5,387,762	1,736,970	589,295	1,000,209	553,909	8,677,855	4,940,554	2,943,885
Compensation of employees	6,799,640	611,186	198,760	148,209	96,672	674,908	505,623	374,135
Net taxes on production	13,815	340,930	225,696	95,632	58,824	761,234	201,226	97,085
Depreciation of fixed assets	333,683	116,807	132,684	34,980	25,062	209,157	132,802	44,067
Operating surplus	0	353,748	310,948	181,808	80,271	672,226	207,414	125,414
Total value added	7,147,138	1,422,671	868,087	460,628	260,830	2,317,525	1,047,065	640,701
Total input	12,534,900	3,159,641	1,457,383	1,460,837	814,738	10,995,380	5,987,619	3,584,586

Sector 9	Sector 10	Sector 11	Sector 12	Sector 13	Sector 14	Sector 15	Sector 16	Sector 17
235,050	151,543	104	485,715	3644	1258	992	1113	690
15,265	29,353	290,203	226,754	475,869	455,890	15,944	35,209	11,053
0	0	2,628,888	482,985	16,740	47,488	0	0	0
0	0	0	76,921	17,474	2,575,259	38,154	30,426	5425
0	0	471	227,568	359,738	30,290	12,986	7237	2289
10,467	12,018	21,735	232,116	23,113	42,000	11,685	33,811	33,400
18,861	52,907	238	97,767	13,351	1322	5067	14,506	16,950
33,637	21,290	1938	31,886	15,914	27,463	10,835	22,158	96,261
565,699	76,829	1432	25,729	47,136	9367	61,311	33,276	106,392
29,846	743,316	2533	214,859	149,765	12,497	33,379	45,148	34,275
14,034	14,196	283,252	853,088	257,856	514,943	35,714	105,575	44,006
177,262	402,990	87,404	5,546,944	481,955	170,710	132,755	372,826	529,719
16,626	8317	15,821	108,529	1,096,033	242,877	38,196	74,217	70,936
44,274	34,188	2985	146,208	173,095	3,339,294	1,178,750	2,170,080	1,158,320
48,608	46,658	15,015	135,319	181,185	110,200	425,017	433,273	190,401
28,225	39,722	59,970	210,478	175,212	382,810	165,893	2,323,803	791,742
8720	24,312	10,522	42,474	34,009	65,779	32,700	155,749	2,734,114
6416	19,352	5283	37,192	25,110	39,413	27,818	565,011	363,391
819	26,389	636	8593	4009	3177	4191	215,288	108,056
1568	39,215	6627	106,970	18,487	40,908	8852	54,011	172,557
16,044	20,690	2059	94,314	54,251	39,945	19,066	71,396	136,847
6963	247,783	920	75,793	120,590	459,367	35,025	122,517	9527
63,511	89,141	77,028	745,743	498,929	757,483	254,684	230,754	159,111
882	840	6833	34,708	7245	15,995	3368	9204	5958
1239	6030	1405	15,170	6333	8121	2927	3446	4410
512	735	2250	5431	2195	3973	661	3568	3336
62,531	70,072	88,952	384,278	265,359	230,812	76,807	295,409	178,038
1312	1498	139	10,492	1620	3486	621	9156	1897
7038	7179	14,365	49,218	13,984	104,952	10,388	32,597	22,329
49,197	86,072	54,826	305,665	172,084	222,098	79,865	264,196	480,280
10,150	16,201	8174	80,173	45,997	43,021	39,677	79,755	36,252
30,554	45,250	27,359	271,166	199,856	203,257	40,100	109,789	94,519
8787	14,399	728	33,342	15,546	3334	15,473	28,809	16,105
17,377	23,717	21,544	200,836	41,768	20,008	13,985	91,104	123,194
1154	1719	1245	34,174	4785	20,032	3020	35,863	77,638
3574	15,176	7668	85,619	24,526	41,607	18,957	61,758	81,588
278	8290	903	13,471	1780	3514	454	2133	2720
2024	5479	3972	28,189	18,192	27,259	5502	12,144	11,475
300	379	181	1986	933	931	446	3278	1702
1715	3438	1168	8584	10,717	10,637	7727	24,508	17,545
2291	2935	1031	15,815	9479	6985	3332	14,468	8601
253	430	208	1688	1019	1208	449	1401	1282
1,543,063	2,410,046	3,758,014	11,793,949	5,086,884	10,340,970	2,872,771	8,199,972	7,944,333
156,591	225,517	151,468	932,158	445,143	753,764	246,654	904,046	562,747
65,019	86,857	500,071	514,806	242,393	364,921	102,685	414,840	373,864
37,400	85,730	112,252	398,976	172,929	449,561	72,025	240,670	163,110
94,471	133,730	166,483	772,178	398,766	515,476	172,756	639,618	463,287
353,481	531,833	930,273	2,618,117	1,259,232	2,083,722	594,120	2,199,174	1,563,007
1,896,545	2,941,879	4,688,287	14,412,066	6,346,115	12,424,692	3,466,891	10,399,146	9,507,340

(continued)

Table 10.8 (continued)

	Sector 18	Sector 19	Sector 20	Sector 21	Sector 22	Sector 23	Sector 24	Sector 25
Sector 1	125	0	31	256,715	177	119	53	0
Sector 2	5789	2380	1184	14,035	815	1,179,813	15,363	345
Sector 3	0	0	0	20	1	41,947	150,528	0
Sector 4	9060	122	966	195	538	198	0	0
Sector 5	2859	52	84	5895	68	1802	12	26
Sector 6	26,188	28,356	2571	28,984	700	20,428	1192	679
Sector 7	12,102	2036	1161	191,763	77	345	37	77
Sector 8	14,623	13,517	2394	31,614	120	22,255	3589	1281
Sector 9	19,361	13,247	2187	64,921	90	4144	375	382
Sector 10	75,735	64,072	7649	38,035	123	12,766	806	678
Sector 11	29,309	20,956	3743	15,549	1203	197,211	6651	2690
Sector 12	550,567	495,959	68,669	162,011	11,751	20,826	2054	12,824
Sector 13	139,248	138,772	28,109	44,507	1447	14,278	738	758
Sector 14	1,784,621	242,784	31,512	202,318	5810	6472	788	300
Sector 15	295,817	179,339	31,576	89,542	609	39,773	3047	6999
Sector 16	336,229	81,381	22,247	8503	359	67,359	2582	2761
Sector 17	18,901	27,344	4885	21,712	124	75,018	7009	3223
Sector 18	961,754	223,617	40,146	13,300	914	371,356	742	873
Sector 19	402,907	3,527,639	185,650	5291	46	4121	168	209
Sector 20	71,237	380,613	99,208	4591	54	90,266	2006	1409
Sector 21	68,655	63,094	11,982	233,052	108	4479	239	268
Sector 22	66,340	9157	98	421	5318	735	0	0
Sector 23	89,948	102,021	9156	28,590	2473	2,398,832	6394	60,865
Sector 24	4022	4104	541	2029	121	7801	17,819	846
Sector 25	2575	3140	561	1442	115	8356	171	5443
Sector 26	1776	1466	221	333	57	2301	237	316
Sector 27	156,944	130,206	17,424	50,458	4218	81,820	8458	2462
Sector 28	4647	2709	914	600	13	793	42	81
Sector 29	21,377	73,425	3092	14,333	168	41,114	456	960
Sector 30	206,847	292,028	19,883	42,294	2583	53,180	5371	1576
Sector 31	45,298	23,400	7319	15,212	516	19,243	1467	2070
Sector 32	95,870	270,060	6140	23,223	2376	278,731	10,604	14,957
Sector 33	20,882	20,007	3649	14,427	180	2941	336	137
Sector 34	152,558	87,663	10,121	17,881	437	29,926	1409	1278
Sector 35	28,786	56,329	4134	2081	45	4467	84	53
Sector 36	38,572	39,464	4912	4557	221	53,849	737	395
Sector 37	1292	910	78	340	32	17,934	140	14,878
Sector 38	5582	5183	660	901	478	51,432	922	2049
Sector 39	963	713	150	354	5	1038	75	199
Sector 40	11,888	3820	1386	332	87	8017	898	57
Sector 41	7139	6534	850	2347	80	15,142	604	780
Sector 42	744	758	105	303	0	959	27	52
Total intermediate input	5,789,138	6,638,378	637,347	1,655,008	44,658	5,253,589	254,234	145,238
Compensation of employees	425,860	592,651	83,008	191,812	6478	537,431	34,221	52,919
Net taxes on production	164,791	120,385	23,409	129,248	3880	326,809	7940	10,429
Depreciation of fixed assets	116,208	186,073	16,932	43,291	2913	673,563	10,626	45,574
Operating surplus	313,395	221,754	42,377	105,649	395,289	218,657	15,535	6110
Total value added	1,020,253	1,120,862	165,727	469,999	408,560	1,756,461	68,322	115,032
Total input	6,809,392	7,759,240	803,073	2,125,007	453,217	7,010,050	322,556	260,271

Sector 26	Sector 27	Sector 28	Sector 29	Sector 30	Sector 31	Sector 32	Sector 33	Sector 34
76,422	178,919	0	0	1630	386,290	0	376	8701
13,800	25,817	278	0	2053	2230	0	972	5170
0	9313	0	0	0	0	0	0	0
0	0	0	0	0	0	0	0	0
175,718	1943	0	0	0	0	0	0	0
69,441	58,971	493	6960	35,026	1,162,625	15,770	12,880	43,384
14,684	9686	300	350	8120	16,453	805	5097	47,398
66,201	53,540	1429	5402	59,480	12,624	14,837	20,677	45,728
304,521	9133	987	2736	11,115	4626	4229	5546	9262
29,068	27,511	5255	88,999	62,369	11,381	109,355	26,883	328,570
224,230	1,397,038	5848	5503	22,537	16,461	35,580	59,906	116,737
536,581	131,868	658	15,126	61,082	25,512	19,095	76,564	131,364
3,941,367	15,877	686	774	1834	6859	904	1358	1946
2,168,412	38,586	0	29	3298	151	0	4112	2801
565,575	41,702	154	5740	2708	3909	5090	27,825	73,624
514,898	180,442	2076	23,093	14,552	3925	23,967	16,246	24,071
71,896	527,424	15,672	20,127	59,021	11,298	16,381	17,504	111,582
657,950	21,069	351	196,553	45,947	23,980	11,571	50,856	102,831
41,931	18,883	1739	170,586	26,330	1365	4211	13,740	211,847
37,984	11,932	235	25,594	2198	105	21,610	8959	10,918
56,621	6857	501	3025	5156	2085	5863	5924	22,225
0	0	0	0	0	0	0	0	0
129,910	89,802	2707	71,186	79,009	71,515	53,079	27,447	34,969
2058	7039	189	1625	1382	12,213	280	483	3541
4176	4513	318	3488	2586	5809	2663	836	2394
184,499	32,234	8379	4838	17,433	7658	8147	88,242	3580
1,620,456	632,449	28,251	87,531	473,643	40,702	139,675	34,604	180,446
3751	2353	8312	3730	9120	1923	24,245	2927	17,245
251,839	53,820	4026	123,700	54,959	10,285	160,627	17,867	13,709
427,948	96,921	14,859	89,682	34,292	70,767	23,885	19,563	85,051
173,933	119,389	1425	56,073	129,747	22,648	207,761	74,694	241,960
164,562	441,976	2321	59,095	230,650	55,390	452,837	196,348	166,397
10,135	36,561	3017	74,653	150,133	35,476	180,625	60,321	61,181
62,686	41,607	7691	143,120	329,399	39,983	343,581	154,751	160,555
13,028	2304	107	4776	10,539	30	7904	965	1831
201,182	8818	35	13,955	26,051	616	3908	8787	2735
855	2260	0	799	1562	1097	3621	1473	31,498
24,357	170,481	316	18,279	62,475	31,231	21,899	33,734	56,791
2165	5594	199	3344	4039	872	10,966	1082	1764
5163	4042	38	851	1086	564	3899	1583	599
24,163	15,422	1482	15,974	26,326	8647	33,745	24,227	15,737
2219	2113	65	1087	1681	515	3217	3034	698
12,876,382	4,536,208	120,401	1,348,383	2,070,569	2,109,822	1,975,831	1,108,393	2,380,842
2,583,447	1,213,161	56,823	383,903	1,393,822	757,637	1,164,075	375,920	703,749
717,444	243,752	5122	112,926	1,382,602	143,876	446,541	541,889	178,477
250,984	539,521	13,391	370,676	323,987	169,840	85,967	1,659,740	342,096
984,150	960,954	164	632,422	1,974,353	179,952	1,969,999	862,818	100,736
4,536,026	2,957,389	75,500	1,499,927	5,074,764	1,251,306	3,666,582	3,440,367	1,325,058
17,412,408	7,493,597	195,901	2,848,309	7,145,333	3,361,127	5,642,413	4,548,760	3,705,900

(continued)

Table 10.8 (continued)

	Sector 35	Sector 36	Sector 37	Sector 38	Sector 39	Sector 40	Sector 41	Sector 42
Sector 1	18,082	6203	17,641	22,303	5473	14,258	4263	0
Sector 2	831	2087	3780	7770	5815	12,023	707	12,546
Sector 3	0	0	0	0	0	0	0	0
Sector 4	4525	969	1273	0	0	0	0	0
Sector 5	74	48	475	100	1057	1522	428	1770
Sector 6	8541	3819	7676	43,323	23,063	24,823	48,982	16,258
Sector 7	9718	1030	3125	28,711	5013	16,563	2565	18,232
Sector 8	3711	2607	9432	13,442	3211	40,830	14,560	59,607
Sector 9	1629	1984	9306	17,987	34,890	22,211	10,464	72,622
Sector 10	5354	21,974	4950	31,842	99,957	37,062	98,317	116,402
Sector 11	4390	31,876	26,799	35,023	22,545	12,150	5679	64,919
Sector 12	52,195	34,349	49,088	166,257	25,705	1,268,629	35,878	45,241
Sector 13	3781	3023	10,158	2028	14,020	16,589	5600	21,798
Sector 14	2408	11,450	5500	4896	2138	142	403	0
Sector 15	12,240	33,313	5819	9474	6131	6213	3474	5291
Sector 16	7222	12,529	15,271	4632	5701	132,907	2446	10,389
Sector 17	4266	22,750	25,681	72,802	19,014	3300	4089	58,071
Sector 18	21,187	64,952	19,035	31,309	8422	9466	4824	10,175
Sector 19	18,934	58,623	10,493	47,047	9233	2189	5218	34,383
Sector 20	16,942	35,354	10,046	18,583	27,807	6935	3276	7860
Sector 21	2191	3447	3826	15,174	4866	3733	1764	10,119
Sector 22	0	0	0	0	0	0	0	0
Sector 23	9782	10,782	20,790	21,169	52,106	34,524	7856	60,710
Sector 24	158	397	668	1231	1838	1381	585	1344
Sector 25	746	1841	1359	3132	3864	1980	692	2557
Sector 26	1709	4707	18,218	14,636	13,731	57,539	7120	62,531
Sector 27	19,924	92,546	16,853	40,098	73,196	14,883	25,716	185,722
Sector 28	807	1859	698	2658	5974	953	1991	47,628
Sector 29	2733	7709	6479	6525	23,163	27,042	7055	83,065
Sector 30	10,017	18,324	10,418	34,130	21,032	88,672	17,951	45,638
Sector 31	18,621	64,014	17,420	30,070	68,535	28,606	27,225	283,666
Sector 32	2069	49,365	23,274	40,482	71,796	16,852	13,076	70,908
Sector 33	2190	15,978	2202	53,710	20,412	5670	11,251	24,104
Sector 34	4548	23,518	10,773	27,225	49,541	21,120	19,176	39,823
Sector 35	11,531	3593	703	52	5717	1347	880	60
Sector 36	6975	97,470	3763	1471	5820	4150	596	6209
Sector 37	301	330	17,356	2770	3610	749	780	10,327
Sector 38	1430	14,019	18,254	74,341	20,072	16,204	8226	63,701
Sector 39	1616	1544	1205	586	12,321	2824	835	33,963
Sector 40	1268	1453	2719	618	4048	7314	1251	12,161
Sector 41	7475	7941	5314	7674	16,506	4258	26,717	55,407
Sector 42	123	498	288	442	533	427	180	1191
Total intermediate input	302,246	770,274	418,128	935,723	797,878	1,968,041	432,095	1,656,397
Compensation of employees	117,394	416,805	159,030	655,130	2,017,707	913,760	252,496	2,537,902
Net taxes on production	7497	69,709	13,054	90,503	12,982	9534	43,532	13,779
Depreciation of fixed assets	22,029	79,286	63,744	53,264	239,814	110,924	67,597	373,630
Operating surplus	25,576	214,171	47,844	184,458	33,850	65,432	79,372	80,317
Total value added	172,496	779,971	283,673	983,355	2,304,353	1,099,649	442,998	3,005,627
Total input	474,742	1,550,246	701,801	1,919,078	3,102,231	3,067,691	875,093	4,662,025

Total intermediate use	Rural household consumption expenditure	Urban household consumption expenditure	Government consumption expenditure	Gross fixed capital formation	Increase in inventories	Export	Import	Total output
9,220,769	1,073,834	1,311,065	96,029	666,673	221,672	344,469	596,011	12,338,500
3,524,839	16,353	11,611	0	32,267	−87,034	16,927	276,877	3,238,086
3,400,306	0	0	0	18,209	−55,601	292,939	2,215,580	1,440,273
2,919,812	0	0	0	71,142	−19,071	7635	1,483,953	1,495,564
898,371	0	0	0	7572	−22,672	20,196	63,726	839,742
5,892,500	1,245,372	3,529,544	0	205,483	255,511	220,825	242,567	11,106,667
3,699,664	40,391	61,957	0	28,508	−27,421	2,701,085	600,434	5,903,748
1,442,781	261,246	1,218,992	0	70,742	90,604	579,653	75,977	3,588,042
1,637,832	23,638	101,002	0	165,339	−1685	269,856	254,914	1,941,068
2,921,538	15,367	106,715	0	33,562	−34,103	226,932	344,145	2,925,865
4,848,019	22,007	202,808	0	−6095	−134,219	104,029	352,180	4,684,369
14,376,648	140,454	457,949	0	168,975	−131,943	944,203	1,543,108	14,413,178
6,283,853	7906	54,151	0	62,086	−135,481	344,354	198,867	6,418,002
13,011,244	0	0	0	144,507	−188,641	365,674	774,558	12,558,227
3,287,623	14,583	78,108	0	82,125	−54,600	345,348	181,976	3,571,211
6,365,496	1055	15,529	0	4,099,031	89,393	1,449,941	1,504,831	10,515,612
4,461,685	123,592	614,794	0	4,056,830	119,509	1,422,790	1,177,570	9,621,630
4,108,047	112,107	430,593	0	1,468,015	34,952	1,418,906	663,516	6,909,105
5,197,556	93,105	373,530	0	956,815	187,744	3,797,575	2,763,553	7,842,771
1,445,878	9946	38,485	0	184,932	−24,607	483,907	1,281,150	857,390
1,182,646	64,554	332,979	0	422,305	9937	212,808	83,185	2,142,044
1,201,553	0	0	0	0	6321	4265	734,118	478,020
7,382,070	60,900	224,622	0	0	−668,114	12,509	1938	7,010,050
184,901	15,507	119,529	0	2116	358	0	0	322,411
131,447	11,094	113,898	0	0	1226	0	0	257,665
579,771	0	234,074	0	16,378,825	72,498	298,671	84,505	17,479,333
6,831,402	115,687	452,768	339,868	53,436	−179,449	369,632	454,030	7,529,314
202,436	1684	11,117	0	0	2855	6772	24,649	200,216
1,382,045	142,948	603,333	0	579,406	43,230	140,723	64,950	2,826,734
4,287,287	425,245	1,291,919	0	481,063	47,359	818,705	0	7,351,577
2,261,054	265,303	1,030,718	0	0	98,640	−37,049	169,930	3,448,736
4,337,515	164,394	853,684	63,074	0	108,023	63,009	57,032	5,532,667
1,034,233	487,890	1,804,860	0	1,127,096	120,162	0	0	4,574,242
2,661,942	23,479	337,482	150,823	0	59,141	736,733	249,350	3,720,251
385,497	0	0	260,084	0	2978	16,177	193,394	471,343
1,122,021	1776	22,731	220,085	165,217	−1537	0	0	1,530,293
202,418	6668	83,192	413,450	0	−2160	0	0	703,567
920,024	150,021	867,697	0	0	35,106	31,478	57,633	1,946,694
109,710	295,331	662,299	2,103,507	0	−27,196	4384	21,823	3,126,212
190,148	117,953	1,323,273	1,445,436	0	−23,344	3828	2677	3,054,617
434,004	29,811	159,790	244,286	0	12,312	42,790	53,539	869,456
37,062	0	0	4,625,594	0	−12,349	14,167	12,530	4,651,946

Table 10.9 China's national input–output table in 2018

	Sector 1	Sector 2	Sector 3	Sector 4	Sector 5	Sector 6	Sector 7	Sector 8
Sector 1	1,657,510	33,953	26	2460	365	4,696,891	1,193,234	174,733
Sector 2	11,476	586,540	11,612	6171	2325	32,395	34,260	15,777
Sector 3	0	0	22,397	0	0	0	0	0
Sector 4	0	0	0	158,308	0	0	0	0
Sector 5	1537	1168	711	1430	60,357	4265	0	0
Sector 6	1,281,833	5735	2431	3593	2491	2,446,401	43,845	188,069
Sector 7	1607	875	1076	829	505	6120	1,947,089	1,213,107
Sector 8	4276	11,529	4793	6409	3730	15,545	93,469	546,784
Sector 9	17,455	27,739	5156	4922	1766	12,712	8092	9564
Sector 10	12,104	4398	2880	2677	2322	221,311	67,062	61,411
Sector 11	87,265	36,802	51,029	84,882	42,686	33,317	14,955	23,340
Sector 12	985,342	76,714	50,479	76,473	92,877	327,329	709,906	228,505
Sector 13	17,956	30,186	12,548	12,352	32,260	81,951	14,728	5226
Sector 14	3166	142,166	66,164	19,180	6257	13,377	5063	5465
Sector 15	52,036	62,732	10,412	30,474	14,449	54,903	14,505	20,213
Sector 16	72,727	175,712	112,235	119,360	76,983	40,557	75,298	24,989
Sector 17	38,071	22,662	9232	16,015	18,386	15,053	12,961	6073
Sector 18	3515	49,008	12,552	19,843	8467	17,284	13,817	7262
Sector 19	1186	5288	1746	2171	989	4372	4636	4193
Sector 20	15,875	25,553	12,805	6906	3283	8362	22,376	7790
Sector 21	45,383	25,020	1866	6143	10,428	18,161	52,830	42,616
Sector 22	1433	370	22	1223	8549	9239	18,390	2924
Sector 23	158,202	150,240	99,046	251,169	104,074	123,055	152,859	30,693
Sector 24	455	1330	1674	7827	4752	7026	3110	1328
Sector 25	1459	2499	1064	1338	952	4807	4810	1713
Sector 26	3264	4326	1532	576	319	3348	1410	1155
Sector 27	227,607	153,732	26,294	65,975	55,587	340,364	111,419	90,231
Sector 28	10,047	877	218	255	261	2678	4594	7915
Sector 29	46,524	10,216	2299	2249	2264	21,080	12,051	18,945
Sector 30	274,511	38,005	12,431	16,395	13,602	286,692	139,612	88,983
Sector 31	35,628	33,216	12,038	16,690	11,775	70,326	27,445	23,191
Sector 32	134,855	84,796	11,117	17,172	17,009	122,122	155,188	54,388
Sector 33	3258	4874	606	899	1264	20,411	20,542	40,391
Sector 34	18,493	17,736	2465	5958	3747	196,065	22,718	80,281
Sector 35	22,504	4793	2479	500	228	9633	3918	2078
Sector 36	111,788	31,253	15,280	19,725	10,137	40,848	14,862	13,382
Sector 37	34,286	4097	1105	980	2293	3640	8404	502
Sector 38	43,946	23,384	5066	3349	3184	14,311	6537	6634
Sector 39	4270	2921	250	576	1051	1632	592	378
Sector 40	6906	7568	486	2832	1759	5913	2639	2411
Sector 41	2831	4626	1299	3622	3666	12,509	6205	6281
Sector 42	3673	800	410	312	165	1660	656	472
Total intermediate input	5,456,258	1,905,441	589,329	1,000,216	627,563	9,347,663	5,046,089	3,059,393
Compensation of employees	6,932,381	660,164	198,771	148,210	105,820	727,001	516,492	386,806
Net taxes on production	14,085	368,250	225,708	95,633	64,390	819,991	205,552	100,373
Depreciation of fixed assets	340,197	126,167	132,692	34,980	27,434	225,301	135,656	45,560
Operating surplus	0	382,096	310,966	181,810	87,866	724,113	211,873	129,661
Total value added	7,286,662	1,536,678	868,137	460,633	285,510	2,496,406	1,069,573	662,400
Total input	12,742,920	3,442,118	1,457,466	1,460,849	913,073	11,844,069	6,115,661	3,721,792

Sector 9	Sector 10	Sector 11	Sector 12	Sector 13	Sector 14	Sector 15	Sector 16	Sector 17
251,465	152,974	108	523,658	4177	1455	1165	1198	851
15,734	29,472	301,120	234,784	509,190	491,365	17,360	37,891	11,956
0	0	2,727,781	564,086	20,832	51,441	0	0	0
0	0	0	82,103	19,637	2,881,743	43,166	32,744	6221
0	0	489	235,346	383,025	32,504	14,044	7788	2455
10,780	12,072	22,552	241,663	24,822	45,129	12,689	36,386	36,594
19,645	53,247	247	102,586	14,741	1468	5734	15,610	19,284
34,705	21,390	2011	33,318	17,193	29,603	11,782	23,846	107,287
583,176	77,166	1486	27,614	51,970	10,000	65,905	35,810	124,772
31,551	751,302	2628	239,545	177,245	14,231	38,654	48,586	44,735
14,499	14,270	293,908	898,147	282,267	558,158	39,133	113,616	49,879
186,225	405,774	90,692	5,777,709	529,943	193,006	152,714	401,220	598,251
17,053	8349	16,416	112,205	1,170,011	259,432	41,138	79,870	76,548
45,445	34,344	3098	155,540	189,467	3,561,243	1,267,105	2,335,346	1,317,675
49,535	46,825	15,580	145,059	197,336	115,265	448,416	466,270	215,971
28,954	39,881	62,226	219,513	187,991	408,251	178,430	2,500,776	871,482
8941	24,396	10,917	43,602	35,897	69,950	34,976	167,610	2,922,768
6581	19,430	5481	38,695	26,851	42,043	29,996	608,041	396,559
840	26,491	660	8944	4290	3386	4506	231,684	118,112
1603	39,266	6876	109,592	19,143	42,506	9248	58,124	180,809
16,592	20,866	2136	101,779	61,393	44,546	21,635	76,833	151,720
7020	248,075	955	77,859	124,352	469,249	35,865	131,848	9876
66,194	89,676	79,926	780,766	542,542	832,741	284,567	248,328	177,032
913	845	7090	36,487	7916	17,426	3712	9905	6764
1311	6083	1457	16,153	7165	9285	3401	3709	5279
531	739	2335	5668	2382	4332	730	3840	3701
64,418	70,413	92,298	403,055	288,292	248,806	83,631	317,906	199,914
1366	1506	144	10,814	1742	3875	704	9853	2045
7495	7248	14,906	52,444	15,927	121,703	12,283	35,079	26,706
50,914	86,475	56,889	316,906	184,913	243,214	88,943	284,316	517,078
10,346	16,247	8481	83,162	48,716	45,112	41,857	85,828	39,144
33,782	45,962	28,389	296,174	241,578	251,466	50,138	118,150	126,041
9265	14,510	755	35,765	17,653	3776	17,848	31,003	19,051
17,992	23,846	22,354	211,304	45,514	21,733	15,359	98,042	139,546
1208	1730	1292	35,821	5290	22,317	3381	38,594	88,713
3781	15,296	7956	90,429	27,695	47,753	22,030	66,461	94,890
290	8345	937	14,053	1947	3935	518	2295	3029
2089	5502	4121	29,276	19,528	29,384	5989	13,069	12,583
309	381	188	2074	1005	1002	486	3527	1884
1779	3457	1212	9006	11,704	11,618	8547	26,374	19,745
2410	2958	1070	16,743	10,614	7923	3842	15,570	10,084
262	432	215	1772	1112	1315	495	1507	1439
1,606,997	2,427,237	3,899,382	12,421,221	5,535,007	11,254,691	3,122,121	8,824,454	8,758,471
161,429	227,940	157,166	970,937	475,189	803,863	260,233	972,895	601,576
67,028	87,790	518,882	536,223	258,754	389,175	108,338	446,433	399,660
38,556	86,651	116,474	415,574	184,601	479,442	75,991	258,999	174,364
97,390	135,167	172,745	804,302	425,682	549,738	182,267	688,329	495,254
364,402	537,547	965,268	2,727,036	1,344,226	2,222,218	626,829	2,366,656	1,670,855
1,971,400	2,964,785	4,864,650	15,148,257	6,879,233	13,476,909	3,748,951	11,191,110	10,429,325

(continued)

Table 10.9 (continued)

	Sector 18	Sector 19	Sector 20	Sector 21	Sector 22	Sector 23	Sector 24	Sector 25
Sector 1	144	0	35	258,026	177	126	55	0
Sector 2	6152	2498	1247	14,107	815	1,248,182	15,870	355
Sector 3	0	0	0	20	1	44,378	155,496	0
Sector 4	10,003	132	1064	196	538	209	0	0
Sector 5	3021	54	88	5925	68	1907	13	27
Sector 6	27,989	29,884	2717	29,132	700	21,612	1232	698
Sector 7	13,331	2186	1255	192,742	77	365	39	80
Sector 8	15,716	14,317	2544	31,775	120	23,545	3707	1317
Sector 9	20,974	14,111	2361	65,252	90	4384	387	393
Sector 10	88,250	71,381	8823	38,229	123	13,505	833	697
Sector 11	31,783	22,292	4011	15,629	1203	208,639	6871	2766
Sector 12	607,182	534,298	74,410	162,838	11,751	22,033	2122	13,184
Sector 13	147,593	144,982	29,465	44,734	1447	15,105	763	780
Sector 14	1,926,222	257,183	33,628	203,351	5810	6847	814	308
Sector 15	317,062	188,696	33,348	89,999	609	42,078	3147	7196
Sector 16	358,408	85,377	23,429	8546	359	71,262	2668	2838
Sector 17	19,881	28,425	5097	21,823	124	79,365	7240	3314
Sector 18	1,024,113	234,246	42,198	13,368	914	392,875	767	898
Sector 19	428,841	3,705,926	195,467	5318	46	4360	173	215
Sector 20	73,339	389,532	103,258	4615	54	95,496	2073	1448
Sector 21	76,339	67,862	13,582	234,242	108	4738	247	276
Sector 22	68,241	9366	100	423	5318	777	0	0
Sector 23	97,895	109,210	9912	28,736	2473	2,537,841	6605	62,573
Sector 24	4373	4366	580	2039	121	8253	18,407	870
Sector 25	2904	3435	625	1449	115	8841	177	5596
Sector 26	1920	1556	236	334	57	2434	245	325
Sector 27	169,222	137,500	18,545	50,715	4218	86,562	8737	2531
Sector 28	5012	2869	970	603	13	839	43	83
Sector 29	24,305	80,896	3477	14,406	168	43,497	471	987
Sector 30	221,899	308,049	21,109	42,510	2583	56,261	5548	1621
Sector 31	47,576	24,270	7612	15,290	516	20,358	1516	2128
Sector 32	114,697	308,897	7474	23,342	2376	294,884	10,954	15,377
Sector 33	23,498	21,939	4057	14,500	180	3112	347	141
Sector 34	165,474	93,286	10,859	17,973	437	31,660	1455	1314
Sector 35	31,457	60,691	4515	2091	45	4726	87	55
Sector 36	43,190	43,252	5452	4580	221	56,969	762	407
Sector 37	1417	976	84	342	32	18,973	144	15,295
Sector 38	5960	5452	698	906	478	54,412	952	2107
Sector 39	1033	756	160	356	5	1098	78	205
Sector 40	12,928	4088	1489	334	87	8482	927	59
Sector 41	7980	7100	935	2359	80	16,019	624	802
Sector 42	807	811	113	304	0	1014	28	53
Total intermediate input	6,248,134	7,022,148	677,029	1,663,456	44,658	5,558,027	262,624	149,316
Compensation of employees	453,472	622,603	86,397	192,791	6478	568,575	35,350	54,404
Net taxes on production	175,475	126,469	24,365	129,908	3880	345,747	8202	10,721
Depreciation of fixed assets	123,743	195,477	17,623	43,512	2913	712,595	10,977	46,854
Operating surplus	333,715	232,961	44,107	106,188	395,289	231,328	16,048	6282
Total value added	1,086,405	1,177,511	172,493	472,399	408,560	1,858,246	70,576	118,261
Total input	7,334,539	8,199,658	849,521	2,135,855	453,217	7,416,272	333,201	267,577

Sector 26	Sector 27	Sector 28	Sector 29	Sector 30	Sector 31	Sector 32	Sector 33	Sector 34
83,090	187,983	0	0	1813	413,832	0	392	9249
15,004	27,125	304	0	2284	2389	0	1015	5496
0	9785	0	0	0	0	0	0	0
0	0	0	0	0	0	0	0	0
191,049	2041	0	0	0	0	0	0	0
75,500	61,959	541	7246	38,972	1,245,519	16,309	13,446	46,116
15,965	10,177	329	364	9035	17,626	832	5321	50,383
71,977	56,252	1565	5624	66,180	13,525	15,345	21,586	48,608
331,090	9596	1082	2848	12,368	4955	4373	5789	9845
31,604	28,905	5758	92,646	69,394	12,193	113,096	28,064	349,262
243,794	1,467,814	6408	5729	25,076	17,635	36,797	62,537	124,089
583,398	138,549	720	15,745	67,962	27,331	19,749	79,927	139,637
4,285,252	16,681	751	806	2040	7349	935	1418	2069
2,357,606	40,541	0	30	3670	161	0	4293	2977
614,921	43,814	169	5975	3013	4188	5264	29,047	78,260
559,822	189,584	2275	24,039	16,192	4205	24,787	16,960	25,587
78,169	554,145	17,172	20,952	65,669	12,104	16,942	18,272	118,609
715,356	22,136	384	204,606	51,123	25,690	11,967	53,090	109,307
45,590	19,840	1905	177,576	29,296	1462	4355	14,343	225,188
41,298	12,536	257	26,643	2446	112	22,349	9353	11,606
61,562	7205	549	3149	5737	2233	6063	6184	23,625
0	0	0	0	0	0	0	0	0
141,245	94,352	2966	74,102	87,908	76,614	54,895	28,652	37,171
2237	7395	207	1691	1538	13,084	289	505	3764
4540	4742	349	3631	2877	6224	2754	873	2545
200,597	33,867	9181	5037	19,396	8204	8426	92,118	3805
1,761,841	664,491	30,954	91,117	526,995	43,604	144,453	36,124	191,810
4078	2472	9107	3882	10,148	2060	25,074	3056	18,331
273,812	56,547	4412	128,769	61,149	11,018	166,122	18,651	14,573
465,286	101,831	16,280	93,356	38,155	75,813	24,702	20,422	90,407
189,109	125,438	1562	58,371	144,361	24,263	214,868	77,975	257,197
178,920	464,368	2544	61,517	256,631	59,339	468,329	204,973	176,875
11,020	38,413	3306	77,712	167,045	38,006	186,804	62,971	65,034
68,155	43,715	8427	148,984	366,503	42,834	355,335	161,548	170,667
14,165	2421	118	4972	11,726	32	8174	1008	1946
218,735	9265	38	14,527	28,985	660	4042	9173	2907
929	2375	0	832	1738	1175	3744	1537	33,482
26,482	179,118	346	19,028	69,512	33,457	22,649	35,216	60,367
2354	5878	218	3481	4494	934	11,341	1129	1875
5614	4246	42	886	1209	604	4033	1652	637
26,271	16,204	1624	16,629	29,292	9264	34,899	25,292	16,728
2412	2220	72	1132	1871	552	3327	3167	742
13,999,848	4,766,022	131,922	1,403,632	2,303,801	2,260,250	2,043,423	1,157,078	2,530,777
2,808,853	1,274,622	62,261	399,633	1,550,824	811,656	1,203,897	392,431	748,068
780,041	256,101	5612	117,553	1,538,340	154,134	461,817	565,692	189,717
272,883	566,854	14,673	385,864	360,481	181,949	88,908	1,732,643	363,640
1,070,018	1,009,638	180	658,335	2,196,747	192,783	2,037,392	900,717	107,080
4,931,795	3,107,216	82,725	1,561,386	5,646,391	1,340,523	3,792,013	3,591,483	1,408,505
18,931,643	7,873,237	214,647	2,965,017	7,950,192	3,600,773	5,835,436	4,748,561	3,939,282

(continued)

Table 10.9 (continued)

	Sector 35	Sector 36	Sector 37	Sector 38	Sector 39	Sector 40	Sector 41	Sector 42
Sector 1	18,840	6449	18,747	24,020	5906	14,978	4475	0
Sector 2	866	2169	4017	8368	6275	12,630	742	13,224
Sector 3	0	0	0	0	0	0	0	0
Sector 4	4714	1007	1353	0	0	0	0	0
Sector 5	77	50	505	108	1140	1599	449	1865
Sector 6	8899	3970	8157	46,660	24,889	26,077	51,410	17,137
Sector 7	10,126	1070	3320	30,923	5410	17,400	2693	19,217
Sector 8	3867	2710	10,024	14,478	3465	42,893	15,281	62,829
Sector 9	1697	2062	9889	19,373	37,652	23,333	10,983	76,547
Sector 10	5579	22,843	5260	34,294	107,869	38,934	103,190	122,694
Sector 11	4574	33,136	28,479	37,720	24,329	12,764	5960	68,428
Sector 12	54,383	35,707	52,165	179,062	27,739	1,332,708	37,656	47,686
Sector 13	3940	3143	10,794	2185	15,130	17,427	5877	22,976
Sector 14	2509	11,903	5845	5273	2307	150	422	0
Sector 15	12,753	34,630	6184	10,204	6616	6527	3646	5577
Sector 16	7525	13,025	16,228	4989	6152	139,621	2567	10,950
Sector 17	4445	23,650	27,290	78,409	20,519	3467	4291	61,210
Sector 18	22,075	67,521	20,228	33,720	9089	9945	5063	10,725
Sector 19	19,727	60,942	11,151	50,671	9964	2300	5476	36,241
Sector 20	17,652	36,752	10,676	20,014	30,008	7286	3439	8284
Sector 21	2282	3583	4066	16,342	5251	3921	1852	10,666
Sector 22	0	0	0	0	0	0	0	0
Sector 23	10,192	11,208	22,093	22,799	56,230	36,268	8245	63,992
Sector 24	165	412	710	1326	1984	1451	614	1416
Sector 25	777	1914	1445	3373	4170	2080	727	2695
Sector 26	1781	4893	19,360	15,763	14,818	60,445	7472	65,912
Sector 27	20,759	96,206	17,910	43,187	78,990	15,635	26,991	195,762
Sector 28	841	1933	741	2863	6447	1001	2090	50,203
Sector 29	2848	8013	6885	7027	24,996	28,408	7405	87,555
Sector 30	10,437	19,049	11,072	36,759	22,697	93,151	18,841	48,105
Sector 31	19,401	66,546	18,512	32,386	73,960	30,051	28,575	299,000
Sector 32	2156	51,318	24,733	43,600	77,479	17,703	13,724	74,741
Sector 33	2282	16,610	2340	57,846	22,028	5957	11,808	25,407
Sector 34	4739	24,448	11,449	29,321	53,462	22,187	20,126	41,975
Sector 35	12,014	3735	747	56	6169	1415	923	63
Sector 36	7267	101,325	3998	1584	6281	4359	625	6545
Sector 37	314	343	18,444	2984	3896	786	818	10,885
Sector 38	1490	14,574	19,398	80,067	21,661	17,022	8634	67,144
Sector 39	1683	1605	1281	631	13,296	2967	876	35,799
Sector 40	1321	1510	2890	665	4368	7683	1313	12,819
Sector 41	7788	8255	5647	8265	17,813	4474	28,041	58,402
Sector 42	128	518	306	476	575	448	189	1256
Total intermediate input	314,911	800,738	444,339	1,007,790	861,031	2,067,448	453,511	1,745,934
Compensation of employees	122,313	433,289	168,999	705,586	2,177,411	959,914	265,011	2,675,088
Net taxes on production	7811	72,466	13,873	97,473	14,010	10,015	45,689	14,523
Depreciation of fixed assets	22,952	82,422	67,740	57,366	258,795	116,527	70,948	393,827
Operating surplus	26,647	222,642	50,844	198,665	36,530	68,737	83,306	84,659
Total value added	179,724	810,818	301,456	1,059,091	2,486,746	1,155,193	464,954	3,168,097
Total input	494,634	1,611,555	745,794	2,066,881	3,347,777	3,222,642	918,465	4,914,032

Total intermediate use	Rural household consumption expenditure	Urban household consumption expenditure	Government consumption expenditure	Gross fixed capital formation	Increase in inventories	Export	Import	Total output
9,744,559	1,139,662	1,399,123	102,653	679,687	221,672	363,968	619,140	13,032,184
3,740,564	17,768	12,738	0	34,852	−87,034	17,328	301,164	3,435,052
3,596,216	0	0	0	18,210	−55,601	312,969	2,320,055	1,551,740
3,243,137	0	0	0	71,142	−19,071	7635	1,672,957	1,629,886
955,105	0	0	0	8289	−22,672	20,819	68,587	892,954
6,223,851	1,318,055	3,763,040	0	221,343	255,511	220,825	242,567	11,760,059
3,814,036	43,051	66,183	0	29,121	−27,421	2,960,594	674,708	6,210,855
1,516,917	276,164	1,297,693	0	73,138	90,604	579,653	75,977	3,758,194
1,736,738	25,055	107,581	0	170,447	−1685	269,856	285,893	2,022,099
3,116,066	16,245	111,924	0	33,922	−34,103	226,932	354,697	3,116,289
5,136,612	23,351	214,032	0	−6324	−134,219	107,225	375,507	4,965,170
15,151,100	149,499	487,428	0	176,005	−131,943	944,203	1,595,489	15,180,802
6,771,872	8685	58,242	0	66,276	−135,481	366,394	219,850	6,916,136
14,041,952	0	0	0	154,112	−188,641	365,674	813,361	13,559,736
3,512,906	15,618	83,864	0	86,646	−54,600	345,348	198,255	3,791,528
6,812,759	1131	16,600	0	4,411,199	89,393	1,554,775	1,584,727	11,301,130
4,748,093	131,613	650,224	0	4,483,396	119,509	1,552,963	1,264,132	10,421,666
4,386,829	118,780	457,639	0	1,563,200	34,952	1,489,063	698,731	7,351,730
5,479,869	99,799	400,017	0	1,005,173	187,744	3,980,011	2,903,629	8,248,983
1,500,641	10,955	40,734	0	192,482	−24,607	494,544	1,400,991	813,758
1,261,611	67,205	352,476	0	424,461	9937	219,300	88,297	2,246,692
1,231,473	0	0	0	0	6321	4396	769,315	472,874
7,855,289	60,900	224,622	0	0	−735,755	13,166	1950	7,416,272
196,358	16,174	124,055	0	2186	358	0	0	339,130
141,340	11,773	117,995	0	0	1226	0	0	272,335
618,367	0	250,883	0	17,807,880	72,498	327,366	91,947	18,985,047
7,304,799	123,585	481,883	371,309	56,143	−179,449	369,632	496,162	8,031,739
213,653	1802	11,980	0	0	2855	7009	27,512	209,787
1,485,816	150,626	638,855	0	603,146	43,230	150,873	70,565	3,001,981
4,545,822	454,467	1,384,759	0	535,250	47,359	838,301	0	7,805,957
2,390,042	283,394	1,102,149	0	0	98,640	−51,319	183,980	3,638,925
4,745,277	173,318	903,589	68,348	0	108,023	69,099	62,275	6,005,379
1,104,232	517,205	1,883,706	0	1,176,603	120,162	0	0	4,801,909
2,839,487	24,782	355,681	161,857	0	59,141	780,848	249,350	3,972,447
417,829	0	0	273,343	0	2978	17,686	208,991	502,846
1,208,716	1776	22,731	235,078	171,751	−1537	0	0	1,638,514
212,204	7050	87,640	435,604	0	−2160	0	0	740,338
975,083	159,089	921,092	0	0	35,106	32,868	62,328	2,060,912
116,059	316,323	708,407	2,248,257	0	−27,196	4510	23,660	3,342,700
203,843	123,201	1,393,968	1,557,046	0	−23,344	3828	2677	3,255,866
463,036	31,616	169,636	261,910	0	12,312	42,790	53,539	927,762
39,219	0	0	4,929,195	0	−12,349	15,272	13,234	4,958,103

Table 10.10 China's national input–output table in 2019

	Sector 1	Sector 2	Sector 3	Sector 4	Sector 5	Sector 6	Sector 7	Sector 8
Sector 1	1,859,330	38,318	26	3331	392	4,946,925	1,325,307	190,489
Sector 2	12,423	674,171	11,633	8339	2525	34,120	37,291	16,725
Sector 3	0	0	22,439	0	0	0	0	0
Sector 4	0	0	0	165,930	0	0	0	0
Sector 5	1691	1383	713	2188	65,499	4492	0	0
Sector 6	1,497,323	6977	2435	5208	2801	2,576,632	51,684	208,678
Sector 7	1740	978	1078	1097	536	6446	2,141,048	1,290,005
Sector 8	4758	13,063	4802	8237	4021	16,372	103,924	587,374
Sector 9	18,859	31,507	5165	6062	1921	13,389	8736	10,037
Sector 10	12,618	4662	2885	3152	2406	233,092	70,063	63,518
Sector 11	97,542	42,478	51,123	109,052	46,796	35,091	16,880	25,060
Sector 12	1,092,282	88,858	50,572	113,317	100,155	344,754	788,403	248,141
Sector 13	19,468	34,094	12,571	15,850	34,670	86,314	16,072	5524
Sector 14	3527	165,355	66,286	25,691	6868	14,089	5678	5858
Sector 15	57,603	73,823	10,431	38,852	16,110	57,825	16,212	21,501
Sector 16	80,100	199,851	112,441	151,306	83,624	42,716	83,460	26,581
Sector 17	41,362	25,561	9249	20,703	19,670	15,854	14,069	6427
Sector 18	3899	56,448	12,575	26,024	9236	18,204	15,478	7773
Sector 19	1326	6095	1750	2792	1083	4605	5217	4503
Sector 20	16,764	31,905	12,828	11,335	3660	8807	23,535	8084
Sector 21	46,184	26,064	1870	7511	10,711	19,128	54,052	43,341
Sector 22	1437	373	22	1256	8582	9731	18,436	2934
Sector 23	183,626	180,480	99,228	370,398	116,913	129,606	179,485	34,096
Sector 24	497	1487	1677	9924	5091	7400	3414	1410
Sector 25	1598	2829	1066	1867	1018	5063	5274	1841
Sector 26	3731	5144	1534	815	355	3526	1627	1264
Sector 27	244,194	169,433	26,342	81,096	59,020	358,483	120,125	94,678
Sector 28	11,866	1158	218	484	302	2821	5490	8951
Sector 29	51,962	11,922	2303	3393	2459	22,202	13,498	20,680
Sector 30	315,966	49,176	12,454	29,209	15,719	301,954	162,875	98,534
Sector 31	37,743	36,395	12,060	19,825	12,460	74,070	29,128	24,073
Sector 32	144,788	95,034	11,138	25,478	17,948	128,623	165,379	58,296
Sector 33	3681	5773	607	1329	1393	21,497	23,290	44,166
Sector 34	21,140	20,839	2469	8103	4150	206,503	26,061	87,572
Sector 35	24,785	5522	2484	685	248	10,145	4306	2229
Sector 36	122,953	35,937	15,308	28,307	10,971	43,022	16,385	14,421
Sector 37	41,058	5223	1108	1669	2632	3834	10,226	571
Sector 38	49,935	28,035	5076	4760	3561	15,073	7456	7216
Sector 39	5106	3564	251	812	1201	1719	712	423
Sector 40	7858	8776	487	3793	1929	6228	3021	2629
Sector 41	3173	5399	1301	5340	3983	13,174	6981	6844
Sector 42	4234	939	411	421	182	1748	762	519
Total intermediate input	6,150,129	2,195,025	590,415	1,324,941	682,798	9,845,275	5,581,039	3,282,966
Compensation of employees	7,776,474	758,795	199,138	155,346	114,835	765,703	567,942	415,520
Net taxes on production	15,800	423,269	226,125	100,237	69,876	863,642	226,028	107,824
Depreciation of fixed assets	381,620	145,017	132,936	36,664	29,771	237,294	149,170	48,942
Operating surplus	0	439,183	311,539	190,564	95,352	762,660	232,979	139,286
Total value added	8,173,893	1,766,264	869,737	482,812	309,833	2,629,299	1,176,118	711,572
Total input	14,324,022	3,961,289	1,460,153	1,807,753	992,631	12,474,574	6,757,157	3,994,538

Sector 9	Sector 10	Sector 11	Sector 12	Sector 13	Sector 14	Sector 15	Sector 16	Sector 17
294,354	182,580	124	593,405	4727	1716	1358	1391	930
17,695	34,241	345,706	264,712	574,296	562,946	19,627	43,995	12,884
0	0	3,131,677	623,401	22,439	54,893	0	0	0
0	0	0	101,969	21,969	3,148,098	45,684	38,019	6718
0	0	561	264,494	444,782	38,519	16,330	9043	2690
12,519	15,552	25,891	293,945	29,484	53,142	14,566	42,249	41,529
21,999	61,592	283	112,594	16,418	1685	6562	18,125	20,520
38,707	25,199	2309	37,870	19,294	33,493	13,113	27,687	116,072
640,312	89,531	1706	31,570	57,736	11,054	71,271	41,579	133,995
33,990	824,641	3017	252,836	187,923	15,654	42,386	56,414	45,837
16,089	17,148	337,426	1,045,176	319,963	631,316	43,404	131,920	54,290
219,854	490,988	104,121	6,512,617	612,457	233,075	182,758	465,861	661,489
18,774	9639	18,847	125,359	1,305,641	291,464	45,541	92,737	81,657
50,580	41,940	3556	180,877	215,741	4,033,496	1,399,857	2,711,598	1,431,709
54,186	57,141	17,887	171,926	224,455	128,054	484,777	541,392	234,449
31,857	46,910	71,440	251,489	210,662	457,571	196,415	2,903,681	940,347
9928	28,226	12,534	48,626	40,031	78,727	38,713	194,615	3,152,990
7329	23,285	6293	44,697	30,434	47,695	33,474	706,003	431,665
934	31,600	758	10,419	4863	3825	4990	269,011	128,806
1969	45,701	7895	127,701	22,363	49,739	10,515	67,489	203,820
17,344	22,340	2452	105,057	63,852	48,211	23,408	89,212	153,304
7102	249,982	1096	78,234	125,404	477,494	36,502	153,090	9711
78,291	115,944	91,760	954,208	641,369	993,629	333,113	288,336	199,935
1009	982	8140	40,910	8784	19,655	4135	11,501	7229
1513	7265	1673	17,978	8092	10,942	3967	4306	5723
612	927	2681	6778	2788	5087	839	4459	4140
70,080	80,123	105,964	443,854	315,563	275,998	91,739	369,125	210,357
1726	2095	165	13,579	2222	5034	892	11,441	2432
8938	9042	17,113	59,952	18,497	147,863	14,744	40,731	29,595
62,103	116,581	65,312	395,654	229,661	304,816	108,679	330,123	594,459
11,079	18,011	9737	90,856	52,763	48,968	44,882	99,656	40,808
40,630	55,089	32,592	319,666	272,020	305,420	60,116	137,186	136,350
10,935	18,169	867	41,908	20,681	4529	21,060	35,998	21,124
20,532	29,809	25,664	252,116	52,641	25,167	17,386	113,838	155,664
1380	2049	1483	40,879	6013	25,894	3833	44,812	96,840
4384	18,158	9135	102,052	31,647	56,754	25,786	77,169	103,173
357	11,551	1076	17,743	2419	4955	637	2665	3556
2417	6880	4731	35,151	22,935	34,400	6839	15,175	14,201
361	496	215	2613	1201	1179	555	4096	2169
2023	4190	1391	10,555	13,423	13,412	9698	30,623	21,732
2820	3671	1229	19,183	12,276	9489	4536	18,078	11,159
300	532	247	2117	1287	1522	561	1750	1598
1,817,013	2,799,803	4,476,755	14,146,726	6,271,217	12,696,581	3,485,247	10,246,181	9,527,653
177,245	250,190	180,438	1,094,438	530,274	910,463	281,335	1,129,640	648,961
73,595	96,360	595,712	604,429	288,750	440,784	117,123	518,359	431,141
42,333	95,110	133,720	468,434	206,000	543,020	82,153	300,727	188,099
106,931	148,361	198,324	906,607	475,027	622,638	197,047	799,227	534,264
400,104	590,021	1,108,193	3,073,908	1,500,051	2,516,904	677,658	2,747,953	1,802,465
2,217,117	3,389,823	5,584,949	17,220,634	7,771,268	15,213,485	4,162,904	12,994,134	11,330,119

(continued)

Table 10.10 (continued)

	Sector 18	Sector 19	Sector 20	Sector 21	Sector 22	Sector 23	Sector 24	Sector 25
Sector 1	159	0	36	311,962	177	144	61	0
Sector 2	6709	2692	1254	17,056	815	1,420,412	17,723	390
Sector 3	0	0	0	24	1	50,502	173,650	0
Sector 4	10,659	138	1071	237	538	238	0	0
Sector 5	3370	59	89	7163	68	2170	14	30
Sector 6	31,686	33,204	2734	35,221	700	24,594	1375	767
Sector 7	14,545	2351	1261	233,032	77	416	43	87
Sector 8	17,121	15,422	2557	38,417	120	26,794	4140	1446
Sector 9	22,540	14,970	2372	78,892	90	4989	432	432
Sector 10	92,804	74,304	8856	46,220	123	15,369	930	765
Sector 11	34,702	23,951	4030	18,896	1203	237,428	7673	3038
Sector 12	684,158	592,512	74,942	196,876	11,751	25,073	2370	14,484
Sector 13	159,467	154,473	29,588	54,085	1447	17,190	852	856
Sector 14	2,110,557	276,747	33,786	245,858	5810	7791	909	339
Sector 15	346,464	201,519	33,493	108,812	609	47,884	3515	7905
Sector 16	388,836	91,152	23,531	10,332	359	81,095	2979	3118
Sector 17	21,513	30,375	5119	26,384	124	90,316	8085	3640
Sector 18	1,123,099	252,101	42,400	16,162	914	447,086	856	986
Sector 19	469,052	3,997,067	196,408	6429	46	4962	194	237
Sector 20	81,577	425,958	104,205	5580	54	108,673	2315	1591
Sector 21	78,893	69,598	13,623	283,206	108	5392	276	303
Sector 22	68,830	9438	100	512	5318	885	0	0
Sector 23	111,378	122,234	9994	34,742	2473	2,888,025	7376	68,743
Sector 24	4735	4659	583	2466	121	9392	20,556	956
Sector 25	3208	3737	629	1752	115	10,061	197	6148
Sector 26	2154	1713	237	404	57	2770	274	357
Sector 27	181,041	145,046	18,615	61,316	4218	98,506	9757	2781
Sector 28	6048	3360	980	730	13	955	48	91
Sector 29	27,445	89,780	3505	17,417	168	49,499	527	1084
Sector 30	260,770	352,204	21,302	51,396	2583	64,025	6196	1780
Sector 31	50,322	25,377	7635	18,486	516	23,167	1693	2338
Sector 32	126,937	337,683	7539	28,221	2376	335,573	12,233	16,893
Sector 33	26,523	24,362	4087	17,531	180	3541	388	155
Sector 34	184,238	101,936	10,925	21,730	437	36,029	1625	1443
Sector 35	34,428	65,963	4543	2529	45	5378	97	60
Sector 36	47,816	47,455	5489	5538	221	64,830	851	447
Sector 37	1677	1126	85	413	32	21,591	161	16,804
Sector 38	6682	5994	703	1095	478	61,920	1064	2314
Sector 39	1174	843	161	431	5	1250	87	225
Sector 40	14,314	4472	1497	404	87	9652	1035	65
Sector 41	8964	7830	942	2853	80	18,230	697	881
Sector 42	898	891	113	368	0	1154	31	59
Total intermediate input	6,867,492	7,614,699	681,017	2,011,175	44,658	6,324,952	293,285	164,039
Compensation of employees	497,302	671,516	87,189	233,091	6478	647,030	39,477	59,769
Net taxes on production	192,436	136,405	24,588	157,063	3880	393,455	9159	11,779
Depreciation of fixed assets	135,703	210,834	17,785	52,607	2913	810,923	12,258	51,474
Operating surplus	365,970	251,263	44,512	128,385	395,289	263,248	17,922	6901
Total value added	1,191,412	1,270,017	174,074	571,146	408,560	2,114,656	78,816	129,922
Total input	8,058,904	8,884,715	855,091	2,582,320	453,217	8,439,607	372,102	293,961

Sector 26	Sector 27	Sector 28	Sector 29	Sector 30	Sector 31	Sector 32	Sector 33	Sector 34
94,329	213,006	0	0	1927	453,351	0	418	10,156
17,034	30,735	316	0	2427	2618	0	1081	6035
0	11,087	0	0	0	0	0	0	0
0	0	0	0	0	0	0	0	0
216,890	2313	0	0	0	0	0	0	0
85,712	70,206	561	8015	41,407	1,364,460	18,245	14,322	50,640
18,125	11,532	341	403	9599	19,310	931	5668	55,326
81,713	63,740	1626	6221	70,316	14,816	17,166	22,993	53,377
375,873	10,873	1123	3150	13,140	5429	4893	6167	10,811
35,878	32,753	5979	102,478	73,730	13,357	126,521	29,894	383,527
276,769	1,663,199	6653	6337	26,643	19,319	41,165	66,614	136,262
662,307	156,991	748	17,416	72,209	29,941	22,093	85,138	153,336
4,864,868	18,901	780	892	2168	8050	1046	1510	2272
2,676,492	45,937	0	34	3899	177	0	4573	3270
698,094	49,647	175	6609	3202	4588	5889	30,941	85,938
635,543	214,820	2362	26,590	17,204	4607	27,730	18,065	28,098
88,742	627,908	17,831	23,175	69,772	13,260	18,953	19,464	130,245
812,114	25,083	399	226,320	54,317	28,143	13,388	56,551	120,031
51,756	22,481	1978	196,421	31,127	1601	4872	15,279	247,281
46,884	14,205	267	29,470	2598	123	25,003	9962	12,744
69,888	8164	570	3483	6095	2446	6783	6587	25,942
0	0	0	0	0	0	0	0	0
160,349	106,911	3080	81,967	93,402	83,931	61,412	30,520	40,818
2540	8380	215	1871	1634	14,333	324	537	4133
5154	5373	362	4016	3057	6818	3081	930	2794
227,729	38,375	9533	5571	20,608	8987	9426	98,124	4179
2,000,145	752,943	32,142	100,787	559,928	47,768	161,601	38,480	210,627
4630	2802	9457	4295	10,782	2256	28,051	3255	20,130
310,847	64,074	4581	142,434	64,971	12,071	185,843	19,867	16,002
528,220	115,386	16,905	103,264	40,539	83,052	27,634	21,753	99,277
214,687	142,135	1622	64,565	153,383	26,580	240,375	83,059	282,430
203,120	526,181	2641	68,045	272,669	65,006	523,924	218,336	194,228
12,510	43,527	3432	85,959	177,484	41,635	208,979	67,076	71,414
77,374	49,534	8751	164,795	389,407	46,925	397,517	172,081	187,410
16,081	2743	122	5500	12,458	35	9144	1073	2137
248,321	10,498	40	16,069	30,796	723	4521	9771	3193
1055	2691	0	920	1847	1288	4189	1637	36,767
30,064	202,961	360	21,047	73,856	36,652	25,337	37,512	66,290
2672	6660	226	3850	4775	1023	12,687	1203	2059
6373	4812	43	980	1284	662	4511	1760	699
29,825	18,361	1687	18,393	31,122	10,148	39,042	26,941	18,369
2739	2515	74	1252	1988	605	3722	3373	815
15,893,445	5,400,440	136,983	1,552,592	2,447,769	2,476,094	2,285,997	1,232,517	2,779,061
3,188,774	1,444,291	64,649	442,044	1,647,737	889,166	1,346,812	418,017	821,458
885,548	290,192	5828	130,028	1,634,474	168,854	516,639	602,573	208,329
309,793	642,310	15,236	426,813	383,008	199,325	99,462	1,845,607	399,315
1,214,746	1,144,034	187	728,201	2,334,025	211,193	2,279,250	959,441	117,585
5,598,861	3,520,826	85,899	1,727,087	5,999,244	1,468,537	4,242,162	3,825,639	1,546,687
21,492,306	8,921,266	222,882	3,279,679	8,447,013	3,944,630	6,528,159	5,058,156	4,325,749

(continued)

Table 10.10 (continued)

	Sector 35	Sector 36	Sector 37	Sector 38	Sector 39	Sector 40	Sector 41	Sector 42
Sector 1	21,052	7450	19,597	25,727	6318	16,248	4834	0
Sector 2	967	2506	4200	8963	6712	13,701	802	14,291
Sector 3	0	0	0	0	0	0	0	0
Sector 4	5268	1164	1414	0	0	0	0	0
Sector 5	86	57	528	116	1220	1734	485	2016
Sector 6	9943	4587	8527	49,976	26,622	28,288	55,532	18,519
Sector 7	11,315	1237	3471	33,120	5787	18,875	2908	20,768
Sector 8	4321	3131	10,478	15,507	3706	46,529	16,507	67,899
Sector 9	1896	2383	10,338	20,750	40,275	25,311	11,864	82,723
Sector 10	6234	26,391	5499	36,731	115,383	42,235	111,464	132,593
Sector 11	5111	38,283	29,771	40,401	26,024	13,846	6438	73,949
Sector 12	60,768	41,254	54,532	191,788	29,672	1,445,689	40,676	51,534
Sector 13	4402	3631	11,284	2340	16,184	18,905	6349	24,830
Sector 14	2804	13,751	6110	5648	2468	162	456	0
Sector 15	14,251	40,009	6465	10,929	7077	7080	3939	6027
Sector 16	8408	15,048	16,964	5344	6581	151,457	2773	11,834
Sector 17	4967	27,323	28,528	83,982	21,948	3761	4635	66,149
Sector 18	24,667	78,008	21,146	36,117	9722	10,788	5469	11,591
Sector 19	22,043	70,407	11,657	54,272	10,658	2495	5915	39,165
Sector 20	19,725	42,461	11,160	21,437	32,098	7903	3715	8953
Sector 21	2550	4140	4250	17,504	5617	4254	2000	11,526
Sector 22	0	0	0	0	0	0	0	0
Sector 23	11,389	12,949	23,095	24,420	60,147	39,343	8906	69,155
Sector 24	184	476	742	1420	2122	1574	664	1531
Sector 25	868	2212	1510	3613	4461	2256	785	2913
Sector 26	1990	5653	20,238	16,883	15,850	65,569	8072	71,230
Sector 27	23,196	111,149	18,722	46,256	84,492	16,960	29,155	211,556
Sector 28	940	2233	775	3066	6896	1086	2258	54,253
Sector 29	3182	9258	7198	7527	26,737	30,816	7998	94,619
Sector 30	11,662	22,007	11,574	39,371	24,278	101,048	20,351	51,986
Sector 31	21,679	76,882	19,352	34,687	79,112	32,599	30,866	323,124
Sector 32	2409	59,288	25,855	46,698	82,876	19,204	14,824	80,772
Sector 33	2550	19,189	2446	61,957	23,562	6462	12,755	27,457
Sector 34	5295	28,245	11,968	31,405	57,186	24,067	21,740	45,362
Sector 35	13,425	4315	781	60	6599	1534	997	68
Sector 36	8120	117,063	4180	1697	6718	4729	675	7073
Sector 37	351	396	19,281	3196	4167	853	884	11,764
Sector 38	1665	16,837	20,278	85,757	23,170	18,465	9326	72,562
Sector 39	1881	1855	1339	676	14,222	3219	946	38,688
Sector 40	1476	1745	3021	712	4673	8335	1418	13,853
Sector 41	8702	9537	5903	8853	19,053	4853	30,289	63,114
Sector 42	143	598	320	509	615	486	205	1357
Total intermediate input	351,886	925,109	464,496	1,079,416	921,011	2,242,717	489,874	1,886,803
Compensation of employees	136,675	500,587	176,666	755,734	2,329,089	1,041,291	286,260	2,890,924
Net taxes on production	8728	83,721	14,502	104,400	14,986	10,864	49,352	15,695
Depreciation of fixed assets	25,647	95,224	70,813	61,444	276,823	126,406	76,637	425,602
Operating surplus	29,776	257,222	53,150	212,785	39,074	74,564	89,986	91,489
Total value added	200,826	936,755	315,131	1,134,363	2,659,973	1,253,125	502,235	3,423,711
Total input	552,712	1,861,863	779,626	2,213,779	3,580,983	3,495,843	992,109	5,310,514

Total intermediate use	Rural household consumption expenditure	Urban household consumption expenditure	Government consumption expenditure	Gross fixed capital formation	Increase in inventories	Export	Import	Total output
10,631,654	1,217,889	1,503,766	109,277	762,447	221,672	383,386	643,972	14,186,119
4,252,770	19,284	13,945	0	40,059	−87,034	17,728	325,505	3,931,249
4,090,113	0	0	0	18,244	−55,601	332,711	2,965,838	1,419,630
3,549,114	0	0	0	74,568	−19,071	7635	1,951,101	1,661,144
1,090,792	0	0	0	8995	−22,672	21,443	73,435	1,025,123
6,866,460	1,389,186	3,991,547	0	233,126	255,511	220,825	242,567	12,714,088
4,173,235	46,111	71,046	0	32,022	−27,421	3,220,254	748,938	6,766,308
1,662,355	292,384	1,383,259	0	78,567	90,604	579,653	75,977	4,010,847
1,926,145	26,717	115,302	0	187,146	−1685	269,856	317,005	2,206,475
3,375,112	17,349	118,470	0	37,234	−34,103	226,932	460,718	3,280,277
5,828,457	24,924	227,175	0	−7260	−134,219	110,403	398,961	5,650,519
17,027,959	159,338	519,495	0	198,392	−131,943	944,203	1,647,955	17,069,489
7,620,591	9512	62,589	0	73,959	−135,481	388,392	240,873	7,778,689
15,810,283	0	0	0	174,549	−188,641	365,674	852,309	15,309,556
3,937,682	16,752	90,171	0	93,672	−54,600	345,348	214,609	4,214,415
7,685,282	1215	17,779	0	5,121,897	89,393	1,659,560	1,664,661	12,910,463
5,193,485	139,774	686,273	0	5,003,955	119,509	1,683,369	1,350,541	11,475,824
4,897,970	125,832	486,227	0	1,714,291	34,952	1,558,897	734,107	8,084,061
5,946,378	106,878	428,031	0	1,084,140	187,744	4,161,980	3,060,691	8,854,460
1,672,769	11,846	42,722	0	194,248	−24,607	505,255	1,491,409	910,824
1,367,241	71,173	381,654	0	513,187	9937	225,609	93,554	2,475,247
1,266,467	0	0	0	0	6321	4505	810,185	467,107
9,147,178	60,900	224,622	0	0	−1,004,953	13,823	1962	8,439,607
219,390	16,935	129,217	0	2441	358	0	0	368,341
158,066	12,538	122,607	0	0	1226	0	0	294,437
682,291	0	268,625	0	20,216,544	72,498	356,020	99,399	21,496,579
8,083,362	132,357	514,221	402,750	63,616	−179,449	369,632	538,708	8,847,783
240,263	1910	12,774	0	0	2855	7247	30,361	234,688
1,672,344	159,026	677,719	0	667,155	43,230	160,987	76,200	3,304,260
5,271,839	482,340	1,473,314	0	568,699	47,359	857,937	0	8,701,489
2,619,184	302,193	1,176,380	0	0	98,640	−65,591	198,029	3,932,778
5,279,284	183,334	959,608	73,623	0	108,023	75,133	67,566	6,611,439
1,222,169	547,830	1,966,075	0	1,253,315	120,162	0	0	5,109,550
3,147,079	26,157	374,899	172,891	0	59,141	824,800	249,350	4,355,619
463,704	0	0	286,601	0	2978	19,190	224,656	547,818
1,362,394	1776	22,731	250,070	198,427	−1537	0	0	1,833,861
244,451	7425	92,002	457,758	0	−2160	0	0	799,476
1,086,232	168,162	974,515	0	0	35,106	34,259	67,019	2,231,254
128,829	337,218	754,303	2,393,006	0	−27,196	4636	25,502	3,565,295
229,653	128,774	1,469,039	1,668,657	0	−23,344	3828	2677	3,473,930
513,302	33,519	180,022	279,534	0	12,312	42,790	53,539	1,007,940
43,961	0	0	5,232,795	0	−12,349	16,378	13,938	5,266,847

Table 10.11 China's national input–output table in 2020

	Sector 1	Sector 2	Sector 3	Sector 4	Sector 5	Sector 6	Sector 7	Sector 8
Sector 1	2,053,479	42,187	26	3377	452	5,579,462	1,471,479	211,569
Sector 2	13,323	753,136	11,652	8453	3001	38,482	40,714	18,086
Sector 3	0	0	22,473	0	0	0	0	0
Sector 4	0	0	0	166,084	0	0	0	0
Sector 5	1821	1569	714	2243	73,821	5066	0	0
Sector 6	1,641,541	7787	2439	5295	3352	2,906,093	57,889	231,029
Sector 7	1870	1076	1080	1117	599	7270	2,340,417	1,398,787
Sector 8	5148	14,251	4810	8335	4628	18,466	113,975	638,412
Sector 9	20,166	34,951	5173	6122	2317	15,101	9488	10,790
Sector 10	13,499	5104	2889	3206	2682	262,896	75,497	68,649
Sector 11	105,882	47,107	51,202	110,972	55,085	39,577	18,689	27,362
Sector 12	1,204,454	100,855	50,650	115,537	116,302	388,835	876,029	277,217
Sector 13	20,776	37,350	12,590	16,041	39,964	97,350	17,443	5927
Sector 14	3818	183,455	66,388	25,954	8235	15,891	6294	6394
Sector 15	61,109	80,336	10,447	39,147	19,441	65,219	17,665	22,861
Sector 16	87,226	222,618	112,616	153,072	100,380	48,178	93,043	29,110
Sector 17	43,561	27,368	9264	20,893	21,975	17,881	14,959	6782
Sector 18	4208	62,174	12,594	26,323	10,908	20,532	17,110	8432
Sector 19	1452	6793	1752	2821	1310	5194	5859	4959
Sector 20	17,064	33,983	12,848	11,521	4015	9933	23,980	8239
Sector 21	48,373	29,051	1873	7897	11,859	21,574	57,097	46,262
Sector 22	1438	374	22	1257	8611	10,975	18,456	2939
Sector 23	206,266	206,447	99,382	376,713	146,501	146,178	207,903	38,908
Sector 24	543	1657	1679	10,082	5964	8346	3796	1553
Sector 25	1732	3135	1068	1907	1149	5710	5742	2016
Sector 26	4136	5829	1537	827	437	3977	1852	1415
Sector 27	261,444	185,362	26,383	82,093	67,633	404,320	130,637	101,944
Sector 28	12,452	1238	218	490	341	3181	5850	9416
Sector 29	56,320	13,210	2307	3458	2813	25,041	14,768	22,639
Sector 30	335,369	53,887	12,473	29,568	18,757	340,563	176,793	105,183
Sector 31	39,717	39,292	12,079	20,009	14,099	83,541	31,004	25,411
Sector 32	152,390	102,147	11,155	26,016	19,224	145,069	172,900	61,858
Sector 33	3950	6286	608	1342	1626	24,246	25,494	47,641
Sector 34	23,056	22,952	2473	8195	4974	232,907	29,102	95,877
Sector 35	26,987	6186	2487	694	297	11,443	4754	2447
Sector 36	136,081	41,187	15,332	28,895	13,090	48,523	18,352	16,180
Sector 37	44,398	5719	1109	1687	3138	4324	11,377	620
Sector 38	54,045	30,985	5084	4805	4316	17,000	8263	7839
Sector 39	5901	4167	251	825	1576	1938	850	497
Sector 40	8854	10,030	488	3863	2355	7024	3480	2999
Sector 41	3378	5815	1303	5399	4488	14,859	7539	7312
Sector 42	4849	1091	412	430	229	1972	895	603
Total intermediate input	6,732,077	2,438,145	591,330	1,342,964	801,949	11,104,140	6,137,433	3,576,163
Compensation of employees	8,588,483	847,671	199,446	155,490	129,426	863,609	620,827	451,626
Net taxes on production	17,450	472,845	226,475	100,330	78,755	974,072	247,075	117,193
Depreciation of fixed assets	421,468	162,003	133,142	36,698	33,554	267,636	163,060	53,194
Operating surplus	0	490,623	312,022	190,741	107,467	860,177	254,673	151,389
Total value added	9,027,401	1,973,143	871,085	483,259	349,201	2,965,494	1,285,635	773,402
Total input	15,759,478	4,411,288	1,462,415	1,826,223	1,151,150	14,069,633	7,423,068	4,349,565

Sector 9	Sector 10	Sector 11	Sector 12	Sector 13	Sector 14	Sector 15	Sector 16	Sector 17
316,173	190,662	141	648,676	5316	1915	1555	1533	1140
18,842	35,617	393,796	288,397	649,361	628,028	22,511	48,493	15,529
0	0	3,567,314	683,844	24,945	59,282	0	0	0
0	0	0	133,074	27,470	3,603,320	49,052	41,906	8865
0	0	639	280,697	510,707	45,329	20,247	9968	3340
13,463	16,266	29,493	324,046	33,572	60,436	17,090	46,568	51,151
23,469	63,802	323	119,833	18,312	1893	7659	19,978	24,281
40,929	26,133	2630	40,987	21,420	36,934	14,780	30,518	137,448
682,051	93,380	1943	34,686	65,117	12,288	80,842	45,830	162,229
36,306	860,988	3437	268,282	208,525	17,521	49,131	62,181	53,559
17,193	17,902	384,364	1,141,916	360,782	712,371	50,263	145,406	65,292
240,176	514,107	118,604	7,090,702	709,395	270,035	222,278	513,485	858,073
19,797	9975	21,469	134,786	1,453,094	321,674	51,511	102,218	95,699
53,728	43,860	4051	198,361	244,410	4,510,850	1,611,209	2,988,800	1,727,611
56,542	59,452	20,375	188,561	247,984	139,468	533,630	596,737	266,963
33,894	49,066	81,378	276,995	239,562	512,510	226,216	3,200,520	1,161,690
10,353	28,984	14,277	51,627	43,375	84,828	42,338	214,510	3,573,096
7757	24,254	7168	48,865	34,151	52,948	38,177	778,177	517,814
994	33,058	863	11,542	5540	4275	5730	296,512	159,340
2090	46,347	8993	133,528	23,765	54,137	11,609	74,388	222,772
18,727	23,461	2794	109,868	71,683	57,576	29,664	98,332	170,587
7115	250,179	1249	78,386	125,765	479,395	36,674	168,740	9343
85,765	123,151	104,525	1,083,205	753,647	1,152,066	402,552	317,813	260,409
1082	1028	9273	44,652	9987	22,258	4844	12,676	8959
1634	7575	1906	19,287	9103	12,450	4668	4747	6960
662	977	3053	7583	3242	5833	997	4914	5302
74,066	83,185	120,704	477,310	351,209	305,390	104,317	406,860	248,349
1801	2151	188	14,539	2400	5454	980	12,611	2696
9638	9438	19,493	64,799	20,827	167,372	17,231	44,895	35,862
65,513	120,921	74,397	432,965	255,130	337,065	122,742	363,871	685,323
11,577	18,523	11,092	96,735	57,686	53,131	49,693	109,844	46,530
43,296	56,833	37,126	333,315	294,349	336,178	67,265	151,210	154,146
11,523	18,874	987	45,629	22,936	4949	23,483	39,678	24,529
21,791	31,193	29,234	278,950	59,223	27,937	19,681	125,476	188,795
1479	2143	1689	44,951	6853	29,009	4381	49,394	119,951
4794	19,099	10,405	112,328	36,974	65,815	31,101	85,058	132,229
377	12,035	1226	19,581	2697	5465	716	2938	4201
2562	7179	5390	38,914	25,889	38,028	7715	16,726	17,314
399	531	245	3035	1454	1393	685	4514	3025
2211	4422	1584	11,806	15,749	15,582	11,751	33,754	28,427
2957	3793	1399	20,541	13,394	10,315	5022	19,926	12,728
332	566	282	2405	1542	1801	697	1929	2166
1,943,056	2,911,110	5,099,501	15,440,188	7,068,544	14,264,505	4,002,689	11,293,631	11,273,723
188,799	261,218	205,538	1,191,584	590,161	1,018,213	309,686	1,245,121	735,429
78,392	100,607	678,579	658,081	321,359	492,949	128,926	571,350	488,586
45,092	99,302	152,322	510,014	229,265	607,285	90,432	331,469	213,161
113,902	154,900	225,912	987,081	528,675	696,325	216,904	880,931	605,450
426,185	616,026	1,262,350	3,346,759	1,669,459	2,814,773	745,948	3,028,871	2,042,627
2,369,241	3,527,136	6,361,851	18,786,948	8,738,003	17,079,278	4,748,637	14,322,502	13,316,350

(continued)

Table 10.11 (continued)

	Sector 18	Sector 19	Sector 20	Sector 21	Sector 22	Sector 23	Sector 24	Sector 25
Sector 1	178	0	41	321,145	177	161	66	0
Sector 2	7454	2862	1423	17,558	815	1,591,160	18,954	427
Sector 3	0	0	0	25	1	56,573	185,714	0
Sector 4	12,302	148	1303	244	538	267	0	0
Sector 5	3856	64	100	7374	68	2431	15	33
Sector 6	35,730	35,734	3096	36,258	700	27,551	1471	840
Sector 7	16,181	2504	1386	239,891	77	466	46	96
Sector 8	18,815	16,319	2827	39,548	120	30,015	4428	1584
Sector 9	24,987	15,850	2727	81,214	90	5589	462	473
Sector 10	102,765	78,530	9903	47,580	123	17,216	995	838
Sector 11	38,665	25,543	4477	19,452	1203	265,969	8206	3327
Sector 12	790,075	645,983	87,001	202,672	11,751	28,087	2534	15,862
Sector 13	174,905	162,722	32,588	55,677	1447	19,256	911	938
Sector 14	2,357,446	293,783	38,612	253,095	5810	8728	973	371
Sector 15	377,135	210,395	37,015	112,015	609	53,641	3759	8658
Sector 16	435,642	96,989	26,732	10,637	359	90,844	3186	3415
Sector 17	23,062	31,616	5509	27,161	124	101,173	8647	3987
Sector 18	1,246,325	266,874	47,449	16,638	914	500,830	916	1080
Sector 19	526,048	4,266,208	224,602	6618	46	5558	207	259
Sector 20	86,279	442,033	111,984	5744	54	121,737	2475	1742
Sector 21	88,720	74,308	15,571	291,543	108	6040	295	332
Sector 22	69,041	9456	100	527	5318	991	0	0
Sector 23	129,700	133,969	12,164	35,765	2473	3,235,194	7888	75,286
Sector 24	5337	4985	659	2538	121	10,521	21,985	1047
Sector 25	3600	4009	703	1804	115	11,270	211	6733
Sector 26	2470	1853	278	416	57	3103	293	391
Sector 27	199,432	152,843	20,614	63,121	4218	110,347	10,435	3045
Sector 28	6494	3510	1049	751	13	1069	51	100
Sector 29	30,771	96,238	3957	17,930	168	55,449	563	1187
Sector 30	286,518	371,957	23,819	52,909	2583	71,721	6627	1950
Sector 31	54,364	26,460	8285	19,030	516	25,952	1810	2560
Sector 32	137,096	354,832	8133	29,052	2376	375,912	13,083	18,501
Sector 33	29,045	25,664	4575	18,047	180	3967	415	169
Sector 34	204,633	108,166	12,358	22,369	437	40,360	1738	1581
Sector 35	38,381	70,515	5241	2603	45	6025	104	66
Sector 36	55,018	51,779	6450	5701	221	72,623	910	489
Sector 37	1851	1192	95	425	32	24,187	172	18,403
Sector 38	7408	6347	801	1127	478	69,364	1138	2535
Sector 39	1399	936	200	443	5	1400	93	247
Sector 40	16,605	4899	1763	415	87	10,813	1107	71
Sector 41	9712	8200	1026	2937	80	20,421	745	965
Sector 42	1063	989	137	379	0	1293	34	64
Total intermediate input	7,656,505	8,107,266	766,755	2,070,378	44,658	7,085,273	313,660	179,651
Compensation of employees	551,866	716,732	93,698	239,952	6478	724,809	42,220	65,457
Net taxes on production	213,550	145,589	26,424	161,686	3880	440,752	9796	12,900
Depreciation of fixed assets	150,592	225,031	19,113	54,156	2913	908,404	13,110	56,373
Operating surplus	406,124	268,181	47,835	132,164	395,289	294,893	19,167	7558
Total value added	1,322,133	1,355,533	187,070	587,959	408,560	2,368,858	84,292	142,287
Total input	8,978,638	9,462,799	953,825	2,658,337	453,217	9,454,131	397,952	321,938

Sector 26	Sector 27	Sector 28	Sector 29	Sector 30	Sector 31	Sector 32	Sector 33	Sector 34
106,123	233,481	0	0	2235	490,812	0	455	11,521
19,163	33,690	383	0	2816	2834	0	1176	6846
0	12,153	0	0	0	0	0	0	0
0	0	0	0	0	0	0	0	0
244,009	2535	0	0	0	0	0	0	0
96,429	76,955	679	9048	48,039	1,477,208	20,893	15,584	57,444
20,391	12,640	413	455	11,136	20,905	1066	6167	62,760
91,930	69,867	1967	7023	81,577	16,040	19,657	25,017	60,548
422,870	11,918	1359	3557	15,245	5877	5603	6710	12,264
40,364	35,901	7235	115,696	85,538	14,461	144,881	32,526	435,059
311,374	1,823,073	8051	7154	30,909	20,915	47,139	72,480	154,571
745,118	172,082	905	19,663	83,774	32,415	25,299	92,635	173,939
5,473,142	20,718	944	1007	2515	8716	1198	1643	2577
3,011,145	50,353	0	38	4523	191	0	4975	3709
785,380	54,419	212	7461	3714	4967	6743	33,666	97,485
715,008	235,469	2859	30,020	19,959	4987	31,754	19,656	31,873
99,837	688,266	21,577	26,165	80,947	14,356	21,703	21,178	147,745
913,656	27,494	483	255,512	63,017	30,468	15,330	61,531	136,159
58,227	24,642	2394	221,756	36,112	1734	5578	16,624	280,506
52,746	15,571	323	33,271	3015	133	28,631	10,840	14,457
78,627	8948	689	3932	7071	2649	7767	7167	29,428
0	0	0	0	0	0	0	0	0
180,398	117,188	3727	92,539	108,361	90,866	70,323	33,208	46,302
2857	9185	260	2112	1896	15,517	371	585	4688
5799	5889	438	4534	3546	7381	3528	1011	3170
256,203	42,064	11,536	6290	23,909	9730	10,794	106,764	4740
2,250,231	825,319	38,895	113,787	649,602	51,716	185,051	41,868	238,928
5209	3071	11,444	4848	12,508	2443	32,121	3542	22,834
349,713	70,233	5544	160,806	75,376	13,068	212,811	21,617	18,152
594,265	126,477	20,457	116,583	47,031	89,915	31,644	23,669	112,616
241,531	155,797	1963	72,893	177,948	28,777	275,257	90,373	320,378
228,517	576,760	3196	76,822	316,337	70,377	599,952	237,562	220,325
14,074	47,711	4154	97,046	205,908	45,076	239,305	72,983	81,010
87,048	54,295	10,589	186,052	451,771	50,802	455,202	187,233	212,591
18,092	3007	148	6209	14,454	38	10,471	1168	2425
279,370	11,507	48	18,141	35,729	783	5177	10,632	3622
1187	2950	0	1039	2142	1394	4797	1782	41,707
33,823	222,471	435	23,762	85,684	39,681	29,014	40,815	75,196
3006	7300	273	4347	5540	1108	14,528	1309	2336
7170	5274	52	1106	1490	716	5166	1915	793
33,554	20,126	2041	20,766	36,107	10,987	44,707	29,313	20,837
3081	2757	90	1413	2306	655	4262	3670	925
17,880,665	5,919,552	165,765	1,752,855	2,839,788	2,680,698	2,617,726	1,341,047	3,152,466
3,587,479	1,583,122	78,233	499,062	1,911,628	962,639	1,542,252	454,826	931,832
996,272	318,086	7052	146,800	1,896,240	182,806	591,610	655,633	236,321
348,527	704,051	18,437	481,867	444,348	215,795	113,895	2,008,124	452,968
1,366,631	1,254,003	226	822,129	2,707,827	228,644	2,609,999	1,043,925	133,384
6,298,909	3,859,262	103,947	1,949,858	6,960,043	1,589,885	4,857,756	4,162,508	1,754,506
24,179,574	9,778,814	269,712	3,702,713	9,799,831	4,270,582	7,475,481	5,503,555	4,906,971

(continued)

Table 10.11 (continued)

	Sector 35	Sector 36	Sector 37	Sector 38	Sector 39	Sector 40	Sector 41	Sector 42
Sector 1	23,125	8111	21,995	27,907	6627	17,204	5430	0
Sector 2	1063	2729	4713	9722	7041	14,507	901	14,991
Sector 3	0	0	0	0	0	0	0	0
Sector 4	5787	1267	1587	0	0	0	0	0
Sector 5	94	63	592	126	1279	1836	545	2115
Sector 6	10,923	4993	9570	54,209	27,927	29,953	62,386	19,427
Sector 7	12,429	1346	3896	35,926	6071	19,986	3267	21,786
Sector 8	4746	3408	11,760	16,820	3888	49,267	18,544	71,227
Sector 9	2083	2594	11,602	22,507	42,248	26,800	13,328	86,778
Sector 10	6848	28,731	6171	39,842	121,037	44,721	125,221	139,092
Sector 11	5614	41,678	33,413	43,823	27,299	14,660	7233	77,573
Sector 12	66,754	44,912	61,203	208,034	31,126	1,530,778	45,696	54,059
Sector 13	4836	3953	12,665	2538	16,977	20,018	7132	26,047
Sector 14	3080	14,971	6857	6126	2588	172	513	0
Sector 15	15,655	43,557	7256	11,855	7424	7497	4425	6323
Sector 16	9236	16,382	19,040	5796	6903	160,371	3115	12,414
Sector 17	5456	29,746	32,019	91,096	23,024	3982	5207	69,391
Sector 18	27,096	84,926	23,733	39,176	10,199	11,423	6144	12,159
Sector 19	24,215	76,651	13,083	58,869	11,181	2641	6646	41,085
Sector 20	21,668	46,226	12,526	23,252	33,671	8368	4173	9392
Sector 21	2802	4507	4770	18,986	5892	4504	2247	12,091
Sector 22	0	0	0	0	0	0	0	0
Sector 23	12,511	14,098	25,920	26,488	63,094	41,658	10,006	72,544
Sector 24	203	519	833	1541	2226	1666	745	1606
Sector 25	954	2408	1695	3919	4679	2389	882	3056
Sector 26	2186	6154	22,714	18,313	16,627	69,429	9068	74,721
Sector 27	25,481	121,005	21,012	50,174	88,632	17,958	32,753	221,925
Sector 28	1032	2431	870	3326	7234	1149	2536	56,912
Sector 29	3496	10,079	8078	8164	28,048	32,630	8986	99,257
Sector 30	12,811	23,959	12,990	42,706	25,468	106,995	22,863	54,534
Sector 31	23,815	83,699	21,719	37,625	82,988	34,517	34,676	338,961
Sector 32	2646	64,546	29,018	50,654	86,937	20,334	16,654	84,730
Sector 33	2801	20,891	2745	67,205	24,717	6842	14,329	28,803
Sector 34	5817	30,750	13,432	34,065	59,989	25,484	24,423	47,585
Sector 35	14,748	4698	876	65	6922	1625	1121	72
Sector 36	8920	127,444	4691	1841	7048	5007	759	7419
Sector 37	385	431	21,640	3466	4371	903	993	12,340
Sector 38	1829	18,330	22,759	93,021	24,305	19,552	10,477	76,118
Sector 39	2066	2019	1503	733	14,919	3408	1063	40,584
Sector 40	1622	1900	3390	773	4902	8825	1593	14,532
Sector 41	9560	10,383	6625	9602	19,987	5138	34,028	66,207
Sector 42	158	651	359	553	645	515	230	1423
Total intermediate input	386,550	1,007,143	521,320	1,170,847	966,141	2,374,716	550,336	1,979,278
Compensation of employees	150,138	544,977	198,278	819,748	2,443,216	1,102,578	321,591	3,032,614
Net taxes on production	9588	91,145	16,276	113,244	15,720	11,504	55,444	16,464
Depreciation of fixed assets	28,173	103,668	79,476	66,648	290,387	133,846	86,095	446,462
Operating surplus	32,709	280,032	59,652	230,808	40,989	78,952	101,092	95,973
Total value added	220,609	1,019,821	353,683	1,230,447	2,790,312	1,326,880	564,222	3,591,513
Total input	607,159	2,026,964	875,003	2,401,294	3,756,453	3,701,596	1,114,558	5,570,791

Total intermediate use	Rural household consumption expenditure	Urban household consumption expenditure	Government consumption expenditure	Gross fixed capital formation	Increase in inventories	Export	Import	Total output
11,805,937	1,297,157	1,609,802	115,901	842,060	221,672	402,786	669,134	15,626,181
4,750,649	20,800	15,153	0	44,751	−87,034	18,129	349,808	4,412,640
4,612,324	0	0	0	18,272	−55,601	352,499	3,657,223	1,270,271
4,053,213	0	0	0	74,637	−19,071	7635	2,172,440	1,943,973
1,223,296	0	0	0	10,138	−22,672	22,067	78,281	1,154,549
7,610,555	1,466,957	4,241,383	0	262,935	255,511	220,825	242,567	13,815,599
4,533,053	49,188	75,938	0	35,003	−27,421	3,478,772	823,495	7,321,037
1,826,747	308,991	1,470,868	0	85,394	90,604	579,653	75,977	4,286,280
2,107,208	28,332	122,803	0	199,345	−1685	269,856	348,065	2,377,793
3,675,625	18,386	124,618	0	38,875	−34,103	226,932	499,552	3,550,781
6,445,168	26,506	240,388	0	−8270	−134,219	113,589	422,360	6,260,803
18,839,091	169,131	551,414	0	216,002	−131,943	944,203	1,700,143	18,887,754
8,512,731	10,346	66,977	0	82,312	−135,481	410,315	261,966	8,685,234
17,767,370	0	0	0	195,206	−188,641	365,674	891,263	17,248,347
4,327,201	17,936	96,756	0	103,112	−54,600	345,348	231,007	4,604,745
8,620,719	1296	18,922	0	5,645,500	89,393	1,764,439	1,744,524	14,395,745
5,809,044	148,464	724,662	0	5,953,862	119,509	1,812,830	1,437,571	13,130,799
5,471,122	133,039	515,440	0	1,902,382	34,952	1,628,856	769,421	8,916,369
6,459,485	113,895	455,797	0	1,157,140	187,744	4,344,268	3,205,370	9,512,959
1,789,523	12,942	45,164	0	208,749	−24,607	515,872	1,642,618	905,024
1,485,672	74,447	405,731	0	528,294	9937	232,074	98,686	2,637,467
1,286,350	0	0	0	0	6321	4545	869,584	427,632
10,353,101	60,900	224,622	0	0	−1,196,998	14,479	1974	9,454,131
242,380	17,667	134,181	0	2611	358	0	0	397,197
174,513	13,310	127,261	0	0	1226	0	0	316,310
752,677	0	286,458	0	22,744,299	72,498	384,742	106,834	24,133,841
8,969,604	140,965	545,955	434,191	69,732	−179,449	369,632	581,371	9,769,259
262,556	2040	13,724	0	0	2855	7482	33,243	255,414
1,862,434	167,729	717,985	0	753,209	43,230	171,086	81,844	3,633,829
5,813,586	513,164	1,571,246	0	659,778	47,359	877,478	0	9,482,611
2,881,853	320,893	1,250,217	0	0	98,640	−79,788	212,150	4,259,666
5,788,830	193,755	1,017,885	78,898	0	108,023	81,163	72,861	7,195,694
1,361,444	579,612	2,051,555	0	1,363,677	120,162	0	0	5,476,450
3,530,587	27,588	394,893	183,925	0	59,141	868,823	249,350	4,815,609
524,261	0	0	299,859	0	2978	20,698	240,264	607,533
1,536,770	1776	22,731	265,063	216,023	−1537	0	0	2,040,825
269,493	7843	96,867	479,912	0	−2160	0	0	851,955
1,198,523	177,462	1,029,278	0	0	35,106	35,649	71,717	2,404,302
142,055	357,564	798,993	2,537,756	0	−27,196	4763	27,335	3,786,599
261,362	134,201	1,542,137	1,780,267	0	−23,344	3828	2677	3,695,773
564,222	35,531	191,003	297,158	0	12,312	42,790	53,539	1,089,479
49,850	0	0	5,536,395	0	−12,349	17,484	14,643	5,576,737

Table 10.12 China's national input–output table in 2021

	Sector 1	Sector 2	Sector 3	Sector 4	Sector 5	Sector 6	Sector 7	Sector 8
Sector 1	2,128,365	46,981	26	3825	520	5,645,041	1,497,080	218,769
Sector 2	13,556	818,059	11,652	9467	3298	38,935	41,076	18,407
Sector 3	0	0	22,474	0	0	0	0	0
Sector 4	0	0	0	167,074	0	0	0	0
Sector 5	1870	1739	714	2702	79,348	5126	0	0
Sector 6	1,701,644	8883	2439	6211	3873	2,940,250	58,934	239,629
Sector 7	1936	1219	1080	1317	682	7356	2,384,958	1,445,056
Sector 8	5301	15,795	4810	9347	5265	18,683	115,672	656,745
Sector 9	20,450	37,597	5173	6668	2565	15,278	9559	10,946
Sector 10	13,740	5447	2890	3663	2843	265,986	76,094	69,836
Sector 11	108,726	51,934	51,202	130,235	59,778	40,043	18,911	28,086
Sector 12	1,238,424	110,662	50,651	134,208	126,478	393,406	886,759	284,532
Sector 13	21,181	40,545	12,590	17,858	43,767	98,494	17,614	6039
Sector 14	3885	197,851	66,389	28,229	9185	16,078	6357	6508
Sector 15	62,116	87,802	10,447	42,737	21,946	65,986	17,837	23,268
Sector 16	88,748	238,440	112,616	168,367	108,717	48,744	93,857	29,609
Sector 17	44,406	29,655	9264	22,839	24,255	18,091	15,104	6908
Sector 18	4294	67,518	12,594	29,206	12,006	20,774	17,294	8604
Sector 19	1479	7303	1752	3080	1437	5255	5917	5051
Sector 20	17,318	39,355	12,848	14,043	4512	10,049	24,122	8362
Sector 21	49,394	32,687	1873	11,533	13,048	21,827	57,711	47,352
Sector 22	1442	382	22	1271	9087	11,104	18,491	2949
Sector 23	211,977	227,279	99,383	433,747	162,584	147,896	210,691	40,017
Sector 24	556	1802	1679	11,502	6473	8444	3838	1589
Sector 25	1803	3603	1068	2319	1321	5777	5850	2094
Sector 26	4244	6412	1537	942	486	4024	1875	1452
Sector 27	266,389	199,727	26,383	91,173	73,698	409,073	131,885	103,809
Sector 28	13,128	1581	218	572	506	3219	6060	9948
Sector 29	58,570	15,231	2307	4148	3296	25,335	15,058	23,520
Sector 30	345,663	63,037	12,473	33,973	23,443	344,566	180,085	108,646
Sector 31	40,284	42,020	12,079	21,719	15,344	84,523	31,233	25,762
Sector 32	160,289	123,813	11,155	32,741	23,964	146,774	177,019	64,958
Sector 33	4025	6781	608	1460	1807	24,531	25,759	48,563
Sector 34	23,768	25,792	2473	9195	5864	235,645	29,602	98,855
Sector 35	27,570	6758	2487	779	331	11,577	4804	2500
Sector 36	139,544	45,100	15,332	33,795	14,244	49,094	18,556	16,577
Sector 37	45,997	6592	1109	1901	3890	4375	11,629	643
Sector 38	54,919	33,154	5084	5207	4718	17,200	8333	7967
Sector 39	6051	4527	251	937	1715	1961	860	510
Sector 40	9138	11,113	488	4496	2603	7107	3530	3092
Sector 41	3486	6572	1303	6060	5358	15,034	7679	7543
Sector 42	5008	1210	412	505	253	1995	908	623
Total intermediate input	6,950,684	2,671,959	591,334	1,511,047	884,506	11,234,654	6,238,600	3,685,322
Compensation of employees	8,901,687	920,744	199,448	156,417	139,115	873,760	632,642	464,595
Net taxes on production	18,086	513,606	226,476	100,928	84,650	985,521	251,777	120,558
Depreciation of fixed assets	436,838	175,968	133,143	36,917	36,066	270,781	166,163	54,722
Operating surplus	0	532,917	312,024	191,877	115,513	870,288	259,520	155,737
Total value added	9,356,611	2,143,235	871,091	486,139	375,344	3,000,349	1,310,102	795,611
Total input	16,307,296	4,815,194	1,462,425	1,997,186	1,259,850	14,235,003	7,548,703	4,480,934

Sector 9	Sector 10	Sector 11	Sector 12	Sector 13	Sector 14	Sector 15	Sector 16	Sector 17
346,302	204,751	149	708,049	5790	2220	1619	1711	1222
20,024	37,058	417,235	302,271	688,970	705,013	23,248	54,113	16,159
0	0	3,779,641	947,352	37,803	103,316	0	0	0
0	0	0	137,452	28,279	3,654,883	49,359	46,763	8924
0	0	678	292,552	549,151	54,472	21,307	11,123	3544
14,954	17,439	31,249	350,901	37,071	73,323	18,052	51,965	55,473
25,772	68,240	342	129,262	20,094	2268	8060	22,294	26,180
44,249	27,719	2787	44,014	23,194	42,618	15,368	34,055	146,379
722,589	97,277	2059	36,417	68,815	13,586	83,115	51,142	168,083
38,514	897,303	3641	277,700	218,614	19,674	50,701	69,388	55,249
18,664	18,758	407,242	1,197,097	387,176	838,407	52,602	162,259	69,033
260,230	538,471	125,664	7,471,878	759,678	314,792	231,903	572,999	908,281
21,020	10,388	22,747	141,279	1,544,844	363,073	53,247	114,065	99,753
56,754	45,947	4292	209,211	259,304	5,011,718	1,659,181	3,335,210	1,790,757
60,176	62,520	21,588	199,495	265,113	157,206	552,818	665,900	278,027
35,743	50,923	86,221	288,943	252,332	569,654	232,965	3,571,468	1,201,526
10,977	30,216	15,127	54,369	46,091	94,787	43,610	239,372	3,744,823
8260	25,361	7595	51,437	36,348	59,883	39,532	868,369	541,132
1050	34,421	915	12,120	5857	4746	5900	330,878	165,276
2424	48,129	9528	141,499	26,038	67,812	12,377	83,010	242,969
20,576	25,290	2960	115,389	77,711	72,751	31,664	109,729	177,740
7197	253,597	1323	80,255	128,190	489,987	36,899	188,297	9325
93,456	130,137	110,746	1,150,866	810,377	1,340,349	420,311	354,648	275,199
1160	1075	9825	46,838	10,630	25,530	5030	14,146	9393
1833	8208	2019	21,000	10,091	15,305	4938	5297	7614
716	1030	3235	8041	3486	6753	1039	5484	5596
78,439	86,759	127,889	499,777	371,748	342,480	107,663	454,016	258,037
2101	2484	199	17,769	2916	7098	1060	14,072	3160
10,817	10,276	20,653	71,141	23,177	204,693	18,213	50,098	39,215
72,923	131,688	78,826	482,830	287,341	409,472	129,949	406,045	745,689
12,145	19,154	11,752	100,854	60,675	58,466	51,018	122,575	47,985
50,577	64,982	39,336	380,694	339,378	428,239	71,634	168,736	175,194
12,217	19,754	1046	48,181	24,348	5493	24,178	44,277	25,499
23,781	33,615	30,974	304,638	64,962	32,555	20,543	140,019	202,912
1583	2249	1790	47,584	7319	32,737	4521	55,118	125,787
5174	19,983	11,025	118,032	39,488	76,042	32,331	94,916	138,840
416	13,160	1299	21,916	3019	6499	753	3278	4584
2702	7450	5710	40,716	27,284	41,911	7927	18,665	17,931
431	556	260	3187	1549	1599	713	5038	3171
2413	4666	1679	12,540	17,006	18,336	12,296	37,666	30,193
3220	4108	1483	22,595	14,741	12,002	5237	22,236	13,733
364	597	298	2552	1666	2132	730	2152	2301
2,091,943	3,055,739	5,403,024	16,590,693	7,587,663	15,783,877	4,143,611	12,602,591	11,841,888
200,020	272,235	217,771	1,255,641	627,424	1,131,272	320,821	1,389,434	770,775
83,051	104,850	718,969	693,457	341,651	547,685	133,562	637,571	512,068
47,772	103,490	161,388	537,431	243,741	674,716	93,683	369,888	223,406
120,671	161,434	239,358	1,040,143	562,056	773,643	224,703	983,033	634,549
451,515	642,009	1,337,486	3,526,672	1,774,872	3,127,316	772,769	3,379,925	2,140,798
2,543,459	3,697,748	6,740,510	20,117,364	9,362,534	18,911,193	4,916,381	15,982,516	13,982,686

(continued)

Table 10.12 (continued)

	Sector 18	Sector 19	Sector 20	Sector 21	Sector 22	Sector 23	Sector 24	Sector 25
Sector 1	193	0	41	323,691	177	172	69	0
Sector 2	7858	3065	1427	17,697	815	1,702,670	20,019	439
Sector 3	0	0	0	25	1	60,537	196,153	0
Sector 4	12,566	151	1307	246	538	286	0	0
Sector 5	4135	71	100	7433	68	2601	16	34
Sector 6	39,253	40,476	3107	36,545	700	29,481	1554	863
Sector 7	17,654	2795	1390	241,793	77	498	49	99
Sector 8	20,229	17,930	2835	39,861	120	32,118	4677	1629
Sector 9	26,253	16,857	2734	81,858	90	5981	488	486
Sector 10	107,678	83,179	9918	47,958	123	18,423	1051	862
Sector 11	41,393	27,962	4485	19,606	1203	284,608	8667	3421
Sector 12	844,902	706,822	87,179	204,278	11,751	30,056	2677	16,308
Sector 13	184,719	174,278	32,652	56,118	1447	20,606	962	964
Sector 14	2,486,677	313,396	38,713	255,102	5810	9340	1027	381
Sector 15	401,643	226,297	37,127	112,903	609	57,400	3970	8901
Sector 16	457,293	103,075	26,782	10,721	359	97,210	3365	3511
Sector 17	24,328	33,798	5521	27,376	124	108,263	9133	4099
Sector 18	1,322,107	287,073	47,556	16,769	914	535,929	967	1110
Sector 19	553,786	4,558,784	225,098	6671	46	5948	219	266
Sector 20	94,045	495,557	112,474	5789	54	130,269	2614	1791
Sector 21	95,985	81,746	15,612	293,854	108	6464	312	341
Sector 22	70,203	9623	100	531	5318	1060	0	0
Sector 23	139,372	147,861	12,200	36,049	2473	3,461,920	8332	77,402
Sector 24	5668	5385	661	2558	121	11,258	23,220	1076
Sector 25	3969	4545	706	1818	115	12,060	223	6922
Sector 26	2646	2028	279	420	57	3321	309	402
Sector 27	210,168	162,876	20,654	63,622	4218	118,081	11,022	3131
Sector 28	7722	4301	1061	757	13	1144	54	103
Sector 29	34,013	109,458	3974	18,072	168	59,335	595	1221
Sector 30	319,428	424,739	23,964	53,328	2583	76,747	6999	2005
Sector 31	56,851	27,920	8300	19,181	516	27,771	1912	2632
Sector 32	155,346	414,862	8183	29,282	2376	402,257	13,818	19,021
Sector 33	30,663	27,486	4588	18,190	180	4245	438	174
Sector 34	222,743	120,082	12,411	22,547	437	43,188	1836	1625
Sector 35	40,609	76,179	5255	2624	45	6447	110	68
Sector 36	58,473	56,393	6464	5746	221	77,713	961	503
Sector 37	2053	1354	96	428	32	25,882	182	18,920
Sector 38	7771	6741	803	1136	478	74,225	1201	2606
Sector 39	1489	1022	201	447	5	1498	98	254
Sector 40	17,882	5432	1768	419	87	11,570	1170	73
Sector 41	10,583	9089	1030	2960	80	21,852	787	992
Sector 42	1147	1102	138	382	0	1384	35	66
Total intermediate input	8,141,494	8,791,788	768,891	2,086,791	44,658	7,581,816	331,292	184,700
Compensation of employees	585,422	765,885	94,108	241,854	6478	775,604	44,593	67,297
Net taxes on production	226,535	155,574	26,539	162,968	3880	471,641	10,346	13,262
Depreciation of fixed assets	159,749	240,463	19,196	54,585	2913	972,065	13,847	57,957
Operating surplus	430,819	286,573	48,044	133,212	395,289	315,559	20,244	7770
Total value added	1,402,524	1,448,495	187,888	592,620	408,560	2,534,870	89,030	146,286
Total input	9,544,018	10,240,284	956,778	2,679,411	453,217	10,116,686	420,322	330,986

Sector 26	Sector 27	Sector 28	Sector 29	Sector 30	Sector 31	Sector 32	Sector 33	Sector 34
114,398	249,854	0	0	2322	523,094	0	485	12,053
20,658	36,052	388	0	2925	3020	0	1254	7162
0	13,005	0	0	0	0	0	0	0
0	0	0	0	0	0	0	0	0
263,035	2713	0	0	0	0	0	0	0
103,948	82,351	689	9380	49,897	1,574,368	21,556	16,619	60,095
21,981	13,527	419	471	11,567	22,280	1100	6577	65,656
99,098	74,766	1996	7280	84,733	17,095	20,281	26,680	63,342
455,843	12,754	1379	3687	15,835	6264	5780	7156	12,830
43,512	38,419	7340	119,937	88,848	15,412	149,477	34,688	455,135
335,654	1,950,915	8168	7417	32,105	22,291	48,634	77,297	161,704
803,219	184,149	918	20,384	87,015	34,547	26,102	98,791	181,965
5,899,908	22,171	958	1043	2612	9289	1236	1752	2696
3,245,938	53,884	0	39	4698	204	0	5306	3880
846,619	58,235	215	7735	3858	5294	6957	35,903	101,983
770,760	251,981	2900	31,120	20,731	5315	32,761	20,963	33,344
107,622	736,530	21,891	27,124	84,078	15,300	22,392	22,585	154,563
984,898	29,422	490	264,879	65,455	32,472	15,817	65,620	142,442
62,768	26,370	2429	229,885	37,509	1848	5755	17,729	293,450
56,858	16,662	328	34,491	3131	142	29,539	11,560	15,124
84,758	9576	699	4076	7345	2823	8014	7644	30,786
0	0	0	0	0	0	0	0	0
194,465	125,406	3781	95,931	112,553	96,843	72,554	35,415	48,439
3080	9829	264	2189	1969	16,538	382	624	4904
6251	6302	445	4701	3683	7867	3640	1079	3316
276,180	45,013	11,704	6520	24,834	10,370	11,136	113,860	4959
2,425,691	883,194	39,461	117,959	674,735	55,117	190,921	44,650	249,953
5615	3286	11,610	5026	12,992	2603	33,140	3777	23,888
376,982	75,158	5624	166,701	78,292	13,927	219,562	23,053	18,990
640,603	135,346	20,754	120,857	48,851	95,829	32,648	25,242	117,813
260,364	166,723	1991	75,565	184,832	30,669	283,989	96,379	335,162
246,336	617,205	3243	79,638	328,576	75,006	618,983	253,350	230,492
15,172	51,056	4214	100,604	213,875	48,040	246,896	77,833	84,748
93,836	58,102	10,743	192,872	469,250	54,143	469,641	199,677	222,401
19,502	3217	150	6437	15,013	40	10,804	1246	2536
301,153	12,314	49	18,806	37,111	835	5342	11,338	3789
1280	3156	0	1077	2225	1486	4949	1900	43,632
36,461	238,071	442	24,633	88,999	42,291	29,934	43,528	78,666
3240	7812	277	4506	5754	1181	14,989	1396	2444
7729	5644	53	1147	1547	764	5330	2042	830
36,170	21,537	2071	21,527	37,504	11,710	46,125	31,261	21,799
3321	2950	91	1465	2395	698	4398	3914	967
19,274,904	6,334,658	168,175	1,817,110	2,949,657	2,857,015	2,700,763	1,430,172	3,297,937
3,867,212	1,694,138	79,370	517,356	1,985,588	1,025,955	1,591,174	485,053	974,832
1,073,955	340,392	7154	152,182	1,969,605	194,830	610,376	699,206	247,226
375,703	753,422	18,705	499,531	461,539	229,989	117,508	2,141,583	473,871
1,473,194	1,341,939	229	852,266	2,812,592	243,683	2,692,791	1,113,304	139,539
6,790,064	4,129,891	105,459	2,021,335	7,229,324	1,694,456	5,011,850	4,439,147	1,835,468
26,064,968	10,464,549	273,633	3,838,445	10,178,981	4,551,470	7,712,613	5,869,319	5,133,404

(continued)

Table 10.12 (continued)

	Sector 35	Sector 36	Sector 37	Sector 38	Sector 39	Sector 40	Sector 41	Sector 42
Sector 1	24,776	8827	22,734	30,083	7110	18,045	5688	0
Sector 2	1139	2969	4872	10,481	7554	15,217	944	15,732
Sector 3	0	0	0	0	0	0	0	0
Sector 4	6200	1379	1641	0	0	0	0	0
Sector 5	101	68	612	135	1373	1926	570	2219
Sector 6	11,702	5434	9892	58,437	29,961	31,417	65,349	20,387
Sector 7	13,316	1465	4027	38,728	6513	20,963	3423	22,862
Sector 8	5085	3709	12,155	18,132	4171	51,677	19,425	74,746
Sector 9	2231	2823	11,992	24,263	45,325	28,111	13,961	91,065
Sector 10	7337	31,267	6379	42,950	129,852	46,908	131,168	145,964
Sector 11	6015	45,356	34,536	47,241	29,287	15,377	7576	81,406
Sector 12	71,517	48,876	63,261	224,260	33,392	1,605,631	47,866	56,730
Sector 13	5181	4302	13,090	2736	18,213	20,996	7471	27,334
Sector 14	3300	16,292	7088	6604	2777	180	537	0
Sector 15	16,772	47,401	7500	12,780	7965	7863	4635	6635
Sector 16	9896	17,828	19,680	6249	7406	168,213	3263	13,027
Sector 17	5846	32,372	33,095	98,201	24,701	4177	5455	72,819
Sector 18	29,030	92,421	24,531	42,232	10,941	11,981	6436	12,760
Sector 19	25,943	83,416	13,522	63,461	11,995	2771	6961	43,115
Sector 20	23,214	50,306	12,947	25,066	36,123	8778	4371	9856
Sector 21	3002	4905	4930	20,467	6322	4725	2354	12,688
Sector 22	0	0	0	0	0	0	0	0
Sector 23	13,404	15,342	26,792	28,554	67,689	43,695	10,481	76,128
Sector 24	217	565	861	1661	2388	1748	781	1685
Sector 25	1022	2620	1752	4225	5020	2506	923	3206
Sector 26	2342	6697	23,477	19,742	17,838	72,824	9498	78,412
Sector 27	27,299	131,685	21,719	54,088	95,087	18,837	34,309	232,889
Sector 28	1106	2646	899	3585	7761	1206	2657	59,724
Sector 29	3745	10,969	8350	8801	30,090	34,226	9412	104,161
Sector 30	13,725	26,073	13,426	46,037	27,322	112,227	23,949	57,228
Sector 31	25,514	91,087	22,450	40,560	89,032	36,205	36,322	355,708
Sector 32	2835	70,242	29,993	54,605	93,268	21,328	17,445	88,917
Sector 33	3001	22,735	2838	72,447	26,517	7177	15,010	30,226
Sector 34	6232	33,464	13,884	36,722	64,357	26,730	25,583	49,936
Sector 35	15,800	5112	906	70	7426	1704	1174	75
Sector 36	9557	138,692	4849	1984	7561	5252	795	7786
Sector 37	413	469	22,367	3737	4689	947	1040	12,950
Sector 38	1960	19,948	23,524	100,277	26,075	20,508	10,974	79,879
Sector 39	2214	2198	1553	790	16,005	3575	1113	42,589
Sector 40	1738	2067	3504	833	5258	9257	1669	15,250
Sector 41	10,242	11,299	6848	10,351	21,443	5390	35,644	69,478
Sector 42	169	708	371	596	692	540	241	1494
Total intermediate input	414,133	1,096,033	538,845	1,262,171	1,036,500	2,490,837	576,471	2,077,066
Compensation of employees	160,852	593,077	204,944	883,687	2,621,144	1,156,493	336,863	3,182,442
Net taxes on production	10,272	99,189	16,823	122,076	16,865	12,066	58,077	17,278
Depreciation of fixed assets	30,183	112,818	82,147	71,846	311,535	140,391	90,184	468,519
Operating surplus	35,043	304,747	61,657	248,811	43,974	82,813	105,893	100,715
Total value added	236,351	1,109,831	365,572	1,326,421	2,993,518	1,391,763	591,017	3,768,954
Total input	650,485	2,205,864	904,417	2,588,592	4,030,018	3,882,600	1,167,488	5,846,019

Total intermediate use	Rural household consumption expenditure	Urban household consumption expenditure	Government consumption expenditure	Gross fixed capital formation	Increase in inventories	Export	Import	Total output
12,156,231	1,369,096	1,706,036	122,525	872,769	221,672	422,271	691,619	16,178,981
5,102,895	22,283	16,333	0	48,609	−87,034	18,530	374,078	4,747,538
5,160,307	0	0	0	18,272	−55,601	372,560	3,707,789	1,787,749
4,117,046	0	0	0	75,082	−19,071	7635	2,479,457	1,701,235
1,311,535	0	0	0	10,897	−22,672	22,692	83,129	1,239,323
7,915,750	1,530,021	4,443,976	0	266,025	255,511	220,825	242,567	14,389,541
4,665,343	51,893	80,237	0	35,670	−27,421	3,739,086	897,539	7,647,269
1,915,773	323,938	1,549,724	0	87,846	90,604	579,653	75,977	4,471,562
2,227,363	29,930	130,227	0	211,194	−1685	269,856	379,125	2,487,759
3,835,062	19,427	130,791	0	40,514	−34,103	226,932	533,140	3,685,483
6,942,478	27,984	252,727	0	−8762	−134,219	116,779	445,729	6,751,257
19,901,610	178,888	583,217	0	227,614	−131,943	944,203	1,752,397	19,951,192
9,141,240	11,143	71,165	0	87,509	−135,481	432,371	282,934	9,325,012
19,168,029	0	0	0	216,881	−188,641	365,674	930,163	18,631,781
4,622,180	19,017	102,770	0	106,819	−54,600	345,348	247,276	4,894,258
9,296,624	1379	20,090	0	6,299,827	89,393	1,869,128	1,824,531	15,751,910
6,131,288	156,405	759,740	0	6,342,156	119,509	1,943,618	1,523,728	13,928,987
5,854,457	139,893	543,220	0	2,018,055	34,952	1,699,132	804,576	9,485,133
6,872,181	120,941	483,680	0	1,236,497	187,744	4,526,515	3,353,782	10,073,776
1,945,487	13,809	47,097	0	209,662	−24,607	526,600	1,731,592	986,455
1,578,769	77,452	427,834	0	532,482	9937	238,565	103,798	2,761,241
1,326,653	0	0	0	0	6321	4668	907,025	430,617
11,172,741	60,900	224,622	0	0	−1,354,729	15,137	1986	10,116,686
259,142	18,390	139,084	0	2758	358	0	0	419,732
189,055	14,007	131,469	0	0	1226	0	0	335,757
801,218	0	303,659	0	24,517,778	72,498	413,491	114,261	25,994,382
9,520,307	149,384	576,991	465,632	74,621	−179,449	369,632	623,579	10,353,540
282,768	2140	14,459	0	0	2855	7721	36,086	273,857
2,010,627	175,557	754,198	0	780,820	43,230	181,239	87,457	3,858,213
6,324,375	540,183	1,657,085	0	685,305	47,359	897,150	0	10,151,458
3,035,221	339,308	1,322,928	0	0	98,640	−94,126	226,135	4,475,837
6,336,090	202,914	1,069,112	84,173	0	108,023	87,247	78,109	7,809,451
1,426,880	610,736	2,135,268	0	1,454,306	120,162	0	0	5,747,352
3,737,634	28,877	412,904	194,960	0	59,141	913,021	249,350	5,097,187
558,046	0	0	313,117	0	2978	22,206	255,883	640,465
1,641,758	1776	22,731	280,055	235,089	−1537	0	0	2,179,871
286,324	8213	101,174	502,066	0	−2160	0	0	895,618
1,268,007	186,766	1,084,060	0	0	35,106	37,038	76,415	2,534,561
149,965	378,851	845,749	2,682,506	0	−27,196	4889	29,174	4,005,590
281,425	139,581	1,614,614	1,891,877	0	−23,344	3828	2677	3,905,305
600,161	37,380	201,092	314,782	0	12,312	42,790	53,539	1,154,979
52,968	0	0	5,839,995	0	−12,349	18,593	15,345	5,883,862

Table 10.13 China's national input–output table in 2022

	Sector 1	Sector 2	Sector 3	Sector 4	Sector 5	Sector 6	Sector 7	Sector 8
Sector 1	2,168,422	47,887	26	3825	559	5,977,729	1,644,937	225,609
Sector 2	13,717	831,991	11,652	9467	3484	41,229	43,992	18,805
Sector 3	0	0	22,475	0	0	0	0	0
Sector 4	0	0	0	167,074	0	0	0	0
Sector 5	1898	1763	714	2702	82,446	5428	0	0
Sector 6	1,717,163	8993	2439	6211	4080	3,113,532	61,973	243,544
Sector 7	1951	1232	1080	1317	713	7789	2,504,998	1,463,935
Sector 8	5369	16,041	4810	9347	5589	19,784	123,812	671,126
Sector 9	20,808	38,509	5174	6668	2751	16,179	10,481	11,295
Sector 10	13,856	5495	2890	3663	2925	281,662	79,850	70,861
Sector 11	109,977	52,521	51,205	130,235	61,910	42,402	20,221	28,660
Sector 12	1,253,758	112,001	50,653	134,208	131,843	416,591	952,959	290,477
Sector 13	21,395	41,065	12,591	17,858	45,754	104,299	18,683	6145
Sector 14	3956	202,112	66,392	28,229	9823	17,025	7049	6725
Sector 15	62,749	89,344	10,448	42,737	23,321	69,875	18,890	23,717
Sector 16	89,960	242,070	112,622	168,367	114,025	51,617	101,413	30,318
Sector 17	44,849	30,045	9264	22,839	25,455	19,157	15,949	7025
Sector 18	4345	68,479	12,595	29,206	12,606	21,998	18,516	8782
Sector 19	1503	7435	1752	3080	1522	5564	6477	5196
Sector 20	17,360	39,635	12,849	14,043	4643	10,642	24,429	8399
Sector 21	49,490	32,845	1873	11,533	13,362	23,114	58,444	47,533
Sector 22	1469	393	22	1271	9727	11,758	20,255	3066
Sector 23	214,481	230,350	99,388	433,747	170,762	156,612	225,308	40,884
Sector 24	562	1822	1679	11,502	6722	8942	4071	1617
Sector 25	1819	3644	1068	2319	1385	6118	6148	2126
Sector 26	4292	6499	1537	942	510	4261	2000	1481
Sector 27	269,798	202,635	26,384	91,173	77,289	433,181	141,972	106,106
Sector 28	13,449	1645	218	572	589	3408	6934	10,385
Sector 29	59,317	15,472	2307	4148	3508	26,829	16,171	24,040
Sector 30	351,167	64,749	12,474	33,973	25,915	364,873	197,397	111,902
Sector 31	40,853	42,757	12,079	21,719	16,183	89,504	33,801	26,392
Sector 32	162,438	126,079	11,156	32,741	25,827	155,424	189,188	66,446
Sector 33	4121	6944	608	1460	1939	25,977	29,372	50,655
Sector 34	24,199	26,385	2473	9195	6369	249,532	32,598	102,024
Sector 35	28,064	6899	2488	779	354	12,259	5279	2579
Sector 36	141,677	45,776	15,332	33,795	14,915	51,987	20,229	17,017
Sector 37	47,094	6795	1109	1901	4339	4633	13,304	671
Sector 38	55,998	33,845	5084	5207	5013	18,213	9284	8247
Sector 39	6121	4582	251	937	1788	2077	916	521
Sector 40	9224	11,229	488	4496	2709	7526	3721	3142
Sector 41	3565	6750	1304	6060	5886	15,920	8706	7840
Sector 42	5055	1222	412	505	263	2113	956	633
Total intermediate input	7,047,288	2,715,935	591,363	1,511,047	928,806	11,896,762	6,680,684	3,755,925
Compensation of employees	9,069,222	936,425	199,457	156,417	144,546	925,254	664,485	474,768
Net taxes on production	18,426	522,353	226,487	100,928	87,955	1,043,602	264,449	123,198
Depreciation of fixed assets	445,059	178,965	133,150	36,917	37,474	286,740	174,527	55,920
Operating surplus	0	541,993	312,039	191,877	120,023	921,577	272,582	159,147
Total value added	9,532,708	2,179,737	871,133	486,139	389,998	3,177,173	1,376,042	813,034
Total input	16,579,996	4,895,671	1,462,496	1,997,186	1,318,804	15,073,936	8,056,726	4,568,959

Sector 9	Sector 10	Sector 11	Sector 12	Sector 13	Sector 14	Sector 15	Sector 16	Sector 17
348,326	217,940	157	747,616	6114	2230	1638	1753	1277
20,107	38,668	438,749	313,053	720,975	707,449	23,515	55,437	16,709
0	0	3,974,532	1,052,921	53,520	105,006	0	0	0
0	0	0	203,712	55,001	3,958,293	65,024	47,906	15,382
0	0	712	300,868	572,528	54,604	21,565	11,395	3667
15,006	18,064	32,860	362,054	38,346	73,532	18,189	53,236	56,682
25,842	70,422	360	132,935	20,681	2273	8109	22,839	26,597
44,437	28,985	2931	45,769	24,216	42,770	15,518	34,888	151,137
727,605	103,560	2165	38,195	73,539	13,661	84,752	52,392	177,995
38,607	926,933	3829	283,285	224,208	19,713	51,062	71,085	56,101
18,718	19,381	428,241	1,226,334	400,157	840,349	53,095	166,228	70,705
261,024	558,531	132,143	7,706,065	786,954	315,544	233,931	587,014	932,884
21,085	10,747	23,919	145,292	1,601,559	364,071	53,726	116,855	102,206
57,080	48,831	4513	218,875	275,565	5,037,241	1,688,721	3,416,783	1,879,747
60,468	65,440	22,701	206,667	277,707	157,916	560,424	682,187	286,846
35,881	53,143	90,667	298,643	263,529	571,573	235,757	3,658,820	1,243,112
11,014	31,286	15,907	56,063	47,807	95,062	43,981	245,227	3,851,937
8291	26,395	7987	53,124	37,836	60,075	39,955	889,608	557,369
1055	36,169	962	12,615	6165	4765	5982	338,971	172,225
2430	48,688	10,019	143,529	26,501	67,911	12,440	85,040	245,836
20,601	25,766	3112	116,684	78,656	72,842	31,757	112,413	178,288
7244	269,365	1391	84,748	139,023	493,246	37,997	192,903	10,126
93,825	135,692	116,456	1,190,369	841,773	1,344,428	424,366	363,322	282,756
1163	1112	10,331	48,070	10,969	25,591	5071	14,492	9596
1838	8512	2123	21,649	10,407	15,342	4971	5426	7767
719	1071	3402	8309	3623	6774	1049	5618	5749
78,729	90,570	134,483	516,304	387,647	343,583	108,866	465,120	266,402
2125	2752	210	19,625	3236	7165	1086	14,416	3444
10,864	10,808	21,718	74,178	24,212	205,389	18,391	51,323	40,498
73,458	141,086	82,890	513,515	308,670	412,065	132,366	415,976	785,696
12,196	20,014	12,358	104,530	63,766	58,686	51,730	125,573	50,077
50,851	69,365	41,364	400,235	356,345	429,888	72,324	172,863	182,001
12,297	21,161	1100	50,768	26,118	5521	24,670	45,360	27,065
23,939	35,971	32,571	322,055	69,210	32,735	20,863	143,443	213,700
1592	2381	1882	49,788	7753	32,879	4585	56,467	131,937
5194	20,877	11,593	122,132	41,223	76,262	32,687	97,237	143,471
420	14,352	1366	23,612	3283	6546	769	3358	4914
2717	7879	6005	42,533	28,988	42,100	8068	19,121	18,940
432	576	273	3281	1604	1604	719	5161	3251
2420	4812	1765	12,878	17,512	18,378	12,380	38,587	30,748
3243	4453	1559	24,130	15,879	12,075	5332	22,780	14,653
365	615	314	2619	1714	2136	735	2205	2341
2,103,204	3,192,372	5,681,623	17,299,629	7,954,520	16,139,271	4,218,169	12,910,830	12,261,833
201,409	281,225	229,000	1,294,995	650,458	1,137,033	325,236	1,423,417	792,822
83,628	108,312	756,041	715,192	354,193	550,474	135,399	653,165	526,715
48,104	106,907	169,710	554,275	252,689	678,152	94,972	378,934	229,796
121,509	166,764	251,700	1,072,744	582,690	777,583	227,795	1,007,076	652,699
454,650	663,209	1,406,451	3,637,206	1,840,031	3,143,242	783,402	3,462,592	2,202,032
2,557,854	3,855,582	7,088,074	20,936,835	9,794,551	19,282,513	5,001,571	16,373,422	14,463,865

(continued)

Table 10.13 (continued)

	Sector 18	Sector 19	Sector 20	Sector 21	Sector 22	Sector 23	Sector 24	Sector 25
Sector 1	199	0	43	352,724	177	177	72	0
Sector 2	8064	3097	1469	19,285	815	1,744,122	20,895	464
Sector 3	0	0	0	27	1	62,011	204,733	0
Sector 4	19,719	182	2325	268	538	293	0	0
Sector 5	4232	71	103	8099	68	2665	17	35
Sector 6	40,066	40,823	3177	39,823	700	30,199	1622	912
Sector 7	17,957	2814	1414	263,481	77	510	51	104
Sector 8	20,755	18,121	2912	43,437	120	32,900	4881	1719
Sector 9	27,323	17,121	2865	89,200	90	6126	509	513
Sector 10	109,335	83,631	10,080	52,259	123	18,871	1096	910
Sector 11	42,205	28,165	4580	21,365	1203	291,537	9046	3612
Sector 12	862,695	712,465	89,227	222,601	11,751	30,787	2794	17,217
Sector 13	188,538	175,635	33,336	61,152	1447	21,107	1004	1018
Sector 14	2,580,293	317,839	40,253	277,983	5810	9567	1072	403
Sector 15	413,736	228,860	38,263	123,029	609	58,797	4144	9397
Sector 16	469,320	104,087	27,497	11,682	359	99,577	3512	3707
Sector 17	24,842	34,073	5644	29,832	124	110,899	9532	4327
Sector 18	1,354,780	289,729	48,753	18,274	914	548,976	1009	1172
Sector 19	571,130	4,614,919	232,454	7269	46	6092	228	281
Sector 20	95,006	497,506	114,674	6309	54	133,440	2729	1891
Sector 21	96,667	81,966	15,747	320,211	108	6621	325	360
Sector 22	74,038	9805	104	579	5318	1086	0	0
Sector 23	142,717	149,326	12,553	39,282	2473	3,546,203	8696	81,717
Sector 24	5775	5423	673	2788	121	11,532	24,236	1136
Sector 25	4042	4578	720	1981	115	12,353	233	7308
Sector 26	2708	2047	286	457	57	3402	323	424
Sector 27	215,421	164,346	21,177	69,328	4218	120,955	11,504	3305
Sector 28	8247	4418	1151	825	13	1172	57	108
Sector 29	34,913	110,600	4099	19,693	168	60,779	621	1289
Sector 30	333,960	432,385	25,320	58,111	2583	78,616	7305	2116
Sector 31	58,508	28,212	8542	20,902	516	28,447	1996	2779
Sector 32	159,762	419,523	8497	31,908	2376	412,050	14,423	20,081
Sector 33	31,991	27,938	4819	19,822	180	4348	457	184
Sector 34	231,658	121,997	12,961	24,569	437	44,239	1916	1716
Sector 35	41,939	77,205	5464	2859	45	6604	114	72
Sector 36	59,917	56,941	6649	6261	221	79,605	1003	531
Sector 37	2165	1383	102	467	32	26,512	190	19,975
Sector 38	8058	6834	837	1238	478	76,032	1254	2751
Sector 39	1521	1031	205	487	5	1535	102	268
Sector 40	18,198	5472	1801	456	87	11,852	1221	77
Sector 41	11,073	9250	1084	3225	80	22,384	822	1047
Sector 42	1166	1110	140	416	0	1417	37	70
Total intermediate input	8,394,638	8,890,929	792,004	2,273,963	44,658	7,766,400	345,782	194,997
Compensation of employees	599,889	775,316	95,949	263,547	6478	794,487	46,543	71,049
Net taxes on production	232,133	157,490	27,058	177,585	3880	483,123	10,799	14,001
Depreciation of fixed assets	163,697	243,424	19,572	59,481	2913	995,731	14,452	61,188
Operating surplus	441,465	290,102	48,984	145,160	395,289	323,242	21,130	8204
Total value added	1,437,184	1,466,331	191,563	645,774	408,560	2,596,583	92,924	154,442
Total input	9,831,822	10,357,261	983,567	2,919,737	453,217	10,362,983	438,706	349,439

Sector 26	Sector 27	Sector 28	Sector 29	Sector 30	Sector 31	Sector 32	Sector 33	Sector 34
119,263	258,336	0	0	2377	541,918	0	499	12,288
21,536	37,276	392	0	2995	3129	0	1290	7302
0	13,447	0	0	0	0	0	0	0
0	0	0	0	0	0	0	0	0
274,221	2805	0	0	0	0	0	0	0
108,368	85,147	696	9795	51,089	1,631,024	22,482	17,095	61,268
22,915	13,986	423	492	11,844	23,082	1147	6765	66,937
103,312	77,305	2016	7602	86,757	17,711	21,152	27,443	64,579
475,227	13,187	1393	3850	16,213	6489	6029	7361	13,080
45,362	39,723	7415	125,240	90,970	15,966	155,900	35,680	464,020
349,927	2,017,143	8252	7744	32,872	23,093	50,724	79,508	164,861
837,375	190,400	928	21,285	89,093	35,790	27,223	101,617	185,518
6,150,796	22,924	967	1090	2675	9623	1289	1802	2749
3,383,968	55,713	0	41	4811	211	0	5458	3956
882,621	60,212	218	8077	3950	5485	7256	36,930	103,974
803,536	260,535	2930	32,496	21,226	5507	34,169	21,562	33,995
112,199	761,533	22,114	28,323	86,087	15,850	23,354	23,231	157,580
1,026,779	30,420	495	276,590	67,018	33,641	16,496	67,497	145,223
65,437	27,265	2454	240,049	38,405	1914	6003	18,236	299,179
59,276	17,228	331	36,016	3206	147	30,808	11,891	15,419
88,362	9901	707	4257	7520	2924	8358	7862	31,387
0	0	0	0	0	0	0	0	0
202,734	129,663	3820	100,173	115,241	100,328	75,672	36,428	49,384
3211	10,163	267	2286	2016	17,133	399	642	5000
6516	6516	449	4909	3771	8150	3796	1110	3381
287,924	46,541	11,823	6809	25,427	10,743	11,615	117,117	5056
2,528,842	913,176	39,864	123,174	690,851	57,101	199,125	45,928	254,833
5853	3398	11,729	5248	13,303	2697	34,564	3885	24,354
393,013	77,709	5682	174,071	80,162	14,429	228,996	23,713	19,361
667,844	139,941	20,966	126,201	50,018	99,278	34,051	25,964	120,113
271,436	172,382	2011	78,906	189,247	31,773	296,192	99,136	341,705
256,811	638,157	3276	83,159	336,424	77,705	645,581	260,598	234,992
15,817	52,789	4257	105,052	218,983	49,769	257,505	80,059	86,403
97,826	60,075	10,853	201,399	480,458	56,092	489,822	205,389	226,743
20,332	3327	151	6721	15,371	42	11,268	1281	2586
313,960	12,732	50	19,638	37,997	865	5571	11,662	3863
1334	3264	0	1124	2279	1539	5162	1954	44,483
38,011	246,153	446	25,722	91,125	43,813	31,221	44,773	80,202
3378	8077	280	4705	5892	1223	15,633	1436	2491
8058	5835	54	1198	1584	791	5559	2101	846
37,708	22,268	2092	22,479	38,399	12,131	48,107	32,155	22,224
3462	3050	92	1530	2453	723	4586	4026	986
20,094,551	6,549,702	169,892	1,897,451	3,020,109	2,959,829	2,816,816	1,471,085	3,362,324
4,031,661	1,751,649	80,181	540,231	2,033,013	1,062,875	1,659,548	498,929	993,864
1,119,624	351,947	7228	158,910	2,016,648	201,841	636,605	719,209	252,053
391,680	778,999	18,896	521,617	472,563	238,265	122,557	2,202,847	483,122
1,535,840	1,387,494	231	889,948	2,879,769	252,452	2,808,502	1,145,152	142,263
7,078,805	4,270,090	106,535	2,110,705	7,401,993	1,755,433	5,227,212	4,566,137	1,871,303
27,173,356	10,819,792	276,427	4,008,157	10,422,102	4,715,262	8,044,028	6,037,222	5,233,627

(continued)

Table 10.13 (continued)

	Sector 35	Sector 36	Sector 37	Sector 38	Sector 39	Sector 40	Sector 41	Sector 42
Sector 1	25,215	9179	23,042	31,115	7458	19,191	5832	0
Sector 2	1159	3088	4938	10,840	7924	16,183	967	16,777
Sector 3	0	0	0	0	0	0	0	0
Sector 4	6310	1434	1663	0	0	0	0	0
Sector 5	103	71	620	140	1440	2048	585	2367
Sector 6	11,910	5651	10,026	60,441	31,429	33,411	66,998	21,741
Sector 7	13,552	1524	4081	40,056	6832	22,294	3509	24,380
Sector 8	5175	3857	12,320	18,754	4375	54,957	19,915	79,710
Sector 9	2271	2936	12,155	25,095	47,545	29,895	14,313	97,114
Sector 10	7467	32,515	6465	44,423	136,213	49,885	134,478	155,659
Sector 11	6121	47,166	35,003	48,862	30,722	16,353	7767	86,813
Sector 12	72,786	50,826	64,117	231,951	35,028	1,707,552	49,074	60,498
Sector 13	5273	4473	13,268	2830	19,106	22,329	7660	29,149
Sector 14	3358	16,942	7184	6831	2913	192	551	0
Sector 15	17,069	49,293	7601	13,218	8355	8362	4752	7076
Sector 16	10,071	18,540	19,946	6463	7769	178,891	3345	13,892
Sector 17	5949	33,663	33,543	101,569	25,911	4442	5592	77,656
Sector 18	29,545	96,110	24,863	43,680	11,477	12,742	6598	13,607
Sector 19	26,403	86,745	13,705	65,637	12,582	2947	7137	45,978
Sector 20	23,626	52,314	13,122	25,926	37,893	9335	4482	10,510
Sector 21	3055	5100	4997	21,169	6631	5024	2413	13,531
Sector 22	0	0	0	0	0	0	0	0
Sector 23	13,642	15,954	27,154	29,533	71,005	46,469	10,745	81,185
Sector 24	221	587	872	1718	2505	1859	801	1797
Sector 25	1040	2725	1776	4369	5266	2665	947	3419
Sector 26	2383	6964	23,795	20,419	18,712	77,446	9738	83,620
Sector 27	27,783	136,940	22,013	55,943	99,745	20,032	35,174	248,358
Sector 28	1126	2751	911	3708	8141	1282	2724	63,691
Sector 29	3812	11,406	8463	9103	31,564	36,398	9650	111,079
Sector 30	13,968	27,114	13,608	47,616	28,661	119,351	24,553	61,029
Sector 31	25,967	94,722	22,753	41,951	93,394	38,503	37,239	379,333
Sector 32	2886	73,046	30,399	56,478	97,838	22,682	17,885	94,822
Sector 33	3055	23,642	2876	74,932	27,816	7632	15,389	32,234
Sector 34	6342	34,800	14,072	37,982	67,510	28,427	26,228	53,253
Sector 35	16,080	5316	918	72	7790	1812	1203	80
Sector 36	9726	144,227	4915	2052	7931	5586	815	8303
Sector 37	420	488	22,670	3865	4919	1008	1066	13,810
Sector 38	1995	20,744	23,843	103,716	27,352	21,810	11,251	85,184
Sector 39	2253	2285	1574	817	16,789	3802	1142	45,418
Sector 40	1768	2150	3552	861	5516	9844	1711	16,263
Sector 41	10,423	11,750	6941	10,706	22,493	5732	36,543	74,093
Sector 42	172	737	376	616	726	574	247	1593
Total intermediate input	421,480	1,139,776	546,141	1,305,458	1,087,280	2,648,948	591,019	2,215,022
Compensation of employees	163,705	616,746	207,719	913,993	2,749,558	1,229,904	345,365	3,393,816
Net taxes on production	10,455	103,148	17,051	126,263	17,691	12,832	59,542	18,425
Depreciation of fixed assets	30,719	117,320	83,260	74,310	326,797	149,302	92,460	499,638
Operating surplus	35,665	316,910	62,492	257,344	46,128	88,070	108,566	107,404
Total value added	240,544	1,154,124	370,522	1,371,911	3,140,176	1,480,108	605,932	4,019,284
Total input	662,024	2,293,900	916,662	2,677,369	4,227,456	4,129,056	1,196,952	6,234,306

Total intermediate use	Rural household consumption expenditure	Urban household consumption expenditure	Government consumption expenditure	Gross fixed capital formation	Increase in inventories	Export	Import	Total output
12,805,148	1,437,032	1,796,913	129,149	889,194	221,672	441,783	714,121	17,006,772
5,243,005	23,551	17,344	0	49,437	−87,034	18,932	398,260	4,866,976
5,488,673	0	0	0	18,273	−55,601	392,376	4,573,200	1,270,521
4,545,123	0	0	0	75,082	−19,071	7635	2,479,457	2,129,312
1,359,980	0	0	0	11,322	−22,672	23,318	87,963	1,283,985
8,211,795	1,603,376	4,679,627	0	281,703	255,511	220,825	242,567	15,010,270
4,843,299	54,872	84,971	0	37,465	−27,421	3,999,518	971,549	8,021,155
1,978,315	338,577	1,626,948	0	89,770	90,604	579,653	75,977	4,627,890
2,303,629	31,238	136,305	0	212,660	−1685	269,856	409,926	2,542,076
3,959,712	20,445	136,824	0	41,852	−34,103	226,932	567,272	3,784,391
7,134,984	29,426	264,775	0	−9214	−134,219	119,976	469,040	6,936,688
20,567,178	188,146	613,391	0	234,748	−131,943	944,203	1,804,718	20,611,004
9,488,491	11,901	75,147	0	90,722	−135,481	454,501	303,830	9,681,450
19,694,017	0	0	0	217,986	−188,641	365,674	968,811	19,120,224
4,762,722	20,023	108,369	0	108,289	−54,600	345,348	263,504	5,026,647
9,556,133	1451	21,093	0	6,453,911	89,393	1,974,660	1,903,898	16,192,743
6,310,739	164,002	793,299	0	6,584,354	119,509	2,074,682	1,609,704	14,436,880
6,023,557	146,195	568,767	0	2,067,926	34,952	1,769,540	839,665	9,771,273
7,003,935	126,934	507,394	0	1,251,722	187,744	4,709,877	3,475,278	10,312,329
1,977,531	14,831	49,377	0	213,763	−24,607	537,278	1,827,870	940,303
1,623,516	81,311	456,212	0	580,243	9937	244,963	108,984	2,887,197
1,374,933	0	0	0	0	6321	4801	941,769	444,286
11,456,617	60,900	224,622	0	0	−1,392,953	15,795	1998	10,362,983
266,471	19,096	143,877	0	2878	358	0	0	432,681
194,797	14,756	135,988	0	0	1226	0	0	346,767
833,673	0	319,971	0	25,560,374	72,498	442,239	121,689	27,107,065
9,849,380	157,251	605,991	497,073	77,155	−179,449	369,632	665,708	10,711,324
296,617	2239	15,181	0	0	2855	7959	38,926	285,925
2,079,947	183,561	791,229	0	815,343	43,230	191,389	93,073	4,011,626
6,588,846	566,266	1,739,955	0	701,673	47,359	916,841	0	10,560,940
3,148,770	356,744	1,391,772	0	0	98,640	−108,555	240,032	4,647,340
6,545,894	212,384	1,122,071	89,448	0	108,023	93,333	83,356	8,087,797
1,483,086	639,518	2,212,682	0	1,495,909	120,162	0	0	5,951,357
3,884,026	30,082	429,738	205,994	0	59,141	957,283	249,350	5,316,915
576,619	0	0	326,375	0	2978	23,720	271,438	658,255
1,692,426	1776	22,731	295,048	244,471	−1537	0	0	2,254,914
298,676	8563	105,245	524,220	0	−2160	0	0	934,544
1,312,084	195,399	1,134,893	0	0	35,106	38,435	81,084	2,634,834
156,446	399,444	890,980	2,827,255	0	−27,196	5016	31,010	4,220,934
288,872	145,149	1,689,613	2,003,488	0	−23,344	3828	2677	4,104,930
625,344	39,141	210,699	332,406	0	12,312	42,790	53,539	1,209,154
54,508	0	0	6,143,595	0	−12,349	19,697	16,050	6,189,401

Table 10.14 China's national input–output table in 2023

	Sector 1	Sector 2	Sector 3	Sector 4	Sector 5	Sector 6	Sector 7	Sector 8
Sector 1	2,353,957	47,988	26	3825	559	6,238,465	1,705,743	236,687
Sector 2	15,982	836,743	11,652	9467	3486	43,027	46,894	20,631
Sector 3	0	0	22,475	0	0	0	0	0
Sector 4	0	0	0	167,074	0	0	0	0
Sector 5	2824	1789	714	2702	82,561	5664	0	0
Sector 6	1,897,106	9016	2439	6211	4081	3,249,338	64,482	258,512
Sector 7	2131	1234	1080	1317	713	8129	2,606,768	1,538,204
Sector 8	5825	16,077	4810	9347	5590	20,647	128,256	704,010
Sector 9	25,685	38,841	5174	6668	2753	16,884	11,399	12,859
Sector 10	16,509	5523	2890	3663	2926	293,948	85,470	78,419
Sector 11	141,968	52,917	51,205	130,235	61,951	44,252	22,389	33,390
Sector 12	1,593,408	112,738	50,653	134,208	131,925	434,762	1,047,873	332,689
Sector 13	25,444	41,289	12,591	17,858	45,776	108,849	20,016	6797
Sector 14	4979	203,706	66,392	28,229	9830	17,768	7773	7746
Sector 15	68,427	89,641	10,448	42,737	23,327	72,922	19,610	25,065
Sector 16	110,576	243,647	112,622	168,367	114,091	53,869	110,286	34,242
Sector 17	50,014	30,148	9264	22,839	25,463	19,993	16,670	7476
Sector 18	5025	68,789	12,595	29,206	12,611	22,957	19,708	9575
Sector 19	1799	7477	1752	3080	1523	5807	6992	5785
Sector 20	22,442	40,271	12,849	14,043	4648	11,106	26,266	9848
Sector 21	51,099	32,885	1873	11,533	13,364	24,122	59,241	48,530
Sector 22	2218	402	22	1271	9747	12,271	23,385	4146
Sector 23	242,388	231,141	99,388	433,747	170,818	163,443	237,492	44,088
Sector 24	670	1830	1679	11,502	6725	9332	4366	1789
Sector 25	2063	3655	1068	2319	1385	6385	6451	2283
Sector 26	4855	6522	1537	942	510	4447	2108	1594
Sector 27	323,046	203,788	26,384	91,173	77,330	452,076	153,143	117,833
Sector 28	14,228	1648	218	572	589	3557	7159	10,771
Sector 29	66,924	15,523	2307	4148	3509	27,999	17,018	25,802
Sector 30	384,440	64,975	12,474	33,973	25,924	380,788	206,236	118,629
Sector 31	51,767	43,130	12,079	21,719	16,196	93,408	37,151	30,306
Sector 32	180,831	126,441	11,156	32,741	25,834	162,204	197,061	70,722
Sector 33	5939	7028	608	1460	1941	27,110	34,060	63,521
Sector 34	26,564	26,460	2473	9195	6371	260,416	34,037	108,022
Sector 35	34,210	6946	2488	779	354	12,794	5726	2906
Sector 36	182,264	46,139	15,332	33,795	14,926	54,254	22,468	19,733
Sector 37	51,803	6817	1109	1901	4340	4835	14,001	714
Sector 38	77,189	34,220	5084	5207	5018	19,008	10,564	10,027
Sector 39	6865	4597	251	937	1789	2167	962	558
Sector 40	10,339	11,262	488	4496	2710	7854	3898	3357
Sector 41	4078	6778	1304	6060	5888	16,614	9283	8499
Sector 42	5682	1226	412	505	263	2205	1002	678
Total intermediate input	8,073,564	2,731,249	591,363	1,511,047	929,346	12,415,674	7,033,411	4,016,441
Compensation of employees	9,845,206	941,774	199,457	156,417	144,749	965,612	691,481	498,031
Net taxes on production	20,003	525,337	226,487	100,928	88,079	1,089,122	275,193	129,235
Depreciation of fixed assets	483,140	179,987	133,150	36,917	37,526	299,247	181,617	58,660
Operating surplus	0	545,089	312,039	191,877	120,191	961,775	283,656	166,945
Total value added	10,348,349	2,192,187	871,133	486,139	390,545	3,315,755	1,431,947	852,871
Total input	18,421,913	4,923,436	1,462,496	1,997,186	1,319,891	15,731,429	8,465,357	4,869,312

Sector 9	Sector 10	Sector 11	Sector 12	Sector 13	Sector 14	Sector 15	Sector 16	Sector 17
348,587	218,902	157	748,498	6145	2231	1652	1762	1292
20,142	39,010	439,636	313,727	729,288	707,970	24,059	55,730	17,283
0	0	3,982,569	1,319,777	117,038	112,853	0	0	0
0	0	0	204,712	56,281	3,967,411	68,729	48,160	16,164
0	0	714	302,547	588,890	54,704	22,811	11,455	4041
15,019	18,147	32,926	362,436	38,558	73,558	18,409	53,518	57,504
25,861	70,723	360	133,066	20,783	2274	8191	22,960	26,890
44,469	29,104	2936	45,817	24,340	42,784	15,668	35,073	152,808
729,841	104,990	2169	38,320	74,853	13,678	88,063	52,670	188,330
38,677	937,463	3837	283,860	226,547	19,728	52,244	71,461	57,907
18,764	19,638	429,106	1,229,879	406,398	841,225	54,910	167,108	74,555
261,586	565,290	132,411	7,728,942	797,794	315,814	240,294	590,121	977,668
21,121	10,842	23,968	145,590	1,619,536	364,349	55,010	117,474	105,951
57,229	49,508	4523	219,561	280,231	5,043,194	1,751,307	3,434,870	1,977,765
60,535	65,798	22,747	206,916	279,662	157,998	570,351	685,798	291,709
35,957	53,739	90,851	299,419	267,148	572,107	242,652	3,678,187	1,299,822
11,026	31,463	15,939	56,143	48,161	95,106	44,608	246,525	3,938,144
8303	26,604	8003	53,224	38,207	60,113	40,789	894,317	573,774
1056	36,520	964	12,645	6241	4769	6131	340,765	178,701
2449	49,425	10,040	144,114	27,122	68,087	13,483	85,490	277,745
20,609	25,841	3119	116,731	78,837	72,856	31,965	113,008	178,880
7300	277,629	1394	85,498	145,346	495,311	42,570	193,924	12,019
93,940	136,598	116,692	1,192,236	848,178	1,345,065	430,979	365,245	288,614
1165	1122	10,352	48,168	11,082	25,608	5181	14,568	9915
1840	8565	2128	21,678	10,475	15,349	5039	5455	7922
720	1079	3409	8322	3652	6777	1066	5648	5872
78,871	91,512	134,755	517,519	392,425	343,859	111,611	467,582	276,848
2126	2762	210	19,645	3252	7167	1097	14,493	3479
10,876	10,880	21,762	74,292	24,388	205,477	18,651	51,595	41,332
73,538	141,891	83,058	514,279	311,057	412,261	134,561	418,178	800,685
12,229	20,282	12,383	104,875	64,915	58,758	53,714	126,238	53,400
50,903	69,813	41,447	400,767	358,554	430,044	73,165	173,778	185,379
12,349	21,608	1102	51,018	26,787	5531	26,031	45,600	29,380
23,961	36,166	32,637	322,486	69,672	32,748	21,134	144,202	217,192
1595	2409	1886	49,928	7865	32,907	4704	56,765	137,772
5207	21,174	11,616	122,555	41,915	76,337	33,688	97,752	150,799
420	14,438	1369	23,650	3308	6548	780	3376	5006
2727	8022	6017	42,720	29,665	42,166	8471	19,222	20,528
433	580	274	3286	1616	1605	729	5188	3312
2423	4838	1769	12,895	17,623	18,385	12,536	38,792	31,252
3247	4489	1562	24,179	16,034	12,082	5428	22,900	15,054
365	619	314	2622	1725	2137	745	2217	2381
2,107,465	3,229,479	5,693,111	17,608,543	8,121,593	16,168,931	4,343,210	12,979,171	12,695,073
202,028	284,419	229,463	1,298,840	657,759	1,138,377	330,996	1,430,951	810,565
83,885	109,543	757,570	717,315	358,169	551,124	137,798	656,622	538,503
48,252	108,122	170,053	555,920	255,526	678,953	96,654	380,940	234,939
121,882	168,659	252,209	1,075,929	589,230	778,502	231,830	1,012,407	667,306
456,047	670,743	1,409,295	3,648,004	1,860,684	3,146,956	797,279	3,480,921	2,251,313
2,563,512	3,900,222	7,102,406	21,256,547	9,982,278	19,315,888	5,140,488	16,460,092	14,946,387

(continued)

Table 10.14 (continued)

	Sector 18	Sector 19	Sector 20	Sector 21	Sector 22	Sector 23	Sector 24	Sector 25
Sector 1	200	0	43	369,415	177	179	73	0
Sector 2	8165	3116	1470	20,197	815	1,769,049	21,146	480
Sector 3	0	0	0	29	1	62,897	207,198	0
Sector 4	20,206	185	2332	280	538	297	0	0
Sector 5	4371	72	103	8483	68	2703	17	37
Sector 6	40,329	40,955	3178	41,708	700	30,631	1641	944
Sector 7	18,059	2821	1414	275,948	77	518	51	108
Sector 8	20,870	18,168	2914	45,492	120	33,370	4940	1780
Sector 9	27,874	17,287	2869	93,421	90	6214	516	531
Sector 10	110,711	84,111	10,087	54,732	123	19,141	1110	942
Sector 11	42,984	28,417	4584	22,376	1203	295,704	9155	3739
Sector 12	876,731	718,131	89,302	233,134	11,751	31,227	2828	17,823
Sector 13	190,962	176,717	33,357	64,045	1447	21,409	1016	1054
Sector 14	2,630,082	320,698	40,301	291,137	5810	9704	1085	417
Sector 15	417,214	229,716	38,283	128,851	609	59,637	4194	9728
Sector 16	476,775	104,867	27,520	12,235	359	101,000	3555	3837
Sector 17	25,041	34,200	5647	31,244	124	112,484	9647	4480
Sector 18	1,370,111	291,229	48,785	19,138	914	556,822	1022	1214
Sector 19	579,006	4,645,404	232,655	7613	46	6179	231	291
Sector 20	98,575	508,215	114,796	6607	54	135,347	2762	1958
Sector 21	96,951	82,078	15,749	335,363	108	6716	329	373
Sector 22	78,617	10,106	104	606	5318	1102	0	0
Sector 23	143,980	149,977	12,562	41,141	2473	3,596,885	8801	84,592
Sector 24	5845	5454	674	2920	121	11,697	24,528	1176
Sector 25	4074	4595	721	2075	115	12,530	235	7566
Sector 26	2732	2056	286	479	57	3450	327	439
Sector 27	218,400	165,366	21,193	72,609	4218	122,684	11,642	3422
Sector 28	8290	4428	1152	864	13	1189	57	112
Sector 29	35,200	111,026	4102	20,625	168	61,648	628	1334
Sector 30	336,823	434,056	25,340	60,861	2583	79,739	7393	2191
Sector 31	59,695	28,475	8550	21,891	516	28,853	2020	2877
Sector 32	160,855	420,851	8502	33,418	2376	417,939	14,596	20,788
Sector 33	32,952	28,351	4828	20,760	180	4410	463	190
Sector 34	233,330	122,393	12,968	25,731	437	44,872	1939	1776
Sector 35	42,560	77,775	5470	2994	45	6699	116	74
Sector 36	60,995	57,448	6657	6558	221	80,743	1015	550
Sector 37	2183	1388	102	489	32	26,891	192	20,678
Sector 38	8275	6923	838	1296	478	77,119	1269	2848
Sector 39	1534	1035	206	510	5	1557	103	277
Sector 40	18,331	5492	1802	478	87	12,022	1236	80
Sector 41	11,187	9292	1085	3378	80	22,704	832	1084
Sector 42	1175	1114	140	436	0	1437	37	72
Total intermediate input	8,522,252	8,953,987	792,670	2,381,568	44,658	7,877,398	349,946	201,859
Compensation of employees	606,678	780,437	96,051	276,018	6478	805,842	47,104	73,549
Net taxes on production	234,760	158,530	27,087	185,989	3880	490,028	10,929	14,494
Depreciation of fixed assets	165,549	245,032	19,593	62,296	2913	1,009,962	14,626	63,341
Operating surplus	446,461	292,018	49,036	152,029	395,289	327,862	21,384	8492
Total value added	1,453,448	1,476,018	191,767	676,332	408,560	2,633,694	94,043	159,876
Total input	9,975,700	10,430,004	984,437	3,057,900	453,217	10,511,092	443,989	361,735

Sector 26	Sector 27	Sector 28	Sector 29	Sector 30	Sector 31	Sector 32	Sector 33	Sector 34
122,234	262,627	0	0	2493	553,352	0	507	13,010
22,073	37,895	423	0	3141	3195	0	1310	7731
0	13,670	0	0	0	0	0	0	0
0	0	0	0	0	0	0	0	0
281,052	2851	0	0	0	0	0	0	0
111,068	86,561	752	10,230	53,578	1,665,437	23,603	17,360	64,866
23,486	14,218	457	514	12,420	23,569	1204	6870	70,868
105,886	78,589	2177	7940	90,983	18,084	22,207	27,869	68,371
487,067	13,406	1504	4021	17,003	6626	6329	7475	13,848
46,492	40,383	8008	130,808	95,401	16,303	163,671	36,234	491,270
358,645	2,050,649	8912	8089	34,473	23,580	53,253	80,743	174,542
858,237	193,563	1002	22,231	93,433	36,545	28,580	103,194	196,412
6,304,035	23,305	1045	1138	2805	9826	1353	1830	2910
3,468,275	56,639	0	43	5045	216	0	5543	4188
904,610	61,212	235	8436	4143	5600	7618	37,503	110,080
823,555	264,863	3164	33,941	22,260	5623	35,872	21,897	35,991
114,994	774,182	23,882	29,582	90,280	16,185	24,518	23,592	166,834
1,052,360	30,926	534	288,888	70,283	34,351	17,319	68,545	153,751
67,067	27,718	2650	250,722	40,276	1955	6302	18,519	316,749
60,753	17,514	358	37,617	3362	150	32,344	12,075	16,325
90,563	10,065	763	4446	7887	2986	8775	7984	33,230
0	0	0	0	0	0	0	0	0
207,785	131,817	4125	104,627	120,855	102,444	79,444	36,994	52,285
3291	10,332	288	2388	2114	17,495	419	651	5294
6679	6624	485	5127	3955	8322	3986	1127	3579
295,098	47,315	12,769	7111	26,665	10,970	12,194	118,935	5353
2,591,844	928,344	43,051	128,650	724,504	58,305	209,052	46,641	269,798
5999	3454	12,667	5482	13,951	2754	36,287	3945	25,785
402,804	79,000	6136	181,811	84,067	14,733	240,412	24,081	20,498
684,482	142,265	22,643	131,812	52,454	101,372	35,748	26,367	127,167
278,198	175,246	2172	82,415	198,465	32,443	310,957	100,675	361,772
263,209	648,757	3538	86,857	352,812	79,345	677,763	264,643	248,792
16,211	53,666	4597	109,723	229,650	50,819	270,342	81,302	91,477
100,263	61,073	11,721	210,354	503,862	57,275	514,239	208,577	240,059
20,838	3382	164	7020	16,120	43	11,829	1301	2738
321,781	12,943	53	20,511	39,848	883	5849	11,843	4089
1367	3318	0	1174	2390	1572	5419	1985	47,096
38,958	250,242	482	26,865	95,564	44,737	32,777	45,468	84,912
3462	8212	303	4914	6179	1249	16,413	1458	2638
8258	5932	58	1251	1662	808	5836	2133	896
38,648	22,638	2259	23,478	40,270	12,387	50,506	32,654	23,529
3549	3101	100	1598	2572	738	4815	4089	1044
20,595,180	6,658,496	183,477	1,981,814	3,167,224	3,022,278	2,957,233	1,493,919	3,559,776
4,132,105	1,780,745	86,592	564,250	2,132,045	1,085,301	1,742,275	506,674	1,052,229
1,147,518	357,793	7805	165,975	2,114,883	206,100	668,339	730,372	266,855
401,438	791,938	20,407	544,808	495,582	243,292	128,667	2,237,040	511,494
1,574,103	1,410,541	250	929,516	3,020,048	257,778	2,948,504	1,162,928	150,618
7,255,164	4,341,018	115,054	2,204,549	7,762,558	1,792,471	5,487,785	4,637,013	1,981,195
27,850,344	10,999,514	298,531	4,186,363	10,929,782	4,814,750	8,445,017	6,130,933	5,540,971

(continued)

Table 10.14 (continued)

	Sector 35	Sector 36	Sector 37	Sector 38	Sector 39	Sector 40	Sector 41	Sector 42
Sector 1	25,465	9248	24,280	31,669	7769	20,014	6060	0
Sector 2	1170	3111	5203	11,033	8254	16,877	1005	17,476
Sector 3	0	0	0	0	0	0	0	0
Sector 4	6372	1445	1752	0	0	0	0	0
Sector 5	104	71	654	142	1500	2136	608	2465
Sector 6	12,028	5693	10,564	61,517	32,737	34,845	69,619	22,646
Sector 7	13,687	1535	4301	40,769	7116	23,251	3646	25,396
Sector 8	5226	3886	12,982	19,088	4557	57,315	20,694	83,029
Sector 9	2294	2958	12,808	25,541	49,524	31,178	14,873	101,157
Sector 10	7541	32,759	6813	45,214	141,883	52,025	139,739	162,140
Sector 11	6182	47,521	36,884	49,731	32,001	17,055	8071	90,427
Sector 12	73,509	51,208	67,562	236,079	36,486	1,780,806	50,994	63,017
Sector 13	5325	4507	13,980	2880	19,901	23,287	7959	30,363
Sector 14	3392	17,070	7570	6952	3034	200	572	0
Sector 15	17,239	49,663	8009	13,453	8703	8721	4938	7370
Sector 16	10,171	18,679	21,018	6578	8092	186,565	3476	14,471
Sector 17	6008	33,916	35,345	103,377	26,989	4633	5811	80,889
Sector 18	29,838	96,832	26,199	44,458	11,955	13,288	6856	14,174
Sector 19	26,665	87,397	14,442	66,806	13,106	3073	7416	47,893
Sector 20	23,860	52,707	13,827	26,387	39,470	9735	4657	10,948
Sector 21	3085	5139	5266	21,546	6907	5240	2508	14,095
Sector 22	0	0	0	0	0	0	0	0
Sector 23	13,777	16,074	28,613	30,059	73,961	48,463	11,166	84,565
Sector 24	223	591	919	1748	2609	1939	832	1872
Sector 25	1050	2745	1871	4447	5485	2779	984	3562
Sector 26	2407	7017	25,074	20,782	19,491	80,769	10,119	87,102
Sector 27	28,059	137,970	23,196	56,938	103,897	20,892	36,551	258,699
Sector 28	1137	2772	960	3774	8480	1337	2830	66,343
Sector 29	3850	11,492	8918	9265	32,878	37,960	10,027	115,704
Sector 30	14,107	27,318	14,339	48,463	29,854	124,471	25,514	63,570
Sector 31	26,224	95,434	23,976	42,698	97,281	40,155	38,696	395,128
Sector 32	2914	73,595	32,032	57,483	101,910	23,655	18,585	98,771
Sector 33	3085	23,820	3031	76,266	28,973	7960	15,991	33,576
Sector 34	6405	35,061	14,828	38,658	70,320	29,646	27,254	55,470
Sector 35	16,240	5356	967	74	8115	1890	1250	84
Sector 36	9823	145,311	5179	2089	8261	5825	847	8649
Sector 37	424	492	23,888	3934	5124	1051	1108	14,385
Sector 38	2014	20,900	25,124	105,562	28,491	22,745	11,692	88,731
Sector 39	2275	2302	1659	832	17,488	3965	1186	47,309
Sector 40	1786	2166	3742	877	5746	10,267	1778	16,940
Sector 41	10,527	11,838	7314	10,897	23,429	5978	37,973	77,178
Sector 42	173	742	396	627	757	599	256	1659
Total intermediate input	425,664	1,148,342	575,485	1,328,693	1,132,534	2,762,589	614,140	2,307,254
Compensation of employees	165,330	621,382	218,879	930,261	2,863,999	1,282,667	358,875	3,535,131
Net taxes on production	10,558	103,923	17,967	128,510	18,428	13,383	61,872	19,193
Depreciation of fixed assets	31,024	118,202	87,733	75,633	340,399	155,707	96,077	520,442
Operating surplus	36,019	319,291	65,850	261,924	48,048	91,848	112,813	111,876
Total value added	242,932	1,162,798	390,430	1,396,329	3,270,875	1,543,605	629,636	4,186,643
Total input	668,595	2,311,140	965,915	2,725,022	4,403,409	4,306,194	1,243,777	6,493,896

Total intermediate use	Rural household consumption expenditure	Urban household consumption expenditure	Government consumption expenditure	Gross fixed capital formation	Increase in inventories	Export	Import	Total output
13,365,291	1,516,234	1,902,860	135,774	965,276	221,672	461,221	737,324	17,831,004
5,299,065	24,698	18,257	0	49,719	−87,034	19,336	422,399	4,901,643
5,838,506	0	0	0	18,273	−55,601	412,473	4,573,200	1,640,451
4,561,938	0	0	0	75,082	−19,071	7635	2,588,229	2,037,354
1,388,854	0	0	0	11,338	−22,672	23,948	92,764	1,308,704
8,604,447	1,675,488	4,911,285	0	293,991	255,511	220,825	242,567	15,718,980
5,042,989	57,805	89,632	0	38,987	−27,421	4,259,103	1,045,801	8,415,293
2,044,100	354,511	1,711,009	0	94,169	90,604	579,653	75,977	4,798,069
2,359,590	32,457	141,966	0	213,313	−1685	269,856	440,712	2,574,784
4,078,712	21,364	142,278	0	42,328	−34,103	226,932	591,834	3,885,677
7,268,811	30,499	273,735	0	−9233	−134,219	123,187	472,615	7,080,165
21,321,962	195,559	637,553	0	235,445	−131,943	944,203	1,856,668	21,346,111
9,688,917	12,573	78,685	0	91,740	−135,481	476,690	324,672	9,888,451
20,042,580	0	0	0	218,243	−188,641	365,674	1,007,407	19,430,450
4,839,458	21,058	114,123	0	110,207	−54,600	345,348	279,727	5,095,868
9,733,946	1513	21,966	0	6,488,073	89,393	2,080,412	1,983,098	16,432,204
6,452,869	171,434	826,127	0	6,779,276	119,509	2,205,987	1,695,521	14,859,681
6,133,592	152,054	592,515	0	2,091,328	34,952	1,840,036	874,710	9,969,766
7,092,189	132,612	529,860	0	1,259,991	187,744	4,893,444	3,597,095	10,498,745
2,049,831	15,524	50,921	0	213,991	−24,607	548,035	1,921,434	932,261
1,652,643	84,927	482,804	0	607,700	9937	251,396	114,142	2,975,265
1,410,306	0	0	0	0	6321	4934	976,445	445,115
11,627,515	60,900	224,622	0	0	−1,416,389	16,453	2010	10,511,092
270,473	19,721	148,114	0	2913	358	0	0	441,579
198,778	15,473	140,310	0	0	1226	0	0	355,787
858,054	0	335,508	0	26,197,176	72,498	471,036	129,104	27,805,168
10,145,678	164,622	633,165	528,514	78,436	−179,449	369,632	707,726	11,032,873
307,034	2361	16,076	0	0	2855	8197	41,786	294,736
2,140,852	191,606	828,451	0	851,594	43,230	201,538	98,688	4,158,583
6,737,881	594,375	1,829,259	0	735,853	47,359	936,504	0	10,881,231
3,267,364	373,355	1,457,364	0	0	98,640	−123,038	253,875	4,819,810
6,704,824	222,047	1,176,114	94,722	0	108,023	99,403	88,616	8,316,517
1,553,696	666,737	2,285,891	0	1,519,129	120,162	0	0	6,145,616
4,002,249	31,429	448,558	217,028	0	59,141	1,001,473	249,350	5,510,529
595,179	0	0	339,633	0	2978	25,237	286,962	676,066
1,767,929	1776	22,731	310,040	246,308	−1537	0	0	2,347,247
311,095	8967	109,950	546,374	0	−2160	0	0	974,226
1,370,167	203,554	1,182,912	0	0	35,106	39,837	85,740	2,745,836
162,219	419,823	935,742	2,972,005	0	−27,196	5143	32,842	4,434,892
294,578	150,548	1,762,331	2,115,098	0	−23,344	3828	2677	4,300,363
644,622	40,991	220,794	350,030	0	12,312	42,790	53,539	1,258,001
56,324	0	0	6,447,195	0	−12,349	20,806	16,752	6,495,225

Table 10.15 China's national input–output table in 2024

	Sector 1	Sector 2	Sector 3	Sector 4	Sector 5	Sector 6	Sector 7	Sector 8
Sector 1	2,385,938	48,087	26	3825	560	6,380,823	1,767,884	243,692
Sector 2	16,219	839,006	11,652	9467	3487	44,009	48,260	21,223
Sector 3	0	0	22,475	0	0	0	0	0
Sector 4	0	0	0	167,074	0	0	0	0
Sector 5	2941	1804	714	2702	82,601	5794	0	0
Sector 6	1,915,542	9027	2439	6211	4081	3,323,486	65,553	263,069
Sector 7	2151	1236	1080	1317	713	8314	2,659,076	1,564,267
Sector 8	5880	16,098	4810	9347	5591	21,118	130,862	717,033
Sector 9	26,085	38,948	5174	6668	2753	17,270	11,676	13,227
Sector 10	16,800	5536	2890	3663	2926	300,655	88,113	80,905
Sector 11	145,463	53,109	51,205	130,235	61,961	45,262	23,416	34,943
Sector 12	1,625,111	113,022	50,653	134,208	131,941	444,683	1,082,660	344,079
Sector 13	25,957	41,426	12,591	17,858	45,783	111,332	20,838	7057
Sector 14	5061	204,189	66,392	28,229	9831	18,173	7976	7976
Sector 15	69,063	89,798	10,448	42,737	23,329	74,587	19,994	25,548
Sector 16	112,631	244,324	112,622	168,367	114,106	55,098	113,972	35,414
Sector 17	50,567	30,198	9264	22,839	25,465	20,449	17,023	7627
Sector 18	5090	68,913	12,595	29,206	12,612	23,481	20,165	9798
Sector 19	1821	7489	1752	3080	1523	5940	7134	5915
Sector 20	24,472	41,573	12,849	14,043	4655	11,359	30,228	11,582
Sector 21	51,444	32,941	1873	11,533	13,365	24,672	60,507	49,312
Sector 22	2334	410	22	1271	9756	12,551	26,058	4701
Sector 23	244,971	231,448	99,388	433,747	170,828	167,173	242,023	44,972
Sector 24	682	1835	1679	11,502	6726	9545	4513	1847
Sector 25	2100	3664	1068	2319	1386	6530	6708	2362
Sector 26	4909	6532	1537	942	510	4548	2150	1627
Sector 27	327,728	204,202	26,384	91,173	77,337	462,392	156,949	120,842
Sector 28	14,302	1650	218	572	589	3638	7271	10,901
Sector 29	67,755	15,549	2307	4148	3509	28,638	17,443	26,405
Sector 30	387,543	65,071	12,474	33,973	25,925	389,477	209,947	120,623
Sector 31	52,802	43,277	12,079	21,719	16,198	95,539	38,404	31,390
Sector 32	183,362	126,711	11,156	32,741	25,837	165,905	203,215	72,735
Sector 33	6104	7059	608	1460	1942	27,729	35,670	66,889
Sector 34	26,778	26,491	2473	9195	6372	266,359	34,623	109,765
Sector 35	34,633	6958	2488	779	354	13,086	5832	2969
Sector 36	184,955	46,225	15,332	33,795	14,928	55,492	22,949	20,224
Sector 37	52,078	6822	1109	1901	4340	4945	14,162	721
Sector 38	78,901	34,335	5084	5207	5018	19,442	10,927	10,430
Sector 39	6920	4601	251	937	1789	2217	975	566
Sector 40	10,448	11,275	488	4496	2710	8033	3968	3419
Sector 41	4109	6785	1304	6060	5889	16,993	9399	8611
Sector 42	5742	1227	412	505	263	2255	1020	690
Total intermediate input	8,187,393	2,738,850	591,363	1,511,047	929,488	12,698,992	7,229,541	4,105,358
Compensation of employees	9,978,961	944,321	199,457	156,417	144,817	987,646	705,356	507,243
Net taxes on production	20,275	526,758	226,487	100,928	88,120	1,113,975	280,715	131,625
Depreciation of fixed assets	489,704	180,474	133,150	36,917	37,544	306,075	185,261	59,745
Operating surplus	0	546,563	312,039	191,877	120,248	983,722	289,348	170,033
Total value added	10,488,939	2,198,115	871,133	486,139	390,729	3,391,418	1,460,680	868,647
Total input	18,676,332	4,936,965	1,462,496	1,997,186	1,320,218	16,090,411	8,690,221	4,974,005

Sector 9	Sector 10	Sector 11	Sector 12	Sector 13	Sector 14	Sector 15	Sector 16	Sector 17
350,127	220,664	158	751,847	6162	2239	1666	1769	1314
20,221	39,279	440,028	314,778	731,496	710,358	24,313	55,945	17,610
0	0	3,986,116	1,320,618	117,132	112,965	0	0	0
0	0	0	205,105	56,343	3,985,045	69,122	48,345	16,249
0	0	715	306,153	594,826	55,263	23,538	11,499	4318
15,044	18,204	32,956	362,943	38,610	73,655	18,504	53,724	57,908
25,904	70,977	361	133,286	20,813	2277	8233	23,048	27,078
44,568	29,224	2939	45,913	24,380	42,868	15,754	35,208	154,167
733,047	105,695	2171	38,431	75,079	13,726	89,142	52,873	192,109
38,820	945,514	3840	284,712	227,191	19,788	52,800	71,736	58,934
18,861	19,836	429,489	1,235,194	408,134	844,759	55,763	167,752	76,759
262,471	569,257	132,529	7,753,977	800,151	316,623	242,707	592,395	997,162
21,221	10,939	23,989	146,210	1,625,928	365,903	55,758	117,926	108,801
57,423	49,815	4527	220,115	280,989	5,057,952	1,770,655	3,448,106	2,010,662
60,715	66,123	22,767	207,371	280,238	158,454	575,504	688,441	294,969
36,102	54,147	90,932	300,433	268,013	574,125	245,569	3,692,361	1,328,207
11,054	31,609	15,954	56,274	48,258	95,320	44,911	247,475	3,989,669
8324	26,736	8010	53,342	38,288	60,250	41,120	897,764	581,118
1059	36,676	965	12,670	6253	4778	6177	342,078	180,693
2570	51,756	10,049	147,925	28,024	69,981	15,354	85,820	350,702
20,666	26,046	3122	117,040	79,001	73,076	32,222	113,443	180,259
7513	289,424	1395	87,741	148,447	511,365	46,098	194,671	13,904
94,146	137,156	116,796	1,194,406	849,526	1,347,333	433,542	366,653	291,191
1169	1130	10,361	48,325	11,114	25,686	5235	14,625	10,108
1848	8642	2130	21,769	10,508	15,401	5091	5476	8093
721	1083	3412	8338	3658	6790	1073	5670	5929
79,096	92,036	134,875	518,780	393,363	344,707	112,596	469,384	280,956
2130	2771	210	19,682	3256	7180	1102	14,549	3501
10,906	10,941	21,782	74,483	24,438	205,921	18,780	51,794	41,847
73,721	142,475	83,132	515,371	311,598	413,195	135,489	419,789	808,258
12,285	20,445	12,394	105,266	65,165	58,993	54,486	126,724	54,895
51,108	70,419	41,484	402,306	359,502	431,359	73,757	174,448	188,754
12,429	21,855	1103	51,272	26,921	5559	26,527	45,776	30,367
24,013	36,304	32,666	323,085	69,772	32,811	21,246	144,758	218,853
1599	2419	1888	50,025	7880	32,970	4735	56,984	139,343
5219	21,276	11,627	122,802	42,003	76,470	33,932	98,129	152,621
421	14,472	1370	23,680	3311	6557	783	3389	5025
2740	8087	6022	42,869	29,778	42,322	8597	19,296	21,080
433	581	274	3290	1617	1606	732	5208	3328
2427	4854	1770	12,915	17,649	18,411	12,600	38,941	31,479
3252	4501	1564	24,208	16,051	12,095	5450	22,989	15,126
366	621	315	2626	1728	2140	749	2225	2397
2,115,740	3,263,992	5,698,182	17,667,577	8,152,594	16,238,276	4,391,408	13,029,186	12,955,743
202,915	286,862	229,668	1,303,047	660,356	1,141,709	333,987	1,436,466	821,170
84,254	110,484	758,245	719,638	359,583	552,737	139,043	659,153	545,549
48,464	109,050	170,204	557,721	256,534	680,940	97,528	382,408	238,013
122,418	170,107	252,434	1,079,414	591,556	780,780	233,925	1,016,308	676,037
458,050	676,504	1,410,550	3,659,820	1,868,029	3,156,166	804,482	3,494,335	2,280,769
2,573,790	3,940,495	7,108,732	21,327,397	10,020,623	19,394,442	5,195,890	16,523,521	15,236,512

(continued)

Table 10.15 (continued)

	Sector 18	Sector 19	Sector 20	Sector 21	Sector 22	Sector 23	Sector 24	Sector 25
Sector 1	203	0	43	370,420	177	180	74	0
Sector 2	8276	3177	1470	20,252	815	1,782,326	21,284	485
Sector 3	0	0	0	29	1	63,369	208,548	0
Sector 4	20,359	188	2332	281	538	299	0	0
Sector 5	4563	77	103	8506	68	2723	17	37
Sector 6	40,580	41,310	3178	41,821	700	30,861	1652	953
Sector 7	18,178	2846	1414	276,700	77	522	52	109
Sector 8	21,034	18,383	2914	45,616	120	33,621	4972	1798
Sector 9	28,258	17,626	2869	93,675	90	6260	519	537
Sector 10	112,143	85,509	10,087	54,881	123	19,285	1117	951
Sector 11	43,799	29,153	4584	22,437	1203	297,923	9215	3777
Sector 12	888,017	730,664	89,302	233,769	11,751	31,462	2846	18,004
Sector 13	194,296	180,978	33,357	64,220	1447	21,570	1023	1065
Sector 14	2,662,234	326,025	40,301	291,929	5810	9776	1092	421
Sector 15	421,620	233,061	38,283	129,202	609	60,085	4221	9827
Sector 16	483,828	107,005	27,520	12,269	359	101,758	3578	3876
Sector 17	25,269	34,636	5647	31,329	124	113,328	9710	4525
Sector 18	1,383,628	295,102	48,785	19,190	914	561,001	1028	1226
Sector 19	584,026	4,702,439	232,655	7634	46	6226	232	294
Sector 20	111,896	594,747	114,797	6625	54	136,363	2780	1978
Sector 21	97,925	83,175	15,749	336,276	108	6766	331	376
Sector 22	87,284	11,737	104	608	5318	1110	0	0
Sector 23	145,078	151,663	12,562	41,253	2473	3,623,879	8858	85,454
Sector 24	5923	5554	674	2928	121	11,785	24,688	1188
Sector 25	4134	4692	721	2081	115	12,624	237	7643
Sector 26	2754	2079	286	480	57	3476	329	443
Sector 27	220,723	167,670	21,193	72,806	4218	123,605	11,718	3457
Sector 28	8347	4474	1152	866	13	1198	58	113
Sector 29	35,538	112,531	4102	20,681	168	62,111	632	1348
Sector 30	339,790	439,644	25,340	61,027	2583	80,338	7441	2213
Sector 31	60,695	29,104	8550	21,950	516	29,070	2033	2906
Sector 32	162,869	428,413	8502	33,509	2376	421,075	14,691	21,000
Sector 33	33,695	29,253	4828	20,816	180	4443	466	192
Sector 34	235,041	123,728	12,968	25,802	437	45,208	1952	1794
Sector 35	42,903	78,706	5470	3003	45	6749	117	75
Sector 36	61,517	58,134	6657	6575	221	81,349	1022	555
Sector 37	2193	1398	102	490	32	27,093	193	20,888
Sector 38	8413	7080	838	1300	478	77,697	1277	2877
Sector 39	1541	1043	206	512	5	1568	104	280
Sector 40	18,449	5543	1802	479	87	12,112	1244	81
Sector 41	11,243	9356	1085	3387	80	22,874	837	1095
Sector 42	1182	1124	140	437	0	1448	38	73
Total intermediate input	8,639,446	9,159,027	792,672	2,388,049	44,658	7,936,517	352,227	203,914
Compensation of employees	612,663	790,019	96,052	276,770	6478	811,889	47,411	74,298
Net taxes on production	237,076	160,476	27,087	186,495	3880	493,706	11,000	14,642
Depreciation of fixed assets	167,183	248,041	19,593	62,465	2913	1,017,542	14,722	63,986
Operating surplus	450,866	295,603	49,037	152,443	395,289	330,322	21,523	8579
Total value added	1,467,788	1,494,140	191,769	678,173	408,560	2,653,459	94,656	161,504
Total input	10,107,234	10,653,166	984,441	3,066,222	453,217	10,589,976	446,882	365,418

Sector 26	Sector 27	Sector 28	Sector 29	Sector 30	Sector 31	Sector 32	Sector 33	Sector 34
124,367	267,395	0	0	2582	564,817	0	515	13,535
22,458	38,583	442	0	3253	3261	0	1332	8043
0	13,918	0	0	0	0	0	0	0
0	0	0	0	0	0	0	0	0
285,956	2903	0	0	0	0	0	0	0
113,006	88,132	785	10,503	55,501	1,699,942	24,206	17,646	67,487
23,896	14,476	477	528	12,866	24,057	1235	6983	73,732
107,733	80,015	2272	8152	94,249	18,459	22,774	28,329	71,134
495,564	13,650	1570	4129	17,613	6763	6491	7598	14,408
47,303	41,116	8357	134,300	98,826	16,641	167,853	36,831	511,120
364,902	2,087,880	9300	8305	35,711	24,069	54,613	82,073	181,595
873,210	197,077	1046	22,825	96,787	37,302	29,310	104,895	204,348
6,414,018	23,728	1090	1168	2906	10,030	1387	1860	3028
3,528,784	57,667	0	44	5226	220	0	5634	4357
920,393	62,323	245	8661	4291	5716	7813	38,121	114,528
837,923	269,672	3302	34,847	23,059	5739	36,788	22,258	37,445
117,000	788,238	24,924	30,372	93,521	16,520	25,144	23,981	173,575
1,070,720	31,487	557	296,599	72,806	35,063	17,761	69,674	159,964
68,237	28,221	2766	257,415	41,722	1995	6463	18,824	329,547
61,813	17,832	373	38,621	3483	153	33,170	12,274	16,984
92,143	10,248	796	4565	8170	3048	8999	8116	34,573
0	0	0	0	0	0	0	0	0
211,410	134,210	4305	107,419	125,193	104,567	81,473	37,603	54,397
3349	10,519	301	2451	2190	17,857	429	662	5508
6795	6745	506	5264	4097	8494	4087	1145	3724
300,246	48,174	13,326	7301	27,623	11,197	12,506	120,894	5569
2,637,063	945,199	44,929	132,084	750,512	59,513	214,393	47,409	280,700
6104	3517	13,219	5628	14,452	2811	37,214	4010	26,826
409,832	80,434	6404	186,664	87,085	15,038	246,554	24,478	21,326
696,424	144,848	23,630	135,330	54,337	103,472	36,662	26,801	132,305
283,052	178,428	2267	84,614	205,590	33,115	318,901	102,333	376,390
267,801	660,536	3692	89,175	365,477	80,989	695,079	269,003	258,845
16,494	54,641	4798	112,651	237,894	51,872	277,249	82,642	95,173
102,012	62,181	12,232	215,968	521,950	58,462	527,377	212,013	249,759
21,202	3443	171	7207	16,699	43	12,132	1322	2848
327,395	13,178	56	21,058	41,279	901	5998	12,039	4255
1391	3378	0	1206	2475	1604	5557	2017	48,999
39,638	254,785	503	27,582	98,994	45,664	33,614	46,217	88,343
3523	8361	316	5046	6401	1275	16,832	1482	2744
8403	6040	61	1284	1721	824	5985	2169	932
39,322	23,049	2358	24,105	41,715	12,644	51,796	33,192	24,480
3611	3157	104	1640	2664	753	4938	4156	1086
20,954,494	6,779,386	191,481	2,034,712	3,280,922	3,084,894	3,032,785	1,518,532	3,703,614
4,204,196	1,813,076	90,369	579,311	2,208,581	1,107,786	1,786,787	515,021	1,094,745
1,167,539	364,289	8146	170,406	2,190,803	210,370	685,414	742,405	277,637
408,442	806,317	21,297	559,350	513,373	248,333	131,954	2,273,896	532,161
1,601,566	1,436,151	261	954,326	3,128,463	263,119	3,023,834	1,182,087	156,704
7,381,742	4,419,832	120,073	2,263,393	8,041,220	1,829,608	5,627,988	4,713,410	2,061,248
28,336,236	11,199,218	311,554	4,298,105	11,322,142	4,914,501	8,660,773	6,231,942	5,764,862

(continued)

Table 10.15 (continued)

	Sector 35	Sector 36	Sector 37	Sector 38	Sector 39	Sector 40	Sector 41	Sector 42
Sector 1	26,028	9483	26,016	32,378	8245	20,758	6418	0
Sector 2	1196	3190	5575	11,280	8760	17,504	1065	18,178
Sector 3	0	0	0	0	0	0	0	0
Sector 4	6513	1481	1877	0	0	0	0	0
Sector 5	106	73	700	146	1592	2216	644	2564
Sector 6	12,294	5838	11,320	62,895	34,743	36,140	73,730	23,556
Sector 7	13,990	1574	4608	41,682	7552	24,115	3862	26,416
Sector 8	5342	3985	13,910	19,515	4837	59,445	21,916	86,367
Sector 9	2344	3033	13,723	26,113	52,559	32,337	15,752	105,224
Sector 10	7708	33,590	7300	46,226	150,578	53,959	147,992	168,658
Sector 11	6319	48,727	39,521	50,845	33,962	17,689	8548	94,062
Sector 12	75,134	52,508	72,392	241,366	38,722	1,847,006	54,005	65,550
Sector 13	5443	4621	14,980	2945	21,121	24,153	8429	31,583
Sector 14	3467	17,503	8111	7108	3220	207	606	0
Sector 15	17,620	50,924	8582	13,754	9236	9045	5229	7667
Sector 16	10,396	19,153	22,521	6725	8588	193,501	3682	15,053
Sector 17	6141	34,777	37,872	105,692	28,643	4805	6154	84,141
Sector 18	30,498	99,290	28,072	45,453	12,688	13,782	7261	14,743
Sector 19	27,255	89,615	15,474	68,302	13,909	3187	7854	49,818
Sector 20	24,388	54,044	14,815	26,978	41,889	10,097	4932	11,388
Sector 21	3153	5269	5642	22,028	7330	5435	2656	14,661
Sector 22	0	0	0	0	0	0	0	0
Sector 23	14,082	16,482	30,659	30,732	78,493	50,264	11,825	87,964
Sector 24	228	606	985	1787	2769	2011	881	1947
Sector 25	1073	2815	2005	4547	5821	2883	1042	3705
Sector 26	2460	7195	26,866	21,247	20,685	83,771	10,717	90,603
Sector 27	28,680	141,471	24,854	58,213	110,264	21,668	38,709	269,098
Sector 28	1162	2842	1029	3859	9000	1387	2997	69,010
Sector 29	3935	11,784	9555	9472	34,893	39,371	10,620	120,355
Sector 30	14,419	28,011	15,364	49,549	31,683	129,098	27,021	66,126
Sector 31	26,804	97,856	25,690	43,654	103,243	41,648	40,981	411,012
Sector 32	2979	75,462	34,322	58,770	108,155	24,535	19,682	102,741
Sector 33	3153	24,424	3247	77,974	30,749	8256	16,935	34,925
Sector 34	6547	35,951	15,888	39,524	74,629	30,749	28,864	57,700
Sector 35	16,599	5492	1037	75	8612	1960	1324	87
Sector 36	10,040	148,999	5549	2136	8768	6042	896	8996
Sector 37	434	504	25,596	4022	5438	1090	1173	14,963
Sector 38	2059	21,431	26,920	107,926	30,237	23,591	12,382	92,298
Sector 39	2326	2361	1778	850	18,560	4112	1256	49,211
Sector 40	1825	2221	4010	896	6098	10,648	1883	17,621
Sector 41	10,760	12,139	7836	11,141	24,865	6200	40,215	80,280
Sector 42	177	761	425	641	803	621	272	1726
Total intermediate input	435,077	1,177,484	616,625	1,358,448	1,201,937	2,865,285	650,410	2,399,999
Compensation of employees	168,987	637,151	234,527	951,094	3,039,507	1,330,349	380,070	3,677,234
Net taxes on production	10,792	106,561	19,252	131,388	19,557	13,880	65,526	19,964
Depreciation of fixed assets	31,710	121,202	94,005	77,327	361,259	161,496	101,751	541,363
Operating surplus	36,816	327,394	70,558	267,790	50,993	95,262	119,475	116,374
Total value added	248,304	1,192,308	418,341	1,427,599	3,471,316	1,600,987	666,821	4,354,935
Total input	683,381	2,369,792	1,034,966	2,786,047	4,673,253	4,466,272	1,317,231	6,754,934

Total intermediate use	Rural household consumption expenditure	Urban household consumption expenditure	Government consumption expenditure	Gross fixed capital formation	Increase in inventories	Export	Import	Total output
13,640,412	1,580,924	1,989,396	142,398	978,390	221,672	480,755	760,080	18,273,868
5,329,556	25,763	19,105	0	49,854	−87,034	19,740	446,511	4,910,474
5,845,171	0	0	0	18,273	−55,601	432,394	4,882,724	1,357,514
4,581,152	0	0	0	75,082	−19,071	7635	2,771,696	1,873,101
1,405,859	0	0	0	11,344	−22,672	24,578	97,563	1,321,546
8,759,733	1,744,345	5,132,488	0	300,699	255,511	220,825	242,567	16,171,035
5,132,079	60,562	94,014	0	39,769	−27,421	4,519,865	1,119,717	8,699,151
2,082,651	368,932	1,787,090	0	95,910	90,604	579,653	75,977	4,928,864
2,392,779	33,719	147,830	0	214,250	−1685	269,856	471,488	2,585,260
4,167,276	22,262	147,600	0	42,691	−34,103	226,932	604,393	3,968,265
7,362,351	31,505	282,139	0	−9241	−134,219	126,402	472,615	7,186,323
21,612,928	203,091	662,101	0	236,207	−131,943	944,203	1,908,593	21,617,994
9,829,964	13,185	81,900	0	92,102	−135,481	498,930	345,464	10,035,136
20,227,815	0	0	0	218,882	−188,641	365,674	1,046,032	19,577,698
4,891,144	22,023	119,493	0	111,203	−54,600	345,348	295,931	5,138,680
9,837,267	1573	22,814	0	6,513,075	89,393	2,186,223	2,062,254	16,588,091
6,550,028	178,486	857,280	0	6,895,781	119,509	2,337,517	1,781,190	15,157,411
6,214,104	157,844	615,984	0	2,111,961	34,952	1,910,622	909,709	10,135,758
7,190,159	138,647	553,742	0	1,275,461	187,744	5,076,867	3,716,680	10,705,940
2,255,417	16,014	52,014	0	213,993	−24,607	558,806	2,019,841	1,051,797
1,668,062	87,694	503,153	0	609,354	9937	257,897	119,246	3,016,850
1,463,822	0	0	0	0	6321	5067	1,011,110	464,099
11,717,597	60,900	224,622	0	0	−1,428,233	17,111	2022	10,589,976
273,424	20,312	152,122	0	2932	358	0	0	449,148
202,090	16,110	144,156	0	0	1226	0	0	363,582
879,714	0	350,547	0	26,654,226	72,498	499,866	136,511	28,320,339
10,322,948	172,072	660,628	559,955	79,860	−179,449	369,632	749,682	11,235,964
314,810	2475	16,917	0	0	2855	8434	44,637	300,855
2,181,555	199,279	863,949	0	874,324	43,230	211,704	104,295	4,269,746
6,821,518	621,701	1,916,077	0	762,268	47,359	956,179	0	11,125,103
3,352,464	389,974	1,522,981	0	0	98,640	−137,524	267,717	4,958,818
6,824,477	231,111	1,226,804	99,997	0	108,023	105,498	93,855	8,502,054
1,596,220	694,157	2,359,640	0	1,544,157	120,162	0	0	6,314,335
4,084,301	32,729	466,722	228,062	0	59,141	1,045,703	249,350	5,667,309
602,262	0	0	352,892	0	2978	26,751	302,517	682,366
1,791,592	1776	22,731	325,033	252,559	−1537	0	0	2,392,154
317,326	9389	114,861	568,528	0	−2160	0	0	1,007,945
1,402,280	211,929	1,232,230	0	0	35,106	41,238	90,398	2,832,385
167,018	441,111	982,502	3,116,754	0	−27,196	5269	34,680	4,650,778
298,302	155,914	1,834,622	2,226,709	0	−23,344	3828	2677	4,493,354
660,438	42,935	231,405	367,654	0	12,312	42,790	53,539	1,303,997
57,257	0	0	6,750,795	0	−12,349	21,914	17,455	6,800,163

Table 10.16 China's national input–output table in 2025

	Sector 1	Sector 2	Sector 3	Sector 4	Sector 5	Sector 6	Sector 7	Sector 8
Sector 1	2,388,519	48,190	26	3825	560	6,652,493	1,921,167	248,800
Sector 2	16,225	840,074	11,652	9467	3488	45,883	49,928	21,389
Sector 3	0	0	22,475	0	0	0	0	0
Sector 4	0	0	0	167,074	0	0	0	0
Sector 5	2943	1809	714	2702	82,657	6040	0	0
Sector 6	1,915,812	9030	2439	6211	4082	3,464,987	66,410	263,848
Sector 7	2152	1236	1080	1317	713	8668	2,709,946	1,569,912
Sector 8	5882	16,110	4810	9347	5592	22,017	134,687	721,867
Sector 9	26,091	38,979	5174	6668	2754	18,005	11,889	13,287
Sector 10	16,804	5540	2890	3663	2927	313,456	89,950	81,271
Sector 11	145,501	53,152	51,205	130,235	61,971	47,189	24,002	35,130
Sector 12	1,625,396	113,078	50,653	134,208	131,956	463,615	1,100,649	345,266
Sector 13	25,967	41,475	12,591	17,858	45,795	116,072	21,579	7110
Sector 14	5062	204,308	66,392	28,229	9832	18,947	8111	8007
Sector 15	69,086	89,898	10,448	42,737	23,335	77,762	20,612	25,739
Sector 16	112,663	244,530	112,622	168,367	114,126	57,444	116,899	35,615
Sector 17	50,581	30,222	9264	22,839	25,470	21,320	17,453	7670
Sector 18	5091	68,952	12,595	29,206	12,614	24,481	20,548	9838
Sector 19	1822	7493	1752	3080	1523	6192	7238	5935
Sector 20	24,476	41,618	12,849	14,043	4656	11,843	30,573	11,612
Sector 21	51,464	32,990	1873	11,533	13,371	25,723	63,163	49,759
Sector 22	2335	411	22	1271	9760	13,086	27,119	4754
Sector 23	245,019	231,562	99,388	433,747	170,848	174,290	246,430	45,161
Sector 24	682	1836	1679	11,502	6727	9951	4625	1857
Sector 25	2102	3669	1068	2319	1386	6808	7020	2386
Sector 26	4910	6535	1537	942	511	4742	2192	1634
Sector 27	327,797	204,329	26,384	91,173	77,348	482,079	160,072	121,364
Sector 28	14,313	1653	218	572	589	3793	7802	11,089
Sector 29	67,778	15,562	2307	4148	3510	29,857	17,997	26,591
Sector 30	387,692	65,150	12,474	33,973	25,933	406,059	217,742	121,677
Sector 31	52,814	43,312	12,079	21,719	16,201	99,607	39,193	31,532
Sector 32	183,505	126,953	11,156	32,741	25,849	172,969	216,936	73,918
Sector 33	6106	7065	608	1460	1942	28,909	36,502	67,243
Sector 34	26,789	26,516	2473	9195	6373	277,699	35,813	110,672
Sector 35	34,640	6962	2488	779	354	13,643	5937	2982
Sector 36	184,980	46,244	15,332	33,795	14,929	57,855	23,252	20,283
Sector 37	52,097	6828	1109	1901	4341	5156	14,640	727
Sector 38	78,913	34,354	5084	5207	5019	20,269	11,087	10,465
Sector 39	6921	4603	251	937	1789	2311	989	568
Sector 40	10,450	11,279	488	4496	2710	8375	4021	3429
Sector 41	4110	6788	1304	6060	5889	17,717	9569	8647
Sector 42	5743	1228	412	505	263	2351	1033	692
Total intermediate input	8,191,230	2,741,522	591,363	1,511,047	929,692	13,239,665	7,504,776	4,129,728
Compensation of Employees	9,989,756	945,523	199,457	156,417	144,916	1,029,697	718,850	510,663
Net taxes on production	20,297	527,428	226,487	100,928	88,181	1,161,403	286,085	132,513
Depreciation of fixed assets	490,233	180,704	133,150	36,917	37,570	319,107	188,806	60,148
Operating surplus	0	547,259	312,039	191,877	120,330	1,025,605	294,883	171,179
Total value added	10,500,286	2,200,914	871,133	486,139	390,996	3,535,811	1,488,625	874,503
Total input	18,691,516	4,942,436	1,462,496	1,997,186	1,320,689	16,775,477	8,993,401	5,004,231

Sector 9	Sector 10	Sector 11	Sector 12	Sector 13	Sector 14	Sector 15	Sector 16	Sector 17
356,769	229,629	159	825,247	6220	2304	1675	1781	1350
20,406	40,003	442,815	328,622	735,239	721,542	24,383	56,306	17,818
0	0	4,011,369	1,503,216	118,024	118,729	0	0	0
0	0	0	219,490	56,812	4,257,263	69,824	48,657	16,617
0	0	719	335,534	601,242	57,078	23,686	11,574	4437
15,083	18,306	33,164	367,567	38,667	73,975	18,521	54,070	58,015
25,986	71,525	363	135,659	20,853	2289	8242	23,197	27,152
44,834	29,598	2958	47,285	24,461	43,307	15,785	35,435	155,362
737,914	106,894	2185	39,404	75,310	13,879	89,325	53,214	193,386
39,012	957,556	3865	291,063	227,766	19,957	52,886	72,199	59,223
18,964	20,069	432,210	1,264,773	409,334	852,832	55,876	168,834	77,270
263,369	573,673	133,368	7,895,804	801,669	318,403	242,983	596,216	1,000,526
21,390	11,124	24,141	151,812	1,633,464	371,252	55,917	118,687	110,030
57,682	50,272	4555	224,480	281,653	5,099,886	1,773,403	3,470,345	2,018,933
61,262	67,271	22,911	215,093	281,567	161,140	577,446	692,881	298,039
36,323	54,842	91,508	308,955	268,905	580,523	246,095	3,716,176	1,338,081
11,119	31,994	16,055	57,941	48,421	96,304	44,996	249,071	4,022,525
8359	26,976	8061	54,414	38,378	60,708	41,182	903,554	583,630
1063	36,932	971	12,873	6265	4808	6184	344,285	181,203
2589	52,136	10,112	150,718	28,103	70,699	15,393	86,373	353,354
20,882	26,971	3141	123,430	79,510	74,648	32,357	114,175	181,985
7654	297,593	1404	93,472	149,754	532,910	46,489	195,927	14,277
94,532	138,346	117,536	1,217,497	851,295	1,356,075	434,103	369,017	292,352
1175	1144	10,427	49,523	11,146	25,918	5244	14,719	10,170
1865	8836	2143	22,830	10,561	15,630	5106	5511	8202
725	1093	3434	8511	3666	6839	1074	5706	5955
79,459	92,990	135,729	530,544	394,388	347,577	112,773	472,411	282,313
2167	2864	212	21,463	3288	7384	1108	14,642	3588
10,980	11,113	21,920	77,122	24,528	208,097	18,819	52,128	42,206
74,448	145,209	83,658	539,945	313,287	420,252	135,946	422,497	818,039
12,353	20,666	12,472	107,908	65,365	59,605	54,595	127,542	55,260
51,954	73,384	41,747	437,523	362,651	441,560	74,077	175,573	193,256
12,509	22,125	1110	52,643	27,005	5620	26,585	46,071	30,581
24,208	36,917	32,873	335,544	70,070	33,259	21,300	145,692	221,120
1607	2444	1900	51,237	7901	33,236	4742	57,352	140,053
5235	21,427	11,700	124,913	42,078	76,859	33,963	98,762	153,004
424	14,706	1379	24,631	3325	6642	785	3411	5074
2751	8149	6061	43,596	29,839	42,623	8609	19,421	21,163
435	585	276	3341	1620	1615	733	5242	3337
2433	4882	1782	13,084	17,674	18,489	12,611	39,192	31,534
3265	4542	1574	24,738	16,088	12,179	5456	23,137	15,188
367	624	317	2659	1730	2149	749	2240	2401
2,133,550	3,315,413	5,734,282	18,336,102	8,189,123	16,656,042	4,401,025	13,113,220	13,048,013
204,262	290,516	231,123	1,326,881	663,416	1,151,174	335,114	1,445,730	827,933
84,813	111,891	763,048	732,801	361,249	557,320	139,512	663,404	550,041
48,786	110,439	171,283	567,922	257,723	686,586	97,857	384,875	239,973
123,230	172,274	254,033	1,099,157	594,298	787,253	234,714	1,022,863	681,604
461,091	685,119	1,419,486	3,726,762	1,876,687	3,182,333	807,197	3,516,872	2,299,551
2,594,642	4,000,533	7,153,768	22,062,864	10,065,810	19,838,374	5,208,222	16,630,093	15,347,565

(continued)

Table 10.16 (continued)

	Sector 18	Sector 19	Sector 20	Sector 21	Sector 22	Sector 23	Sector 24	Sector 25
Sector 1	214	0	43	370,579	177	182	75	0
Sector 2	8509	3349	1480	20,261	815	1,800,792	21,526	488
Sector 3	0	0	0	29	1	64,026	210,919	0
Sector 4	21,658	219	2351	281	538	302	0	0
Sector 5	4830	85	104	8509	68	2751	17	37
Sector 6	40,927	41,983	3184	41,839	700	31,180	1671	960
Sector 7	18,377	2902	1417	276,818	77	527	52	110
Sector 8	21,437	19,062	2925	45,635	120	33,969	5029	1810
Sector 9	28,768	18,246	2880	93,716	90	6325	525	540
Sector 10	113,816	87,763	10,115	54,904	123	19,485	1130	958
Sector 11	44,551	30,095	4597	22,446	1203	301,010	9320	3803
Sector 12	897,799	746,001	89,483	233,869	11,751	31,788	2878	18,128
Sector 13	199,288	189,537	33,508	64,247	1447	21,793	1035	1072
Sector 14	2,699,299	334,606	40,416	292,055	5810	9878	1105	424
Sector 15	433,862	245,113	38,526	129,257	609	60,708	4269	9895
Sector 16	493,242	110,904	27,615	12,274	359	102,812	3619	3903
Sector 17	25,741	35,832	5669	31,342	124	114,502	9820	4557
Sector 18	1,402,871	302,619	48,911	19,199	914	566,813	1040	1234
Sector 19	590,397	4,802,927	233,139	7637	46	6290	235	296
Sector 20	113,611	611,348	115,107	6628	54	137,776	2811	1991
Sector 21	101,423	88,115	15,912	336,420	108	6836	335	379
Sector 22	92,172	12,978	105	608	5318	1121	0	0
Sector 23	146,908	155,462	12,598	41,270	2473	3,661,425	8959	86,043
Sector 24	6021	5724	676	2929	121	11,907	24,968	1196
Sector 25	4256	4944	725	2082	115	12,755	240	7695
Sector 26	2792	2133	287	480	57	3512	332	446
Sector 27	224,012	172,157	21,254	72,838	4218	124,885	11,851	3480
Sector 28	8827	4948	1174	867	13	1210	58	114
Sector 29	36,293	116,965	4123	20,690	168	62,754	640	1357
Sector 30	350,730	466,018	25,571	61,053	2583	81,170	7526	2228
Sector 31	61,749	30,026	8575	21,960	516	29,371	2056	2926
Sector 32	170,660	465,375	8629	33,523	2376	425,438	14,858	21,144
Sector 33	34,328	30,335	4844	20,825	180	4490	471	193
Sector 34	241,033	129,688	13,047	25,813	437	45,677	1974	1807
Sector 35	43,519	81,009	5490	3004	45	6819	118	76
Sector 36	62,113	59,257	6673	6578	221	82,191	1033	559
Sector 37	2249	1466	103	491	32	27,373	196	21,032
Sector 38	8515	7247	840	1301	478	78,502	1292	2897
Sector 39	1557	1064	206	512	5	1585	105	282
Sector 40	18,602	5634	1805	479	87	12,237	1258	81
Sector 41	11,393	9586	1089	3389	80	23,111	847	1103
Sector 42	1192	1142	141	437	0	1463	38	73
Total intermediate input	8,789,541	9,433,865	795,338	2,389,074	44,658	8,018,743	356,231	205,320
Compensation of employees	621,184	806,902	96,311	276,888	6478	820,301	47,950	74,810
Net taxes on production	240,373	163,905	27,160	186,575	3880	498,821	11,125	14,743
Depreciation of fixed assets	169,508	253,341	19,646	62,492	2913	1,028,084	14,889	64,427
Operating surplus	457,136	301,920	49,169	152,509	395,289	333,745	21,768	8638
Total value added	1,488,201	1,526,069	192,286	678,464	408,560	2,680,950	95,732	162,618
Total input	10,277,742	10,959,934	987,624	3,067,538	453,217	10,699,694	451,963	367,938

Sector 26	Sector 27	Sector 28	Sector 29	Sector 30	Sector 31	Sector 32	Sector 33	Sector 34
126,270	273,269	0	0	2612	580,368	0	529	13,658
22,801	39,431	443	0	3290	3351	0	1368	8116
0	14,224	0	0	0	0	0	0	0
0	0	0	0	0	0	0	0	0
290,332	2967	0	0	0	0	0	0	0
114,735	90,068	787	10,673	56,128	1,746,747	24,477	18,129	68,102
24,262	14,794	479	536	13,012	24,720	1249	7174	74,403
109,382	81,773	2279	8284	95,314	18,967	23,029	29,104	71,782
503,149	13,949	1575	4195	17,812	6950	6564	7806	14,539
48,027	42,019	8384	136,469	99,943	17,099	169,733	37,839	515,775
370,487	2,133,746	9330	8439	36,115	24,732	55,225	84,319	183,249
886,575	201,407	1049	23,193	97,881	38,329	29,639	107,766	206,209
6,512,185	24,249	1094	1187	2939	10,306	1403	1911	3056
3,582,793	58,934	0	45	5285	226	0	5788	4397
934,479	63,692	246	8801	4340	5874	7900	39,165	115,571
850,748	275,596	3313	35,410	23,320	5897	37,201	22,867	37,786
118,791	805,554	25,003	30,863	94,578	16,975	25,426	24,637	175,156
1,087,108	32,179	559	301,388	73,628	36,028	17,960	71,582	161,420
69,281	28,841	2774	261,572	42,193	2050	6535	19,339	332,548
62,759	18,224	375	39,245	3522	158	33,542	12,610	17,139
93,554	10,473	799	4638	8262	3132	9100	8338	34,888
0	0	0	0	0	0	0	0	0
214,646	137,158	4319	109,154	126,608	107,446	82,386	38,632	54,893
3400	10,750	302	2491	2215	18,349	434	680	5558
6899	6893	508	5349	4143	8728	4133	1177	3758
304,841	49,232	13,368	7419	27,935	11,505	12,646	124,203	5620
2,677,423	965,963	45,072	134,217	758,992	61,152	216,795	48,707	283,256
6197	3594	13,261	5719	14,615	2888	37,631	4120	27,071
416,104	82,201	6424	189,678	88,069	15,452	249,316	25,148	21,520
707,083	148,030	23,705	137,515	54,951	106,321	37,072	27,535	133,510
287,384	182,347	2274	85,981	207,913	34,027	322,474	105,134	379,818
271,900	675,046	3704	90,615	369,607	83,219	702,866	276,366	261,202
16,746	55,841	4813	114,471	240,582	53,300	280,355	84,904	96,040
103,574	63,547	12,271	219,456	527,848	60,072	533,286	217,817	252,033
21,526	3519	171	7324	16,887	45	12,268	1359	2874
332,406	13,468	56	21,398	41,745	926	6066	12,368	4293
1412	3452	0	1225	2503	1648	5620	2073	49,445
40,245	260,382	504	28,028	100,113	46,921	33,991	47,482	89,148
3577	8544	317	5127	6473	1310	17,020	1522	2769
8531	6173	61	1305	1741	847	6052	2228	941
39,924	23,555	2365	24,494	42,187	12,992	52,376	34,101	24,703
3666	3227	104	1667	2694	774	4993	4270	1096
21,275,203	6,928,314	192,089	2,067,571	3,317,994	3,169,832	3,066,762	1,560,098	3,737,341
4,268,541	1,852,905	90,656	588,666	2,233,537	1,138,287	1,806,805	529,119	1,104,715
1,185,408	372,292	8172	173,157	2,215,558	216,162	693,093	762,727	280,166
414,693	824,030	21,365	568,383	519,174	255,170	133,432	2,336,138	537,007
1,626,078	1,467,700	262	969,738	3,163,812	270,364	3,057,711	1,214,444	158,131
7,494,719	4,516,926	120,454	2,299,944	8,132,081	1,879,983	5,691,041	4,842,427	2,080,018
28,769,922	11,445,240	312,543	4,367,515	11,450,074	5,049,815	8,757,803	6,402,526	5,817,360

(continued)

Table 10.16 (continued)

	Sector 35	Sector 36	Sector 37	Sector 38	Sector 39	Sector 40	Sector 41	Sector 42
Sector 1	26,539	9927	26,339	33,700	8607	21,840	6657	0
Sector 2	1220	3340	5644	11,741	9145	18,417	1105	19,214
Sector 3	0	0	0	0	0	0	0	0
Sector 4	6641	1551	1901	0	0	0	0	0
Sector 5	108	77	709	152	1662	2331	668	2710
Sector 6	12,535	6112	11,460	65,462	36,271	38,024	76,486	24,899
Sector 7	14,264	1648	4665	43,384	7884	25,372	4006	27,922
Sector 8	5447	4172	14,083	20,312	5049	62,543	22,735	91,288
Sector 9	2390	3175	13,894	27,179	54,871	34,022	16,340	111,220
Sector 10	7859	35,165	7390	48,113	157,200	56,771	153,523	178,268
Sector 11	6443	51,012	40,012	52,921	35,456	18,611	8867	99,422
Sector 12	76,607	54,970	73,291	251,219	40,425	1,943,259	56,024	69,286
Sector 13	5550	4838	15,166	3065	22,049	25,411	8744	33,383
Sector 14	3535	18,324	8212	7398	3362	218	629	0
Sector 15	17,965	53,312	8689	14,316	9642	9517	5425	8104
Sector 16	10,600	20,051	22,800	7000	8966	203,585	3819	15,910
Sector 17	6262	36,408	38,342	110,006	29,903	5056	6384	88,935
Sector 18	31,096	103,946	28,420	47,309	13,246	14,501	7532	15,584
Sector 19	27,789	93,817	15,666	71,090	14,521	3353	8147	52,657
Sector 20	24,866	56,579	14,999	28,079	43,731	10,623	5116	12,037
Sector 21	3215	5516	5712	22,928	7653	5718	2755	15,496
Sector 22	0	0	0	0	0	0	0	0
Sector 23	14,358	17,255	31,040	31,987	81,945	52,883	12,267	92,977
Sector 24	232	635	997	1860	2891	2115	914	2058
Sector 25	1094	2947	2030	4732	6077	3033	1081	3916
Sector 26	2508	7532	27,200	22,115	21,595	88,137	11,117	95,766
Sector 27	29,242	148,105	25,162	60,590	115,113	22,798	40,156	284,432
Sector 28	1185	2976	1041	4017	9396	1459	3109	72,942
Sector 29	4012	12,336	9674	9859	36,427	41,423	11,016	127,213
Sector 30	14,702	29,325	15,555	51,571	33,077	135,826	28,031	69,894
Sector 31	27,330	102,445	26,009	45,436	107,783	43,818	42,513	434,432
Sector 32	3037	79,001	34,749	61,169	112,911	25,813	20,418	108,595
Sector 33	3215	25,570	3288	81,157	32,101	8686	17,568	36,916
Sector 34	6675	37,637	16,085	41,137	77,911	32,351	29,943	60,988
Sector 35	16,924	5750	1049	78	8991	2063	1374	92
Sector 36	10,237	155,986	5618	2223	9153	6357	930	9509
Sector 37	442	528	25,914	4186	5677	1147	1217	15,816
Sector 38	2099	22,436	27,254	112,331	31,567	24,820	12,845	97,557
Sector 39	2371	2472	1800	885	19,376	4326	1303	52,015
Sector 40	1861	2325	4060	933	6366	11,203	1953	18,625
Sector 41	10,971	12,708	7934	11,596	25,959	6523	41,718	84,855
Sector 42	181	797	430	667	838	653	282	1824
Total intermediate input	443,607	1,232,703	624,281	1,413,903	1,254,796	3,014,605	674,718	2,536,757
Compensation of employees	172,300	667,031	237,438	989,919	3,173,179	1,399,678	394,274	3,886,773
Net taxes on production	11,004	111,558	19,491	136,752	20,417	14,603	67,975	21,102
Depreciation of fixed assets	32,331	126,886	95,172	80,483	377,147	169,912	105,554	572,211
Operating surplus	37,537	342,748	71,434	278,722	53,235	100,227	123,940	123,005
Total value added	253,172	1,248,222	423,535	1,485,876	3,623,978	1,684,420	691,743	4,603,090
Total input	696,779	2,480,925	1,047,816	2,899,779	4,878,773	4,699,025	1,366,460	7,139,847

Total intermediate use	Rural household consumption expenditure	Urban household consumption expenditure	Government consumption expenditure	Gross fixed capital formation	Increase in inventories	Export	Import	Total output
14,190,299	1,633,681	2,059,969	149,022	979,448	221,672	500,324	784,335	18,950,082
5,391,088	26,753	19,894	0	49,917	−87,034	20,144	470,608	4,950,155
6,063,013	0	0	0	18,273	−55,601	452,447	4,966,454	1,511,677
4,871,178	0	0	0	75,082	−19,071	7635	2,771,696	2,163,127
1,449,252	0	0	0	11,351	−22,672	25,209	102,362	1,360,778
8,973,725	1,817,983	5,369,047	0	313,502	255,511	220,825	242,567	16,708,025
5,200,346	63,311	98,384	0	40,530	−27,421	4,781,121	1,193,492	8,962,779
2,114,866	382,131	1,856,717	0	96,557	90,604	579,653	75,977	5,044,552
2,425,078	35,037	153,949	0	215,673	−1685	269,856	502,297	2,595,610
4,247,898	23,196	153,138	0	43,235	−34,103	226,932	632,328	4,027,968
7,483,952	32,703	292,140	0	−9300	−134,219	129,610	486,477	7,308,409
21,979,657	212,068	691,362	0	240,528	−131,943	944,203	1,960,735	21,975,140
9,974,727	13,806	85,165	0	92,529	−135,481	521,172	366,255	10,185,662
20,412,835	0	0	0	220,697	−188,641	365,674	1,084,695	19,725,870
4,966,501	22,902	124,380	0	111,578	−54,600	345,348	312,091	5,204,019
9,943,267	1637	23,706	0	6,555,082	89,393	2,292,030	2,141,412	16,763,704
6,629,112	185,240	887,114	0	6,970,072	119,509	2,469,224	1,866,743	15,393,528
6,295,676	163,838	640,281	0	2,141,333	34,952	1,981,206	944,710	10,312,576
7,322,760	145,046	579,065	0	1,302,716	187,744	5,260,076	3,835,976	10,961,433
2,294,079	16,862	53,904	0	214,570	−24,607	569,547	2,109,743	1,014,611
1,698,678	90,040	520,407	0	609,615	9937	264,403	124,346	3,068,733
1,510,541	0	0	0	0	6321	5199	1,045,804	476,257
11,842,291	60,900	224,622	0	0	−1,443,856	17,770	2034	10,699,694
277,748	20,935	156,346	0	2965	358	0	0	458,352
206,752	16,726	147,870	0	0	1226	0	0	372,575
902,785	0	365,419	0	27,062,169	72,498	528,686	143,920	28,787,636
10,490,603	179,682	688,680	591,396	81,615	−179,449	369,632	791,735	11,430,425
325,179	2565	17,574	0	0	2855	8674	47,474	309,372
2,223,525	206,604	897,838	0	888,444	43,230	221,879	109,896	4,371,623
6,974,567	646,173	1,993,824	0	770,882	47,359	975,897	0	11,408,701
3,426,518	407,104	1,590,620	0	0	98,640	−151,987	281,580	5,089,315
7,017,975	239,497	1,273,706	105,272	0	108,023	111,599	99,089	8,756,984
1,625,303	723,033	2,437,305	0	1,586,424	120,162	0	0	6,492,227
4,168,618	33,862	482,546	239,096	0	59,141	1,090,043	249,350	5,823,956
610,630	0	0	366,150	0	2978	28,265	318,066	689,957
1,815,977	1776	22,731	340,025	264,403	−1537	0	0	2,443,375
322,424	9750	119,058	590,682	0	−2160	0	0	1,039,754
1,431,404	220,862	1,284,829	0	0	35,106	42,634	95,070	2,919,766
172,078	461,818	1,027,984	3,261,504	0	−27,196	5396	36,513	4,865,070
302,318	161,503	1,909,911	2,338,319	0	−23,344	3828	2677	4,689,859
675,807	44,805	241,610	385,278	0	12,312	42,790	53,539	1,349,064
58,110	0	0	7,054,395	0	−12,349	23,022	18,157	7,105,021

10.2.2 Analysis

In this section, we focus on the trends of the influence coefficients and reaction coefficients of various sectors in the BAU scenario. Figure 10.3 illustrates the trends of the influence coefficients of Sector 2, Sector 3, Sector 4, and Sector 5 during the 2011–2025 period in BAU scenario. Sector 2 is Mining and Washing of Coal; Sector 3 is Extraction of Petroleum and Natural Gas; Sector 4 is Mining of Metal Ores; Sector 5 is Mining and Processing of Nonmetal Ores and Other Ores. It is clear that the influence coefficients of most sectors are quite stable, because no mutation will happen in China's economy during this period. The obvious change only occurs in Sector 4, Mining and Dressing of Metal Ores. Its influence coefficient increases slightly from 1.0475 in 2011 to 1.1570 in 2025, indicating the enhancement of pulling effect on other sectors.

Figure 10.4 shows the trends of the influence coefficients of Sector 12, Sector 13, Sector 14, and Sector 15 during the 2011–2025 period in BAU scenario. Sector 12 is Chemical Industry; Sector 13 is Manufacture of Nonmetallic Mineral Products; Sector 14 is Smelting and Rolling of Metals; Sector 15 is Manufacture of Metal Products. There are some changes for the influence coefficients of these sectors. The coefficient of Chemical Industry decreases slenderly from 1.2231 to 1.2083, which reflects that the pulling effects of Chemical Industry will decline in the future. Oppositely, the coefficients of Manufacture of Nonmetallic Mineral

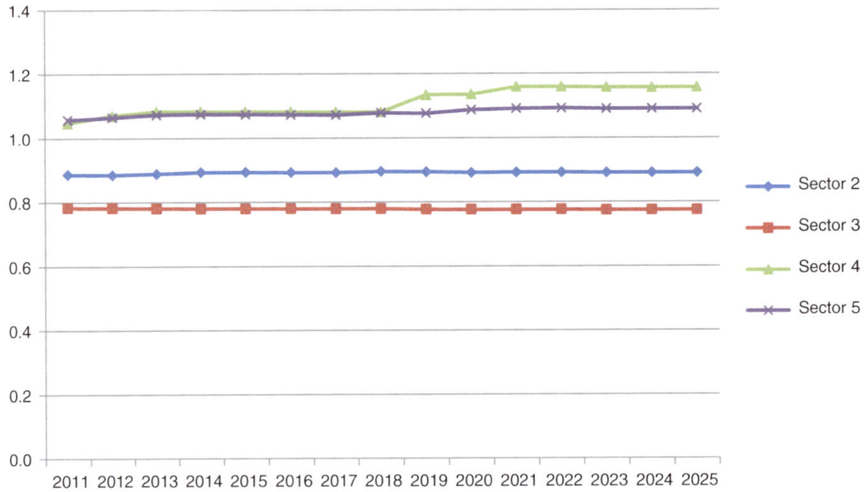

Fig. 10.3 The influence coefficients of Sector 2, Sector 3, Sector 4, and Sector 5 during the 2011–2025 period in BAU scenario

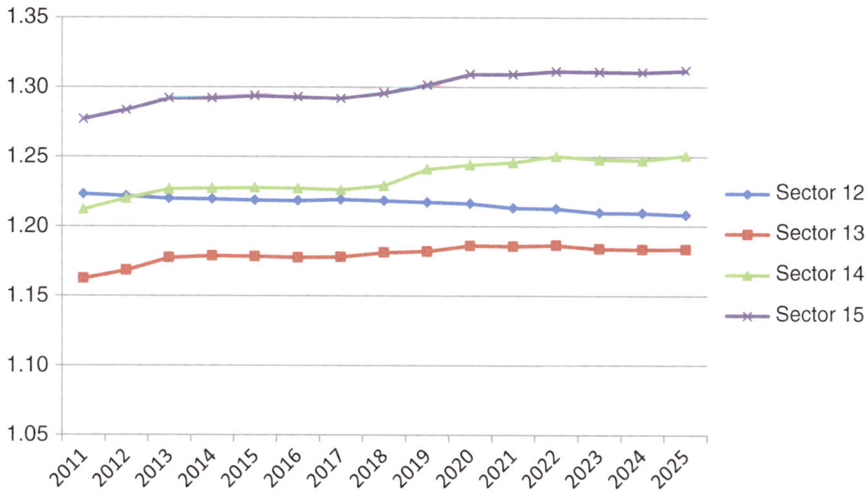

Fig. 10.4 The influence coefficients of Sector 12, Sector 13, Sector 14, and Sector 15 during the 2011–2025 period in BAU scenario

Products (Sector 13), Smelting and Rolling of Metals (Sector 14), and Manufacture of Metal Products (Sector 15) increase more or less during this period.

Figure 10.5 depicts the trends of the influence coefficients of Sector 17, Sector 18, Sector 19, and Sector 20 during the 2011–2025 period in BAU scenario. Sector 17 is Manufacture of Transport Equipment; Sector 18 is Manufacture of Electrical Machinery and Equipment; Sector 19 is Manufacture of Communication Equipment, Computers, and Other Electronic Equipment; and Sector 20 is Manufacture of Instruments, Meters, and Machinery for Cultural and Office Use. The influence coefficient of Manufacture of Transport Equipment (Sector 17) increases a little bit, which verifies the increase in its pulling effect on China's economy. On the contrary, the influence coefficient of Manufacture of Communication Equipment, Computers, and Other Electronic Equipment (Sector 19) decreases from 1.4047 to 1.3841, demonstrating the decline of its pulling effect.

Figure 10.6 illustrates the trends of the influence coefficients of Sector 30, Sector 31, Sector 32, and Sector 33 during the 2011–2025 period in BAU scenario. Sector 30 is Wholesale and Retail Trades; Sector 31 is Hotels and Catering Services; Sector 32 is Financial Intermediation; Sector 33 is Real Estate. Evidently, the influence coefficients of all these sectors are quite stable like most other sectors in China's economy. The reason is that there will be no mutation within the fifteen-year period according to our research.

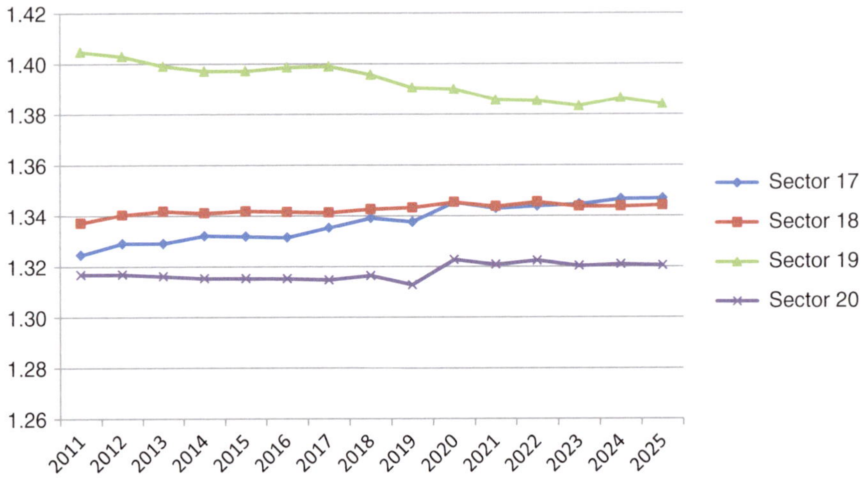

Fig. 10.5 The influence coefficients of Sector 17, Sector 18, Sector 19, and Sector 20 during the 2011–2025 period in BAU scenario

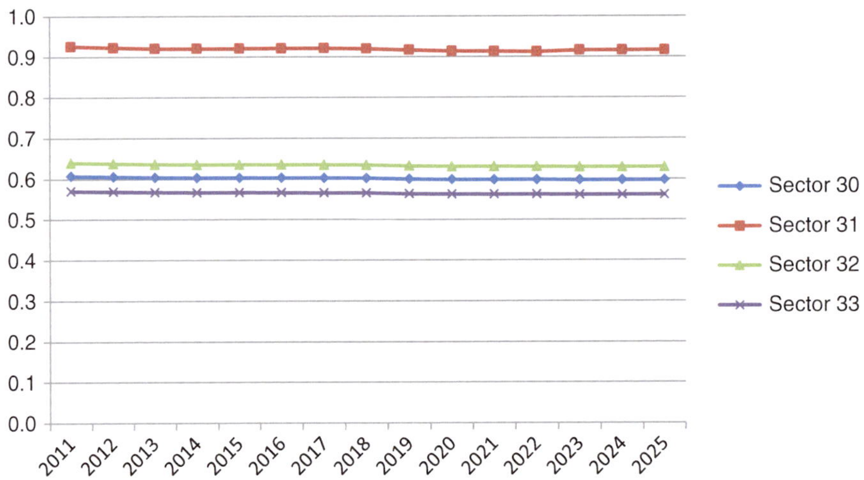

Fig. 10.6 The influence coefficients of Sector 30, Sector 31, Sector 32, and Sector 33 during the 2011–2025 period in BAU scenario

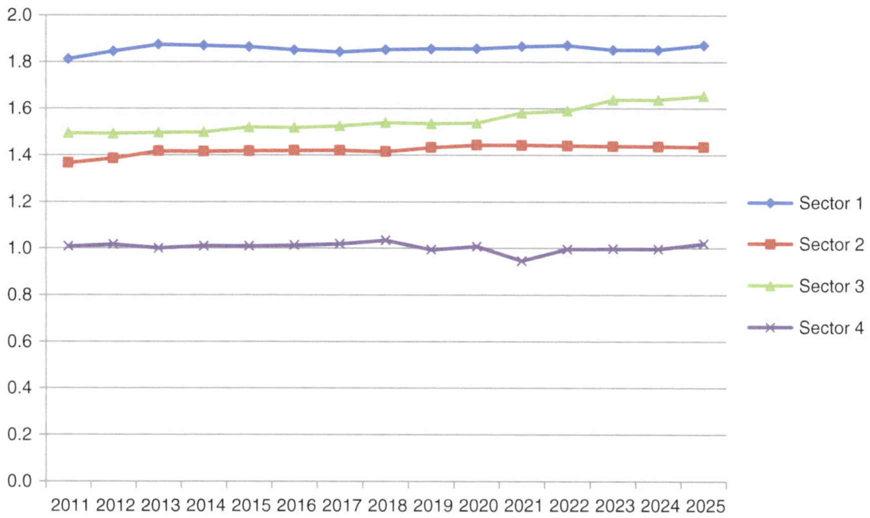

Fig. 10.7 The reaction coefficients of Sector 1, Sector 2, Sector 3, and Sector 4 during the 2011–2025 period in BAU scenario

Figure 10.7 shows the trends of the reaction coefficients of Sector 1, Sector 2, Sector 3, and Sector 4 during the 2011–2025 period in BAU scenario. Sector 1 is Agriculture, Forestry, Animal Husbandry, and Fishery; Sector 2 is Mining and Washing of Coal; Sector 3 is Extraction of Petroleum and Natural Gas; Sector 4 is Mining of Metal Ores. Most reaction coefficients are relatively stable within the 15 years. The most obvious change happens in Sector 3, Extraction of Petroleum and Natural Gas, whose reaction coefficient increases from 1.4936 to 1.6529, which evidences that this sector will play an increasingly important supporting role in the national economy within the 15-year period.

Figure 10.8 expresses the trends of the reaction coefficients of Sector 6, Sector 7, Sector 8, and Sector 9 during the 2011–2025 period in BAU scenario. Sector 6 is Manufacture of Foods and Tobacco; Sector 7 is Manufacture of Textile; Sector 8 is Manufacture of Textile Wearing Apparel, Footwear, Caps, Leather, Fur, Feather (Down) and Its Products; Sector 9 is Processing of Timbers and Manufacture of Furniture. The reaction coefficient of Sector 7, Manufacture of Textile, decreases within the 15 years, which means that the pushing effects of this labor-intensive sector will decline in China's economy in the future. The reaction coefficients of the other three sectors are far more stable.

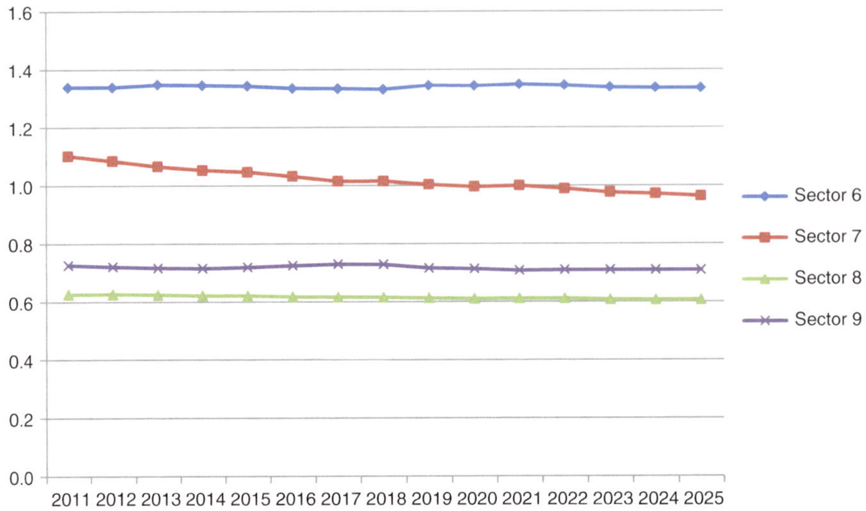

Fig. 10.8 The reaction coefficients of Sector 6, Sector 7, Sector 8, and Sector 9 during the 2011–2025 period in BAU scenario

Figure 10.9 depicts the trends of the reaction coefficients of Sector 16, Sector 17, Sector 18, and Sector 19 during the 2011–2025 period in BAU scenario. Sector 16 is Manufacture of General and Special Purpose Machinery; Sector 17 is Manufacture of Transport Equipment; Sector 18 is Manufacture of Electrical Machinery and Equipment; Sector 19 is Manufacture of Communication Equipment, Computer, and Other Electronic Equipment. The reaction coefficients of these four sectors are relatively steady. What they have in common is that their reaction coefficients gradually decline somewhat. Since these sectors all belong to manufacturing, we may safely draw the conclusion that the foundation role of manufacturing in China's economy will decrease in the next period.

Figure 10.10 shows the trends of the reaction coefficients of Sector 29, Sector 30, Sector 31, and Sector 32 during the 2011–2025 period in BAU scenario. Sector 29 is Information Transmission, Computer Services, and Software; Sector 30 is Wholesale and Retail Trades; Sector 31 is Hotels and Catering Services; and Sector 32 is Financial Intermediation. The reaction coefficients of Sector 29 and Sector 31 remain at the same level for 15 years, respectively. The reaction coefficients of Sector 30 and Sector 32 increase more or less, which reflects the increase in the pushing effect of Wholesale and Retail Trades and Financial Intermediation in China's national economy.

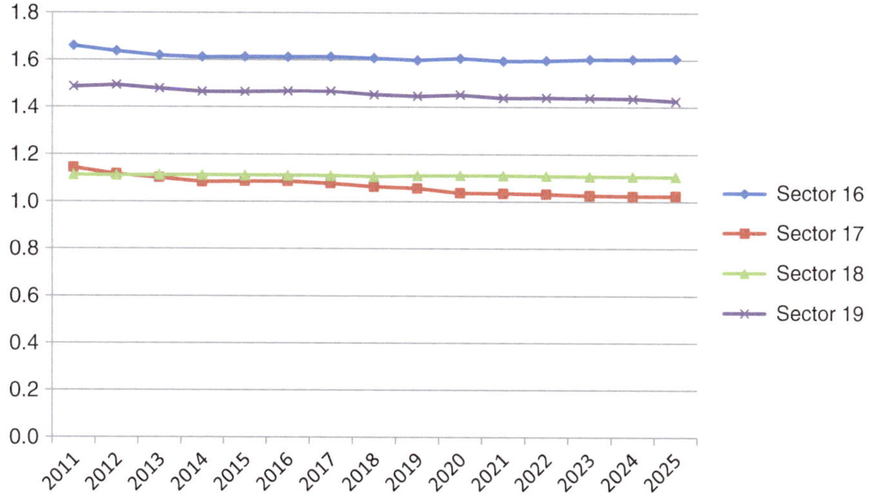

Fig. 10.9 The reaction coefficients of Sector 16, Sector 17, Sector 18, and Sector 19 during the 2011–2025 period in BAU scenario

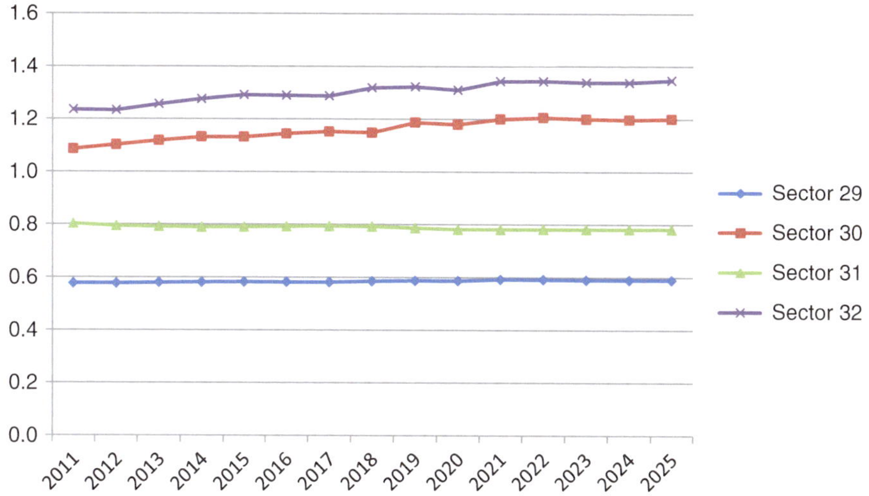

Fig. 10.10 The reaction coefficients of Sector 29, Sector 30, Sector 31, and Sector 32 during the 2011–2025 period in BAU scenario

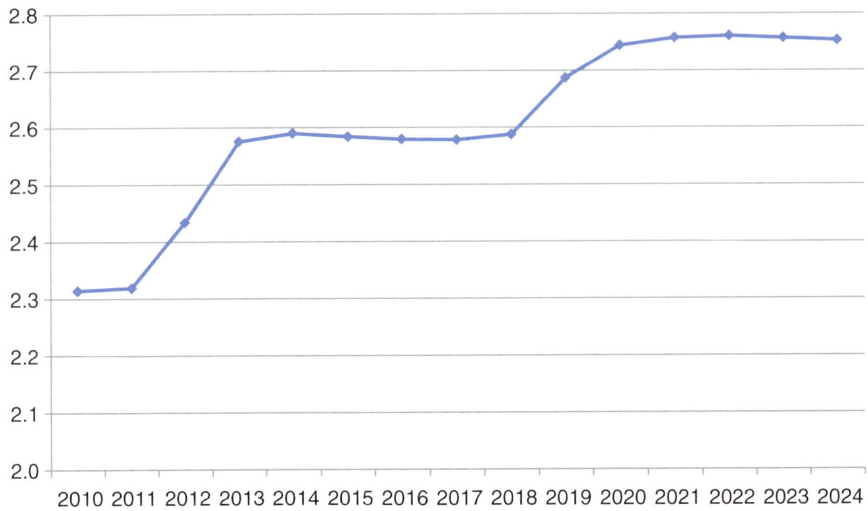

Fig. 10.11 The reaction coefficients of Sector 23 (Production and Supply of Electric Power and Heat Power) during the 2011–2025 period in BAU scenario

Based on the electricity economics, we pay attention to the situation of Sector 23 (Production and Supply of Electric Power and Heat Power). Figure 10.11 pictures the trends of the reaction coefficients of the sector during the 2011–2025 period in BAU scenario. It is clear that the supporting role of this sector will be greater and greater.

10.3 Input–Output Tables of 2015–2025 (Policy Scenario)

10.3.1 Input–Output Tables

In this section, we pay attention to the period of 2015–2025 in the policy scenario in which the monetary policies and fiscal policies are used in the ARE simulation. The input–output table in each year within the period in this scenario is going to be displayed. The unit of the numbers in all the tables is million CNY in constant 2010 CNY.

China's national input–output table in 2015 in the policy scenario is shown in Table 10.17.

China's national input–output table in 2016 in the policy scenario is exhibited in Table 10.18.

China's national input–output table in 2017 in the policy scenario is shown in Table 10.19.

China's national input–output table in 2018 in the policy scenario is shown in Table 10.20.

China's national input–output table in 2019 in the policy scenario is displayed in Table 10.21.

China's national input–output table in 2020 in the policy scenario is shown in Table 10.22.

China's national input–output table in 2021 in the policy scenario is shown in Table 10.23.

China's national input–output table in 2022 in the policy scenario is displayed in Table 10.24.

China's national input–output table in 2023 in the policy scenario is shown in Table 10.25.

China's national input–output table in 2024 in the policy scenario is shown in Table 10.26.

China's national input–output table in 2025 in the policy scenario is exhibited in Table 10.27.

Table 10.17 China's national input–output table in 2015 in the policy scenario

	Sector 1	Sector 2	Sector 3	Sector 4	Sector 5	Sector 6	Sector 7	Sector 8
Sector 1	1,457,232	29,243	26	2460	294	3,916,274	1,031,322	147,888
Sector 2	9033	538,476	11,600	6171	2056	27,011	25,405	12,167
Sector 3	0	0	22,375	0	0	0	0	0
Sector 4	0	0	0	158,307	0	0	0	0
Sector 5	1260	1083	710	1430	54,080	3556	0	0
Sector 6	1,144,950	5280	2428	3593	2217	2,039,812	37,540	167,990
Sector 7	1484	782	1075	829	432	5103	1,726,241	1,089,009
Sector 8	3859	10,557	4789	6409	3291	12,961	82,175	490,567
Sector 9	11,380	24,334	5151	4922	1457	10,599	4895	6294
Sector 10	10,552	3577	2877	2677	1756	184,529	55,487	52,659
Sector 11	78,028	33,023	50,978	84,881	36,766	27,780	12,882	20,585
Sector 12	835,963	69,100	50,428	76,473	79,717	272,927	576,920	190,590
Sector 13	15,439	27,951	12,535	12,352	28,868	68,331	12,118	4456
Sector 14	2614	127,132	66,098	19,180	5338	11,154	3903	4554
Sector 15	36,261	55,759	10,401	30,474	12,215	45,778	8613	14,811
Sector 16	61,717	161,517	112,122	119,359	68,234	33,817	60,879	21,243
Sector 17	32,243	21,292	9223	16,015	16,773	12,551	10,760	5042
Sector 18	2983	45,264	12,539	19,843	7553	14,412	11,039	6190
Sector 19	1010	4880	1745	2171	881	3645	3771	3564
Sector 20	5835	23,812	12,792	6906	2726	6972	7357	2627
Sector 21	40,202	21,272	1864	6143	8226	15,143	42,757	37,529
Sector 22	737	353	22	1223	7751	7703	8485	1593
Sector 23	138,473	136,659	98,947	251,168	90,645	102,603	127,895	26,408
Sector 24	406	1201	1672	7827	4112	5858	2677	1177
Sector 25	1307	2171	1063	1338	782	4008	4180	1492
Sector 26	2867	3943	1530	576	279	2791	1190	1004
Sector 27	198,670	139,493	26,267	65,975	48,429	283,796	93,074	78,543
Sector 28	7864	807	217	255	230	2233	3302	6093
Sector 29	41,001	8815	2297	2249	1839	17,577	10,233	16,158
Sector 30	206,240	34,759	12,419	16,395	11,877	239,044	93,937	65,669
Sector 31	26,754	30,883	12,026	16,690	10,542	58,638	19,249	17,680
Sector 32	120,482	67,913	11,106	17,172	12,580	101,825	137,107	45,663
Sector 33	2657	4208	605	899	1023	17,019	15,666	32,677
Sector 34	16,168	16,064	2462	5958	3252	163,480	19,111	69,866
Sector 35	19,235	4253	2476	500	192	8032	3318	1691
Sector 36	97,266	27,132	15,264	19,725	8300	34,059	12,601	11,118
Sector 37	30,006	3703	1104	980	1984	3035	6983	432
Sector 38	35,586	21,527	5061	3349	2815	11,932	5049	5302
Sector 39	3794	2685	250	576	932	1361	507	335
Sector 40	6263	6850	486	2832	1527	4930	2317	2153
Sector 41	2531	4062	1298	3622	3058	10,430	5348	5490
Sector 42	3322	725	410	312	143	1384	574	421
Total intermediate input	4,713,677	1,722,537	588,740	1,000,212	545,169	7,794,094	4,286,864	2,668,729
Compensation of employees	6,094,738	606,067	198,573	148,209	94,815	606,175	457,909	347,037
Net taxes on production	12,383	338,074	225,483	95,632	57,694	683,710	182,237	90,053
Depreciation of fixed assets	299,091	115,828	132,559	34,980	24,581	187,856	120,270	40,876
Operating surplus	0	350,785	310,655	181,809	78,728	603,766	187,841	116,330
Total value added	6,406,212	1,410,754	867,270	460,630	255,818	2,081,506	948,256	594,296
Total input	11,119,889	3,133,291	1,456,010	1,460,842	800,987	9,875,600	5,235,120	3,263,025

Sector 9	Sector 10	Sector 11	Sector 12	Sector 13	Sector 14	Sector 15	Sector 16	Sector 17
234,920	144,781	98	461,532	3552	1255	995	1086	659
15,250	25,540	272,866	212,214	452,125	450,301	15,999	34,327	10,358
0	0	2,471,834	424,121	15,511	47,661	0	0	0
0	0	0	67,533	15,829	2,518,713	38,145	29,664	4798
0	0	443	212,914	342,876	30,019	13,024	7056	2146
10,462	11,592	20,436	221,282	22,482	41,908	11,729	32,964	32,016
18,855	50,968	224	94,570	13,095	1319	5078	14,142	16,551
33,623	20,611	1823	30,564	15,545	27,415	10,875	21,603	92,751
564,418	60,925	1346	22,165	42,507	9157	61,449	32,442	89,264
29,815	712,360	2381	194,467	142,207	12,466	33,607	44,016	30,835
14,027	13,566	266,331	811,738	250,413	513,450	35,868	102,930	42,045
177,162	372,003	82,182	5,318,389	468,134	169,584	132,995	363,485	510,907
16,616	7697	14,876	103,713	1,060,596	241,311	38,299	72,358	68,107
44,227	31,628	2807	133,987	164,166	3,314,304	1,180,904	2,115,705	1,064,525
48,503	40,035	14,118	115,183	163,263	108,308	424,350	422,417	161,538
28,205	36,945	56,388	198,730	168,908	380,475	166,305	2,265,577	749,168
8717	22,019	9893	40,994	33,072	65,380	32,909	151,847	2,637,813
6411	18,012	4967	35,096	24,199	39,167	27,864	550,854	344,258
818	24,659	598	8143	3876	3159	4200	209,894	102,915
1554	15,107	6231	83,994	14,232	36,260	8663	52,658	128,636
16,024	19,304	1936	85,441	50,999	39,599	19,052	69,607	129,281
6943	186,670	865	64,757	105,968	443,321	34,794	119,447	8100
63,472	84,167	72,426	709,120	484,747	754,209	255,334	224,972	152,651
882	805	6425	33,067	7047	15,951	3380	8973	5699
1239	5749	1321	14,481	6168	8106	2946	3360	4217
512	694	2116	5184	2132	3956	663	3479	3205
62,493	66,024	83,637	363,984	256,555	229,831	77,114	288,007	168,925
1311	1295	130	9917	1548	3438	621	8927	1807
7035	6757	13,507	46,827	13,576	104,663	10,455	31,780	21,299
49,135	71,850	51,551	282,119	161,562	218,360	79,910	257,576	448,439
10,139	14,008	7685	74,059	43,350	42,428	39,708	77,756	33,218
30,545	42,333	25,725	258,818	194,348	203,602	40,779	107,038	89,218
8778	13,267	684	30,676	14,769	3311	15,516	28,087	14,894
17,367	22,536	20,257	190,089	40,477	19,942	14,041	88,821	116,907
1153	1558	1171	32,847	4647	19,930	3048	34,965	74,609
3573	14,031	7210	82,208	23,852	41,401	19,060	60,210	78,801
277	7784	849	12,899	1732	3498	456	2080	2625
2023	4887	3735	26,620	17,457	27,025	5522	11,840	10,834
299	366	170	1892	908	929	448	3196	1630
1714	3330	1098	8242	10,479	10,621	7754	23,894	16,970
2290	2800	970	15,111	9231	6972	3351	14,105	8244
253	417	195	1617	996	1206	451	1365	1237
1,541,039	2,179,078	3,533,505	11,141,303	4,869,135	10,213,913	2,877,660	7,994,511	7,482,102
156,237	216,125	142,419	893,749	430,751	748,123	246,267	881,393	542,926
64,872	83,240	470,196	493,594	234,556	362,190	102,524	404,446	360,695
37,315	82,160	105,546	382,536	167,338	446,197	71,912	234,640	157,365
94,257	128,161	156,537	740,361	385,873	511,618	172,485	623,591	446,969
352,681	509,685	874,698	2,510,241	1,218,518	2,068,128	593,188	2,144,071	1,507,955
1,893,721	2,688,763	4,408,203	13,651,543	6,087,654	12,282,041	3,470,848	10,138,582	8,990,057

(continued)

Table 10.17 (continued)

	Sector 18	Sector 19	Sector 20	Sector 21	Sector 22	Sector 23	Sector 24	Sector 25
Sector 1	124	0	31	253,174	177	115	50	0
Sector 2	5619	2302	1183	13,842	815	1,137,077	14,406	315
Sector 3	0	0	0	20	1	40,428	141,150	0
Sector 4	8618	115	965	192	538	191	0	0
Sector 5	2790	50	84	5813	68	1737	12	24
Sector 6	25,894	27,927	2570	28,584	700	19,688	1118	620
Sector 7	11,988	2014	1161	189,118	77	333	35	71
Sector 8	14,486	13,342	2394	31,177	120	21,449	3365	1169
Sector 9	18,270	12,405	2185	64,025	90	3994	351	349
Sector 10	74,262	62,578	7644	37,510	123	12,303	756	619
Sector 11	28,942	20,632	3743	15,335	1203	190,067	6237	2457
Sector 12	541,238	486,428	68,659	159,776	11,751	20,072	1926	11,710
Sector 13	136,852	136,100	28,104	43,893	1447	13,761	692	692
Sector 14	1,739,854	235,800	31,500	199,527	5810	6237	739	274
Sector 15	280,201	169,056	31,557	88,306	609	38,333	2857	6392
Sector 16	329,979	79,648	22,242	8385	359	64,919	2422	2521
Sector 17	18,608	26,845	4885	21,412	124	72,300	6572	2943
Sector 18	943,327	218,774	40,138	13,116	914	357,904	696	797
Sector 19	395,859	3,453,410	185,612	5218	46	3972	157	191
Sector 20	58,638	310,946	99,154	4528	54	86,996	1881	1286
Sector 21	66,786	61,286	11,973	229,837	108	4317	224	245
Sector 22	61,036	8247	98	415	5318	708	0	0
Sector 23	88,619	100,132	9154	28,195	2473	2,311,939	5995	55,581
Sector 24	3974	4043	541	2001	121	7519	16,709	773
Sector 25	2549	3096	561	1422	115	8054	160	4971
Sector 26	1751	1441	221	328	57	2218	222	288
Sector 27	154,560	127,880	17,420	49,761	4218	78,857	7931	2248
Sector 28	4510	2622	914	592	13	764	39	74
Sector 29	21,113	72,235	3091	14,135	168	39,625	428	877
Sector 30	199,022	279,767	19,879	41,710	2583	51,253	5036	1439
Sector 31	43,784	22,554	7317	15,002	516	18,546	1376	1890
Sector 32	95,374	266,817	6138	22,903	2376	268,635	9943	13,659
Sector 33	20,383	19,421	3648	14,228	180	2835	315	125
Sector 34	150,466	86,121	10,119	17,635	437	28,842	1321	1167
Sector 35	28,363	55,267	4133	2052	45	4306	79	49
Sector 36	38,019	38,772	4911	4494	221	51,898	691	361
Sector 37	1274	894	78	335	32	17,284	131	13,586
Sector 38	5456	5048	659	889	478	49,569	865	1871
Sector 39	953	703	150	350	5	1001	70	182
Sector 40	11,784	3772	1385	327	87	7727	842	52
Sector 41	7066	6443	850	2315	80	14,593	567	712
Sector 42	737	748	105	298	0	924	25	47
Total intermediate input	5,643,127	6,425,683	637,158	1,632,176	44,658	5,063,287	238,395	132,629
Compensation of employees	417,700	580,180	82,963	189,165	6478	517,964	32,089	48,325
Net taxes on production	161,633	117,852	23,396	127,465	3880	314,971	7445	9523
Depreciation of fixed assets	113,981	182,158	16,923	42,694	2913	649,165	9964	41,618
Operating surplus	307,390	217,087	42,354	104,191	395,289	210,737	14,568	5580
Total value added	1,000,705	1,097,277	165,637	463,515	408,560	1,692,836	64,065	105,045
Total input	6,643,831	7,522,960	802,794	2,095,691	453,217	6,756,123	302,460	237,675

Sector 26	Sector 27	Sector 28	Sector 29	Sector 30	Sector 31	Sector 32	Sector 33	Sector 34
71,810	168,870	0	0	1571	368,899	0	352	7961
12,967	24,367	266	0	1980	2130	0	910	4731
0	8790	0	0	0	0	0	0	0
0	0	0	0	0	0	0	0	0
165,113	1833	0	0	0	0	0	0	0
65,250	55,659	473	6246	33,771	1,110,283	14,010	12,064	39,692
13,798	9142	288	314	7829	15,712	715	4774	43,365
62,206	50,533	1369	4848	57,349	12,056	13,181	19,367	41,837
286,143	8620	946	2455	10,717	4417	3757	5195	8474
27,313	25,966	5036	79,860	60,134	10,869	97,150	25,180	300,612
210,697	1,318,577	5605	4938	21,730	15,720	31,609	56,111	106,804
504,198	124,462	630	13,572	58,894	24,363	16,964	71,713	120,186
3,703,504	14,985	657	695	1768	6551	803	1272	1781
2,037,548	36,419	0	26	3180	144	0	3852	2563
531,442	39,360	148	5150	2611	3733	4522	26,062	67,359
483,823	170,308	1990	20,721	14,031	3749	21,292	15,217	22,023
67,557	497,803	15,020	18,060	56,906	10,790	14,553	16,395	102,087
618,242	19,885	336	176,369	44,301	22,900	10,280	47,634	94,082
39,401	17,823	1667	153,069	25,387	1303	3741	12,869	193,821
35,691	11,262	225	22,966	2119	100	19,198	8392	9989
53,204	6472	480	2714	4971	1991	5208	5549	20,334
0	0	0	0	0	0	0	0	0
122,070	84,759	2595	63,876	76,178	68,296	47,155	25,708	31,993
1933	6643	181	1458	1333	11,663	248	453	3239
3924	4260	305	3130	2493	5548	2366	783	2190
173,364	30,423	8031	4342	16,808	7313	7238	82,652	3275
1,522,661	596,930	27,076	78,543	456,675	38,870	124,086	32,412	165,092
3524	2221	7966	3347	8794	1836	21,539	2742	15,778
236,640	50,797	3859	110,998	52,990	9822	142,700	16,735	12,543
402,121	91,477	14,241	80,473	33,063	67,581	21,219	18,323	77,814
163,436	112,684	1366	50,315	125,098	21,629	184,573	69,962	221,372
154,631	417,154	2225	53,027	222,387	52,896	402,296	183,909	152,238
9524	34,508	2891	66,987	144,755	33,879	160,465	56,500	55,975
58,903	39,270	7371	128,423	317,598	38,183	305,234	144,947	146,894
12,242	2175	103	4286	10,161	28	7022	904	1675
189,041	8323	34	12,522	25,117	589	3472	8230	2502
803	2133	0	717	1506	1048	3216	1379	28,818
22,887	160,907	303	16,402	60,237	29,825	19,455	31,597	51,958
2034	5280	190	3000	3895	833	9742	1013	1614
4852	3815	36	764	1047	538	3464	1483	548
22,705	14,556	1421	14,334	25,383	8258	29,978	22,693	14,398
2085	1994	63	975	1621	492	2858	2841	639
12,099,289	4,281,444	115,394	1,209,921	1,996,390	2,014,836	1,755,310	1,038,174	2,178,259
2,427,535	1,145,027	54,460	344,481	1,343,888	723,528	1,034,153	352,104	643,868
674,146	230,063	4909	101,330	1,333,070	137,399	396,703	507,560	163,291
235,837	509,220	12,834	332,612	312,380	162,194	76,372	1,554,593	312,987
924,757	906,985	157	567,481	1,903,621	171,851	1,750,129	808,157	92,164
4,262,275	2,791,294	72,361	1,345,904	4,892,958	1,194,971	3,257,357	3,222,414	1,212,311
16,361,564	7,072,738	187,755	2,555,825	6,889,348	3,209,807	5,012,666	4,260,588	3,390,569

(continued)

Table 10.17 (continued)

	Sector 35	Sector 36	Sector 37	Sector 38	Sector 39	Sector 40	Sector 41	Sector 42
Sector 1	17,048	5646	16,125	20,782	4730	12,349	3776	0
Sector 2	783	1899	3455	7240	5026	10,414	626	10,818
Sector 3	0	0	0	0	0	0	0	0
Sector 4	4266	882	1164	0	0	0	0	0
Sector 5	70	44	434	93	913	1318	379	1526
Sector 6	8052	3476	7016	40,369	19,933	21,501	43,384	14,019
Sector 7	9163	937	2856	26,754	4333	14,346	2272	15,721
Sector 8	3499	2373	8622	12,526	2775	35,365	12,896	51,400
Sector 9	1535	1806	8506	16,761	30,155	19,238	9269	62,623
Sector 10	5048	19,999	4524	29,670	86,391	32,102	87,081	100,375
Sector 11	4139	29,011	24,496	32,635	19,485	10,524	5030	55,980
Sector 12	49,210	31,263	44,869	154,920	22,216	1,098,824	31,778	39,012
Sector 13	3565	2752	9285	1890	12,117	14,369	4960	18,797
Sector 14	2271	10,421	5027	4562	1848	123	357	0
Sector 15	11,540	30,319	5319	8828	5299	5381	3077	4563
Sector 16	6809	11,403	13,959	4317	4927	115,118	2166	8958
Sector 17	4022	20,706	23,474	67,838	16,433	2859	3621	50,075
Sector 18	19,975	59,116	17,399	29,174	7279	8199	4273	8774
Sector 19	17,851	53,356	9591	43,839	7980	1896	4621	29,649
Sector 20	15,973	32,177	9183	17,316	24,033	6007	2902	6777
Sector 21	2065	3137	3497	14,139	4206	3233	1563	8725
Sector 22	0	0	0	0	0	0	0	0
Sector 23	9223	9813	19,003	19,725	45,034	29,903	6958	52,351
Sector 24	149	361	610	1147	1589	1196	518	1159
Sector 25	703	1676	1243	2918	3340	1715	613	2205
Sector 26	1611	4284	16,652	13,638	11,868	49,837	6306	53,922
Sector 27	18,784	84,230	15,405	37,364	63,261	12,891	22,777	160,151
Sector 28	761	1692	638	2477	5163	825	1764	41,070
Sector 29	2577	7016	5922	6080	20,019	23,423	6249	71,628
Sector 30	9444	16,677	9523	31,803	18,178	76,803	15,899	39,354
Sector 31	17,556	58,262	15,923	28,019	59,233	24,777	24,114	244,609
Sector 32	1951	44,929	21,273	37,721	62,052	14,596	11,582	61,145
Sector 33	2065	14,542	2013	50,047	17,641	4911	9965	20,785
Sector 34	4288	21,405	9847	25,368	42,817	18,293	16,984	34,340
Sector 35	10,872	3270	642	48	4941	1166	779	52
Sector 36	6576	88,712	3439	1371	5030	3594	528	5354
Sector 37	284	300	15,865	2581	3120	648	690	8905
Sector 38	1349	12,760	16,685	69,272	17,348	14,035	7286	54,930
Sector 39	1523	1406	1102	546	10,648	2446	739	29,287
Sector 40	1196	1322	2485	575	3498	6335	1108	10,487
Sector 41	7047	7227	4857	7151	14,266	3688	23,664	47,778
Sector 42	116	453	263	411	461	370	160	1027
Total intermediate input	284,961	701,059	382,191	871,915	689,586	1,704,620	382,714	1,428,333
Compensation of employees	110,681	379,352	145,362	610,456	1,743,853	791,453	223,640	2,188,465
Net taxes on production	7068	63,445	11,932	84,331	11,220	8258	38,557	11,881
Depreciation of fixed assets	20,769	72,162	58,265	49,632	207,265	96,077	59,872	322,186
Operating surplus	24,113	194,926	43,732	171,880	29,256	56,674	70,301	69,258
Total value added	162,631	709,885	259,293	916,299	1,991,594	952,462	392,371	2,591,791
Total input	447,591	1,410,944	641,484	1,788,215	2,681,180	2,657,082	775,084	4,020,123

Total intermediate use	Rural household consumption expenditure	Urban household consumption expenditure	Government consumption expenditure	Gross fixed capital formation	Increase in inventories	Export	Import	Total output
8,387,205	930,261	1,119,009	82,780	597,560	221,672	305,589	548,481	11,095,595
3,384,072	14,403	10,057	0	31,996	−87,034	16,120	228,620	3,140,995
3,171,890	0	0	0	18,192	−55,601	253,009	1,835,772	1,551,717
2,849,920	0	0	0	71,142	−19,071	7635	1,495,437	1,414,189
852,898	0	0	0	7427	−22,672	18,939	54,099	802,493
5,410,980	1,109,733	3,093,810	0	184,556	255,511	220,825	242,567	10,032,848
3,416,871	34,728	52,955	0	25,818	−27,421	2,181,813	451,960	5,232,804
1,349,122	231,413	1,061,609	0	65,618	90,604	579,653	75,977	3,302,043
1,534,991	21,537	91,241	0	164,964	−1685	269,856	193,390	1,887,514
2,709,373	13,512	95,716	0	32,164	−34,103	226,932	295,349	2,748,246
4,616,992	19,388	180,939	0	−5730	−134,219	97,644	307,273	4,467,741
13,455,595	124,143	404,785	0	162,013	−131,943	944,203	1,438,929	13,519,866
5,966,916	6580	47,176	0	60,078	−135,481	300,061	157,105	6,088,225
12,620,307	0	0	0	143,426	−188,641	365,674	697,172	12,243,595
3,084,053	12,988	69,238	0	81,996	−54,600	345,348	149,619	3,389,404
6,120,874	926	13,724	0	3,996,324	89,393	1,238,891	1,346,086	10,114,045
4,268,432	110,278	555,978	0	3,839,082	119,509	1,160,082	1,006,000	9,047,359
3,934,537	100,934	385,302	0	1,439,887	34,952	1,278,012	593,377	6,580,246
5,042,258	81,623	328,097	0	936,682	187,744	3,431,306	2,509,144	7,498,567
1,204,158	8945	36,252	0	184,832	−24,607	462,374	1,086,254	785,700
1,120,643	59,203	293,627	0	416,479	9937	199,882	72,917	2,026,853
1,074,555	0	0	0	0	6321	4000	664,597	420,278
7,090,627	60,900	224,622	0	0	−629,307	11,195	1914	6,756,123
176,721	14,182	110,542	0	1984	358	0	0	303,788
124,267	9688	105,413	0	0	1226	0	0	240,593
534,211	0	204,064	0	15,390,359	72,498	241,164	69,652	16,372,645
6,430,898	100,767	397,767	276,986	50,435	−179,449	369,632	369,697	7,077,339
181,659	1487	9670	0	0	2855	6295	18,956	183,010
1,287,735	127,353	531,184	0	519,908	43,230	120,454	53,702	2,576,162
3,925,526	377,103	1,138,968	0	463,829	47,359	779,320	0	6,732,104
2,068,726	232,967	903,040	0	0	98,640	−8109	142,217	3,153,047
4,088,113	145,857	750,010	52,524	0	108,023	50,880	46,501	5,148,905
953,325	430,957	1,651,730	0	1,055,693	120,162	0	0	4,211,866
2,483,073	20,973	302,464	128,755	0	59,141	648,505	249,350	3,393,561
368,285	0	0	233,568	0	2978	13,150	162,271	455,711
1,059,612	1776	22,731	190,100	150,371	−1537	0	0	1,423,052
186,056	5939	74,713	369,142	0	−2160	0	0	633,690
852,634	133,412	769,896	0	0	35,106	28,682	48,297	1,771,433
98,990	254,219	571,997	1,814,008	0	−27,196	4132	18,137	2,698,012
180,990	107,243	1,179,011	1,222,215	0	−23,344	3828	2677	2,667,266
400,941	26,181	139,977	209,038	0	12,312	42,790	53,539	777,702
34,741	0	0	4,018,394	0	−12,349	11,962	11,118	4,041,631

Table 10.18 China's national input–output table in 2016 in the policy scenario

	Sector 1	Sector 2	Sector 3	Sector 4	Sector 5	Sector 6	Sector 7	Sector 8
Sector 1	1,567,556	29,309	26	2460	297	4,308,584	1,090,602	160,636
Sector 2	10,120	541,254	11,647	6171	2110	29,717	28,105	14,058
Sector 3	0	0	22,466	0	0	0	0	0
Sector 4	0	0	0	158,307	0	0	0	0
Sector 5	1383	1088	713	1430	55,230	3912	0	0
Sector 6	1,208,410	5293	2438	3593	2240	2,244,149	39,623	177,668
Sector 7	1563	783	1080	829	437	5614	1,817,756	1,161,648
Sector 8	4082	10,580	4808	6409	3323	14,260	86,423	519,856
Sector 9	13,842	24,541	5172	4922	1520	11,661	5787	7922
Sector 10	11,019	3585	2889	2677	1772	203,014	57,911	55,070
Sector 11	82,349	33,101	51,184	84,881	37,169	30,563	13,555	21,871
Sector 12	921,401	69,361	50,633	76,473	81,203	300,267	630,413	212,428
Sector 13	16,733	28,048	12,586	12,352	29,372	75,176	13,047	4895
Sector 14	2840	127,614	66,365	19,180	5431	12,271	4235	4981
Sector 15	42,697	56,226	10,443	30,474	12,658	50,364	10,327	17,566
Sector 16	67,255	162,098	112,577	119,359	69,396	37,204	65,968	23,333
Sector 17	35,931	21,379	9260	16,015	17,142	13,808	11,731	5745
Sector 18	3263	45,438	12,590	19,843	7687	15,855	12,071	6821
Sector 19	1104	4898	1752	2171	896	4011	4094	3933
Sector 20	9479	24,597	12,844	6906	3154	7670	11,528	4682
Sector 21	41,894	21,349	1872	6143	8348	16,660	45,186	39,240
Sector 22	932	357	22	1223	8114	8475	10,790	1990
Sector 23	147,460	136,989	99,348	251,168	91,728	112,882	135,715	28,450
Sector 24	432	1204	1679	7827	4160	6445	2839	1255
Sector 25	1395	2177	1067	1338	792	4410	4433	1606
Sector 26	3064	3954	1536	576	283	3071	1266	1082
Sector 27	214,409	139,974	26,374	65,975	49,185	312,225	100,178	85,134
Sector 28	9310	813	218	255	239	2457	3925	7387
Sector 29	44,357	8846	2306	2249	1869	19,337	10,999	17,689
Sector 30	254,175	35,152	12,469	16,395	12,509	262,990	117,169	83,930
Sector 31	33,162	31,197	12,075	16,690	11,029	64,512	23,618	22,083
Sector 32	129,231	68,141	11,151	17,172	12,783	112,026	144,758	49,777
Sector 33	3060	4236	608	899	1053	18,723	18,051	37,798
Sector 34	17,487	16,116	2472	5958	3300	179,856	20,565	75,821
Sector 35	21,099	4268	2487	500	196	8836	3556	1911
Sector 36	106,292	27,231	15,326	19,725	8462	37,471	13,566	12,437
Sector 37	33,245	3719	1109	980	2025	3339	7749	483
Sector 38	42,612	21,700	5082	3349	2922	13,128	5966	6461
Sector 39	4075	2693	251	576	943	1497	543	361
Sector 40	6692	6867	488	2832	1545	5424	2462	2315
Sector 41	2753	4078	1303	3622	3109	11,474	5798	6004
Sector 42	3549	727	412	312	145	1523	610	453
Total intermediate input	5,121,710	1,730,979	591,126	1,000,212	555,776	8,574,863	4,582,916	2,886,780
Compensation of employees	6,556,157	609,194	199,377	148,209	96,831	666,898	482,185	367,757
Net taxes on production	13,320	339,818	226,397	95,632	58,921	752,200	191,898	95,430
Depreciation of fixed assets	321,734	116,426	133,096	34,980	25,103	206,674	126,646	43,316
Operating surplus	0	352,595	311,914	181,809	80,402	664,248	197,800	123,276
Total value added	6,891,211	1,418,034	870,784	460,630	261,257	2,290,020	998,528	629,778
Total input	12,012,921	3,149,013	1,461,910	1,460,842	817,033	10,864,883	5,581,443	3,516,558

Sector 9	Sector 10	Sector 11	Sector 12	Sector 13	Sector 14	Sector 15	Sector 16	Sector 17
235,135	155,911	104	465,897	3611	1268	997	1110	664
15,276	29,688	289,947	215,884	468,385	460,691	16,083	35,085	10,513
0	0	2,626,572	426,723	15,696	48,171	0	0	0
0	0	0	68,938	16,603	2,593,793	38,377	30,320	4894
0	0	471	215,830	353,109	30,551	13,074	7212	2172
10,470	12,229	21,715	222,884	22,819	42,233	11,749	33,693	32,161
18,868	54,556	238	95,292	13,291	1333	5090	14,455	16,598
33,647	21,745	1937	30,786	15,770	27,630	10,893	22,080	93,139
566,155	77,809	1431	22,831	45,096	9498	61,941	33,159	92,416
29,835	753,024	2530	195,735	144,139	12,559	33,661	44,989	30,957
14,037	14,433	283,003	817,909	254,339	518,143	35,942	105,204	42,221
177,390	415,515	87,327	5,390,411	480,608	172,290	133,463	371,517	516,913
16,635	8497	15,807	104,918	1,086,770	244,945	38,428	73,957	68,701
44,278	34,821	2983	135,578	168,295	3,361,086	1,184,474	2,162,458	1,075,356
48,620	48,209	15,002	118,006	172,024	111,339	426,945	431,752	166,062
28,237	40,700	59,918	201,109	173,043	385,892	166,823	2,315,641	756,762
8730	25,218	10,512	41,636	34,082	66,707	33,057	155,202	2,675,414
6419	19,934	5278	35,539	24,827	39,740	27,955	563,027	348,099
819	27,109	636	8242	3969	3203	4213	214,532	103,938
1576	32,153	6621	92,169	17,388	40,783	8872	53,821	146,676
16,038	21,002	2057	86,127	52,033	40,060	19,097	71,145	129,984
6967	223,678	919	67,072	114,115	456,924	34,977	122,087	8507
63,528	90,685	76,960	715,444	492,890	761,861	255,929	229,944	153,561
883	861	6827	33,346	7168	16,103	3387	9172	5732
1240	6229	1403	14,618	6283	8197	2953	3434	4247
512	750	2248	5232	2173	4000	665	3556	3225
62,559	72,140	88,873	367,941	262,447	232,767	77,327	294,371	170,499
1314	1594	139	10,172	1631	3546	626	9124	1854
7043	7470	14,353	47,373	13,892	106,124	10,488	32,482	21,508
49,305	94,854	54,778	291,883	173,577	227,885	80,680	263,268	463,692
10,170	17,613	8167	76,370	46,277	43,960	40,008	79,474	34,584
30,584	47,304	27,335	261,698	198,697	206,521	40,904	109,404	90,147
8796	15,352	727	31,298	15,411	3383	15,587	28,708	15,226
17,385	24,491	21,525	192,123	41,344	20,174	14,076	90,784	118,002
1155	1763	1244	33,304	4774	20,309	3060	35,737	75,421
3577	15,716	7661	83,295	24,480	42,126	19,134	61,541	79,569
278	8767	903	13,090	1783	3558	457	2125	2661
2028	5990	3968	27,325	18,415	27,854	5559	12,102	11,169
300	391	181	1910	925	938	449	3266	1642
1716	3556	1167	8314	10,659	10,725	7771	24,422	17,066
2293	3082	1030	15,292	9457	7067	3361	14,417	8336
253	445	207	1631	1013	1218	451	1396	1244
1,544,053	2,435,284	3,754,704	11,291,174	5,013,307	10,417,154	2,888,984	8,171,173	7,601,532
156,718	228,462	151,335	905,852	441,382	758,683	247,773	900,870	550,665
65,072	87,991	499,630	500,278	240,345	367,302	103,151	413,383	365,837
37,430	86,850	112,153	387,717	171,467	452,495	72,352	239,825	159,608
94,547	135,476	166,336	750,387	395,396	518,840	173,540	637,372	453,341
353,767	538,779	929,454	2,544,234	1,248,590	2,097,320	596,816	2,191,450	1,529,450
1,897,820	2,974,064	4,684,158	13,835,408	6,261,898	12,514,474	3,485,800	10,362,623	9,130,982

(continued)

Table 10.18 (continued)

	Sector 18	Sector 19	Sector 20	Sector 21	Sector 22	Sector 23	Sector 24	Sector 25
Sector 1	125	0	31	258,165	177	118	53	0
Sector 2	5747	2357	1183	14,115	815	1,162,578	15,176	337
Sector 3	0	0	0	20	1	41,335	148,694	0
Sector 4	8849	119	965	196	538	195	0	0
Sector 5	2838	51	84	5928	68	1776	12	26
Sector 6	26,111	28,190	2570	29,147	700	20,130	1178	663
Sector 7	12,113	2035	1161	192,846	77	340	37	76
Sector 8	14,608	13,468	2394	31,792	120	21,930	3545	1250
Sector 9	18,950	12,897	2185	65,287	90	4084	370	373
Sector 10	74,838	63,060	7644	38,249	123	12,579	796	661
Sector 11	29,215	20,835	3743	15,637	1203	194,330	6570	2625
Sector 12	550,292	494,983	68,660	162,925	11,751	20,522	2029	12,514
Sector 13	138,905	138,175	28,104	44,758	1447	14,069	729	740
Sector 14	1,764,900	239,390	31,500	203,460	5810	6377	779	293
Sector 15	288,580	174,326	31,557	90,047	609	39,192	3010	6830
Sector 16	334,848	80,869	22,242	8551	359	66,375	2551	2694
Sector 17	18,984	27,393	4885	21,835	124	73,922	6923	3145
Sector 18	958,024	222,306	40,138	13,375	914	365,931	733	852
Sector 19	401,714	3,509,326	185,614	5320	46	4061	166	204
Sector 20	65,502	351,557	99,163	4617	54	88,947	1982	1375
Sector 21	67,432	61,904	11,973	234,368	108	4413	236	262
Sector 22	63,243	8641	98	424	5318	724	0	0
Sector 23	89,584	101,341	9154	28,751	2473	2,363,789	6316	59,396
Sector 24	4014	4086	541	2040	121	7687	17,602	826
Sector 25	2578	3133	561	1450	115	8234	169	5312
Sector 26	1771	1459	221	335	57	2267	234	308
Sector 27	156,610	129,607	17,420	50,742	4218	80,625	8355	2403
Sector 28	4655	2712	914	604	13	781	41	79
Sector 29	21,418	73,315	3091	14,414	168	40,514	451	937
Sector 30	207,727	292,863	19,879	42,533	2583	52,403	5306	1538
Sector 31	45,415	23,441	7317	15,298	516	18,962	1450	2020
Sector 32	96,691	270,503	6138	23,354	2376	274,660	10,475	14,596
Sector 33	20,856	19,933	3648	14,508	180	2898	332	133
Sector 34	152,344	87,291	10,119	17,982	437	29,489	1391	1247
Sector 35	28,868	56,338	4133	2093	45	4402	83	52
Sector 36	38,662	39,475	4911	4583	221	53,062	728	386
Sector 37	1297	912	78	342	32	17,672	138	14,519
Sector 38	5629	5220	659	906	478	50,680	911	2000
Sector 39	963	711	150	357	5	1023	74	195
Sector 40	11,907	3816	1385	334	87	7900	887	56
Sector 41	7167	6540	850	2361	80	14,921	597	761
Sector 42	745	757	105	304	0	945	27	51
Total intermediate input	5,744,717	6,575,336	637,171	1,664,354	44,658	5,176,841	251,136	141,734
Compensation of employees	424,208	589,574	82,970	192,895	6478	529,580	33,804	51,642
Net taxes on production	164,152	119,760	23,398	129,978	3880	322,035	7843	10,177
Depreciation of fixed assets	115,757	185,107	16,925	43,535	2913	663,723	10,496	44,475
Operating surplus	312,179	220,602	42,358	106,245	395,289	215,463	15,346	5963
Total value added	1,016,296	1,115,043	165,651	472,653	408,560	1,730,802	67,489	112,256
Total input	6,761,014	7,690,379	802,822	2,137,007	453,217	6,907,643	318,625	253,990

Sector 26	Sector 27	Sector 28	Sector 29	Sector 30	Sector 31	Sector 32	Sector 33	Sector 34
75,171	177,686	0	0	1600	382,677	0	372	8519
13,574	25,639	271	0	2016	2209	0	962	5063
0	9249	0	0	0	0	0	0	0
0	0	0	0	0	0	0	0	0
172,840	1929	0	0	0	0	0	0	0
68,304	58,565	481	6733	34,385	1,151,751	15,424	12,753	42,478
14,444	9620	293	338	7971	16,299	787	5047	46,409
65,117	53,171	1394	5226	58,391	12,506	14,511	20,474	44,774
299,533	9070	963	2647	10,912	4582	4136	5491	9069
28,592	27,322	5126	86,089	61,227	11,275	106,955	26,619	321,715
220,557	1,387,412	5705	5323	22,124	16,307	34,799	59,316	114,301
527,793	130,959	641	14,631	59,964	25,273	18,676	75,810	128,623
3,876,818	15,767	669	749	1800	6795	884	1345	1906
2,132,899	38,320	0	28	3238	149	0	4072	2743
556,312	41,414	150	5552	2659	3873	4978	27,551	72,088
506,465	179,199	2026	22,338	14,286	3889	23,441	16,086	23,569
70,718	523,790	15,289	19,469	57,940	11,193	16,022	17,331	109,254
647,174	20,924	342	190,126	45,106	23,756	11,317	50,356	100,686
41,244	18,753	1697	165,008	25,848	1352	4118	13,605	207,427
37,362	11,850	229	24,757	2158	104	21,136	8871	10,690
55,694	6810	489	2926	5062	2065	5734	5866	21,761
0	0	0	0	0	0	0	0	0
127,783	89,183	2641	68,858	77,562	70,847	51,914	27,177	34,239
2024	6990	184	1571	1357	12,099	274	479	3467
4107	4482	311	3374	2538	5755	2605	828	2344
181,477	32,012	8175	4680	17,113	7586	7968	87,374	3505
1,593,917	628,092	27,561	84,669	464,973	40,322	136,610	34,264	176,681
3689	2337	8109	3608	8953	1905	23,713	2898	16,885
247,714	53,449	3928	119,655	53,953	10,189	157,103	17,691	13,423
420,939	96,253	14,496	86,749	33,664	70,105	23,361	19,370	83,277
171,085	118,566	1391	54,240	127,371	22,436	203,202	73,959	236,912
161,867	438,931	2265	57,163	226,428	54,872	442,900	194,417	162,925
9969	36,309	2943	72,212	147,385	35,145	176,661	59,728	59,905
61,659	41,320	7504	138,440	323,369	39,609	336,042	153,228	157,206
12,815	2288	105	4620	10,346	29	7730	956	1793
197,887	8757	34	13,499	25,574	611	3822	8701	2678
841	2245	0	773	1534	1087	3541	1458	30,841
23,958	169,307	308	17,681	61,331	30,939	21,419	33,403	55,606
2129	5556	194	3234	3966	864	10,725	1071	1727
5079	4014	37	823	1066	559	3814	1567	587
23,767	15,316	1446	15,452	25,844	8566	33,004	23,989	15,409
2182	2098	64	1052	1651	510	3147	3004	684
12,665,501	4,504,954	117,461	1,304,295	2,032,664	2,090,088	1,932,472	1,097,489	2,331,170
2,541,137	1,204,803	55,436	371,351	1,368,306	750,551	1,138,530	372,221	689,067
705,694	242,073	4997	109,234	1,357,291	142,530	436,742	536,559	174,754
246,874	535,804	13,064	358,556	318,055	168,251	84,080	1,643,413	334,959
968,033	954,333	160	611,744	1,938,209	178,269	1,926,768	854,330	98,634
4,461,738	2,937,012	73,657	1,450,884	4,981,862	1,239,602	3,586,120	3,406,523	1,297,413
17,127,239	7,441,966	191,117	2,755,178	7,014,526	3,329,691	5,518,592	4,504,012	3,628,584

(continued)

Table 10.18 (continued)

	Sector 35	Sector 36	Sector 37	Sector 38	Sector 39	Sector 40	Sector 41	Sector 42
Sector 1	17,602	5905	16,857	21,755	5225	13,421	4102	0
Sector 2	809	1986	3612	7579	5551	11,317	680	11,821
Sector 3	0	0	0	0	0	0	0	0
Sector 4	4405	922	1216	0	0	0	0	0
Sector 5	72	46	454	98	1009	1432	411	1667
Sector 6	8314	3635	7335	42,260	22,018	23,366	47,122	15,318
Sector 7	9461	980	2986	28,007	4786	15,591	2468	17,178
Sector 8	3613	2481	9013	13,112	3065	38,433	14,007	56,162
Sector 9	1585	1888	8892	17,546	33,309	20,907	10,067	68,424
Sector 10	5212	20,917	4730	31,060	95,427	34,886	94,583	109,674
Sector 11	4273	30,342	25,608	34,163	21,523	11,436	5463	61,166
Sector 12	50,811	32,697	46,907	162,177	24,540	1,194,145	34,515	42,626
Sector 13	3681	2878	9706	1979	13,385	15,615	5387	20,538
Sector 14	2344	10,899	5255	4776	2041	134	387	0
Sector 15	11,916	31,710	5561	9242	5853	5848	3342	4985
Sector 16	7030	11,927	14,592	4519	5443	125,104	2353	9788
Sector 17	4153	21,656	24,539	71,015	18,152	3107	3933	54,715
Sector 18	20,625	61,828	18,189	30,540	8041	8911	4641	9587
Sector 19	18,431	55,803	10,027	45,893	8815	2061	5020	32,395
Sector 20	16,493	33,653	9600	18,127	26,547	6528	3152	7405
Sector 21	2132	3281	3656	14,801	4646	3514	1697	9534
Sector 22	0	0	0	0	0	0	0	0
Sector 23	9523	10,263	19,866	20,649	49,744	32,497	7557	57,201
Sector 24	154	378	638	1201	1755	1300	563	1266
Sector 25	726	1753	1299	3055	3689	1864	666	2409
Sector 26	1664	4480	17,408	14,276	13,109	54,161	6849	58,917
Sector 27	19,395	88,094	16,104	39,114	69,878	14,009	24,739	174,988
Sector 28	786	1770	667	2593	5703	897	1916	44,875
Sector 29	2661	7338	6191	6365	22,113	25,455	6787	78,264
Sector 30	9751	17,442	9955	33,292	20,079	83,466	17,269	43,000
Sector 31	18,127	60,935	16,646	29,332	65,429	26,927	26,191	267,270
Sector 32	2014	46,991	22,239	39,488	68,542	15,862	12,579	66,810
Sector 33	2132	15,209	2104	52,391	19,487	5337	10,823	22,711
Sector 34	4427	22,387	10,295	26,556	47,296	19,880	18,447	37,521
Sector 35	11,225	3420	672	51	5458	1267	846	57
Sector 36	6790	92,782	3595	1435	5556	3906	573	5850
Sector 37	293	314	16,585	2702	3446	705	750	9730
Sector 38	1392	13,345	17,443	72,516	19,162	15,252	7913	60,019
Sector 39	1573	1470	1152	571	11,762	2659	803	32,000
Sector 40	1234	1383	2598	602	3864	6885	1203	11,459
Sector 41	7276	7559	5078	7486	15,758	4008	25,702	52,204
Sector 42	120	474	275	431	509	402	174	1122
Total intermediate input	294,226	733,221	399,545	912,756	761,715	1,852,493	415,683	1,560,658
Compensation of employees	114,279	396,755	151,962	639,050	1,926,255	860,111	242,906	2,391,212
Net taxes on production	7298	66,355	12,474	88,281	12,394	8974	41,878	12,982
Depreciation of fixed assets	21,444	75,472	60,911	51,957	228,944	104,412	65,030	352,034
Operating surplus	24,897	203,869	45,718	179,931	32,316	61,590	76,358	75,675
Total value added	167,919	742,451	271,066	959,218	2,199,909	1,035,086	426,172	2,831,903
Total input	462,145	1,475,672	670,611	1,871,974	2,961,624	2,887,579	841,854	4,392,561

Total intermediate use	Rural household consumption expenditure	Urban household consumption expenditure	Government consumption expenditure	Gross fixed capital formation	Increase in inventories	Export	Import	Total output
9,013,725	1,004,915	1,218,873	89,404	642,800	221,672	325,043	572,113	11,944,320
3,480,135	15,458	10,898	0	32,161	−87,034	16,523	252,751	3,215,391
3,338,926	0	0	0	18,266	−55,601	272,866	2,214,607	1,359,850
2,928,637	0	0	0	71,142	−19,071	7635	1,495,437	1,492,906
876,916	0	0	0	7585	−22,672	19,568	58,916	822,481
5,762,295	1,183,413	3,330,504	0	203,044	255,511	220,825	242,567	10,713,025
3,600,779	37,587	57,500	0	27,186	−27,421	2,442,520	525,892	5,612,260
1,411,924	247,198	1,144,883	0	69,536	90,604	579,653	75,977	3,467,823
1,598,960	22,704	96,662	0	165,472	−1685	269,856	224,175	1,927,794
2,854,725	14,560	101,929	0	34,000	−34,103	226,932	329,824	2,868,219
4,773,884	20,819	192,890	0	−6089	−134,219	100,835	330,633	4,617,487
13,982,077	132,288	431,333	0	164,207	−131,943	944,203	1,490,967	14,031,197
6,197,698	7283	50,872	0	61,561	−135,481	322,245	177,950	6,326,227
12,872,041	0	0	0	145,450	−188,641	365,674	735,898	12,458,628
3,195,900	13,865	74,116	0	82,497	−54,600	345,348	165,795	3,491,331
6,275,859	994	14,684	0	4,084,634	89,393	1,344,438	1,425,442	10,384,560
4,381,058	117,055	585,917	0	3,924,102	119,509	1,291,641	1,091,651	9,327,631
4,048,809	106,777	408,987	0	1,462,322	34,952	1,348,455	628,448	6,781,853
5,153,468	87,652	351,951	0	951,848	187,744	3,614,596	2,635,526	7,711,732
1,332,779	9478	37,441	0	184,848	−24,607	473,142	1,184,291	828,791
1,148,598	62,242	315,977	0	424,690	9937	206,339	78,056	2,089,726
1,145,598	0	0	0	0	6321	4132	699,407	456,643
7,262,854	60,900	224,622	0	0	−650,659	11,852	1926	6,907,643
181,935	14,884	115,304	0	2091	358	0	0	314,572
129,178	10,428	109,878	0	0	1226	0	0	250,710
564,600	0	219,866	0	16,110,584	72,498	269,907	77,081	17,160,373
6,705,768	108,723	427,096	308,427	53,068	−179,449	369,632	411,728	7,381,537
195,706	1587	10,399	0	0	2855	6534	21,797	195,284
1,354,969	135,502	568,885	0	560,461	43,230	130,601	59,319	2,734,329
4,202,013	401,432	1,216,263	0	472,256	47,359	799,038	0	7,138,361
2,205,422	249,974	970,190	0	0	98,640	−22,589	156,063	3,345,573
4,272,718	155,734	805,250	57,799	0	108,023	56,953	51,758	5,404,718
1,011,858	461,052	1,732,674	0	1,116,009	120,162	0	0	4,441,755
2,606,968	22,290	320,866	139,789	0	59,141	692,615	249,350	3,592,318
379,314	0	0	246,826	0	2978	14,663	177,834	465,947
1,101,687	1776	22,731	205,093	157,269	−1537	0	0	1,487,018
198,105	6303	78,950	391,296	0	−2160	0	0	672,494
903,108	141,994	820,431	0	0	35,106	30,083	52,957	1,877,765
105,873	275,587	618,932	1,958,758	0	−27,196	4258	19,976	2,916,234
187,166	112,702	1,252,537	1,333,825	0	−23,344	3828	2677	2,864,038
423,616	28,062	150,245	226,662	0	12,312	42,790	53,539	830,149
36,417	0	0	4,321,994	0	−12,349	13,066	11,823	4,347,307

Table 10.19 China's national input–output table in 2017 in the policy scenario

	Sector 1	Sector 2	Sector 3	Sector 4	Sector 5	Sector 6	Sector 7	Sector 8
Sector 1	1,645,563	31,485	26	2460	318	4,371,393	1,188,412	164,919
Sector 2	10,467	570,291	11,651	6171	2204	30,150	30,186	14,293
Sector 3	0	0	22,472	0	0	0	0	0
Sector 4	0	0	0	158,308	0	0	0	0
Sector 5	1440	1156	714	1430	58,088	3969	0	0
Sector 6	1,265,744	5667	2439	3593	2373	2,276,864	43,642	181,943
Sector 7	1604	824	1080	829	457	5696	1,933,341	1,178,533
Sector 8	4280	11,331	4809	6409	3526	14,468	94,658	532,500
Sector 9	14,734	26,595	5173	4922	1623	11,831	6544	8148
Sector 10	11,826	4199	2889	2677	2034	205,974	66,734	57,456
Sector 11	86,023	35,802	51,198	84,882	39,699	31,008	14,851	22,387
Sector 12	957,462	72,821	50,646	76,473	84,745	304,645	682,057	216,316
Sector 13	17,306	29,582	12,589	12,352	30,735	76,272	14,016	4977
Sector 14	2993	138,089	66,383	19,180	5804	12,450	4731	5111
Sector 15	46,478	63,037	10,446	30,474	13,871	51,098	12,491	18,273
Sector 16	70,354	172,695	112,606	119,360	73,184	37,747	72,412	23,841
Sector 17	37,050	22,239	9263	16,015	17,727	14,009	12,453	5827
Sector 18	3419	48,378	12,593	19,844	8100	16,087	13,351	6973
Sector 19	1156	5215	1752	2171	945	4069	4498	4021
Sector 20	10,477	28,005	12,847	6906	3490	7782	14,056	4890
Sector 21	44,067	24,582	1872	6143	9475	16,902	51,386	40,492
Sector 22	1062	400	22	1223	8931	8599	13,631	2130
Sector 23	152,512	143,530	99,374	251,169	95,586	114,527	145,728	28,957
Sector 24	455	1308	1679	7827	4466	6539	3150	1288
Sector 25	1472	2402	1068	1338	863	4474	4927	1653
Sector 26	3175	4166	1537	576	296	3116	1364	1102
Sector 27	226,548	153,263	26,381	65,975	53,129	316,777	112,248	87,543
Sector 28	9831	871	218	255	253	2493	4429	7575
Sector 29	46,888	9786	2307	2249	2042	19,619	12,309	18,239
Sector 30	272,599	38,690	12,472	16,395	13,526	266,824	137,102	86,734
Sector 31	36,669	36,139	12,078	16,690	12,446	65,452	28,884	23,137
Sector 32	138,737	81,215	11,154	17,172	14,916	113,659	163,951	52,076
Sector 33	3458	5282	608	899	1279	18,996	23,043	40,459
Sector 34	18,566	17,605	2473	5958	3560	182,478	23,153	78,163
Sector 35	21,897	4582	2487	500	209	8965	3812	1950
Sector 36	110,475	29,232	15,330	19,725	9001	38,017	14,653	12,699
Sector 37	35,134	4024	1109	980	2165	3388	8734	497
Sector 38	46,576	24,856	5083	3349	3279	13,319	7124	6750
Sector 39	4340	2935	251	576	1016	1519	614	372
Sector 40	7014	7362	488	2832	1641	5503	2707	2372
Sector 41	2961	4650	1303	3622	3483	11,642	6718	6247
Sector 42	3730	783	412	312	154	1545	674	465
Total intermediate input	5,376,544	1,865,074	591,281	1,000,220	590,638	8,699,864	4,978,775	2,951,311
Compensation of employees	6,882,413	641,875	199,430	148,210	101,841	676,620	512,845	376,702
Net taxes on production	13,983	358,049	226,456	95,633	61,970	763,165	204,100	97,751
Depreciation of fixed assets	337,745	122,672	133,131	34,980	26,402	209,687	134,698	44,370
Operating surplus	0	371,511	311,996	181,811	84,563	673,931	210,377	126,274
Total value added	7,234,141	1,494,107	871,012	460,634	274,776	2,323,403	1,062,020	645,096
Total input	12,610,685	3,359,180	1,462,293	1,460,853	865,414	11,023,267	6,040,796	3,596,407

Sector 9	Sector 10	Sector 11	Sector 12	Sector 13	Sector 14	Sector 15	Sector 16	Sector 17
240,661	156,911	106	507,977	3867	1283	1006	1147	769
15,456	29,860	295,667	226,577	492,268	464,144	16,164	36,287	11,319
0	0	2,678,387	575,588	21,230	51,072	0	0	0
0	0	0	73,784	17,883	2,627,051	38,710	31,357	5434
0	0	480	228,635	375,038	30,819	13,146	7459	2377
10,641	12,315	22,144	239,365	24,374	42,652	11,830	34,846	36,115
19,119	54,778	242	100,307	13,922	1343	5123	14,950	17,957
34,237	21,889	1975	33,167	16,846	27,913	10,972	22,836	105,827
576,473	78,640	1459	24,804	49,010	9611	62,366	34,294	107,766
31,204	763,186	2580	229,545	165,837	12,884	34,341	46,529	41,717
14,302	14,532	288,586	883,644	273,272	524,016	36,239	108,806	48,232
179,556	418,061	89,050	5,685,699	505,188	173,632	134,174	384,234	561,744
16,836	8544	16,119	110,483	1,144,744	246,879	38,637	76,488	74,501
45,066	35,130	3042	146,985	181,923	3,401,048	1,193,756	2,236,482	1,227,487
49,719	48,945	15,298	132,549	193,466	113,239	431,447	446,531	201,507
28,643	40,992	61,100	214,787	184,219	389,543	167,875	2,394,909	845,057
8811	25,342	10,720	43,368	35,477	67,106	33,182	160,515	2,856,010
6509	20,086	5382	37,969	26,448	40,115	28,131	582,300	387,661
831	27,295	648	8809	4223	3233	4240	221,876	115,766
1623	32,876	6752	104,443	20,081	41,636	8950	55,664	186,109
16,536	21,370	2097	97,398	59,077	41,128	19,468	73,581	147,636
7115	229,108	937	77,197	131,572	466,791	35,312	126,266	10,602
64,346	91,185	78,478	754,103	516,125	767,429	257,283	237,815	165,494
901	868	6962	36,396	7754	16,307	3418	9486	6697
1274	6284	1431	16,196	6877	8324	2987	3552	5126
519	754	2293	5531	2288	4032	669	3677	3511
63,943	72,807	90,626	403,826	286,456	236,011	78,088	304,448	201,730
1334	1610	141	10,922	1753	3584	630	9436	2064
7245	7542	14,636	52,629	15,261	107,823	10,614	33,594	25,888
50,251	96,155	55,859	320,889	191,568	231,278	81,397	272,280	531,887
10,505	17,901	8328	88,535	53,512	44,934	40,574	82,195	45,988
32,403	47,985	27,875	309,816	230,994	212,408	41,766	113,149	126,930
9365	15,660	741	39,373	18,988	3506	16,017	29,691	22,375
17,784	24,722	21,949	211,610	45,052	20,449	14,210	93,892	139,925
1176	1774	1268	35,617	5086	20,518	3082	36,961	84,816
3642	15,812	7812	89,250	26,117	42,575	19,280	63,648	88,880
283	8850	920	14,238	1927	3602	461	2198	3035
2099	6078	4047	31,452	21,025	28,435	5637	12,516	14,440
306	394	184	2103	1007	950	453	3378	1932
1747	3579	1190	8967	11,396	10,837	7830	25,258	19,293
2374	3119	1051	17,520	10,669	7214	3413	14,911	10,684
258	448	211	1767	1087	1231	455	1443	1416
1,575,092	2,463,388	3,828,773	12,233,818	5,394,909	10,548,585	2,913,332	8,450,883	8,493,703
159,574	231,545	154,320	955,475	464,927	767,703	250,385	931,708	587,836
66,257	89,178	509,486	527,684	253,166	371,669	104,238	427,534	390,532
38,112	88,022	114,365	408,956	180,614	457,875	73,115	248,034	170,382
96,270	137,305	169,617	791,493	416,489	525,009	175,370	659,190	483,942
360,214	546,050	947,789	2,683,608	1,315,197	2,122,257	603,109	2,266,467	1,632,691
1,935,306	3,009,438	4,776,562	14,917,426	6,710,105	12,670,842	3,516,440	10,717,350	10,126,394

(continued)

Table 10.19 (continued)

	Sector 18	Sector 19	Sector 20	Sector 21	Sector 22	Sector 23	Sector 24	Sector 25
Sector 1	129	0	31	258,619	177	121	54	0
Sector 2	5862	2400	1188	14,140	815	1,196,404	15,511	349
Sector 3	0	0	0	20	1	42,537	151,980	0
Sector 4	9121	122	973	196	538	201	0	0
Sector 5	2904	52	85	5938	68	1828	13	27
Sector 6	26,863	28,921	2585	29,199	700	20,716	1204	687
Sector 7	12,393	2071	1167	193,185	77	350	38	78
Sector 8	15,043	13,827	2408	31,848	120	22,568	3624	1295
Sector 9	19,551	13,297	2197	65,402	90	4202	378	386
Sector 10	80,349	66,476	7784	38,317	123	12,945	814	685
Sector 11	30,161	21,403	3768	15,665	1203	199,984	6715	2720
Sector 12	562,363	504,555	68,956	163,212	11,751	21,119	2074	12,967
Sector 13	141,899	140,733	28,223	44,837	1447	14,479	746	767
Sector 14	1,822,690	246,255	31,683	203,818	5810	6563	796	303
Sector 15	302,301	181,790	31,752	90,206	609	40,333	3076	7077
Sector 16	343,743	82,746	22,346	8566	359	68,306	2607	2792
Sector 17	19,290	27,795	4900	21,873	124	76,073	7076	3259
Sector 18	983,790	227,525	40,317	13,398	914	376,577	749	883
Sector 19	412,446	3,595,116	186,534	5330	46	4179	169	212
Sector 20	68,623	370,328	99,746	4626	54	91,535	2026	1424
Sector 21	71,351	64,649	12,160	234,780	108	4542	242	271
Sector 22	66,838	9188	98	424	5318	745	0	0
Sector 23	91,496	103,317	9199	28,802	2473	2,432,564	6455	61,544
Sector 24	4162	4209	545	2044	121	7911	17,991	856
Sector 25	2692	3245	567	1453	115	8474	173	5504
Sector 26	1812	1488	222	335	57	2333	240	319
Sector 27	162,772	133,763	17,553	50,832	4218	82,971	8539	2489
Sector 28	4791	2786	918	605	13	804	42	82
Sector 29	22,395	76,057	3125	14,439	168	41,692	461	971
Sector 30	215,772	303,800	19,993	42,607	2583	53,928	5423	1594
Sector 31	48,180	24,705	7390	15,325	516	19,513	1482	2093
Sector 32	104,131	286,608	6293	23,395	2376	282,651	10,706	15,124
Sector 33	23,036	21,708	3737	14,534	180	2983	339	138
Sector 34	158,350	90,271	10,200	18,014	437	30,347	1422	1292
Sector 35	29,653	57,752	4163	2096	45	4530	85	54
Sector 36	39,775	40,486	4946	4591	221	54,606	744	400
Sector 37	1342	940	78	342	32	18,186	141	15,044
Sector 38	5956	5491	667	908	478	52,155	931	2072
Sector 39	999	736	151	357	5	1053	76	202
Sector 40	12,273	3922	1395	335	87	8130	906	58
Sector 41	7593	6856	862	2365	80	15,355	610	789
Sector 42	769	779	106	305	0	972	27	52
Total intermediate input	5,935,660	6,768,170	641,010	1,667,280	44,658	5,327,464	256,685	146,859
Compensation of employees	435,617	603,987	83,458	193,234	6478	544,989	34,551	53,509
Net taxes on production	168,566	122,688	23,536	130,206	3880	331,405	8016	10,545
Depreciation of fixed assets	118,871	189,632	17,024	43,612	2913	683,035	10,728	46,083
Operating surplus	320,575	225,995	42,607	106,432	395,289	221,732	15,685	6178
Total value added	1,043,629	1,142,302	166,626	473,484	408,560	1,781,160	68,980	116,316
Total input	6,979,290	7,910,472	807,636	2,140,764	453,217	7,108,625	325,666	263,174

Sector 26	Sector 27	Sector 28	Sector 29	Sector 30	Sector 31	Sector 32	Sector 33	Sector 34
80,337	184,018	0	0	1701	399,769	0	385	8833
14,507	26,553	289	0	2143	2308	0	996	5249
0	9579	0	0	0	0	0	0	0
0	0	0	0	0	0	0	0	0
184,719	1998	0	0	0	0	0	0	0
72,998	60,652	512	7077	36,554	1,203,194	16,247	13,194	44,040
15,436	9963	312	356	8474	17,027	829	5221	48,116
69,593	55,066	1484	5493	62,075	13,065	15,286	21,182	46,420
320,120	9394	1025	2782	11,600	4787	4357	5681	9402
30,557	28,296	5458	90,492	65,089	11,778	112,663	27,539	333,545
235,716	1,436,858	6074	5596	23,520	17,036	36,656	61,367	118,504
564,068	135,627	683	15,379	63,747	26,402	19,673	78,431	133,353
4,143,266	16,329	712	787	1914	7099	931	1391	1976
2,279,490	39,686	0	30	3442	156	0	4213	2843
594,546	42,890	160	5836	2826	4046	5244	28,504	74,738
541,274	185,585	2156	23,480	15,187	4062	24,692	16,642	24,436
75,579	542,458	16,277	20,465	61,595	11,693	16,877	17,931	113,271
691,654	21,669	364	199,850	47,952	24,817	11,921	52,097	104,388
44,079	19,422	1806	173,448	27,479	1412	4338	14,075	215,055
39,929	12,272	244	26,023	2294	109	22,264	9178	11,084
59,522	7053	520	3076	5381	2157	6040	6068	22,561
0	0	0	0	0	0	0	0	0
136,565	92,362	2812	72,380	82,456	74,011	54,685	28,116	35,498
2163	7239	196	1652	1443	12,639	288	495	3594
4390	4642	331	3547	2698	6012	2744	856	2430
193,950	33,153	8702	4920	18,193	7925	8394	90,395	3634
1,703,465	650,476	29,341	88,999	494,307	42,123	143,901	35,449	183,178
3943	2420	8633	3792	9518	1990	24,978	2999	17,506
264,739	55,354	4182	125,775	57,356	10,644	165,487	18,302	13,917
449,869	99,683	15,432	91,186	35,788	73,236	24,607	20,040	86,339
182,843	122,792	1480	57,014	135,407	23,439	214,047	76,516	245,623
172,992	454,574	2411	60,087	240,713	57,323	466,538	201,138	168,916
10,655	37,603	3133	75,905	156,683	36,714	186,090	61,793	62,107
65,897	42,793	7988	145,521	343,770	41,378	353,977	158,526	162,986
13,696	2370	111	4856	10,998	31	8143	989	1859
211,488	9069	36	14,189	27,187	638	4026	9001	2777
899	2325	0	812	1630	1135	3730	1508	31,975
25,605	175,341	328	18,585	65,200	32,320	22,562	34,557	57,651
2276	5754	206	3400	4216	902	11,298	1108	1791
5428	4157	40	865	1134	584	4017	1622	608
25,401	15,862	1540	16,242	27,475	8949	34,766	24,818	15,975
2332	2173	68	1105	1755	533	3314	3108	709
13,535,982	4,665,505	125,046	1,371,004	2,160,900	2,183,443	2,035,611	1,135,433	2,416,888
2,715,786	1,247,740	59,015	390,344	1,454,629	784,075	1,199,295	385,090	714,404
754,195	250,700	5320	114,821	1,442,920	148,897	460,051	555,109	181,179
263,841	554,899	13,908	376,894	338,121	175,766	88,568	1,700,231	347,275
1,034,564	988,344	170	643,032	2,060,487	186,232	2,029,603	883,867	102,261
4,768,386	3,041,684	78,413	1,525,091	5,296,157	1,294,970	3,777,517	3,524,298	1,345,120
18,304,369	7,707,189	203,459	2,896,096	7,457,057	3,478,412	5,813,129	4,659,731	3,762,008

(continued)

Table 10.19 (continued)

	Sector 35	Sector 36	Sector 37	Sector 38	Sector 39	Sector 40	Sector 41	Sector 42
Sector 1	18,591	6335	17,748	22,908	5439	14,184	4305	0
Sector 2	854	2131	3803	7981	5779	11,961	714	12,468
Sector 3	0	0	0	0	0	0	0	0
Sector 4	4652	990	1281	0	0	0	0	0
Sector 5	76	49	478	103	1050	1514	432	1759
Sector 6	8781	3900	7722	44,499	22,921	24,695	49,454	16,157
Sector 7	9992	1052	3144	29,491	4983	16,478	2590	18,118
Sector 8	3816	2662	9489	13,807	3191	40,620	14,700	59,237
Sector 9	1674	2026	9362	18,476	34,675	22,097	10,565	72,171
Sector 10	5505	22,442	4980	32,706	99,341	36,872	99,264	115,679
Sector 11	4513	32,554	26,961	35,974	22,406	12,087	5733	64,515
Sector 12	53,665	35,080	49,386	170,771	25,546	1,262,103	36,224	44,960
Sector 13	3888	3088	10,219	2084	13,934	16,504	5654	21,662
Sector 14	2476	11,694	5533	5029	2124	142	406	0
Sector 15	12,585	34,022	5855	9732	6093	6181	3507	5258
Sector 16	7425	12,796	15,364	4758	5666	132,224	2469	10,324
Sector 17	4386	23,235	25,836	74,779	18,897	3283	4128	57,710
Sector 18	21,783	66,335	19,150	32,159	8370	9418	4870	10,112
Sector 19	19,467	59,872	10,557	48,325	9176	2178	5268	34,169
Sector 20	17,419	36,107	10,107	19,088	27,635	6900	3308	7811
Sector 21	2252	3520	3849	15,586	4836	3714	1781	10,056
Sector 22	0	0	0	0	0	0	0	0
Sector 23	10,058	11,012	20,915	21,744	51,785	34,347	7932	60,333
Sector 24	163	405	672	1265	1827	1374	591	1335
Sector 25	767	1881	1368	3217	3841	1970	699	2541
Sector 26	1757	4807	18,328	15,033	13,647	57,243	7188	62,143
Sector 27	20,484	94,517	16,955	41,187	72,745	14,806	25,964	184,569
Sector 28	830	1899	702	2730	5937	948	2011	47,332
Sector 29	2810	7873	6519	6702	23,020	26,903	7123	82,549
Sector 30	10,299	18,714	10,482	35,057	20,902	88,216	18,124	45,354
Sector 31	19,145	65,377	17,526	30,886	68,113	28,459	27,488	281,905
Sector 32	2127	50,416	23,415	41,581	71,353	16,765	13,202	70,468
Sector 33	2252	16,318	2215	55,168	20,286	5641	11,359	23,955
Sector 34	4676	24,019	10,839	27,964	49,236	21,011	19,360	39,575
Sector 35	11,856	3669	707	53	5681	1340	888	60
Sector 36	7171	99,546	3785	1511	5784	4128	601	6170
Sector 37	310	337	17,461	2845	3588	745	787	10,263
Sector 38	1471	14,318	18,365	76,359	19,948	16,120	8305	63,305
Sector 39	1661	1577	1213	602	12,245	2810	843	33,753
Sector 40	1304	1484	2736	634	4023	7276	1263	12,086
Sector 41	7685	8110	5346	7882	16,404	4237	26,974	55,063
Sector 42	127	508	290	454	530	424	182	1184
Total intermediate input	310,753	786,676	420,660	961,128	792,958	1,957,918	436,258	1,646,111
Compensation of employees	120,699	425,680	159,993	672,917	2,005,265	909,059	254,929	2,522,141
Net taxes on production	7708	71,193	13,133	92,960	12,902	9485	43,951	13,693
Depreciation of fixed assets	22,649	80,975	64,130	54,710	238,335	110,354	68,249	371,310
Operating surplus	26,296	218,732	48,134	189,467	33,642	65,095	80,137	79,818
Total value added	177,351	796,580	285,391	1,010,054	2,290,144	1,093,993	447,266	2,986,963
Total input	488,104	1,583,256	706,051	1,971,182	3,083,102	3,051,911	883,524	4,633,074

Total intermediate use	Rural household consumption expenditure	Urban household consumption expenditure	Government consumption expenditure	Gross fixed capital formation	Increase in inventories	Export	Import	Total output
9,341,986	1,076,748	1,314,963	96,029	674,788	221,672	344,546	595,098	12,475,634
3,597,556	16,814	11,978	0	33,887	−87,034	16,926	276,960	3,313,167
3,552,864	0	0	0	18,271	−55,601	292,894	2,303,144	1,505,284
2,970,601	0	0	0	71,142	−19,071	7635	1,592,732	1,437,575
927,843	0	0	0	7977	−22,672	20,195	63,737	869,606
5,960,019	1,245,759	3,530,789	0	206,004	255,511	220,825	242,567	11,176,340
3,751,045	40,528	62,174	0	28,915	−27,421	2,702,797	599,946	5,958,092
1,475,562	261,738	1,221,589	0	71,227	90,604	579,653	75,977	3,624,397
1,669,691	24,081	103,059	0	168,488	−1685	269,856	254,992	1,978,497
3,011,309	15,469	107,321	0	34,459	−34,103	226,932	333,744	3,027,643
4,985,167	22,101	203,593	0	−6209	−134,219	104,038	347,203	4,827,267
14,638,598	141,650	461,844	0	173,202	−131,943	944,203	1,543,055	14,684,499
6,555,628	8047	54,891	0	64,845	−135,481	344,338	198,882	6,693,386
13,400,842	0	0	0	147,180	−188,641	365,674	774,583	12,950,472
3,378,038	14,780	79,202	0	83,367	−54,600	345,348	181,991	3,664,143
6,569,331	1066	15,686	0	4,224,457	89,393	1,449,955	1,504,820	10,845,067
4,623,902	125,057	621,265	0	4,332,449	119,509	1,422,660	1,177,656	10,067,186
4,234,412	112,921	433,891	0	1,501,651	34,952	1,418,977	663,480	7,073,323
5,304,907	93,940	376,834	0	975,117	187,744	3,797,737	2,762,445	7,973,834
1,436,720	10,219	39,093	0	185,936	−24,607	483,889	1,278,413	852,836
1,219,286	64,725	334,239	0	425,436	9937	212,844	83,156	2,183,311
1,203,511	0	0	0	0	6321	4264	734,211	479,885
7,496,503	60,900	224,622	0	0	−683,971	12,509	1938	7,108,625
190,682	15,530	119,689	0	2137	358	0	0	328,395
136,400	11,126	114,087	0	0	1226	0	0	262,839
594,821	0	236,383	0	17,217,840	72,498	298,682	84,502	18,335,722
7,085,403	116,425	455,488	339,868	54,959	−179,449	369,632	453,838	7,788,488
205,609	1699	11,224	0	0	2855	6772	24,645	203,513
1,427,235	143,370	605,287	0	589,126	43,230	140,764	64,927	2,884,085
4,464,935	428,704	1,302,907	0	502,050	47,359	818,717	0	7,564,671
2,341,226	267,345	1,038,780	0	0	98,640	−37,084	169,896	3,539,012
4,581,510	165,075	857,489	63,074	0	108,023	63,044	57,001	5,781,213
1,084,272	489,849	1,810,127	0	1,154,593	120,162	0	0	4,659,002
2,751,399	23,535	338,259	150,823	0	59,141	736,920	249,350	3,810,726
400,388	0	0	260,084	0	2978	16,177	193,389	486,239
1,159,030	1776	22,731	220,085	168,735	−1537	0	0	1,570,819
208,001	6677	83,303	413,450	0	−2160	0	0	709,272
955,062	150,739	871,925	0	0	35,106	31,483	57,616	1,986,700
111,562	295,315	662,263	2,103,507	0	−27,196	4385	21,808	3,128,027
196,380	117,997	1,323,859	1,445,436	0	−23,344	3828	2677	3,061,479
448,819	29,863	160,070	244,286	0	12,312	42,790	53,539	884,601
38,197	0	0	4,625,594	0	−12,349	14,176	12,524	4,653,095

Table 10.20 China's national input–output table in 2018 in the policy scenario

	Sector 1	Sector 2	Sector 3	Sector 4	Sector 5	Sector 6	Sector 7	Sector 8
Sector 1	1,680,983	32,748	26	2554	354	4,767,968	1,215,503	180,308
Sector 2	10,574	586,498	11,655	6378	2360	32,885	30,569	14,972
Sector 3	0	0	22,481	0	0	0	0	0
Sector 4	0	0	0	159,176	0	0	0	0
Sector 5	1455	1185	714	1487	61,615	4329	0	0
Sector 6	1,276,296	5780	2440	3671	2495	2,483,422	44,087	188,501
Sector 7	1627	851	1080	858	496	6213	1,970,891	1,250,244
Sector 8	4325	11,592	4811	6545	3744	15,780	95,785	555,683
Sector 9	14,842	27,114	5175	5065	1698	12,905	6597	8405
Sector 10	12,017	4403	2890	2744	2320	224,660	68,122	61,457
Sector 11	86,766	36,653	51,218	86,475	42,222	33,821	15,016	23,210
Sector 12	971,346	75,052	50,667	79,786	91,303	332,282	693,591	230,123
Sector 13	17,456	30,253	12,594	12,649	32,543	83,191	14,168	5164
Sector 14	3014	140,714	66,410	19,605	6088	13,580	4773	5267
Sector 15	46,727	64,042	10,450	31,152	14,399	55,734	12,576	18,689
Sector 16	70,882	175,935	112,652	121,768	76,728	41,171	73,075	24,596
Sector 17	37,353	22,662	9266	16,411	18,560	15,280	12,567	6030
Sector 18	3446	49,287	12,598	20,311	8482	17,546	13,480	7206
Sector 19	1165	5321	1753	2215	995	4438	4544	4166
Sector 20	10,514	28,385	12,852	7179	3579	8488	14,107	4969
Sector 21	44,316	25,282	1873	6294	10,292	18,436	51,863	41,603
Sector 22	1064	402	22	1225	9063	9379	13,652	2143
Sector 23	154,008	146,587	99,414	258,173	100,947	124,917	147,483	30,264
Sector 24	459	1342	1680	7989	4784	7132	3190	1344
Sector 25	1495	2498	1068	1386	959	4880	5022	1770
Sector 26	3205	4257	1537	591	313	3399	1379	1147
Sector 27	228,710	157,275	26,391	67,394	56,914	345,515	113,598	91,073
Sector 28	9954	898	218	270	271	2719	4496	8004
Sector 29	47,726	10,227	2308	2352	2293	21,399	12,575	19,715
Sector 30	275,458	39,848	12,477	17,465	14,431	291,030	138,830	90,992
Sector 31	37,063	37,302	12,083	17,184	13,502	71,390	29,253	24,194
Sector 32	142,149	87,190	11,159	18,034	18,032	123,970	168,434	57,960
Sector 33	3575	5726	608	983	1577	20,720	23,945	46,807
Sector 34	18,761	18,052	2474	6106	3803	199,032	23,442	81,669
Sector 35	22,148	4727	2488	515	226	9778	3861	2051
Sector 36	112,012	30,293	15,337	20,513	9864	41,466	14,895	13,496
Sector 37	35,590	4148	1110	1025	2331	3695	8872	526
Sector 38	47,320	25,911	5085	3514	3645	14,527	7255	7226
Sector 39	4387	3007	251	590	1082	1656	622	390
Sector 40	7100	7561	488	2910	1759	6003	2747	2498
Sector 41	3019	4881	1304	3789	3954	12,698	6877	6784
Sector 42	3777	804	412	321	166	1685	684	490
Total intermediate input	5,454,082	1,916,693	591,520	1,024,653	630,189	9,489,119	5,072,427	3,121,135
Compensation of employees	7,030,555	660,117	199,510	149,023	108,025	738,003	522,806	393,102
Net taxes on production	14,284	368,224	226,547	96,157	65,733	832,400	208,064	102,006
Depreciation of fixed assets	345,015	126,158	133,185	35,172	28,006	228,710	137,315	46,301
Operating surplus	0	382,069	312,122	182,807	89,698	735,070	214,463	131,771
Total value added	7,389,854	1,536,568	871,364	463,158	291,462	2,534,183	1,082,647	673,181
Total input	12,843,936	3,453,261	1,462,884	1,487,811	921,650	12,023,302	6,155,075	3,794,315

Sector 9	Sector 10	Sector 11	Sector 12	Sector 13	Sector 14	Sector 15	Sector 16	Sector 17
280,309	161,262	114	525,701	4291	1641	1316	1271	896
16,857	30,373	318,892	230,598	527,780	548,133	19,388	40,206	12,244
0	0	2,888,774	611,753	24,454	65,391	0	0	0
0	0	0	75,187	18,745	2,902,538	43,805	34,745	5770
0	0	518	232,124	400,477	36,005	15,429	8264	2538
11,301	12,468	23,883	242,936	25,568	47,477	13,326	38,610	38,102
21,126	55,979	261	102,660	15,139	1632	6406	16,564	19,933
36,835	22,190	2130	33,771	17,818	31,706	12,618	25,303	113,720
612,858	79,689	1574	25,125	51,402	10,756	70,038	37,998	113,629
36,582	780,207	2783	238,756	183,889	16,432	45,784	51,555	48,691
15,339	14,704	311,254	899,377	288,341	590,700	41,539	120,559	51,560
198,360	428,339	96,045	5,808,493	548,687	212,088	166,325	425,739	622,023
18,001	8660	17,385	112,188	1,211,421	281,408	44,577	84,750	79,287
47,606	35,548	3280	149,014	190,301	3,774,525	1,336,576	2,478,063	1,286,030
51,893	49,425	16,499	134,018	200,505	123,039	473,983	494,765	207,923
30,289	41,464	65,899	217,744	192,785	432,981	188,560	2,653,603	888,721
9354	25,684	11,562	43,926	37,277	75,417	37,438	177,853	3,025,639
6884	20,347	5805	38,481	27,717	44,791	31,795	645,199	407,278
885	27,625	699	8943	4436	3624	4811	245,843	122,501
1703	33,190	7282	105,161	20,703	45,137	9674	61,676	191,253
17,891	21,643	2262	99,388	62,516	48,314	23,746	81,529	154,744
7227	229,388	1011	77,588	132,883	477,641	36,320	139,905	10,632
69,317	92,684	84,643	766,461	545,427	876,396	298,495	263,504	176,213
977	880	7509	37,150	8240	18,774	4013	10,510	7268
1457	6437	1543	16,706	7566	10,439	3847	3935	5873
556	766	2473	5618	2420	4596	771	4075	3736
69,301	73,822	97,745	411,988	304,467	271,825	91,709	337,334	218,388
1460	1650	152	11,131	1899	4376	780	10,456	2240
8438	7764	15,785	54,434	17,013	139,804	14,184	37,223	29,985
54,733	98,416	60,246	326,511	206,236	279,318	99,794	301,691	569,203
11,596	18,179	8982	90,629	57,499	53,357	49,280	91,074	50,609
42,123	49,904	30,064	327,316	270,242	307,251	62,257	125,371	160,940
12,413	16,506	800	41,890	23,081	5578	27,021	32,898	28,370
19,309	25,106	23,673	215,912	47,854	23,519	16,614	104,034	151,498
1300	1804	1368	36,402	5472	24,283	3668	40,953	93,375
4076	16,153	8426	91,499	28,505	52,293	24,008	70,523	98,841
311	9059	993	14,547	2081	4346	567	2436	3314
2413	6234	4364	32,426	23,239	36,128	7343	13,868	16,469
332	400	199	2144	1068	1084	525	3743	2080
1904	3639	1283	9155	12,185	12,625	9282	27,986	20,972
2783	3213	1133	18,190	11,964	9490	4646	16,521	12,530
282	455	228	1805	1163	1437	541	1599	1541
1,726,379	2,511,256	4,129,523	12,524,847	5,764,756	11,908,296	3,342,799	9,363,735	9,056,558
169,645	236,709	166,442	976,111	492,008	852,006	275,071	1,032,350	622,750
70,439	91,167	549,507	539,080	267,912	412,483	114,515	473,716	413,727
40,518	89,985	123,349	417,788	191,134	508,156	80,324	274,827	180,501
102,346	140,367	182,941	808,587	440,748	582,661	192,660	730,394	512,685
382,949	558,228	1,022,238	2,741,566	1,391,802	2,355,306	662,569	2,511,287	1,729,663
2,109,328	3,069,485	5,151,761	15,266,414	7,156,558	14,263,602	4,005,368	11,875,021	10,786,221

(continued)

Table 10.20 (continued)

	Sector 18	Sector 19	Sector 20	Sector 21	Sector 22	Sector 23	Sector 24	Sector 25
Sector 1	154	0	38	264,229	177	128	56	0
Sector 2	6514	2592	1319	14,446	815	1,267,950	16,286	361
Sector 3	0	0	0	20	1	45,081	159,576	0
Sector 4	9834	127	1095	201	538	213	0	0
Sector 5	3192	56	92	6067	68	1937	13	28
Sector 6	28,968	30,512	2787	29,832	700	21,954	1264	711
Sector 7	14,177	2269	1317	197,375	77	371	40	81
Sector 8	16,457	14,737	2638	32,539	120	23,918	3805	1341
Sector 9	21,007	14,030	2357	66,821	90	4454	397	400
Sector 10	96,200	73,852	9771	39,148	123	13,719	855	709
Sector 11	32,854	22,642	4128	16,004	1203	211,943	7051	2816
Sector 12	641,140	555,012	77,203	166,752	11,751	22,382	2178	13,423
Sector 13	154,605	149,217	30,579	45,809	1447	15,345	783	794
Sector 14	1,955,387	258,308	33,924	208,239	5810	6955	836	314
Sector 15	320,241	188,927	33,454	92,162	609	42,745	3230	7326
Sector 16	369,468	86,887	23,917	8751	359	72,391	2738	2890
Sector 17	20,782	29,333	5241	22,347	124	80,622	7430	3374
Sector 18	1,060,487	239,511	43,090	13,689	914	399,097	787	914
Sector 19	446,324	3,802,516	201,354	5445	46	4429	178	219
Sector 20	71,764	384,119	105,155	4726	54	97,009	2127	1475
Sector 21	78,607	68,187	14,239	239,873	108	4813	254	281
Sector 22	68,174	9298	100	434	5318	790	0	0
Sector 23	100,218	110,543	10,118	29,426	2473	2,578,035	6778	63,709
Sector 24	4604	4492	605	2088	121	8384	18,890	886
Sector 25	3153	3602	674	1484	115	8981	181	5698
Sector 26	1978	1583	241	343	57	2473	251	331
Sector 27	179,840	142,422	19,469	51,934	4218	87,933	8966	2577
Sector 28	5429	3059	1015	618	13	852	44	85
Sector 29	26,773	85,881	3806	14,752	168	44,186	484	1005
Sector 30	242,085	331,433	22,061	43,532	2583	57,153	5694	1650
Sector 31	54,244	26,699	8352	15,658	516	20,680	1556	2166
Sector 32	134,368	338,085	9074	23,903	2376	299,554	11,241	15,656
Sector 33	31,830	27,396	5360	14,849	180	3161	356	143
Sector 34	174,939	96,585	11,333	18,405	437	32,161	1493	1338
Sector 35	33,055	62,633	4734	2142	45	4801	89	56
Sector 36	45,439	44,654	5694	4690	221	57,872	782	414
Sector 37	1521	1029	88	350	32	19,274	148	15,573
Sector 38	7004	6154	793	927	478	55,274	977	2145
Sector 39	1100	789	167	365	5	1116	80	209
Sector 40	13,664	4242	1552	342	87	8616	952	60
Sector 41	9200	7793	1058	2416	80	16,273	641	817
Sector 42	858	845	118	312	0	1030	29	54
Total intermediate input	6,487,640	7,232,052	700,109	1,703,445	44,658	5,646,053	269,514	152,027
Compensation of employees	469,578	638,830	87,985	197,425	6478	577,580	36,277	55,392
Net taxes on production	181,708	129,765	24,812	133,031	3880	351,223	8417	10,916
Depreciation of fixed assets	128,138	200,572	17,947	44,558	2913	723,881	11,265	47,705
Operating surplus	345,568	239,033	44,918	108,741	395,289	234,992	16,469	6396
Total value added	1,124,991	1,208,201	175,662	483,755	408,560	1,887,676	72,428	120,409
Total input	7,612,631	8,440,253	875,771	2,187,200	453,217	7,533,729	341,942	272,436

Sector 26	Sector 27	Sector 28	Sector 29	Sector 30	Sector 31	Sector 32	Sector 33	Sector 34
87,515	197,156	0	0	1878	429,768	0	398	9603
15,803	28,448	308	0	2366	2481	0	1030	5707
0	10,262	0	0	0	0	0	0	0
0	0	0	0	0	0	0	0	0
201,222	2141	0	0	0	0	0	0	0
79,520	64,982	547	7382	40,360	1,293,482	17,012	13,647	47,883
16,815	10,674	333	371	9356	18,305	868	5401	52,314
75,810	58,997	1584	5730	68,538	14,045	16,006	21,908	50,471
348,720	10,064	1094	2902	12,808	5146	4562	5876	10,222
33,287	30,316	5824	94,390	71,866	12,662	117,971	28,484	362,647
256,775	1,539,441	6482	5837	25,969	18,314	38,383	63,473	128,844
614,463	145,310	729	16,042	70,383	28,383	20,600	81,122	144,988
4,513,433	17,495	760	821	2113	7632	975	1439	2148
2,483,144	42,519	0	31	3800	168	0	4357	3092
647,664	45,952	171	6087	3121	4349	5491	29,482	81,259
589,632	198,835	2301	24,491	16,769	4367	25,856	17,213	26,568
82,331	581,186	17,371	21,346	68,008	12,570	17,672	18,546	123,154
753,447	23,216	389	208,457	52,944	26,679	12,483	53,884	113,496
48,017	20,808	1927	180,918	30,340	1518	4542	14,558	233,818
43,497	13,148	260	27,144	2533	117	23,313	9493	12,051
64,840	7556	555	3208	5941	2319	6325	6277	24,530
0	0	0	0	0	0	0	0	0
148,766	98,956	3001	75,497	91,040	79,565	57,261	29,081	38,595
2356	7756	209	1723	1593	13,587	302	512	3908
4782	4973	353	3699	2979	6463	2873	886	2642
211,278	35,519	9287	5131	20,087	8520	8789	93,496	3951
1,855,655	696,916	31,313	92,832	545,769	45,284	150,680	36,665	199,160
4295	2593	9213	3956	10,509	2139	26,155	3102	19,034
288,392	59,306	4463	131,192	63,328	11,443	173,284	18,930	15,131
490,062	106,800	16,469	95,114	39,514	78,732	25,767	20,727	93,872
199,178	131,559	1580	59,469	149,504	25,197	224,131	79,141	267,054
188,447	487,028	2573	62,675	265,773	61,624	488,517	208,039	183,654
11,606	40,288	3344	79,174	172,996	39,469	194,857	63,912	67,526
71,784	45,848	8525	151,788	379,560	44,484	370,653	163,964	177,207
14,919	2539	119	5066	12,143	33	8526	1023	2021
230,382	9717	39	14,800	30,018	686	4216	9310	3019
979	2491	0	847	1800	1221	3906	1560	34,765
27,892	187,859	350	19,386	71,988	34,746	23,625	35,743	62,681
2479	6164	220	3546	4655	970	11,830	1146	1947
5913	4454	42	903	1252	627	4207	1677	661
27,670	16,994	1643	16,942	30,335	9621	36,403	25,670	17,369
2541	2328	72	1153	1938	573	3471	3214	771
14,745,312	4,998,593	133,450	1,430,050	2,385,871	2,347,288	2,131,511	1,174,385	2,627,762
2,958,419	1,336,821	62,982	407,155	1,606,070	842,912	1,255,795	398,301	776,736
821,577	268,599	5677	119,766	1,593,142	160,070	481,725	574,153	196,987
287,413	594,515	14,843	393,126	373,323	188,956	92,740	1,758,560	377,575
1,126,994	1,058,906	182	670,726	2,275,004	200,207	2,125,220	914,189	111,183
5,194,403	3,258,841	83,684	1,590,773	5,847,539	1,392,144	3,955,480	3,645,204	1,462,482
19,939,715	8,257,434	217,134	3,020,824	8,233,410	3,739,432	6,086,992	4,819,589	4,090,244

(continued)

Table 10.20 (continued)

	Sector 35	Sector 36	Sector 37	Sector 38	Sector 39	Sector 40	Sector 41	Sector 42
Sector 1	19,515	6624	18,881	24,462	5948	15,110	4530	0
Sector 2	897	2228	4046	8522	6319	12,742	751	13,295
Sector 3	0	0	0	0	0	0	0	0
Sector 4	4883	1035	1362	0	0	0	0	0
Sector 5	80	51	508	110	1148	1613	454	1875
Sector 6	9218	4078	8215	47,517	25,064	26,308	52,041	17,229
Sector 7	10,489	1100	3344	31,491	5448	17,554	2726	19,321
Sector 8	4005	2783	10,095	14,744	3489	43,272	15,469	63,168
Sector 9	1758	2118	9960	19,729	37,917	23,539	11,118	76,960
Sector 10	5779	23,463	5298	34,924	108,629	39,279	104,456	123,355
Sector 11	4738	34,036	28,682	38,413	24,501	12,876	6033	68,797
Sector 12	56,332	36,677	52,538	182,352	27,935	1,344,502	38,118	47,943
Sector 13	4081	3228	10,871	2225	15,237	17,582	5950	23,100
Sector 14	2599	12,226	5886	5370	2323	151	428	0
Sector 15	13,211	35,570	6228	10,391	6663	6584	3691	5607
Sector 16	7794	13,378	16,344	5081	6196	140,856	2599	11,009
Sector 17	4604	24,292	27,485	79,850	20,664	3498	4344	61,540
Sector 18	22,866	69,353	20,373	34,340	9153	10,033	5125	10,783
Sector 19	20,434	62,596	11,230	51,602	10,034	2320	5544	36,436
Sector 20	18,285	37,750	10,752	20,382	30,219	7350	3481	8329
Sector 21	2364	3680	4095	16,642	5288	3956	1875	10,723
Sector 22	0	0	0	0	0	0	0	0
Sector 23	10,558	11,513	22,250	23,218	56,626	36,589	8346	64,337
Sector 24	171	424	715	1350	1998	1464	622	1424
Sector 25	805	1966	1455	3435	4200	2098	735	2710
Sector 26	1845	5026	19,498	16,052	14,922	60,980	7564	66,267
Sector 27	21,503	98,817	18,037	43,980	79,546	15,773	27,322	196,817
Sector 28	871	1985	747	2915	6493	1010	2116	50,473
Sector 29	2950	8231	6935	7156	25,172	28,660	7496	88,027
Sector 30	10,811	19,566	11,151	37,434	22,857	93,975	19,072	48,364
Sector 31	20,097	68,352	18,644	32,980	74,481	30,317	28,925	300,611
Sector 32	2233	52,710	24,909	44,401	78,024	17,860	13,892	75,144
Sector 33	2364	17,060	2357	58,909	22,183	6010	11,953	25,544
Sector 34	4908	25,112	11,530	29,860	53,839	22,383	20,373	42,202
Sector 35	12,445	3836	752	57	6213	1427	935	64
Sector 36	7528	104,075	4027	1613	6325	4398	633	6580
Sector 37	325	352	18,576	3038	3923	793	828	10,944
Sector 38	1544	14,969	19,537	81,538	21,813	17,173	8740	67,506
Sector 39	1744	1649	1290	642	13,389	2993	887	35,992
Sector 40	1369	1551	2910	677	4399	7751	1329	12,888
Sector 41	8067	8479	5687	8417	17,938	4513	28,385	58,716
Sector 42	133	532	308	484	579	452	192	1262
Total intermediate input	326,200	822,469	447,510	1,026,305	867,093	2,085,745	459,075	1,755,345
Compensation of employees	126,698	445,048	170,206	718,549	2,192,740	968,409	268,262	2,689,507
Net taxes on production	8091	74,432	13,972	99,264	14,109	10,104	46,250	14,602
Depreciation of fixed assets	23,774	84,659	68,223	58,420	260,617	117,559	71,818	395,949
Operating surplus	27,603	228,684	51,206	202,315	36,787	69,345	84,328	85,115
Total value added	186,167	832,823	303,607	1,078,548	2,504,252	1,165,416	470,659	3,185,173
Total input	512,367	1,655,292	751,117	2,104,853	3,371,345	3,251,161	929,734	4,940,517

Total intermediate use	Rural household consumption expenditure	Urban household consumption expenditure	Government consumption expenditure	Gross fixed capital formation	Increase in inventories	Export	Import	Total output
9,943,414	1,143,572	1,404,353	102,653	689,313	221,672	364,069	618,480	13,250,565
3,855,590	18,103	13,004	0	34,850	−87,034	17,328	301,143	3,550,699
3,827,792	0	0	0	18,278	−55,601	312,868	2,518,968	1,584,368
3,259,252	0	0	0	71,532	−19,071	7635	1,784,713	1,534,635
986,796	0	0	0	8462	−22,672	20,821	68,569	924,838
6,331,574	1,320,335	3,770,362	0	224,693	255,511	220,825	242,567	11,880,733
3,893,577	43,183	66,393	0	29,477	−27,421	2,963,336	673,926	6,294,618
1,560,022	277,138	1,302,831	0	74,328	90,604	579,653	75,977	3,808,601
1,778,917	25,675	110,464	0	179,122	−1685	269,856	285,992	2,076,357
3,220,295	16,417	112,944	0	35,227	−34,103	226,932	345,143	3,232,571
5,308,988	23,551	215,701	0	−6697	−134,219	107,233	370,531	5,144,025
15,428,506	150,232	489,818	0	176,943	−131,943	944,203	1,595,224	15,462,535
7,093,363	8822	58,966	0	68,622	−135,481	366,464	219,784	7,240,971
14,596,262	0	0	0	163,342	−188,641	365,674	813,415	14,123,223
3,606,036	15,947	85,698	0	91,586	−54,600	345,348	198,308	3,891,707
7,085,542	1146	16,816	0	4,680,776	89,393	1,554,923	1,584,615	11,843,981
4,919,993	133,043	656,539	0	4,716,000	119,509	1,553,580	1,263,726	10,834,937
4,546,161	119,829	461,892	0	1,618,720	34,952	1,489,262	698,631	7,572,184
5,641,122	100,802	403,989	0	1,031,371	187,744	3,980,558	2,892,004	8,453,582
1,500,386	11,250	41,392	0	196,020	−24,607	494,557	1,382,078	836,918
1,288,528	67,765	356,593	0	434,665	9937	219,327	88,275	2,288,540
1,233,659	0	0	0	0	6321	4393	770,069	474,303
7,991,432	60,900	224,622	0	0	−754,441	13,166	1950	7,533,729
203,426	16,226	124,410	0	2244	358	0	0	346,664
147,824	11,823	118,296	0	0	1226	0	0	279,170
636,308	0	253,459	0	18,756,114	72,498	327,415	91,934	19,953,861
7,616,877	124,665	485,865	371,309	58,883	−179,449	369,632	495,896	8,351,886
219,695	1813	12,062	0	0	2855	7010	27,494	215,941
1,560,373	151,139	641,231	0	614,499	43,230	150,931	70,532	3,090,871
4,813,157	457,904	1,395,680	0	554,318	47,359	838,375	0	8,106,793
2,514,267	285,722	1,111,344	0	0	98,640	−51,506	183,799	3,774,669
5,094,158	174,340	909,309	68,348	0	108,023	69,138	62,240	6,361,076
1,195,327	518,715	1,887,766	0	1,194,203	120,162	0	0	4,916,172
2,941,571	24,891	357,213	161,857	0	59,141	781,099	249,350	4,076,423
434,093	0	0	273,343	0	2978	17,689	208,961	519,142
1,249,330	1776	22,731	235,078	176,412	−1537	0	0	1,683,788
219,313	7063	87,793	435,604	0	−2160	0	0	747,614
1,027,810	159,760	925,044	0	0	35,106	32,882	62,281	2,118,322
118,935	316,784	709,421	2,248,257	0	−27,196	4511	23,641	3,347,072
212,220	123,410	1,396,780	1,557,046	0	−23,344	3828	2677	3,267,264
484,901	31,688	170,030	261,910	0	12,312	42,790	53,539	950,094
40,639	0	0	4,929,195	0	−12,349	15,284	13,227	4,959,542

Table 10.21 China's national input–output table in 2019 in the policy scenario

	Sector 1	Sector 2	Sector 3	Sector 4	Sector 5	Sector 6	Sector 7	Sector 8
Sector 1	1,923,153	36,667	26	2775	394	4,990,051	1,397,450	192,555
Sector 2	11,723	663,806	11,672	6878	2692	34,417	35,036	15,747
Sector 3	0	0	22,514	0	0	0	0	0
Sector 4	0	0	0	161,446	0	0	0	0
Sector 5	1626	1381	715	1656	68,943	4531	0	0
Sector 6	1,508,623	6830	2443	3940	2928	2,599,095	53,771	203,528
Sector 7	1785	950	1082	932	539	6502	2,219,934	1,312,192
Sector 8	4985	13,203	4818	6957	4258	16,515	112,323	592,397
Sector 9	17,468	33,244	5183	5577	2024	13,506	7888	9066
Sector 10	12,799	4671	2895	2875	2492	235,124	73,829	63,374
Sector 11	100,522	42,534	51,294	91,582	49,162	35,396	18,004	24,755
Sector 12	1,101,366	86,752	50,741	88,818	101,849	347,759	795,548	245,607
Sector 13	19,484	34,241	12,613	13,479	36,823	87,066	16,281	5433
Sector 14	3453	163,241	66,508	21,045	7053	14,212	5632	5609
Sector 15	54,579	77,586	10,466	33,864	17,157	58,330	15,212	20,084
Sector 16	81,571	202,903	112,818	129,460	89,185	43,089	86,946	26,175
Sector 17	41,674	25,703	9280	17,587	20,655	15,992	14,179	6346
Sector 18	3973	57,413	12,617	21,831	9809	18,363	16,068	7691
Sector 19	1342	6109	1755	2353	1151	4645	5408	4440
Sector 20	12,077	38,090	12,871	8435	4345	8883	16,189	5328
Sector 21	45,520	26,297	1876	6609	10,874	19,295	54,200	42,231
Sector 22	1066	404	22	1229	9087	9816	13,684	2147
Sector 23	178,483	168,996	99,561	278,837	117,057	130,736	177,383	32,507
Sector 24	523	1512	1682	8438	5488	7465	3756	1423
Sector 25	1685	2823	1070	1505	1077	5107	5762	1876
Sector 26	3736	4982	1540	639	366	3557	1662	1232
Sector 27	257,957	177,459	26,430	71,404	65,032	361,608	132,077	96,116
Sector 28	11,919	1152	219	319	324	2845	5511	8779
Sector 29	54,571	11,826	2311	2611	2611	22,396	14,663	21,085
Sector 30	341,797	54,810	12,496	21,246	18,457	304,586	178,490	101,590
Sector 31	41,826	42,975	12,100	18,595	15,456	74,715	33,590	25,593
Sector 32	155,957	96,540	11,175	19,837	19,762	129,744	185,342	60,833
Sector 33	4407	7230	609	1200	1863	21,685	29,569	52,185
Sector 34	22,616	21,502	2478	6583	4571	208,303	29,157	88,782
Sector 35	25,582	5531	2492	558	266	10,234	4551	2193
Sector 36	128,451	35,379	15,359	22,607	11,398	43,397	17,468	14,428
Sector 37	44,085	5265	1111	1169	2871	3867	11,432	583
Sector 38	57,124	32,105	5093	3997	4415	15,204	8939	7910
Sector 39	5546	3673	252	638	1365	1734	823	433
Sector 40	8514	8914	489	3142	2093	6282	3418	2713
Sector 41	3475	5662	1306	4214	4499	13,289	8053	7270
Sector 42	4587	956	412	347	200	1764	865	536
Total intermediate input	6,301,629	2,211,317	592,393	1,097,212	720,591	9,931,105	5,810,094	3,312,772
Compensation of employees	8,043,409	747,129	199,805	151,148	120,872	772,378	588,868	419,074
Net taxes on production	16,342	416,761	226,882	97,528	73,550	871,171	234,355	108,746
Depreciation of fixed assets	394,719	142,788	133,381	35,673	31,336	239,363	154,666	49,360
Operating surplus	0	432,431	312,582	185,414	100,365	769,309	241,563	140,478
Total value added	8,454,470	1,739,109	872,650	469,764	326,123	2,652,221	1,219,452	717,658
Total input	14,756,099	3,950,426	1,465,043	1,566,976	1,046,714	12,583,326	7,029,546	4,030,430

Sector 9	Sector 10	Sector 11	Sector 12	Sector 13	Sector 14	Sector 15	Sector 16	Sector 17
286,019	192,628	124	596,992	4717	1727	1341	1383	968
17,185	35,208	345,642	262,399	580,940	576,735	19,769	43,731	13,219
0	0	3,131,100	901,646	30,343	73,371	0	0	0
0	0	0	101,326	21,566	3,153,091	45,309	37,791	6770
0	0	561	258,209	448,852	38,314	15,784	8989	2733
11,576	15,690	25,887	287,932	29,090	50,812	13,663	41,995	42,480
21,436	65,294	283	112,008	16,491	1709	6521	18,016	20,974
37,568	26,667	2309	38,860	19,765	33,498	12,872	27,521	124,012
631,715	101,996	1706	30,183	59,738	11,674	72,135	41,329	130,450
36,992	869,938	3017	257,614	193,458	16,929	46,301	56,075	50,201
15,667	18,004	337,364	1,054,413	323,370	627,128	42,451	131,128	56,773
202,473	524,135	104,101	6,524,909	614,130	225,580	170,241	463,062	675,840
18,320	10,142	18,843	126,884	1,336,022	296,264	45,435	92,180	85,328
48,647	43,650	3556	173,144	213,772	4,025,912	1,368,592	2,695,305	1,410,141
53,284	63,104	17,884	160,457	231,089	133,433	489,074	538,138	232,942
30,945	50,199	71,427	255,018	215,827	460,167	192,872	2,886,233	982,376
9519	30,205	12,532	49,099	41,043	79,376	38,127	193,445	3,271,189
7036	25,024	6292	44,897	31,210	47,767	32,580	701,761	449,867
904	33,170	758	10,452	4951	3844	4918	267,395	134,746
1789	43,721	7893	127,226	24,857	50,522	10,076	67,083	224,343
18,013	23,369	2452	104,100	64,619	49,506	23,974	88,676	156,782
7230	230,440	1096	77,817	133,240	478,580	36,347	152,170	10,412
71,009	114,535	91,743	902,536	611,712	932,745	305,736	286,604	194,334
997	1049	8139	43,067	9091	19,806	4094	11,432	7919
1485	7786	1673	18,932	8337	11,009	3923	4280	6332
569	952	2680	6628	2743	4913	791	4432	4154
70,644	87,914	105,944	473,777	335,936	286,861	93,568	366,907	236,903
1507	2242	165	13,359	2268	4830	810	11,372	2556
8631	9710	17,109	62,612	19,060	148,720	14,513	40,486	32,726
57,120	142,812	65,300	418,827	257,766	317,322	104,873	328,139	671,782
11,846	21,711	9735	104,222	64,032	56,699	50,407	99,058	55,140
42,863	60,050	32,586	361,055	292,507	320,977	63,209	136,362	170,243
12,836	23,520	867	50,462	27,446	6115	27,947	35,782	31,857
19,864	32,432	25,659	264,329	55,015	25,369	17,084	113,154	172,626
1332	2205	1483	42,860	6170	25,856	3749	44,543	103,755
4169	19,763	9133	106,042	32,098	55,692	24,559	76,705	108,269
322	12,584	1076	18,230	2489	4796	589	2649	3850
2495	8280	4731	39,778	27,327	39,464	7588	15,084	18,883
344	532	215	2768	1259	1184	543	4071	2451
1955	4561	1391	11,070	13,917	13,538	9529	30,439	23,621
2845	4037	1228	20,934	13,414	10,108	4756	17,970	13,659
290	578	247	2216	1338	1547	556	1739	1750
1,769,442	2,959,840	4,475,930	14,519,289	6,423,018	12,723,489	3,427,203	10,184,614	9,945,357
174,865	263,933	180,404	1,096,504	542,613	908,751	283,828	1,122,852	673,290
72,607	101,653	595,602	605,570	295,468	439,955	118,161	515,244	447,303
41,765	100,334	133,696	469,318	210,794	541,999	82,881	298,920	195,150
105,495	156,510	198,287	908,318	486,081	621,467	198,794	794,425	554,293
394,732	622,430	1,107,989	3,079,709	1,534,955	2,512,172	683,664	2,731,441	1,870,036
2,164,174	3,582,269	5,583,920	17,598,998	7,957,974	15,235,661	4,110,867	12,916,055	11,815,393

(continued)

Table 10.21 (continued)

	Sector 18	Sector 19	Sector 20	Sector 21	Sector 22	Sector 23	Sector 24	Sector 25
Sector 1	161	0	38	313,872	177	140	62	0
Sector 2	6810	2729	1324	17,160	815	1,381,960	17,835	403
Sector 3	0	0	0	24	1	49,135	174,749	0
Sector 4	10,568	135	1107	238	538	232	0	0
Sector 5	3359	60	92	7207	68	2111	14	31
Sector 6	30,786	32,781	2797	35,437	700	23,928	1384	792
Sector 7	14,765	2384	1319	234,458	77	404	43	90
Sector 8	17,285	15,631	2646	38,652	120	26,069	4166	1494
Sector 9	22,535	15,262	2367	79,375	90	4854	435	446
Sector 10	98,705	75,889	9792	46,503	123	14,953	936	790
Sector 11	34,641	24,026	4142	19,011	1203	231,000	7722	3138
Sector 12	676,946	594,162	77,413	198,081	11,751	24,394	2385	14,958
Sector 13	161,638	156,889	30,667	54,416	1447	16,724	857	885
Sector 14	2,066,558	274,815	34,046	247,363	5810	7580	915	350
Sector 15	343,813	204,057	33,590	109,478	609	46,588	3537	8164
Sector 16	390,026	92,183	24,005	10,396	359	78,900	2998	3220
Sector 17	21,712	30,915	5254	26,546	124	87,871	8136	3760
Sector 18	1,123,486	255,400	43,239	16,261	914	434,983	862	1019
Sector 19	470,583	4,041,560	202,099	6468	46	4827	195	244
Sector 20	78,394	428,475	105,797	5614	54	105,732	2329	1643
Sector 21	79,943	69,307	14,267	284,940	108	5246	278	313
Sector 22	68,282	9317	100	515	5318	861	0	0
Sector 23	106,125	118,357	10,162	34,955	2473	2,809,843	7423	70,995
Sector 24	4831	4731	608	2481	121	9138	20,687	987
Sector 25	3303	3805	677	1763	115	9788	199	6350
Sector 26	2100	1696	242	407	57	2695	275	368
Sector 27	188,594	149,784	19,536	61,692	4218	95,839	9819	2872
Sector 28	5917	3403	1019	734	13	929	48	94
Sector 29	28,275	91,749	3820	17,524	168	48,159	530	1120
Sector 30	270,228	378,342	22,192	51,710	2583	62,292	6235	1839
Sector 31	57,161	28,337	8380	18,599	516	22,540	1704	2414
Sector 32	139,364	353,759	9100	28,394	2376	326,489	12,310	17,447
Sector 33	34,546	30,628	5381	17,639	180	3445	390	160
Sector 34	187,336	104,329	11,388	21,863	437	35,053	1635	1491
Sector 35	34,849	66,872	4754	2544	45	5233	98	62
Sector 36	47,950	47,762	5714	5572	221	63,075	856	461
Sector 37	1662	1144	88	415	32	21,007	162	17,354
Sector 38	7573	6763	797	1102	478	60,244	1070	2390
Sector 39	1194	867	168	433	5	1216	87	233
Sector 40	14,569	4580	1559	406	87	9391	1042	67
Sector 41	9725	8335	1062	2870	80	17,736	701	910
Sector 42	918	917	119	370	0	1123	32	61
Total intermediate input	6,867,217	7,732,136	702,869	2,023,486	44,658	6,153,727	295,142	169,413
Compensation of employees	497,473	678,990	88,522	234,518	6478	629,514	39,727	61,727
Net taxes on production	192,502	137,923	24,964	158,024	3880	382,804	9217	12,164
Depreciation of fixed assets	135,750	213,181	18,057	52,929	2913	788,970	12,336	53,160
Operating surplus	366,096	254,059	45,192	129,171	395,289	256,122	18,035	7127
Total value added	1,191,822	1,284,154	176,735	574,642	408,560	2,057,409	79,315	134,179
Total input	8,059,039	9,016,290	879,604	2,598,129	453,217	8,211,137	374,457	303,592

Sector 26	Sector 27	Sector 28	Sector 29	Sector 30	Sector 31	Sector 32	Sector 33	Sector 34
95,806	218,765	0	0	1983	470,587	0	427	10,374
17,300	31,566	328	0	2499	2717	0	1104	6165
0	11,387	0	0	0	0	0	0	0
0	0	0	0	0	0	0	0	0
220,287	2375	0	0	0	0	0	0	0
87,054	72,104	583	8390	42,625	1,416,336	19,829	14,632	51,724
18,408	11,844	355	422	9881	20,044	1012	5791	56,511
82,993	65,463	1689	6512	72,383	15,379	18,657	23,490	54,520
381,760	11,167	1167	3298	13,527	5635	5317	6300	11,043
36,440	33,638	6212	107,282	75,898	13,865	137,507	30,541	391,741
281,104	1,708,165	6913	6634	27,426	20,053	44,740	68,056	139,181
672,681	161,236	777	18,233	74,333	31,079	24,011	86,980	156,620
4,941,065	19,413	810	933	2232	8356	1137	1543	2321
2,718,413	47,179	0	35	4014	184	0	4672	3340
709,028	50,989	182	6919	3296	4763	6400	31,611	87,779
645,497	220,627	2455	27,836	17,709	4782	30,137	18,456	28,699
90,132	644,884	18,527	24,262	71,824	13,764	20,599	19,885	133,034
824,834	25,761	414	236,930	55,915	29,213	14,550	57,775	122,602
52,567	23,089	2056	205,629	32,042	1662	5295	15,609	252,577
47,618	14,589	278	30,852	2675	128	27,173	10,178	13,017
70,983	8384	592	3646	6274	2539	7372	6730	26,498
0	0	0	0	0	0	0	0	0
162,861	109,802	3200	85,809	96,149	87,122	66,744	31,181	41,692
2580	8606	223	1958	1682	14,878	352	549	4221
5235	5518	376	4205	3146	7077	3348	950	2854
231,296	39,412	9905	5832	21,214	9329	10,245	100,247	4268
2,031,472	773,299	33,397	105,512	576,393	49,585	175,633	39,312	215,138
4702	2877	9826	4496	11,099	2342	30,486	3326	20,561
315,716	65,806	4760	149,111	66,881	12,529	201,979	20,297	16,345
536,493	118,505	17,565	108,105	41,731	86,210	30,034	22,224	101,403
218,050	145,978	1685	67,592	157,893	27,591	261,247	84,856	288,479
206,302	540,407	2744	71,235	280,686	67,477	569,416	223,061	198,388
12,706	44,703	3566	89,989	182,703	43,218	227,125	68,528	72,944
78,586	50,873	9092	172,521	400,857	48,709	432,033	175,805	191,424
16,333	2817	127	5757	12,825	36	9938	1097	2183
252,210	10,782	41	16,822	31,702	751	4914	9983	3261
1072	2764	0	963	1901	1336	4553	1673	37,554
30,535	208,448	374	22,033	76,028	38,046	27,537	38,324	67,709
2714	6840	235	4031	4916	1062	13,789	1229	2103
6473	4942	45	1026	1322	687	4903	1798	714
30,292	18,857	1753	19,256	32,037	10,534	42,432	27,524	18,762
2781	2583	77	1310	2046	628	4045	3446	832
16,142,379	5,546,444	142,330	1,625,378	2,519,746	2,570,232	2,484,490	1,259,189	2,838,582
3,238,719	1,483,338	67,173	462,768	1,696,189	922,971	1,463,755	427,063	839,052
899,418	298,037	6055	136,124	1,682,536	175,273	561,498	615,614	212,791
314,645	659,675	15,830	446,823	394,270	206,903	108,098	1,885,548	407,867
1,233,773	1,174,963	194	762,339	2,402,658	219,222	2,477,156	980,204	120,103
5,686,554	3,616,014	89,252	1,808,054	6,175,653	1,524,369	4,610,508	3,908,429	1,579,814
21,828,934	9,162,458	231,582	3,433,432	8,695,399	4,094,601	7,094,997	5,167,619	4,418,395

(continued)

Table 10.21 (continued)

	Sector 35	Sector 36	Sector 37	Sector 38	Sector 39	Sector 40	Sector 41	Sector 42
Sector 1	21,333	7463	19,954	26,498	6329	16,298	5049	0
Sector 2	980	2511	4276	9232	6724	13,744	838	14,295
Sector 3	0	0	0	0	0	0	0	0
Sector 4	5338	1166	1440	0	0	0	0	0
Sector 5	87	58	537	119	1222	1740	506	2016
Sector 6	10,076	4595	8682	51,472	26,669	28,376	58,009	18,524
Sector 7	11,466	1239	3534	34,112	5797	18,934	3038	20,773
Sector 8	4378	3136	10,669	15,971	3713	46,674	17,243	67,915
Sector 9	1921	2387	10,526	21,371	40,345	25,390	12,393	82,744
Sector 10	6317	26,437	5599	37,831	115,586	42,367	116,436	132,626
Sector 11	5179	38,350	30,313	41,611	26,070	13,889	6725	73,967
Sector 12	61,582	41,326	55,525	197,531	29,724	1,450,209	42,490	51,546
Sector 13	4461	3637	11,490	2410	16,213	18,964	6632	24,836
Sector 14	2841	13,776	6221	5817	2472	163	477	0
Sector 15	14,442	40,079	6582	11,256	7090	7102	4114	6029
Sector 16	8521	15,074	17,273	5504	6592	151,931	2897	11,837
Sector 17	5033	27,371	29,048	86,497	21,987	3773	4842	66,165
Sector 18	24,997	78,145	21,531	37,198	9739	10,821	5713	11,594
Sector 19	22,339	70,531	11,869	55,897	10,677	2502	6179	39,175
Sector 20	19,989	42,535	11,364	22,078	32,155	7928	3880	8955
Sector 21	2585	4147	4327	18,028	5627	4267	2090	11,529
Sector 22	0	0	0	0	0	0	0	0
Sector 23	11,542	12,972	23,516	25,151	60,253	39,466	9304	69,172
Sector 24	187	477	755	1463	2126	1579	693	1531
Sector 25	880	2215	1538	3721	4469	2263	820	2913
Sector 26	2016	5663	20,606	17,389	15,878	65,774	8432	71,247
Sector 27	23,507	111,344	19,063	47,641	84,640	17,013	30,455	211,609
Sector 28	952	2237	789	3158	6908	1089	2358	54,267
Sector 29	3225	9274	7329	7752	26,784	30,913	8355	94,643
Sector 30	11,818	22,046	11,785	40,550	24,321	101,364	21,259	51,999
Sector 31	21,969	77,016	19,704	35,726	79,251	32,700	32,243	323,204
Sector 32	2441	59,392	26,325	48,097	83,022	19,264	15,486	80,792
Sector 33	2584	19,223	2491	63,812	23,603	6482	13,324	27,464
Sector 34	5366	28,295	12,186	32,346	57,287	24,143	22,710	45,373
Sector 35	13,605	4322	795	62	6611	1539	1042	68
Sector 36	8229	117,268	4256	1748	6730	4744	705	7074
Sector 37	355	397	19,632	3291	4174	856	923	11,767
Sector 38	1688	16,867	20,648	88,325	23,210	18,523	9742	72,580
Sector 39	1906	1858	1363	696	14,247	3229	988	38,697
Sector 40	1496	1748	3076	734	4681	8361	1481	13,857
Sector 41	8819	9554	6011	9118	19,087	4868	31,640	63,129
Sector 42	145	599	326	525	616	488	214	1357
Total intermediate input	356,598	926,729	472,954	1,111,734	922,629	2,249,730	511,726	1,887,270
Compensation of employees	138,505	501,464	179,883	778,361	2,333,181	1,044,547	299,029	2,891,640
Net taxes on production	8845	83,868	14,766	107,526	15,012	10,898	51,554	15,699
Depreciation of fixed assets	25,990	95,391	72,102	63,283	277,309	126,801	80,055	425,708
Operating surplus	30,175	257,673	54,118	219,155	39,143	74,797	94,000	91,512
Total value added	203,515	938,396	320,869	1,168,325	2,664,645	1,257,044	524,638	3,424,559
Total input	560,113	1,865,124	793,823	2,280,059	3,587,273	3,506,773	1,036,364	5,311,828

Total intermediate use	Rural household consumption expenditure	Urban household consumption expenditure	Government consumption expenditure	Gross fixed capital formation	Increase in inventories	Export	Import	Total output
10,844,285	1,224,701	1,512,879	109,277	788,618	221,672	383,523	641,542	14,443,412
4,220,113	19,615	14,208	0	39,444	−87,034	17,729	325,415	3,898,660
4,394,270	0	0	0	18,305	−55,601	332,740	3,021,263	1,668,450
3,548,061	0	0	0	72,553	−19,071	7635	1,989,499	1,619,678
1,094,184	0	0	0	9468	−22,672	21,447	73,407	1,029,019
6,948,571	1,391,445	3,998,802	0	235,158	255,511	220,825	242,567	12,807,746
4,283,353	46,356	71,437	0	33,201	−27,421	3,223,572	747,992	6,882,506
1,695,379	293,351	1,388,363	0	79,239	90,604	579,653	75,977	4,050,614
1,936,538	27,169	117,403	0	184,634	−1685	269,856	316,855	2,217,060
3,492,599	17,561	119,728	0	39,279	−34,103	226,932	400,906	3,461,090
5,882,805	25,056	228,271	0	−7259	−134,219	110,426	393,880	5,711,199
17,078,886	160,276	522,553	0	198,767	−131,943	944,203	1,647,567	17,125,174
7,744,813	9648	63,309	0	75,680	−135,481	388,561	240,712	7,905,817
15,716,512	0	0	0	174,221	−188,641	365,674	852,206	15,215,560
3,954,180	17,008	91,598	0	94,502	−54,600	345,348	214,542	4,233,494
7,735,126	1226	17,936	0	5,091,120	89,393	1,660,244	1,664,142	12,930,903
5,341,895	141,309	693,056	0	5,271,217	119,509	1,684,534	1,349,774	11,901,746
4,938,093	126,607	489,366	0	1,714,881	34,952	1,559,489	733,810	8,129,577
6,024,057	107,778	431,591	0	1,096,208	187,744	4,163,284	3,023,268	8,987,394
1,687,158	12,087	43,258	0	197,216	−24,607	505,298	1,473,472	946,938
1,378,410	71,775	386,080	0	516,329	9937	225,696	93,484	2,494,743
1,249,179	0	0	0	0	6321	4494	813,174	446,819
8,790,780	60,900	224,622	0	0	−877,027	13,823	1962	8,211,137
223,324	16,980	129,526	0	2457	358	0	0	372,645
162,190	12,616	123,075	0	0	1226	0	0	299,107
692,170	0	270,788	0	20,533,190	72,498	356,136	99,369	21,825,413
8,323,905	133,378	517,984	402,750	65,336	−179,449	369,632	538,090	9,095,447
243,808	1927	12,900	0	0	2855	7248	30,342	238,397
1,718,686	159,912	681,817	0	698,432	43,230	161,066	76,156	3,386,986
5,540,246	485,636	1,483,785	0	585,422	47,359	858,060	0	9,000,508
2,752,536	304,697	1,186,266	0	0	98,640	−65,937	197,694	4,078,509
5,542,815	184,973	968,773	73,623	0	108,023	75,190	67,517	6,885,881
1,332,415	549,858	1,971,530	0	1,280,438	120,162	0	0	5,254,403
3,260,659	26,254	376,250	172,891	0	59,141	825,246	249,350	4,471,092
476,975	0	0	286,601	0	2978	19,198	224,566	561,186
1,377,751	1776	22,731	250,070	198,775	−1537	0	0	1,849,566
252,045	7449	92,279	457,758	0	−2160	0	0	807,372
1,139,448	169,039	979,677	0	0	35,106	34,279	66,952	2,290,599
131,937	337,636	755,220	2,393,006	0	−27,196	4638	25,476	3,569,764
234,631	128,981	1,471,820	1,668,657	0	−23,344	3828	2677	3,481,896
531,810	33,681	180,906	279,534	0	12,312	42,790	53,539	1,027,495
45,486	0	0	5,232,795	0	−12,349	16,393	13,929	5,268,396

Table 10.22 China's national input–output table in 2020 in the policy scenario

	Sector 1	Sector 2	Sector 3	Sector 4	Sector 5	Sector 6	Sector 7	Sector 8
Sector 1	2,031,738	38,487	26	2796	419	5,519,426	1,480,656	213,230
Sector 2	12,189	695,215	11,674	6921	2884	38,068	36,974	16,976
Sector 3	0	0	22,516	0	0	0	0	0
Sector 4	0	0	0	161,599	0	0	0	0
Sector 5	1702	1480	715	1680	72,888	5012	0	0
Sector 6	1,561,406	7089	2444	3968	3077	2,874,823	56,135	217,677
Sector 7	1863	1005	1082	942	565	7192	2,329,034	1,425,752
Sector 8	5168	13,676	4819	6999	4470	18,267	117,267	633,695
Sector 9	17,910	34,266	5183	5609	2118	14,938	8145	9500
Sector 10	13,439	4946	2895	2909	2643	260,068	77,785	69,822
Sector 11	103,847	44,186	51,299	92,281	51,551	39,152	18,744	26,341
Sector 12	1,157,719	92,095	50,747	89,776	108,054	384,651	839,157	270,708
Sector 13	20,187	35,675	12,614	13,565	38,962	96,303	17,055	5810
Sector 14	3570	168,883	66,515	21,150	7474	15,720	5901	5986
Sector 15	55,809	79,669	10,467	34,044	17,909	64,518	15,710	20,953
Sector 16	84,615	210,385	112,830	130,135	94,717	47,660	91,418	28,029
Sector 17	42,717	26,449	9281	17,671	21,441	17,689	14,608	6641
Sector 18	4102	59,424	12,618	21,958	10,318	20,311	16,777	8170
Sector 19	1398	6350	1756	2365	1225	5138	5707	4790
Sector 20	12,209	39,144	12,873	8515	4442	9826	16,363	5445
Sector 21	46,878	27,938	1876	6779	11,475	21,341	56,146	45,278
Sector 22	1068	405	22	1230	9103	10,857	13,700	2156
Sector 23	186,161	176,155	99,571	280,762	124,555	144,605	187,544	35,349
Sector 24	547	1579	1682	8495	5851	8256	3973	1545
Sector 25	1775	3005	1070	1524	1146	5649	6088	2078
Sector 26	3893	5201	1540	644	389	3934	1753	1334
Sector 27	268,772	185,287	26,433	71,896	69,294	399,970	139,265	103,917
Sector 28	12,264	1204	219	322	337	3147	5689	9292
Sector 29	57,689	12,679	2311	2650	2790	24,772	15,537	23,548
Sector 30	351,104	56,994	12,497	21,455	19,405	336,898	184,867	107,271
Sector 31	43,486	45,147	12,102	18,770	16,399	82,642	35,122	27,585
Sector 32	166,822	106,126	11,176	20,330	21,145	143,508	195,786	69,717
Sector 33	4544	7457	609	1208	1947	23,985	30,702	55,280
Sector 34	23,513	22,286	2478	6622	4861	230,401	30,703	95,498
Sector 35	26,581	5766	2492	562	283	11,319	4775	2360
Sector 36	134,584	37,429	15,361	22,842	12,186	48,001	18,438	15,817
Sector 37	45,814	5497	1111	1180	3048	4278	12,019	630
Sector 38	59,727	33,801	5093	4037	4725	16,817	9425	8628
Sector 39	5943	3905	252	643	1508	1918	898	497
Sector 40	9032	9441	489	3171	2265	6949	3673	3043
Sector 41	3658	6026	1306	4272	4780	14,699	8498	8037
Sector 42	4894	1017	413	350	218	1951	935	609
Total intermediate input	6,590,338	2,312,771	592,456	1,104,630	762,866	10,984,656	6,112,971	3,588,996
Compensation of employees	8,497,554	782,481	199,826	151,291	127,789	854,316	617,808	448,289
Net taxes on production	17,265	436,481	226,906	97,621	77,758	963,591	245,873	116,327
Depreciation of fixed assets	417,006	149,544	133,396	35,707	33,129	264,756	162,267	52,801
Operating surplus	0	452,892	312,616	185,590	106,108	850,922	253,434	150,271
Total value added	8,931,824	1,821,397	872,743	470,209	344,784	2,933,584	1,279,382	767,688
Total input	15,522,162	4,134,168	1,465,199	1,574,839	1,107,650	13,918,241	7,392,353	4,356,684

Sector 9	Sector 10	Sector 11	Sector 12	Sector 13	Sector 14	Sector 15	Sector 16	Sector 17
333,661	196,819	137	621,611	5097	1946	1550	1506	1182
19,587	35,870	383,878	272,535	624,933	639,218	22,716	47,634	15,742
0	0	3,477,470	922,266	32,667	90,086	0	0	0
0	0	0	107,257	22,962	3,348,943	49,547	41,163	7685
0	0	623	266,323	493,029	45,548	19,029	9791	3382
12,826	15,927	28,750	296,387	30,827	56,132	15,337	45,743	48,381
24,419	66,504	315	115,221	17,835	1963	7606	19,624	25,088
41,525	27,069	2564	40,018	20,912	36,728	14,342	29,977	141,727
692,263	103,167	1894	30,980	62,618	12,661	79,183	45,018	145,787
43,267	894,802	3350	266,921	210,662	19,611	54,986	61,079	60,756
17,242	18,274	374,684	1,083,545	342,371	692,901	47,382	142,830	64,200
238,117	535,089	115,617	6,767,438	670,345	261,917	203,115	504,389	845,623
20,467	10,308	20,928	130,925	1,426,835	327,849	51,579	100,407	99,368
54,215	44,367	3949	178,856	226,917	4,412,503	1,551,576	2,935,853	1,618,787
57,421	63,802	19,862	164,381	241,068	143,897	537,297	586,166	252,987
34,547	51,051	79,328	264,009	229,694	505,632	218,722	3,143,822	1,146,321
10,295	30,530	13,918	50,176	42,893	85,323	41,429	210,710	3,630,115
7764	25,392	6988	46,244	33,003	52,347	36,573	764,391	512,306
1018	33,756	841	10,841	5290	4245	5619	291,259	158,738
1935	43,963	8766	128,830	25,629	54,824	10,789	73,070	238,544
20,654	24,037	2723	106,866	70,154	59,892	29,643	96,590	175,049
7281	230,640	1217	77,906	133,809	484,588	36,700	165,751	10,165
81,244	116,779	101,892	937,287	655,306	1,039,543	352,528	312,183	229,504
1132	1071	9039	44,731	9767	22,147	4727	12,452	9456
1752	7981	1858	19,669	9124	12,874	4672	4662	7865
648	970	2977	6874	2946	5497	913	4827	4932
80,293	89,706	117,664	491,487	361,063	321,209	108,295	399,652	282,074
1671	2270	183	13,683	2402	5429	915	12,387	2880
10,375	9969	19,002	65,222	21,031	176,707	17,564	44,099	41,162
63,636	144,623	72,524	430,922	272,547	353,834	119,005	357,425	752,753
13,582	22,100	10,812	107,850	68,911	64,463	58,735	107,898	65,586
54,144	62,358	36,191	376,624	330,200	399,434	78,223	148,531	222,825
14,092	23,842	963	51,869	28,785	6590	30,559	38,975	35,049
22,286	33,020	28,498	274,235	58,518	27,939	19,326	123,253	200,759
1507	2245	1647	44,426	6595	28,553	4220	48,519	121,823
4854	20,178	10,143	110,190	34,910	63,781	28,787	83,551	131,313
365	12,805	1195	18,893	2661	5386	679	2886	4493
2920	8448	5254	41,414	29,577	45,052	8909	16,430	22,804
426	548	239	2939	1402	1393	679	4434	3230
2326	4674	1544	11,611	15,288	15,677	11,489	33,156	29,826
3337	4134	1364	21,732	14,641	11,817	5644	19,573	16,692
352	593	275	2335	1483	1816	683	1895	2255
1,999,446	3,019,680	4,971,069	15,023,528	6,896,709	13,947,899	3,891,273	11,093,562	11,389,214
191,626	271,477	200,361	1,137,260	579,496	996,014	311,814	1,223,063	747,165
79,566	104,558	661,489	628,079	315,552	482,202	129,812	561,228	496,383
45,768	103,202	148,485	486,762	225,122	594,045	91,053	325,597	216,563
115,607	160,984	220,222	942,080	519,121	681,144	218,394	865,325	615,112
432,566	640,220	1,230,558	3,194,181	1,639,291	2,753,405	751,073	2,975,214	2,075,223
2,432,012	3,659,900	6,201,627	18,217,710	8,536,000	16,701,304	4,642,346	14,068,776	13,464,437

(continued)

Table 10.22 (continued)

	Sector 18	Sector 19	Sector 20	Sector 21	Sector 22	Sector 23	Sector 24	Sector 25
Sector 1	182	0	47	317,099	177	149	65	0
Sector 2	7557	2878	1593	17,337	815	1,476,017	18,733	424
Sector 3	0	0	0	25	1	52,479	183,554	0
Sector 4	11,400	139	1299	241	538	248	0	0
Sector 5	3849	64	110	7281	68	2255	15	32
Sector 6	33,608	34,423	3183	35,801	700	25,557	1454	833
Sector 7	16,595	2547	1520	236,869	77	432	45	95
Sector 8	18,808	16,378	3007	39,050	120	27,843	4376	1572
Sector 9	24,183	15,857	2670	80,191	90	5184	457	469
Sector 10	112,317	81,218	11,872	46,981	123	15,970	983	832
Sector 11	37,658	25,170	4651	19,207	1203	246,722	8111	3302
Sector 12	774,811	640,166	93,709	200,118	11,751	26,055	2505	15,742
Sector 13	177,653	164,618	35,435	54,975	1447	17,863	900	931
Sector 14	2,264,440	287,131	39,809	249,906	5810	8096	961	368
Sector 15	367,752	211,006	37,177	110,603	609	49,759	3715	8592
Sector 16	428,687	96,521	27,975	10,503	359	84,270	3149	3389
Sector 17	23,146	31,984	5791	26,819	124	93,852	8546	3957
Sector 18	1,224,904	266,757	49,252	16,428	914	464,588	905	1072
Sector 19	520,535	4,255,614	238,632	6535	46	5156	205	257
Sector 20	82,221	442,133	114,533	5671	54	112,928	2447	1729
Sector 21	90,536	73,949	17,553	287,869	108	5603	292	329
Sector 22	68,788	9364	101	520	5318	919	0	0
Sector 23	118,244	125,416	12,315	35,314	2473	3,001,082	7797	74,716
Sector 24	5390	4997	721	2506	121	9760	21,729	1039
Sector 25	3782	4106	822	1781	115	10,454	209	6682
Sector 26	2339	1793	288	411	57	2879	289	388
Sector 27	210,348	158,005	23,194	62,326	4218	102,362	10,314	3022
Sector 28	6473	3589	1143	742	13	992	51	99
Sector 29	32,783	99,929	4737	17,704	168	51,436	557	1178
Sector 30	295,553	397,640	25,576	52,242	2583	66,531	6550	1935
Sector 31	63,992	30,088	9899	18,790	516	24,074	1789	2541
Sector 32	165,292	393,352	11,639	28,685	2376	348,710	12,931	18,361
Sector 33	37,123	31,905	6074	17,820	180	3680	410	168
Sector 34	206,086	109,535	13,392	22,087	437	37,439	1718	1569
Sector 35	38,214	70,395	5641	2570	45	5589	103	66
Sector 36	54,106	51,167	6930	5629	221	67,368	899	486
Sector 37	1849	1211	105	420	32	22,437	170	18,264
Sector 38	8555	7228	975	1113	478	64,344	1124	2515
Sector 39	1414	948	227	438	5	1299	92	245
Sector 40	16,817	4957	1959	410	87	10,030	1095	70
Sector 41	11,107	8998	1277	2900	80	18,943	737	958
Sector 42	1075	1001	153	374	0	1199	33	64
Total intermediate input	7,570,172	8,164,181	816,982	2,044,291	44,658	6,572,553	310,012	178,292
Compensation of employees	542,381	714,952	95,831	236,929	6478	672,359	41,729	64,962
Net taxes on production	209,880	145,228	27,025	159,649	3880	408,858	9682	12,802
Depreciation of fixed assets	148,004	224,472	19,548	53,473	2913	842,668	12,957	55,946
Operating surplus	399,144	267,515	48,924	130,499	395,289	273,553	18,944	7501
Total value added	1,299,409	1,352,167	191,327	580,550	408,560	2,197,438	83,311	141,211
Total input	8,869,581	9,516,348	1,008,309	2,624,842	453,217	8,769,991	393,324	319,504

Sector 26	Sector 27	Sector 28	Sector 29	Sector 30	Sector 31	Sector 32	Sector 33	Sector 34
103,878	233,769	0	0	2282	506,672	0	475	11,577
18,758	33,731	374	0	2875	2925	0	1227	6879
0	12,168	0	0	0	0	0	0	0
0	0	0	0	0	0	0	0	0
238,845	2538	0	0	0	0	0	0	0
94,388	77,049	664	8957	49,054	1,524,942	21,112	16,260	57,723
19,959	12,656	404	450	11,372	21,581	1077	6435	63,065
89,985	69,953	1922	6952	83,300	16,559	19,863	26,104	60,843
413,921	11,933	1328	3521	15,567	6067	5661	7001	12,323
39,510	35,945	7069	114,531	87,346	14,928	146,400	33,939	437,174
304,785	1,825,317	7867	7082	31,563	21,591	47,633	75,629	155,322
729,351	172,294	885	19,465	85,544	33,462	25,564	96,658	174,784
5,357,323	20,744	922	996	2568	8997	1210	1714	2590
2,947,425	50,415	0	38	4619	198	0	5192	3727
768,760	54,486	207	7386	3793	5128	6814	35,128	97,959
699,877	235,759	2793	29,717	20,380	5149	32,087	20,510	32,028
97,725	689,113	21,083	25,901	82,657	14,820	21,931	22,098	148,463
894,321	27,528	472	252,940	64,348	31,453	15,491	64,204	136,821
56,995	24,672	2339	219,524	36,875	1790	5637	17,346	281,870
51,629	15,590	316	32,936	3078	138	28,931	11,310	14,527
76,963	8959	674	3893	7221	2734	7849	7479	29,571
0	0	0	0	0	0	0	0	0
176,581	117,332	3642	91,607	110,650	93,802	71,061	34,650	46,527
2797	9196	254	2091	1936	16,019	374	610	4711
5676	5897	428	4489	3621	7620	3565	1055	3185
250,781	42,115	11,272	6226	24,414	10,044	10,907	111,402	4763
2,202,613	826,335	38,006	112,642	663,327	53,387	186,992	43,687	240,089
5098	3075	11,182	4800	12,773	2522	32,458	3696	22,945
342,313	70,319	5417	159,187	76,968	13,490	215,043	22,556	18,241
581,690	126,633	19,989	115,410	48,025	92,821	31,976	24,697	113,164
236,419	155,989	1918	72,159	181,707	29,706	278,143	94,298	321,935
223,682	577,470	3123	76,049	323,021	72,651	606,243	247,881	221,396
13,776	47,769	4059	96,069	210,259	46,532	241,814	76,153	81,404
85,206	54,362	10,347	184,178	461,316	52,444	459,975	195,366	213,624
17,709	3010	144	6146	14,759	39	10,581	1219	2436
273,458	11,521	47	17,959	36,483	809	5232	11,093	3639
1162	2953	0	1028	2188	1439	4847	1859	41,910
33,108	222,744	425	23,522	87,494	40,963	29,318	42,588	75,562
2942	7309	267	4303	5657	1144	14,681	1366	2347
7018	5281	51	1095	1521	740	5220	1998	797
32,844	20,150	1994	20,557	36,869	11,342	45,176	30,586	20,938
3016	2760	88	1399	2355	676	4307	3830	929
17,502,286	5,926,841	161,973	1,735,208	2,899,785	2,767,321	2,645,175	1,399,300	3,167,790
3,511,564	1,585,071	76,443	494,038	1,952,016	993,746	1,558,424	474,583	936,362
975,189	318,478	6891	145,322	1,936,303	188,713	597,813	684,113	237,470
341,152	704,918	18,015	477,015	453,736	222,768	115,089	2,095,353	455,170
1,337,711	1,255,547	221	813,852	2,765,037	236,032	2,637,367	1,089,272	134,032
6,165,616	3,864,014	101,570	1,930,227	7,107,092	1,641,260	4,908,694	4,343,321	1,763,034
23,667,902	9,790,854	263,543	3,665,435	10,006,878	4,408,581	7,553,869	5,742,620	4,930,824

(continued)

Table 10.22 (continued)

	Sector 35	Sector 36	Sector 37	Sector 38	Sector 39	Sector 40	Sector 41	Sector 42
Sector 1	23,268	8000	21,849	28,736	6590	17,099	5445	0
Sector 2	1069	2691	4682	10,011	7002	14,419	903	14,915
Sector 3	0	0	0	0	0	0	0	0
Sector 4	5823	1250	1577	0	0	0	0	0
Sector 5	95	62	588	129	1272	1825	546	2104
Sector 6	10,990	4925	9507	55,820	27,770	29,770	62,553	19,328
Sector 7	12,506	1328	3870	36,993	6037	19,864	3276	21,675
Sector 8	4775	3362	11,682	17,320	3866	48,968	18,594	70,865
Sector 9	2096	2559	11,525	23,176	42,010	26,637	13,364	86,337
Sector 10	6890	28,340	6131	41,026	120,356	44,449	125,557	138,386
Sector 11	5649	41,111	33,191	45,126	27,146	14,571	7252	77,179
Sector 12	67,165	44,301	60,798	214,215	30,950	1,521,459	45,818	53,785
Sector 13	4866	3899	12,581	2614	16,882	19,896	7151	25,915
Sector 14	3099	14,767	6812	6308	2574	171	514	0
Sector 15	15,751	42,964	7208	12,207	7382	7451	4437	6291
Sector 16	9293	16,159	18,914	5969	6864	159,395	3124	12,351
Sector 17	5490	29,341	31,806	93,803	22,894	3958	5221	69,038
Sector 18	27,263	83,770	23,576	40,340	10,141	11,353	6160	12,097
Sector 19	24,364	75,608	12,996	60,619	11,118	2625	6663	40,876
Sector 20	21,801	45,597	12,443	23,943	33,481	8317	4184	9344
Sector 21	2819	4445	4738	19,550	5859	4477	2253	12,030
Sector 22	0	0	0	0	0	0	0	0
Sector 23	12,588	13,906	25,749	27,275	62,739	41,405	10,032	72,176
Sector 24	204	512	827	1586	2213	1656	747	1598
Sector 25	959	2375	1684	4035	4653	2375	884	3040
Sector 26	2199	6070	22,563	18,857	16,533	69,006	9092	74,341
Sector 27	25,638	119,359	20,873	51,665	88,133	17,849	32,841	220,798
Sector 28	1039	2398	864	3425	7193	1142	2543	56,623
Sector 29	3517	9942	8025	8407	27,890	32,432	9010	98,753
Sector 30	12,890	23,633	12,904	43,975	25,324	106,344	22,925	54,257
Sector 31	23,961	82,560	21,575	38,743	82,521	34,307	34,769	337,240
Sector 32	2663	63,667	28,825	52,159	86,448	20,210	16,699	84,300
Sector 33	2819	20,607	2727	69,202	24,577	6800	14,368	28,657
Sector 34	5852	30,332	13,343	35,078	59,651	25,329	24,488	47,344
Sector 35	14,838	4634	871	67	6883	1615	1124	71
Sector 36	8975	125,710	4660	1895	7008	4977	761	7382
Sector 37	388	425	21,496	3569	4346	898	995	12,278
Sector 38	1841	18,081	22,608	95,785	24,168	19,433	10,505	75,731
Sector 39	2079	1992	1493	755	14,835	3387	1066	40,378
Sector 40	1632	1874	3368	796	4874	8772	1597	14,458
Sector 41	9618	10,242	6581	9888	19,874	5107	34,119	65,871
Sector 42	158	642	357	569	642	512	230	1416
Total intermediate input	388,930	993,440	517,866	1,205,638	960,700	2,360,261	551,811	1,969,228
Compensation of employees	151,063	537,562	196,965	844,106	2,429,458	1,095,867	322,453	3,017,215
Net taxes on production	9647	89,905	16,168	116,609	15,632	11,434	55,592	16,381
Depreciation of fixed assets	28,346	102,257	78,949	68,628	288,752	133,031	86,326	444,195
Operating surplus	32,911	276,222	59,257	237,667	40,758	78,472	101,363	95,486
Total value added	221,967	1,005,946	351,339	1,267,010	2,774,600	1,318,803	565,735	3,573,276
Total input	610,897	1,999,386	869,206	2,472,649	3,735,300	3,679,064	1,117,546	5,542,504

Total intermediate use	Rural household consumption expenditure	Urban household consumption expenditure	Government consumption expenditure	Gross fixed capital formation	Increase in inventories	Export	Import	Total output
11,737,647	1,300,144	1,613,799	115,901	833,145	221,672	403,011	664,292	15,561,027
4,530,831	21,009	15,319	0	41,310	−87,034	18,132	349,611	4,189,956
4,793,232	0	0	0	18,307	−55,601	352,521	3,838,063	1,270,396
3,761,670	0	0	0	72,621	−19,071	7635	2,224,679	1,598,176
1,182,882	0	0	0	10,010	−22,672	22,074	78,227	1,114,067
7,483,301	1,469,029	4,248,039	0	260,106	255,511	220,825	242,567	13,694,244
4,546,740	49,262	76,055	0	34,833	−27,421	3,484,091	821,978	7,341,581
1,825,292	309,746	1,474,851	0	84,763	90,604	579,653	75,977	4,288,933
2,085,372	28,851	125,218	0	202,330	−1685	269,856	347,849	2,362,092
3,748,213	18,571	125,715	0	40,402	−34,103	226,932	429,910	3,695,819
6,284,867	26,623	241,360	0	−8062	−134,219	113,620	417,212	6,106,978
18,275,911	169,395	552,275	0	206,155	−131,943	944,203	1,699,594	18,316,401
8,373,526	10,449	67,519	0	80,824	−135,481	410,652	261,646	8,545,843
17,224,600	0	0	0	190,950	−188,641	365,674	891,034	16,701,550
4,238,524	18,202	98,240	0	103,820	−54,600	345,348	230,866	4,518,668
8,438,137	1306	19,066	0	5,545,489	89,393	1,765,492	1,743,725	14,115,159
5,851,449	149,921	731,097	0	6,082,789	119,509	1,815,052	1,436,109	13,313,708
5,365,785	133,717	518,185	0	1,869,685	34,952	1,629,728	768,984	8,783,069
6,449,233	114,673	458,877	0	1,154,267	187,744	4,346,241	3,149,194	9,561,841
1,774,449	13,189	45,715	0	213,500	−24,607	515,957	1,588,692	949,511
1,481,075	74,812	408,415	0	521,638	9937	232,188	98,595	2,629,469
1,271,609	0	0	0	0	6321	4598	855,588	426,939
9,450,048	60,900	224,622	0	0	−978,086	14,481	1974	8,769,991
240,044	17,698	134,392	0	2581	358	0	0	395,072
176,285	13,354	127,530	0	0	1226	0	0	318,395
748,305	0	288,070	0	22,262,999	72,498	384,913	106,789	23,649,997
9,034,300	141,789	548,992	434,191	69,817	−179,449	369,632	580,305	9,838,967
261,169	2052	13,811	0	0	2855	7486	33,198	254,174
1,877,110	168,162	719,989	0	745,626	43,230	171,223	81,767	3,643,573
5,963,025	516,553	1,582,013	0	673,718	47,359	877,685	0	9,660,353
2,980,833	323,435	1,260,254	0	0	98,640	−80,351	211,604	4,371,206
6,130,974	194,764	1,023,529	78,898	0	108,023	81,282	72,758	7,544,712
1,441,410	582,718	2,059,911	0	1,422,912	120,162	0	0	5,627,114
3,554,694	27,679	396,164	183,925	0	59,141	869,432	249,350	4,841,685
522,441	0	0	299,859	0	2978	20,711	240,134	605,855
1,500,779	1776	22,731	265,063	213,083	−1537	0	0	2,001,894
269,208	7859	97,052	479,912	0	−2160	0	0	851,872
1,233,272	178,413	1,034,878	0	0	35,106	35,677	71,621	2,445,725
141,521	357,800	799,511	2,537,756	0	−27,196	4765	27,306	3,786,851
260,172	134,354	1,544,200	1,780,267	0	−23,344	3828	2677	3,696,801
576,968	35,621	191,492	297,158	0	12,312	42,790	53,539	1,102,803
49,859	0	0	5,536,395	0	−12,349	17,503	14,631	5,576,777

Table 10.23 China's national input–output table in 2021 in the policy scenario

	Sector 1	Sector 2	Sector 3	Sector 4	Sector 5	Sector 6	Sector 7	Sector 8
Sector 1	2,118,961	42,219	26	2843	457	5,586,668	1,498,866	217,988
Sector 2	12,472	745,912	11,674	7005	3123	38,532	37,294	17,189
Sector 3	0	0	22,516	0	0	0	0	0
Sector 4	0	0	0	161,898	0	0	0	0
Sector 5	1759	1667	715	1727	77,384	5073	0	0
Sector 6	1,623,309	7867	2444	4041	3384	2,909,846	56,834	222,673
Sector 7	1953	1154	1082	967	623	7279	2,366,032	1,463,898
Sector 8	5390	15,106	4819	7109	4974	18,490	118,865	648,090
Sector 9	18,704	38,599	5183	5709	2477	15,120	8274	9722
Sector 10	13,836	5406	2895	2977	2789	263,236	78,393	71,044
Sector 11	107,333	48,792	51,299	94,083	55,382	39,629	18,926	26,868
Sector 12	1,200,494	102,346	50,747	91,726	116,047	389,338	847,939	276,432
Sector 13	20,716	38,571	12,614	13,752	41,872	97,476	17,204	5894
Sector 14	3666	180,970	66,515	21,387	8171	15,911	5959	6071
Sector 15	57,465	87,112	10,467	34,582	19,688	65,304	15,874	21,310
Sector 16	86,536	223,848	112,830	131,567	100,908	48,240	92,113	28,380
Sector 17	43,890	28,533	9281	17,882	23,172	17,904	14,739	6737
Sector 18	4223	64,464	12,618	22,270	11,155	20,559	16,942	8304
Sector 19	1434	6789	1756	2392	1316	5200	5757	4858
Sector 20	12,706	47,041	12,873	8856	5144	9946	16,553	5582
Sector 21	48,004	31,332	1876	7137	12,267	21,601	56,570	46,044
Sector 22	1085	421	22	1236	9775	10,989	13,776	2176
Sector 23	191,454	189,841	99,571	284,861	133,421	146,366	189,264	35,939
Sector 24	561	1697	1682	8614	6248	8357	4007	1569
Sector 25	1860	3428	1070	1568	1261	5718	6169	2134
Sector 26	4009	5647	1540	654	418	3982	1769	1357
Sector 27	275,844	199,413	26,433	72,919	74,315	404,843	140,453	105,434
Sector 28	13,247	1490	219	334	433	3185	5858	9695
Sector 29	60,879	14,801	2311	2742	3120	25,073	15,780	24,296
Sector 30	371,809	67,842	12,497	22,150	23,928	341,003	188,931	110,891
Sector 31	45,334	51,003	12,102	19,197	18,570	83,649	35,604	28,216
Sector 32	177,207	127,746	11,176	21,401	23,433	145,257	198,604	72,296
Sector 33	4778	8321	609	1229	2295	24,277	31,300	56,723
Sector 34	24,574	24,702	2478	6730	5552	233,208	31,186	97,777
Sector 35	27,484	6335	2492	571	314	11,457	4827	2404
Sector 36	139,327	41,593	15,361	23,346	13,240	48,586	18,630	16,140
Sector 37	48,294	6300	1111	1208	3567	4330	12,252	649
Sector 38	62,688	38,546	5093	4138	5443	17,022	9578	8862
Sector 39	6117	4218	252	654	1602	1941	905	506
Sector 40	9354	10,371	489	3231	2432	7033	3710	3106
Sector 41	3903	7145	1306	4421	5568	14,878	8674	8329
Sector 42	5075	1122	413	358	234	1975	945	622
Total intermediate input	6,857,731	2,529,708	592,456	1,121,472	825,505	11,118,481	6,195,357	3,676,207
Compensation of employees	8,862,355	839,541	199,826	151,571	135,671	864,724	627,622	458,472
Net taxes on production	18,006	468,310	226,906	97,801	82,555	975,330	249,779	118,969
Depreciation of fixed assets	434,908	160,449	133,396	35,773	35,173	267,981	164,845	54,001
Operating surplus	0	485,917	312,616	185,933	112,653	861,288	257,460	153,684
Total value added	9,315,269	1,954,217	872,743	471,079	366,052	2,969,324	1,299,706	785,126
Total input	16,173,000	4,483,925	1,465,199	1,592,551	1,191,557	14,087,805	7,495,063	4,461,333

Sector 9	Sector 10	Sector 11	Sector 12	Sector 13	Sector 14	Sector 15	Sector 16	Sector 17
337,469	205,301	141	646,104	5448	2110	1580	1650	1244
19,727	36,971	395,018	280,275	657,154	675,968	23,040	52,173	16,223
0	0	3,578,383	1,103,335	62,817	216,225	0	0	0
0	0	0	109,953	23,644	3,395,693	49,935	45,086	7750
0	0	641	272,931	528,937	50,978	19,490	10,724	3567
12,983	16,609	29,585	307,463	33,241	61,789	15,711	50,102	51,200
24,783	69,738	324	120,006	19,514	2208	7822	21,495	26,939
41,992	28,344	2638	41,721	22,559	40,105	14,666	32,834	150,206
702,297	109,140	1949	32,813	68,920	14,035	81,537	49,308	157,941
43,715	928,041	3447	273,067	222,634	21,356	55,975	66,900	63,007
17,438	18,915	385,557	1,114,231	365,907	760,416	48,421	156,442	67,398
241,155	553,401	118,972	6,978,512	717,319	288,597	207,584	552,457	891,922
20,636	10,619	21,535	134,419	1,507,623	351,237	52,433	109,976	102,906
54,618	45,980	4064	184,637	239,871	4,697,678	1,577,791	3,215,641	1,674,733
58,012	66,223	20,438	170,136	257,743	156,810	550,385	642,028	264,917
34,786	52,422	81,630	270,309	240,421	535,816	221,868	3,443,429	1,178,308
10,382	31,478	14,322	51,657	45,337	91,155	42,103	230,790	3,787,158
7836	26,250	7191	47,645	35,014	56,466	37,284	837,238	534,014
1026	34,766	866	11,140	5562	4511	5706	319,016	163,807
1998	46,140	9021	136,174	28,857	65,534	11,323	80,034	266,969
20,906	25,193	2802	109,417	75,121	68,124	30,485	105,795	180,834
7311	236,517	1253	79,978	138,715	503,769	37,134	181,547	10,354
82,002	120,556	104,849	963,531	691,219	1,115,151	358,646	341,934	238,066
1141	1105	9302	45,889	10,284	23,716	4804	13,639	9777
1779	8401	1912	20,474	9948	14,481	4798	5107	8410
654	1002	3063	7070	3123	5937	930	5287	5134
80,955	92,649	121,079	504,671	380,716	344,205	110,099	437,739	291,699
1713	2491	189	15,090	2826	6497	965	13,568	3291
10,569	10,579	19,554	68,395	23,269	202,324	18,121	48,302	44,505
65,006	155,722	74,628	465,865	311,436	412,935	124,756	391,488	831,069
13,769	23,239	11,126	113,046	75,205	71,802	60,367	118,181	70,022
55,393	67,049	37,241	394,338	368,689	467,135	80,796	162,687	242,508
14,272	25,512	990	55,336	31,537	7176	31,337	42,690	37,264
22,551	34,972	29,325	288,540	63,678	30,607	19,824	134,999	213,834
1522	2340	1695	46,100	7064	30,855	4300	53,142	127,865
4912	20,933	10,437	113,806	37,413	69,930	29,390	91,513	137,826
371	13,662	1230	20,086	2955	6077	702	3161	4866
2966	8954	5406	43,758	32,590	50,496	9179	17,996	24,525
430	565	246	3012	1478	1504	691	4857	3346
2352	4832	1589	11,956	16,291	17,117	11,723	36,316	31,191
3402	4463	1404	23,148	16,481	13,654	5848	21,439	18,289
356	614	283	2404	1584	1991	698	2075	2361
2,025,187	3,141,684	5,115,325	15,682,442	7,390,144	14,954,168	3,970,244	12,150,783	11,947,244
194,403	281,561	206,175	1,172,731	612,307	1,060,385	319,409	1,339,622	779,489
80,719	108,442	680,685	647,668	333,419	513,366	132,974	614,714	517,857
46,431	107,035	152,794	501,944	237,868	632,437	93,271	356,627	225,931
117,283	166,964	226,613	971,463	548,514	725,165	223,714	947,791	641,722
438,836	664,002	1,266,267	3,293,807	1,732,108	2,931,354	769,368	3,258,754	2,165,000
2,464,023	3,805,686	6,381,592	18,976,249	9,122,252	17,885,522	4,739,613	15,409,537	14,112,244

(continued)

Table 10.23 (continued)

	Sector 18	Sector 19	Sector 20	Sector 21	Sector 22	Sector 23	Sector 24	Sector 25
Sector 1	189	0	47	319,950	177	156	69	0
Sector 2	7784	3025	1593	17,493	815	1,537,157	19,804	439
Sector 3	0	0	0	25	1	54,653	194,042	0
Sector 4	11,592	142	1300	243	538	258	0	0
Sector 5	4025	70	110	7347	68	2348	16	34
Sector 6	35,290	37,429	3184	36,123	700	26,616	1537	862
Sector 7	17,567	2806	1521	238,999	77	450	48	98
Sector 8	19,727	17,771	3009	39,401	120	28,996	4626	1627
Sector 9	25,705	17,479	2672	80,912	90	5399	483	485
Sector 10	116,378	86,364	11,875	47,403	123	16,632	1039	861
Sector 11	39,298	27,084	4652	19,380	1203	256,942	8574	3417
Sector 12	809,823	691,689	93,742	201,917	11,751	27,134	2648	16,289
Sector 13	183,668	173,966	35,444	55,470	1447	18,603	952	963
Sector 14	2,344,573	303,382	39,822	252,153	5810	8432	1016	381
Sector 15	385,046	226,480	37,192	111,598	609	51,820	3928	8891
Sector 16	441,365	101,218	27,982	10,597	359	87,761	3329	3507
Sector 17	23,945	33,816	5792	27,060	124	97,739	9035	4094
Sector 18	1,273,357	283,986	49,268	16,576	914	483,832	957	1109
Sector 19	537,325	4,491,340	238,699	6594	46	5369	216	266
Sector 20	89,471	507,639	114,674	5722	54	117,605	2586	1789
Sector 21	94,634	79,521	17,560	290,458	108	5835	308	341
Sector 22	70,599	9743	101	525	5318	957	0	0
Sector 23	122,487	133,464	12,319	35,632	2473	3,125,394	8242	77,314
Sector 24	5571	5282	721	2529	121	10,164	22,970	1075
Sector 25	3996	4521	822	1797	115	10,888	220	6915
Sector 26	2428	1911	288	415	57	2998	306	401
Sector 27	217,538	166,875	23,200	62,886	4218	106,602	10,903	3127
Sector 28	7216	4257	1144	748	13	1033	54	103
Sector 29	34,970	112,019	4740	17,863	168	53,567	588	1219
Sector 30	323,855	459,097	25,601	52,712	2583	69,287	6924	2002
Sector 31	67,678	32,980	9904	18,959	516	25,071	1892	2629
Sector 32	176,796	445,820	11,645	28,943	2376	363,154	13,669	18,999
Sector 33	39,313	35,068	6078	17,980	180	3832	433	174
Sector 34	217,653	119,639	13,399	22,286	437	38,990	1816	1623
Sector 35	39,774	75,631	5643	2593	45	5821	108	68
Sector 36	56,441	55,266	6933	5679	221	70,159	950	502
Sector 37	1983	1358	105	424	32	23,366	180	18,898
Sector 38	9107	8031	975	1123	478	67,010	1189	2603
Sector 39	1465	1012	227	442	5	1353	97	253
Sector 40	17,523	5347	1959	414	87	10,446	1157	73
Sector 41	11,979	10,241	1278	2926	80	19,728	779	991
Sector 42	1122	1084	153	377	0	1249	35	66
Total intermediate input	7,890,257	8,773,854	817,375	2,062,672	44,658	6,844,804	327,727	184,490
Compensation of employees	563,836	754,554	95,949	239,059	6478	700,209	44,113	67,221
Net taxes on production	218,182	153,272	27,058	161,085	3880	425,794	10,235	13,247
Depreciation of fixed assets	153,859	236,906	19,572	53,954	2913	877,573	13,698	57,891
Operating surplus	414,933	282,333	48,984	131,673	395,289	284,885	20,026	7762
Total value added	1,350,809	1,427,066	191,564	585,770	408,560	2,288,461	88,072	146,120
Total input	9,241,066	10,200,920	1,008,939	2,648,443	453,217	9,133,264	415,798	330,611

Sector 26	Sector 27	Sector 28	Sector 29	Sector 30	Sector 31	Sector 32	Sector 33	Sector 34
110,821	250,209	0	0	2341	533,223	0	499	12,055
20,012	36,104	379	0	2950	3079	0	1291	7164
0	13,024	0	0	0	0	0	0	0
0	0	0	0	0	0	0	0	0
254,811	2717	0	0	0	0	0	0	0
100,698	82,468	673	9299	50,322	1,604,854	22,171	17,102	60,107
21,294	13,546	409	467	11,666	22,712	1131	6768	65,668
96,000	74,873	1949	7217	85,455	17,426	20,859	27,456	63,355
441,590	12,772	1346	3655	15,970	6385	5945	7364	12,832
42,151	38,473	7167	118,899	89,605	15,710	153,741	35,696	455,223
325,159	1,953,691	7976	7352	32,379	22,723	50,022	79,544	161,735
778,105	184,411	897	20,207	87,756	35,216	26,846	101,663	182,000
5,715,440	22,203	935	1034	2635	9469	1271	1803	2697
3,144,449	53,961	0	39	4738	208	0	5460	3881
820,149	58,318	210	7668	3891	5397	7156	36,947	102,003
746,661	252,340	2832	30,851	20,908	5418	33,695	21,572	33,350
104,257	737,578	21,374	26,889	84,795	15,596	23,030	23,242	154,593
954,103	29,464	478	262,586	66,012	33,101	16,268	67,528	142,470
60,805	26,407	2372	227,896	37,829	1884	5920	18,244	293,508
55,081	16,686	320	34,192	3158	145	30,382	11,896	15,127
82,108	9589	683	4041	7407	2877	8242	7866	30,792
0	0	0	0	0	0	0	0	0
188,385	125,584	3692	95,101	113,512	98,718	74,624	36,444	48,448
2984	9843	258	2170	1986	16,858	393	642	4905
6055	6311	434	4660	3715	8019	3744	1110	3317
267,545	45,077	11,428	6464	25,045	10,571	11,454	117,169	4960
2,349,849	884,451	38,529	116,938	680,484	56,184	196,368	45,948	250,002
5439	3291	11,336	4983	13,103	2654	34,086	3887	23,893
365,195	75,265	5491	165,258	78,959	14,197	225,825	23,724	18,994
620,574	135,539	20,264	119,811	49,267	97,685	33,580	25,976	117,836
252,223	166,960	1944	74,911	186,407	31,263	292,090	99,180	335,227
238,634	618,083	3166	78,949	331,376	76,459	636,642	260,715	230,537
14,697	51,129	4114	99,733	215,697	48,971	253,940	80,095	84,765
90,902	58,185	10,490	191,203	473,248	55,192	483,039	205,481	222,445
18,893	3222	146	6381	15,141	41	11,112	1282	2537
291,737	12,331	48	18,643	37,427	851	5494	11,668	3789
1240	3161	0	1067	2244	1514	5090	1955	43,640
35,321	238,410	431	24,419	89,757	43,110	30,788	44,793	78,682
3139	7823	271	4467	5804	1203	15,417	1436	2444
7487	5652	52	1137	1561	778	5482	2102	830
35,039	21,568	2022	21,341	37,823	11,936	47,441	32,170	21,803
3217	2954	89	1452	2416	711	4523	4028	967
18,672,249	6,343,673	164,204	1,801,384	2,974,788	2,912,338	2,777,812	1,471,744	3,298,580
3,746,298	1,696,549	77,496	512,879	2,002,505	1,045,821	1,636,568	499,153	975,022
1,040,377	340,876	6986	150,864	1,986,385	198,603	627,790	719,531	247,274
363,957	754,494	18,263	495,207	465,472	234,442	120,860	2,203,834	473,963
1,427,132	1,343,849	224	844,890	2,836,554	248,401	2,769,613	1,145,666	139,566
6,577,764	4,135,768	102,969	2,003,841	7,290,916	1,727,267	5,154,831	4,568,184	1,835,826
25,250,013	10,479,441	267,173	3,805,225	10,265,703	4,639,605	7,932,643	6,039,928	5,134,406

(continued)

Table 10.23 (continued)

	Sector 35	Sector 36	Sector 37	Sector 38	Sector 39	Sector 40	Sector 41	Sector 42
Sector 1	24,411	8480	22,513	30,273	7148	18,140	5655	0
Sector 2	1122	2853	4824	10,547	7595	15,297	938	15,813
Sector 3	0	0	0	0	0	0	0	0
Sector 4	6109	1324	1625	0	0	0	0	0
Sector 5	100	65	606	136	1380	1936	567	2231
Sector 6	11,530	5220	9796	58,805	30,122	31,583	64,974	20,492
Sector 7	13,120	1408	3988	38,972	6548	21,074	3403	22,980
Sector 8	5010	3563	12,037	18,246	4193	51,949	19,313	75,131
Sector 9	2199	2712	11,876	24,416	45,569	28,259	13,881	91,535
Sector 10	7229	30,037	6317	43,221	130,551	47,155	130,416	146,716
Sector 11	5926	43,572	34,200	47,539	29,445	15,458	7533	81,825
Sector 12	70,466	46,953	62,645	225,673	33,572	1,614,087	47,592	57,023
Sector 13	5105	4132	12,963	2753	18,311	21,107	7428	27,475
Sector 14	3251	15,651	7019	6646	2792	181	534	0
Sector 15	16,525	45,537	7427	12,860	8008	7905	4608	6669
Sector 16	9750	17,127	19,489	6288	7446	169,099	3244	13,094
Sector 17	5760	31,098	32,773	98,820	24,834	4199	5423	73,195
Sector 18	28,603	88,786	24,292	42,498	11,000	12,044	6399	12,825
Sector 19	25,561	80,135	13,391	63,861	12,059	2785	6921	43,337
Sector 20	22,872	48,327	12,821	25,224	36,318	8824	4346	9906
Sector 21	2957	4712	4882	20,596	6356	4749	2340	12,754
Sector 22	0	0	0	0	0	0	0	0
Sector 23	13,207	14,738	26,531	28,734	68,054	43,925	10,421	76,521
Sector 24	214	542	852	1671	2401	1757	776	1694
Sector 25	1007	2517	1735	4251	5047	2519	918	3223
Sector 26	2307	6434	23,249	19,866	17,934	73,207	9444	78,817
Sector 27	26,898	126,506	21,508	54,428	95,599	18,936	34,112	234,090
Sector 28	1090	2542	890	3608	7803	1212	2641	60,032
Sector 29	3690	10,537	8269	8856	30,252	34,406	9358	104,698
Sector 30	13,523	25,048	13,296	46,327	27,469	112,818	23,812	57,523
Sector 31	25,139	87,504	22,231	40,816	89,511	36,396	36,114	357,541
Sector 32	2794	67,480	29,701	54,949	93,770	21,441	17,345	89,375
Sector 33	2957	21,841	2810	72,904	26,659	7215	14,924	30,382
Sector 34	6140	32,148	13,749	36,954	64,704	26,871	25,436	50,194
Sector 35	15,567	4911	897	70	7466	1713	1167	76
Sector 36	9416	133,237	4802	1997	7602	5280	790	7826
Sector 37	407	451	22,150	3760	4715	952	1034	13,017
Sector 38	1931	19,164	23,296	100,908	26,215	20,616	10,912	80,290
Sector 39	2181	2111	1538	795	16,092	3594	1107	42,809
Sector 40	1712	1986	3470	838	5287	9306	1659	15,329
Sector 41	10,091	10,855	6781	10,417	21,558	5418	35,439	69,836
Sector 42	166	681	368	599	696	543	239	1501
Total intermediate input	408,042	1,052,924	533,606	1,270,123	1,042,079	2,503,956	573,167	2,087,774
Compensation of employees	158,486	569,750	202,951	889,254	2,635,252	1,162,584	334,933	3,198,848
Net taxes on production	10,121	95,288	16,660	122,845	16,956	12,130	57,744	17,367
Depreciation of fixed assets	29,739	108,380	81,349	72,299	313,212	141,130	89,667	470,935
Operating surplus	34,528	292,761	61,058	250,379	44,211	83,249	105,286	101,234
Total value added	232,875	1,066,180	362,018	1,334,777	3,009,630	1,399,093	587,630	3,788,384
Total input	640,917	2,119,104	895,624	2,604,900	4,051,709	3,903,049	1,160,797	5,876,158

Total intermediate use	Rural household consumption expenditure	Urban household consumption expenditure	Government consumption expenditure	Gross fixed capital formation	Increase in inventories	Export	Import	Total output
12,015,433	1,374,131	1,712,770	122,525	868,912	221,672	422,521	686,745	16,051,220
4,747,826	22,458	16,473	0	44,322	−87,034	18,534	373,818	4,388,761
5,245,020	0	0	0	18,307	−55,601	372,609	3,838,063	1,742,272
3,817,088	0	0	0	72,756	−19,071	7635	2,485,802	1,392,605
1,254,160	0	0	0	10,627	−22,672	22,701	83,054	1,181,762
7,731,039	1,532,874	4,453,142	0	263,274	255,511	220,825	242,567	14,214,099
4,652,536	51,937	80,306	0	35,386	−27,421	3,745,254	895,779	7,642,219
1,897,217	324,451	1,552,427	0	86,689	90,604	579,653	75,977	4,455,065
2,183,262	30,241	131,670	0	205,263	−1685	269,856	378,676	2,439,930
3,897,701	19,611	131,884	0	41,902	−34,103	226,932	458,281	3,825,646
6,643,666	28,023	253,052	0	−8296	−134,219	116,821	440,498	6,458,549
19,055,093	178,613	582,320	0	212,585	−131,943	944,203	1,751,677	19,089,193
8,886,757	11,240	71,677	0	85,400	−135,481	432,791	282,535	9,069,850
18,208,043	0	0	0	203,291	−188,641	365,674	929,793	17,658,575
4,475,331	19,252	104,080	0	106,349	−54,600	345,348	247,073	4,748,687
8,952,693	1388	20,213	0	6,073,976	89,393	1,870,634	1,823,388	15,184,910
6,135,583	157,846	766,102	0	6,437,881	119,509	1,946,348	1,521,932	14,041,337
5,657,927	140,315	544,933	0	1,943,643	34,952	1,700,229	804,026	9,217,973
6,778,670	121,615	486,344	0	1,218,203	187,744	4,529,159	3,276,363	10,045,372
1,939,918	13,967	47,451	0	213,764	−24,607	526,713	1,681,910	1,035,296
1,545,227	77,847	430,740	0	526,328	9937	238,685	103,702	2,725,062
1,323,300	0	0	0	0	6321	4729	890,749	443,601
9,870,636	60,900	224,622	0	0	−1,036,047	15,138	1986	9,133,264
250,771	18,428	139,343	0	2728	358	0	0	411,628
186,373	14,070	131,849	0	0	1226	0	0	333,518
791,351	0	305,093	0	23,751,198	72,498	413,701	114,207	25,219,634
9,489,645	150,274	580,271	465,632	74,728	−179,449	369,632	622,356	10,328,377
278,147	2153	14,552	0	0	2855	7725	36,037	269,393
2,002,733	176,060	756,528	0	774,062	43,230	181,394	87,371	3,846,635
6,454,365	542,851	1,665,564	0	691,143	47,359	897,398	0	10,298,680
3,145,519	341,606	1,332,002	0	0	98,640	−94,826	225,456	4,597,485
6,565,773	204,413	1,077,495	84,173	0	108,023	87,372	78,001	8,049,249
1,511,537	613,360	2,142,327	0	1,496,580	120,162	0	0	5,883,965
3,726,761	28,965	414,131	194,960	0	59,141	913,725	249,350	5,088,333
551,073	0	0	313,117	0	2978	22,224	255,696	633,696
1,581,473	1776	22,731	280,055	225,842	−1537	0	0	2,110,340
283,563	8229	101,362	502,066	0	−2160	0	0	893,060
1,306,798	187,445	1,088,059	0	0	35,106	37,077	76,282	2,578,203
149,406	379,604	847,403	2,682,506	0	−27,196	4892	29,140	4,007,474
274,768	139,885	1,618,708	1,891,877	0	−23,344	3828	2677	3,903,046
611,891	37,454	201,494	314,782	0	12,312	42,790	53,539	1,167,185
52,347	0	0	5,839,995	0	−12,349	18,613	15,332	5,883,274

Table 10.24 China's national input–output table in 2022 in the policy scenario

	Sector 1	Sector 2	Sector 3	Sector 4	Sector 5	Sector 6	Sector 7	Sector 8
Sector 1	2,177,374	42,359	26	2843	488	5,881,133	1,683,595	222,767
Sector 2	12,838	749,523	11,674	7005	3378	40,563	43,024	17,640
Sector 3	0	0	22,516	0	0	0	0	0
Sector 4	0	0	0	161,898	0	0	0	0
Sector 5	1797	1671	715	1727	80,725	5340	0	0
Sector 6	1,647,503	7886	2444	4041	3560	3,063,220	60,801	225,485
Sector 7	1976	1156	1082	967	650	7663	2,520,855	1,477,691
Sector 8	5519	15,158	4819	7109	5331	19,464	131,841	660,772
Sector 9	19,418	38,846	5183	5709	2784	15,917	9801	10,053
Sector 10	14,018	5416	2895	2977	2877	277,111	83,391	71,863
Sector 11	108,821	48,888	51,299	94,083	57,479	41,717	20,217	27,209
Sector 12	1,219,737	102,557	50,747	91,726	120,817	409,859	917,840	280,198
Sector 13	21,092	38,680	12,614	13,752	44,119	102,614	18,810	5986
Sector 14	3774	181,752	66,515	21,387	8816	16,750	6853	6235
Sector 15	58,426	87,372	10,467	34,582	20,887	68,746	17,215	21,635
Sector 16	88,650	224,591	112,830	131,567	106,725	50,783	103,486	29,013
Sector 17	44,746	28,617	9281	17,882	24,501	18,848	16,137	6848
Sector 18	4302	64,645	12,618	22,270	11,765	21,642	18,579	8442
Sector 19	1474	6814	1756	2392	1403	5474	6570	4981
Sector 20	12,825	47,152	12,873	8856	5431	10,470	17,257	5630
Sector 21	48,179	31,362	1876	7137	12,556	22,740	57,681	46,194
Sector 22	1109	423	22	1236	10,498	11,568	15,172	2231
Sector 23	194,766	190,302	99,571	284,861	139,837	154,081	205,694	36,499
Sector 24	573	1702	1682	8614	6579	8798	4432	1598
Sector 25	1893	3436	1070	1568	1326	6019	6694	2166
Sector 26	4077	5661	1540	654	439	4192	1919	1377
Sector 27	283,158	200,131	26,433	72,919	78,967	426,181	159,247	107,957
Sector 28	13,642	1499	219	334	489	3353	6730	9927
Sector 29	62,016	14,841	2311	2742	3298	26,395	17,203	24,676
Sector 30	382,489	68,228	12,497	22,150	26,893	358,976	216,341	113,724
Sector 31	46,939	51,275	12,102	19,197	20,385	88,058	41,745	29,099
Sector 32	180,405	128,051	11,176	21,401	24,546	152,913	214,499	73,420
Sector 33	5244	8416	609	1229	2719	25,557	46,644	62,004
Sector 34	25,420	24,831	2478	6730	6126	245,500	36,151	100,667
Sector 35	28,326	6364	2492	571	340	12,061	5515	2470
Sector 36	142,479	41,726	15,361	23,346	14,020	51,147	20,726	16,468
Sector 37	49,851	6332	1111	1208	3947	4558	14,226	667
Sector 38	65,442	38,784	5093	4138	6061	17,919	11,632	9218
Sector 39	6255	4230	252	654	1680	2043	1001	517
Sector 40	9530	10,396	489	3231	2544	7404	4049	3158
Sector 41	4012	7174	1306	4421	6052	15,662	9827	8518
Sector 42	5169	1125	413	358	244	2079	1029	632
Total intermediate input	7,005,264	2,539,373	592,456	1,121,472	871,280	11,704,520	6,774,427	3,735,636
Compensation of employees	9,106,664	843,605	199,826	151,571	141,529	910,303	668,691	467,443
Net taxes on production	18,502	470,577	226,906	97,801	86,119	1,026,738	266,123	121,298
Depreciation of fixed assets	446,897	161,226	133,396	35,773	36,692	282,106	175,631	55,058
Operating surplus	0	488,270	312,616	185,933	117,517	906,685	274,307	156,691
Total value added	9,572,064	1,963,677	872,743	471,079	381,857	3,125,833	1,384,753	800,490
Total input	16,577,327	4,503,051	1,465,199	1,592,551	1,253,137	14,830,353	8,159,180	4,536,125

Sector 9	Sector 10	Sector 11	Sector 12	Sector 13	Sector 14	Sector 15	Sector 16	Sector 17
338,129	219,374	150	697,525	5658	2116	1608	1680	1277
19,778	39,936	418,992	305,373	692,761	678,811	23,661	53,136	16,831
0	0	3,795,564	1,133,195	64,366	216,479	0	0	0
0	0	0	202,205	38,569	3,587,755	68,523	45,918	11,134
0	0	680	288,182	547,576	51,097	19,874	10,922	3645
12,999	17,312	31,380	322,265	34,111	61,935	15,917	51,027	51,942
24,805	72,161	344	124,802	19,913	2212	7896	21,891	27,170
42,070	30,153	2799	44,771	23,437	40,241	14,943	33,440	154,236
705,233	120,716	2068	37,142	73,989	14,127	84,875	50,218	167,898
43,760	968,113	3657	282,786	227,207	21,391	56,560	68,135	63,585
17,458	19,604	408,957	1,156,894	374,422	761,869	49,027	159,330	68,180
241,453	575,090	126,193	7,306,095	735,078	289,134	210,200	562,655	904,267
20,667	11,142	22,842	141,788	1,557,373	352,169	53,294	112,006	104,888
54,759	49,688	4310	200,576	252,410	4,718,899	1,623,723	3,274,999	1,736,114
58,112	69,546	21,679	180,056	266,420	157,318	561,730	653,879	270,413
34,857	55,641	86,585	288,321	250,275	537,608	226,906	3,506,992	1,211,687
10,400	33,075	15,191	54,765	46,892	91,412	42,823	235,050	3,881,712
7848	27,588	7627	50,388	36,172	56,626	37,962	852,693	545,218
1028	37,154	918	11,999	5817	4528	5847	324,905	169,089
2002	47,582	9568	142,652	29,560	65,699	11,503	81,511	271,851
20,915	25,782	2972	111,241	75,839	68,193	30,645	107,748	181,222
7328	253,015	1329	86,801	145,128	506,201	38,241	184,898	10,719
82,123	126,297	111,213	1,013,983	710,864	1,117,816	364,028	348,245	242,162
1143	1169	9866	48,635	10,644	23,784	4893	13,890	9992
1782	8842	2028	21,584	10,231	14,514	4865	5201	8549
655	1048	3249	7431	3214	5951	944	5385	5220
81,132	99,345	128,427	541,510	397,718	345,442	112,769	445,819	300,504
1718	2721	200	16,968	2992	6533	995	13,818	3434
10,585	11,174	20,740	72,475	23,975	202,811	18,386	49,194	45,311
65,226	169,655	79,158	521,168	330,253	415,366	129,266	398,714	866,279
13,814	25,441	11,801	125,430	79,892	72,175	62,450	120,362	73,263
55,474	71,115	39,501	415,021	378,719	468,050	81,797	165,690	246,329
14,394	31,451	1051	69,605	36,051	7254	33,850	43,478	41,229
22,619	38,461	31,104	319,976	67,448	30,773	20,440	137,491	223,381
1527	2534	1797	50,274	7443	30,990	4411	54,123	132,825
4921	22,151	11,071	121,223	38,817	70,130	29,939	93,202	141,049
372	14,941	1304	22,317	3125	6109	724	3219	5068
2979	9981	5734	49,579	34,992	50,800	9553	18,328	25,879
430	596	261	3188	1529	1508	704	4947	3417
2355	5054	1686	12,573	16,763	17,157	11,897	36,986	31,711
3410	4842	1489	25,247	17,223	13,705	5977	21,834	18,840
357	641	300	2526	1629	1995	708	2114	2399
2,030,617	3,320,130	5,425,786	16,630,537	7,676,496	15,188,680	4,084,355	12,375,075	12,279,918
195,216	293,718	218,689	1,227,781	632,513	1,065,176	325,994	1,364,350	798,950
81,057	113,124	721,997	678,071	344,421	515,685	135,715	626,061	530,786
46,625	111,657	162,068	525,506	245,718	635,294	95,194	363,210	231,572
117,773	174,173	240,366	1,017,065	566,614	728,441	228,326	965,286	657,744
440,670	692,673	1,343,120	3,448,423	1,789,266	2,944,596	785,228	3,318,907	2,219,053
2,471,288	4,012,803	6,768,906	20,078,960	9,465,761	18,133,277	4,869,583	15,693,981	14,498,971

(continued)

Table 10.24 (continued)

	Sector 18	Sector 19	Sector 20	Sector 21	Sector 22	Sector 23	Sector 24	Sector 25
Sector 1	195	0	47	348,240	177	160	70	0
Sector 2	8065	3063	1604	19,039	815	1,577,800	20,321	462
Sector 3	0	0	0	27	1	56,098	199,107	0
Sector 4	16,634	162	1418	264	538	265	0	0
Sector 5	4122	70	111	7996	68	2410	17	35
Sector 6	35,955	37,702	3194	39,317	700	27,319	1577	909
Sector 7	17,814	2820	1524	260,131	77	462	49	104
Sector 8	20,273	17,959	3022	42,884	120	29,763	4747	1714
Sector 9	27,102	17,814	2702	88,066	90	5542	495	511
Sector 10	118,030	86,773	11,906	51,595	123	17,072	1066	907
Sector 11	39,945	27,237	4666	21,093	1203	263,736	8798	3600
Sector 12	823,959	695,804	94,049	219,771	11,751	27,851	2717	17,161
Sector 13	187,746	175,297	35,580	60,374	1447	19,094	977	1015
Sector 14	2,432,772	307,194	40,094	274,449	5810	8655	1043	401
Sector 15	395,134	228,488	37,358	121,465	609	53,190	4030	9367
Sector 16	454,339	102,186	28,132	11,534	359	90,081	3416	3695
Sector 17	24,512	34,093	5817	29,453	124	100,324	9270	4313
Sector 18	1,304,289	286,314	49,474	18,041	914	496,625	982	1169
Sector 19	554,936	4,542,146	240,121	7177	46	5511	222	280
Sector 20	91,214	511,163	115,369	6228	54	120,715	2654	1885
Sector 21	95,251	79,713	17,581	316,140	108	5990	316	359
Sector 22	73,371	9872	101	571	5318	983	0	0
Sector 23	125,032	134,489	12,369	38,783	2473	3,208,030	8457	81,450
Sector 24	5710	5327	724	2752	121	10,433	23,570	1132
Sector 25	4076	4553	825	1956	115	11,175	226	7285
Sector 26	2478	1925	289	451	57	3077	314	423
Sector 27	224,421	168,574	23,334	68,447	4218	109,421	11,187	3295
Sector 28	7539	4330	1153	814	13	1060	55	108
Sector 29	35,712	112,897	4758	19,443	168	54,983	604	1285
Sector 30	339,275	467,082	25,834	57,373	2583	71,119	7104	2109
Sector 31	70,742	33,482	9987	20,636	516	25,734	1941	2770
Sector 32	179,965	448,712	11,690	31,503	2376	372,756	14,026	20,016
Sector 33	43,476	36,309	6222	19,570	180	3934	445	183
Sector 34	227,235	121,589	13,500	24,257	437	40,021	1863	1710
Sector 35	41,193	76,647	5684	2823	45	5974	111	72
Sector 36	57,868	55,766	6968	6182	221	72,014	975	529
Sector 37	2070	1380	105	461	32	23,984	185	19,910
Sector 38	9600	8181	986	1222	478	68,781	1220	2742
Sector 39	1502	1021	228	481	5	1388	99	267
Sector 40	17,883	5388	1967	450	87	10,722	1187	77
Sector 41	12,365	10,365	1285	3184	80	20,250	799	1044
Sector 42	1144	1092	154	411	0	1282	36	69
Total intermediate input	8,134,942	8,864,981	821,930	2,245,057	44,658	7,025,782	336,280	194,360
Compensation of employees	577,532	763,090	96,530	260,197	6478	718,723	45,264	70,817
Net taxes on production	223,482	155,006	27,222	175,328	3880	437,052	10,502	13,956
Depreciation of fixed assets	157,596	239,586	19,690	58,725	2913	900,776	14,055	60,988
Operating surplus	425,012	285,527	49,281	143,315	395,289	292,417	20,549	8177
Total value added	1,383,622	1,443,209	192,723	637,565	408,560	2,348,968	90,370	153,938
Total input	9,518,564	10,308,190	1,014,653	2,882,622	453,217	9,374,750	426,651	348,298

Sector 26	Sector 27	Sector 28	Sector 29	Sector 30	Sector 31	Sector 32	Sector 33	Sector 34
116,301	257,662	0	0	2405	552,859	0	508	12,266
21,001	37,179	388	0	3030	3192	0	1314	7289
0	13,412	0	0	0	0	0	0	0
0	0	0	0	0	0	0	0	0
267,411	2797	0	0	0	0	0	0	0
105,677	84,925	690	9922	51,697	1,663,952	23,825	17,406	61,158
22,346	13,950	420	499	11,985	23,548	1216	6888	66,818
100,747	77,103	1998	7701	87,790	18,068	22,416	27,944	64,463
463,426	13,153	1380	3900	16,406	6620	6389	7495	13,057
44,236	39,619	7347	126,862	92,053	16,289	165,216	36,331	463,189
341,237	2,011,884	8176	7845	33,263	23,559	53,755	80,958	164,566
816,581	189,904	919	21,561	90,154	36,513	28,850	103,469	185,185
5,998,060	22,864	958	1104	2707	9817	1366	1835	2744
3,299,938	55,568	0	42	4868	216	0	5557	3949
860,704	60,055	216	8181	3997	5595	7690	37,603	103,788
783,583	259,856	2903	32,917	21,479	5618	36,210	21,955	33,934
109,413	759,548	21,911	28,690	87,111	16,170	24,749	23,655	157,298
1,001,282	30,341	490	280,172	67,816	34,320	17,482	68,728	144,963
63,812	27,194	2431	243,158	38,862	1953	6361	18,568	298,644
57,804	17,183	328	36,482	3244	150	32,649	12,107	15,392
86,168	9875	700	4312	7610	2983	8858	8006	31,331
0	0	0	0	0	0	0	0	0
197,700	129,325	3785	101,470	116,613	102,353	80,194	37,092	49,296
3131	10,136	264	2316	2040	17,479	423	653	4991
6355	6499	445	4972	3816	8315	4023	1130	3375
280,775	46,420	11,715	6897	25,730	10,960	12,309	119,251	5047
2,466,046	910,795	39,498	124,769	699,075	58,253	211,025	46,765	254,377
5708	3389	11,621	5316	13,461	2752	36,630	3956	24,311
383,254	77,506	5629	176,326	81,116	14,720	242,681	24,145	19,326
651,260	139,576	20,774	127,835	50,613	101,282	36,086	26,437	119,898
264,695	171,933	1993	79,928	191,500	32,414	313,892	100,943	341,094
250,434	636,493	3246	84,236	340,429	79,274	684,160	265,348	234,571
15,424	52,652	4218	106,412	221,590	50,774	272,894	81,519	86,248
95,397	59,918	10,753	204,008	486,177	57,224	519,093	209,132	226,337
19,827	3318	150	6808	15,554	42	11,941	1305	2581
306,163	12,699	49	19,892	38,450	882	5904	11,875	3856
1301	3255	0	1139	2306	1570	5470	1990	44,404
37,067	245,511	442	26,055	92,210	44,697	33,086	45,589	80,059
3294	8056	278	4766	5962	1248	16,567	1462	2487
7858	5820	53	1213	1603	807	5891	2139	845
36,772	22,210	2073	22,770	38,856	12,376	50,982	32,741	22,184
3376	3042	91	1550	2482	737	4861	4100	984
19,595,565	6,532,627	168,333	1,922,025	3,056,060	3,019,582	2,985,145	1,497,897	3,356,303
3,931,547	1,747,083	79,445	547,227	2,057,214	1,084,333	1,758,720	508,023	992,084
1,091,822	351,030	7161	160,968	2,040,654	205,916	674,647	732,317	251,601
381,954	776,968	18,722	528,372	478,188	243,075	129,881	2,242,996	482,257
1,497,702	1,383,877	229	901,474	2,914,050	257,548	2,976,334	1,166,024	142,009
6,903,025	4,258,957	105,557	2,138,041	7,490,106	1,790,872	5,539,581	4,649,359	1,867,952
26,498,590	10,791,585	273,890	4,060,066	10,546,166	4,810,454	8,524,726	6,147,256	5,224,255

(continued)

Table 10.24 (continued)

	Sector 35	Sector 36	Sector 37	Sector 38	Sector 39	Sector 40	Sector 41	Sector 42
Sector 1	24,795	8830	23,041	31,301	7466	19,180	5918	0
Sector 2	1139	2970	4937	10,905	7933	16,174	982	16,774
Sector 3	0	0	0	0	0	0	0	0
Sector 4	6205	1379	1663	0	0	0	0	0
Sector 5	101	68	620	141	1441	2047	594	2366
Sector 6	11,711	5436	10,025	60,802	31,462	33,393	67,995	21,736
Sector 7	13,326	1466	4081	40,295	6839	22,282	3561	24,375
Sector 8	5089	3711	12,319	18,866	4380	54,926	20,211	79,693
Sector 9	2233	2824	12,154	25,245	47,596	29,879	14,526	97,093
Sector 10	7342	31,277	6465	44,688	136,359	49,857	136,480	155,626
Sector 11	6019	45,371	35,002	49,154	30,755	16,344	7883	86,794
Sector 12	71,573	48,892	64,114	233,337	35,066	1,706,591	49,804	60,485
Sector 13	5185	4303	13,267	2847	19,126	22,317	7774	29,143
Sector 14	3302	16,298	7183	6872	2916	191	559	0
Sector 15	16,785	47,417	7601	13,297	8364	8358	4822	7074
Sector 16	9903	17,834	19,945	6501	7777	178,790	3395	13,890
Sector 17	5850	32,382	33,541	102,176	25,938	4440	5676	77,639
Sector 18	29,052	92,452	24,862	43,941	11,490	12,735	6696	13,604
Sector 19	25,963	83,444	13,705	66,030	12,596	2945	7243	45,968
Sector 20	23,232	50,323	13,121	26,081	37,933	9330	4548	10,508
Sector 21	3004	4906	4997	21,296	6638	5022	2449	13,528
Sector 22	0	0	0	0	0	0	0	0
Sector 23	13,414	15,347	27,153	29,710	71,081	46,443	10,905	81,168
Sector 24	217	565	872	1728	2508	1858	813	1797
Sector 25	1022	2621	1776	4396	5272	2664	961	3419
Sector 26	2344	6699	23,794	20,541	18,732	77,403	9883	83,603
Sector 27	27,320	131,729	22,012	56,277	99,852	20,021	35,698	248,305
Sector 28	1107	2647	911	3731	8150	1281	2764	63,678
Sector 29	3748	10,972	8463	9157	31,598	36,378	9794	111,055
Sector 30	13,736	26,082	13,607	47,900	28,691	119,284	24,919	61,016
Sector 31	25,534	91,117	22,752	42,202	93,494	38,482	37,793	379,253
Sector 32	2837	70,266	30,398	56,815	97,942	22,669	18,151	94,802
Sector 33	3004	22,742	2876	75,380	27,845	7628	15,618	32,227
Sector 34	6236	33,475	14,071	38,209	67,582	28,411	26,619	53,242
Sector 35	15,812	5114	918	73	7799	1811	1221	80
Sector 36	9564	138,738	4914	2065	7940	5582	827	8301
Sector 37	413	469	22,669	3888	4924	1007	1082	13,807
Sector 38	1961	19,955	23,842	104,335	27,382	21,797	11,419	85,166
Sector 39	2215	2198	1574	822	16,807	3799	1159	45,408
Sector 40	1739	2068	3551	867	5522	9839	1736	16,260
Sector 41	10,250	11,303	6940	10,770	22,517	5729	37,087	74,077
Sector 42	169	709	376	620	727	574	250	1593
Total intermediate input	414,453	1,096,396	546,112	1,313,256	1,088,441	2,647,457	599,817	2,214,554
Compensation of employees	160,976	593,273	207,708	919,453	2,752,495	1,229,212	350,506	3,393,099
Net taxes on production	10,280	99,222	17,050	127,017	17,710	12,825	60,429	18,421
Depreciation of fixed assets	30,207	112,855	83,255	74,754	327,146	149,218	93,836	499,532
Operating surplus	35,070	304,848	62,489	258,881	46,178	88,020	110,182	107,382
Total value added	236,534	1,110,199	370,502	1,380,106	3,143,529	1,479,275	614,952	4,018,434
Total input	650,987	2,206,595	916,614	2,693,362	4,231,970	4,126,732	1,214,769	6,232,989

Total intermediate use	Rural household consumption expenditure	Urban household consumption expenditure	Government consumption expenditure	Gross fixed capital formation	Increase in inventories	Export	Import	Total output
12,689,523	1,445,250	1,807,906	129,149	892,866	221,672	442,045	709,439	16,918,972
4,901,301	23,590	17,374	0	44,537	−87,034	18,938	397,937	4,520,769
5,500,763	0	0	0	18,307	−55,601	392,420	4,565,805	1,290,084
4,144,530	0	0	0	72,756	−19,071	7635	2,485,802	1,720,048
1,306,368	0	0	0	11,086	−22,672	23,329	87,877	1,230,234
8,022,313	1,605,878	4,687,666	0	277,151	255,511	220,825	242,567	14,826,777
4,860,110	55,028	85,219	0	37,702	−27,421	4,006,106	969,670	8,047,074
1,964,009	339,000	1,629,180	0	88,385	90,604	579,653	75,977	4,614,855
2,269,677	31,496	137,501	0	206,121	−1685	269,856	409,434	2,503,532
4,042,449	20,673	138,179	0	43,712	−34,103	226,932	488,259	3,949,583
6,838,298	29,537	265,704	0	−8799	−134,219	120,022	463,789	6,646,754
19,779,705	188,319	613,956	0	222,564	−131,943	944,203	1,803,941	19,812,862
9,260,781	11,995	75,642	0	88,219	−135,481	454,978	303,378	9,452,754
18,705,436	0	0	0	204,209	−188,641	365,674	968,403	18,118,276
4,609,672	20,293	109,866	0	108,541	−54,600	345,348	263,282	4,875,838
9,196,748	1458	21,193	0	6,186,096	89,393	1,976,368	1,902,602	15,568,652
6,302,276	165,397	799,461	0	6,651,677	119,509	2,077,835	1,607,630	14,508,525
5,824,619	146,638	570,563	0	1,990,857	34,952	1,770,759	839,053	9,499,335
6,901,466	127,583	509,962	0	1,231,984	187,744	4,712,789	3,399,330	10,272,198
1,982,120	14,885	49,498	0	215,058	−24,607	537,449	1,770,572	1,003,832
1,589,423	81,710	459,145	0	572,867	9937	245,119	108,860	2,849,339
1,366,135	0	0	0	0	6321	4862	925,478	451,839
10,146,474	60,900	224,622	0	0	−1,071,045	15,796	1998	9,374,750
259,556	19,106	143,943	0	2799	358	0	0	425,763
193,037	14,821	136,380	0	0	1226	0	0	345,464
824,069	0	321,716	0	24,925,660	72,498	442,478	121,627	26,464,794
9,872,373	158,101	609,123	497,073	76,954	−179,449	369,632	664,342	10,739,465
292,056	2259	15,334	0	0	2855	7963	38,879	281,589
2,083,852	184,459	795,385	0	825,902	43,230	191,553	92,981	4,031,400
6,757,858	569,420	1,749,974	0	710,026	47,359	917,108	0	10,751,744
3,288,250	359,243	1,401,640	0	0	98,640	−109,315	239,295	4,799,164
6,781,223	214,534	1,134,093	89,448	0	108,023	93,461	83,244	8,337,537
1,620,154	641,138	2,217,037	0	1,523,173	120,162	0	0	6,121,665
3,906,092	30,166	430,907	205,994	0	59,141	958,060	249,350	5,341,009
570,967	0	0	326,375	0	2978	23,740	271,227	652,834
1,635,998	1776	22,731	295,048	235,167	−1537	0	0	2,189,182
296,901	8597	105,641	524,220	0	−2160	0	0	933,199
1,359,895	196,138	1,139,248	0	0	35,106	38,478	80,939	2,687,927
156,307	400,180	892,597	2,827,255	0	−27,196	5018	30,972	4,223,189
282,557	145,430	1,693,391	2,003,488	0	−23,344	3828	2677	4,102,673
638,548	39,334	211,752	332,406	0	12,312	42,790	53,539	1,223,604
54,147	0	0	6,143,595	0	−12,349	19,722	16,034	6,189,080

Table 10.25 China's national input–output table in 2023 in the policy scenario

	Sector 1	Sector 2	Sector 3	Sector 4	Sector 5	Sector 6	Sector 7	Sector 8
Sector 1	2,371,545	42,474	26	2843	489	6,191,124	1,730,023	235,712
Sector 2	14,820	754,471	11,674	7005	3387	42,701	45,189	19,526
Sector 3	0	0	22,516	0	0	0	0	0
Sector 4	0	0	0	161,898	0	0	0	0
Sector 5	2300	1689	715	1727	81,113	5622	0	0
Sector 6	1,798,661	7912	2444	4041	3564	3,224,680	62,337	239,903
Sector 7	2188	1160	1082	967	651	8067	2,604,818	1,577,074
Sector 8	5971	15,200	4819	7109	5337	20,490	135,138	697,848
Sector 9	21,850	39,032	5183	5709	2790	16,756	10,215	10,987
Sector 10	16,224	5447	2895	2977	2883	291,717	86,845	79,094
Sector 11	128,046	49,222	51,299	94,083	57,623	43,916	21,153	30,466
Sector 12	1,435,167	103,199	50,747	91,726	121,123	431,462	963,019	311,202
Sector 13	23,737	38,877	12,614	13,752	44,206	108,022	19,523	6483
Sector 14	4439	182,961	66,515	21,387	8840	17,633	7231	7012
Sector 15	64,257	87,748	10,467	34,582	20,921	72,369	17,740	23,204
Sector 16	107,877	226,456	112,830	131,567	107,064	53,460	109,761	33,242
Sector 17	49,932	28,743	9281	17,882	24,545	19,841	16,702	7373
Sector 18	4849	64,955	12,618	22,270	11,788	22,783	19,299	9176
Sector 19	1767	6863	1756	2392	1407	5763	6952	5654
Sector 20	14,402	47,550	12,873	8856	5446	11,022	17,735	6131
Sector 21	49,916	31,432	1876	7137	12,567	23,939	58,298	47,410
Sector 22	1343	427	22	1236	10,536	12,178	15,895	2616
Sector 23	215,566	191,013	99,571	284,861	140,051	162,203	212,489	39,246
Sector 24	658	1711	1682	8614	6593	9261	4622	1757
Sector 25	2197	3456	1070	1568	1329	6336	6975	2389
Sector 26	4530	5684	1540	654	439	4413	1983	1482
Sector 27	337,917	201,557	26,433	72,919	79,193	448,645	168,421	122,057
Sector 28	14,514	1503	219	334	490	3530	6878	10,389
Sector 29	71,866	14,927	2311	2742	3306	27,786	17,932	27,239
Sector 30	413,576	68,469	12,497	22,150	26,930	377,898	222,386	120,864
Sector 31	58,059	51,756	12,102	19,197	20,458	92,699	44,503	33,768
Sector 32	235,039	129,528	11,176	21,401	24,646	160,973	228,442	87,390
Sector 33	8631	8604	609	1229	2742	26,904	54,069	90,383
Sector 34	27,649	24,906	2478	6730	6133	258,440	37,212	107,164
Sector 35	32,928	6404	2492	571	341	12,697	5775	2755
Sector 36	168,655	42,013	15,361	23,346	14,058	53,843	21,824	18,507
Sector 37	54,429	6353	1111	1208	3952	4798	14,686	713
Sector 38	83,587	39,179	5093	4138	6085	18,864	12,532	11,013
Sector 39	7161	4250	252	654	1684	2151	1043	570
Sector 40	10,906	10,446	489	3231	2549	7794	4216	3465
Sector 41	4522	7206	1306	4421	6062	16,488	10,197	9247
Sector 42	5935	1130	413	358	245	2188	1072	696
Total intermediate input	7,877,614	2,555,914	592,456	1,121,472	873,565	12,321,457	7,025,128	4,041,204
Compensation of employees	9,918,764	849,174	199,826	151,571	142,210	958,284	690,963	493,672
Net taxes on production	20,152	473,683	226,906	97,801	86,533	1,080,857	274,987	128,104
Depreciation of fixed assets	486,750	162,290	133,396	35,773	36,868	296,976	181,481	58,147
Operating surplus	0	491,493	312,616	185,933	118,082	954,476	283,444	165,484
Total value added	10,425,666	1,976,641	872,743	471,079	383,693	3,290,593	1,430,876	845,406
Total input	18,303,281	4,532,555	1,465,199	1,592,551	1,257,258	15,612,050	8,456,004	4,886,610

Sector 9	Sector 10	Sector 11	Sector 12	Sector 13	Sector 14	Sector 15	Sector 16	Sector 17
340,805	221,303	151	701,057	5741	2120	1626	1690	1293
20,087	40,611	422,370	308,260	715,812	681,193	24,288	53,453	17,399
0	0	3,826,164	1,312,492	99,098	227,225	0	0	0
0	0	0	204,178	39,903	3,602,803	71,126	46,192	11,472
0	0	686	292,751	577,773	51,395	20,741	10,987	3873
13,129	17,454	31,633	323,646	34,661	62,062	16,141	51,330	52,677
25,060	72,864	346	125,444	20,276	2216	8011	22,022	27,612
42,408	30,391	2821	44,965	23,779	40,310	15,127	33,639	156,018
715,815	122,388	2084	37,459	76,064	14,171	87,134	50,517	173,212
44,432	986,282	3686	284,903	233,373	21,462	57,759	68,541	65,270
17,796	19,920	412,254	1,166,121	386,292	764,973	50,340	160,278	70,641
245,643	584,113	127,211	7,371,524	757,101	290,134	215,175	566,004	934,982
20,933	11,280	23,026	142,738	1,596,497	353,199	54,366	112,673	107,463
55,734	50,633	4345	202,616	261,548	4,739,046	1,676,018	3,294,494	1,802,968
58,865	70,263	21,853	181,079	272,089	157,773	573,982	657,771	275,950
35,628	56,801	87,283	291,516	260,594	540,345	234,884	3,527,868	1,271,537
10,524	33,449	15,313	55,124	47,962	91,648	43,616	236,450	3,980,114
7953	27,934	7689	50,734	37,059	56,794	38,792	857,769	559,079
1047	37,858	926	12,126	6033	4548	6031	326,839	176,146
2090	48,134	9645	143,720	30,460	66,070	11,892	81,996	284,300
21,033	25,952	2996	111,472	76,518	68,293	30,911	108,389	182,079
7550	258,423	1339	87,849	151,596	510,176	39,975	185,999	11,365
83,025	127,671	112,109	1,019,882	724,724	1,120,346	370,125	350,318	246,615
1159	1187	9946	49,007	10,933	23,861	5003	13,973	10,271
1812	8981	2044	21,748	10,504	14,561	4968	5232	8795
663	1059	3275	7475	3281	5966	960	5417	5323
82,639	101,430	129,463	547,265	412,805	346,941	116,375	448,473	312,891
1733	2738	202	17,036	3030	6543	1008	13,900	3471
10,768	11,349	20,908	73,035	24,616	203,461	18,780	49,486	46,593
66,004	171,129	79,796	524,004	336,229	416,320	131,786	401,088	881,020
14,177	26,040	11,896	127,092	83,610	72,579	65,038	121,079	77,241
57,392	73,445	39,820	420,446	396,644	470,771	84,559	166,676	259,670
15,371	33,337	1059	71,917	40,061	7348	37,185	43,737	46,432
22,823	38,836	31,355	321,671	68,616	30,834	20,765	138,309	226,761
1552	2577	1812	50,735	7684	31,101	4524	54,446	137,441
5010	22,551	11,160	122,425	40,156	70,395	30,747	93,757	146,079
376	15,091	1315	22,447	3183	6121	737	3238	5152
3071	10,252	5781	50,345	36,827	51,108	10,000	18,437	27,442
436	604	263	3213	1569	1512	719	4976	3502
2387	5120	1699	12,664	17,179	17,205	12,138	37,206	32,478
3453	4905	1501	25,416	17,599	13,740	6087	21,964	19,239
362	650	302	2544	1670	2001	722	2126	2458
2,060,744	3,375,005	5,469,529	16,942,141	7,951,150	15,260,669	4,200,159	12,448,740	12,664,326
198,145	299,231	220,452	1,238,776	648,402	1,069,723	333,104	1,372,472	819,204
82,273	115,247	727,818	684,143	353,074	517,887	138,675	629,788	544,242
47,325	113,752	163,374	530,212	251,891	638,007	97,270	365,372	237,443
119,540	177,442	242,304	1,026,173	580,848	731,551	233,306	971,032	674,418
447,283	705,673	1,353,949	3,479,305	1,834,215	2,957,168	802,354	3,338,663	2,275,306
2,508,027	4,080,677	6,823,478	20,421,446	9,785,365	18,217,837	5,002,513	15,787,404	14,939,632

(continued)

Table 10.25 (continued)

	Sector 18	Sector 19	Sector 20	Sector 21	Sector 22	Sector 23	Sector 24	Sector 25
Sector 1	197	0	47	363,832	177	162	72	0
Sector 2	8250	3081	1608	19,892	815	1,604,789	20,710	477
Sector 3	0	0	0	28	1	57,057	202,924	0
Sector 4	17,080	163	1427	276	538	269	0	0
Sector 5	4279	71	111	8354	68	2452	17	36
Sector 6	36,396	37,820	3197	41,077	700	27,787	1607	937
Sector 7	18,049	2830	1526	271,778	77	470	50	107
Sector 8	20,494	18,009	3026	44,805	120	30,272	4838	1768
Sector 9	27,687	17,904	2709	92,009	90	5637	505	528
Sector 10	120,328	87,200	11,924	53,905	123	17,364	1087	935
Sector 11	40,881	27,400	4674	22,038	1203	268,247	8966	3714
Sector 12	842,056	699,702	94,214	229,611	11,751	28,328	2769	17,703
Sector 13	191,053	176,061	35,633	63,077	1447	19,421	995	1047
Sector 14	2,498,348	309,173	40,201	286,737	5810	8803	1063	414
Sector 15	401,980	229,409	37,415	126,904	609	54,100	4108	9663
Sector 16	468,247	102,960	28,203	12,050	359	91,622	3481	3811
Sector 17	24,912	34,227	5826	30,771	124	102,040	9448	4450
Sector 18	1,328,726	287,596	49,555	18,849	914	505,120	1001	1205
Sector 19	570,191	4,573,894	240,726	7498	46	5606	226	289
Sector 20	93,618	514,677	115,685	6507	54	122,780	2705	1945
Sector 21	95,874	79,843	17,590	330,295	108	6092	322	370
Sector 22	76,282	9975	101	597	5318	999	0	0
Sector 23	126,882	135,002	12,387	40,519	2473	3,262,905	8619	84,026
Sector 24	5825	5353	725	2876	121	10,611	24,022	1168
Sector 25	4156	4576	827	2044	115	11,367	231	7515
Sector 26	2516	1933	289	472	57	3130	320	436
Sector 27	230,643	169,689	23,393	71,511	4218	111,293	11,402	3399
Sector 28	7617	4340	1154	851	13	1078	56	112
Sector 29	36,417	113,478	4766	20,313	168	55,924	615	1325
Sector 30	344,379	468,728	25,882	59,941	2583	72,335	7241	2176
Sector 31	73,235	33,771	10,020	21,560	516	26,174	1978	2857
Sector 32	185,893	452,823	11,719	32,913	2376	379,132	14,295	20,649
Sector 33	47,222	37,132	6286	20,446	180	4001	453	189
Sector 34	230,292	121,982	13,521	25,343	437	40,705	1899	1764
Sector 35	42,118	77,101	5698	2949	45	6077	113	74
Sector 36	59,262	56,118	6984	6458	221	73,245	994	546
Sector 37	2101	1385	106	482	32	24,394	188	20,539
Sector 38	9986	8263	990	1277	478	69,958	1243	2829
Sector 39	1531	1026	229	502	5	1412	101	275
Sector 40	18,208	5414	1970	471	87	10,905	1210	79
Sector 41	12,568	10,407	1287	3327	80	20,596	815	1077
Sector 42	1165	1098	154	429	0	1304	37	72
Total intermediate input	8,326,943	8,921,615	823,786	2,345,574	44,658	7,145,961	342,728	200,507
Compensation of employees	588,353	768,424	96,795	271,847	6478	731,017	46,132	73,056
Net taxes on production	227,669	156,089	27,297	183,178	3880	444,528	10,703	14,397
Depreciation of fixed assets	160,549	241,260	19,744	61,354	2913	916,184	14,325	62,917
Operating surplus	432,975	287,523	49,416	149,732	395,289	297,419	20,943	8435
Total value added	1,409,546	1,453,296	193,251	666,110	408,560	2,389,148	92,103	158,806
Total input	9,736,489	10,374,911	1,017,037	3,011,684	453,217	9,535,109	434,831	359,313

Sector 26	Sector 27	Sector 28	Sector 29	Sector 30	Sector 31	Sector 32	Sector 33	Sector 34
120,559	262,983	0	0	2573	570,916	0	515	13,202
21,770	37,947	424	0	3242	3296	0	1333	7845
0	13,689	0	0	0	0	0	0	0
0	0	0	0	0	0	0	0	0
277,201	2855	0	0	0	0	0	0	0
109,546	86,678	752	10,341	55,307	1,718,300	24,581	17,659	65,825
23,165	14,238	457	520	12,821	24,317	1254	6988	71,916
104,435	78,695	2178	8027	93,919	18,658	23,127	28,349	69,382
480,392	13,424	1505	4065	17,551	6836	6591	7603	14,053
45,855	40,437	8011	132,231	98,480	16,821	170,454	36,858	498,534
353,730	2,053,433	8916	8177	35,586	24,329	55,459	82,132	177,123
846,476	193,826	1002	22,473	96,449	37,705	29,765	104,970	199,316
6,217,649	23,336	1045	1150	2896	10,138	1409	1862	2953
3,420,749	56,715	0	44	5208	223	0	5638	4250
892,214	61,295	235	8528	4276	5778	7934	38,149	111,708
812,270	265,222	3166	34,310	22,978	5801	37,358	22,274	36,523
113,418	775,233	23,894	29,904	93,194	16,699	25,534	23,998	169,301
1,037,939	30,968	534	292,029	72,551	35,441	18,036	69,725	156,024
66,148	27,756	2651	253,449	41,576	2017	6563	18,838	321,432
59,920	17,538	358	38,026	3471	155	33,684	12,283	16,566
89,322	10,079	763	4494	8141	3081	9138	8122	33,722
0	0	0	0	0	0	0	0	0
204,938	131,995	4127	105,764	124,756	105,696	82,736	37,630	53,058
3246	10,346	288	2414	2183	18,050	436	663	5372
6587	6633	486	5183	4083	8586	4151	1146	3632
291,054	47,379	12,775	7189	27,526	11,318	12,699	120,981	5432
2,556,328	929,604	43,072	130,050	747,886	60,156	217,715	47,443	273,788
5917	3459	12,673	5541	14,401	2841	37,791	4013	26,166
397,285	79,107	6139	183,788	86,780	15,201	250,374	24,495	20,801
675,103	142,458	22,653	133,245	54,147	104,590	37,230	26,821	129,047
274,386	175,484	2173	83,311	204,871	33,473	323,842	102,407	367,121
259,602	649,638	3539	87,801	364,199	81,863	705,849	269,197	252,471
15,989	53,739	4600	110,916	237,062	52,432	281,545	82,701	92,829
98,889	61,156	11,726	212,641	520,124	59,093	535,549	212,166	243,608
20,553	3387	164	7096	16,640	44	12,320	1323	2778
317,372	12,961	54	20,734	41,134	911	6091	12,047	4150
1348	3322	0	1187	2467	1621	5643	2019	47,792
38,424	250,582	482	27,157	98,648	46,157	34,135	46,250	86,168
3415	8223	303	4968	6378	1289	17,093	1483	2677
8145	5940	58	1264	1715	833	6078	2170	909
38,118	22,669	2260	23,734	41,569	12,780	52,598	33,216	23,877
3500	3105	100	1615	2655	761	5015	4159	1059
20,312,959	6,667,535	183,563	2,003,366	3,269,444	3,118,209	3,079,778	1,519,624	3,612,412
4,075,482	1,783,162	86,633	570,386	2,200,855	1,119,750	1,814,474	515,392	1,067,787
1,131,794	358,279	7809	167,780	2,183,139	212,642	696,034	742,939	270,800
395,937	793,013	20,416	550,733	511,577	251,015	133,998	2,275,531	519,057
1,552,533	1,412,456	250	939,625	3,117,518	265,961	3,070,688	1,182,937	152,845
7,155,745	4,346,911	115,108	2,228,524	8,013,089	1,849,367	5,715,195	4,716,800	2,010,489
27,468,704	11,014,446	298,671	4,231,891	11,282,533	4,967,576	8,794,973	6,236,424	5,622,901

(continued)

Table 10.25 (continued)

	Sector 35	Sector 36	Sector 37	Sector 38	Sector 39	Sector 40	Sector 41	Sector 42
Sector 1	25,283	9015	24,600	32,270	7734	19,898	6266	0
Sector 2	1162	3033	5272	11,243	8217	16,780	1040	17,394
Sector 3	0	0	0	0	0	0	0	0
Sector 4	6327	1408	1775	0	0	0	0	0
Sector 5	103	70	662	145	1493	2124	628	2454
Sector 6	11,942	5550	10,704	62,685	32,591	34,644	71,986	22,540
Sector 7	13,589	1496	4357	41,543	7085	23,116	3770	25,277
Sector 8	5189	3788	13,153	19,450	4537	56,984	21,398	82,641
Sector 9	2277	2883	12,977	26,027	49,303	30,998	15,379	100,685
Sector 10	7487	31,932	6902	46,073	141,250	51,725	144,491	161,383
Sector 11	6138	46,322	37,370	50,676	31,858	16,957	8346	90,005
Sector 12	72,982	49,916	68,452	240,564	36,323	1,770,528	52,728	62,723
Sector 13	5287	4393	14,165	2935	19,812	23,153	8230	30,221
Sector 14	3368	16,639	7669	7084	3021	199	592	0
Sector 15	17,115	48,410	8115	13,709	8664	8671	5106	7336
Sector 16	10,098	18,208	21,295	6703	8056	185,489	3595	14,403
Sector 17	5965	33,061	35,811	105,340	26,869	4606	6009	80,511
Sector 18	29,625	94,389	26,544	45,302	11,902	13,212	7089	14,108
Sector 19	26,474	85,191	14,632	68,075	13,048	3055	7668	47,669
Sector 20	23,689	51,377	14,009	26,888	39,294	9679	4815	10,897
Sector 21	3063	5009	5335	21,955	6876	5210	2593	14,029
Sector 22	0	0	0	0	0	0	0	0
Sector 23	13,678	15,668	28,990	30,630	73,631	48,183	11,545	84,170
Sector 24	221	577	931	1782	2597	1927	860	1863
Sector 25	1043	2676	1896	4532	5461	2763	1017	3545
Sector 26	2390	6840	25,404	21,177	19,404	80,302	10,463	86,695
Sector 27	27,858	134,488	23,501	58,020	103,433	20,771	37,793	257,491
Sector 28	1129	2702	973	3846	8442	1330	2927	66,033
Sector 29	3822	11,202	9035	9441	32,731	37,741	10,368	115,164
Sector 30	14,006	26,628	14,528	49,384	29,721	123,753	26,382	63,273
Sector 31	26,037	93,025	24,292	43,509	96,847	39,923	40,012	393,283
Sector 32	2893	71,738	32,454	58,575	101,455	23,519	19,217	98,309
Sector 33	3063	23,219	3071	77,714	28,844	7914	16,535	33,419
Sector 34	6359	34,176	15,023	39,392	70,007	29,475	28,181	55,211
Sector 35	16,123	5221	980	75	8078	1879	1293	83
Sector 36	9753	141,644	5247	2129	8225	5792	875	8608
Sector 37	421	479	24,203	4008	5101	1045	1146	14,318
Sector 38	2000	20,373	25,455	107,567	28,364	22,614	12,089	88,317
Sector 39	2259	2244	1681	848	17,410	3942	1227	47,088
Sector 40	1773	2111	3792	893	5720	10,207	1838	16,861
Sector 41	10,451	11,540	7410	11,104	23,325	5943	39,264	76,817
Sector 42	172	723	402	639	753	595	265	1651
Total intermediate input	422,615	1,119,363	583,066	1,353,931	1,127,483	2,746,644	635,024	2,296,475
Compensation of employees	164,146	605,701	221,763	947,931	2,851,226	1,275,264	371,079	3,518,616
Net taxes on production	10,483	101,301	18,204	130,951	18,346	13,305	63,976	19,103
Depreciation of fixed assets	30,801	115,219	88,889	77,070	338,881	154,809	99,344	518,011
Operating surplus	35,761	311,234	66,717	266,900	47,834	91,318	116,649	111,354
Total value added	241,192	1,133,454	395,573	1,422,851	3,256,286	1,534,696	651,047	4,167,084
Total input	663,807	2,252,817	978,639	2,776,782	4,383,770	4,281,341	1,286,071	6,463,559

Total intermediate use	Rural household consumption expenditure	Urban household consumption expenditure	Government consumption expenditure	Gross fixed capital formation	Increase in inventories	Export	Import	Total output
13,310,322	1,525,384	1,915,101	135,774	972,488	221,672	461,525	731,606	17,810,660
4,981,863	24,759	18,306	0	44,831	−87,034	19,341	422,058	4,580,008
5,761,193	0	0	0	18,307	−55,601	412,517	4,565,805	1,570,612
4,166,835	0	0	0	72,756	−19,071	7635	2,640,661	1,587,492
1,354,495	0	0	0	11,139	−22,672	23,959	92,680	1,274,241
8,437,429	1,680,030	4,925,876	0	291,760	255,511	220,825	242,567	15,568,864
5,070,656	57,935	89,838	0	38,958	−27,421	4,266,769	1,043,614	8,453,120
2,038,613	355,267	1,714,999	0	93,344	90,604	579,653	75,977	4,796,504
2,330,985	32,910	144,070	0	209,214	−1685	269,856	440,257	2,545,092
4,185,581	21,655	143,999	0	44,532	−34,103	226,932	505,037	4,083,558
6,988,103	30,771	276,011	0	−8870	−134,219	123,231	472,730	6,802,297
20,411,864	196,591	640,917	0	224,557	−131,943	944,203	1,855,906	20,430,283
9,548,759	12,731	79,512	0	90,435	−135,481	477,161	324,225	9,748,892
19,085,366	0	0	0	205,081	−188,641	365,674	1,007,010	18,460,471
4,712,344	21,341	115,699	0	110,909	−54,600	345,348	279,477	4,971,563
9,407,196	1521	22,078	0	6,222,920	89,393	2,082,203	1,981,739	15,843,572
6,469,643	172,979	832,952	0	6,874,175	119,509	2,209,390	1,693,282	14,985,366
5,959,924	152,815	595,600	0	2,028,158	34,952	1,841,292	874,080	9,738,661
7,008,824	133,310	532,622	0	1,240,595	187,744	4,896,535	3,523,559	10,476,071
2,022,943	15,708	51,333	0	215,647	−24,607	548,201	1,862,033	967,192
1,621,644	85,313	485,646	0	598,515	9937	251,578	113,997	2,938,637
1,391,799	0	0	0	0	6321	4995	960,196	442,918
10,329,824	60,900	224,622	0	0	−1,094,682	16,455	2010	9,535,109
264,695	19,767	148,425	0	2853	358	0	0	436,098
198,212	15,536	140,690	0	0	1226	0	0	355,663
852,323	0	337,927	0	25,838,190	72,498	471,288	129,039	27,443,187
10,250,369	165,689	637,097	528,514	78,543	−179,449	369,632	706,295	11,144,100
302,921	2383	16,240	0	0	2855	8201	41,730	290,871
2,153,587	192,514	832,651	0	860,855	43,230	201,722	98,586	4,185,973
6,928,378	598,858	1,843,502	0	759,602	47,359	936,789	0	11,114,488
3,425,399	376,735	1,470,709	0	0	98,640	−123,812	253,125	4,994,546
7,050,137	223,861	1,186,258	94,722	0	108,023	99,557	88,481	8,674,077
1,731,117	668,446	2,290,486	0	1,545,268	120,162	0	0	6,355,479
4,039,405	31,561	450,392	217,028	0	59,141	1,002,310	249,350	5,550,486
588,050	0	0	339,633	0	2978	25,257	286,750	669,169
1,701,442	1776	22,731	310,040	240,093	−1537	0	0	2,274,545
310,257	9010	110,449	546,374	0	−2160	0	0	973,931
1,423,561	204,797	1,190,232	0	0	35,106	39,880	85,593	2,807,983
162,168	420,451	937,121	2,972,005	0	−27,196	5145	32,800	4,436,894
289,825	150,794	1,765,658	2,115,098	0	−23,344	3828	2677	4,299,183
660,186	41,273	222,336	350,030	0	12,312	42,790	53,539	1,275,389
56,302	0	0	6,447,195	0	−12,349	20,833	16,735	6,495,246

Table 10.26 China's national input–output table in 2024 in the policy scenario

	Sector 1	Sector 2	Sector 3	Sector 4	Sector 5	Sector 6	Sector 7	Sector 8
Sector 1	2,415,031	44,149	26	2843	502	6,330,564	1,804,821	240,951
Sector 2	15,003	771,332	11,674	7005	3424	43,663	45,986	19,788
Sector 3	0	0	22,516	0	0	0	0	0
Sector 4	0	0	0	161,898	0	0	0	0
Sector 5	2347	1739	715	1727	82,463	5748	0	0
Sector 6	1,821,585	8055	2444	4041	3601	3,297,308	63,543	243,394
Sector 7	2219	1183	1082	967	658	8249	2,667,837	1,600,014
Sector 8	6058	15,619	4819	7109	5429	20,952	139,040	709,969
Sector 9	22,039	39,647	5183	5709	2814	17,134	10,346	11,096
Sector 10	16,530	5596	2895	2977	2926	298,287	89,203	80,671
Sector 11	131,008	50,900	51,299	94,083	58,687	44,905	21,874	31,259
Sector 12	1,465,034	106,202	50,747	91,726	123,195	441,180	993,355	317,948
Sector 13	24,114	39,994	12,614	13,752	44,927	110,455	20,047	6597
Sector 14	4498	186,539	66,515	21,387	8930	18,030	7356	7114
Sector 15	65,155	90,132	10,467	34,582	21,253	73,999	18,181	23,598
Sector 16	110,191	233,995	112,830	131,567	108,919	54,664	113,044	34,023
Sector 17	50,540	29,311	9281	17,882	24,824	20,288	17,010	7469
Sector 18	4920	66,469	12,618	22,270	11,944	23,296	19,747	9325
Sector 19	1796	7021	1756	2392	1424	5893	7100	5754
Sector 20	15,168	54,728	12,873	8856	5917	11,270	19,543	6571
Sector 21	50,413	32,402	1876	7137	12,847	24,478	59,884	48,046
Sector 22	1396	456	22	1236	10,868	12,452	16,644	2754
Sector 23	218,123	194,369	99,571	284,861	141,480	165,856	216,437	39,775
Sector 24	669	1751	1682	8614	6677	9470	4727	1786
Sector 25	2252	3601	1070	1568	1362	6479	7280	2456
Sector 26	4590	5804	1540	654	445	4512	2025	1503
Sector 27	343,086	206,170	26,433	72,919	80,148	458,750	171,875	124,055
Sector 28	14,732	1564	219	334	503	3610	7175	10,602
Sector 29	73,684	15,591	2311	2742	3396	28,412	18,775	28,032
Sector 30	419,108	70,897	12,497	22,150	27,524	386,409	229,510	123,073
Sector 31	59,303	53,696	12,102	19,197	20,858	94,787	45,857	34,571
Sector 32	246,697	141,156	11,176	21,401	25,963	164,598	247,865	92,401
Sector 33	8890	9008	609	1229	2795	27,510	55,823	93,474
Sector 34	27,988	25,612	2478	6730	6240	264,261	38,215	108,864
Sector 35	33,389	6562	2492	571	346	12,983	5890	2799
Sector 36	171,370	43,049	15,361	23,346	14,249	55,055	22,317	18,830
Sector 37	55,018	6500	1111	1208	4009	4906	15,024	723
Sector 38	85,316	40,437	5093	4138	6183	19,288	12,865	11,269
Sector 39	7244	4317	252	654	1699	2199	1060	577
Sector 40	11,068	10,654	489	3231	2577	7970	4304	3520
Sector 41	4603	7438	1306	4421	6180	16,859	10,542	9439
Sector 42	6025	1152	413	358	247	2238	1094	707
Total intermediate input	8,018,195	2,634,796	592,456	1,121,472	888,430	12,598,969	7,253,219	4,114,798
Compensation of employees	10,100,640	868,152	199,826	151,571	144,576	979,867	707,680	502,247
Net taxes on production	20,522	484,269	226,906	97,801	87,973	1,105,200	281,640	130,329
Depreciation of fixed assets	495,675	165,917	133,396	35,773	37,481	303,665	185,872	59,157
Operating surplus	0	502,477	312,616	185,933	120,047	975,974	290,301	168,358
Total value added	10,616,837	2,020,815	872,743	471,079	390,077	3,364,706	1,465,493	860,090
Total input	18,635,032	4,655,611	1,465,199	1,592,551	1,278,507	15,963,675	8,718,713	4,974,888

Sector 9	Sector 10	Sector 11	Sector 12	Sector 13	Sector 14	Sector 15	Sector 16	Sector 17
366,841	225,899	152	709,482	5876	2227	1647	1707	1371
20,920	40,980	424,352	309,715	725,404	697,670	24,482	53,968	17,857
0	0	3,844,116	1,325,974	102,035	234,308	0	0	0
0	0	0	205,098	40,428	3,822,944	71,816	46,637	11,812
0	0	689	294,853	589,611	52,999	21,001	11,093	4041
13,661	17,609	31,782	325,118	35,091	63,477	16,273	51,826	53,998
26,032	73,591	348	126,094	20,544	2264	8075	22,234	28,331
44,759	30,799	2834	45,296	24,191	41,796	15,288	33,964	162,600
742,899	123,187	2094	37,601	76,817	14,470	87,735	51,005	176,843
46,872	1,003,542	3703	286,850	237,602	22,144	58,382	69,202	67,663
19,117	20,255	414,189	1,175,585	395,359	796,596	51,105	161,824	74,564
260,855	592,566	127,807	7,430,479	772,081	299,647	217,740	571,464	978,216
22,065	11,426	23,135	143,700	1,625,956	365,374	54,968	113,759	111,853
58,073	51,093	4365	203,525	264,986	4,857,910	1,690,832	3,326,271	1,849,111
62,393	71,119	21,956	182,273	276,887	163,852	581,685	664,116	286,926
38,103	57,668	87,692	293,754	266,352	562,593	238,251	3,561,896	1,339,007
10,957	33,745	15,385	55,393	48,583	93,866	43,959	238,730	4,100,148
8345	28,241	7725	51,027	37,635	58,491	39,195	866,042	578,405
1099	38,255	930	12,191	6126	4674	6091	329,991	181,960
3247	50,813	9691	149,042	33,335	80,014	12,841	82,787	365,756
22,117	26,522	3010	112,316	77,960	71,380	31,327	109,435	186,942
8757	266,158	1346	89,228	158,519	561,058	41,410	187,793	12,805
86,283	128,822	112,635	1,024,595	733,234	1,145,064	372,917	353,697	252,423
1215	1201	9992	49,286	11,103	24,554	5052	14,108	10,610
1969	9189	2054	21,993	10,788	15,284	5047	5283	9365
692	1070	3290	7516	3327	6124	969	5469	5479
86,547	102,574	130,070	550,190	419,113	356,679	117,509	452,799	322,725
1887	2788	203	17,236	3103	6919	1025	14,034	3689
11,780	11,627	21,006	73,937	25,317	214,104	19,100	49,964	49,712
71,402	173,664	80,170	528,878	343,305	437,551	133,935	404,956	924,726
15,325	26,466	11,952	128,202	85,576	75,903	66,064	122,247	81,551
68,404	77,752	40,007	430,730	419,011	515,855	87,165	168,284	293,589
17,193	33,997	1064	72,623	41,204	7730	37,914	44,158	49,393
24,147	39,368	31,502	324,032	69,779	32,090	20,996	139,643	236,024
1631	2605	1820	51,016	7801	32,007	4565	54,971	141,987
5263	22,810	11,213	123,156	40,801	72,339	31,040	94,662	150,864
395	15,254	1321	22,585	3227	6317	744	3269	5314
3310	10,399	5808	50,726	37,588	53,069	10,140	18,615	28,737
452	609	264	3226	1586	1542	724	5024	3570
2489	5170	1707	12,728	17,412	17,612	12,239	37,565	33,350
3683	4990	1508	25,640	17,951	14,283	6162	22,176	20,122
377	656	304	2557	1692	2047	728	2147	2522
2,181,555	3,434,478	5,495,192	17,085,447	8,114,298	15,906,827	4,248,138	12,568,814	13,215,963
205,642	304,467	221,486	1,248,683	660,367	1,096,554	337,574	1,385,710	843,910
85,386	117,264	731,233	689,615	359,589	530,876	140,536	635,862	560,655
49,115	115,743	164,141	534,453	256,538	654,009	98,575	368,896	244,604
124,063	180,547	243,441	1,034,380	591,566	749,900	236,437	980,398	694,757
464,206	718,022	1,360,301	3,507,132	1,868,060	3,031,339	813,122	3,370,866	2,343,926
2,645,762	4,152,500	6,855,494	20,592,578	9,982,358	18,938,166	5,061,260	15,939,680	15,559,889

(continued)

Table 10.26 (continued)

	Sector 18	Sector 19	Sector 20	Sector 21	Sector 22	Sector 23	Sector 24	Sector 25
Sector 1	202	0	47	365,079	177	165	73	0
Sector 2	8372	3133	1608	19,960	815	1,629,291	21,200	484
Sector 3	0	0	0	28	1	57,928	207,720	0
Sector 4	17,392	169	1428	277	538	273	0	0
Sector 5	4374	72	111	8383	68	2489	17	37
Sector 6	36,936	38,431	3197	41,218	700	28,211	1645	951
Sector 7	18,321	2874	1526	272,709	77	477	51	109
Sector 8	20,929	18,485	3026	44,958	120	30,734	4953	1794
Sector 9	28,028	18,157	2709	92,325	90	5723	517	535
Sector 10	122,781	89,083	11,924	54,090	123	17,629	1112	949
Sector 11	41,987	28,172	4674	22,113	1203	272,343	9178	3769
Sector 12	861,192	716,687	94,215	230,398	11,751	28,760	2835	17,967
Sector 13	195,009	180,111	35,633	63,294	1447	19,718	1019	1062
Sector 14	2,536,854	314,190	40,201	287,720	5810	8937	1088	420
Sector 15	411,227	234,974	37,416	127,339	609	54,926	4205	9806
Sector 16	480,321	105,765	28,204	12,092	359	93,021	3564	3868
Sector 17	25,282	34,783	5827	30,877	124	103,598	9672	4516
Sector 18	1,353,774	293,353	49,555	18,914	914	512,832	1024	1223
Sector 19	580,427	4,663,623	240,728	7524	46	5691	232	293
Sector 20	105,483	580,531	115,693	6530	54	124,654	2769	1974
Sector 21	98,009	81,828	17,590	331,427	108	6185	330	376
Sector 22	80,994	10,588	101	599	5318	1015	0	0
Sector 23	128,681	137,215	12,387	40,658	2473	3,312,725	8823	85,275
Sector 24	5931	5459	725	2885	121	10,773	24,590	1186
Sector 25	4287	4741	827	2051	115	11,540	236	7627
Sector 26	2558	1969	289	473	57	3178	327	443
Sector 27	234,710	172,830	23,393	71,756	4218	112,992	11,671	3449
Sector 28	7871	4515	1154	854	13	1095	57	113
Sector 29	37,644	117,921	4766	20,383	168	56,778	630	1345
Sector 30	354,390	484,932	25,883	60,147	2583	73,440	7412	2209
Sector 31	75,308	34,815	10,020	21,634	516	26,574	2025	2900
Sector 32	197,472	485,761	11,719	33,026	2376	384,921	14,633	20,956
Sector 33	48,846	38,551	6286	20,516	180	4062	464	192
Sector 34	235,391	125,425	13,521	25,429	437	41,327	1944	1791
Sector 35	42,829	78,696	5698	2959	45	6169	116	75
Sector 36	60,312	57,257	6984	6481	221	74,364	1017	554
Sector 37	2140	1416	106	483	32	24,766	193	20,845
Sector 38	10,235	8490	990	1281	478	71,026	1272	2871
Sector 39	1551	1042	229	504	5	1434	104	279
Sector 40	18,494	5512	1970	472	87	11,072	1239	80
Sector 41	12,877	10,710	1287	3338	80	20,910	834	1093
Sector 42	1184	1118	154	430	0	1324	37	73
Total intermediate input	8,510,601	9,193,384	823,800	2,353,612	44,658	7,255,069	350,828	203,489
Compensation of employees	599,444	783,498	96,802	272,778	6478	742,179	47,222	74,143
Net taxes on production	231,961	159,152	27,299	183,806	3880	451,315	10,956	14,611
Depreciation of fixed assets	163,575	245,993	19,746	61,565	2913	930,173	14,663	63,853
Operating surplus	441,137	293,163	49,419	150,245	395,289	301,960	21,438	8561
Total value added	1,436,117	1,481,807	193,266	668,393	408,560	2,425,627	94,280	161,168
Total input	9,946,718	10,675,191	1,017,066	3,022,006	453,217	9,680,696	445,107	364,656

Sector 26	Sector 27	Sector 28	Sector 29	Sector 30	Sector 31	Sector 32	Sector 33	Sector 34
124,098	273,061	0	0	2654	591,915	0	531	13,689
22,409	39,401	433	0	3344	3418	0	1373	8135
0	14,213	0	0	0	0	0	0	0
0	0	0	0	0	0	0	0	0
285,337	2965	0	0	0	0	0	0	0
112,761	90,000	768	10,531	57,047	1,781,500	24,976	18,187	68,256
23,844	14,783	467	529	13,225	25,211	1275	7197	74,572
107,500	81,711	2225	8174	96,875	19,345	23,499	29,197	71,944
494,492	13,939	1537	4140	18,104	7088	6697	7831	14,572
47,201	41,987	8183	134,654	101,580	17,439	173,196	37,960	516,941
364,112	2,132,123	9107	8327	36,706	25,224	56,351	84,589	183,663
871,320	201,254	1024	22,885	99,484	39,092	30,243	108,111	206,675
6,400,138	24,231	1068	1171	2987	10,511	1432	1917	3062
3,521,148	58,889	0	44	5372	231	0	5807	4407
918,401	63,644	240	8684	4411	5991	8061	39,290	115,832
836,110	275,386	3233	34,939	23,702	6015	37,959	22,940	37,872
116,747	804,941	24,405	30,452	96,127	17,313	25,945	24,716	175,552
1,068,403	32,154	546	297,381	74,834	36,745	18,326	71,811	161,785
68,089	28,819	2708	258,093	42,884	2091	6669	19,401	333,300
61,679	18,210	366	38,723	3580	161	34,226	12,651	17,178
91,944	10,465	780	4577	8397	3194	9285	8365	34,967
0	0	0	0	0	0	0	0	0
210,953	137,054	4216	107,703	128,682	109,584	84,067	38,756	55,017
3341	10,742	294	2458	2251	18,714	443	683	5570
6781	6888	496	5277	4211	8902	4218	1180	3766
299,596	49,194	13,048	7320	28,392	11,734	12,904	124,601	5632
2,631,357	965,228	43,994	132,433	771,424	62,369	221,217	48,863	283,896
6091	3591	12,944	5643	14,854	2946	38,399	4133	27,132
408,945	82,139	6270	187,156	89,511	15,760	254,401	25,228	21,569
694,917	147,918	23,138	135,687	55,851	108,437	37,829	27,623	133,811
282,439	182,209	2220	84,837	211,318	34,704	329,052	105,471	380,676
267,222	674,533	3615	89,410	375,661	84,874	717,203	277,251	261,792
16,458	55,799	4698	112,948	244,523	54,361	286,073	85,175	96,257
101,792	63,499	11,977	216,538	536,493	61,267	544,164	218,514	252,603
21,156	3516	167	7226	17,164	45	12,518	1363	2881
326,687	13,458	55	21,114	42,429	945	6189	12,408	4303
1388	3450	0	1209	2544	1681	5734	2079	49,557
39,552	260,184	492	27,655	101,753	47,855	34,684	47,634	89,349
3515	8538	309	5059	6579	1336	17,368	1527	2776
8384	6168	59	1288	1769	864	6176	2235	943
39,237	23,537	2309	24,169	42,878	13,250	53,445	34,210	24,759
3603	3224	102	1645	2739	789	5095	4283	1099
20,909,149	6,923,044	187,493	2,040,078	3,372,339	3,232,897	3,129,318	1,565,090	3,745,789
4,195,098	1,851,495	88,487	580,838	2,270,120	1,160,934	1,843,660	530,812	1,107,212
1,165,012	372,009	7976	170,855	2,251,846	220,463	707,230	765,167	280,799
407,558	823,403	20,853	560,825	527,677	260,247	136,154	2,343,613	538,221
1,598,100	1,466,583	255	956,843	3,215,632	275,743	3,120,082	1,218,330	158,488
7,365,768	4,513,490	117,573	2,269,361	8,265,276	1,917,387	5,807,127	4,857,922	2,084,720
28,274,917	11,436,534	305,066	4,309,439	11,637,616	5,150,284	8,936,445	6,423,012	5,830,509

(continued)

Table 10.26 (continued)

	Sector 35	Sector 36	Sector 37	Sector 38	Sector 39	Sector 40	Sector 41	Sector 42
Sector 1	26,128	9366	25,814	33,683	8282	20,836	6512	0
Sector 2	1201	3151	5532	11,735	8800	17,570	1080	18,241
Sector 3	0	0	0	0	0	0	0	0
Sector 4	6538	1463	1863	0	0	0	0	0
Sector 5	107	72	695	151	1599	2224	653	2573
Sector 6	12,341	5766	11,232	65,429	34,901	36,276	74,817	23,638
Sector 7	14,043	1555	4572	43,362	7587	24,206	3919	26,508
Sector 8	5362	3936	13,802	20,302	4859	59,669	22,239	86,667
Sector 9	2353	2995	13,617	27,166	52,799	32,459	15,984	105,590
Sector 10	7737	33,176	7243	48,089	151,263	54,163	150,172	169,245
Sector 11	6343	48,126	39,214	52,894	34,117	17,756	8674	94,390
Sector 12	75,420	51,860	71,830	251,093	38,898	1,853,972	54,801	65,778
Sector 13	5464	4564	14,864	3063	21,217	24,244	8553	31,693
Sector 14	3480	17,287	8048	7394	3235	208	615	0
Sector 15	17,687	50,296	8515	14,309	9278	9079	5306	7693
Sector 16	10,436	18,917	22,346	6996	8627	194,231	3736	15,105
Sector 17	6165	34,348	37,578	109,951	28,774	4823	6245	84,434
Sector 18	30,614	98,065	27,854	47,285	12,745	13,834	7368	14,795
Sector 19	27,358	88,510	15,354	71,054	13,973	3199	7970	49,991
Sector 20	24,481	53,378	14,700	28,065	42,080	10,135	5005	11,428
Sector 21	3165	5204	5598	22,916	7364	5455	2695	14,712
Sector 22	0	0	0	0	0	0	0	0
Sector 23	14,135	16,279	30,421	31,971	78,851	50,454	11,999	88,271
Sector 24	229	599	977	1859	2782	2018	894	1954
Sector 25	1077	2780	1989	4730	5848	2894	1057	3718
Sector 26	2470	7106	26,657	22,104	20,779	84,087	10,875	90,919
Sector 27	28,789	139,726	24,661	60,559	110,766	21,750	39,279	270,034
Sector 28	1166	2807	1021	4014	9041	1392	3042	69,250
Sector 29	3950	11,638	9481	9854	35,052	39,519	10,776	120,774
Sector 30	14,474	27,665	15,245	51,545	31,827	129,585	27,419	66,356
Sector 31	26,906	96,649	25,490	45,413	103,713	41,805	41,585	412,442
Sector 32	2990	74,532	34,056	61,138	108,648	24,627	19,972	103,099
Sector 33	3165	24,123	3222	81,116	30,889	8287	17,185	35,047
Sector 34	6572	35,508	15,764	41,116	74,969	30,864	29,289	57,901
Sector 35	16,662	5424	1029	78	8651	1968	1344	87
Sector 36	10,078	147,161	5506	2222	8808	6065	910	9028
Sector 37	435	498	25,397	4184	5463	1094	1191	15,015
Sector 38	2067	21,166	26,711	112,275	30,375	23,680	12,564	92,619
Sector 39	2334	2332	1764	885	18,645	4128	1275	49,382
Sector 40	1832	2194	3979	933	6126	10,689	1910	17,682
Sector 41	10,801	11,989	7776	11,590	24,978	6223	40,808	80,559
Sector 42	178	752	422	667	807	623	276	1732
Total intermediate input	436,733	1,162,962	611,836	1,413,191	1,207,411	2,876,093	659,994	2,408,350
Compensation of employees	169,630	629,293	232,705	989,421	3,053,350	1,335,367	385,670	3,690,029
Net taxes on production	10,833	105,246	19,102	136,683	19,646	13,932	66,491	20,034
Depreciation of fixed assets	31,830	119,707	93,275	80,443	362,904	162,105	103,250	543,246
Operating surplus	36,956	323,356	70,010	278,581	51,225	95,622	121,236	116,778
Total value added	249,249	1,177,603	415,092	1,485,128	3,487,125	1,607,026	676,647	4,370,088
Total input	685,982	2,340,565	1,026,928	2,898,319	4,694,536	4,483,118	1,336,640	6,778,437

Total intermediate use	Rural household consumption expenditure	Urban household consumption expenditure	Government consumption expenditure	Gross fixed capital formation	Increase in inventories	Export	Import	Total output
13,656,398	1,592,966	2,005,505	142,398	990,320	221,672	481,068	754,598	18,335,729
5,063,907	26,091	19,367	0	45,833	−87,034	19,745	446,207	4,641,702
5,808,841	0	0	0	18,307	−55,601	432,575	4,639,191	1,564,931
4,390,575	0	0	0	72,756	−19,071	7635	2,784,071	1,667,823
1,380,954	0	0	0	11,325	−22,672	24,588	97,490	1,296,705
8,632,519	1,749,443	5,148,864	0	298,331	255,511	220,825	242,567	16,062,927
5,172,720	60,776	94,353	0	39,900	−27,421	4,527,960	1,117,408	8,750,880
2,092,825	369,711	1,791,199	0	94,966	90,604	579,653	75,977	4,942,982
2,396,064	34,482	151,372	0	217,130	−1685	269,856	471,106	2,596,112
4,292,967	22,632	149,796	0	45,311	−34,103	226,932	521,913	4,181,622
7,177,763	31,955	285,892	0	−8912	−134,219	126,443	476,302	7,002,620
20,847,785	204,807	667,695	0	226,353	−131,943	944,203	1,907,836	20,851,065
9,802,208	13,447	83,276	0	92,103	−135,481	499,375	345,041	10,009,887
19,457,919	0	0	0	210,225	−188,641	365,674	1,045,697	18,799,481
4,815,791	22,343	121,269	0	112,397	−54,600	345,348	295,666	5,066,881
9,630,289	1587	23,004	0	6,282,943	89,393	2,188,003	2,060,903	16,154,316
6,660,565	180,715	867,123	0	7,145,586	119,509	2,340,868	1,778,986	15,535,379
6,091,828	159,007	620,699	0	2,066,391	34,952	1,911,891	909,073	9,975,695
7,152,481	139,636	557,658	0	1,264,932	187,744	5,080,059	3,643,933	10,738,577
2,236,183	16,304	52,661	0	215,663	−24,607	558,971	1,959,640	1,095,534
1,653,029	88,157	506,561	0	600,567	9937	258,081	119,099	2,997,232
1,471,518	0	0	0	0	6321	5128	994,911	488,055
10,496,788	60,900	224,622	0	0	−1,116,705	17,113	2022	9,680,696
269,975	20,442	153,008	0	2920	358	0	0	446,703
204,275	16,203	144,712	0	0	1226	0	0	366,416
881,722	0	353,843	0	26,596,546	72,498	500,124	136,444	28,268,289
10,513,006	173,775	666,904	559,955	81,553	−179,449	369,632	748,254	11,437,123
312,761	2489	17,021	0	0	2855	8440	44,570	298,996
2,221,149	199,974	867,165	0	876,630	43,230	211,905	104,183	4,315,869
7,130,080	626,054	1,929,907	0	783,508	47,359	956,496	0	11,473,404
3,534,378	394,492	1,540,823	0	0	98,640	−138,304	266,961	5,163,070
7,383,541	232,645	1,235,382	99,997	0	108,023	105,661	93,712	9,071,537
1,783,038	697,781	2,369,385	0	1,591,501	120,162	0	0	6,561,867
4,142,063	32,855	468,483	228,062	0	59,141	1,046,623	249,350	5,727,878
601,302	0	0	352,892	0	2978	26,772	302,295	681,649
1,740,279	1776	22,731	325,033	249,444	−1537	0	0	2,337,725
317,824	9414	115,143	568,528	0	−2160	0	0	1,008,749
1,470,241	213,797	1,243,230	0	0	35,106	41,281	90,252	2,913,403
167,897	442,271	985,050	3,116,754	0	−27,196	5272	34,634	4,655,414
296,243	156,320	1,840,087	2,226,709	0	−23,344	3828	2677	4,497,165
680,948	43,151	232,581	367,654	0	12,312	42,790	53,539	1,325,898
57,619	0	0	6,750,795	0	−12,349	21,943	17,436	6,800,571

Table 10.27 China's national input–output table in 2025 in the policy scenario

	Sector 1	Sector 2	Sector 3	Sector 4	Sector 5	Sector 6	Sector 7	Sector 8
Sector 1	2,422,509	46,393	26	2843	577	6,639,242	1,976,999	245,017
Sector 2	15,014	785,126	11,674	7005	3586	45,792	47,110	19,872
Sector 3	0	0	22,516	0	0	0	0	0
Sector 4	0	0	0	161,898	0	0	0	0
Sector 5	2349	1761	715	1727	85,978	6028	0	0
Sector 6	1,822,647	8144	2444	4041	3718	3,458,085	64,844	244,251
Sector 7	2220	1198	1082	967	686	8651	2,743,647	1,606,350
Sector 8	6066	16,029	4819	7109	5835	21,973	145,599	715,835
Sector 9	22,066	40,604	5183	5709	3004	17,969	10,700	11,178
Sector 10	16,542	5681	2895	2977	3055	312,832	91,504	80,997
Sector 11	131,076	51,516	51,299	94,083	60,681	47,095	22,318	31,356
Sector 12	1,465,962	107,638	50,747	91,726	128,298	462,692	1,017,920	319,069
Sector 13	24,134	40,758	12,614	13,752	47,358	115,841	20,660	6628
Sector 14	4501	189,646	66,515	21,387	9358	18,909	7544	7147
Sector 15	65,206	91,846	10,467	34,582	22,405	77,607	18,723	23,713
Sector 16	110,255	237,392	112,830	131,567	113,253	57,329	115,541	34,141
Sector 17	50,576	29,761	9281	17,882	25,937	21,277	17,429	7499
Sector 18	4923	67,496	12,618	22,270	12,478	24,432	20,267	9364
Sector 19	1797	7115	1756	2392	1477	6180	7249	5775
Sector 20	15,174	55,506	12,873	8856	6139	11,819	19,857	6585
Sector 21	50,468	33,415	1876	7137	14,169	25,671	62,738	48,386
Sector 22	1396	459	22	1236	11,006	13,060	16,743	2756
Sector 23	218,270	196,908	99,571	284,861	146,945	173,943	221,564	39,937
Sector 24	669	1778	1682	8614	6962	9932	4845	1794
Sector 25	2254	3693	1070	1568	1464	6795	7609	2473
Sector 26	4593	5897	1540	654	466	4732	2081	1511
Sector 27	343,357	209,649	26,433	72,919	84,030	481,118	176,363	124,631
Sector 28	14,778	1661	219	334	592	3786	7978	10,806
Sector 29	73,793	16,069	2311	2742	3707	29,797	19,813	28,272
Sector 30	419,887	73,937	12,497	22,150	30,819	405,250	244,854	124,551
Sector 31	59,364	55,028	12,102	19,197	22,287	99,409	47,451	34,783
Sector 32	247,061	146,344	11,176	21,401	28,756	172,624	262,401	93,198
Sector 33	8894	9134	609	1229	2889	28,852	56,736	93,741
Sector 34	28,046	26,622	2478	6730	6909	277,146	40,693	110,238
Sector 35	33,422	6708	2492	571	367	13,616	6074	2816
Sector 36	171,522	43,865	15,361	23,346	15,049	57,740	22,985	18,929
Sector 37	55,117	6719	1111	1208	4393	5146	15,897	731
Sector 38	85,387	41,216	5093	4138	6509	20,229	13,217	11,326
Sector 39	7250	4385	252	654	1776	2307	1087	580
Sector 40	11,076	10,797	489	3231	2680	8359	4408	3535
Sector 41	4611	7677	1306	4421	6753	17,681	11,155	9533
Sector 42	6029	1167	413	358	257	2347	1121	710
Total intermediate input	8,030,262	2,686,737	592,456	1,121,472	932,604	13,213,293	7,595,724	4,140,015
Compensation of employees	10,131,916	883,678	199,826	151,571	150,740	1,027,645	727,790	506,396
Net taxes on production	20,586	492,930	226,906	97,801	91,724	1,159,090	289,643	131,406
Depreciation of fixed assets	497,210	168,884	133,396	35,773	39,079	318,471	191,154	59,646
Operating surplus	0	511,463	312,616	185,933	125,165	1,023,562	298,551	169,749
Total value added	10,649,711	2,056,955	872,743	471,079	406,708	3,528,768	1,507,137	867,196
Total input	18,679,973	4,743,692	1,465,199	1,592,551	1,339,312	16,742,061	9,102,860	5,007,211

Sector 9	Sector 10	Sector 11	Sector 12	Sector 13	Sector 14	Sector 15	Sector 16	Sector 17
374,482	242,280	162	817,826	6269	2328	1701	1786	1512
21,043	41,811	451,644	323,183	741,363	707,912	24,755	56,465	18,309
0	0	4,091,347	1,333,582	102,213	234,597	0	0	0
0	0	0	207,400	40,818	3,895,876	72,275	48,795	11,954
0	0	733	305,796	599,705	53,596	21,185	11,606	4111
13,725	17,868	33,826	334,977	35,622	64,172	16,410	54,223	54,861
26,154	74,994	370	131,279	20,925	2289	8152	23,263	28,893
45,243	31,855	3017	48,466	25,027	42,865	15,569	35,535	171,037
749,845	126,946	2229	40,423	79,573	14,791	89,638	53,364	185,961
47,139	1,029,233	3942	298,316	242,144	22,456	58,940	72,403	69,074
19,216	20,553	440,827	1,207,374	401,049	806,437	51,501	169,311	75,710
262,296	602,751	136,027	7,705,318	784,979	303,377	219,575	597,901	997,515
22,219	11,684	24,622	150,293	1,664,964	371,852	55,639	119,022	115,199
58,415	52,220	4646	212,980	271,101	4,936,346	1,712,921	3,480,154	1,898,563
62,909	72,709	23,368	190,673	283,560	167,232	590,734	694,840	295,956
38,324	58,637	93,332	303,369	270,903	570,922	240,451	3,726,680	1,365,924
11,021	34,379	16,375	57,690	49,574	95,202	44,422	249,775	4,213,533
8396	28,794	8221	53,098	38,397	59,386	39,643	906,108	592,838
1104	38,882	990	12,617	6232	4733	6148	345,258	185,391
3281	51,360	10,314	152,062	33,740	81,469	12,950	86,617	373,128
22,355	28,111	3204	121,063	81,145	73,725	32,121	114,498	193,475
8781	267,385	1432	89,900	159,228	565,298	41,527	196,481	12,814
86,748	131,170	119,879	1,062,934	746,150	1,159,357	376,498	370,061	257,570
1222	1225	10,635	51,220	11,320	24,912	5104	14,760	10,851
1989	9521	2186	23,445	11,124	15,646	5127	5527	9770
696	1093	3502	7847	3399	6217	980	5722	5622
87,056	104,994	138,436	576,245	428,846	362,293	118,981	473,747	331,394
1938	3004	216	20,409	3365	7322	1081	14,684	4167
11,931	12,144	22,357	80,197	26,295	220,022	19,478	52,275	52,340
72,819	182,203	85,326	589,721	362,447	456,459	138,991	423,691	997,489
15,471	27,252	12,721	136,488	88,239	77,674	67,232	127,902	84,928
69,495	82,251	42,580	468,479	435,871	532,609	88,796	176,069	309,693
17,300	34,500	1132	74,781	41,787	7835	38,215	46,201	50,227
24,541	41,455	33,528	358,590	73,497	33,344	21,641	146,104	255,067
1645	2676	1938	53,947	8026	32,633	4631	57,514	147,326
5298	23,378	11,934	129,865	41,841	73,502	31,438	99,041	155,357
400	15,949	1406	24,881	3381	6518	765	3421	5667
3336	10,646	6181	53,372	38,509	54,024	10,282	19,476	29,625
454	623	281	3379	1622	1564	733	5257	3661
2502	5264	1817	13,226	17,732	17,828	12,356	39,303	34,019
3730	5229	1605	28,072	18,721	14,692	6304	23,202	21,302
379	668	323	2657	1723	2072	735	2246	2572
2,204,898	3,527,697	5,848,611	17,857,441	8,302,425	16,183,386	4,305,628	13,150,286	13,634,406
207,565	312,262	235,731	1,294,870	676,210	1,114,259	342,826	1,449,817	867,247
86,184	120,266	778,262	715,122	368,216	539,448	142,722	665,279	576,160
49,574	118,706	174,698	554,221	262,693	664,569	100,109	385,962	251,368
125,223	185,169	259,098	1,072,640	605,759	762,008	240,115	1,025,755	713,970
468,547	736,404	1,447,788	3,636,854	1,912,877	3,080,283	825,772	3,526,813	2,408,744
2,673,445	4,264,101	7,296,399	21,494,294	10,215,302	19,263,669	5,131,400	16,677,100	16,043,150

(continued)

Table 10.27 (continued)

	Sector 18	Sector 19	Sector 20	Sector 21	Sector 22	Sector 23	Sector 24	Sector 25
Sector 1	219	0	59	365,573	177	169	76	0
Sector 2	8588	3263	1753	19,987	815	1,665,371	21,887	494
Sector 3	0	0	0	28	1	59,211	214,456	0
Sector 4	17,629	173	1479	278	538	279	0	0
Sector 5	4462	75	119	8394	68	2544	18	38
Sector 6	37,651	39,577	3397	41,274	700	28,836	1699	972
Sector 7	18,733	2968	1639	273,079	77	487	53	111
Sector 8	21,849	19,779	3410	45,019	120	31,415	5113	1833
Sector 9	29,316	19,379	3324	92,450	90	5850	534	547
Sector 10	125,649	92,297	12,730	54,163	123	18,019	1149	970
Sector 11	42,756	28,990	4897	22,143	1203	278,374	9476	3850
Sector 12	879,275	740,549	99,703	230,710	11,751	29,397	2927	18,353
Sector 13	200,624	188,142	38,461	63,379	1447	20,154	1052	1085
Sector 14	2,610,569	327,951	44,344	288,109	5810	9135	1123	429
Sector 15	424,974	246,416	40,957	127,511	609	56,142	4341	10,017
Sector 16	490,877	109,474	29,774	12,108	359	95,081	3679	3951
Sector 17	25,910	36,091	6298	30,919	124	105,892	9985	4613
Sector 18	1,389,396	304,894	53,399	18,939	914	524,189	1057	1250
Sector 19	593,028	4,826,731	257,277	7534	46	5817	239	300
Sector 20	107,412	597,964	119,387	6538	54	127,415	2858	2016
Sector 21	102,885	87,598	20,655	331,876	108	6322	341	384
Sector 22	81,569	10,722	102	600	5318	1037	0	0
Sector 23	131,595	142,196	13,350	40,713	2473	3,386,082	9109	87,108
Sector 24	6076	5668	777	2889	121	11,012	25,387	1211
Sector 25	4455	5037	914	2054	115	11,796	244	7791
Sector 26	2627	2051	314	474	57	3248	338	452
Sector 27	241,442	180,233	25,699	71,854	4218	115,494	12,050	3523
Sector 28	8780	5264	1652	855	13	1119	59	116
Sector 29	39,470	126,993	5462	20,410	168	58,035	650	1374
Sector 30	382,169	540,956	33,561	60,228	2583	75,066	7652	2256
Sector 31	78,328	36,978	11,373	21,663	516	27,162	2091	2962
Sector 32	207,092	526,091	13,170	33,071	2376	393,445	15,108	21,406
Sector 33	49,789	39,912	6704	20,544	180	4152	479	196
Sector 34	251,954	139,064	16,641	25,464	437	42,242	2007	1829
Sector 35	44,283	83,155	6433	2963	45	6306	120	76
Sector 36	62,126	60,034	7770	6489	221	76,011	1050	566
Sector 37	2276	1551	132	484	32	25,315	199	21,293
Sector 38	10,561	8923	1112	1283	478	72,599	1314	2933
Sector 39	1595	1090	252	505	5	1465	107	285
Sector 40	18,916	5717	2118	473	87	11,317	1279	82
Sector 41	13,580	11,608	1524	3343	80	21,373	861	1117
Sector 42	1210	1159	166	431	0	1353	39	74
Total intermediate input	8,771,696	9,606,714	892,285	2,356,801	44,658	7,415,727	362,205	207,862
Compensation of employees	615,217	810,901	99,892	273,148	6478	758,614	48,754	75,736
Net taxes on production	238,064	164,718	28,170	184,055	3880	461,309	11,312	14,925
Depreciation of fixed assets	167,880	254,597	20,376	61,648	2913	950,771	15,139	65,225
Operating surplus	452,745	303,417	50,997	150,448	395,289	308,647	22,133	8745
Total value added	1,473,906	1,533,632	199,436	669,299	408,560	2,479,340	97,337	164,631
Total input	10,245,602	11,140,346	1,091,721	3,026,100	453,217	9,895,067	459,542	372,493

Sector 26	Sector 27	Sector 28	Sector 29	Sector 30	Sector 31	Sector 32	Sector 33	Sector 34
127,482	283,362	0	0	2686	614,797	0	564	13,792
23,020	40,887	435	0	3384	3550	0	1459	8196
0	14,750	0	0	0	0	0	0	0
0	0	0	0	0	0	0	0	0
293,118	3076	0	0	0	0	0	0	0
115,836	93,395	772	10,745	57,737	1,850,370	25,747	19,332	68,766
24,495	15,341	469	540	13,385	26,186	1314	7651	75,129
110,432	84,794	2235	8340	98,047	20,092	24,224	31,036	72,482
507,976	14,465	1544	4224	18,323	7362	6904	8324	14,681
48,488	43,571	8219	137,387	102,808	18,114	178,542	40,351	520,807
374,041	2,212,557	9147	8496	37,150	26,199	58,091	89,916	185,036
895,080	208,846	1028	23,349	100,687	40,603	31,177	114,919	208,221
6,574,663	25,145	1072	1195	3023	10,917	1476	2038	3085
3,617,166	61,110	0	45	5437	240	0	6172	4440
943,445	66,045	241	8860	4464	6222	8310	41,764	116,699
858,910	285,775	3248	35,648	23,988	6247	39,131	24,385	38,155
119,931	835,308	24,514	31,070	97,289	17,982	26,746	26,272	176,865
1,097,537	33,367	548	303,417	75,739	38,165	18,892	76,333	162,995
69,946	29,906	2720	263,332	43,403	2172	6875	20,623	335,793
63,361	18,897	367	39,509	3623	167	35,283	13,447	17,306
94,451	10,860	783	4669	8499	3318	9572	8892	35,228
0	0	0	0	0	0	0	0	0
216,705	142,224	4235	109,889	130,238	113,820	86,662	41,196	55,428
3432	11,147	296	2508	2279	19,437	457	726	5612
6966	7148	498	5385	4262	9246	4348	1255	3795
307,766	51,050	13,106	7469	28,736	12,188	13,302	132,447	5675
2,703,111	1,001,641	44,189	135,121	780,752	64,780	228,046	51,940	286,019
6257	3727	13,001	5757	15,034	3060	39,584	4394	27,335
420,096	85,237	6298	190,954	90,594	16,369	262,255	26,817	21,730
713,867	153,498	23,241	138,441	56,527	112,629	38,997	29,363	134,812
290,141	189,082	2230	86,559	213,874	36,046	339,210	112,112	383,523
274,509	699,980	3631	91,225	380,204	88,156	739,344	294,710	263,750
16,907	57,903	4719	115,241	247,480	56,462	294,905	90,539	96,976
104,567	65,895	12,031	220,933	542,981	63,635	560,963	232,274	254,492
21,733	3649	168	7373	17,372	47	12,904	1449	2902
335,595	13,965	55	21,542	42,942	981	6380	13,189	4335
1426	3580	0	1233	2575	1746	5911	2210	49,927
40,631	270,000	494	28,216	102,983	49,705	35,755	50,634	90,017
3611	8860	311	5162	6659	1388	17,904	1624	2796
8613	6401	60	1314	1791	897	6366	2376	950
40,307	24,425	2319	24,659	43,396	13,763	55,094	36,364	24,944
3701	3346	102	1678	2772	820	5253	4553	1107
21,479,317	7,184,214	188,327	2,081,485	3,413,119	3,357,876	3,225,926	1,663,648	3,773,801
4,309,493	1,921,343	88,881	592,627	2,297,571	1,205,814	1,900,578	564,238	1,115,492
1,196,781	386,043	8012	174,323	2,279,077	228,985	729,064	813,352	282,899
418,671	854,465	20,946	572,208	534,058	270,308	140,357	2,491,196	542,246
1,641,678	1,521,910	256	976,264	3,254,517	286,402	3,216,404	1,295,051	159,673
7,566,624	4,683,761	118,095	2,315,422	8,365,224	1,991,510	5,986,403	5,163,837	2,100,310
29,045,941	11,867,975	306,422	4,396,907	11,778,343	5,349,386	9,212,328	6,827,485	5,874,111

(continued)

Table 10.27 (continued)

	Sector 35	Sector 36	Sector 37	Sector 38	Sector 39	Sector 40	Sector 41	Sector 42
Sector 1	26,691	9698	26,169	35,375	8633	21,961	6683	0
Sector 2	1227	3263	5608	12,324	9172	18,519	1109	19,257
Sector 3	0	0	0	0	0	0	0	0
Sector 4	6679	1515	1888	0	0	0	0	0
Sector 5	109	75	705	159	1667	2344	670	2717
Sector 6	12,607	5971	11,386	68,716	36,379	38,235	76,785	24,955
Sector 7	14,346	1610	4635	45,540	7908	25,512	4022	27,985
Sector 8	5478	4076	13,992	21,322	5064	62,890	22,824	91,494
Sector 9	2404	3102	13,804	28,531	55,035	34,211	16,404	111,470
Sector 10	7904	34,355	7343	50,505	157,670	57,087	154,123	178,670
Sector 11	6480	49,836	39,754	55,551	35,562	18,714	8902	99,646
Sector 12	77,047	53,703	72,819	263,708	40,546	1,954,051	56,243	69,442
Sector 13	5582	4727	15,068	3217	22,115	25,553	8778	33,458
Sector 14	3555	17,901	8159	7766	3372	219	631	0
Sector 15	18,069	52,083	8633	15,028	9671	9569	5446	8122
Sector 16	10,661	19,589	22,653	7348	8993	204,715	3834	15,946
Sector 17	6298	35,569	38,095	115,475	29,992	5084	6409	89,136
Sector 18	31,275	101,550	28,237	49,660	13,285	14,581	7562	15,619
Sector 19	27,949	91,655	15,565	74,624	14,564	3372	8179	52,775
Sector 20	25,009	55,275	14,903	29,475	43,862	10,682	5136	12,064
Sector 21	3234	5389	5675	24,067	7676	5750	2766	15,531
Sector 22	0	0	0	0	0	0	0	0
Sector 23	14,440	16,857	30,840	33,577	82,191	53,177	12,315	93,186
Sector 24	234	620	991	1953	2899	2127	918	2063
Sector 25	1101	2879	2017	4968	6096	3050	1085	3925
Sector 26	2523	7359	27,024	23,214	21,659	88,626	11,161	95,982
Sector 27	29,410	144,692	25,000	63,602	115,458	22,924	40,313	285,073
Sector 28	1191	2907	1035	4216	9424	1467	3122	73,107
Sector 29	4035	12,052	9612	10,349	36,537	41,653	11,060	127,500
Sector 30	14,786	28,649	15,455	54,135	33,176	136,580	28,140	70,051
Sector 31	27,487	100,083	25,841	47,695	108,106	44,061	42,679	435,411
Sector 32	3054	77,180	34,525	64,210	113,250	25,957	20,498	108,840
Sector 33	3233	24,980	3266	85,191	32,197	8734	17,637	36,999
Sector 34	6714	36,769	15,981	43,182	78,145	32,531	30,060	61,125
Sector 35	17,022	5617	1043	82	9017	2074	1379	92
Sector 36	10,296	152,391	5582	2333	9181	6392	934	9530
Sector 37	445	515	25,747	4394	5694	1153	1222	15,852
Sector 38	2111	21,919	27,079	117,915	31,661	24,958	12,895	97,777
Sector 39	2385	2415	1788	929	19,434	4350	1308	52,132
Sector 40	1872	2272	4034	979	6385	11,266	1961	18,667
Sector 41	11,034	12,415	7883	12,172	26,036	6559	41,882	85,046
Sector 42	182	778	427	700	841	657	283	1828
Total intermediate input	446,155	1,204,290	620,259	1,484,189	1,258,553	3,031,346	677,356	2,542,472
Compensation of employees	173,290	651,656	235,909	1,039,129	3,182,681	1,407,451	395,816	3,895,529
Net taxes on production	11,067	108,987	19,365	143,550	20,478	14,685	68,240	21,149
Depreciation of fixed assets	32,517	123,961	94,559	84,484	378,276	170,855	105,966	573,500
Operating surplus	37,753	334,848	70,973	292,577	53,395	100,783	124,425	123,282
Total value added	254,627	1,219,451	420,806	1,559,740	3,634,829	1,693,774	694,447	4,613,460
Total input	700,782	2,423,741	1,041,065	3,043,929	4,893,383	4,725,120	1,371,803	7,155,932

Total intermediate use	Rural household consumption expenditure	Urban household consumption expenditure	Government consumption expenditure	Gross fixed capital formation	Increase in inventories	Export	Import	Total output
14,326,116	1,651,605	2,083,945	149,022	993,387	221,672	500,634	778,677	19,147,703
5,195,232	27,398	20,407	0	46,652	−87,034	20,148	470,348	4,752,455
6,072,702	0	0	0	18,307	−55,601	452,465	5,026,650	1,461,223
4,469,475	0	0	0	72,756	−19,071	7635	3,013,339	1,517,456
1,415,649	0	0	0	11,807	−22,672	25,215	102,312	1,327,688
8,905,697	1,824,731	5,390,726	0	312,877	255,511	220,825	242,567	16,667,800
5,274,334	63,671	98,956	0	41,034	−27,421	4,789,395	1,191,132	9,048,837
2,149,238	383,285	1,862,807	0	95,750	90,604	579,653	75,977	5,085,361
2,459,435	35,838	157,672	0	219,160	−1685	269,856	501,892	2,638,384
4,411,172	23,650	155,829	0	46,471	−34,103	226,932	549,761	4,280,190
7,388,169	33,487	298,691	0	−9485	−134,219	129,643	499,602	7,206,684
21,477,923	214,380	698,897	0	234,725	−131,943	944,203	1,959,963	21,478,222
10,072,698	14,182	87,142	0	94,313	−135,481	521,586	365,862	10,288,578
19,976,037	0	0	0	213,619	−188,641	365,674	1,084,348	19,282,342
4,960,166	23,359	126,921	0	114,146	−54,600	345,348	311,845	5,203,495
9,935,381	1664	24,086	0	6,573,611	89,393	2,293,619	2,140,207	16,777,547
6,853,478	188,406	901,100	0	7,401,961	119,509	2,472,417	1,864,642	16,072,229
6,271,532	165,418	646,686	0	2,120,765	34,952	1,982,487	944,067	10,277,773
7,389,519	146,397	584,410	0	1,309,173	187,744	5,263,367	3,762,855	11,117,755
2,293,741	17,343	54,977	0	222,548	−24,607	569,673	2,061,916	1,071,759
1,710,087	90,785	525,886	0	601,380	9937	264,586	124,200	3,078,462
1,488,872	0	0	0	0	6321	5258	1,030,388	470,063
10,741,973	60,900	224,622	0	0	−1,148,166	17,772	2034	9,895,067
278,363	21,142	157,751	0	3015	358	0	0	460,629
211,684	16,893	148,880	0	0	1226	0	0	378,683
915,400	0	369,689	0	27,321,803	72,498	528,949	143,852	29,064,488
10,817,074	181,883	696,794	591,396	84,629	−179,449	369,632	790,279	11,771,681
328,812	2582	17,703	0	0	2855	8679	47,406	313,226
2,299,256	207,519	902,071	0	894,422	43,230	222,086	109,780	4,458,803
7,529,907	651,062	2,009,359	0	792,982	47,359	976,224	0	12,006,893
3,652,741	412,405	1,611,551	0	0	98,640	−152,790	280,802	5,341,746
7,683,583	241,996	1,287,680	105,272	0	108,023	111,759	98,949	9,439,364
1,829,390	729,566	2,454,877	0	1,691,721	120,162	0	0	6,825,717
4,328,544	33,983	484,242	239,096	0	59,141	1,091,009	249,350	5,986,666
623,708	0	0	366,150	0	2978	28,288	317,827	703,297
1,790,341	1776	22,731	340,025	258,309	−1537	0	0	2,411,645
327,619	9777	119,366	590,682	0	−2160	0	0	1,045,284
1,516,519	222,977	1,297,284	0	0	35,106	42,681	94,913	3,019,654
174,224	463,105	1,030,809	3,261,504	0	−27,196	5399	36,464	4,871,381
304,812	161,965	1,916,127	2,338,319	0	−23,344	3828	2677	4,699,030
707,778	44,971	242,513	385,278	0	12,312	42,790	53,539	1,382,104
59,238	0	0	7,054,395	0	−12,349	23,053	18,137	7,106,200

10.3.2 Analysis

In this section, the trends of the influence coefficients and reaction coefficients of various sectors in the policy scenario are analyzed. Figure 10.12 shows the trends of the influence coefficients of Sector 2, Sector 3, Sector 4, and Sector 5 during the 2011–2025 period in the policy adjustment scenario. As stated before, Sector 2 is Mining and Washing of Coal; Sector 3 is Extraction of Petroleum and Natural Gas; Sector 4 is Mining of Metal Ores; Sector 5 is Mining and Processing of Nonmetal Ores and Other Ores. It can be seen from the figure that the influence coefficients of these sectors are quite stable, even more stable than that in the BAU scenario.

Figure 10.13 depicts the trends of the influence coefficients of Sector 12, Sector 13, Sector 14, and Sector 15 during the 2011–2025 period in the policy scenario. As introduced above, Sector 12 is Chemical Industry; Sector 13 is Manufacture of Nonmetallic Mineral Products; Sector 14 is Smelting and Rolling of Metals; Sector 15 is Manufacture of Metal Products. The trend of the influence coefficients of Sector 12 is going down slightly, while the trends of the other three sectors are going up. The situation is very similar to that in the BAU scenario.

Figure 10.14 shows the trends of the influence coefficients of Sector 17, Sector 18, Sector 19, and Sector 20 during the 2011–2025 period in the policy scenario. As described before, Sector 17 is Manufacture of Transport Equipment; Sector 18 is Manufacture of Electrical Machinery and Equipment; Sector 19 is Manufacture of Communication Equipment, Computers, and Other Electronic Equipment; and Sector 20 is Manufacture of Instruments, Meters, and Machinery for Cultural and Office Use. The influence coefficients of Sector 19, Manufacture of

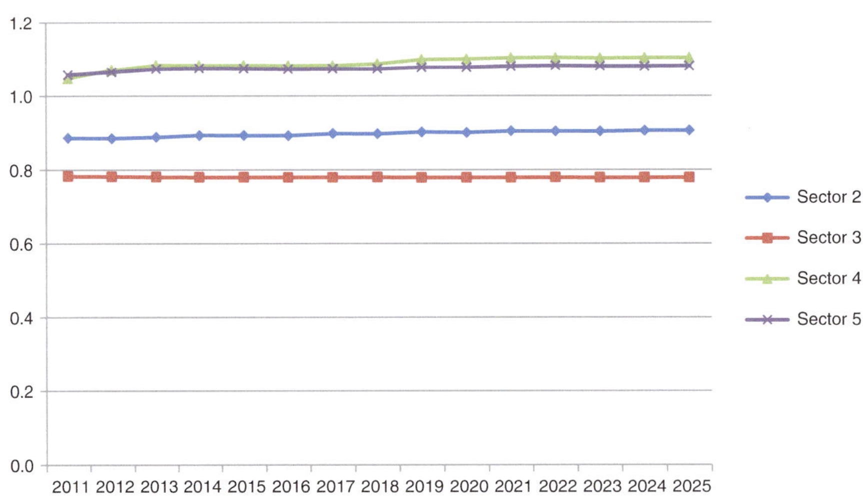

Fig. 10.12 The influence coefficients of Sector 2, Sector 3, Sector 4, and Sector 5 during the 2011–2025 period in policy scenario

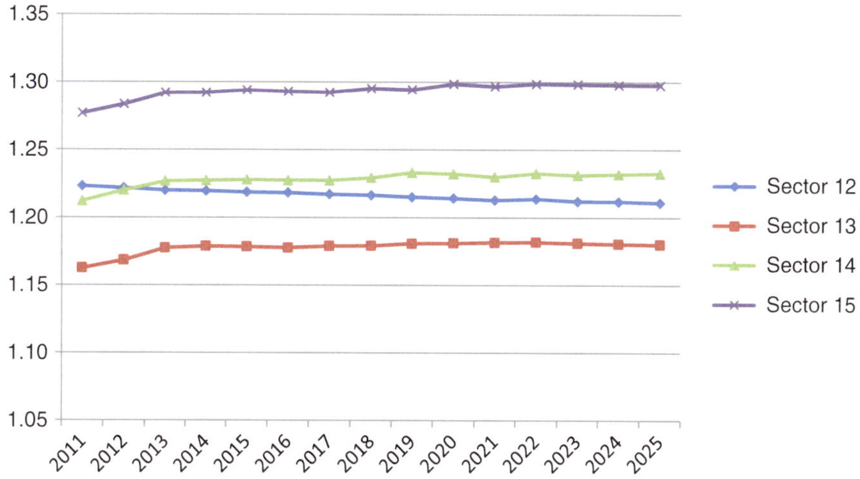

Fig. 10.13 The influence coefficients of Sector 12, Sector 13, Sector 14, and Sector 15 during the 2011–2025 period in policy scenario

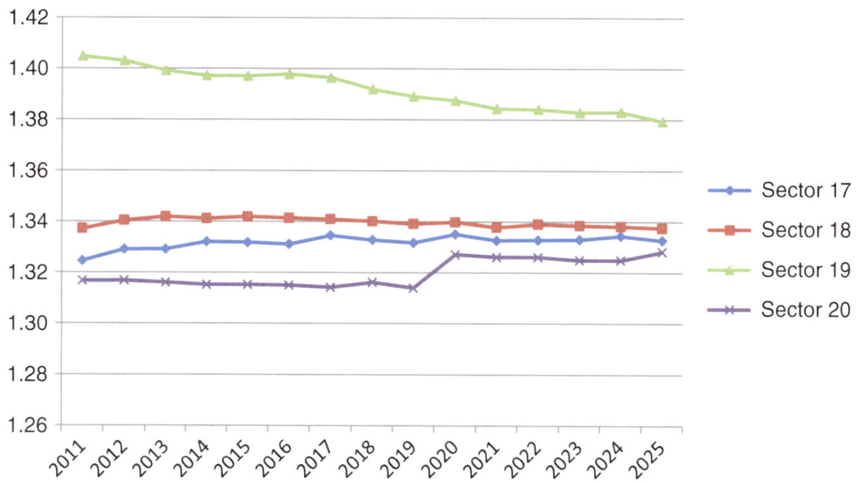

Fig. 10.14 The influence coefficients of Sector 17, Sector 18, Sector 19, and Sector 20 during the 2011–2025 period in policy scenario

Communication Equipment, Computers, and Other Electronic Equipment, decrease from 1.4047 to 1.3794. The decrease rate is larger than that in the BAU scenario, which means the pulling effect of the sector will be undermined more obviously in this scenario. The influence coefficients of the other three sectors are relatively stable.

Figure 10.15 illustrates the trends of the influence coefficients of Sector 30, Sector 31, Sector 32, and Sector 33 during the 2011–2025 period in the policy scenario.

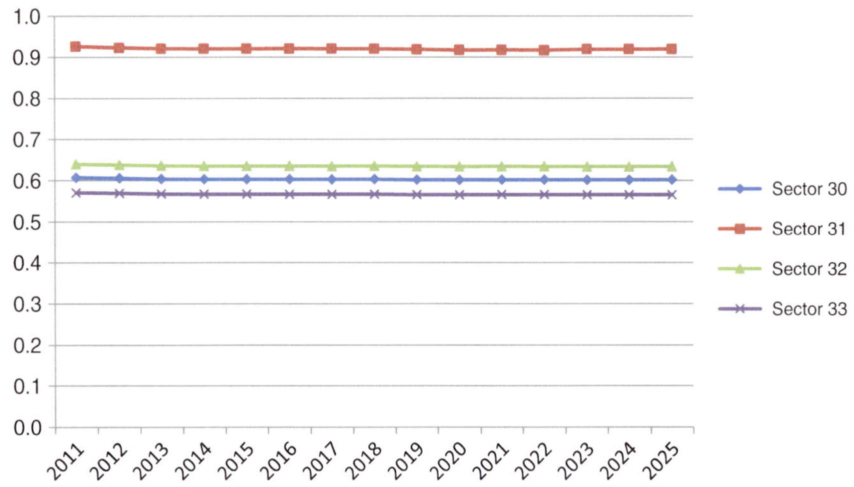

Fig. 10.15 The influence coefficients of Sector 30, Sector 31, Sector 32, and Sector 33 during the 2011–2025 period in policy scenario

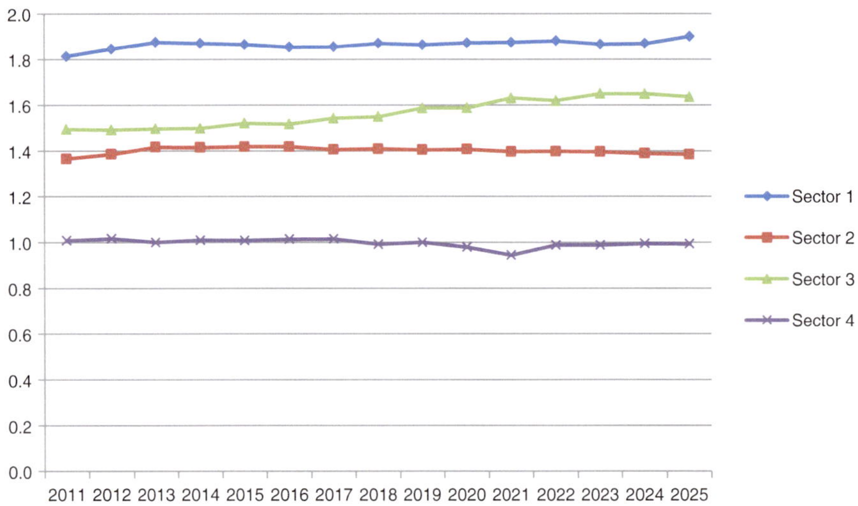

Fig. 10.16 The reaction coefficients of Sector 1, Sector 2, Sector 3, and Sector 4 during the 2011–2025 period in policy scenario

As shown above, Sector 30 is Wholesale and Retail Trades; Sector 31 is Hotels and Catering Services; Sector 32 is Financial Intermediation; Sector 33 is Real Estate. All these influence coefficients maintain at their original level during the whole 15-year period. The reason is that there will be no mutation in China's economy in this period.

Figure 10.16 shows the trends of the reaction coefficients of Sector 1, Sector 2, Sector 3, and Sector 4 during the 2011–2025 period in policy scenario. As stated

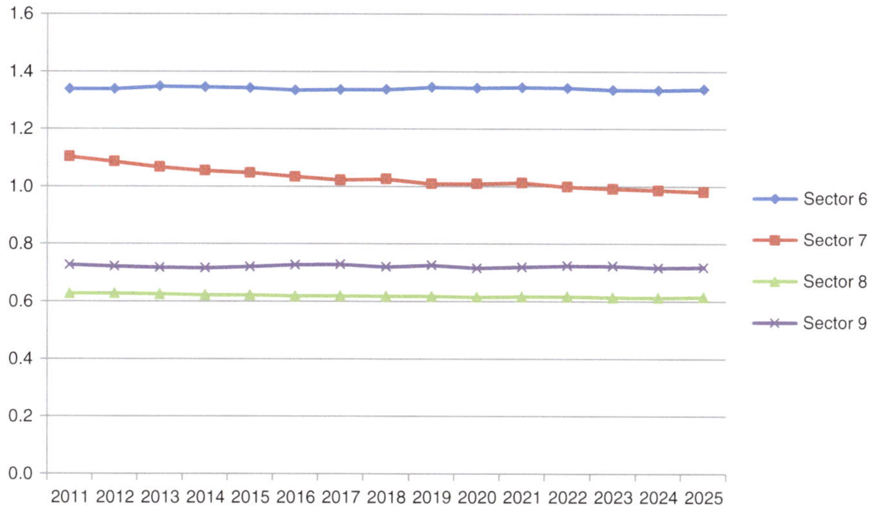

Fig. 10.17 The reaction coefficients of Sector 6, Sector 7, Sector 8, and Sector 9 during the 2011–2025 period in policy adjustment scenario

before, Sector 1 is Agriculture, Forestry, Animal Husbandry, and Fishery; Sector 2 is Mining and Washing of Coal; Sector 3 is Extraction of Petroleum and Natural Gas; Sector 4 is Mining of Metal Ores. There are no obvious changes in the reaction coefficients of these sectors. Only Sector 3 has some little change in its coefficient. The situation is similar to that in the BAU scenario.

Figure 10.17 pictures the trends of the reaction coefficients of Sector 6, Sector 7, Sector 8, and Sector 9 during the 2011–2025 period in the policy scenario. As mentioned above, Sector 6 is Manufacture of Foods and Tobacco; Sector 7 is Manufacture of Textile; Sector 8 is Manufacture of Textile Wearing Apparel, Footwear, Caps, Leather, Fur, Feather (Down) and Its Products; and Sector 9 is Processing of Timbers and Manufacture of Furniture. The reaction coefficients of Sector 6, Sector 8, and Sector 9 remain in their original level in 2011 during the whole period. The reaction coefficient of Sector 7, Textile Industry, decreases obviously from 1.1022 to 0.9803. The decrease rate is less than that in the BAU scenario, which is because the new policy settings provide better environment for the companies in Textile Industry.

Figure 10.18 shows the trends of the reaction coefficients of Sector 16, Sector 17, Sector 18, and Sector 19 during the 2011–2025 period in the policy adjustment scenario. As introduced above, Sector 16 is Manufacture of General and Special Purpose Machinery; Sector 17 is Manufacture of Transport Equipment; Sector 18 is Manufacture of Electrical Machinery and Equipment; Sector 19 is Manufacture of Communication Equipment, Computer and Other Electronic Equipment. These four sectors are all typical manufacturing. There are slight decreases in these coefficients, which is similar to the situation in the BAU scenario, as a result of the economic restructure.

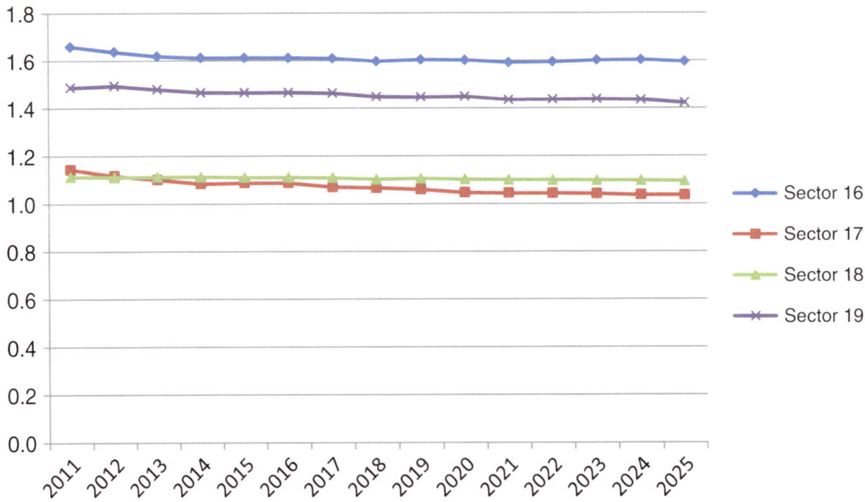

Fig. 10.18 The reaction coefficients of Sector 16, Sector 17, Sector 18, and Sector 19 during the 2011–2025 period in policy scenario

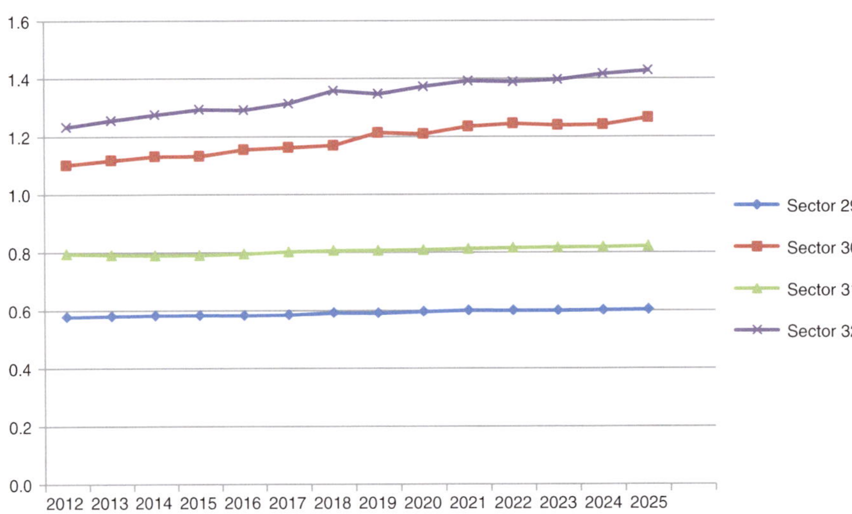

Fig. 10.19 The reaction coefficients of Sector 29, Sector 30, Sector 31, and Sector 32 during the 2011–2025 period in policy scenario

Figure 10.19 illustrates the trends of the reaction coefficients of Sector 29, Sector 30, Sector 31, and Sector 32 during the 2011–2025 period in the policy scenario. As described before, Sector 29 is Information Transmission, Computer Services, and Software; Sector 30 is Wholesale and Retail Trades; Sector 31 is Hotels and Catering Services; Sector 32 is Financial Intermediation. These sectors all belong to the tertiary industry. The reaction coefficients of these four sectors all increase more or less, which verifies that the tertiary industry will play a more fundamental role in China's economy in the future. Besides, the increase is more obvious than that in the BAU scenario, which means that the new policy settings have better impacts on the upgrading of China's national economy.

MIX
Papier aus verantwortungsvollen Quellen
Paper from responsible sources
FSC® C105338

If you have any concerns about our products,
you can contact us on
ProductSafety@springernature.com

In case Publisher is established outside the EU,
the EU authorized representative is:
Springer Nature Customer Service Center GmbH
Europaplatz 3, 69115 Heidelberg, Germany

Printed by Libri Plureos GmbH
in Hamburg, Germany